D1485818

Geology, Resources, and Society

A Series of Books in Geology

EDITORS: *James Gilluly*
A. O. Woodford

Geology, Resources, and Society

AN INTRODUCTION TO EARTH SCIENCE

H. W. Menard

INSTITUTE OF MARINE RESOURCES
UNIVERSITY OF CALIFORNIA, SAN DIEGO

W. H. FREEMAN AND COMPANY
San Francisco

Library of Congress Cataloging in Publication Data

Menard, Henry W
Geology, resources, and society.

Includes bibliographies.
1. Earth sciences. I. Title.
QE28.M48 550 73–17151
ISBN 0–7167–0260-6

Copyright © 1974 by W. H. Freeman and Company

No part of this book may be reproduced by any mechanical, photographic, or electronic process, or in the form of a phonographic recording, nor may it be stored in a retrieval system, transmitted, or otherwise copied for public or private use, without written permission from the publisher.

Printed in the United States of America

9 8 7 6 5 4 3 2 1

Dedicated to
the United States Geological Survey,
which has been deeply concerned with
geology, resources, and society
for most of a century.

CONTENTS

NOT SO SOLID EARTH

SEDIMENT IN MOTION

RESOURCES

PREFACE

This book is one of the results of a broadening concern with man's future on earth that I have come to share with most other fortunate residents of affluent countries. The residents of have-not nations have relatively little control over their fate, for our pollution and our waste of resources will determine much of what happens to the rest of the world. My aim here is not to attempt to influence the future, for that is a role properly reserved for the politician or for the conservationist writing for the popular press. My aim is to provide a broad introduction to the earth sciences in a way that will aid the citizen, young or old, in making rational judgments on environmental matters.

There are several possible approaches to the writing of science texts, and each has its advantages for a particular audience. This book is intended primarily for students who are not majoring in the earth sciences. For this reason, it focuses on comprehensive syntheses and only rarely attempts to lead the student through the details of scientific evidence to a scientific conclusion. Instead, it states the scientific conclusions, if they are uncontroversial themselves, and attempts to lead the reader to a consideration of social problems that have a direct bearing on his life regardless of his occupation.

The book begins with the big picture, the whole earth, before it focuses on the rocks and minerals of which it is composed. I believe that this approach provides the best conceptual framework for understanding, remembering, and relating the subject matter of the whole of the earth sciences. This was the format followed by the great geology textbooks of the nineteenth and early twentieth centuries, although more recent ones have tended to begin with the small and build up to the large—or neglect it.

Introductory texts are all too often collections of small facts and big names packaged in unrelated chapters. The small facts are readily forgotten and, moreover, may not be as important as big "facts" that are less certain. For example, it can be shown in exquisite detail that earthquakes are associated with faults that offset rocks. This is an important fact, but, in itself, it leads to little understanding. It does not tell us whether earthquakes will continue or which faults will be active. It can also be shown that a small, diverse collection of only partially documented data support the idea that the earth's crust is broken into vast, moving plates. Studying the earth with this idea in mind leads to predictions about earthquakes and fault activity, and to realistic evaluations of their social implications.

It appears logical to start with the plates and to follow with details of local effects and social consequences. The same approach is used, wherever possible, elsewhere in the book. The text deals with the properties and circulation of air and water in some detail before turning to climatic change in the past and the social consequences of future changes. Soil formation and the movement of unstable soil are followed by an analysis of the transport of sediment by air and water. Finally, mineral deposits are discussed, both in terms of the major geologic cycles and in terms of our resource needs.

A few remarks about the text may be useful as a guide to the reader. Few complete references are given within the text because they tend to distract all but experienced professionals. However, many seemingly casual references to scientists in the text correspond to entries in the annotated bibliographies at the ends of chapters. The bibliographies themselves include a wide range of material, from articles in popular magazines, which the beginning student should be able to read with ease, to advanced source material for those who wish to pursue a subject in depth. The glossary and appendixes will be of assistance in these collateral activites as well as in understanding the text.

The "boxes" in the text are intended to isolate extended notes and asides that would otherwise break the continuity of the main text. Some consist of relatively advanced material that will aid the reader with a scientific background but that is not essential to understanding the remainder of the text. Others contain important information from nongeologic sources that may already be familiar to some readers; the box on the population explosion is an example. Other boxes, directed to the reader who is an amateur geologist or is interested in becoming one, discuss some of the many science-related activities of rock hounds, cave explorers, and gold prospectors.

The earth sciences are in a dynamic phase: information is proliferating and certain ideas are changing rapidly. This book was as current as possible when it was completed, but some new facts will have been discovered during the time necessary for its publication and distribution. Nothing can be done about this; but a concentration upon broad concepts, of the sort I have attempted to provide, can give the reader the satisfaction of being plunged into lively investigations, discussions, and controversies, rather than into settled and stagnant matters. The reader can expect a continuing review of the earth sciences in the popular media as new ideas are published—or as new disasters strike.

I am indebted to many people for assistance in preparing and reviewing this book. Laurie Rogers, Warren Smith, and Isabel Taylor helped with every stage of the preparation. All or parts of the manuscript were reviewed by Jon Galehouse, Robert Giegengack, James Gilluly, Leigh Mintz, George Sharman, and A. O. Woodford. None of them, however, share the responsibility for the text. Ritchie Mayes, then a recent graduate from high school, identified technical and nontechnical words in the original manuscript that might have proven unfamiliar to college freshmen.

I am also generally indebted to my colleagues at the Scripps Institution of Oceanography for guidance when I was puzzled by unfamiliar material far from my own research specialty. In writing an introductory text, there is no substitute for working in a closely knit group that includes renowned experts on astrophysics, the moon, tides, waves, beaches, sand transport, isotopic tracers and dating, stratigraphy, earthquakes, crustal plates, continental drift, magnetism, gravity, heat flow, atmospheric circulation, weather, modern and ancient climates, vulcanism, rocks, and minerals.

For an even wider range of information, I am indebted to many individuals and organizations for the opportunity to participate in many interdisciplinary investigations. I have served on official committees in Washington concerned with the past, present, and future of oceanography, meteorology, geology, and geophysics. I have been a member of various official groups, including those that investigated the offshore oil spill in Santa Barbara Channel in California, the environmental impact of a proposed jetport north of the Everglades National Park in Florida, and the environmental impact of the proposed extension of Kennedy Airport into Jamaica Bay in New York City. I have been exposed to problems in the earth sciences from the viewpoints of many people: lawyers, economists, businessmen, physicians, biologists, conservationists, sport fishermen, political scientists, politicians, airline pilots, and urban activists. I have learned things from all of them that have helped to shape this book.

June 1973 *H. W. Menard*

Geology, Resources, and Society

1

THE PLANET OF MAN

There is nothing like geology. The pleasure of the first day's partridge shooting or the first day's hunting cannot be compared to finding a fine group of fossil bones, which tell their story of former times with almost a living tongue. . . .

Charles Darwin,
in a letter to his sister Catherine (1833)

Concern about the Earth

The purpose of this book is to provide broad and generalized information about the earth because it is increasingly necessary for understanding matters of everyday interest, matters that appear in newspapers, on television, and on the ballot. Will an underground hydrogen bomb test in the Aleutian Islands trigger a major earthquake? If so, will it generate a wave that will spread damage throughout the Pacific? How will the frozen ground of the Arctic be affected by an oil pipeline across Alaska? What are the risks that the pipeline will be broken where it crosses active earthquake faults? Is it possible to predict or control earthquakes, volcanic eruptions, hurricanes, or other natural disasters? What of climatic change and future ice ages? How soon will the mineral resources of the earth be exhausted? These are examples of the kinds of questions that cannot be answered without an understanding of the history of the earth and the nature of the processes that affect it—which are the subject matter of the earth sciences and of this book.

Until fairly recently, few people asked such questions. Some scholars mused about them, but there was surprisingly little interest in the earth when it was still sparsely populated. Not until the beginning of the nineteenth century did people commonly begin to concern themselves with the world except as a thing to exploit and control. A new appreciation of nature began in England as Constable painted the marshes of Norfolk, Wordsworth went to write poetry in the Lake District, and Byron followed many an English mountain climber to the Alps. What strange things to do at that time! Is it entirely a coincidence that England was the most industrialized nation of the world, and that the quality of the environmental experience of the average citizen began to plummet as his standard of living went up? It is no coincidence that interest in the earth is most intense, at present, in the United States, the most industrially advanced and richest nation of the age. Here, we can afford to concern ourselves with the geologic environment—and we cannot afford *not* to, if we want to preserve what we have. In preserving, however, we shall need to tread carefully, because the earth itself changes. It may be as much of a violation of nature to attempt to stabilize a changing world as to change a stable one.

The Concept of a Changing Earth

"Geology is the science which investigates the successive changes that have taken place in the organic and inorganic kingdoms of nature; it inquires into the causes of these changes, and the influence which they have exerted in modifying the surface and external structure of our planet." So Charles Lyell began the fourth edition of his immensely influential *Principles of Geology*, which was published in 1835. His words show that the fact that the environment changes naturally had already been recognized almost a century and a half ago. This was a revolutionary principle that, shortly before, had begun to displace conventional ideas of an immutable earth. It soon combined with the emerging concept of organic evolution to tear apart the fabric of western philosophy and religion. Organic and inorganic change were intimately linked—not only by Lyell, but by the young Charles Darwin, who had a copy of Lyell's book with him as he voyaged around the world on H.M.S. *Beagle* and began the work that utimately became *The Origin of Species* (1859).

Continuing research in the earth sciences has uncovered evidence of changes undreamed of by Lyell. When we explore the past, we shall find that the day was once only a few hours long, the air was unbreathable, and our magnetic compass sometimes pointed south instead of north. Canada and Scandinavia were more than once covered with glaciers, and some of the deserts of the American west were lakes.

BOX 1.1 THE POPULATION EXPLOSION

The peculiar shapes of these illustrations make the problems obvious. The population is expanding ever faster. Between 1850 and 1930, a period of 80 years, the population doubled from one billion to two billion. It doubled again to four billion in the 45-year period from 1930 to 1975, and it is expected to reach eight billion in 35 years (by 2010) unless it is controlled. Thus, the arithmetical plot of population (figure at left) increases so rapidly it is almost unreadable. The log–log plot (bottom) compresses larger numbers and shows a very different phenomenon: The population increased very rapidly as a result of the cultural revolution 1,000,000 years ago. The increased quantities of food produced in the agricultural revolution 10,000 years ago supported the next expansion. Then, about 500 years ago, the industrial-medical revolution began the expansion that still continues. The population of scientists and engineers has doubled every fifteen years and will increase much faster as China expands in these fields. Thus, technologists become an ever larger, but still small, fraction of the total.

The area of the continents is 57,300,000 square miles; after removing lakes, glaciers, and deserts, however, it is only 33,800,000 square miles. The graph on the right shows how the population density has been growing from 3.8 toward 180 per square mile in a span of 2000 years.

Source: Data from Deevey (1960).

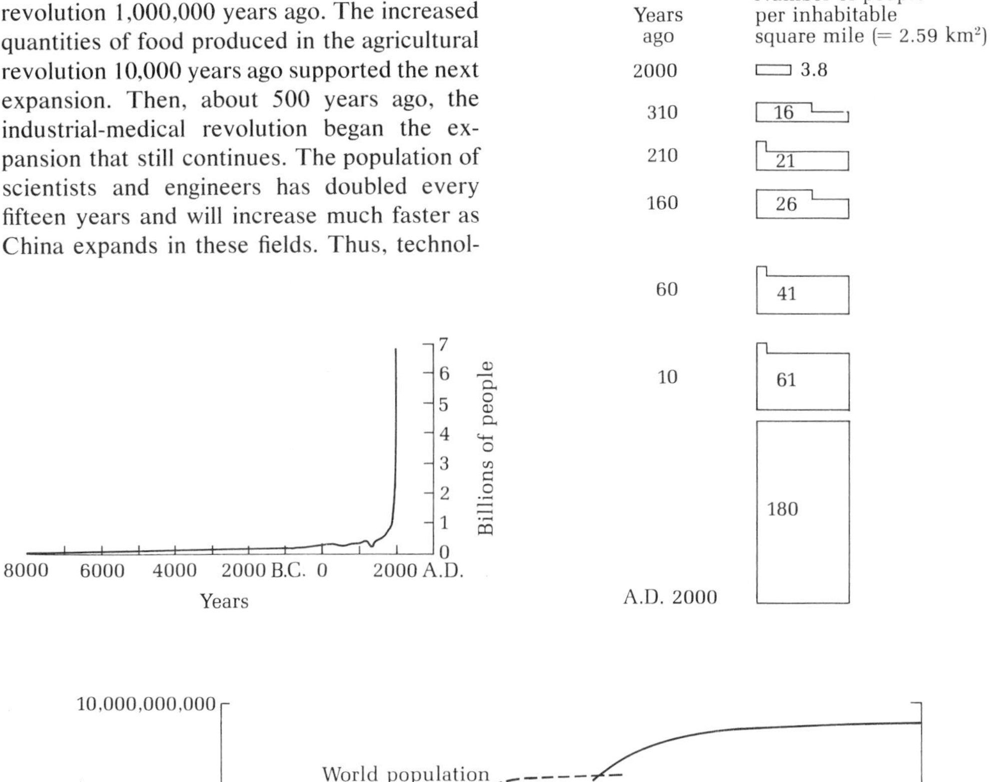

Very recently, many of the changes that have occurred during the long life of the earth have been explained in terms of a revolutionary concept—namely, that the continents drift across the surface of the earth. Southern India, which is now near the equator, shows evidence of having undergone widespread glaciation roughly 200 million years ago, and the reason is that it was once near the south pole. Drifting north, it collided with Asia and pushed up the Himalaya Mountains. The mighty crustal plates continue to crush together, which causes the earthquakes that are such destructive killers in the heavily populated Indian and Burmese valleys at the foot of the mountains. The concept of continental drift clarifies the difficulties that lie in the way of controlling these earthquakes. No action that we can now even imagine can stop the movement of the plates. Knowing this, however, we can concentrate on the earthquakes themselves, and there have already been suggestions about how to control some of them.

In the Ganges Plain, we have a notable example of the reason for the growing interest in the earth sciences and environmental changes. The population explosion is filling this area, as well as the rest of the earth with people who eat, drink, use mineral resources, and, of necessity, live somewhere, no matter how unsuitable the spot. Hordes live on islands of mud near the mouth of the Ganges River, barely above sea level. They have the certain knowledge that flooding periodically will drown the unluckiest among them by the tens of thousands, and that even more will starve before crops can be grown again in their flood-ravaged fields. The islands are the best places available to them. Affluent Americans who build mansions on beaches and mountainsides have other choices, but most of them do not realize that steep slopes are prone to slump and that beaches tend to shift. Thus, some of them also lose their homes to natural changes.

More and more people are concentrated in cities, where they are able to derive little joy from nature but must still face its hazards. Los Angeles, San Francisco, Tokyo, and Mexico City were seriously damaged by earthquakes in times past. Beyond question, other earthquakes will strike them in the future; meanwhile, urban sprawl, high-rise buildings, and the warp and woof of roads and pipelines have vastly increased the potential for damage. Do these problems concern only those who live in regions where earthquakes are a frequent hazard? Memories are short. Charleston, South Carolina, was struck by a quake in 1886, and the whole southeastern seaboard shook. Seventy years earlier, the sparsely populated area around New Madrid, in southern Missouri, suffered the worst earthquakes ever felt in the continental United States. Forests were toppled, new lakes filled sunken ground, and chimneys fell in Cincinnati, Ohio, 640 kilometers away. The region is now rather densely populated, and the potential for earthquake damage has vastly increased.

The geological, archeological, and historical records show that the dramatic climatic changes of the distant past have continued to the present. It would be foolhardy to assume that they will not persist in the future. In general, the changes are gradual and can be accommodated for centuries. In places, however, cities have grown and farmlands have been cultivated to the limit of available water in good years. A temporary decline in rainfall of the sort that has occurred in the same regions in historical time would be very serious, but should hardly come as a surprise. This is not to imply that everyone will be prepared when water shortages occur.

Scale and Duration of Changes

Much of the difficulty in understanding changes in the environment arises because there are usually several different things occurring at the same time. Consider the level of the sea, whose motions are commonly reported in daily newspapers in those coastal cities where the water is warm enough for surfing. In most places, sea level fluctuates up and down 1–10 meters twice a day in a tidal motion caused by the attraction of the sun and moon. In addition, it moves up and down 1–5 meters every 5–15 seconds as the ordinary waves of surf and swell roll in and break. Surf and swell are generated by local and distant winds.

The changes in sea level due to these phenomena are superimposed; thus, the highest level occurs when large surf and swell occur at high tide. Sometimes, engineers are called in to design coastal structures to resist damage when the sea is at the highest common level. Proper design becomes more important as the coastal population increases and high-rise apartments are built along the shoreline. However, the level of the sea and the width of the zone subject to wave damage fluctuate for additional reasons that are more difficult to take into account in structural design. Some natural events, such as hurricanes, temporarily cause great changes in sea level. If they are frequent, their effects must be considered in construction. Structures designed to resist hurricane waves, however, are relatively expensive. If the hurricanes are infrequent, it may be economically more sensible to build a less durable building and take a chance that it will naturally decay long before a hurricane affects it.

Such considerations require accurate predictions of the frequency of natural events, which may require more understanding of their causes than we now possess. Other events, especially those that occur very slowly, may be unpredictable, even though their causes are known. Let us continue with factors affecting sea level: Large quantities of water are normally stored in continental glaciers. If the climate warms, glacial melting will increase, water will flow to the ocean, and sea level will go up. The reverse will occur if the climate cools. This interchange is clearly associated with the global climate, and it appears to produce changes in sea level as great as 120 meters in periods of 10,000–20,000 years. Smaller fluctuations attributable to this cause occur at intervals of 100–1000 years. Thus, some of the largest known changes in sea level are slow ones associated with unpredictable climatic fluctuations.

BOX 1.2 GEOLOGIC HAZARDS AND URBAN EXPANSION OF SAN FRANCISCO

The maps show the San Andreas fault and the two associated zones of maximum intensity of shaking during the great earthquake in 1906. The reason for shaking around the fault is obvious, but the parallel zone of intense shaking is unexplained. A minor fault may be centered in this zone.

The maps and photographs also show the encroachment of the San Francisco urban area upon the fault zone. In 1906, few roads and buildings existed here to be affected by the earthquake. The maximum damage was off to the northeast in downtown San Francisco, and was most intense on filled land.

By 1961, the great earthquake was half forgotten, and the population had spilled over the gently rolling land that once formed the logical limit of the city. Urbanization had overreached the margin of the bay. Thus, San Francisco International Airport, the major highways,

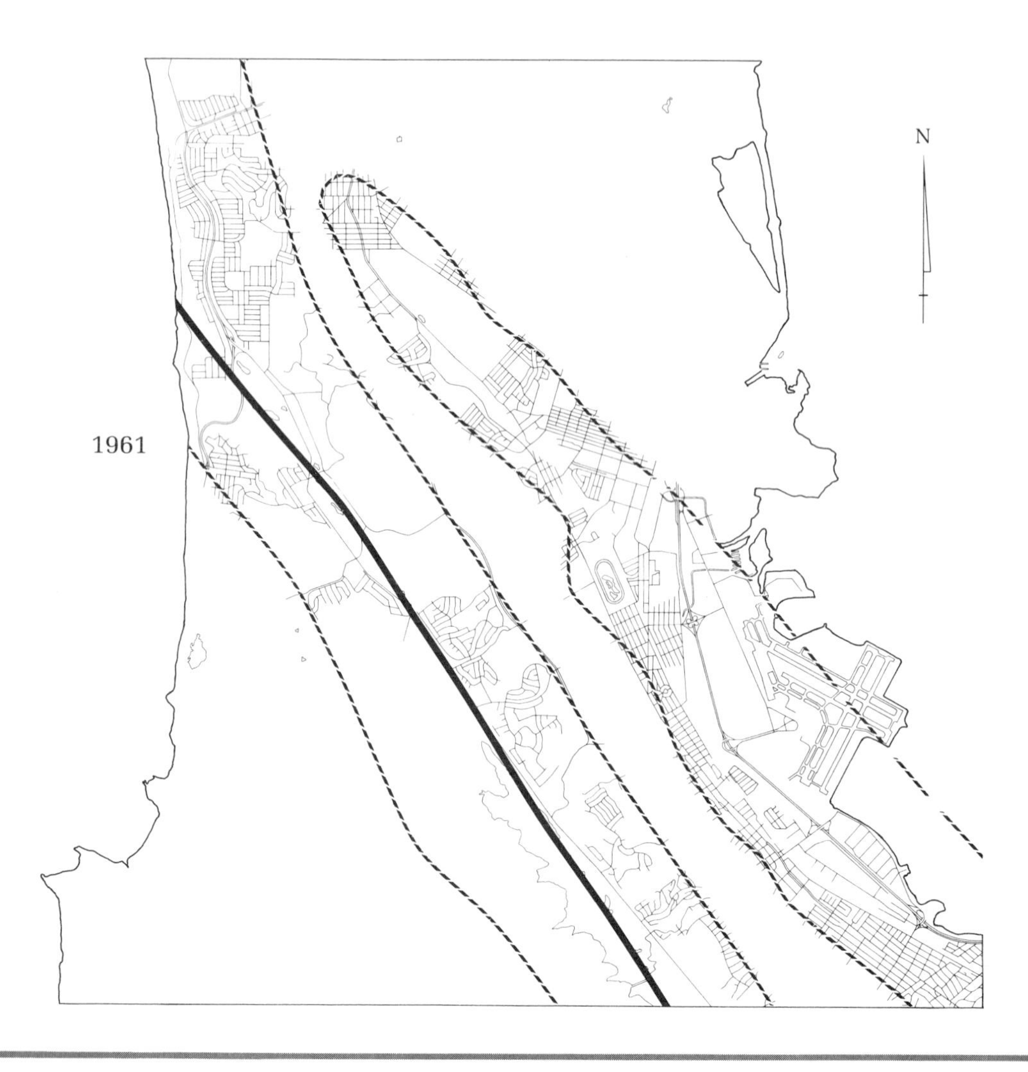

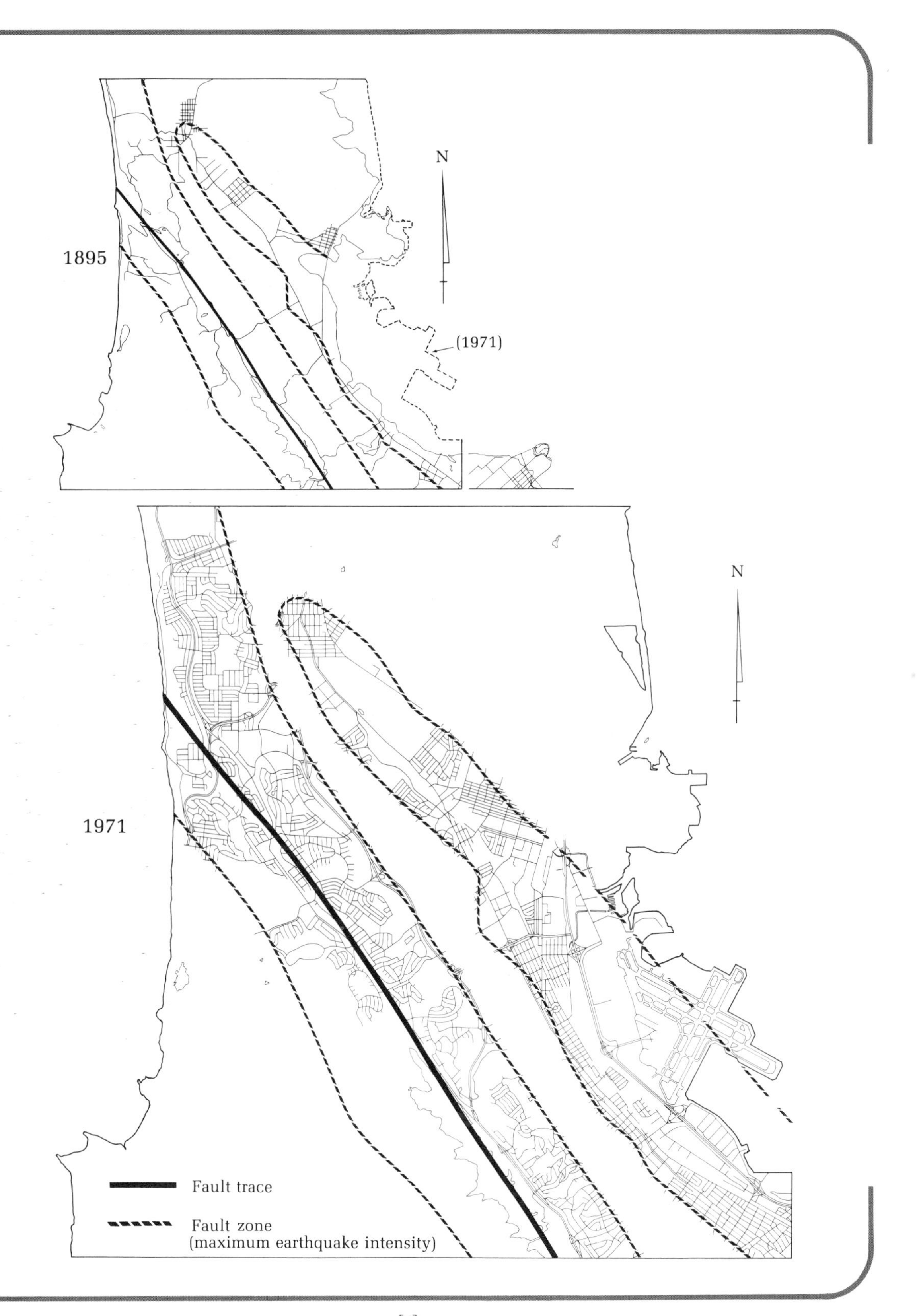
N
1895
(1971)
N
1971
Fault trace
Fault zone
(maximum earthquake intensity)

and some industrial complexes were built on filled land within what was the zone of intense damage in 1906. While industry went to the filled marshlands, housing subdivisions spread onto the adjacent steep slopes of the coastal mountains. The San Andreas fault zone began to be a popular building site.

In the next decade, the urbanization intensified as the population grew. The airport was lengthened, on filled ground, to accommodate jets. The concept of homes around marinas caused subdivisions to spread, on filled land, into the bay. Meanwhile, the mountains also exerted their attractions. Housing roughly

View east across the San Francisco Peninsula showing the San Andreas fault and the extent of urban development in 1953. The Line represents the approximate location of the 1906 fault trace. [Photo by Pacific Resources Incorporated, from Cluff (1968).]

doubled in the main fault zone during the decade (see photographs).

Earthquakes will occur again along this fault. Research is now being directed toward the prediction and modification—and, conceivably, the control—of earthquakes. These maps illustrate why the research is socially as well as scientifically important.

Source: Data from Lawson (1908).

View southeast along the San Andreas fault showing extent of urban development in the Daly City–Pacifica area of the San Francisco Peninsula in 1966. The line represents the 1906 fault trace. [Photo by Robert E. Wallace, U.S. Geological Survey, from Cluff (1968).]

This brings us to the problems of a fluctuating environment and a population that explodes to utilize the optimum. If present trends continue, society will tend to occupy all those areas that are available at a given time. If sea level slowly fluctuates downward, new houses will follow the receding shoreline. When the fluctuation inevitably reverses, the shoreline will be inundated. It is important to realize, as we occupy the shore, that we do not know the long-term trend of sea level or how to predict or control it.

This simplified abstract of the complex fluctuations of sea level is presented here merely to illustrate some very widespread environmental problems. Similar fluctuations occur in the weather and climate, the frequency and magnitude of floods, the eruption of volcanoes, and the generation of earthquakes. Moreover, many of these fluctuations appear to be interrelated in complex

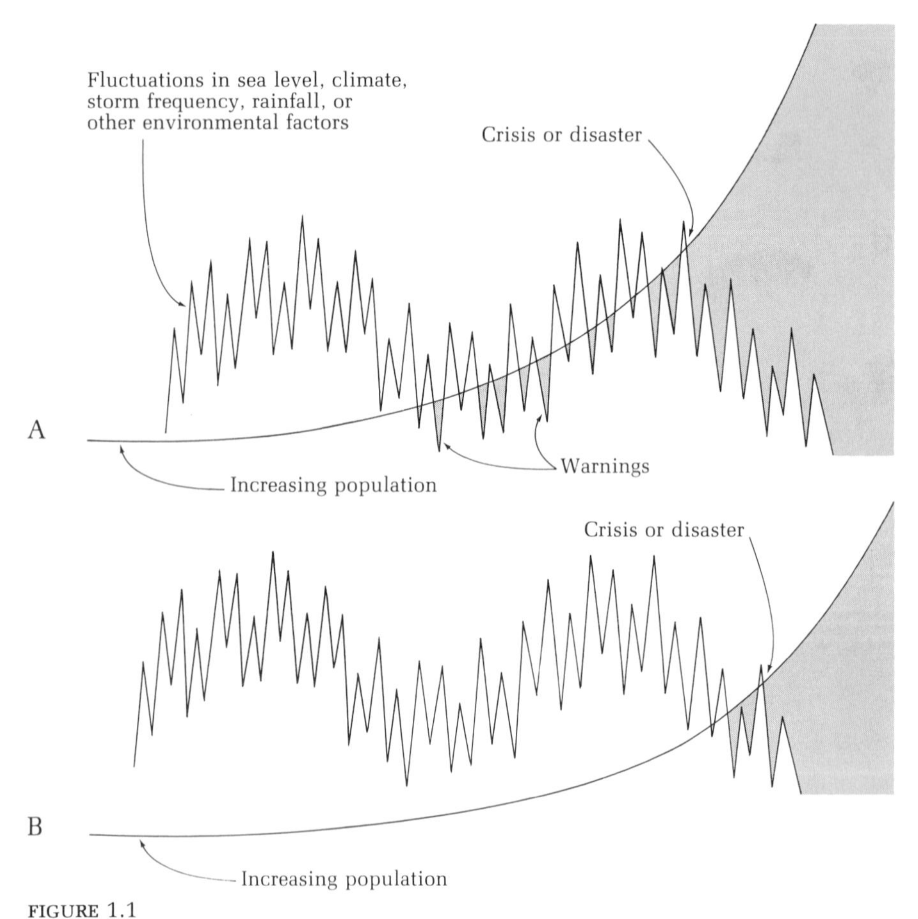

FIGURE 1.1
The problem of a fluctuating environment and a population that explodes to utilize the optimum. Shaded portions represent times when the environment is generally unfavorable or inadequate for people. For population curve *A*, the crisis situation is heralded by many episodic warnings. For curve *B*, however, the crisis occurs after little or no warning.

ways that are still obscure. If present trends continue, it may be anticipated that society will expand to the limit of every aspect of the environment. That is, the population will push north to the limit of tolerable climate, crop production will be increased to the limit of available water, and all other activities will expand to their limits. Consequently, the longer-term fluctuations of the environment and their interactions will be increasingly important. One of the major concerns of the earth sciences is to attempt to understand the history of environmental changes and their causes. This book has been designed to provide, among other things, an introduction to some of our present knowledge of these important matters. In much of the book, emphasis is placed on those aspects of the earth sciences that are related to socially important problems. It should be realized, however, that the apparently diverse earth sciences have an underlying unity, and that each aspect is somewhat related to all of the others.

The Earth Sciences

The earth sciences are concerned with the application of all the appropriate physical and biological sciences to the study of our planet and nearby matter in space. Thus, some earth scientists use satellites to study the radiation belts around the earth, some use ships to follow ocean currents, and others study explosive shocks that reveal the characteristics of the interior of the globe. All of these scientists are mainly concerned with different types of geophysics—that is, the application of physics to the study of the earth. Others, geochemists, use chemistry to study the same sorts of phenomena and attempt to solve the same sorts of problems. So do paleontologists (or geobiologists) and mineralogists and petrologists (who study minerals and rocks, respectively). Geologists—which is merely another way of saying "earth scientists"—now traditionally study surface topography and rocks and the processes that form them.

There are two major questions in the earth sciences: What processes now affect the earth? and, What is the history of the earth since it was born about 4.6 billion years ago? All kinds of earth scientists use all available techniques to try to solve these problems, and thus, all of the scientists are more or less interdependent. For example, stratigraphers, who unravel the history of sedimentary rocks, may obtain useful information about the age of the rocks from geochemical or geophysical dating techniques. Likewise, they may obtain insight about fossil organisms from biologists and ecologists who study modern ones.

These interdependencies promote what appear to be strange collaborations. Contributions to the history of the rotation of the earth are made by classical historians, astronomers, geophysicists, and paleontologists. The history of the magnetic field is studied by mineralogists, physicists, geophysicists, geochemists, petrologists, paleontologists, stratigraphers, and oceanographers. We could continue to compile similarly long lists for every other important problem. However, the point is amply clear: it is only by a broad advance of many types of research that progress in our understanding of the earth can be made.

In the past, the most fruitful scientific research generally originated when a scientist identified an interesting problem and went ahead to solve it. New people have always been available to expand promising lines of research. As a consequence, the volume of scientific literature and the number of scientists have roughly doubled every fifteen years for the last few centuries.

Although the broad advance of individual research has led to great integrating hypotheses in the earth sciences, it has left many puzzles unsolved. Further progress will depend on the efforts of professional earth scientists working alone or (increasingly) in teams. Much of what they do will not be immediately related to social problems, but it should be remembered that much of the scientific knowledge that is now eminently useful was acquired mainly to satisfy the curiosity of scientists. Even so, the earth scientists engaged in self-directed research are few, relative to those in industry and government who use scientific methods directly for social purposes. Year in and year out, the federal government spends about one percent of its budget on measurements and analyses of the physical environment because of their importance to society. The research budget in the earth sciences is but a tiny fraction of that amount. It appears that neither is enough.

The Earth Sciences—Assumptions and Proofs

Knowledge of the earth is increasing rapidly as the need for it grows, but much remains unknown. This reflects the difficulties inherent in research in the earth sciences and other fields concerned with the complex and uncontrolled environment. Some knowledge of these difficulties is essential to an appreciation of the limits of our understanding of the earth.

The great law of the earth sciences, that "the present is the key to the past," was formulated late in the eighteenth century by James Hutton and John Playfair. Charles Lyell later demonstrated convincingly that the phenomena active on the earth today would have been capable, given the great span of geologic time, of producing the effects now observable in ancient rocks. If a river erodes a few sand grains a day, it will eventually wear away a mountain.

This is a powerful and useful law that supports the whole structure of the earth sciences, but it rests on assumptions that we should examine with care. First and most important, it is assumed that the physical relations of the present apply to the past. This assumption cannot be tested on earth, but it can be in space. In astronomy, we are able to look great distances; and, because the velocity of light is limited, we can look into the past. The light arriving from a star 1000 million light years away left it 1000 million years ago, and we can see now what that star was like then. Repeated observations of such distant and ancient objects show that many physical laws have not changed: small stars are held in orbits around big ones by the attraction of gravity; and older stars radiate light that shows that they are made of the same elements as younger ones nearby, and that their atoms are put together in the same way. All seems well—but for the fact that the light emitted by atoms in old and distant objects is redder than that emitted by the same atoms in stars that are young and nearby. However, this striking observation does not show that physical laws change with time. It is elegantly explained by the hypothesis of the American astronomer Edwin P. Hubble (1889–1953) that the universe is expanding. The color of starlight depends on the relative motion of stars, just as the tone of noise from a jet plane changes when it flies past a listener. This phenomenon is known as the Doppler effect (see Figure 1.2).

Observation thus gives grounds for believing the assumption that physical *laws* do not change—but that physical *constants,* which are the ratios or proportions between various physical measurements, do. P. A. M. Dirac, a British physicist of startling imagination, suggested in 1937 that some of these "constants" only appear to be constant, that they actually may vary with time. Such changes appear to be required to explain some astronomical observations that are otherwise puzzling. At the present moment, therefore, it appears safe to view scientific laws as correct at all times and places, but physical constants as not entirely above suspicion for calculations about the earth when it was very young. We should, for example, accept with confidence that, if the earth captured the moon, the interaction was in accordance with the law of gravitation. We should, however, have some reservations about computations that purport to give the exact distance between the two at a time 2000 million years ago.

FORCES THAT SHAPE THE EARTH ARE RECOGNIZABLE FROM THE END PRODUCTS

If we visit the foot of a glacier in the Alps, we find that the streams of meltwater have beds of gravel and boulders. If we walk into a cave in the ice, we find that the bedrock is polished smooth, except for long scratches that mar the finish. Around us, we see piles of similar gravel obviously deposited directly by the glacier, and the nearby bedrock is polished and scratched. In geology, the assumption is made that the existence of a glacier can be deduced from such deposits and marks, even though the ice has long since melted away completely.

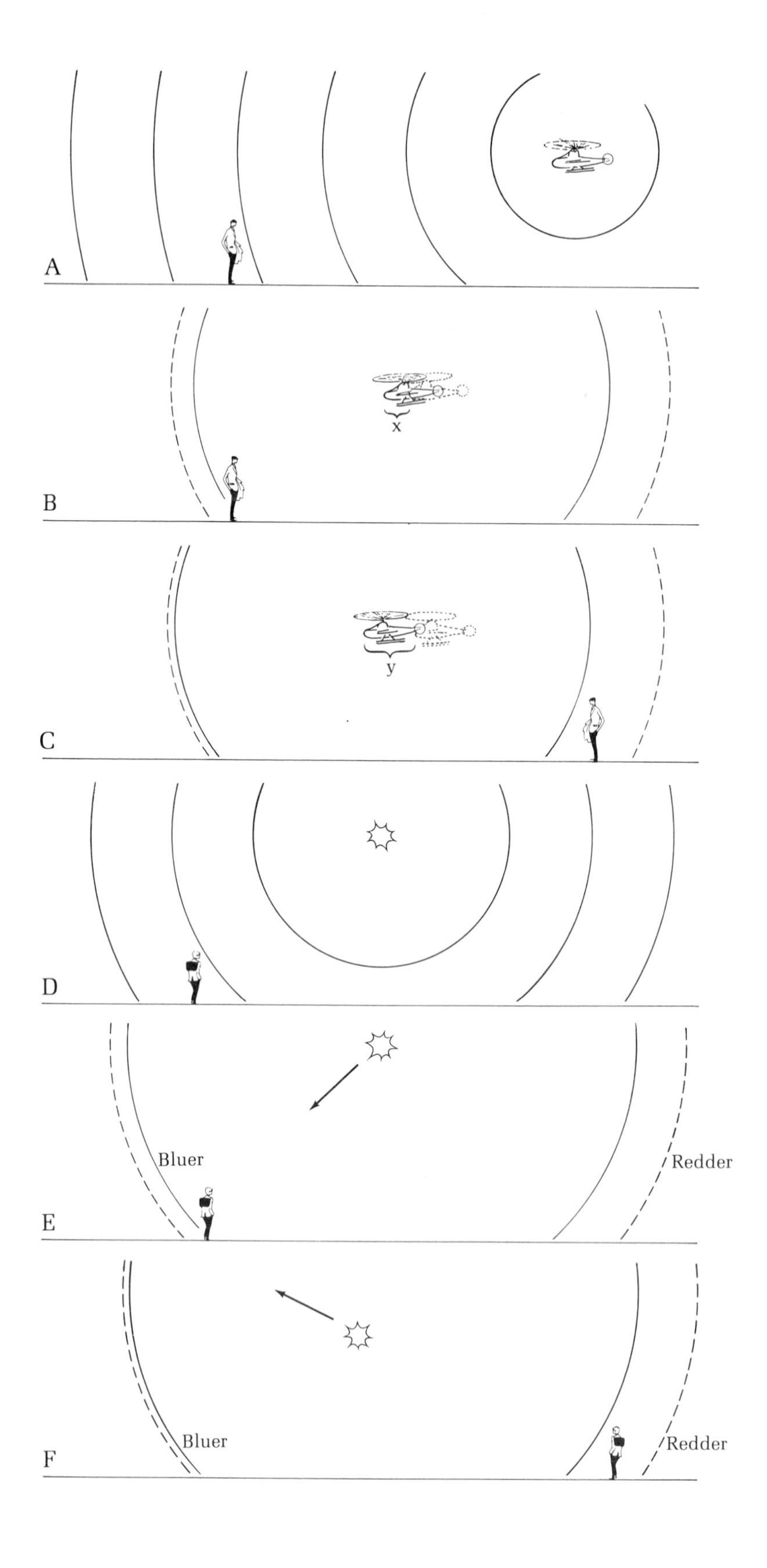
A
B
x
C
y
D
E
Bluer
Redder
F
Bluer
Redder

Given enough distinctive features, we can be confident that a glacier once existed in what is now an empty valley. However, we can be more certain if we have a theory that explains why a glacier produces all the features that are associated with it: The moving glacier scoops up great pieces of rock that it carries with ease in a frozen embrace. With these rocks, it scours away the surrounding bedrock, thereby producing rock flour, which serves as a polishing agent for the bedrock. Pointed or edged rocks are dragged across the polished surface, which is thereby scratched. Broken rocks and sand collect in mounds, called moraines, at the melting glacier edge—or, because of flow patterns, even in the middle of the ice (Figure 1.3). In these and related ways, all the features associated with glaciers can be explained by glacial action.

But what if the characteristics are less distinctive, fewer, or unrelated to the environment by comparable observation or theory? Then interpretations of the ancient environment become increasingly questionable. It may happen that a particular type of sediment—perhaps black, stinking mud—has been found by marine geologists only in deep, narrow fjords. The mud would be altered by temperature and pressure to black rock containing yellow crystals of iron sulfide. Such rocks are widespread in the geologic record. Should all such rocks be interpreted as indicating a narrow, deep environment of deposition? Not at all. They indicate an environment lacking oxygen, and under the proper conditions, they can form in basins of any shape or depth.

ENVIRONMENTS ARE RECOGNIZABLE FROM THE ORGANISMS THAT LIVED IN THEM

Charles Darwin established that reef-dwelling corals live only in warm tropical waters, and the ranges of other organisms are likewise restricted by the environment. A biologist confronted with a species brought to him in a museum can often describe the habitat in which it was collected. If the same species were also found as a fossil in an ancient rock, it is a reasonable assumption that it lived in a similar environment in ancient times. However, species do not endure unchanged for millions of years. Thus, the geologist who seeks to identify ancient environments may assume, for want of better evidence, that

---

FIGURE 1.2

The Doppler effect. In *A*, the helicopter is standing still, and sound waves from the engine reach the observer at their normal frequency. In *B*, the helicopter approaches the observer moving a distance *x* between two successive waves; to the observer, the wavelength seems shorter and the frequency higher. In *C*, the helicopter recedes from the observer, now moving a distance *y* between successive sound waves; to the observer, the wavelength seems longer and the frequency lower. The helicopter's speed is greater in *C*, hence the distance *y* is greater than *x*, and the wavelength and frequency changes are correspondingly greater. Similarly, the light emitted by a star (*D*) appears bluer (*E*) or redder (*F*), depending on whether it moves toward or away from the earth.

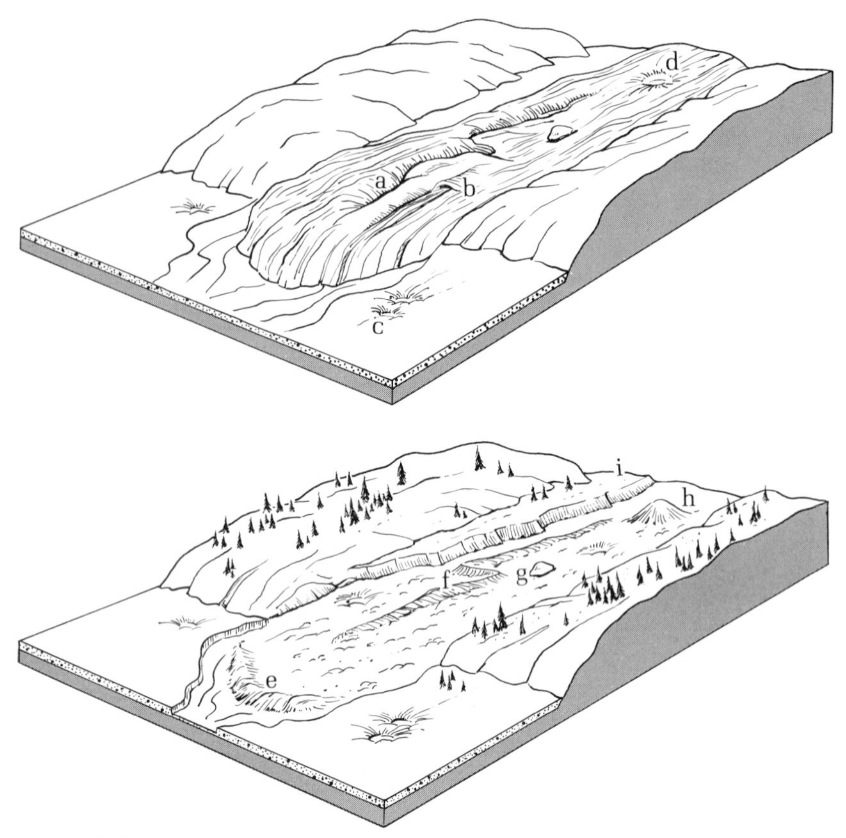

FIGURE 1.3

Nomenclature and origin of some glacial features. *Above*: Glacial ice in a valley has streams draining from crevasses (*a*) and tunnels (*b*) in the ice and from the base of the ice sheet itself; gravel fans are deposited where these discharge at the ice front. Beyond the ice is a plain of stream-deposited outwash where blocks of ice washed from the glacier melt and leave depressions, called kettle holes (*c*), in the plain. In and on the ice are boulders from up the valley. Shafts (*d*) with funnel-shaped openings collect gravel and sand. Lakes ponded against the valley walls contain small deltas. *Below*: When the ice melts, the gravel fans in front of the ice collapse and form an end moraine (*e*); the stream channels on and under the ice are marked by ridges of gravel, known as eskers (*f*). The valley floor is hummocky ground moraine on which boulders rest as glacial erratics (*g*). The gravel and sand deposited in the shafts form hills, or kames (*h*). Collapsed deltas in the marginal lakes form kame terraces (*i*). [After Hunt, *Geology of Soils*, W. H. Freeman and Company, Copyright © 1972.]

a now-vanished species lived in an environment similar to that of a modern species that is only vaguely related. Fortunately, organisms are extraordinarily sensitive and are adapted to their environment by evolutionary selection. They can be used with confidence as environmental indicators, provided only that the mode of adaptation is fully understood. For example, it appears that the ancestral horses that lived 60 million years ago did not graze on grass like modern ones; rather, they browsed on leaves like modern deer. This is

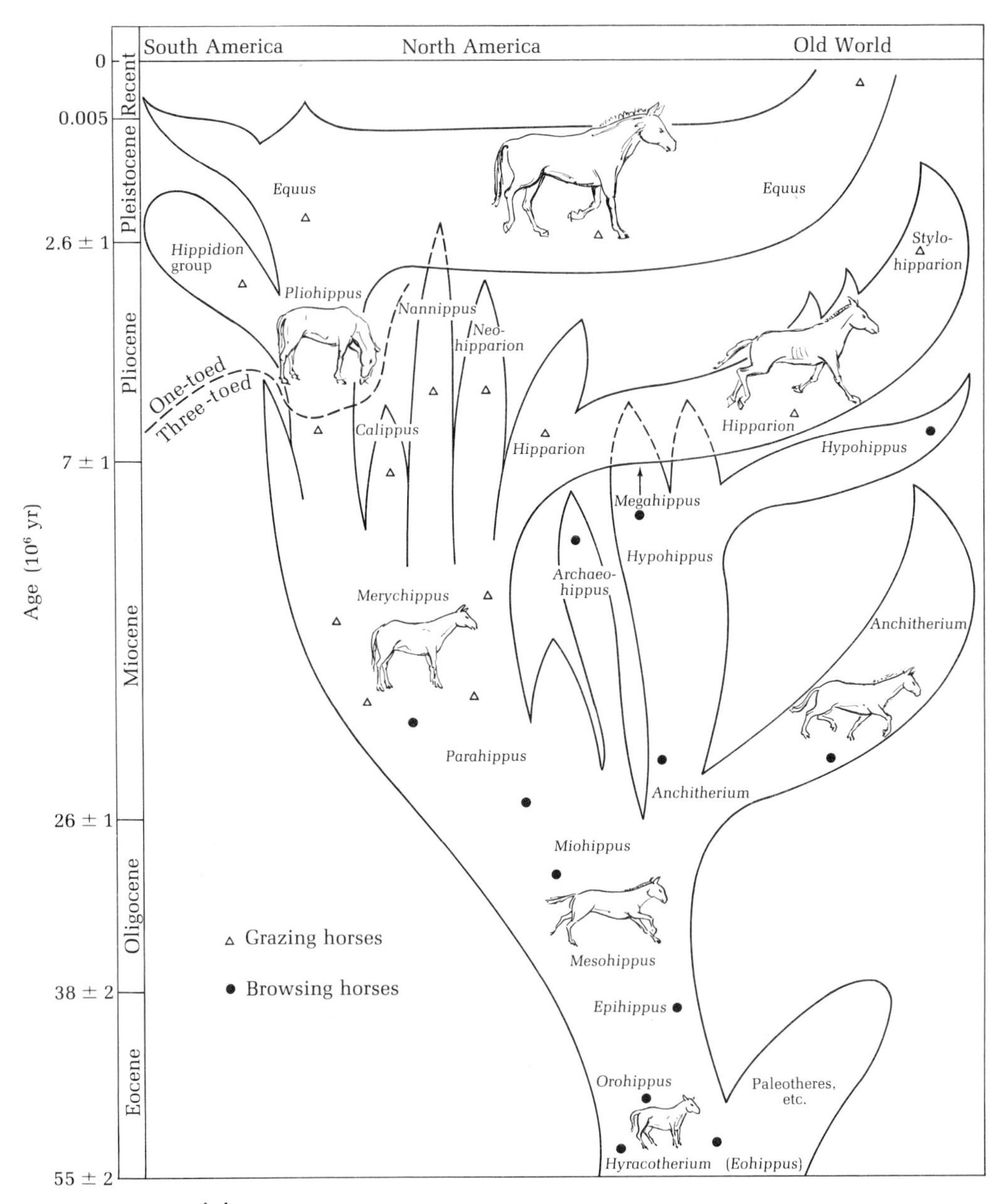

FIGURE 1.4
The evolution of the horse family, drawn to scale. *Equus* is the modern horse, which is large, grazes on grass, and runs on one toe on each foot. Earlier horses were much smaller, ran on three toes, and browsed on leaves. [After Simpson (1951).]

evident in their teeth, which the munching of grass would rapidly have worn away. Modern horses, by contrast, have quite different, "long crowned" teeth that continue to grow throughout life. Other indicators of life style are equally convincing.

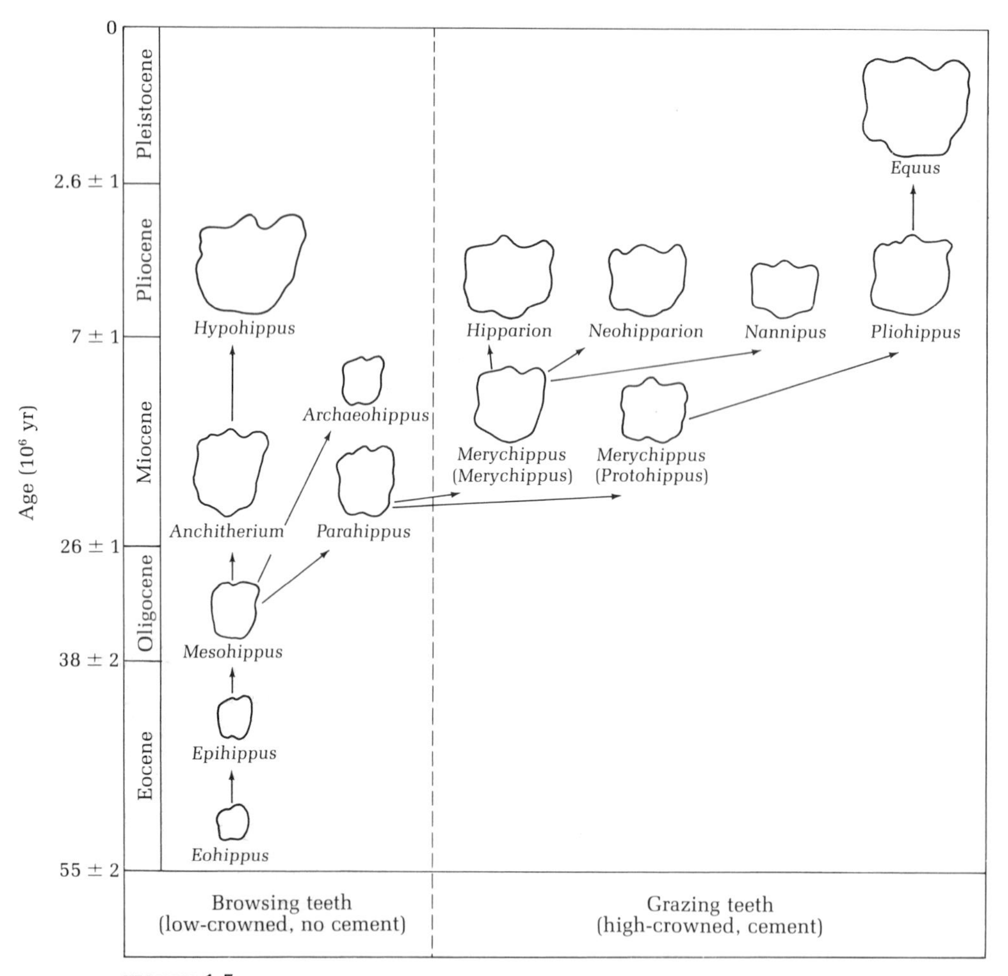

FIGURE 1.5
Evolution of horse teeth, drawn to scale. Grazing teeth grow continuously during the lifetime of the animal because harsh grass wears them down rapidly. All teeth are viewed from above. [From Simpson (1951).]

Even though organisms are sensitive to temperature, salinity, and pressure, among other things, reliable biological indicators of such important environmental characteristics are often difficult to identify. Some bottom-dwelling marine animals are excellent local indicators of water depth—and, thus, of pressure—but they may live at markedly different depths in other localities. Perhaps their distribution is not a function of water depth but of temperature, turbidity, or salinity; or perhaps they are highly insensitive to the physical environment but are affected by the presence or absence of predators that *are* sensitive to it.

NECESSITY FOR COMPLEX ANALYSIS

Physics and chemistry are relatively straightforward sciences. Questions are asked, hypotheses are advanced, critical tests are devised and executed, and new laws are established with comparative ease. The environmental sciences are much more complex, because it is generally impossible to make controlled, critical tests of hypotheses. Nature does the experiments, and there are always uncertainties in evaluating a host of variables. In the earth sciences, there is the further complication that the natural experiments, whose results have become rocks, were finished ages ago. The observer knows that he is not seeing all aspects of the experiment; the bones of a dinosaur may still exist, but rarely his skin, and never the air he was breathing when he died. The earth scientist's hope is to reconstruct the pertinent and important aspects of the experiment despite the fact that most of the evidence has vanished.

The geologist uses what is available: the fragments that have not been eroded away, or buried, or melted and recrystallized. Of necessity, he usually presents complex arguments to test hypotheses. The evidence for a vanished glacier is a typical example. Other arguments may draw on physical, chemical, biological, and geologic evidence, all of it fragmentary. If every aspect of the evidence supports a single hypothesis, it finds ready acceptance and is incorporated into the earth sciences. This rarely occurs. Usually, the fragments are too few to be convincing or—even worse—the evidence is contradictory. Then, each specialist tends to place different emphasis on the conflicting evidence. An unresolvable confusion may result, and the hypothesis is neither rejected nor accepted but left in limbo to vex us. Nothing can be done about this. Everyone would like to resolve such matters, but nature often does not cooperate. Fortunately, the earth has been generous in yielding its secrets in recent years, and most (but not all) earth scientists have accepted the unifying hypothesis that great crustal plates drift slowly across the surface of the earth, and that the continents and the works of man drift with them.

A Model of a Dynamic Earth

After a long period of groping and wandering into dead ends, earth scientists have emerged into the sunlight—at least for the moment. The pattern of the maze of earth history is revealed. Many details remain to be explored, and it is quite likely that future wandering will show that this maze is only a corner of a far vaster and more elegant structure that is now mostly concealed from us. Nevertheless, we now have a gross model of the earth that explains the major mysteries of the past research and seems capable of predicting an extraordinary range of geologic phenomena. It has two great virtues:

the broad outline makes sense, and it is also easy to remember. We may imagine a closed system of cycles with several component cycles, the whole not unlike a gravel works. Consider such a works idealized for our purposes: Coarse stone is assembled at the base of a tower and hauled up by an engine to the top. There it is broken into small fragments of different sizes. Water is also pumped to the top of the tower, and the fragments and water flow by gravity through various screens, which separate the fragments by sizes and drop them on conveyor belts that are also driven by engines. These belts lead the fragments to piles of sorted gravel, which is offered for sale. We must modify the plant, however, in order to make the system *closed*, as it is in the dynamic earth: The water is recycled, and the separate conveyor belts converge and dump the various sorts of gravel together into a furnace, where it is mixed and melted together to form coarse stone like that with which we began; and this stone is once again assembled at the base of the tower, and the cycle is repeated.

The earth can be visualized as working in a similar way. Consider the system of interacting global cycles, beginning with the water cycle, which is the easiest to follow. Water is pumped to the sky by the energy of the sun harnessed through the mechanism of evaporation and atmospheric circulation (Chapter 10). Powered by gravity, it falls on the land and flows back to the sea, perhaps pausing temporarily in lakes or where it seeps into the ground (Chapter 15).

The closely related sedimentary cycle is best visualized by beginning not at the bottom, as in the gravel works or water cycle, but at the top. The rocks on the tops of mountains are broken up by ice and gravity. The pieces move to lower levels, and some are gradually transformed into soil mainly by chemical interaction with water (Chapter 13). Other pieces are broken into tiny grains or dissolved, and are transported by the energy of rivers toward the seashore. The material may pause temporarily on land or at the edge of the sea, but ultimately it is drawn down to the deep-sea floor by the action of gravity (Chapter 16).

The sea floor is broken into a system of moving conveyor belts driven by the energy of radioactivity in the interior of the earth. Thus, it participates in the third of the great global cycles—the tectonic cycle, which physically deforms the surface rocks (Chapters 5–8). The conveyors consist of enormous, rigid crustal plates. Where they move apart, new rock wells up, solidifies, splits, and moves apart in turn. Where they come together, the plates build mountains, plunge back into the interior, and melt. The conveyors, therefore, have a lower limb that is more plastic than the solid rock of the sea floor.

The sediment that falls on the sea floor is ultimately transported by the moving plates to the edge or bottom of a continent, where it is scraped from the plunging oceanic crust. It then enters into the igneous and metamorphic cycles, which are also active in other regions (Chapters 3, 4, 9). The energy

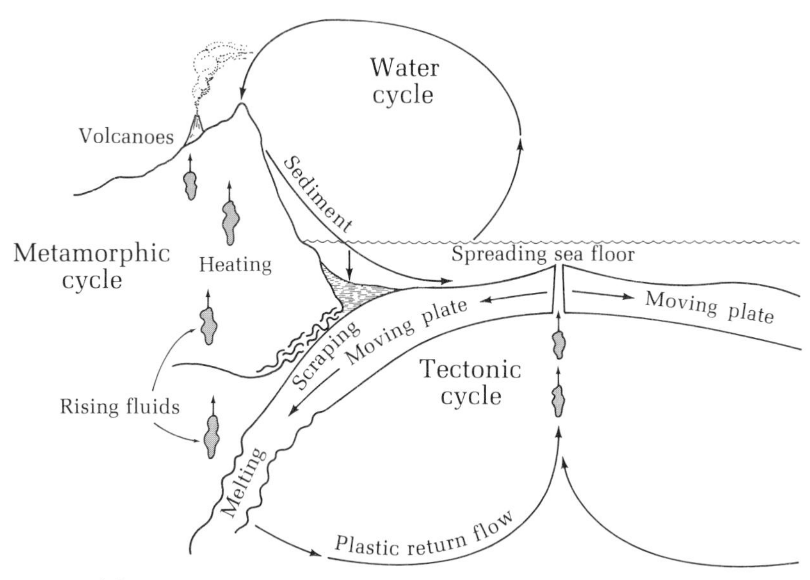

FIGURE 1.6
The major inorganic cycles of the earth.

from radioactive heating of the interior transforms (and may melt) the sediment. The less dense fractions of the melt rise through the overlying rocks, solidify, and form new rocks in continental mountains. Meanwhile, the sedimentary cycle erodes the mountains away. In due course, the sediments scraped onto the bottom of the continental plate emerge as hard rock on the top, where they start to be recycled.

This, of course, is an idealization of the complex events that may occur. However, it is a path actually followed by large quantities of the materials at the surface of the earth under the influence of the external energy of the sun and the internal energy of radioactivity. Other masses of surface rock and sediment have followed much more complex paths and have stalled at some place in the cycle for millions, or even billions, of years. Yet, at any time, they may be reactivated, and no part of the surface of the earth is permanently removed from these cycles. We find rocks in Canada and South Africa that have been parts of continents for more than 3000 million years. It may not seem very useful, therefore, to view them as participating in an active global cycle. Still, many of these ancient rocks apparently started through the whole cycle when the earth was young; they were eroded and transported as sediments; they appear to have been through the tectonic cycle not long after; and the remainder of their 3000 million years has been occupied in the igneous and metamorphic cycle, but it is now completed. If we find the rocks at the surface, it means that they are ready to enter the sedimentary cycle again. Then they will move to the deep-sea floor, and sooner or later, the conveyor belt will restore them to a continent.

If we allow enough time, the global cycle for all surface rocks is realistic. It is also a convenient concept into which we can tie the enormous diversity of materials and processes we shall explore in this book. If the reader prefers, he can consider rocks that are inactive for some arbitrary period—say, 1000 million years—as out of the global cycle entirely and subject to different influences. Such a model of the earth would have both static and dynamic elements. I see little advantage in this division. Living things die, but life goes on. The whole "dead" earth eternally changes.

> We soon cease to lament waste and death, and rather rejoice and exult in the imperishable, unspendable wealth of the universe, and faithfully watch and wait the reappearance of everything that melts and fades and dies about us, feeling sure that its next appearance will be better and more beautiful than the last.
>
> —John Muir

Summary

1. The earth changes in ways and at rates that affect society.
2. The population explosion is causing people to live in locations and in environments that are subject to greater risks from these changes.
3. The various changes in the environment act in different directions and for different periods; for this and other reasons, it is beyond our abilities to predict most of them, despite their importance.
4. The earth sciences are concerned with the application of basic sciences to the study and use of the earth. Major problems in the earth sciences are so complex that many specialists usually collaborate to solve them.
5. One of the major problems of the earth sciences is the comprehension of processes that are now active in shaping the earth. Another problem is unravelling the history of the earth during the last 4.6 billion years.
6. The basic law of the earth sciences is that "the present is the key to the past." Physical laws, apparently, are the same as they were when the earth was born; but it is possible that the proportions of various physical measurements have changed by small but significant amounts.
7. Physical, chemical, and biological phenomena produce distinctive rocks, minerals, and fossils, which may be preserved after the phenomena cease. The geologist can reconstruct the history of the transient phenomena from the material that is preserved.
8. The dynamic processes of the earth are divisible into the interrelated water, sedimentary, tectonic, and igneous–metamorphic cycles. All materials on the surface of the earth and in the crust are ultimately involved in these cycles, although some may be inactive for a billion years or more.

Discussion Questions

1. What are some of the major problems of the earth sciences?
2. The text mentions environmental changes that occur quickly and others that take very long times. Can you think of some that occur during decades or centuries and, thus, have an intermediate duration?
3. What evidence supports the hypothesis that physical laws do not change with time?
4. How can geologists reconstruct the history of the earth?
5. The water and sedimentary cycles are interrelated in several ways. For example, moving river water transports sediment. Can you identify interactions between the two members of the following pairs of cycles: sedimentary and tectonic, sedimentary and igneous, igneous and tectonic, water and igneous?
6. Why has a broad concern with the earth as a limited environment developed only recently?
7. Why do people in the San Francisco area live in houses built in earthquake fault zones?

References

Adams, F. D., 1954. *The Birth and Development of the Geological Sciences.* New York: Dover. (Reprint of 1938 edition.) [Concerned largely with early history.]

Albritton, C. C., Jr., ed., 1963. *Fabric of Geology.* San Francisco: Freeman, Cooper. [Essays on the theory, philosophy, and history of geological thinking.]

Bullard, E., 1969. The origin of the oceans. *Sci. Amer.* 221(3):66–75. (Available as *Sci. Amer.* Offprint 880.) [A lively account of the origin of the basins—not the water.]

Cluff, L. S., 1968. Urban development within the San Andreas fault system. *In* W. R. Dickinson and A. Grantz., *Proceedings of Conference on Geologic Problems of San Andreas Fault System* (Stanford Univ. Publ. Geol. Sci., v. 11), pp. 55–69. Stanford, Calif.: School of Earth Sciences, Stanford University.

Committee on Resources and Man, National Academy of Sciences—National Research Council, 1969. *Resources and Man.* San Francisco: W. H. Freeman and Company. [A distinguished group of scientists makes realistic estimates of available resources.]

Darwin, C., 1860. *The Origin of Species.* London: John Murray. (Reprinted in 1958 by New American Library, New York.) [The most influential scientific book ever written.]

Deevey, E. S., Jr., 1960. The human population. *Sci. Amer.* 203(3):194–204. (Available as *Sci. Amer.* Offprint 608.)

Dietz, R. S., and J. C. Holden, 1970. The breakup of Pangaea. *Sci. Amer.* 223 (4): 30–41. (Available as *Sci. Amer.* Offprint 892.) [The broad sweep of continental drift during the last 200 million years.]

Hoyle, F., and J. V. Narlikar, 1971. On the nature of mass. *Nature* 233:41–44. [Arguments for a small change in the gravitational constant.]

Hunt, C. B., 1972. *Geology of Soils: Their Evolution, Classification, and Uses.* San Francisco: W. H. Freeman and Company.

Krauskopf, K., and A. Beiser, 1966. *Fundamentals of Physical Science.* New York: McGraw-Hill.

Lawson, A. C., chairman, 1908. *The California Earthquake of April 18, 1906* (Report of the State Earthquake Investigation Commission; Carnegie Inst. Wash. Publ. 87). Washington, D.C.: Carnegie Institution of Washington.

Lyell, C., 1835. *Principles of Geology.* London: John Murray. [The best textbook on geology but regrettably out of date.]

Menard, H. W., 1969. The deep-ocean floor. *Sci. Amer.* 221(3):126–142. (Available as *Sci. Amer.* Offprint 883.)

——, 1971. *Science: Growth and Change.* Cambridge, Mass.: Harvard University Press. [The effects of different growth rates on the careers of scientists.]

Moorehead, A., 1969. *Darwin and the Beagle.* London: Hamish Hamilton. [A fascinating account of the scientific awakening of Charles Darwin on the round-the-world voyage of H.M.S. *Beagle*.]

Price, D. J. d. S., 1963. *Little Science, Big Science.* New York: Columbia University Press. [A charming scientific analysis of the structure of the scientific enterprise.]

Simpson, G. G., 1951. *Horses.* New York: Oxford University Press.

Zittel, K. A. von, 1901. *History of Geology and Paleontology.* London: Walter Scott. (Reprinted in 1962 by Hafner, New York.) [The classic book on the subject.]

PLANET EARTH

JOHN S. SHELTON

2

ORIGIN AND EARLY HISTORY

If God had consulted me before embarking on the Creation,
I would have suggested something simpler.

Alfonso of Castile (thirteenth century)

Here about the beach I wander, nourishing a youth sublime
With the fairy tales of science, and the long result of time.

Tennyson, LOCKSLEY HALL

Origin

The rocks of the earth, moon, and meteoroids contain radioactive elements and their decay products in proportions that indicate that all these bodies assumed their present forms about 4600 million years ago. Inasmuch as all are in orbit around the sun, it seems likely that they all formed in the solar system. Thus, the birth certificate of the earth gives the time and the place, and the fact that the birth was multiple, but the remainder of the record is covered with obscuring dust. "None of us was there at the time," remarked the Nobel laureate Harold Urey, when discussing the matter in 1952.

In addition to the rocks, we have the astronomical characteristics of the solar system—the orbits, tilts, spins, and densities of the bodies—from which to deduce the conditions of their birth. Here, the problem is not that facts are too few but, rather, that they are too plentiful. Moreover, it is not certain which of the characteristics are the original ones and which have been changed since birth by the interactions of the bodies.

No one has yet imagined a history of the earth that takes into account all the known facts of the solar system with mathematical rigor. The early scientists can be excused on the grounds that many of the facts were unknown to them. The later scientists have avoided the problem in the only manner possible—namely, by regarding some of the facts as crucial and disregarding the others.

According to most modern hypotheses, the solar system began as a cold, dark dust cloud of the sort observable today in interstellar space. This was driven inward upon itself by the unimaginably gentle pressure of starlight, until it became dense enough to collapse under the attraction of its own gravity. Under pressure, the core heated up to become the sun, but fragments remained behind and condensed into the planets.

In 1970, Hannes Alfvén and Gustaf Arrhenius introduced many novel and attractive insights into the dust-cloud hypothesis, including evidence that the planets formed relatively quickly. This hypothesis of the origin of the earth is only one of several possibilities, and it is still too new to have found general acceptance. It is premature to believe that it will ever by accepted without some modifications based on critical evaluations, and the reader should maintain a scientific aloofness toward it, pending criticism. However, it seems desirable to present a modern hypothesis about the origin of the earth in order to set the scene for the relation of geological history.

Alfvén and Arrhenius regard the origin of the sun as a puzzle that need not be solved in order to decipher the origin of the earth. In its final stage of development, the sun was enveloped by hot ionized gas (plasma) that cooled

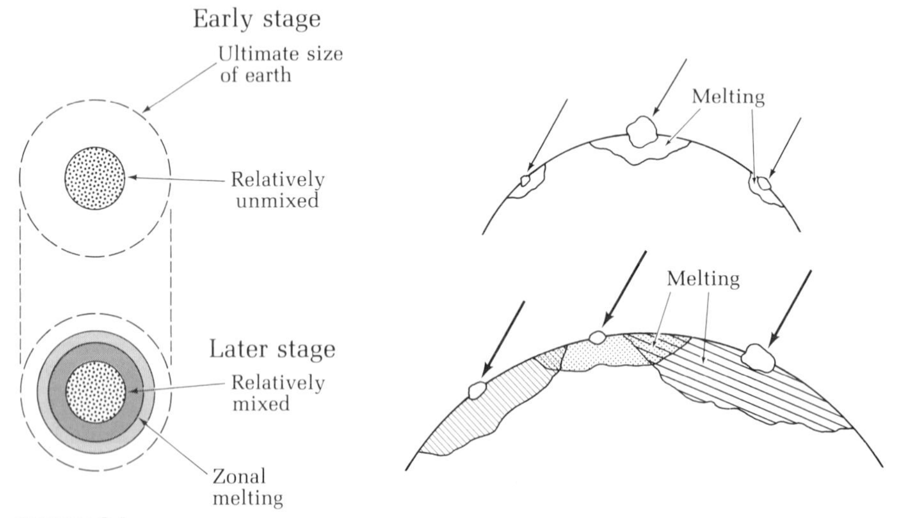

FIGURE 2.1

Accretion of the earth according to the Alfvén–Arrhenius hypothesis. In the early stage, the gravitational attraction is small and little melting occurs. In the later stage, the attraction is greater and impacts cause extensive melting that overlaps to form a zone that migrates outward as accumulation continues.

and condensed into grains. It was in the plasma-grain stages that the solar system mostly differentiated chemically. The former existence of grains and small aggregates floating in orbit in space has been demonstrated elegantly by the study of small grainlike structures, called chondrules, in meteorites. Devendra Lal and R. S. Rajan found, in 1969, that these chondrules have been exposed on all sides to equally intense radiation damage. This must have occurred in space before they were aggregated, because the interiors of meteroids are protected from most radiation.

The grains collected into embryonic planets without shattering upon impact because the small ones tended to melt partially and to stick to the larger ones. Eventually, most of the mass of planets aggregated from embryos. This took only a few million years—which is quick, indeed, by astronomical standards.

The energy of meteoric impact on large embryonic planets generates enough heat to melt a sizable volume of the surrounding rock. The floods of fluid rock that filled the dark maria or "seas" of the moon probably formed in this way. This heating influences the state and temperature of planetary interiors. The earth has a solid inner core, a liquid outer core extending to about half the radius, and a solid mantle reaching almost to the surface (see Figure 3.4). This structure is explained as follows:

1. In the early stage of accumulation, the embryo earth had only a small gravitational attraction, all impacts were at low velocity, and what little rock was melted quickly cooled and solidified. Thus, the interior is a cool solid.

2. Surficial melting became common as accumulation continued, because the heating due to impact increased.

3. If the earth accumulated in two million years, the maximum heating occurred when the radius was roughly 20% to 50% of the present value. This corresponds to the liquid region of the present core. The mantle above is solid and relatively cool, according to this hypothesis. However, it is important to note (a) that most parts of the mantle were melted by superimposed impacts more than once; (b) that they solidified and cooled between impacts; and (c) that, eventually, parts of them were buried too deep for further melting. The low-density components were swept upward through the mantle as it accumulated. Perhaps we live on some of them—namely, the continents.

Matter from Space

Although most of the earth may have been amassed in only a few million years, the addition of solid matter from space is still continuing. Relatively small solid particles in space are collectively called meteoroids. Some of them enter the earth's atmosphere and become meteors or "shooting stars," which are particles that burn up in the air and, thus, are merely seen as streaks of light. If a meteoroid is bigger than 20 kilograms, however, it may be able to punch through the atmosphere, ablating, or shedding droplets of itself, much as the nose cone of a spaceship does at re-entry of the atmosphere. Such meteoroids hit the earth, blast holes in land and sea, and, not as rarely as might be thought, kill people and damage buildings. The fragments of meteoroids that are found on the surface of the earth, which are called meteorites, are of two types: irons, which are nickel-iron alloys with distinctive internal structures, and stones, which are rocks with unusual silicate mineralogy and structures. Individual meteorite fragments show the surface markings of ablation.

Meteoroids are an obvious danger to astronauts; early in the space program, the distribution of micrometeoroids in space around the earth was measured by the frequency of impacts upon spacecraft. It appears that cosmic dust is now raining on the earth at the rate of roughly a million tons per year. Much of this vaporizes in the atmosphere, but then cools and rains down to the surface. The distinctive nickel-iron composition can be identified in dust in the snow layers of the Antarctic, and analysis confirms that the rate of accumulation may be roughly a million tons per year (and suggests that it may be even as much as three times that amount). This may seem a large addition to the earth, but, even if it continued at the same rate for 4000 million years, it would increase the mass of the earth by only 0.0001%.

Most cosmic material vanishes like a needle in a haystack in areas of rapid deposition of sediment, but it can be detected in the sediments of the deep-sea floor because they accumulate very slowly. Ablation yields microscopic spheres whose concentration in sediment is an indication of their rate of accumulation. Unfortunately, similar spheres are produced by volcanoes and by welding, and very painstaking work is required to make a separation of the various kinds of spheres. This puts a limitation on measurements, but it appears that a few thousand tons of cosmic spheres reach the sea floor each year. This is definitely not cosmic dust that has slipped unscathed through the atmosphere. The distribution of particle sizes indicates that they are fragments of large bodies that exploded in their descent.

A mid-air explosion is also indicated by analysis of a fall of meteorites. At Holbrook, Arizona, on July 19, 1912, 16,000 stones fell in an area 3 by 5 kilometers. The fall was accompanied by an explosion, a fireball, and a dust cloud. The meteorites weighed up to 20 kilograms, but half of them weighed less than a gram. It is evident from the small area of the fall that a sizable mass exploded low in the atmosphere. It was, of course, larger than the fragments on the ground, but it was not large enough to reach the ground intact like a spaceship.

Meteoroids of the next larger size are capable of penetrating the whole atmosphere while preserving a significant fraction of their mass intact, but they are unable to retain enough velocity to blast a crater in the earth and vaporize themselves in the encounter. Great irons, weighing 20–60 tons, have been found on the surface in southern Africa, in Greenland, and in Mexico; in fact, irons weighing more than a ton are quite numerous. No stones weighing a ton are known. Presumably, they break easily, and also quickly weather to resemble ordinary rocks, if not found soon after they fall.

Larger masses can form craters, or astroblemes, upon impact. These are still being discovered in remote places, and the thirteen known recent craters, presumably, are but a fraction of the whole. The Barringer Crater in Arizona is typical, and can be seen on many commercial flight routes. It is almost circular and has a diameter of 1200 meters and a depth of 150 meters. The lip is slightly raised, and the sedimentary layers in which the crater occurs have been tilted upward. Shattered rock 300 meters deep fills the bottom of the crater, and rocks and a small number of meteorite fragments have been splashed to the surrounding areas. The crater was created by a high-velocity impact about 25,000 years ago, and the shock waves generated by impact created extremely high pressures in the surrounding rocks.

FIGURE 2.2

Barringer Crater, Arizona: an astrobleme, 1200 meters in diameter and 150 meters deep, created 25,000 years ago. Coarse ejected debris can be seen on the slightly elevated lip. [From Shelton, *Geology Illustrated*, W. H. Freeman and Company, copyright © 1966.]

When minerals are subjected to sufficiently high pressures in a laboratory, the normal low-density phases are changed into denser ones with more closely packed atoms. By this means, any ordinary, low-density, carbon-rich material can be converted to a hard, dense diamond. Either coal or peanut butter will do. The pressure required depends on the temperature and the presence of possible impurities, but it is roughly equivalent to the pressure at a depth in the earth of perhaps 60–80 kilometers, Silica, either in the form of glass or in the form of crystalline quartz, also has a high-pressure phase, which is known as coesite (after the American scientist Loring Coes, Jr., who first made it in the laboratory in 1953). A phase of silica produced at even higher pressures is stishovite (after the Russian scientist S. M. Stishov, who, with his colleague S. V. Popova, first made it in 1961).

These minerals—diamond, coesite, and stishovite—have all been found around Barringer Crater, which demonstrates that the area was once subjected to a pressure that could occur at the surface only as the result of a powerful explosion. Another indication of an explosion is the presence of shatter cones, which are conical fragments of rock with striations that radiate from the apex. These have been produced experimentally only by hypervelocity impacts.

GREATER EVENTS

In the 1950s, the American geologist R. S. Dietz began to demonstrate that several large geologic structures on the surface of the earth are probably the result of meteoritic or asteroidal impacts. He called them astroblemes. His early work has now found general acceptance, both because of the discovery of high-pressure mineral phases at some of the sites and because of the recognition of meteorite craters covering the faces of the moon and Mars.

One of the supposed great astroblemes is the Vredefort ring near Pretoria in South Africa. More that 250 million years ago, a large meteorite-asteroid, perhaps one kilometer in diameter, may have struck and created a crater so great that horizontal strata 15 kilometers deep "peeled back like a flower spreading its petals to the sun," in the words Dietz used in 1961. The present structure of the Vredefort ring is the erosionally exposed root of the original. It is a large fragment of a circle with a diameter of 210 kilometers.

Another astrobleme is the Sudbury structure in Canada, which contains much of the minable nickel of the western world. It was created about 1700 million years ago and has subsequently been modified and eroded. The original impact crater was roughly 50 kilometers in diameter and 3 kilometers deep, with much deeper fracturing of rock. The ore deposits give an indication that the meteorite-asteroid may possibly have been 4 kilometers in diameter.

Two-thirds of the earth is covered by water, in which meteoritic impacts leave but a transient record. Even on land, much of the older rock is buried, which could conceivably mean that many such features as the Barringer Crater are concealed under soil or younger rock. Indeed, most known astroblemes have been found in desert or arctic regions, where they are not obscured by vegetation or soils. Consequently, the frequency of crater-forming events during geologic time is a matter of conjecture.

The local effect of a cratering impact is total annihilation. The regional effect of an impact like that of the Barringer event would be relatively minor on land—little more than a small earthquake and a thunderous noise. At sea, the event would be more of a regional disaster. If an equivalent of the Barringer meteorite hit the ocean, an enormous wave would be generated that would become even higher in shallow water, and the damage near the shoreline would be overwhelming. Numerous events of this sort almost certainly have occurred during geologic time, and the records of some have been tentatively identified in rocks almost 400 million years old.

TEKTITES

Sprinkled on the surface of the earth in many regions are countless thousands of shiny, black, glassy stones called tektites. They have highly distinctive shapes—such as teardrops, disks, dumbbells, and rods—and they look like bits of chilled liquid rock. There seems to be little doubt that they are droplets from a splash of molten rock that have been shaped by cooling while moving almost horizontally for long distances through the atmosphere. Because these droplets are too localized for the splashes to have originated beyond the moon, and because the moon's surface has the wrong composition, the splashes must by elimination, have originated on the earth.

The best preserved record of a tektite event shows that tektites were strewn over the Indian Ocean, western Australia, Indonesia, and southeast Asia roughly 700,000 years ago. Other tektite-generating events occurred 800,000 years ago in Ghana, 15 million years ago in Czechoslovakia, and 35 million years ago in the eastern United States and Libya. But what causes tektites to be formed and scattered? A possible cause is an explosion just above the ground, such as the one that devastated the Tungus region of Siberia on June 30, 1908. Russian scientists I. T. Zotkin and M. A. Cikulin examined 40,000 trees downed in a semicircular area of roughly 40 by 60 kilometers and found them radiating outward from a point near the edge of the semicircle. From model experiments, they estimated that the trees were leveled by an enormous explosion occurring in a meteoroid moving almost horizontally some distance above the ground. There was no crater, but the American oceanographer B. P. Glass has identified glass spherules at the site that are similar to tektites. Perhaps an explosion nearer the ground or over bare sand could produce a true tektite-generating explosion instead of just leveling the trees.

Tektites older than 35 million years are unknown. Presumably, they exist but are difficult to find because they have been buried. Tektite-generating explosions may have occurred from time to time through the history of the earth and, presumably, will occur again.

BOX 2.1 GRAVITY AND GEOLOGY

Newton and Gravity

Contemporaries of Sir Isaac Newton (1642–1727) regarded themselves as fortunate to live in the same age with him, and, if anything, his reputation has improved with time. Wordsworth wrote of Newton's "mind for ever Voyaging through strange seas of Thought, alone." According to Lord Byron, "Sir Isaac Newton could disclose / Through the then unpaved stars the turnpike road." Pope wrote an epitaph intended for his tomb in Westminster Abbey: "Nature and Nature's Laws lay hid in Night; / God said, *Let Newton be!* and all was Light."

Newton, while still in his twenties, isolated himself in the country for a long period to avoid the bubonic plague, which was then raging in the cities. In a burst of unmatched genius, he synthesized and extended the work of Kepler and Galileo by means of the calculus – which he invented – and arrived at the laws of motion that bear his name and the universal law of gravitation, which explains the motions of celestial bodies. The laws, originally written in Latin, may be expressed as follows:

1. If a body in linear motion is not acted upon by any external force, its linear momentum is constant.
2. The acceleration that a force can give an object is in the direction of the applied force and is directly proportional to the force and inversely proportional to the mass of the object.
3. For every force acting on a body, there is an equal and opposite force exerted by the body.

The first law, which is similar to one of Galileo's, is a qualitative definition of force. Force is something that changes the momentum of a body. Momentum (M) is the quantity of motion of a body, a quantity that Newton found useful to identify in his analysis. It is defined as equal to the mass (m) of the body times its velocity (v).

$$M = mv$$

Mass is the quantity of matter in the body, and velocity is the rate of motion.

These laws, conceived only by genius, can now be visualized with ease by anyone who has watched the television coverage of the U.S. manned space program or given any thought to how rockets work.

The first law says that a rocket in space continues on a straight line, unless some force, such as firing the rocket or the attraction of the moon, acts upon it.

The second law says that the force exerted by the ignited rocket fuel is in the direction of the exhaust nozzles, and that, other things being equal, the resulting acceleration is proportional to the amount of rocket fuel fired and inversely proportional to the mass of the rocket.

The third law says that the reason that the rocket moves forward in the vacuum of space is that the gases from the burning rocket fuel move backward and the two forces balance.

With his laws, Newton could visualize the solar system in a way not previously possible. Consider the earth circling the sun: The first law tells us that it would move in a straight line if no outside force were acting on it. Its tendency to do so is balanced by the attraction of gravity, which holds it in orbit around the sun as a string might hold a ball that is twirled in orbit around a child's hand. Newton found in Kepler's laws all the information he needed to exploit this picture of balancing forces.

If the planets had circular orbits and constant orbital speeds, he would have been in difficulty: gravitational attraction depends both on mass and on distance, and, with unknown planetary masses, the effects of distance would be undeterminable. Kepler's first law, however, says

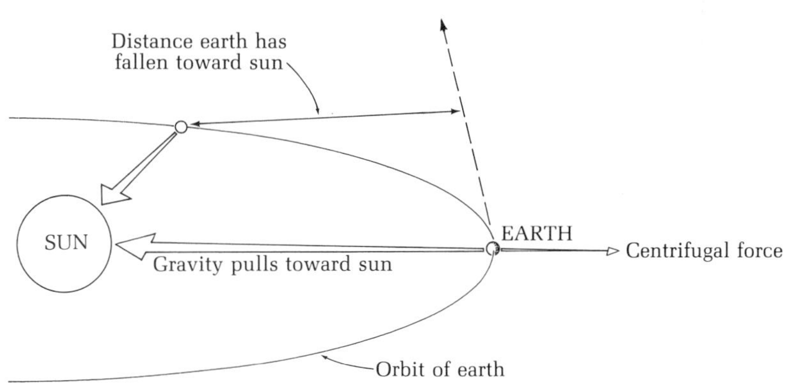

Centripetal motion of earth caused by gravitational attraction between sun and earth. Without this attraction, it would follow the straight path indicated by the dashed line.

that the orbits are ellipses, and his second law says that the orbital speed varies according to the distance from the sun. With the calculus, Newton could analyze the different parts of the orbit of each planet and thereby determine the distance effect for a constant mass. He established that the acceleration of gravity (g) is independent of mass, and for bodies attracted by the mass of the sun, it depends on the inverse square of the distance (r) measured from center to center of more or less spherical objects like planets and the moon.

$$g = \frac{\text{constant}}{r^2}$$

Newton was able to show that this relation applies to falling objects on the earth as well as to celestial objects, and he generalized that all bodies in the universe attract all others. Consider the moon: With a knowledge of its orbit and speed of revolution, Newton could calculate the acceleration of the moon caused by gravity. He compared this with measurements by Galileo of the acceleration of a falling object on the surface of the earth and determined that the accelerations were proportional to the square of the distance of the object and the moon from the center of the earth.

From the acceleration of gravity and his second law, Newton obtained his universal law of gravitation: the force of gravitational attraction (g) equals a universal constant (G) multiplied by the product of the masses (m_1m_2) of two objects divided by the square of the distance (r) between them.

$$g = G\,\frac{m_1m_2}{r^2}$$

Mass of the Earth

The universal gravitational constant was not measured in Newton's lifetime, so all of his work was comparative. Henry Cavendish made the first accurate measurement in 1798 by bringing large metal spheres near small ones attached to a balanced cross bar suspended from a wire. He measured the twisting of the wire from a distance by looking through a telescope. Once the attraction between two masses had been determined, the law of gravitation permitted the calculation of the masses of all objects with known attractive forces or with known orbits. Knowing the acceleration of gravity at the surface of the earth and the mass of an object, Cavendish could calculate the mass of the earth. The size and shape of the earth had been determined in order to test Newton's laws, and, with this information, Cavendish obtained a density of 5.448 grams per cubic centimeter, which is only slightly less than modern measurements.

Given the mass of the earth, the mass of the moon could be calculated; likewise, the mass of the sun around which the earth revolves. Given the mass of the sun, the masses of the planets were known; and given the masses of the planets, those of the moons of Mars and Jupiter could be determined. A very powerful law and a very remarkable measurement!

Gravitational Changes and Geologic Effects

Many properties of the earth are determined by its astronomical characteristics. For example, the spin axis varies in orientation, and the long-term climate varies with it (Chapters 3 and 11). But, if the distance to the sun were very different, the seas would boil or freeze. The persistence of life on earth shows that this has not happened; however, it does not rule out the possibility that there have been smaller variations. Have they occurred?

To answer this question, we need merely think of the gravitational equation. If the mass of the earth or sun changes, then the attraction and distance between them also change. The mass of the sun constantly decreases because it radiates energy that it creates by the conversion of mass in thermonuclear reactions. We can compute the mass lost, assuming present radiation, by Albert Einstein's famous equation that energy equals mass times the square of the velocity of light ($E = mc^2$). Since the earth formed, the sun has converted ten masses equivalent to the earth to energy. During this period, the mass of the earth has varied only slightly, but it has increased somewhat by meteorite impact and decreased by a smaller amount through the loss of hydrogen to space.

The constant called "big G" also appears in the gravitational equation. It seems strange to even consider the possibility of variations in a "constant" that applies to all matter in the universe. But, as discussed in Chapter 1, we only assume for simplicity that it is universal at present and has been constant in the past.

The value of G affects not only the earth–sun relation but the distance from the center of the earth to the surface. Thus, if G is decreasing, as calculated by Hoyle and Narlikar in 1971, the radius of the earth may be expanding at the rate of 10 kilometers every 100 million years. This is a small effect, but it could be important if long persistent.

In sum, when we consider geologic history, we cannot assume that the earth is an island in the sky. It is affected by many astronomical forces, and all of them have, or may have, varied—at least by small amounts—during geologic time.

Earth–Moon System

TIDES

The earth has a unique triad of characteristics for a planet: its surface is largely covered with water; its period of rotation is long, compared to that of many other planets and asteroids; and the mass of its moon is an exceptionally large fraction of its own. These are all related, because the large mass of the moon exerts a large attraction on the water, which forms moving tidal bulges that slow the rotation of the earth by friction. Slowing is accompanied by dynamic changes in the earth–moon system that suggest that the moon and earth were once very close together. Thus, to attempt to understand the probable origin of the system, we should begin with modern tides.

It is sufficient for calculating orbits to assume that the gravitationally attracting masses of celestial bodies are located at points in their centers. However, the different parts of a body may be at significantly different distances from another body and, thereby, may experience appreciably different gravitational attractions. This is the cause of the tides visible to anyone who has spent a day by the seashore.

If the moon and earth were immobile, each would produce two fixed bulges in the other. Consider only the dry, rocky moon: because the region nearest the earth is attracted more than the center, it is pulled into a bulge on a line toward the earth. The opposite side of the moon, of course, is at a greater distance from the earth than the center. Thus, the center is more attracted, and it pulls toward the earth and leaves a bulge behind. The moon revolves around the earth, but it rotates slowly on its axis and keeps almost the same face toward the earth. Consequently, the almost spherical surface is distorted by two fixed tidal bulges in the solid rock, one in the center of the face toward the earth and one diametrically opposed.

Similar tidal bulges occur in the solid earth, but they move as the earth rotates. They are small, so none of us notice them, but they can be detected by sensitive instruments, and they distort the brittle surface rocks twice a day. Tides also occur in the air, but they, too, can be detected only with instruments.

Even primitive men realized that the obvious tides of ocean are associated with the moon. The relation of the tides to the distant but massive sun is not so obvious, because it causes smaller tidal bulges, but the observed tides are the resultant of the sun's attraction as well. The details relating the gravitational attraction of the sun and moon to the tides are complex. Modern tide theory suggests that the periodic nature of the small forces of the sun and moon cause the tides, just as small pushes, applied at the proper times to a child in a swing, can cause significant motion. This is called resonance. The tides, then, are the oceans sloshing back and forth in response to the gentle periodic forces of gravity from the sun and moon. About twice a month, the earth, moon, and sun are in a straight line, and their attractions reinforce to produce the high spring tides. These correspond to the new and full moons. The greatest spring tides occur at the new moon, when the maximum attractions are from the same direction. At the first and last quarters, the moon is perpendicular to the line from the earth to the sun; this causes the neap tides, which are correspondingly low.

As the earth spins, friction between the ocean and the rock below tends to displace the oceanic tidal bulges. If there were no continents and the ocean depths were uniform, the tidal bulges would be displaced symmetrically and the offset mass would tend to accelerate the moon in its orbit and cause it to move away from the earth. However, the real tidal bulges move in a very complicated way through the real ocean basins, and the earth–moon interaction is by no means so simple. Even so, the bottom friction between the sea floor and the complex tidal bulge certainly retards the spin of the earth. This causes the moon to move away from the earth in accordance with the laws of planetary motion. The moon that shone on the dinosaur appeared bigger than the moon we see today.

It is known that the day was once much shorter than it is now. The change in the length of the day can be determined from astronomical and historical data. Astronomers can calculate exactly where and when eclipses should have occurred, assuming that the length of the day has been constant. Eclipses were recorded by the early civilizations of the Middle East, and neither the times nor the locations agree with the calculations. They can be brought into agreement by making the simple assumption that the rotation of the earth has slowed enough in 100,000 years to make the day one second longer. The microseconds per day add up to hours through the centuries.

Tracing the length of the day in prehistoric times is less reliable. It appears, however, that certain ancient coral polyps added new ridges of material to their skeletons each day, and that the ridges varied in size according to an annual cycle. On this assumption, the American paleontologist John Wells has determined that, about 370 million years ago, the year had 400 days, each about 22 hours long. This suggests that tidal friction has had a relatively constant effect in slowing the spin of the earth for several hundred million years.

What of even earlier time? If we simply extrapolate backward for some billions of years, we find that the moon would have been a part of the earth—which is most unlikely. Thus, from the tides moving a few grains of sand at the seashore, we inevitably arrive at the origin of the moon.

CAPTURE OF THE MOON

Sir George Darwin, son of Charles Darwin, proposed that the moon was once, in fact, a part of the earth, and that the two were pulled apart by tidal forces. The moon was born from the surface of the older earth, leaving behind a vast circular scar that is now the Pacific Basin. This hypothesis has the virtue that it can produce the moon at whatever time is indicated by extrapolated measurements of the length of the day. It is now generally discredited for many reasons, however. The moon, now sampled by astronauts, is at least as old as its supposed parent; and the shape of the Pacific Basin has not always been circular.

Where was the moon, if it was neither born of the earth nor moving independently in its present orbit, which is in the same direction (prograde) as the earth's orbit around the sun? The only remaining possibility is that it was

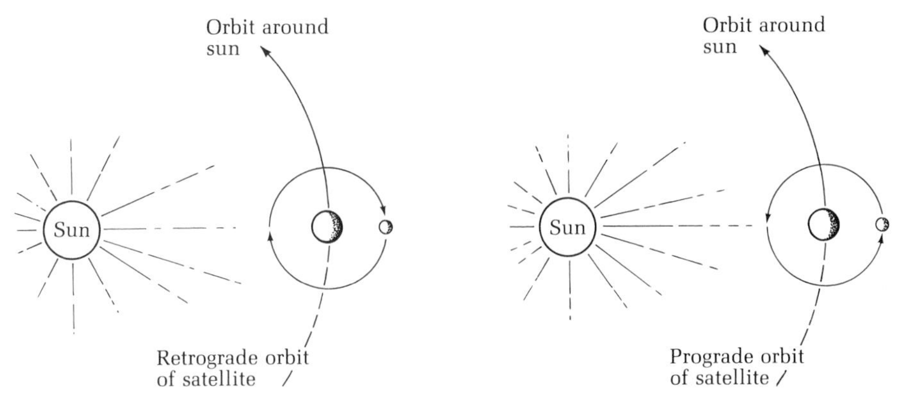

FIGURE 2.3
Satellite orbits.

captured by the earth some time after it formed. How it came to be in a position in which it could be captured is unknown, but moon captures by other planets are not rare. Capture is the only way to explain the orbits of the four moons of Jupiter and the one of Saturn whose motions are opposite in direction (retrograde) to the orbits of their respective planets. Thus, it is reasonable to assume that the moon was captured in a retrograde orbit by some process that is normal to the solar system.

The earth was then very different from what it is now. It was spinning faster; the equatorial bulge produced by the spinning was much larger and the polar regions were more flattened. The Alfvén–Arrhenius hypothesis proposes that the ocean and atmosphere already existed at the time of the moon capture. If so, there was ocean tidal friction just as now. However, the moon was then in a retrograde orbit, and the same laws of planetary motion that cause it to move away from the earth in its present prograde orbit then caused it to move toward the earth.

Thus, a simple extrapolation of the dynamic response of the moon to tidal friction while in retrograde orbit is that it would be driven into the earth. We have returned to the one-body hypothesis of Sir George Darwin; but, having already rejected that hypothesis, we require now a mechanism for holding the moon away from the earth as its orbit changes from retrograde to prograde. Two hypotheses have been offered. H. Gerstenkorn proposed that the orbit slowly decreased until tidal forces began to break up the moon. For 100 to 1000 years the moon was shedding fragments into space that may have become meteoroids. During this period, the earth suffered catastrophic deformation and heating caused by tidal friction. If the oceans existed, they boiled. Earth tides were several kilometers high. After thus using up a large amount of energy, the moon began its prograde motion and tidal friction began to lengthen the day and move the moon away from the earth.

The rocks deposited during at least the last 3600 million years do not appear to contain the record of any such dramatic events. Moreover, stromatolites, which are large calcareous structures produced in shallow seas by living algae, have a height related to the range of tidal movement where they grow. Geologist Preston Cloud finds that the maximum identifiable tidal range occurred about 500–1100 million years ago, and that it was only about 6 meters. Although this observation eliminates the possibility that there was a catastrophe, it requires the moon to have been in orbit by that time, because the tidal range is too large to be accounted for by the sun alone. Indeed, the wide-spread occurrence of stromatolites suggests that the tides were higher than the present average during the whole period from 2000 million years to 500 million years ago. The record of more ancient times is even more questionable, but the existence of ancient sediments of a type thought to be produced only in the intertidal zone suggest that the moon was already in orbit 3200 million years ago.

We require a mode of capturing the moon that did not bring it too close to the earth, but stored it much closer than now, at least for the period from 2000 million to about 500 million years ago. A hypothesis that meets these requirements was advanced by Alfvén and Arrhenius in 1969. They proposed that the moon was captured in a rectrograde orbit more or less according to the hypothesis of Gerstenkorn. It moved ever closer to the earth because of tidal friction; but it did not come close enough to be destroyed, because it locked into a synchronous orbit (like one of the artificial satellites we put in a fixed position in space to transmit television programs across the oceans). This is a phenomenon observed elsewhere in the solar system that occurs whenever a satellite orbits at the same rate that the central body turns—that is, the month and the day are equal. This condition might have existed when the moon was at its closest approach, only about 6–8 times the radius of the earth away. If the earth had a mountain range from pole to pole, at that time, it would have exerted a force on the moon with a component parallel to the lunar orbit. Visualize the moon in a fixed position over the mountain range: if the earth

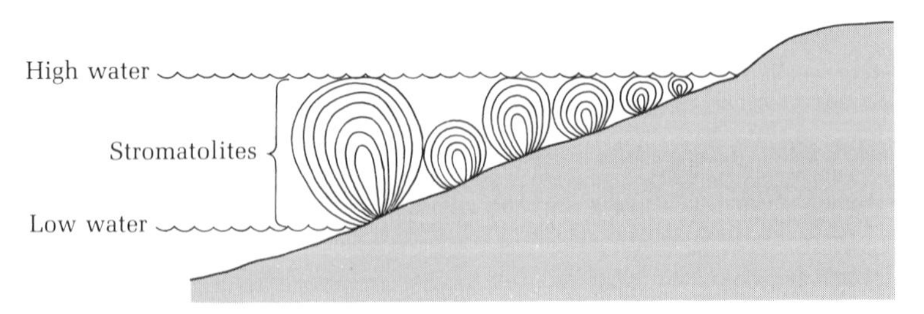

FIGURE 2.4
Stromatolites (structures produced by algae) are normally restricted to the intertidal zone and, thus, indicate the range of the tide.

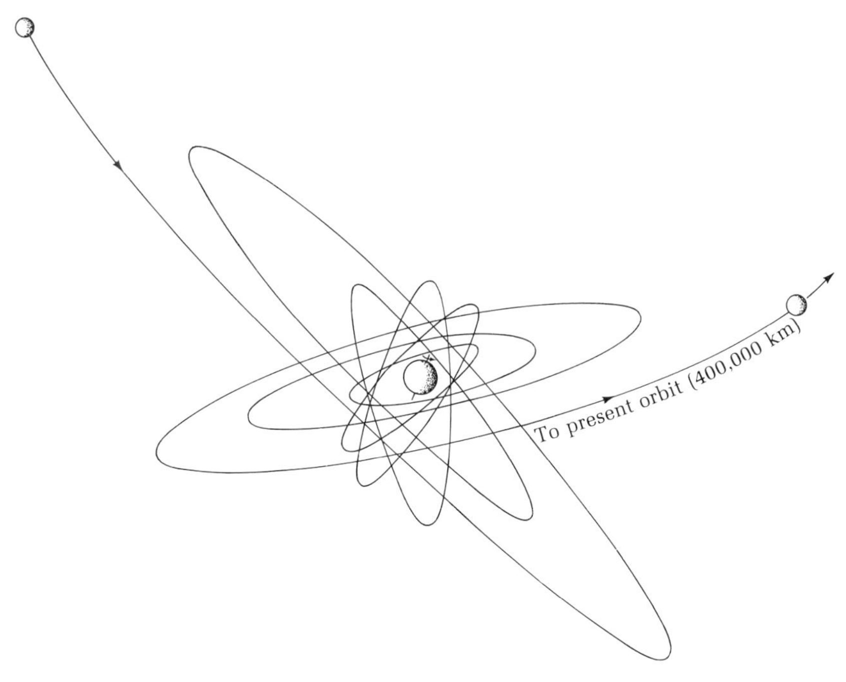

FIGURE 2.5
The capture of the moon in a retrograde orbit that degrades by tidal friction until it is locked in a synchronous orbit near the earth. Tidal friction then causes it to move toward its present orbit.

spins just a trifle slower, the attraction of the mountain range tends to slow the moon as well, and, in this way, they are locked together.

To use a technological analogy, the moon came in from space and entered a parking orbit near the earth perhaps 2000–3000 million years ago. This orbit changed during a very long period. An artificial satellite slows because of atmospheric drag, so it makes a burning plunge into the earth. Tidal friction caused the moon to go the opposite way when it left the synchronous, parking orbit about 500 million years ago. The day and the month, once brief and equal, gradually lengthened to their present much greater and quite unequal values.

The most novel feature of the Alfvén–Arrhenius hypothesis is that it includes a prolonged and intense interaction between earth and moon that is something short of an obliterating catastrophe. The higher tides of that period are a logical consequence of the interaction. Likewise, the melting history of rocks on the moon is explained. The moon has not melted completely in more than 4000 million years, but many younger local melting events have occurred. This history is readily explained by the hypothesis of a moon capture in a parking orbit. The moon may have melted when it formed, but not when it was captured.

Geologic Time

We turn now to the endless span of time. In the following chapters, we shall discuss events that took place "1300 million years ago," or "when dinosaurs roamed the earth," or "early in Cenozoic time." Without an understanding of the means of measuring geologic time and the assumptions and simplifications inherent in the methods, such statements are meaningless, and geology is hardly a science at all. It is a mere jumble of events, like a collection of newspaper clippings without any dates.

If we are to comprehend time in the past, we should first understand it in the present. A little consideration will show that time takes the measure of all the world only in a relative way. Both the starting times and intervals in any system are so arbitrary that different religions and different countries use different calendars. It is equally clear that division of time into hours and shorter intervals is arbitrary. The larger time units that we use, such as days and years, are measures of astronomical events and, thus, are not arbitrary for earthlings—but they might seem so if we lived on another planet. Mars, for example, has a year, but it does not have the same number of days as the earth year. The moon has the same year as the earth, if measured in earth days, but the moon rotates only once an earth month, and, in our sense, it has no natural interval comparable in time to our day.

Moreover, we have found that both the day and the month have been much briefer during the evolution of the earth–moon system than they are now. The year may also have varied, but probably not much, so it is a suitable unit for measuring geologic time. But when does time start, and how do we identify a given year or assign a date to a given event?

It might appear that we should measure the history of the earth forward from the time that the earth was formed—but we cannot, because that time is unknown. By elimination, therefore, we measure geologic time backward from the present, and we designate particular events in the past as having occurred so many years B.P. (Before the present, just as B.C. means Before the birth of Christ).

The present moves ever forward, which may make the B.P. system seem absurd; but, in fact, no appreciable error is introduced, because the geologic "clocks" or "calendars" are not accurate to a year, or even to a century. Some are accurate to only a million years, but, for dating an event 600 million years ago, that accuracy is equivalent to timing an eight-furlong horse race to a tenth of a second.

A few prehistoric events can be dated to an accuracy of a year, because they can be related to annual tree rings, to the alternating layers of dark and light sediment that result from the annual melting cycle of glaciers, or to the records of other annual events. It is appropriate to tie these events into our normal calendar of B.C. or A.D. However, once a gap occurs in the continuous record, the calendars from tree rings and sedimentary layers are referred to B.P. time.

BOX 2.2 PHYSICAL DETERMINATIONS OF AGE

The chemical behavior of an element, which is how we identify it *as* an element, is determined only by the number of protons in the nuclei of its atoms. The atomic weight of an element, however, depends on the number of neutrons as well as on the number of protons; thus, an element may have atoms of several different weights, which are called isotopes. These may or may not be radioactive. Those that are decay naturally to form other isotopes and elements. From the proportions of the decay products, the age of the material that contains them can often be determined. Different decay series give a wide range of ages of effective dating. The following table shows some of the characteristics of radioactive isotopes of importance in dating in the earth sciences.

Highly energetic radiation damages any insulating solid that it encounters, and, in some circumstances, the incidence of damage can also be used to date materials. The radiation damage to moon rocks, earth rocks, and meteorites that is caused by the decay of uranium is theoretically effective for dating materials of any age up to 100 million years. Damage to moon rocks and meteorites caused by cosmic rays is even more useful: the dating method based on such damage is theoretically effective for materials of any age up to 4.5 billion years. Compare these ranges with those for the isotopes in the table.

Origin	*Parent isotope*	*Daughter isotope*	*Half life of parent isotope (yr)*	*Effective dating range (yr)*
Decay of various radioactive elements in earth rocks, moon rocks, and meteorites	Uranium-238	Lead-206	4.51×10^9	10^7–4.5×10^9
	Uranium-235	Lead-207	0.71×10^9	10^7–4.5×10^9
	Thorium-232	Lead-208	14.1×10^9	10^7–4.5×10^9
	Rubidium-87	Strontium-87	47×10^9	10^7–4.5×10^9
	Potassium-40	Argon-40	1.3×10^9	10^4–4.5×10^9
Cosmic-ray bombardment of elements in atmosphere	Nitrogen-14	Carbon-14	5730	50–50,000
	Hydrogen-1	Hydrogen-3	12.3	0–100

Source: After Faul (1966).

Radioactive Clocks

Each radioactive isotope changes into another one, or decays, at a rate that is independent of temperature and pressure within the range of conditions producible in a laboratory. Indeed, the energies within the atomic nucleus are so large and the distances so small that nothing but a nuclear collision can change the rate of decay. Consequently, radioactive decay provides the basis for nearly ideal geologic clocks. The nature of radioactivity is such that half of the original amount of an unstable "parent" isotope decays to "daughter" isotopes in a given time that is known as that isotope's half-life, and half of the remainder decays in the next half-life, and so on. Thus, after two half-lives, one-fourth of the parent isotope remains.

If a pure sample of a radioactive isotope has been isolated in a sealed box that captures all the decay products, and we know the half-life, a measurement of the ratio of daughter isotopes to the remaining parent can be used to calculate the time when the sample was isolated. If the ratio is 3:1 (that is, if only one-fourth of the parent remains), we know that the sample has been isolated for two half-lives. If the half-life is 1000 years, then the sample was isolated 2000 years ago.

Because radioactive isotopes are present in minute amounts almost everywhere in nature, they can be used as radioactive "clocks" to measure geologic ages, provided that no daughter elements were initially present and that the system (box, mineral, cell) was closed—meaning that nothing was added or allowed to escape. The ratios of various stable and radioactive isotopes give clues about the degree to which any system approximates these conditions. Moreover, several different radioactive clocks may operate at once in a system, and, if they give the same age, it is probably correct. The age, however, is the time since the clocks were last set, which, in the case of rocks, may be when they were *last* cooled after reheating, rather than when they *first* solidified.

The range of ages in which a given radioisotope is useful as a geologic clock depends on the length of its half-life and the accuracy of the method employed in the determination. An isotope with a half-life of 10 minutes, for example, would long since have vanished from older rocks, unless it were constantly being formed by decay of yet another isotope. At the opposite extreme, an isotope with a half-life of 1000 million years probably would not be usable as a clock for young rocks, because not enough time would have elapsed for daughter elements to accumulate in measurable amounts.

The utility of radioactive clocks also depends on the chemistry of the system. Potassium, uranium, and most other radioactive elements are relatively abundant in the granitic rocks of the continents, which are thus particularly suitable for dating. The most common rocks, however, are the basalts of the sea floor, and they contain hardly any radioactive elements. Nature is kind to geologists in this matter, because the ancient datable rocks are preserved on the continents, whereas the ancient undatable rocks of the sea floor are disposed of when they plunge into the underlying hot mantle.

Radioactive elements on earth come from two sources: the interior rocks and the outer atmosphere. The elements in the rocks are the products of successive parent–daughter decays of the elements that were originally in the earth or, to a minor extent, that arrived in the rain of meteorites from space. A very different suite of short-lived isotopes is formed by cosmic-ray bombardment of the atmosphere. Cosmic rays are nuclear and atomic particles with extremely high energies, and, when they hit molecules of atmospheric gases, the nuclear changes exceed those in the most powerful laboratory atom-smasher. The most energetic cosmic rays commonly hit atomic nuclei in the upper atmosphere, and a cascade of secondary cosmic rays and fission products rains down to the earth below. At sea level, an average of one cosmic ray hits each square centimeter per minute. If you are out in the open air, several struck you while you were reading these paragraphs.

Some of the fission products are neutrons, which are soon captured by nuclei of nitrogen (by far the most common element in the atmosphere). The nuclei then emit protons and become nuclei of radioactive carbon (or radiocarbon) which has an atomic mass of 14, instead of the 12 of ordinary carbon. This decays with a half-life of 5730 years. The atoms of carbon combine with oxygen to form carbon dioxide, which mixes rapidly and uniformly through the atmosphere. The radioactive carbon dioxide is then taken up by plants and, thus, enters into the food chain. All *living* things on earth contain the various isotopes of carbon in a constant ratio. After they die, the radiocarbon they contain decays; thus, from the change in the ratio of carbon-14 to carbon-12, the time of death can be established. The American chemist Willard Libby was awarded the Nobel Prize for discovering this radioactive clock, which has a distribution and a half-life that are almost perfect for dating human prehistory. It has found widespread use in archeology, as well as geology, for dating events of the last 50,000 years or so. The accuracy of the method has been tested by comparing its results with historical dates for mummies from the early Egyptian dynasties and with tree rings from the bristlecone pines in the White Mountains of California. One of these trees is 7000 years old, quite possibly the oldest living thing in the world. The radiocarbon ages of ancient objects are less than the actual ages by 10–15%. Part of the error results from the fact that, for convenience, all dates are calculated using the original laboratory determinations of the half-life, 5570 years, although later measurements give a value of 5730 years. Correcting for this artifact removes 3% of the apparent error. The remainder may result from the fact that the present ratio of carbon-12 to carbon-14 is not normal. Carbon-12 is now more common than usual because of the burning of fossil fuels such as coal and oil, which, being organic matter long dead, contain no carbon-14. On the other hand, the testing of nuclear weapons creates carbon-14 but no carbon-12.

Radioactive isotopes are also useful as tracers, because different ones are produced in the earth's interior and in the atmosphere. A notable example of an isotope produced in the atmosphere is tritium, which is hydrogen with an atomic mass of 3 instead of the normal 1. It has a half-life of 12.3 years, and its presence in hot springs and volcanic steam is sure proof that the water is recycled rain rather than ancient water from the interior of the earth. Tritium would hardly be detectable in any ancient material because it decays so rapidly.

The constant bombardment of cosmic rays also provides a means of geologic dating. On a nuclear level, cosmic rays damage what they hit; and if the material is an insulator, the damage is preserved in fission tracks. In space or on the moon, wherever there is no shielding by the atmosphere, silicate minerals act as insensitive recorders of bombardment, just like photographic film in a laboratory. Fission tracks can be seen with an electron microscope—or even with an ordinary microscope, if they are enlarged by etching. A constant average flux of cosmic rays is suggested by the fact that isotopically dated rocks from space have a proportional number of fission tracks. Thus, the number of fission tracks may give an independent age for rocks that are otherwise undatable.

Cosmic rays do not cause a recordable flux of fission tracks in rocks protected by the atmosphere. However, the ordinary isotope of uranium, with an atomic mass of 238, decays with sufficient energy to make fission tracks in minerals that contain it. These tracks can be used in a geologic dating method that depends on the half-life of uranium-238 rather than on the cosmic-ray flux.

Relative Time

For precision, we may specify the time of a historical event by giving the year, day, hour, or whatever seems appropriate. For many purposes, however, it may be sufficient to say "in the reign of Queen Elizabeth," and, with regard to the development of the drama or the course of naval warfare, to designate time thus may be even more informative than to specify it in years.

Let us carry this thought further: Were we to memorize the sequence of British monarchs, we would have a useful scale of relative time even without knowing when any of them reigned. Inasmuch as few were monarchs for more than a decade or two, we could usefully consider all events in a reign as contemporary for most purposes. Were we also to memorize the sequence of French monarchs and premiers, we probably would recognize even shorter contemporaneous intervals. Thus, an undated newspaper that refers to living rulers in two countries was doubtless printed at a time when their two reigns overlapped. We could be fairly certain that the resolution of contemporaneity derivable from the reigns of monarchs in a dozen countries probably would rarely exceed a year, even if we had no calendar to indicate which one.

Most rocks, particularly sediments, cannot be dated with radioactive isotopes; and even those that can be may not be, because of the effort and expense involved. Consequently, geology is dependent on relative time scales, although these can be integrated with the radioactive time scale in some places. This permits analysis of rates and intensities of processes in a way that is impossible with a wholly relative time standard.

A relative time scale can be based on any change that is preserved in the geologic record. Its utility depends on the uniqueness of each individual change and the frequency, brevity, and global extent of the type of change.

Some phenomena, such as magnetic reversals, occur over the whole earth in a period of only a few thousand years. They are also distinctive, but they have the disadvantage of being preserved only in certain kinds of rocks. Organic evolutionary changes occur locally, spread elsewhere at different rates, and are preserved only in sedimentary rocks, but each is unique. Climatic changes may be global or regional and tend to be manifest in different phenomena, depending on the latitude. Sedimentary and erosional changes are local or regional and generally occur at different times, even in a single locality. Thus, each base has its advantages and disadvantages, and the use of relative time scales requires the utmost discrimination and care.

Magnetic Epochs

Most major physical characteristics of the earth, such as the length of the day or the height of the lunar tides, are variable, and it is possible that the changes can someday be used as the base of a relative time scale. At present, however, the only physical changes known to have much utility for relative dating are the reversals of the magnetic field discussed in Chapter 3. Let it suffice, for the moment, that a geologist with a machine for time travel would have found that his magnetic compass pointed north at some times in the past, and south at others. The history of these reversals is preserved in lavas and in many sedimentary rocks that are deposited slowly enough so the record is not obscured.

Magnetic reversals have been closely correlated with radioactive dates for the last several million years and, because they are inherently global and so widely preserved, they are ideal for many purposes. They are essentially an areal extension of absolute dating during this period.

Radioactive dates associated with reversals are rarer for earlier times; thus, the relative time scale is vaguer, but the magnetic history preserved in sea floor rocks provides a time standard for 150–200 million years. Reversals are preserved in the still older rocks of the continents; but they are not integrated into a continuous time scale, for lack of identifying dates, and thus are less useful.

Paleontologic–Stratigraphic Time

The absolute time scales based on radioactivity and magnetic reversals have found ready acceptance because they are merely different methods of doing an accepted thing. The development of a scale of relative time began in the eighteenth century, and was conceptually much more difficult. At the beginning, western European scientists were firmly committed to the biblical account of Genesis and the interpretation that the earth was only a few thousand years old, that all live plants and animals were created at the same time, and that man is not an animal. Gradually, they had to reject the common wisdom and the fundamental beliefs of their religion and accept the evidence of their eyes and the inevitable conclusions to which it led.

In 1669, Nicolaus Steno, a Dane, formulated the principle of the law of superposition: namely, that in a sequence of undeformed sedimentary rocks, the youngest rocks are on top. This seems obvious, but so do most fundamental ideas after they are once conceived. At the time, Steno had to make several daring generalizations to formulate the law. Rocks can be deformed and overturned such that the youngest are on the bottom; but this occurs relatively rarely and, in any event, does not violate the law. From the law of superposition alone we can construct a relative time scale wherever we can see a sequence of sedimentary rocks in a sea cliff or river canyon.

Geology found its next great theoretical advance in the work of the British geologist James Hutton, who published his *Theory of the Earth* in 1788. This work established the principle that the present is the key to the past—that ancient rocks were formed by phenomena of the same kinds that are observable today but that have been acting through the immensity of geologic time. His ideas were largely disregarded until his colleague John Playfair expounded them with exceptional clarity in *Illustrations of the Huttonian Theory of the Earth* in 1802. They gained general acceptance only when yet another British geologist, Charles Lyell, published his *Principles of Geology* in several volumes between 1830 and 1833.

The application of geology was mainly beneath the dignity of the theorists. William "Strata" Smith was a humble man of little education, but he rediscovered Steno's law in the course of surveying for the canals that were being dug in Britain around the beginning of the nineteenth century. He could recognize distinctive layers of sandstone and limestone from place to place, and he found that each contained characteristic fossils in an invariable sequence.

It is well to remember that Linnaeus did not establish the now-accepted system for classifying organisms until 1753, only a few decades before Smith made his discovery. Moreover, the now obvious fact that a fossil seashell in a sandstone on land was once an animal living on sand under the ocean was not generally accepted at the time. Indeed, it seemed to be in direct conflict with the biblical account of Genesis.

Smith's work gained little recognition for some years, chiefly for social reasons. Scientists and members of scientific societies were then generally from upper classes; and the laboring Smith was not even a member of the Geological Society of London when it awarded him its first Wollaston Medal for achievement in geology in 1831.

When Charles Darwin published that immediate best-seller *The Origin of Species* in 1859, he established the principle of gradual evolution through natural selection. He recognized fossils as evidence of evolution, and the theoretical fabric for a relative time scale was complete.

More and more paleontologists studied the fossils, and stratigraphers unraveled the sequence of sedimentary rocks in which they occurred. Stratigraphy flowered for the remainder of the century. Fossils were traced from country to country and from continent to continent, until geologists had a relative time scale spanning what we know to be about 600 million years. In the course of this work, many problems were identified and solved. Foremost among these was the problem of facies, which are different types of sediment deposited at the same time. At present, for example, sand is deposited on a beach at the same time that mud is deposited in an adjacent lagoon, and each contains a distinctive fauna (group of animals) that prefers a particular type of sediment. If all this is changed to rock, a stratigrapher may find two faunas, which might be interpreted as indicating two ages instead of two facies of the same age.

Facies relationships may persist for long periods but shift location. Sea level has risen more than 100 meters during the last 25,000 years (Chapter 10), and, while it rose, the shoreline migrated inward across the gently sloping continental shelf. Beaches and lagoons migrated as well. Consider the beach sand as the ocean transgresses onto the shore: At any given time, it forms a long, thin ribbon; as time goes on, however, the advancing ocean leaves behind a sheet of sand, of which the active beach is the shoreward edge. The stratigrapher finds a bed of sandstone deposited on a bed of shale, and the age relationships are quite different from those suggested by a simplistic interpretation of the law of superposition.

Facies complicate stratigraphy, but deformation and erosion can be equally troublesome, especially where the sediments are sparsely populated with fossils. For this and other reasons, stratigraphy based on paleontology has found its greatest utility when applied to undeformed rocks that were initially deposited in quiet, relatively shallow waters teeming with life. It is almost useless for dating rocks deposited during the first 80% of geologic time, either because life was nonexistent or, later, because the remains of whatever life did exist were rarely preserved.

Fossils of microscopic size are particularly useful. Perhaps the most important reason for this is that much of stratigraphy is known from drilling for oil, and only microfossils are abundant enough in drill-core samples to be of much use. The deep sea is another environment in which stratigraphy is studied largely on the basis of samples collected through long pipes. The program of deep-ocean drilling called by its acronym JOIDES (Joint Oceanographic Institutions for Deep Earth Sampling) began to penetrate the whole thickness

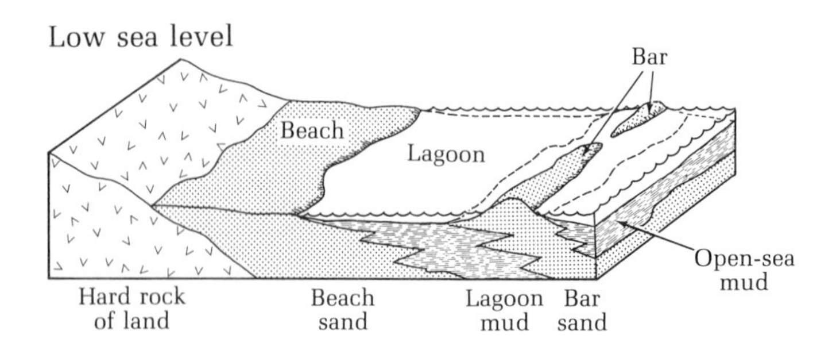

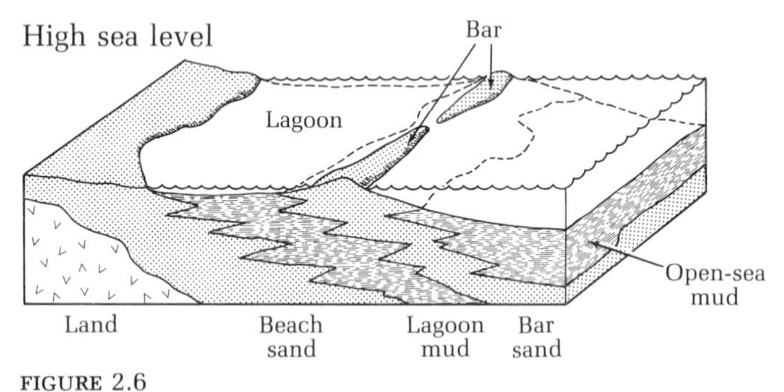

FIGURE 2.6
The geometry and sediments of an offshore bar and lagoon may remain the same even though their locations change as relative sea level rises and the shoreline transgresses the land.

of oceanic sediment for the first time in the late 1960s. Very detailed and relatively continuous stratigraphic sequences are preserved in the sediments of the deep sea. They can be tied to magnetic reversals; and, as a consequence, a new degree of precision is appearing in geologic interpretations of the history of the last 200 million years.

General History of the Earth

What is the geologic history that has emerged from the vast effort of dating and ordering the events of the past? In this section, we will take a very broad overview, in order to place the most important events in a historical context and to emphasize the interdependence of factors that will be dealt with separately in more detail.

Before discussing the evidence and the conclusions derived from it, it may be helpful to make a bold comparison of the earth as it apparently was 4000 million years ago with the earth as it is today, in order to emphasize the magnitude of the changes. The comparison is speculative because, as we shall see, most of the great changes occurred very early and the factual record has been largely destroyed. But, as Preston Cloud has said of what remains, "from such fragile threads we can weave a strong, if coarse-textured, tapestry of events on the primitive earth."

The main mass of the earth had accumulated by 4000 million years ago, although meteorites continued to impact frequently. There was probably no moon, and a day lasted only a few hours on the rapidly spinning earth. The primitive ocean and atmosphere contained juvenile liquids and gases, such as water, carbon dioxide, and nitrogen, but no free oxygen. There was no life. If we could travel in a time machine to such a world, we would find the sun about the same, the climate perhaps not much different, normal-looking clouds, rivers, and lakes, an abundance of dust, and waves breaking at the edge of an almost tideless sea. We could not breathe the air, and we would need a space suit to protect us from the ultraviolet radiation from which we are now shielded by the ozone (oxygen molecules consisting of three atoms) of the upper air. This alien and inhospitable world has evolved into our modern one by an interlocking system of gradual, irreversible changes.

Let us begin with the whole solar system. Lead isotopes in meteorites and in some earth rocks indicate that the solar system came into existence about 4600 million years ago. Presumably, the earth is about the same age, although the oldest rocks known at present are only about 4000 million years old. They have been found in Greenland alone. The oldest preserved rocks that occur on *several* continents have ages of only 3500–3600 million years. Presumably, the gaps represent a period when surface rocks were melted and pulverized repeatedly by meteorite impacts during the later stages of the active accumulation of the earth. The moon's history is more speculative, but it is believed that, perhaps 3000 million years ago, the moon was captured in a retrograde orbit. It then locked into a synchronous orbit with the earth, such that the day and the month were only a few hours each. It stayed in a gradually evolving parking orbit until roughly 600 million years ago, when it broke away and entered its present prograde orbit. While the moon was in the parking orbit, there were high tides in both land and sea.

It is likely that the ocean and the atmosphere both evolved during the accumulation–melting–differentiation stage. African rocks 3200 million years old display evidence of weathering and transportation that would have required an atmosphere and running water, but the presence of minerals in the sediments that combine easily with oxygen (oxidize) shows that there was no free oxygen. Instead, the oxygen-free atmosphere consisted of primitive gases, including hydrogen, derived from the interior of the earth.

Most of the great iron ores of the world are in distinctively thin-bedded deposits of oxides called banded iron formations. They were deposited during the interval from 3200 to 1800 million years ago. During this period, therefore, a balance was achieved on an enormous scale between the output of oxygen and the deposition of iron. The oldest known fossils are 3200 million years old, so life appeared at an even earlier date. According to the analysis of Preston Cloud, the first life was primitive indeed and dependent on fairly complex chemicals in the environment for food. It was replaced, perhaps 3000 million years ago, by autotrophs, which are cells that manufacture their own food substances from fairly simple building blocks. An obvious source of energy for life is sunlight, and these primitive organisms may have used photosynthesis directly; however, the remains of the earliest blue-green algae definitely known to photosynthesize are found in rocks a few hundred million years younger.

If there was photosynthesis, oxygen was released into the atmosphere; and oxygen was a deadly poison to living things before the development of suitable oxygen-mediating enzymes. Cloud surmises that ferrous iron was common in the ocean at that time, and that it combined with the waste oxygen before it could poison the photosynthesizers. This balance produced the banded iron formations. The balance collapsed when the photosynthesizers developed oxygen-mediating enzymes, which, in effect, made oxygen non-poisonous. They flourished, then, and their increased production of oxygen swept the seas free of iron by about 1800 million years ago, when the last major banded iron formations were deposited.

Not long after, free oxygen appeared in the air. This is suggested by the first extensive formations of subaerial sediments rich in iron, which were laid down 1800–2000 million years ago. Presumably, these reflect the retention of iron oxides in soils. Gradually, oxygen increased in the atmosphere, carbon dioxide declined, and ozone increased.

The paleontological record implies that the early life forms were single cells, lacking a nuclear wall and incapable of the kind of cell division that permits genetic exchange. Cells of this type are referred to as procaryotic. On the other hand, multicellular organisms, which abound in the paleontological record starting 600 million years ago, have eucaryotic cells with a nuclear wall and the capacity for genetic recombination by replication of DNA. These organisms led a precarious existence in protected environments until an ozone screen accumulated that was capable of shielding DNA from high-intensity ultraviolet radiation, which inactivates it. The oldest eucaryotic fossils are in rocks aged 1200–1400 million years, which fits well into this chronology of the development of the atmosphere and the evolution of the earth.

Had our time machine materialized a mere 600 million years ago, we would not have been in a familiar world. The land was bare—except, perhaps, for algae and lichens. There were no reptiles, mammals, trees, or grass. Still, the air was breathable, the moon was in the sky, the day had perhaps twenty hours, and life abounded in the sea. The greatest changes were over, and the stage was set for the expansion and diversification of life that still continues.

What of the physical history of the earth during all these ages? Did moun-

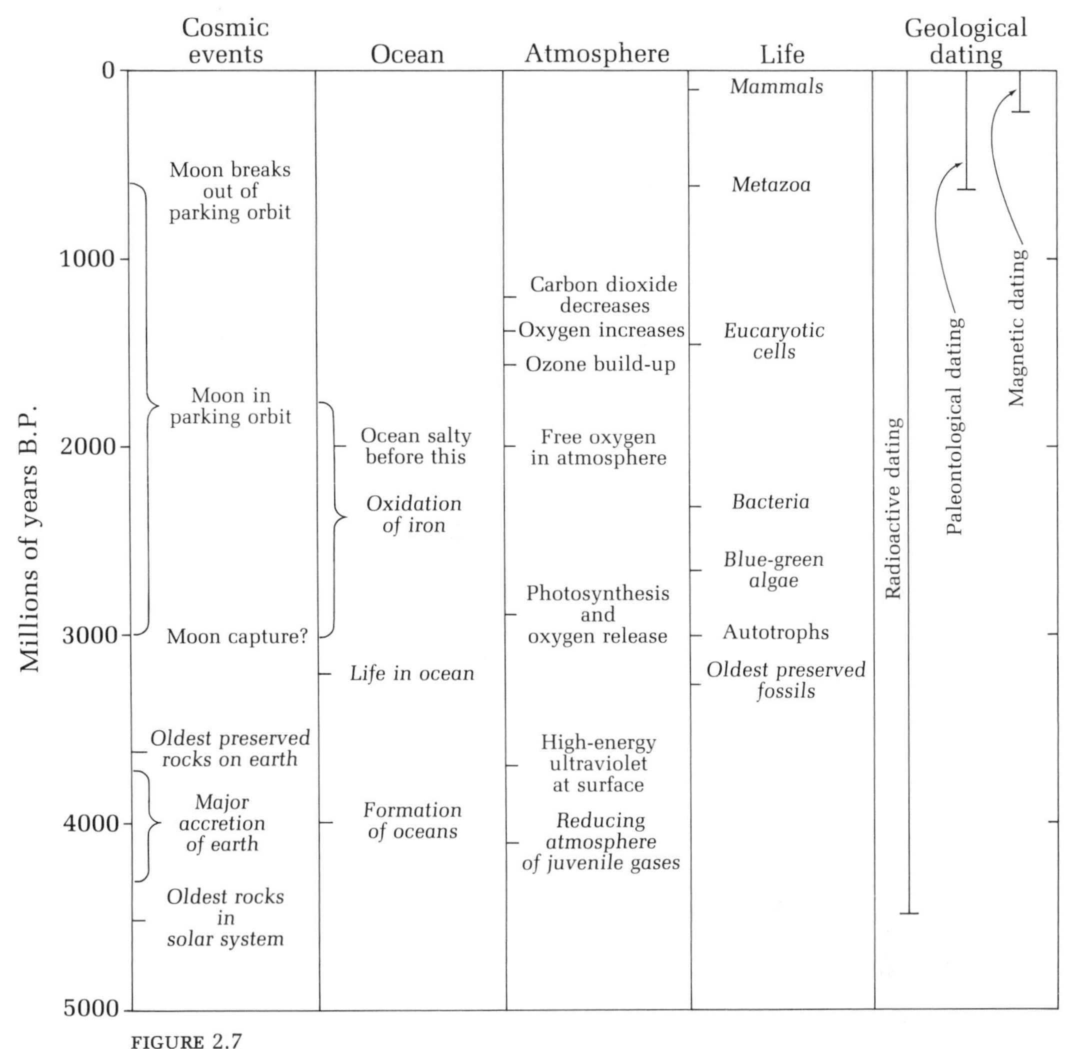

FIGURE 2.7

Diagram of the chronology of the major dated events (*in italics*) and suspected events in the history of the earth.

tains form, volcanoes erupt, continents drift? Did erosion and sedimentation level the mountains and valleys? The answer seems to be a uniform "Yes." The oldest datable rocks are patches of granite, which indicates that geologic processes had long since produced continents. The chemical composition of younger rocks seems to change as time passes, presumably because of repeated erosion and tectonic recycling; but the same rock types—volcanic lavas, sandstones, and shales—occur throughout. The American geologist Albert Engel, long a student of these rocks, has observed that the oldest types of rocks are similar to those that form where moving tectonic plates converge, which suggests continental drift was taking place at the time they were deposited.

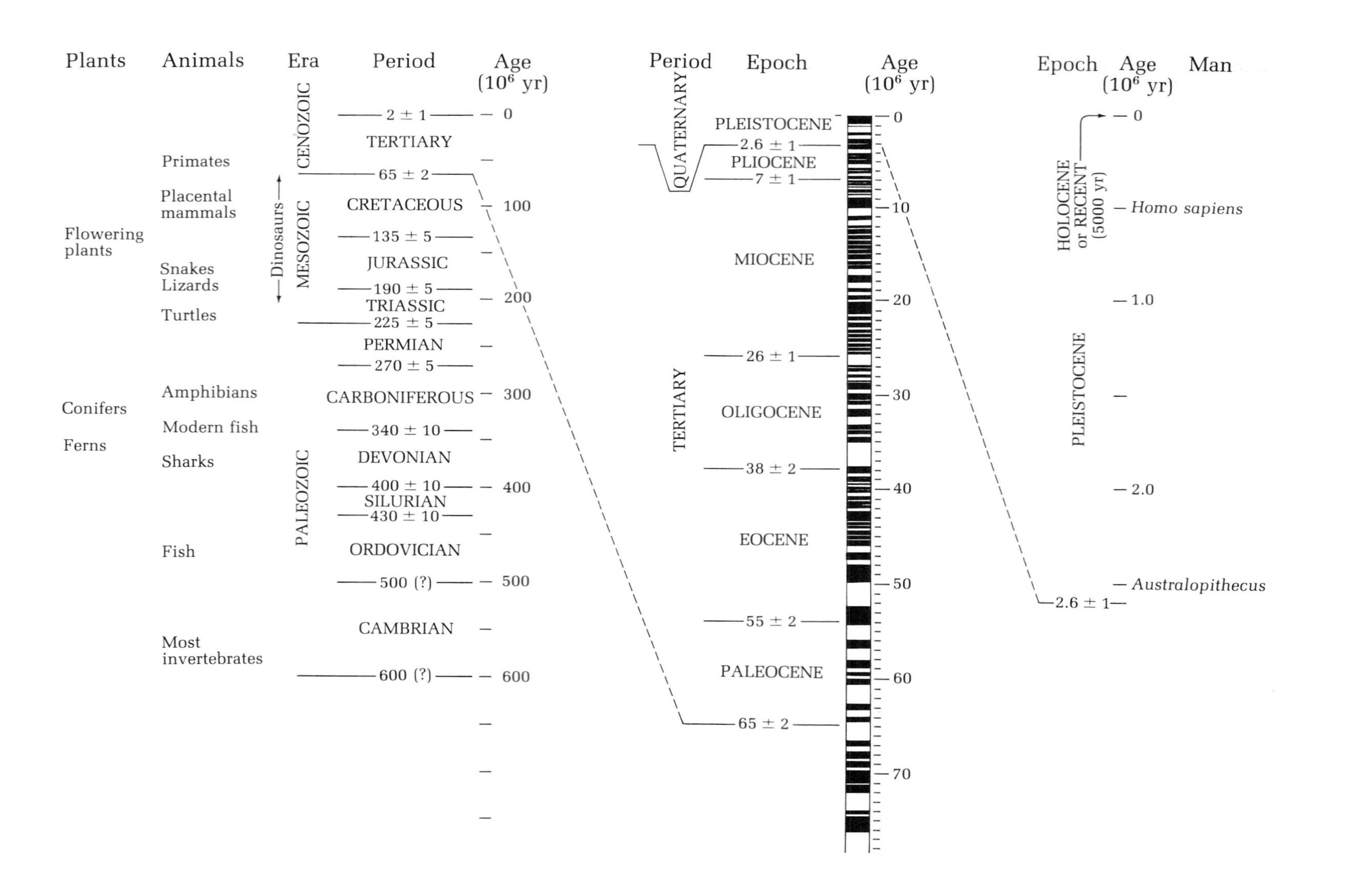

Plants
Animals
Era
Period
Age (10^6 yr)
2 ± 1
0
TERTIARY
CENOZOIC
Primates
65 ± 2
Placental mammals
Dinosaurs
MESOZOIC
CRETACEOUS
100
Flowering plants
135 ± 5
JURASSIC
Snakes
Lizards
190 ± 5
TRIASSIC
200
Turtles
225 ± 5
PERMIAN
270 ± 5
Amphibians
CARBONIFEROUS
300
Conifers
Modern fish
340 ± 10
Ferns
Sharks
DEVONIAN
PALEOZOIC
400 ± 10
400
SILURIAN
430 ± 10
Fish
ORDOVICIAN
500 (?)
500
CAMBRIAN
Most invertebrates
600 (?)
600
Period
Epoch
Age (10^6 yr)
QUATERNARY
PLEISTOCENE
0
2.6 ± 1
PLIOCENE
7 ± 1
10
MIOCENE
20
26 ± 1
30
TERTIARY
OLIGOCENE
38 ± 2
40
EOCENE
50
55 ± 2
PALEOCENE
60
65 ± 2
70
Epoch
Age (10^6 yr)
Man
0
HOLOCENE or RECENT (5000 yr)
Homo sapiens
1.0
PLEISTOCENE
2.0
Australopithecus
2.6 ± 1

As to erosion, vulcanism, and other geologic processes, their products are spread through the whole geologic record. They vary in intensity from time to time, but there is no persistent trend from the first preserved rocks to the present. It is possible that further study will identify physical events of unparalleled intensity in the past, but of this there is no hint. There is nothing unique in the geologic cycles like the change from a reducing to an oxidizing atmosphere. The cycles are, indeed, cycles; the present is the key to the past; and what's past is prologue.

Summary

1. The origin of the earth is uncertain, although we know many ways in which it was not formed. It probably originated by accretion of cold matter ranging in size from dust to asteroids, and, possibly, in a relatively short time.
2. Matter from space is still hitting the earth. Dust tends to burn up in the upper air. Large meteroids may explode either in the lower air or at the surface, where they make craters.
3. The earth and moon interact by gravitational attraction, which causes: (a) the tides; (b) the moon to move away from the earth; and (c) the earth to spin more slowly. At one time, the moon was very close, and the day and month probably had the same duration.
4. Radioactive elements can be used, in favorable circumstances, to date geologic events or to trace geologic phenomena.
5. Relative time scales can be based on various natural phenomena, of which evolutionary changes in organisms are the most useful.
6. In the past, sand and mud facies appeared on the earth's surface at the same time—just as they do at present. Because they often contain different organisms, they complicate the use of the evolutionary time scale.
7. The air, water, and rocks, in addition to life, have evolved during the approximately 4500 million years since the earth was formed: the air has changed from oxygen-free to oxidizing, ocean water has become salty, and the rocks have released gases and have been redistributed.

FIGURE 2.8

The geological time scale for the last 600 million years, and some notable biological developments. The divisions are based on evolutionary changes and are subject to constant minor revision. There are many minor stratigraphic subdivisions that are not shown. Ages in years are based on radioactive clocks and new measurements may change them slightly. The magnetic-reversal pattern (*dark and light bands*) provides another means of measuring relative ages for the last 75 million years or more. [After Faul (1966); and Heirtzler, "Sea-Floor Spreading," copyright © 1968 by Scientific American, Inc. (all rights reserved).]

Discussion Questions

1. How does a scientist make a hypothesis when some of the related facts seem contradictory?
2. What is the evidence for meteoroid explosions in the lower air and at the earth's surface?
3. What are the reasons for believing that the moon was captured and is receding from the earth?
4. A farmer once said. "I know this rock is 4,000,003 years old, because three years ago Professor Lyell told me it was 4,000,000 years old." Where did he err?
5. Why is the radiocarbon method so useful?
6. Compare the merits of evolutionary changes and magnetic reversals for relative dating.
7. Accept the fact that organisms lived in a sequence beginning with *A* and ending with *Z*. What would be your explanation if you found layers of rock with fossils of *A* at the top and of *Z* at the bottom?
8. What would happen to the atmosphere if nuclear war somehow destroyed all life?

References

Alfvén, H., and G. Arrhenius, 1969. Two alternatives for the history of the moon. *Science* 165:11–17. [Suggests that the moon was captured and remained in a parking orbit for more than a billion years.]

——, and ——, 1970. Structures and evolutionary history of the solar system, I and II. *Astrophys. Space Sci.* 8:338–421. [An imaginative synthesis that proposes rapid accretion of the earth accompanied by zonal melting.]

Cloud, P. E., Jr., 1968. Atmospheric and hydrospheric evolution on the primitive earth. *Science* 160:729–736. [Biological and chemical evidence for changes from a reducing to an oxidizing atmosphere.]

——, 1969. The primitive earth. *New Sci.* 43(662):325–327.

Cox, A., G. B. Dalrymple, and R. R. Doell, 1967. Reversals of the earth's magnetic field. *Sci. Amer.* 216(2):44–54.

Dietz, R. S., 1961. Astroblemes. *Sci. Amer.* 205(2):50–58. (Available as *Sci. Amer.* Offprint 801.) [Meteorite impacts described in an imaginative summary.]

——, 1964. Sudbury structure as an astrobleme. *J. Geol.* 72(4):412–434. [Some of the great nickel ores of Canada appear to be of meteoritic origin.]

Engel, A. E. J., and C. G. Engel, 1964. Continental accretion and the evolution of North America. *In* A. P. Subramanian and S. Balakrishna, eds., *Advancing Frontiers in Geology and Geophysics: A Volume in Honour of M. S. Krishnan,* pp. 17–37e. Hyderabad: Indian Geophysical Union. (Reprinted in P. E. Cloud, ed., 1970. *Adventures in Earth History,* pp. 293–312. San Francisco: W. H. Freeman and Company.) [The continent has slowly grown.]

Faul, H., 1966. *Ages of Rocks, Planets and Stars.* New York: McGraw-Hill.
Gerstenkorn, H., 1955. Über Gezeitenreibung beim Zweikörperproblem. *Z. Astrophys.* 36:245–274. [Proposes that the moon was very close to the earth after being captured.]
Glass, B. P., and B. C. Heezen, 1967. Tektites and geomagnetic reversals. *Sci. Amer.* 217(1):32–38.
Heirtzler, J. R., 1968. Sea-floor spreading. *Sci. Amer.* 219(6):60–70. (Available as *Sci. Amer.* Offprint 875.)
Hutton, J., 1795. *Theory of the Earth; with Proofs and Illustrations,* 2 vols. Edinburgh: William Creech. Reprinted in 1959 by Hafner Publishing Company, New York. [A poorly written classic, but a classic nonetheless.]
Lal, D., and R. S. Rajan, 1969. Observations on space irradiation of individual crystals of gas-rich meteorites. *Nature* 223:260–271.
McLaren, D. J., 1970. Presidential address: time, life and boundaries. *J. Paleontol.* 44(5):801–815.
Nininger, H. H., 1952. *Out of the Sky: An Introduction to Meteorites.* New York: Dover. [A popular, descriptive, and intriguing account.]
Playfair, J., 1802. *Illustrations of the Huttonian Theory of the Earth.* Edinburgh: William Creech. (Reprinted in 1956 by Dover, New York.) [A readable classic.]
Runcorn, S. K., 1966. Corals as paleontological clocks. *Sci. Amer.* 215(4): 26–33. (Available as *Sci. Amer.* Offprint 871.) [The Devonian year had 400 days.]
Shelton, J. S., 1966. *Geology Illustrated.* San Francisco: W. H. Freeman and Company. [Remarkable photographs, most of them taken from the air, that illuminate geology.]
Urey, H. C., 1952. The origin of the earth. *Sci. Amer.* 187(4):53–57. (Available as *Sci. Amer.* Offprint 833.) [The earth is an agglomeration of small bodies that were cold when they accreted.]
Zotkin, I. T., and M. A. Cikulin, 1966. Modeling the Tungas meteorite explosion. *Dokl. Akad. Nauk. SSSR* 167(1):59–62. [Thousands of trees fell in a radial pattern after a natural explosion above Siberia.]

3

VISIT TO A SMALL PLANET

The party having dined, Ward and I had retired to another room that we might examine under the microscope some of his volcanic rocks, and compare them with the Paleozoic volcanic series of Scotland. We had been engaged on this task for an hour or two when Ramsay joined us. He sat rather impatiently watching us for a while, and then starting up, left the room, after exclaiming, "I cannot see of what use these slides can be to a field-man. I don't believe in looking at a mountain with a microscope."

Archibald Geikie, MEMOIR OF A. C. RAMSAY

In this chapter, we shall study the earth in the manner that the space projects study the moon and Mars; namely, we shall look at the big things before turning to the small. This seems to be the logical way to study any planet, but it is impossible for its residents until they can escape from it. Geologists began by studying small features visible on the surface of land in little-deformed younger rocks. It was almost inevitable that the small approach led to the concept of a relatively static earth. Indeed, it is a credit to the imagination of many of the early scientists that a more dynamic earth was considered at all.

With modern tools operating at a distance in space or probing the ocean depths, mantle, and core, it is impossible to detect small features; attention, inevitably, is directed to the large ones. A dynamic model of the earth emerges, and the small-scale features examined in the earlier geology appear different when seen in a new light. For example, the ridges on the skeletons of corals that died several hundred million years ago are very small indeed, but, seen against the right background, they can be used to determine the ancient tidal friction between the earth and the moon.

We shall consider the earth as seen from space—as seen, perhaps, by a scientific expedition approaching from the constellation Aquarius and seeking a desirable planet for permanent colonization. A changeable planet makes for high upkeep in a colony; therefore, if changes in the environment of the new colony are very large, the project may have to be abandoned. Thus, stability is a major consideration in picking the best among a myriad of planets in the universe. Is the earth a planet stable enough to be selected by anyone who was in a position to choose?

Gross Properties of the Earth

Many of the characteristics of the earth are familiar to all of us. The major ones are discussed here, however, because most of them have changed and may do so again.

SOLAR ORBIT

The earth orbits the sun in what is called "the plane of the ecliptic," along an elliptical path, at a distance that varies from 147 to 152 million kilometers, and with a period of one year. Both the period and the distances have changed very slightly over the ages as the sun has converted its mass into energy and the mass of the earth has grown by meteorite accretion. The changes have not affected the physical environment in any significant way. If the earth and sun were the only bodies in the solar system, the orbit of the earth would not vary. Gravitational interactions with the moon and planets, however, cause minor but very complex variations in its orbit. The shape of the elliptical orbit varies; thus, the average distance from the earth to the moon changes slightly. This cycle of change has a period of about 90,000 years. The location of the orbit also varies in a way that causes the time of the year when the earth is closest to the sun to fluctuate over a period of about 21,000 years. Both of these very slow effects influence the earth's climate and may influence the occurrence of glacial periods.

SIZE AND SHAPE

The earth has an approximately spherical shape because its gravitational attraction overwhelms the puny strength of the rocks. Its average radius is 6380 kilometers, and this affects many important characteristics: If the earth were like a tiny martian moon, for example, it could be of any shape. If it were of the size and mass of our moon, it could not hold an atmosphere or an ocean. If it were as massive as Jupiter, its gravity would grossly alter weather and ocean currents.

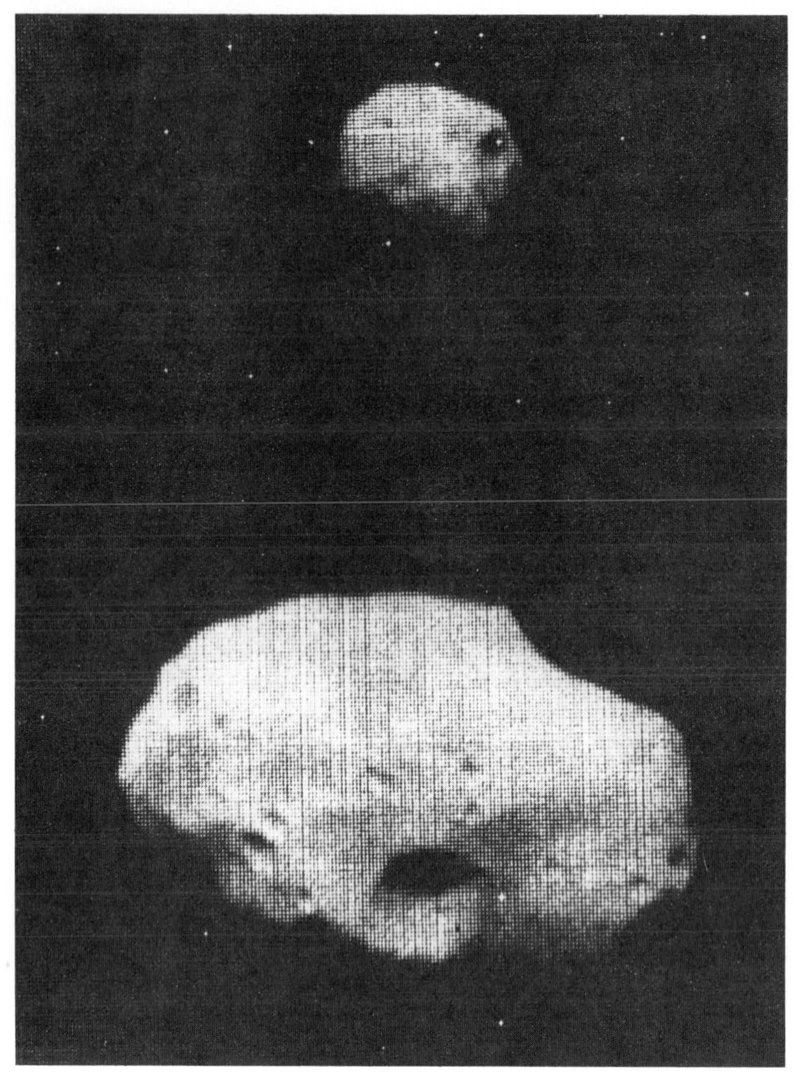

FIGURE 3.1

Phobos (*below*) and Deimos, the small moons of Mars photographed by the Mariner 9 flyby in 1971. They are only 24 kilometers and 13 kilometers, respectively, in average diameter; thus, they are comparable in size to the planetesimals that aggregated together to form the primitive earth. They are not spherical because their gravitational self-attraction is too small to overcome the strength of the rocks. [NASA photo.]

The earth actually resembles an ellipsoid of revolution more than a sphere because of its spin, which bulges the equatorial radius by 7.2 kilometers and pulls in each pole by 14.8 kilometers. However, it is not a perfect ellipsoid because compositional variations in the interior and tectonic forces cause additional small deviation in its shape.

Since the earth finished its initial accumulation, its size has probably remained relatively constant, but it may be increasing very slowly because of a decrease in the gravitational constant. On the other hand, its shape has changed significantly, because the equatorial bulge was much greater when the spin was faster. The rate of change, however, is very slow at present, and changes in shape are probably not important contributors to deformation.

SPIN AND TILT

The earth spins in a day, and the axis of rotation on the average, is tilted at an angle of 23°.5 to a perpendicular to the plane of the ecliptic. The length of the day has increased enormously (as discussed in Chapter 2), and continues to do so, because of tidal friction. The tilt of the axis is a matter of chance, rather than an inherent characteristic of a planet. Jupiter has hardly any tilt, and Uranus spins on an axis that lies almost in its orbital plane. The variations in tilt can be explained as the result of planetary accumulation of small bodies with different angular momenta. A series of balls thrown in the air and struck by small bullets will spin at speeds and around axes that are determined, among other things, by the speed and location of the impacts.

MINOR MOTIONS OF THE EARTH AND ITS AXIS OF ROTATION

The orientation of the spin axis moves, and the earth also moves relative to the spin axis. The largest of these motions is the precession of the equinoxes, which is caused by the attraction of the sun and moon for the equatorial bulge. As the axis moves, it describes a cone, whose apex is at the center of the earth, with an apical angle of 47° centered on a perpendicular to the plane of the ecliptic. The motion takes 26,000 years; during most of that time, the north pole does not point toward Polaris, which we call the "pole star."

Another important component of the motion of the spin axis is the nutation, which has the same cause as the precession of the equinoxes but has an apical angle of only 18 seconds and a period of 18.6 years.

The earth also wobbles around its axis: If an observer could remain suspended above the north pole (the spin axis), he would see the earth beneath him move through an elliptical path with a mean diameter of 6 meters during a 14 month period. The American geophysicists Walter Munk and Gordon MacDonald attribute the wobble to regional variations in the atmosphere.

In the preceding discussion, the angle of the spin axis to the plane of the ecliptic has been treated as though it were constant. However, because of the gravitational attraction of the planets, it varies by a small amount during a period of about 41,000 years.

SEASONS

The orientation of the spin axis in a given year is relatively constant as the earth orbits the sun. As a result, the northern and southern hemispheres

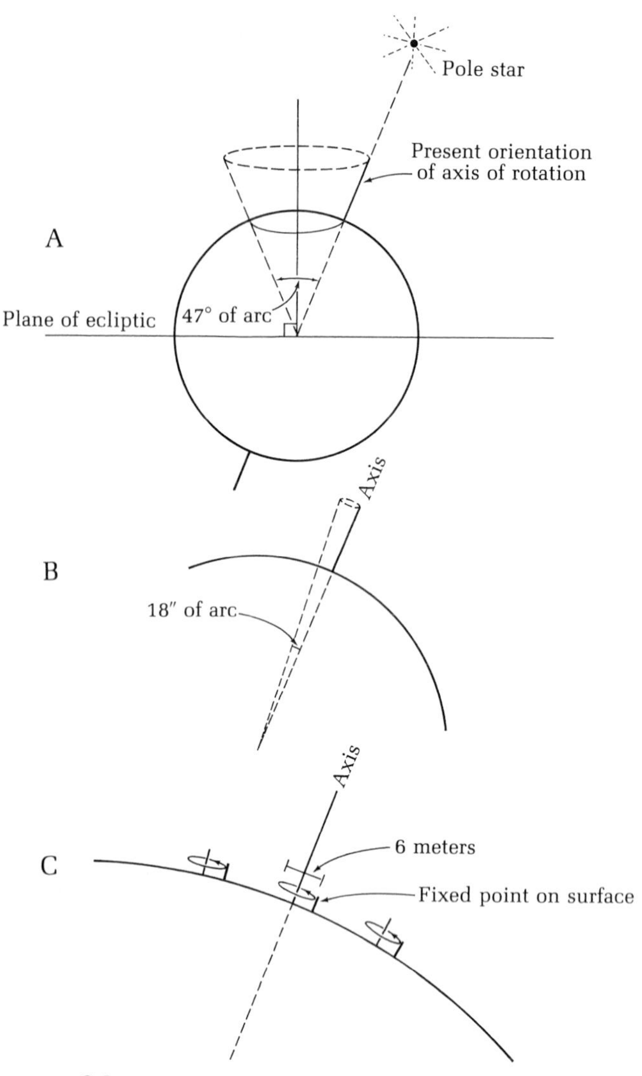

FIGURE 3.2
The earth precesses like a spinning top but with a more complex motion with two different components: *A*, the precession of the equinoxes, with a period of 26,000 years in which the axis of the earth moves through 47 degrees of arc; and *B*, the nutation, with a period of 18.6 years in which the axis moves through 18 seconds of arc. An additional motion, known as the wobble (*C*), has a period of 14 months.

alternate in their exposure to sunlight. The seasons result from this and the variable distance of the elliptical orbit from the sun.

It is evident that the seasons change in a complex but regular cycle as a result of the various motions of the spin axis and changes in the earth's orbit.

The effect on the climate is uncertain, but major climatic fluctuations have occurred, and changes in the axis may have influenced them.

The orientation of the spin axis also changes slightly every time a very large meteorite hits, just as it did under similar circumstances while the earth was forming.

STRENGTH

While looking at a mountain, it is hard to think of anything stronger than rock; and yet if we were to examine many of the rocks of the peaks, it would be obvious that they once flowed like molasses (Chapter 4). Once-rounded cobbles have been drawn out to resemble pencils, or even threads. Sedimentary rocks, originally horizontal, have intricate contortions. The strength of rocks clearly varies, and the factors that affect it include the type and duration of stress, the temperature, the pressure, the water content of the rock, and its mineralogy.

Some rocks, such as common salt, are particularly prone to uniform flow because of the ease with which they recrystallize. Because salt is less dense than most sedimentary rocks, it sometimes moves upward through denser rock, after it has been buried, like a balloon rising through air, and deforms the surrounding rocks into domes. Such domes are commonly the sites of oil fields. Other types of rocks flow because of different stresses; and gravity plays an important role in separating rocks vertically, if they are able to flow.

The lithosphere, which is composed of tectonic plates, forms the outer most rigid layer of the earth. (The lithosphere should not be thought of as equivalent to the crust; in fact, the American geologist Reginald Daly coined the former term in an effort to avoid some of the meanings inherent in the latter one.) The lithosphere is strong enough to remain rigid and, possibly to transmit horizontal stresses for enormous distances, even around the curve of the earth. The interaction of these great tectonic plates squeezes up high mountains, tears continents apart, and molds the deepest trenches under the oceans—all by horizontal stressing.

The response of the lithosphere to vertical stresses is like that of thin ice in a pond. Such ice is capable of transmitting horizontal stresses for long distances despite its relative thinness. If the wind blows, it may break free and gouge the mud on the bank toward which it is blown. However, if any weight, such as a rock, is placed on the ice, it will bend downward; if the weight is too great, the ice will break. The plates that make up the lithosphere behave in this way because, like sheets of ice, they are thin compared to their area and are, essentially, afloat. (The lithosphere "floats" on a layer called the asthenosphere, which, although very weak, is not liquid.) The lithosphere is 50–100 kilometers thick in most places, so it can easily support local loads. It may be surprising, however, that precise leveling before and after the construction of Boulder Dam has shown that the load of Lake Mead, a small and shallow affair, has depressed a dimple in the top of the lithosphere.

Larger loads, such as oceanic volcanoes, river deltas, and glaciers, depress the lithosphere even more. For example, the great continental glaciers that covered Scandinavia, eastern Canada, and New England more than 20,000 years ago depressed the regions over which they spread and formed encircling moats that filled with seas and lakes. The removal of the load of melting ice was accompanied by the slow elevation of the depressed lithosphere. This elevation, which is still continuing, has affected sea and lake levels in the regions involved.

On a large scale, the response of the earth to vertical stresses depends on their duration. In a large earthquake, it is as though the earth had been struck with a celestial hammer. The earth responds elastically by ringing like a gigantic bell. If the earth were equally rigid in response to the centrifugal force of the rotation of the earth, it would be almost spherical instead of having an equatorial bulge. In that case, the strengthless, fluid seawater would tend to pile up at the equator several kilometers deeper than at the poles. In fact, the depth does not appear to vary with latitude; therefore, the solid rock has been deformed as much as the water.

The earth is not a *perfect* ellipsoid of revolution, the figure assumed by a self-gravitating, rotating body without strength. It has continents with high relief and low ocean basins, and large, gently undulating swells and swales with areas of about 10% of the surface and elevations of tens to hundreds of meters. These latter are most easily measured by observing their effect upon the orbits of artificial satellites.

ISOSTASY

Continents rise high above the ocean basins. How can this be, if the lithosphere has hardly any strength to resist vertical stresses? It is because the continents have roots of relatively low-density rocks, and continents and ocean basins alike float on the asthenosphere.

The idea of low-density rocks under high regions was formalized by the American geologist C. E. Dutton when, in 1899, he coined the term isostasy for the principle that columns of unit area have the same mass above a certain level in the earth. Gravity measurements indicate that the level is about 100 kilometers, which corresponds, in part by coincidence, with the thickness of the thickest lithosphere.

The principle of isostasy is extremely useful in the earth sciences. It tells us, for example, that if a mountain is eroded away, the land will rise; that if sediment is deposited on the sea, the floor will sink; and so on. Equilibrium is not perfect. The lithosphere has some strength, and flow in the asthenosphere is not instantaneous. Deviations from perfect hydrostatic equilibrium result in a deficiency or excess of mass and a gravity "anomaly," or deviation from the calculated or average value in a region. The largest anomalies, however, are only about 0.03% of the value of gravity; and even much smaller ones persist only because of such dynamic effects as plate motions.

Surface of the Earth

Some of the surface of the earth is coated with air and some with water, and our impressions are highly colored by the fact that we can breathe in and see through only one of them. Consider the impressions of hypothetical Aquarian explorers, who remain prudently sealed in their metallic space ship, sensing the earth with sonic probes that penetrate water as easily as air: Most of the surface is at one of two levels, and the logical Aquarians, not making special allowances for the water, would measure heights from the more extensive of these—the level that we call the deep sea floor.

Neither the ocean basin nor the continental "levels" are uniform everywhere. There are great, gently-sloping undulations with as much as 1–2 kilometers of relief that accompany the horizontal drifting of crustal plates. An example is the Mid-Atlantic Ridge. Rising about 5–6 kilometers above the undulating base "levels" are the great, steep-sided plateaus that earth scientists call "continents." The fringes of the tops, however, are covered with shallow water, so they are more extensive than what we call "land." The covered fringe is the continental shelf and the steep sides are the continental slope. Several kilometers above and below the two broad levels are elongate regions whose relief is the result of dynamic forces. These are the great continental mountain ranges, which are mainly made of folded and faulted sedimentary rocks, and the oceanic mountain ranges—which are the great straight lines of oceanic volcanoes, such as the Hawaiian group, and the curved volcanic chains of the western Pacific. Mountains of this last type have a relief of about 10 kilometers, because there are long trenches below the base level as well as mountains above.

Superimposed on the primary and secondary relief are the relatively small features produced by volcanoes, local folding and faulting, and erosion and deposition, but few of these would be noteworthy from space.

Interior

It might seem that all we know of the interior of the earth is speculative because we cannot enter it or even drill down any significant distance. However, we can probe through the earth with geophysical techniques and measure, thereby, the physical properties of the unseen materials. We can at least eliminate the possibility that the interior of the earth is made up of any of those

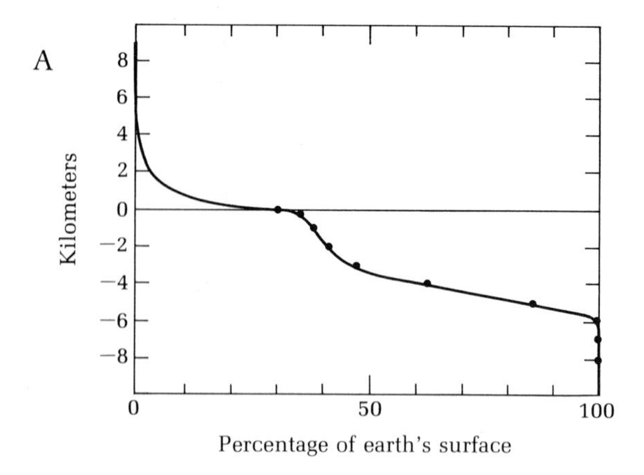

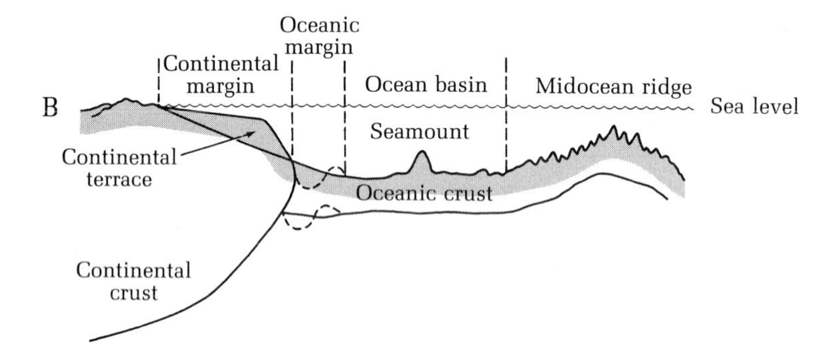

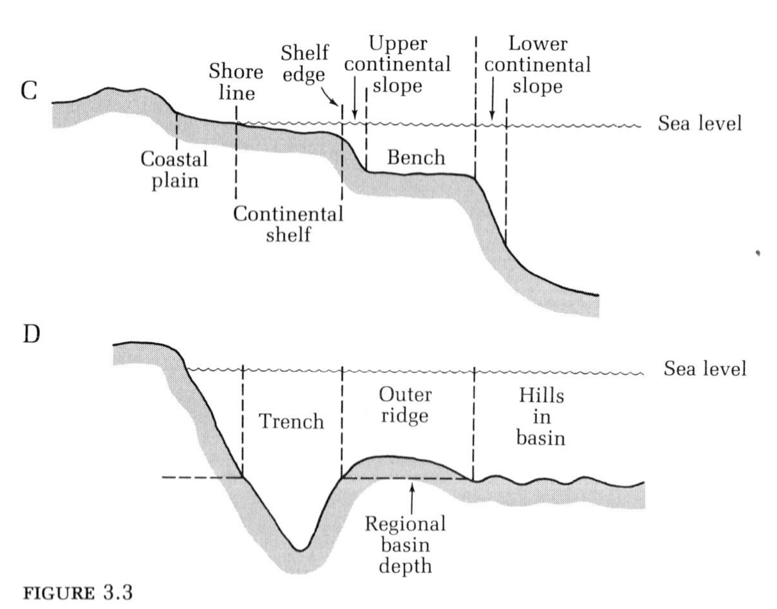

FIGURE 3.3

Distribution of levels (*A*) and major structural and geographical provinces (*B*, *C*, *D*): *A*, most of the surface of the earth is under water, and all but a tiny fraction is either near sea level or else at the level of the deep sea; *B*, generalized cross section of a continent and ocean basin; *C*, features of continental margins whose existence depends on the relative intensity of sedimentation and deformation; *D*, features of oceanic margins whose existence depends mainly on whether the margin is a plate boundary.

materials that do not have the observed properties. Likewise, we can map the distribution of the various properties, even though we cannot be certain about the materials.

The surface rocks themselves contain information about the interior. For example, continental and oceanic rocks are generally different (Chapter 4), and this suggests that they remain so for at least a small distance into the interior. The difference in elevation of continents and ocean basins also implies a difference in density that extends down for tens of kilometers into the interior. Moreover, samples of even deeper rocks are available at the surface because the interior rocks penetrate surface ones in a few places, and small fragments are commonly carried up by lava rising in volcanoes. The fragments are largely composed of dense, dark-colored, coarsely crystalline rocks that are relatively rich in magnesium and deficient in silica and aluminum compared to surface ones. They are also notably deficient in uranium, thorium, and potassium, which are the common radioactive elements that are capable of generating heat in the interior.

Various geophysical studies can help us to map the physical properties of the interior although samples are beyond reach. The most informative study by far is seismology, which analyzes earthquake waves after they penetrate the interior.

On the average, continental crust is about 30–35 kilometers thick, whereas the oceanic crust is only 5–6 kilometers thick. Both are embedded in the lithosphere, which is generally 50–100 kilometers thick. The asthenosphere is a thin layer with its top generally at a depth of 50–200 kilometers. Down to the bottom of the asthenosphere, the interior at a given depth seems to differ from place to place. The deeper interior is divided into the mantle and a core (Figure 3.4) that has about half the diameter—and, thus, an eighth of the volume—of the earth. The mantle is mainly solid, but the core is liquid, except for a small, dense inner core.

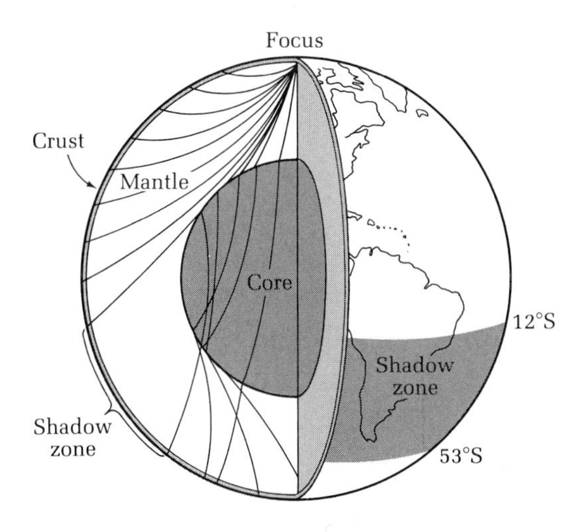

FIGURE 3.4
Different paths followed by energy radiated by an earthquake near the North Pole. By analysis of these paths, the interior can be divided into a core and a mantle overlain by a thin crust. Refraction in the core causes a shadow zone in which the quake is not detectable by body waves (but is by waves that are channeled along the surface). [Modified from Gilluly, Waters, and Woodford, *Principles of Geology*, W. H. Freeman and Company, copyright © 1968.]

The mantle and core seem to have relatively constant properties at any given depth. How is this compatible with the model of an earth accumulating from diverse planetesimals and meteorites? Perhaps the geophysical probes are incapable of detecting the plums deep in the pudding, or else the mantle has been stirred until it is relatively uniform at different levels.

COMPOSITION

Any attempt to estimate the composition of the deep interior of the earth "entails enormous difficulties and more or less questionable assumptions," in the words of the American physicist L. H. Aller in 1967. Despite the uncertainties, most earth scientists accept the working hypothesis that the mantle consists of high-pressure phases of silicate rocks that are not unlike those brought up from the upper mantle by volcanoes, and that have the average composition of stony meteorites.

Many meteorites, however, consist of nickel-iron; and whatever the proportion of stony to metallic meteorites, it is likely that a similar proportion of stony to metallic materials exists in the earth. The metals melt at lower temperatures than do the silicate minerals, so it is possible that, at the high temperatures in the mantle, they have become liquid and have trickled downward to form the core. In any event, the hypothesis that the core is slightly impure iron is very attractive for several reasons, apart from the fact that it provides us with a place to put the iron that presumably accumulated in meteoritic accretion. It is mainly liquid, as iron would be and silicates would not; it is a conductor, as is required to generate the magnetic field; and, with the addition or reasonable impurities, it has the proper density, considering the temperature and pressure.

HEAT

The most important characteristic of the interior of the earth, with regard to surficial tectonic processes, is the distribution of temperature. It is certainly hot in the interior, but we are woefully ignorant of most of the details of the temperature distribution in the earth. The interior heat flows out to the surface where we can measure it, but we don't know how it flows. The possibilities are (1) flow by conduction, (2) radiation, and (3) convection. Conduction, in which heat moves from atom to atom in a material without the material itself moving, is a very passive process that might not cause tectonic plates to move. Moreover, it is a slow process, and one that is incapable of bringing heat to the surface at the observed rate if the earth consists of meteoritic material. In the hot interior, it is possible that heat radiates through silicate minerals, like sunlight through air. This is a much more effective process than conduction, and it could possibly transmit the observed heat, but it is difficult to see how it would cause plates to move as they do.

In volcanoes, we can observe the third mode of heat flow, convection, in which the hot material itself moves and, thereby, carries heat with it. For example, glowing lava flows up into the crater of Kilauea volcano, in Hawaii, and forms a lake of boiling rock. A crust of solid rock forms occasionally, and motion pictures show that it breaks into small plates that behave exactly like the great tectonic plates of the earth's surface. Similar but slower convection cells may convey heat up from the interior of the earth and, in doing so, may move the tectonic plates.

Magnetic Field

The average magnetic field of the earth has the following characteristics:

1. It resembles the field that would result if a large bar magnet in the core were aligned along the axis of rotation. However, other characteristics suggest it is the result of a dynamo produced by the motion of a conductor in the core.
2. The intensity and direction of the field vary with time.
3. The field reverses episodically; that is, the north and south poles change places.

FIELD VARIATIONS

The direction of the field is described by two components: the first, the magnetic dip, is the angle from the horizontal assumed by a freely suspended magnetic needle; and the second, the magnetic declination, is the horizontal angle between the field and a meridian of longitude. Fluctuations with periods of hundreds of years have been recorded at the older observatories. In London, in 1576, a compass needle pointed 8° east of true north and dipped 69° north. By 1600, it pointed 10° east and dipped 72° north; in 1700, it pointed 6° west and dipped 74° north; in 1800, 24° west and 71° north; and, at present, it points 8° west and dips 66° north. Full many a mariner has blessed the magnetic observatory at Greenwich while entering the foggy Thames for London. Fluctuations of the sort observed at Greenwich occur as widespread patterns that drift westward. They seem to be the result of some phenomenon in the dynamo in the core.

The magnetic poles are the places where the lines of force are perpendicular to the surface. Thus, the north-seeking end of a compass needle points vertically downward at the north magnetic pole. This pole moves hundreds of kilometers in only a few decades, and it has been as much as 2000 kilometers from the pole of rotation during historical time.

We can examine longer time spans by studying magnetism of the past, or paleomagnetism. Many rocks contain crystals of iron or iron oxides that are made up of magnetic domains that behave like little magnets. When they are heated above a temperature called the Curie point, they lose their magnetism. This occurs in lava flows. When cooled in a magnetic field, they become magnetized more or less in the direction of the field, just as a bar of iron can be converted to a bar magnet by placing it in a magnetic field. Thus, almost all lava flows contain a record of the direction of the field when they cooled. It appears from paleomagnetic studies that the magnetic pole has been confined for a very long time to a limited area with the pole of rotation in the center. The *average* position of the two poles has been the same for millions of years.

The intensity of the earth's magnetic field was measured in 1835 by Karl F. Gauss, a German mathematician, astronomer, and physicist. Since that time, it has decreased throughout the world by 6%. Paleomagnetic techniques developed by the French physicist Emile Thellier can recapture the intensity of the field when ferromagnetic materials cooled through the Curie point. Because orientation is immaterial, the intensity can be measured using fragments of clay pots, which are superabundant in dated archeological sites. These show that the field was about 1.5 times the present value 2000 years ago and again 8000 years ago, but that it dropped to only half of the present intensity 6000 years ago. If it can fluctuate so much so quickly, might it not drop to zero and even reverse itself as the magnetic field of the sun does every 11 years? We now turn our attention to the remarkable record that shows that reversals have indeed occurred, although with a frequency of hundreds of thousands of years.

REVERSALS

Theory says that, if a dynamo can maintain a certain magnetic field, it can also maintain an exactly opposite one, but it need not do so. In the words of the British geophysicist E. C. Bullard, writing on this subject, "a man can stand on his head, but it does not follow that he will."

In the early 1960s, the American earth scientists Allan Cox and Richard Doell began to apply accurate radioactive dating to samples whose orientation of the magnetic field they also measured. They found that rocks of the same age everywhere in the world have the same polarity, although it is *normal* at some times and reversed at others (these times are called polarity epochs). This was a clear demonstration that most reversals are the consequence of a global phenomenon, which can only be the reversal of the field itself.

This work has subsequently been elaborated and confirmed in special studies in such places as Iceland and Hawaii, where frequent lava flows are stacked in eroded but undistorted piles. It is now evident that the present polarity epoch began about 0.7 million years ago and that it was preceded by

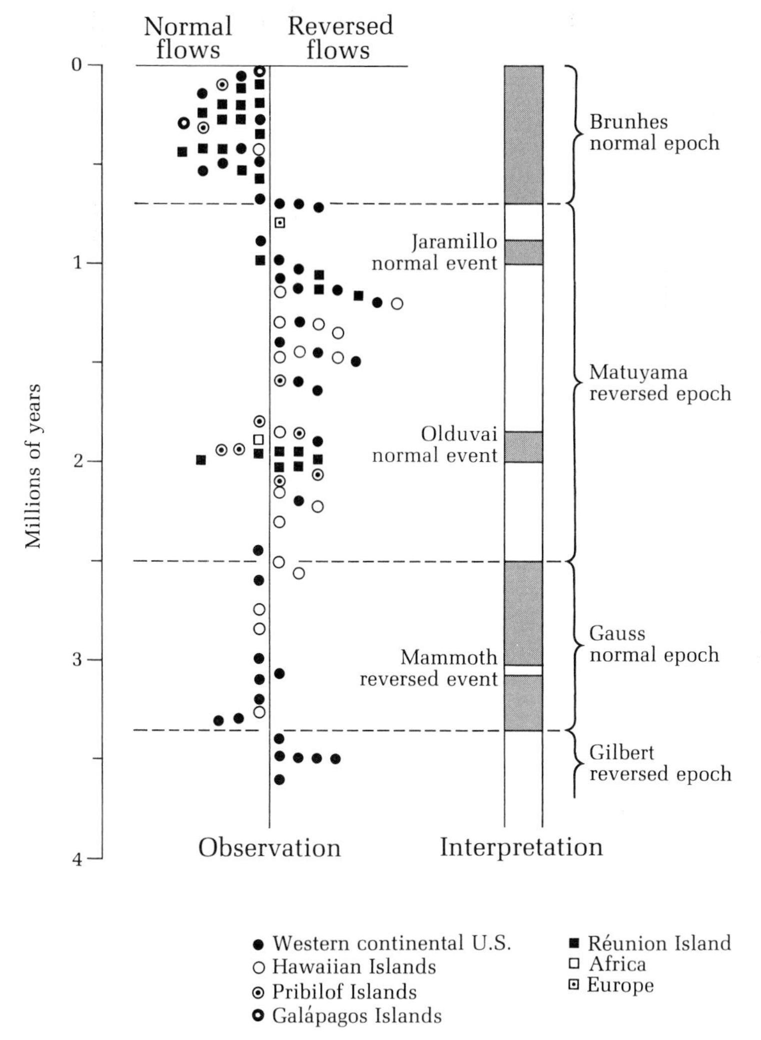

FIGURE 3.5

Time scale for recent reversals of the earth's magnetic field was established on the basis of paleomagnetic data and radiometric age obtained for nearly 100 volcanic formations in both hemispheres. Here, the flows with "normal" and "reversed" magnetism are arranged according to their ages. It is clear that the data fall into four principal time groupings, or geomagnetic polarity "epochs," during which the field was entirely or predominantly of one polarity. Superimposed on the epochs are shorter polarity "events." [From Cox, Dalrymple, and Doell, "Reversals of the Earth's Magnetic Field," copyright © 1967 by Scientific American, Inc. (all rights reserved).]

an epoch of reversed polarity with a duration of 1.8 million years; before that, there were a normal and a reversed epoch with durations of 0.8 million years and 0.6 million years, respectively.

These longer epochs were interrupted by at least three shorter events lasting only about 0.1 to 0.2 million years. The Matuyama reversed epoch, for

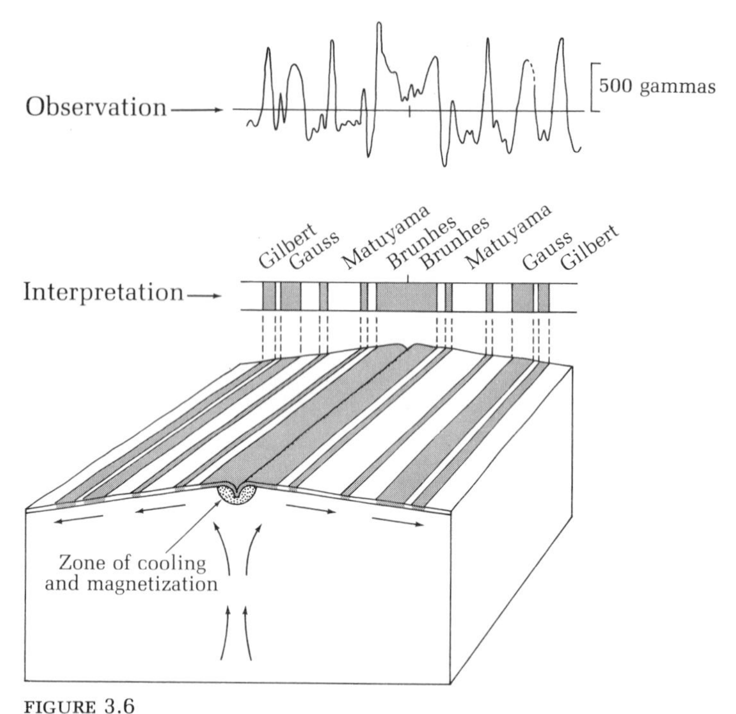

FIGURE 3.6
The observed, bilaterally symmetrical, magnetic-anomaly stripes of the sea floor can be explained by the same magnetic reversals being recorded in the cooling lavas of a spreading center. [After Pitman and Heirtzler (1966); and Cox, Dalrymple, and Doell, "Reversals of the Earth's Magnetic Field," copyright © 1967 by Scientific American, Inc. (all rights reserved).]

example, was interrupted by the Olduvai normal event about 1.8 million years ago. The dating method cannot identify short events older than 4 million years. However, the method can distinguish epochs of a million years duration tens of millions of years in the past, and longer epochs far back into geologic time. Reversals have occurred from time to time throughout the decipherable history of the earth. This implies that the core formed early.

We shall return to paleomagnetism more than once, because it is the principal means for mapping continental drift and plate tectonics.

Energy Balance of the Earth

We have examined the earth as though we were astronauts or starniks approaching from the constellation of Aquarius. Following the rational program of investigation outlined by the Aquarian Academy of Sciences, we now know all the major characteristics of the planet earth and the forces that affect it. The next step is to evaluate the relative magnitudes of the various forces. Any further study would be hampered without the insight that such a comparison provides. We might, for example, be misled into thinking that the forces that build mountains and pour lava from volcanoes must be more potent than the light of the sun, which is powerless to penetrate a thin layer of suntan lotion. But we shall see!

MAJOR SOURCES OF ENERGY

Work expended on the surface of the earth comes from several major sources: radiation from the sun, tidal attraction by the moon and the sun, and heat flow from the interior. Only a tiny fraction of the radiant energy of the sun reaches the earth, but it is unequivocally the most powerful of all the sources. If we think of the earth's internal heat flow (30×10^{12} watts) as one "earth-power unit," the tidal power of the moon and sun is 0.1 units and the solar radiation striking the earth is almost 6000 units.

The ratios of power that directly affect the surface are slightly different, because 77% of the solar power is reradiated into space as light and heat. Even so, an overwhelming 1500 units remains.

In addition to the active sources of power, the earth has many important sources of energy that have the capacity of doing work if circumstances change. Preeminent by far is the kinetic energy of the orbital motion, some 2.7×10^{33} joules. If this were converted into power, it would be equivalent to 1000 earth-power units for 3000 million years, or for most of the age of the earth. If there were some astronomical way—perhaps a retrograde collision with an asteroid—to quickly change the orbital velocity, the effects on the earth would be cataclysmic.

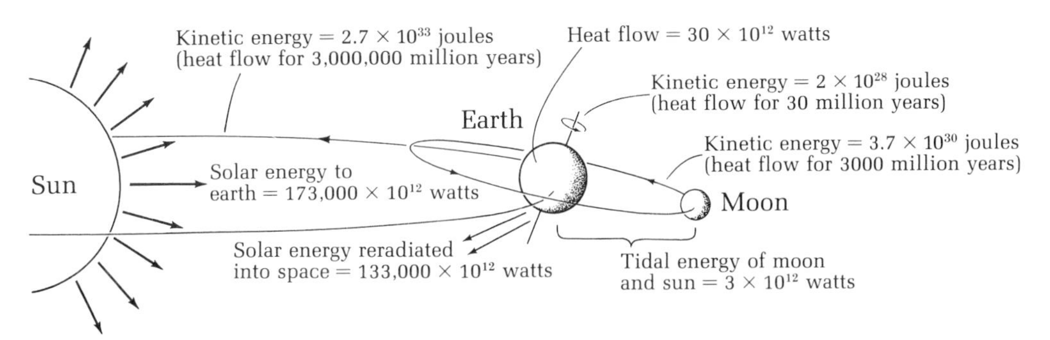

FIGURE 3.7
The motions of the earth are the result of the delicate balance of enormous forces that have changed during geologic time.

BOX 3.1 ENERGY, POWER, AND WORK

In order to compare the effectiveness of forces, we must first consider the appropriate ways to measure energy, power, and work. We begin with energy, which is the capacity for doing work and is of two kinds, potential and kinetic. Consider potential energy: We build a dam across a mountain river and impound water in a lake. The water does nothing in the lake, but it has the potential to ravage in an uncontrolled flood or to flow through penstocks and generate power. Its potential energy (PE) is defined as equal to the product of mass (m) times gravity (g) times height (h) above a certain level (usually sea level).

$$PE = mgh$$

Energy may be expressed in dyne-centimeters (= dyne × cm). The dyne is a unit of force, and force is equal to mass times acceleration ($F = ma$). Specifically, the dyne is equal to a mass of one gram accelerated one centimeter per second per second.

$$\text{dyne} = 1\ \text{g}\ (\text{cm}/\text{sec}^2)$$

The lake water is in an unstable situation: sooner or later, as the lake overspills the dam or the basin fills with mud, the water moves downhill and potential energy becomes kinetic. Kinetic energy (KE) is defined as equal to half the product of the mass times the square of the velocity (v).

$$KE = \frac{1}{2} mv^2$$

It may also be expressed in dyne-centimeters.

The kinetic energy of the flowing water does work if, for example, it transports a mass of sediment for a distance. Work, thus, is defined as force acting for a distance.

$$\text{work} = \text{force} \times \text{distance}$$

$$= \frac{\text{gm cm}}{\text{sec}^2} \times \text{cm}$$

$$= \frac{\text{gm cm}^2}{\text{sec}^2}$$

Work may be expressed in ergs or joules. An erg is a dyne times a centimeter, and a joule is 10 million ergs.

$$\text{erg} = \text{dyne} \times \text{cm}$$

$$\text{joule} = 10^7\ \text{erg}$$

The river can transport a mass of sediment a certain distance quickly in a flood or slowly a grain at a time. The work is the same, but the power, which is defined as the rate of work and is measured in watts, is different.

$$\text{power} = \frac{\text{work}}{\text{time}} = \frac{\text{joule}}{\text{sec}} = \text{watt}$$

We can complete this background by considering a river—in Los Angeles, perhaps—flowing in a conduit of concrete. It is losing potential energy as it flows down toward the sea. It may not be accelerating. If not, where is the energy going? It is dissipated in friction with the concrete and in internal friction in the water. Friction, as we all know from the experience of sliding on a rough surface or braking a car to avoid a crash, generates heat. Energy, therefore, can be equated in some way to heat. Heat, which is defined as the capacity to elevate the temperature of a mass, may be expressed in calories. The calorie is defined as the heat required to elevate the temperature of a gram of water by one degree Celsius, from 14.5°C to 15.5°C, at normal pressure.

Careful experiments by Count Rumford in 1798 elucidated the following relation between the work done drilling the bore of a cannon and the heating that occurred:

$$4.185\ \text{joules} = 1\ \text{calorie}$$

This is the mechanical equivalent of heat.

The orbital motion of the moon has a kinetic energy equivalent to one earth-power unit for 3000 million years, and the energy of the rotation of the earth is equivalent to one unit for 30 million years. As we know, both of these motions change gradually, and the changes have been large in the geologic past.

In sum, the earth is subject to enormous, delicately balanced, changing forces. The environment that seems eternal is eternally changing.

ENERGY BALANCE

Most of the solar energy radiated to the earth is reradiated after it heats the air, sea, and land (Figure 3.8). Most of the remainder is expended in evaporating water. If this continued without interruption, the oceans would evaporate into the air in only 3000 years! A trifling amount of water is stored as vapor in the air, and another dollop is stored in snow and ice. Most of the water outside the ocean is in constant motion as vapor going up and rain coming down, vapor going up and rain coming down, in a never-ending labor. It is evident on a global scale, not only that evaporation is matched by precipitation, but that that balance is extremely delicate. If there is a slight increase in the amount of water stored as ice, the world will become covered with glaciers. How fortunate for us the balance is where it is at the moment.

The water is transported by currents in the atmosphere and ocean. The sun expends 370×10^{12} watts on this work, or little more than ten times the internal heat flow. Most of the work goes into mass convection, which is caused by differential heating in the tropics and polar regions. We shall examine this in more detail in Chapter 10. The winds of the air stress the surface of the ocean and produce waves (2.5×10^{12} watts) and currents (2.4×10^{12} watts). Even though the interaction is inefficient and not much energy is transferred, it is of great geologic importance. By this mechanism, the energy of the sun is concentrated at sea level to erode cliffs and transport sand in a vigorous environment. Much of the tidal energy (3×10^{12} watts) is also dissipated in very shallow water, which further intensifies these geologic processes.

Next in importance in the energy balance is photosynthesis, by which plants convert solar energy into tissue, which then serves to feed entire food chains of animals—herbivores directly and carnivores indirectly. Just as water can be stored as ice, animal and vegetable tissues can be stored geologically as coal and oil. This energy of fossil sunlight, which drives our civilization, is trifling in the energy balance. It is equivalent to only 3600 years of photosynthesis. Unfortunately, man can exhaust the earth's supply of fossil fuels in only a fraction of that time (see Chapter 18). This only serves to emphasize that, in the whole history of the earth, only a negligible amount of solar energy has been stored for more than a few years at a time.

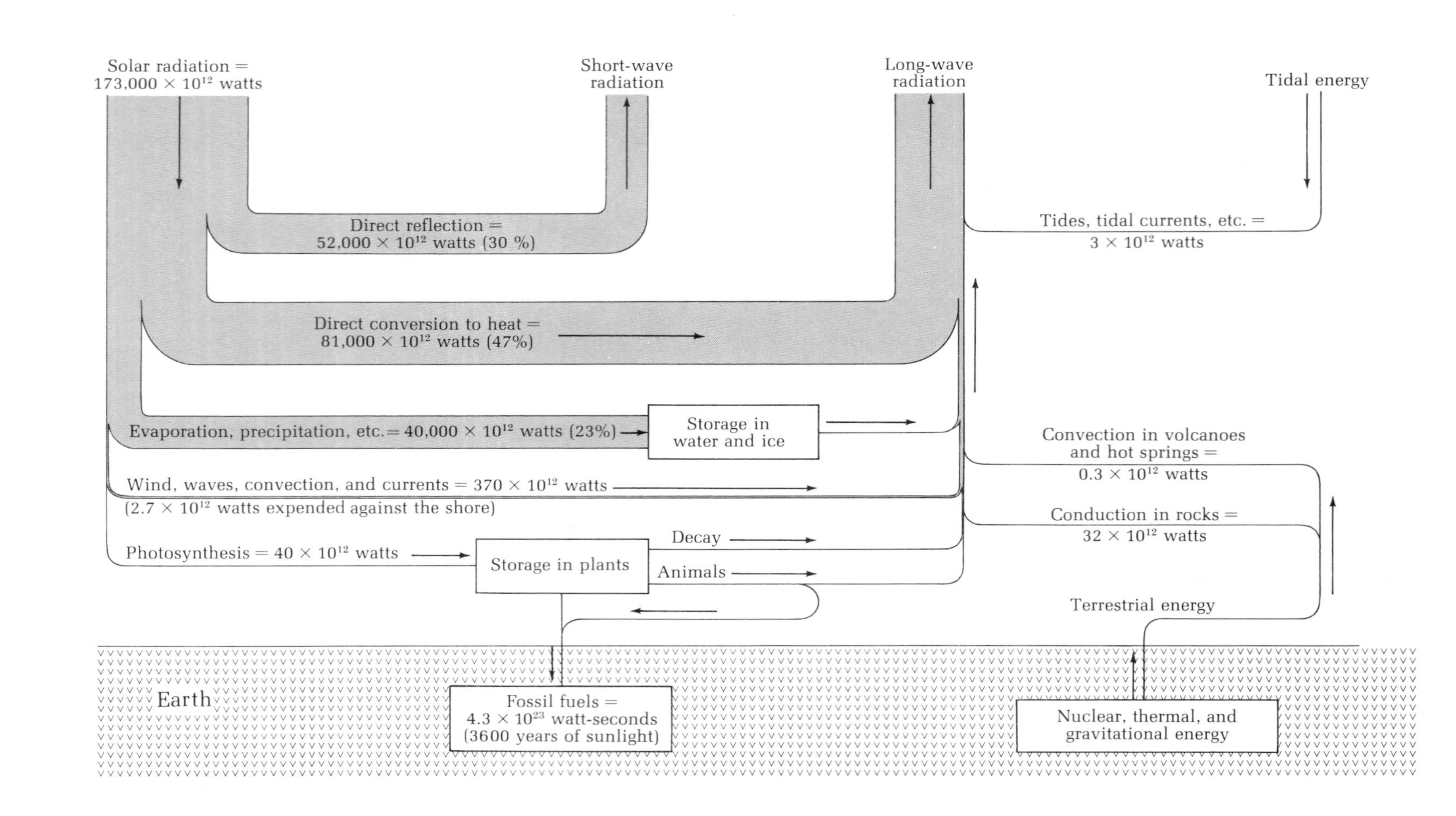
Solar radiation = 173,000 × 10^12 watts
Short-wave radiation
Long-wave radiation
Tidal energy
Direct reflection = 52,000 × 10^12 watts (30 %)
Tides, tidal currents, etc. = 3 × 10^12 watts
Direct conversion to heat = 81,000 × 10^12 watts (47%)
Evaporation, precipitation, etc. = 40,000 × 10^12 watts (23%)
Storage in water and ice
Convection in volcanoes and hot springs = 0.3 × 10^12 watts
Wind, waves, convection, and currents = 370 × 10^12 watts
(2.7 × 10^12 watts expended against the shore)
Conduction in rocks = 32 × 10^12 watts
Photosynthesis = 40 × 10^12 watts
Storage in plants
Decay
Animals
Terrestrial energy
Earth
Fossil fuels = 4.3 × 10^23 watt-seconds (3600 years of sunlight)
Nuclear, thermal, and gravitational energy

We have followed the enormous energy of the sun down to the level of photosynthesis. Now we turn to energy from the interior of the earth. The total amount of energy from the initial heat of the formation of the earth, from gravitational sorting, and from nuclear decay is unknown. All we know is that about 30×10^{12} watts escapes from the interior as general heat flow and 0.3×10^{12} watts is localized in volcanoes and hot springs. The additional energy of the interior drives tectonic plates, elevates mountains, and generally deforms the surface of the earth. That it is small compared to the other forces we have discussed is indicated by the fact that it is incapable of distorting the surface of the earth very much from the equilibrium shape for a rotating body.

There are many mountains and oceanic ridges but, even so, their development has required little energy on the global scale. Thus, the measured heat flow is a reasonable approximation of the energy emanating from the interior. Moreover, the supply of energy grossly exceeds that expended in deforming the surface. The earth appears quite capable of much more mountain building, especially if the energy is favorably focused at the boundaries of moving tectonic plates.

FOCUSING ENERGY

We have all noticed the importance of focusing energy. A nail penetrates a board, but the broader hammer head that drives it does not. Geologically, focusing occurs in both space and in time. Gravitational energy, for example, is important on steep slopes, but it is not on level plains. Similarly, sea waves are important only at sea level, tectonic effects occur at plate boundaries, and so on. Likewise, for long periods, heat flow dominates tectonics, but if some of the potential energy of rotation or of the earth–moon tidal interaction is briefly released, it may take an important or preeminent role for a while.

The dynamic processes of the earth, and the temporary products that are land forms, sediments, and rocks, are the results of the focusing of different forms of energy. The energy available to atmospheric, oceanic, and solid-earth processes, or to life, is much greater than necessary for geologic processes on the surface. Different forms of energy may be focused in sequence.

FIGURE 3.8

Flow of energy to and from the earth is depicted by means of bands and lines that suggest, by their width, the contribution of each item to the earth's energy budget. The principal inputs are solar radiation, tidal energy, and the energy from nuclear, thermal, and gravitational sources. More than 99% of the input is solar radiation. The apportionment of incoming solar radiation is indicated by the horizontal bands beginning with "Direct reflection" and reading downward. The smallest portion goes to photosynthesis. Dead plants and animals buried in the earth give rise to fossil fuels, which may contain stored energy from millions of years past but are equivalent to only a few thousand years of plant production. [From Hubbert, "The Energy Resources of the Earth," copyright © 1971 by Scientific American, Inc. (All rights reserved).]

For example, internal heat builds a volcanic island, and solar energy later drives waves to plane it flat. Solar energy then makes it possible for life to cover it with a coral reef; and, even later, internal heat destroys the volcano by driving the crustal plate, on which it sits, back into the mantle whence it came.

The Aquarian expedition might be very dubious about founding a permanent colony here because of the obvious risks and cost of living on such a dynamic planet. But we have no choice.

Summary

1. The earth is almost spherical because the attraction of its gravity far exceeds the strength of its materials.
2. The earth bulges at the equator because it spins. The size of the bulge has decreased as the rate of spinning has decreased in geologic time.
3. The axis of rotation wobbles complexly like the axis of a spinning top. The changing orientation of the axis influences the seasons and climate.
4. Although the lithosphere is relatively strong and rigid with regard to plate motions, which are horizontal, it is unable to support large vertical loads without sinking.
5. The principle of isostasy says that columns of unit area have the same mass above a certain level in the earth. Thus, mountains are high because they have a low density.
6. The surface of the earth has two dominant levels: one near sea level and one about 5 kilometers deeper at the level of the deep-sea floor.
7. The earth is layered with a crust at the surface, a mantle below it, and a core in the center. The crust and mantle are solid and consist predominately of silicate rocks. The core is mainly liquid, although its very center is solid. It is probably a nickel-iron alloy.
8. The interior of the earth is heated by radioactivity. The heat probably escapes by convective overturn of the mantle.
9. It is probable that the magnetic field of the earth is produced by motions of the core. The intensity and direction of the field change with time, and the field often reverses at intervals of about a million years.
10. The history of the magnetic reversals is commonly preserved, even for a billion years, in the orientation of magnetically sensitive minerals.
11. Enormous forces hold the earth in balance, and none of them are constant.
12. The sun is by far the largest source of energy to the earth, even though much of it is radiated back to space.
13. Heat from the interior of the earth powers the movement of tectonic plates.

Discussion Questions

1. Why do the size and mass of the earth control its shape?
2. Will the North Star (Polaris) be a useful indicator of the north direction 13,000 years from now?
3. Visualize a high mountain connected to a nearby deep lake by a river. What vertical movements occur in the lithosphere if the mountain is eroded away and deposited as sand in the lake?
4. How can the earth be composed of uniform layers if it is an accumulation of meteoroids and asteroids?
5. What is the evidence that the magnetic field varies in intensity and direction? What is the evidence that it reverses?
6. What would happen if radioactive heat could not escape from the mantle? Considering what was discussed in Chapter 2, do you believe that this has ever happened?
7. How has the energy of ancient sunlight been stored?

References

Ahrens, L. H., 1965. *Distribution of the Elements in Our Planet.* New York: McGraw-Hill.

Aller, L. H., 1967. Earth, chemical composition of and its comparison with that of the moon and planets. *In* J. K. Runcorn, ed., *International Dictionary of Geophysics,* p. 285. New York: Pergamon Press. [A most useful dictionary for the professional.]

Bullard, E. C., 1965. Historical introduction to terrestrial heat flow. *In* W. Lee, ed., *Terrestrial Heat Flow* (Amer. Geophys. Union Geophys. Monogr. 8), pp. 1–6. Washington, D.C.: American Geophysical Union.

———, 1966. Solar and terrestrial dynamos. *In* G. Barbera, ed., *Atti del convegno sui campi magnetici solari,* pp. 1–8. Firenze: Comitato nazionale per le manifestazioni celebrative del IV centenario della nascita di Galileo Galilei. [They appear similar in many respects except that the solar dynamo is easier to study.]

Clark, S. P., Jr., 1966. Composition of rocks. *In* S. P. Clark, Jr., ed., *Handbook of physical constants* (Geol. Soc. Amer. Mem. 97), pp. 2–5. New York: Geological Society of America.

Cox, A., G. B. Dalrymple, and R. R. Doell, 1967. Reversals of the earth's magnetic field. *Sci. Amer.* 216(2):44–54.

Daly, R. A., 1940. *Strength and Structure of the Earth.* New York: Prentice-Hall. [The existence of a lithosphere and asthenosphere is indicated by gravity measurements.]

Dutton, C. E., 1889. On some of the greater problems of physical geology. *Bull. Phil. Soc. Wash.* 11:51–64. (Reprinted in 1925 in *J. Wash. Acad. Sci.* 15:359–369.) [The definition of isostasy.]

Gilluly, J., A. C. Waters, and A. O. Woodford, 1968. *Principles of Geology* (3rd ed.). San Francisco: W. H. Freeman and Company.

Hubbert, M. K., 1971. The energy resources of the earth. *Sci. Amer.* 224(3): 60–84. (Available as *Sci. Amer.* Offprint 663.) [The end is foreseeable.]

Munk, W. H., and G. J. F. MacDonald, 1960. *The Rotation of the Earth: A Geophysical Discussion.* London: Cambridge University Press. [A lucid, prize-winning account.]

Oort, A. H., 1970. The energy cycle of the earth. *Sci. Amer.* 223(3):54–63. (Available as *Sci. Amer.* Offprint 1189.)

Pitman, W. C., III, and J. R. Heirtzler, 1966. Magnetic anomalies over the Pacific-Antarctic Ridge. *Science* 154(3753):1164–1171.

Ronov, A. B., and A. A. Yaroshevskiy, 1969. Chemical structure of the earth's crust. *Geochem. Int.* 4(6):1041–1066.

Thellier, E., and O. Thellier, 1959. Sur l'intensité du champ magnétique terrestre dans le passé historique et geologique. *Ann. Géophys.* 15:285–376. [Measuring magnetic field intensity at dated archeological sites.]

4

MATERIALS OF THE CRUST

What stuff 'tis made of, whereof it is born,
I am to learn.

Shakespeare, THE MERCHANT OF VENICE

I hate definitions.

Disraeli

Measurements from a space ship are relatively tidy, in that they are not obscured by a mass of apparently confusing detail. We may surmise, therefore, that many scientists of the Aquarian expedition to the earth would face the first sampling of the surface materials of the crust with mixed feelings. They would be excited because the materials would contain not only valuable resources but also all that is preserved of the history of the planet; and they would be chilled at the thought of the difficulties that might arise in interpreting the meaning of the millions of samples they might be required to collect.

The most cursory sort of world-wide sampling would confirm their gloomy expectations. Under thin layers of air, soil, water, or ice, they would find solid rock composed mainly of small particles of inorganic minerals, although with important traces of life as well. Thus, any understanding of the origin and history of the surface materials requires study not only of the rocks but also of the minerals. The minerals, moreover, are mainly compounds of silicon and oxygen with other elements, which, as we shall see, are relatively complex compared to other possible materials, such as elemental (uncombined) metals or compounds of carbon and oxygen with other elements.

Like the Aquarians, we must try to understand the nature of the complex materials of the crust of the earth. This is because the composition and structure of these materials affect their interactions in the various geological cycles. To view it in another way, the materials *are* the products of the activities of the various cycles, and their properties and distribution tell us much of what is known about the cycles.

Fortunately for our purposes, the necessary millions of samples have been collected and partially studied. In the following brief introduction to some aspects of these studies, we shall begin with the nature of oxides of silicon and proceed to minerals and then to the rocks that are composed of them. We shall continue with information about the average composition of the crust, and conclude with the distribution of materials in the various geologic subdivisions of the crust.

Chemistry of Silicates

Silicon combines with oxygen to form an oxide with the formula SiO_2 (see Appendix 3 for symbols of chemical elements). This oxide, which is called silicon dioxide or silica, occurs as the common mineral quartz, as well as in other forms. If silica is mixed in just the right proportions with oxides of various metals and then heated, new compounds are formed. This singular ability of silica to combine with different amounts of metallic oxides makes possible the numerous and complex silicates that characterize the crust of the earth.

The silicates of magnesium and sodium illustrate both the complexities of silicates and the underlying simplicity of different proportions of oxides. Both $MgSiO_3$ and Mg_2SiO_4 are widespread minerals. They may be remembered more easily if they are written as ratios of oxides—that is, as $MgO \cdot SiO_2$ and $2MgO \cdot SiO_2$, respectively. Likewise, the various silicates of sodium may be considered as merely differing in the ratio of oxides. Thus, Na_2SiO_3 may also be written as $Na_2O \cdot SiO_2$, Na_4SiO_4 as $2Na_2O \cdot SiO_2$, Na_4SiO_6 as $2Na_2O \cdot 2SiO_2$, and so on. Many of the most common minerals are combinations of more than one metallic oxide with silica. Because of the relative complexity of the chemical formula of such a mineral, expressing the formula as a ratio of oxides can be particularly helpful. For example, the mineral formula $CaAl_2Si_2O_8$ can be expressed as $CaO \cdot Al_2O_3 \cdot 2SiO_2$.

The silicates vary greatly, but most of the important ones are relatively insoluble crystalline solids that melt at high temperatures to form viscous liquids. Otherwise, their various compositions and structures are accompanied by correspondingly variable properties.

Structure of Silicates

A crystal is a solid whose atoms or molecules are arranged in a definite three-dimensional pattern, or lattice. The simple chemical formulas of silicates do not tell us anything about the arrangement or structural pattern of their lattices. Inasmuch as the lattices determine several of their properties, it is important to determine the crystalline structure of the silicates as well as their chemistry.

Just as the chemistry of the silicates can be understood in terms of ratios of oxides, their structure can be visualized in terms of the geometrical forms of tetrahedra. The chemical formula of a silicate tetrahedron is SiO_4^{4-}, that is, it includes one atom of silicon with an electrical charge of +4 and four atoms of oxygen, each of which has a charge of −2. The four (spherical) oxygen atoms are disposed at the corners of the tetrahedron. The pattern can be duplicated by placing three baseballs in a triangle on a table and then piling one on top. The (spherical) silicon atom fits into the small cavity in the center between the four oxygen atoms.

The silicate tetrahedron is held together because each oxygen atom is bonded to the silicon atom by sharing an electron with it. However, each oxygen atom has *two* electrons to share; thus, the four corners of the tetrahedron must be bonded to other atoms to give a neutral electrical balance. In the simplest arrangement, each oxygen atom is bonded to another silicon atom in a tightly bound lattice of interconnected tetrahdera to form SiO_2 (quartz). Because of the interconnections, the lattice is difficult to break. Thus, quartz has a high melting temperature and fractures with an irregualr surface.

Most oxygen atoms in silicates are bonded to positively charged ions of metallic elements. Some simple silicates have each oxygen atom in the tetrahedron bonded to ions of the same metal. In others, some oxygen atoms are bonded to silicon atoms and others to one silicon and one metallic ion. The tetrahedra form chains and sheets in more complex silicates, and these larger structures are bonded to metallic ions at the sides of the chains or the tops and bottoms of the sheets. These larger structures determine many physical properties of the silicates. Some, such as asbestos, are long fibers, and others—mica, for example—break into very thin sheets.

Silicon is an element with many intermediate characteristics. It is neither a poor electrical conductor like carbon nor a good one like the metals. Instead, it is a semiconductor—one of the materials whose electrical properties are utilized in transistors. Chemically, silicon is also intermediate between the strongly metallic and strongly nonmetallic elements, and it does not readily form ions. It tends, rather, to accommodate to elements that do. Thus, the silicate structure adapts to leave enough oxygen bonds to satisfy any metallic ions that are present. The silicates, accordingly, exhibit a wide variety of structures, as well as a variety of oxide ratios.

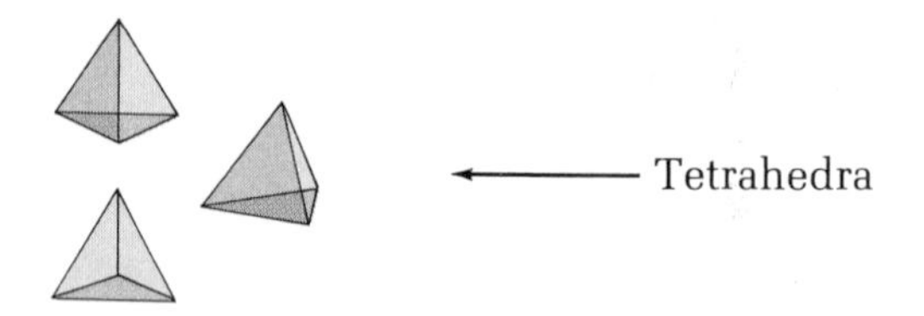

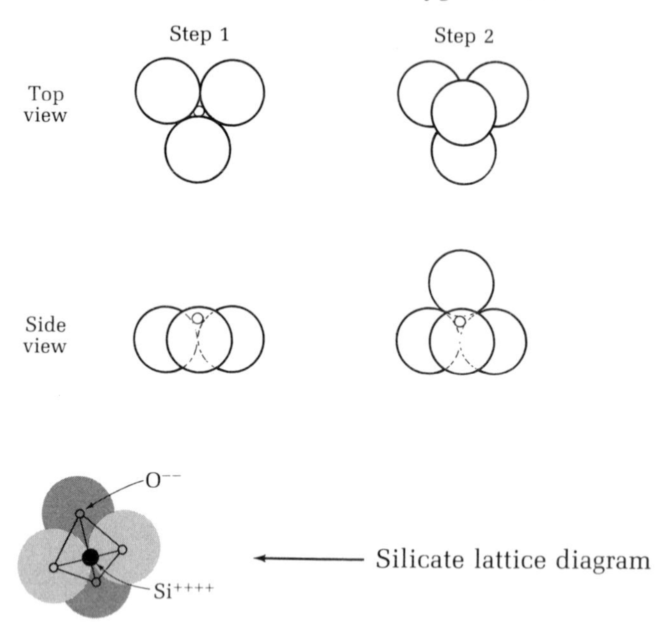

FIGURE 4.1
The centers of four spheres stacked in a particular way mark the apices of a tetrahedron. So do the oxygen atoms of silicate molecules.

The accommodating nature of silicon permits a ready substitution of one metallic ion for another in a silicate structure, provided that the ions have a similar size and electrical charge. For this reason, a number of important minerals are solid solutions containing variable quantities of each of two metallic ions. Such a mineral is intermediate between two other minerals, known as the end members, each of which contains only one of the metallic ions. Olivine, for example, whose chemical formula is $(Mg,Fe)_2SiO_4$, is a group of minerals grading between fayalite, Fe_2SiO_4, and forsterite, Mg_2SiO_4.

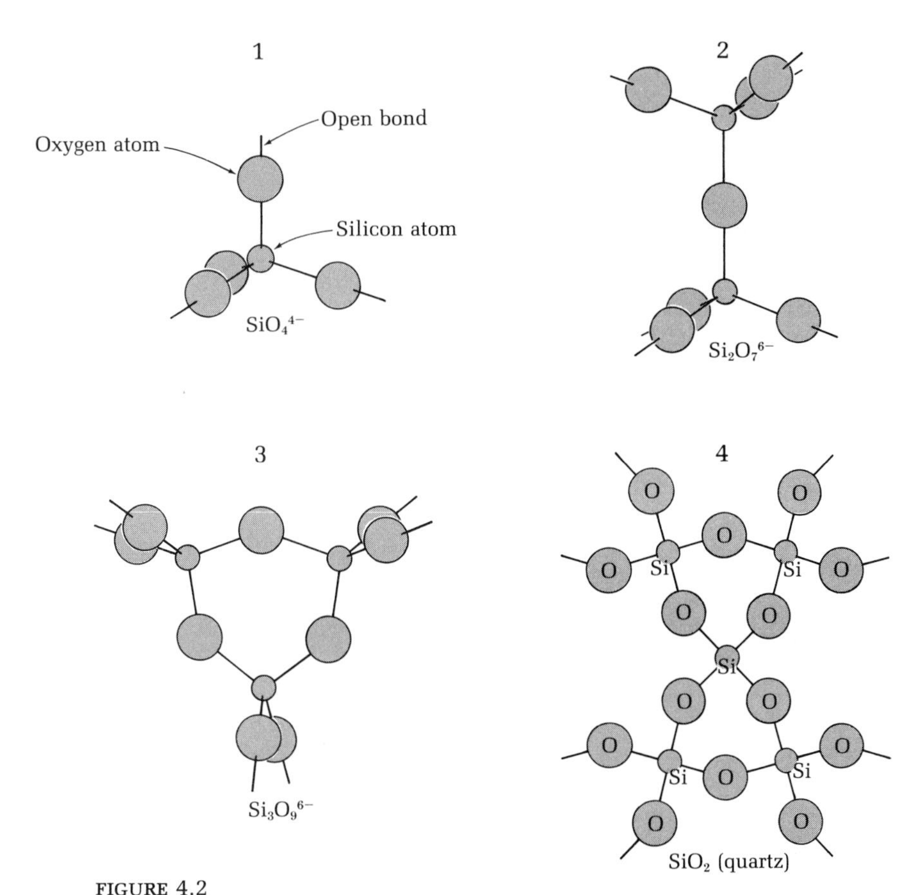

FIGURE 4.2

A silicate tetrahedron (*1*) may be bonded to other tetrahedra, by sharing one or more oxygen atoms (*2, 3, 4*), or to metallic ions by ionic bonds. These options result in a wide range of possible structures. The metallic ions may bond with oxygen atoms with an open link.

Common Minerals

We have been using the word "mineral," until now, in its common sense: "any chemical element or compound occurring naturally." The word is used in various ways, however, by different scientists—even by different *earth* scientists. A distinction between natural and artificial diamonds that makes one a mineral and the other not seems unnecessarily arbitrary. For these reasons, the definition may be broadened, for some purposes; it may also be even more restricted, for certain other purposes, to include only inorganic, crystalline solids.

Many people study minerals as a hobby (See Box 4.1) because they are interesting, valuable, and beautiful, but others, on first acquaintance, are repelled by the necessity of learning an imposing list of names. The reaction of a student of chemistry might be that $CaCO_3$ is "calcium carbonate" and it is unnecessary to call it "calcite." This, however, is not correct, because $CaCO_3$ occurs in two mineral forms that differ in crystalline structure and other properties: one is calcite and the other is aragonite, and, for many purposes, it is necessary to distinguish between them.

By far the purest of the solid, crystalline compounds known are produced by man to make lasers and transistors. Naturally occurring minerals are rarely very pure, whether they are elements, compounds, or even solid solutions. Thus, there is another reason for applying mineral rather than chemical names. This is particularly true of the silicates, because of the ease of substitution of various ions. Calcite also commonly contains minor quantities of iron and magnesium; when the chemical compounds $CaCO_3$, $FeCO_3$, and $MgCO_3$ all occur in the same crystal lattice, it is highly convenient to be able to call them by a single mineral name.

There are about 1500 known minerals, but most of them are rare and can be disregarded for many purposes. Other minerals, including many ores and semiprecious gems, are common enough to be represented in most mineral collections but are seldom found except in local concentrations in veins or thin layers. They are economically important, in many cases, simply because they *are* locally concentrated. If they were spread throughout a large volume of ordinary rock, it would cost more to obtain them than they are worth.

The minerals that do make up ordinary rock are few; some of them are members of solid solutions, and can be grouped under collective names. Quartz and the silicate known as feldspar together make up 63% of the crust of the earth. Hydrated silicates—amphibole, mica, and clay—total 14.6%; the magnesium and iron silicates—pyroxene and olivine—add another 14%; the carbonates—calcite, aragonite, and dolomite—amount to 2%; and all the other minerals add up to only 6.4% of the volume of the crust.

In sum, it is necessary to study only half a dozen minerals or groups of minerals in order to understand the important ones in common rocks. These minerals or groups are relatively easy to distinguish from one another by such familiar properties as color, density, and hardness, and by such less familiar properties as crystal form and cleavage.

The "form" of a crystal refers to the external shape that a perfect one would have if it solidified in a liquid or gas without being restricted or hindered by neighboring solids. The atoms within the crystal are arranged in a lattice that has definite fixed geometrical patterns that can be expressed as a number of intersecting planes. The atoms at the surface of the crystal are also in the lattice and, therefore, are in the same geometrical patterns; they form smooth planes that intersect at sharp angles and determine the crystal's form.

The forms of many perfect crystals are distinctive and, in combination with other properties, they facilitate identification. But perfect crystals of common minerals are rare because they usually interfere with each other as they crystallize out of molten rock (magma). They do not all crystallize out at the same time, but, rather, in a definite sequence in each magma. Those that crystallize first are least impeded; thus, they are the ones most apt to exhibit good crystal form. They produce a network of touching and intersecting solid crystals within the magma, and the remaining minerals must fill in the diminishing spaces between them.

Although the external crystalline form of common minerals is rarely evident in rocks, the internal lattice may often be seen. This is because of the property known as cleavage, which is the tendency to split along planes that are determined by the crystal lattice. When a crystal is broken, cleavage planes may appear; they can be recognized as such because they are mirror-smooth and reflect light if held at the proper angles. A mineral may also display cleavage as a system of fine, parallel cracks that are the surface expression of planes that extend into the crystal. Both the ease of cleaving and the angles between cleavage planes are diagnostic properties of minerals.

Quartz, SiO_2, is abundant not only in mixtures with other minerals in most common rocks but also in a relatively pure state in veins and cavities. In these, it displays a form with six-sided prisms and pyramids; it often occurs as large, clear "rock crystals," which are used in optical instruments and for decorations. Quartz is normally clear or milky white, but impurities produce such varieties as rose quartz and amethyst, which are used in jewelry. Quartz normally crystallizes late in magmas; thus, it often has an irregular form. It also lacks cleavage; this fact alone often serves to distinguish it from feldspar and calcite, which are the other common light-colored minerals with which it might be confused. Also, it is harder than either of these minerals and, thus, will scratch them. Because quartz is chemically and physically resistant to alteration, it is the most persistent of the common minerals. For this reason, it appears throughout the sedimentary cycle.

The feldspar minerals include (1) $KAlSi_3O_8$—which comprises the minerals orthoclase, microcline, and sanidine—and (2) the solid solution grading from $NaAlSi_3O_8$ to $CaAl_2Si_2O_8$—which comprises several minerals that are collectively called plagioclase. Feldspar is normally white or light shades of grey or pink. The crystal forms are readily seen both in veins and in common rocks. They are almost-rectangular prisms with blunt-pointed ends. There are two cleavage directions almost at right angles, and many feldspars display not only smooth cleavage faces but also the parallel cracks of the intersecting cleavage planes. Because feldspars tend to become hydrated to clay when exposed to air and water, they usually disappear from the sedimentary cycle.

Clay minerals are essentially hydrous aluminum silicates, some with calcium, sodium, and potassium, but with variable composition because of the

A
B
C
D
E
F

easy exchange of magnesium and iron in the crystal lattices. The crystals are microscopic in size, but they generally have a lattice of layered sheets like mica. The most important clay minerals are kaolinite, montmorillonite, illite, and chlorite, which commonly form from the alteration of various other minerals in different climates. Although crystalline, the clays appear as soft, earthy masses, which are white unless colored by impurities.

Mica occurs in several varieties, of which the most important are the white mica, muscovite, and the black one, biotite. Both have variable compositions, but muscovite approximates $KAl_3Si_3O_{10}(OH)_2$ and biotite is $K(Fe,Mg)_3AlSi_3O_{10}(OH)_2$. Mica occurs as crystals, but it is distinguished mostly by its perfect cleavage in one plane. It is soft and can be broken into extremely thin cleavage sheets with a fingernail. Micas are fragmented easily in the sedimentary cycle, but the tiny flakes may persist for long periods. On the West Coast, wave action sifts these flakes from beach sands and deposits them in dark streaks along the shore.

Ferromagnesian minerals are a diverse group of iron and magnesium silicates that are conveniently lumped together for purposes of identification because they are all dark green to black. One important subgroup is olivine, the solid solution between Fe_2SiO_4 and Mg_2SiO_4. It is dark green, but it weathers to form a red mineral, iddingsite. Other important ferromagnesian minerals are pyroxene, which is a solid-solution group of minerals that generally contain calcium, magnesium, iron and aluminum, and amphibole, a group of hydrated solid solutions that usually contain the same metallic ions.

Calcite, $CaCO_3$, forms white or colorless hexagonal crystals that appear similar to those of quartz and are common in veins and cavities. Calcite is readily distinguishable from quartz, however, because it has perfect cleavages in three directions at angles of about 75° and tends to break into rhombohedrons. Calcite is easily scratched by quartz and feldspar and, unlike them, readily dissolves and effervesces in dilute acid. A drop of lemon juice will cause calcite to generate bubbles of carbon dioxide.

Common Rocks

Rocks can be classified in a host of ways: by color, hardness, chemistry, mineralogy, fossil content, age, origin, and so on. Because all the classifi-

---

FIGURE 4.3

Common rock-forming minerals, most in the form of individual crystals: *A*, black obsidian, a glassy form of the mineral quartz; *B*, a transparent crystal of quartz; *C*, orthoclase feldspar with intergrown crystals; *D*, olivine, which appears as small grains in center portion of this volcanic rock; *E*, a muscovite mica crystal, viewed from above toward a cleavage plane (it appears as a stack of paper-thin sheets when viewed from side); and *F*, calcite, both pyramidal crystals coated with sand and a smooth-sided, blocky cleavage fragment showing double refraction (double image). [A, D, E, courtesy John Sinkankas; B, courtesy Smithsonian Institution, from Sinkankas (1970); C, courtesy Smithsonian Institution; F, courtesy U.S. National Museum, from Sinkankas (1955).]

BOX 4.1 ROCKHOUNDS AND GEMSTONES

Many people in the United States study rocks and minerals for pleasure. Sometimes called rockhounds, they collect and prepare specimens for themselves, for trading with other amateurs, or for sale. This is an example of the fact that an introduction to the earth sciences opens many possibilities for further enjoyment of nature.

Gems have been collected by rich and mighty men—generally for beautiful women—since the dawn of history. The magnificent collection of ancient gems in the British Museum in London contains examples of every type of precious and semiprecious stone. A few of the very large stones have had long and lurid histories. The Kohinoor diamond had motivated theft and murder in India for 650 years before the British East India Company "seized" it in 1849 and presented it to Queen Victoria.

The nineteenth century saw the rise of rock and gem collecting for new purposes. Minerals were gathered for science, and the fields of mineralogy and crystallography were developed by scientists who specialized in their study. Other collections were made by amateurs and hobbyists who were motivated by the thrill of discovery, admiration of natural beauty, and the pleasures of personally fabricating beautiful objects from precious materials. The processes of shaping hard silicate rocks and minerals had been greatly advanced by the new machine technology then developing. The aspirations of the amateur may have derived from the custom of the rich and mighty of the times to make monuments of unusual rocks instead of conventional marble, which is easy to cut. Les Invalides, the tomb of Napoleon, for example, was built of the most striking and beautiful stones obtainable in the world. This was available for all to see. Meanwhile, in the splendor of the palace of the czars, there were whole rooms paneled in slabs of the brilliant copper minerals azurite (blue) and malachite (green), with statuary and decorations to match. These are preserved in the Hermitage Museum in Leningrad.

The first amateur gem club was founded in New York City in 1886. Perhaps the founding members had varied motives, but they certainly struck a popular vein. There are now more than 600 rock and gem clubs in the United States, and they are the most widespread type of science-oriented group in the country. Tens of thousands of rockhounds collect in the field and grind and polish minerals at home.

How does someone who is so inclined become a rockhound? Most of the population of America is not far from a mineral collecting site. One need merely read the appropriate books and magazines, buy the right detailed maps, and head into the field. However, there are rockhound clubs in all parts of the country, and virtually anyone interested in rocks and minerals can find a group of people eager to share his enthusiasm. These clubs are listed in the phone book or in various other directories, and they commonly exhibit their work at fairs and hobby shows in cities.

Almost any city or university museum has a mineral collection that can be viewed with pleasure by interested amateurs. The outstanding collections of the American Museum of Natural History in New York, The Field Museum of Natural History in Chicago, the Smithsonian Institution in Washington, D.C., and the California Division of Mines in San Francisco are open to the public.

Diamonds are among the most valuable and beautiful of gems, and we can cite them as an unusual example of the lures that draw the rockhound. They are not an inappropriate example of a mineral hunted in the United States because diamonds, although very rare, are surprisingly widespread. They constitute a highly appropriate example for this book because their distribution depends on so many basic aspects of the earth sciences.

A diamond is crystalline carbon with a high density—about 3.52 grams per cubic centimeter—which indicates that it was formed at high pressure. Diamonds sometimes occur as octahedral crystals, but these are rarely seen except in museums. Although diamond is the hardest natural substance known, it has an excellent cleavage, so it is easily broken if struck in the right way. Professional gem cutters utilize the cleavage to make the initial "cuts" in shaping diamonds.

Small commercial diamonds can be manufactured, but large gems form naturally at a depth of a few hundred kilometers in the interior of the earth. They reach the surface mainly in almost vertical structures known as diamond pipes which, in the principal African mines, are 200–300 meters in diameter. They are filled with serpentine rock, a mass of the hydrated equivalent of the ultrabasic rock periodotite, which has become stirred and mixed in the process of penetrating upward from the depths.

The only diamond-bearing pipe known in the United States is near Murfreesboro, Arkansas. Diamonds were discovered there by a farmer in 1906 and, since then, perhaps 100,000 have been found—the largest weighing 40 carats. Normal commercial mining, however, has never been very successful. Consequently, parts of the area were opened to the public in 1950. Visitors pay a fee to search, and may keep diamonds that they find by scratching at the sediment and rock. About 1000 diamonds, one of them weighing 15 carats, were found in the first decade by the rockhounds, some of whom came on special tours.

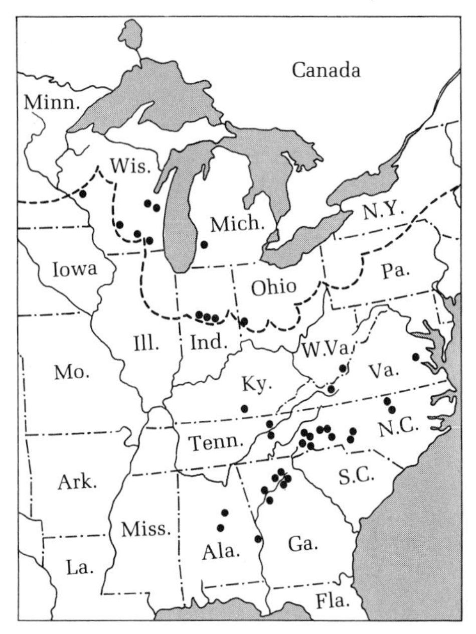

Dots show where diamonds have been discovered in the eastern United States. Dashed line indicates limit of Pleistocene glaciers. [After MacFall (1963).]

Although diamonds come up from the interior mainly in narrow pipes that are quite rare, they are widespread on the surface of the continents. This is because they are hard and relatively inert chemically; thus, they are preserved during the sedimentary cycle. Moreover, they are denser than most minerals, and the processes of sediment transport in rivers are such that they are usually buried by less dense sand and ordinarily protected from wear.

Hardly a state in the union lacks diamond finds in soil, sediment, or sedimentary rock, even though the source pipes are unknown. Many diamonds have been found accidentally during mining for gold in California and in the mountains from Alabama to West Virginia. In Peterstown, West Virginia, William P. Jones uncovered a 34-carat diamond while pitching horseshoes in a vacant lot. That was in 1854; the odd find was preserved but not identified as a diamond until 1943.

By far the most intriguing occurrences of diamonds are in the moraines left in the northern plains states by Pleistocene glaciers. Only fifty diamonds have been reported from these moraines, but they are dispersed and few people look for them. A search has been made for the source rocks, which must have been eroded by the glaciers somewhere near the glacial center in Canada, because the diamonds are in moraines deposited by different lobes of the glaciers. The history of ice movements has been unraveled and the lobes traced toward their source, but to no avail.

Diamonds are mined by professionals but often discovered by amateurs who are not seeking them. The chance that a rockhound will find one remains very slim, but most of the thrill of the search is in seeking rather than finding. Perhaps that is why there are so many rockhounds seeking beautiful and valuable treasures in the earth.

cations try to put transitional and gradational features into separate pigeonholes, all of them are arbitrary. There is no logical basis for preferring one classification over another except on the grounds of utility. Consequently, a sculptor may classify rocks in different ways from a building contractor or a chemist.

The classification system that earth scientists generally find most useful is one based on the origin of rocks. Accordingly, they are divided into three groups: igneous rocks, which cool from the molten state; sedimentary rocks, which are eroded from others and redeposited; and metamorphic ones, which are transformed in the solid state by increased pressure and temperature. It is evident that these divisions are as arbitrary as any others, because the geologic cycles produce a small proportion of transitional types.

The chemistry and mineralogy of rocks provide a basis for further distinctions, which are discussed below. The identification of rock specimens, particularly in the field, may also require consideration of texture, which is described in Appendix 5.

IGNEOUS ROCKS

Igneous rocks can usually be distinguished from the two other types because the minerals are randomly oriented, intertwined, and generally lacking crystal faces because of mutual interference during growth. Useful subclassifications of the igneous rocks can be based on color, grain size, density, chemistry, mineralogy, association with other rocks, and relation to tectonic plates. Fortunately, most of the classifications give very similar results because most of the properties of the rocks are related. Thus, the light-colored rocks tend to have a low density, to be rich in silicon, and to occur on continents. Geologists place many of these in a group called silicic or acidic rocks. With regard to color, the rocks at the opposite extreme are relatively rich in ferromagnesian minerals and are dark green or black. Such rocks most commonly have

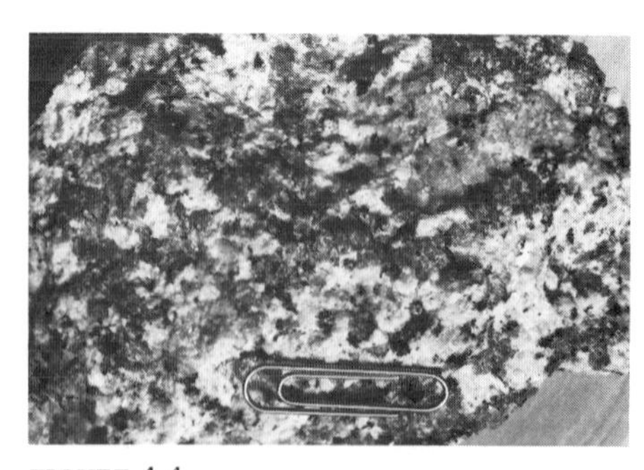

FIGURE 4.4

A specimen of granite: *left*, the surface of the specimen, with a paperclip shown for scale; *right*, the same rock, sliced thin enough to transmit light, × 10 (note interlocking crystals). [From Shelton, *Geology Illustrated*, W. H. Freeman and Company, copyright © 1966.]

higher density, are less rich in silicon, and occur on the sea floor. They are collectively called mafic (from *ma*gnesium, *f*erric) or basic rocks. Between light and dark are the rocks of less intense hue that are referred to as intermediate, and they are exactly that in almost every characteristic.

From color alone, therefore, much can be inferred about igneous rocks. The two other most useful properties for classifying rocks by eye are their grain size and their mineralogy. Magmas that chill very rapidly may explode (Chapter 9) or merely form a natural glass, obsidian, that lacks crystalline minerals. Those that cool a little more slowly allow time for crystals to start to grow but not to become very large. Such rocks typically solidify at the surface of the earth or near it, where the heat of the molten rock radiates rapidly. Those at the surface are called volcanic or extrusive rocks. Molten rock that solidifies deeper in the earth (intrusive rock) is insulated by surrounding rocks and cools very slowly. Therefore, large mineral crystals have time to grow.

Mineralogy provides the basis for very detailed classification of igneous rocks by specialists. However, the few common mineral groups, which are generally identifiable by eye, can provide the basis for a simple and useful classification. The minerals involved are quartz, the feldspars, and the ferromagnesian group (Table 4.1), and the proportions of these determine to which of the three types of igneous rock (silicic, intermediate, or mafic) a particular rock belongs.

The chemistry of the igneous rocks varies systematically with the three types (Table 4.2), and it serves to emphasize the existence of a fourth chemical type—namely, ultramafic (or ultrabasic) rocks, of which peridotite rock is an example. As Table 4.2 shows, SiO_2 decreases from the silicic to the ultramafic rocks, and so do many other oxides. Others, such as FeO and MgO, increase from silicic to mafic rocks, and MgO is by far the most abundant in ultramafic rocks.

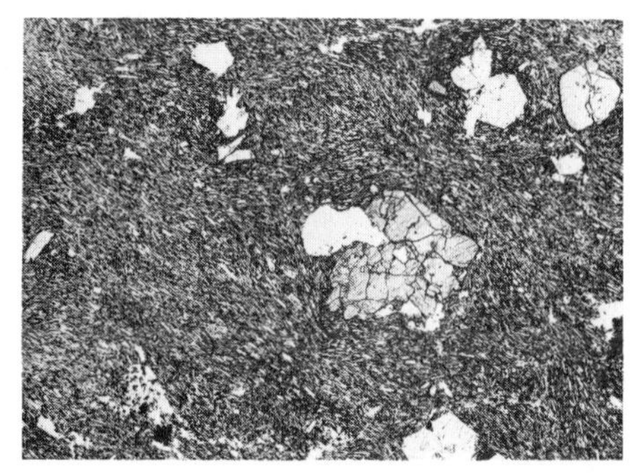

FIGURE 4.5

A fine-grained basalt: *left*, a weathered surface of the rock with a paperclip shown for scale; *right*, the same rock, sliced thin, × 10 (note the large crystals revealed among the fine grains). [From Shelton, *Geology Illustrated*, W. H. Freeman and Company, copyright © 1966.]

TABLE 4.1
A simple classification of the igneous rocks

Type	*Color*	*Minerals*	*Coarse-grained rocks*	*Fine-grained rocks*
Silicic	Light	Quartz, feldspar, ferromagnesian	Granite	Rhyolite
Intermediate	Intermediate	No quartz, feldspar predominant, ferromagnesian	Diorite	Andesite
Mafic	Dark	No quartz, feldspar, ferromagnesian	Gabbro	Basalt

TABLE 4.2
Percentages of oxides in various types of rock

	Type of rock					
Constituent	*Peridotite (ultramafic)*	*Basaltic (mafic)*	*Intermediate*	*Granitic (silicic)*	*Crust*	*Shale*
SiO_2	43.5	48.5	54.5	69.1	58.7	58.1
TiO_2	0.8	1.8	1.5	0.5	1.2	0.7
Al_2O_3	2.0	15.5	16.4	14.5	15.0	15.4
Fe_2O_3	2.5	2.8	3.3	1.7	2.3	4.0
FeO	9.9	8.1	5.2	2.2	5.2	2.5
MnO	0.2	0.17	0.15	0.07	0.12	
MgO	37.0	8.6	3.8	1.1	4.9	2.4
CaO	3.0	10.7	6.5	2.6	6.7	3.1
Na_2O	0.4	2.3	4.2	3.9	3.1	1.3
K_2O	0.1	0.7	3.2	3.8	2.3	3.2

Source: After Ahrens (1965).

SEDIMENTARY ROCKS

When igneous (or any other) rocks are exposed to the elements, they slowly become weathered—that is, they are dissolved, chemically altered, or fragmented (Chapter 13). The sedimentary rocks are derived from these products of weathering. They can easily be identified as sedimentary if they contain fossils of plants or animals, because these would be destroyed during the formation of most igneous or metamorphic rocks. They are often recognizable by the layering or stratification that they exhibit, which results from variations in the intensity of the depositional process. Dissolved materials are ultimately precipitated or evaporated out of solution, and may form different rocks depending on the materials and chemistry involved. Dissolved calcium carbonate forms the mineral calcite, which makes up the rock limestone. Dissolved silica is deposited as microcrystalline quartz to form the rock known as chert. Other, less common, dissolved elements may also form layers of rock in favorable circumstances. Thick layers of common salt (NaCl) and various other salts have formed by evaporation of sea water in unusual circumstances. These layers of salts are usually buried. If they return to the surface in the sedimentary cycle, they are readily dissolved by rainwater. Limestone and chert, however, are much less soluble.

The rock fragments that result from weathering are of various sizes, and the sedimentary cycle tends to sort them out. The coarser ones remain near the source and the finer ones are transported long distances. Thus, the fragments that are collectively called sediment can be divided by size into the rough but commonly understood categories of gravel, sand, and mud. These are unconsolidated sediments, not rocks. Neither are they soil, which is a complex, layered material, only partially composed of inorganic sediment.

TABLE 4.3
A simple classification of the sedimentary rocks

Type	*Grain size*	*Rock name*	*Common constituents*
Fragmental	Coarse	Conglomerate	Rock fragments
	Medium	Sandstone	Quartz most abundant
	Fine	Shale	Clay minerals
Precipitates	Fine	Chert	Microcrystalline quartz
	Fine	Limestone	Calcite

FIGURE 4.6
Wind-deposited Coconino sandstone from the upper wall of the Grand Canyon: *left,* a close view of a vertical break across the sandstone, showing fine layering that was produced during deposition; *right,* the same rock, sliced thin, × 10 (note the rounded shape and relatively uniform size of the sand grains). [From Shelton, *Geology Illustrated,* W. H. Freeman and Company, copyright © 1966.]

FIGURE 4.7
A sedimentary breccia of sandstone and limestone fragments exposed on the lower Kaibab trail, Grand Canyon. The width of this view is about five feet. The boulders indicate energetic transports, perhaps by a flood, and the range of small and large sizes indicates sudden deposition. [From Shelton, *Geology Illustrated,* W. H. Freeman and Company, copyright © 1966.]

FIGURE 4.8
An outcrop of conglomerate and associated cross-bedded sandstone, Puente Hills, California. The conditions of transport and deposition may change markedly from one sedimentary layer to the next above it. The time between deposition of the two layers may be days or centuries. [From Shelton, *Geology Illustrated,* W. H. Freeman and Company, copyright © 1966.]

Sediment usually has water in the pore spaces between particles, and the water usually contains dissolved matter. This is generally dissolved silica and calcium carbonate that is precipitated, under favorable circumstances, to form layers of chert and limestone, respectively. Likewise, if circumstances are suitable, SiO_2 or $CaCO_3$ are deposited from solution within the pore spaces. Thus, the grains bounding a pore are covered with layer on layer of silica or calcite cement until the pores are more or less filled. The sediment is transformed, in this way, to solid rock. Its hardness depends on the cementing material and the degree to which the pores are filled. It is worth emphasizing that the "cement" is not similar to the commercial "cement" that is used to make concrete; under most circumstances, nature makes a less resistant product.) Sedimentary rocks can be classified on the basis of fragment size, just as sediments can be. Thus, gravel becomes conglomerate, sand becomes sandstone, and mud becomes shale.

Some sedimentary rocks are biochemical precipitates—that is, they are precipitated as the shells of dead organisms. These may be intermediate in character between fragmental and precipitated rocks because the shells, especially those of microorganisms, may be transported before they finally come to rest to form rock. A common example of a biochemical precipitate is chalk, which is a poorly consolidated limestone made of the shells of marine microorganisms.

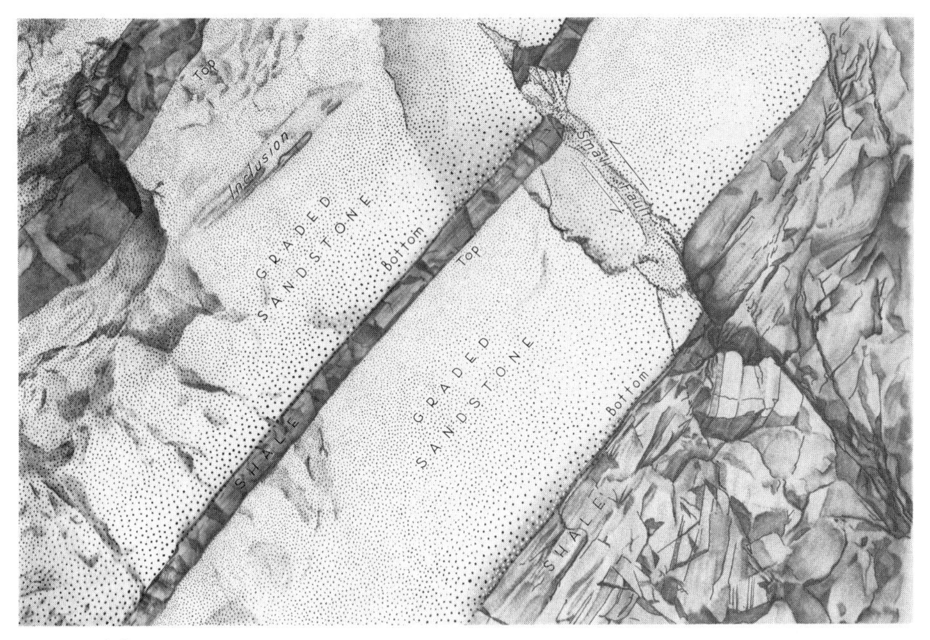

FIGURE 4.9

Exposed strata north of Duarte, California: shown are two 8-inch beds of graded marine sandstone separated by a thin parting of shale and offset a few centimeters, near the top, by a small fault. The grains in the graded sandstone become finer from bottom toward the top. This indicates a gradual change in deposition as the transporting current waned. [From Shelton, *Geology Illustrated*, W. H. Freeman and Company, copyright © 1966.]

METAMORPHIC ROCKS

If chalk becomes buried deeply in the earth, it is subjected to great heat and pressure. The fragile calcareous shells of the microorganisms slowly dissolve and may recrystallize into a metamorphic rock, marble, with coarse intergrown crystals of calcite. All other types of rock, whether igneous or sedimentary, can also be metamorphosed; thus, metamorphic rocks are a highly diverse group. An important consideration in the formation of these rocks is the existence of differential pressure or stress during metamorphosis. The new minerals tend to grow elongate in a direction perpendicular to the maximum pressure. Thus, the metamorphic rock often develops a foliation with flat or elongate grains in layers. This property provides a useful basis for classifying the metamorphic rocks, both because it is usually easily recognizable and because it is diagnostic of the conditions of formation. The size of the mineral crystals in the rock is also useful, but it is noteworthy that the crystals grow and that their size, therefore, is indicative of the intensity of metamorphism rather than of the grain size of the rock in its unmetamorphosed state. Likewise, the minerals present are generally the products of metamorphism, and are not remnants of the original rock.

FIGURE 4.10

An outcrop of Vishnu schist at the mouth of Pipe Creek in the Grand Canyon. Note the intense contortions produced by metamorphism of a sedimentary rock. [From Shelton, *Geology Illustrated*, W. H. Freeman and Company, copyright © 1966.]

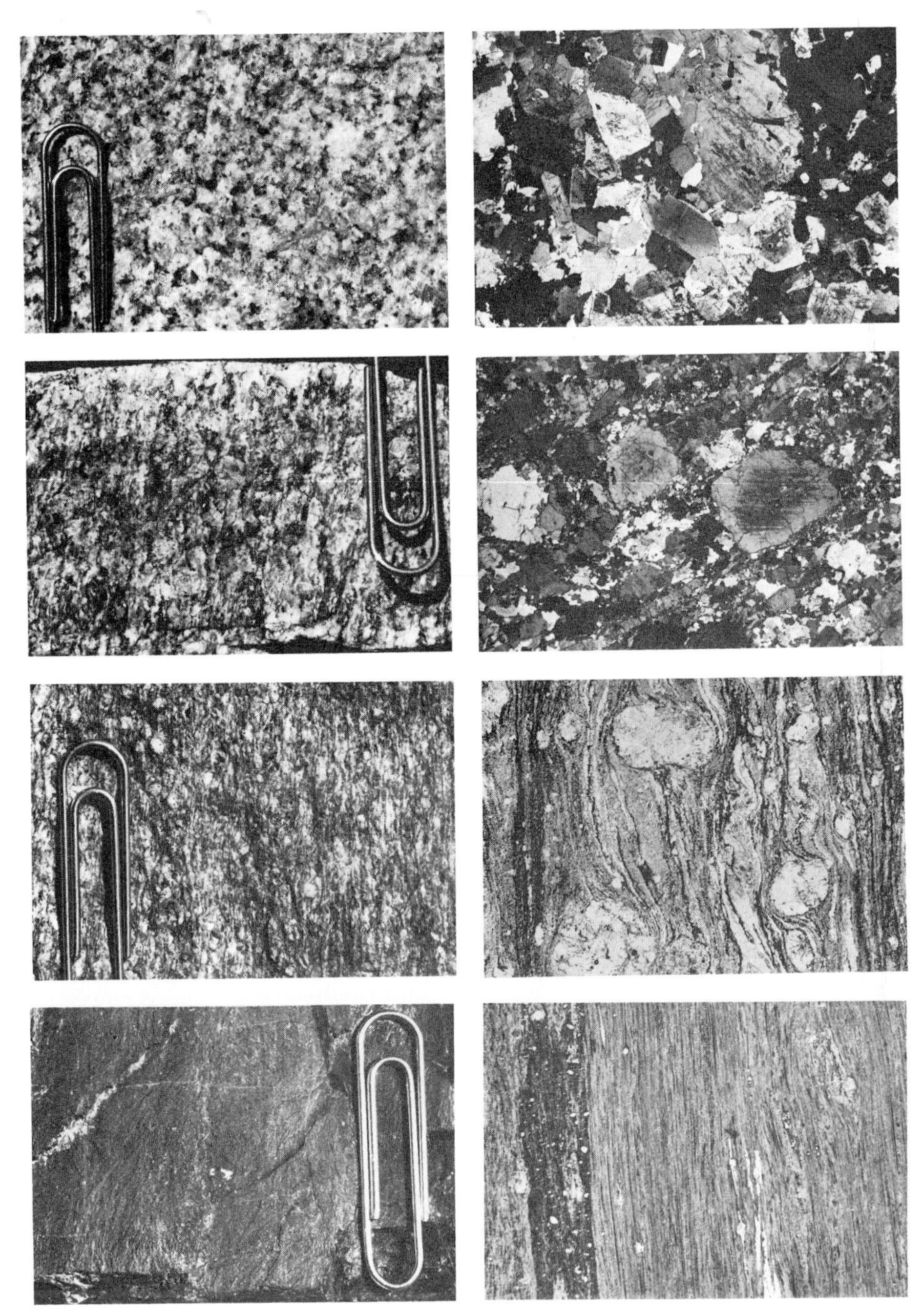

FIGURE 4.11

A suite of four specimens (*left*) and corresponding photomicrographs, × 10 (*right*) showing progressive grinding and fracturing by faulting of granitic rock in the southeastern San Gabriel Mountains, California. [From Shelton, *Geology Illustrated*, W. H. Freeman and Company, copyright © 1966.]

FIGURE 4.12
A light-colored vein of igneous rock that has been closely crumpled by metamorphism. The specimen, a gneiss, is from west of Helsinki, Finland. [From Fenton and Fenton (1940).]

TABLE 4.4
A simple classification of metamorphic rocks

Type	*Grain size*	*Rock name*	*Common constituents*
Foliated	Fine	Slate	Mica and usually quartz
	Medium	Schist	Mica, ferromagnesian minerals, quartz
Unfoliated	Coarse	Gneiss	Quartz, feldspar, mica
	Medium-coarse	Marble	Calcite
	Medium-coarse	Quartzite	Quartz

Abundance and Distribution of Rocks

Now that we have identified most of the common rock types that occur in the crust of the earth, we are prepared to consider their abundance and distribution in the crust. These are important matters because they have a fundamental bearing on most of the geologic cycles and the forces that drive them. Estimates of abundances are full of uncertainties, because they involve the whole surface and depth of the crust and the millions of samples upon which they are based come almost entirely from the surface of the continents. Thus, only the broadest generalizations about abundances are meaningful. Further, it should be remembered that the abundances we are now concerned with are those of minerals *in the crust,* which consists of the diverse surface or near-surface layers of the earth. We are not concerned with the minerals of the deep lithosphere, which is the rigid layer involved in the tectonic cycle.

Even with all these warnings, it is significant that less than 10% of the crust is sedimentary rock and that two-thirds of the remainder is igneous and one-third metamorphic rock. If most of the crust flows through the various geologic cycles, as it may do, it appears that it resides most of the time as igneous and metamorphic rock. An average crustal rock spends only 8% of its time in the stages of the cycle between when it starts to be weathered and when it is again

TABLE 4.5
Proportions of rocks and minerals in the crust

Constituent	Volume of crust (%)	Constituent	Volume of crust (%)
IGNEOUS ROCKS			
Mafic and ultramafic	44	MINERALS	
Intermediate	11	Feldspar	51
Silicic	10	Quartz	12
Total igneous	65	Pyroxenes	11
SEDIMENTARY ROCKS		Amphiboles	5
Clays and shales	4	Micas	5
Carbonates (including salt-bearing deposits)	2	Clay minerals	4.6
Sands and sandstone	2	Olivines	3
Total sedimentary	8	Magnetite	1.5
METAMORPHIC ROCKS		Calcite	1.5
Gneisses	21	Dolomite	0.5
Schists	5	Others	4.9
Marbles	1		
Total metamorphic	27		

Source: Data from Ronov and Yaroshevskiy (1967), rounded off.

metamorphosed or melted. This is another way of saying that most of the rock in the crust is deep under the surface, and all sediment is very near the surface.

Among the igneous rocks, the silicic ones, such as granite, amount to about 10% of the whole crust; so do the intermediate ones, such as granodiorite. The mafic and ultramafic rocks are by far the dominant ones, as they constitute 44% of the crust. This contrasts with the distribution of igneous rocks on the surface, where rocks more like granite are most abundant. It accords with the general downward increase in density in the earth. The low-density silicic and intermediate rocks are relatively abundant at the surface; the more dense mafic rocks dominate the interior.

Among the sedimentary rocks, the fragmental ones predominate, and the finer-grained of these are the most abundant. The limestones and other precipitates from sea water are not very abundant. Limestone, however, is more abundant than marble, its metamorphic equivalent. This contrasts with the foliated rocks, gneiss and schist, which are more than four times as abundant as the sandstone and shale from which they might once have been derived. This can be explained in at least two ways: Perhaps sandstone and shale were relatively more abundant in the past. It is more likely, however, that many of the foliated metamorphic rocks were derived from igneous rather than sedimentary rocks.

Distribution in Continents and Ocean Basins

The continental and oceanic crusts are made up of different types of rocks with different chemistry. With regard to nonsedimentary rocks, the continents generally are composed half of silicic and half of mafic rocks (Table 4.6). The crust of the ocean basins, by contrast, is almost exclusively mafic. Moreover, because it is very low in potassium and other elements that may be radioactive, there is a major difference between its heat-generating capacity and that of the continental crust. The comparison can be made even sharper by considering the most common types of rocks in each region. The continents are typified by old, light-colored, coarse-grained, relatively radioactive granite, and the ocean basins by young, dark, fined-grained, relatively nonradioactive basalt. There are also simple rules of exclusion. For example, islands of granite—such as the Seychelles Islands in the Indian Ocean—can be taken as small fragments of continent that have broken off during drifting.

It is noteworthy that more sedimentary rock is perched on the continents than is spread out on the deep-sea floor. Presumably, this is mainly because

TABLE 4.6
Distribution of rock types in parts of the crust

Crustal and unit layer	*Average thickness (km)*	*Volume (km^3)*	*Mass (10^{24} g)*	*Type of rock and abundance in layer (%)*
CONTINENTAL				
Igneous	48	7340	20.72	Acid igneous and metamorphic rocks, 50.0 Basic igneous and metamorphic rocks, 50.0
Sedimentary			1.77	
Thin, interior facies	1.8	138	0.35	Sands, 23.6 Clays, 49.5 Carbonates, 21.0 Evaporites, 2.0 Igneous, 3.9
Geosynclinal belts	10	372	0.94	Sands, 18.7 Clays and Shales, 39.4 Carbonates, 16.3 Evaporites, 0.3 Igneous, 25.3
Shelf and slope	2.9	192	0.48	Similar to above groups
OCEANIC				
Igneous	6.4	1960	5.62	Basalt
Sedimentary	1.0	242	0.49	

Source: Data from Ronov and Yaroshevиskiy (1967), modified.

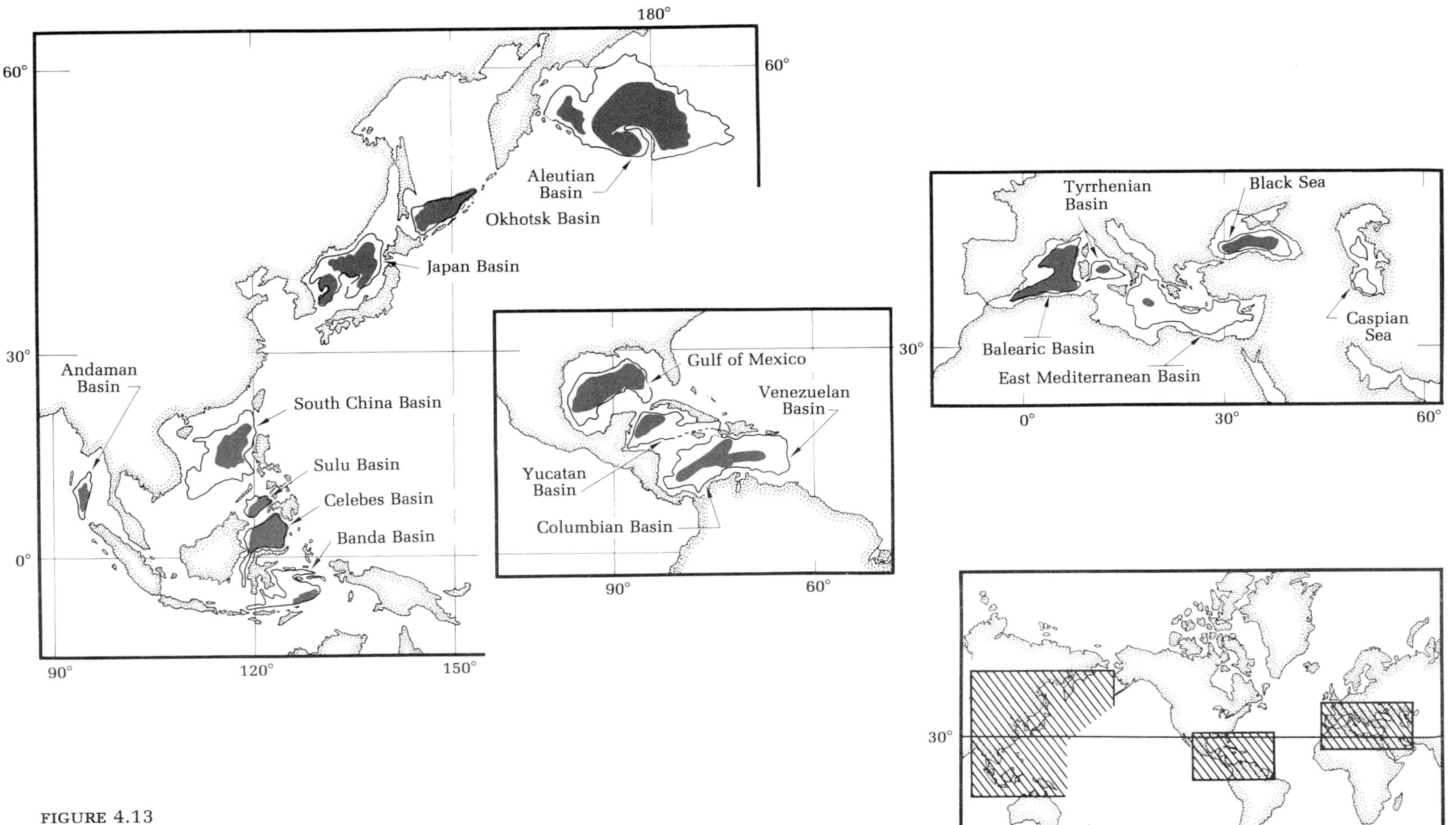

FIGURE 4.13
Small ocean basins. Boundaries shown by solid lines and abyssal plains are shown in black. [From Menard, *J. Geophys. Res.* 72(12):3062, fig. 1, 1967.]

plate motions sweep the oceanic sediments back into the continents in a relatively short time. It is striking, however, that most continental sedimentary rock, and about half of *all* such rock, is concentrated in great thicknesses in the long, narrow traps called geosynclines. The only apparently similar structures in the ocean basins are long, narrow, deep trenches; but sediment does not have time to accumulate thickly in them because it is constantly being drawn thence into the interior by the motions of the plates.

The thickest modern accumulations of sediment and sedimentary rock are in small, rather circular, ocean basins that are enclosed by or marginal to continents. The Caspian Sea and the Black Sea are underlain by 20 kilometers and 15 kilometers of sediment, respectively. This sediment has the same thickness and volume as sedimentary rock in most ancient geosynclines. One reason for this extensive accumulation of sediment is that the great Volga, Dnepr, and Danube rivers empty sediment into these semi-enclosed basins, where it is effectively trapped. For the same reason, the Mississippi River has piled sediment to a depth of 10 kilometers on the floor of the semienclosed Gulf of Mexico.

Summary

1. The commonest rock-forming minerals are compounds of silicon.
2. The minerals are crystalline solids—that is, their atoms or molecules are arranged in a three-dimensional lattice. The properties of the lattice determine many of the physical properties of the minerals.
3. Silicate crystals consist of various linkages between tetrahedra composed of one silicon atom surrounded by four oxygen atoms.
4. Ordinary rocks are composed largely of the following minerals or groups of minerals: feldspar, quartz, amphibole, mica, clay, pyroxine, olivine, and calcite. These can be readily identified in many rocks.
5. Rocks are classified in a multitude of ways for different purposes. A classification according to origin is in most common use.
6. Igneous rocks generally have randomly oriented, intertwined minerals and are classified on the basis of color, mineralogy, and grain size.
7. Sedimentary rocks are those derived by weathering of other rocks. They consist of either chemical precipitates or rock and mineral fragments, and they are classified on the basis of chemical composition and grain size.
8. Metamorphic rocks are a diverse group that have been significantly altered from other types by heating, pressure, or stress.
9. The order of abundance of rocks in the crust is (1) igneous, (2) metamorphic, and (3) sedimentary.

Discussion Questions

1. Why is it necessary to have mineral names in addition to the names of chemical compounds?
2. Why does mica separate easily into flakes?
3. How can you distinguish an igneous rock from a sedimentary one by observation of a hand specimen?
4. What types of igneous rock are most abundant in the crust?
5. Are granites equally common in continental and oceanic crust?
6. Is there more sedimentary rock on the continents or in the ocean basins? Why?

References

Ahrens, L. H., 1965. *Distribution of the Elements in Our Planet.* New York: McGraw-Hill.

Deer, W. A., R. A. Howie, and J. Zussman, 1966. *An Introduction to the Rock-forming Minerals.* New York: John Wiley & Sons. [By the authors of the great modern five-volume work on this subject.]

Fenton, C. L., and M. A. Fenton, 1940. *The Rock Book.* Garden City, N. Y.: Doubleday.

MacFall, R., 1963. Gem hunter's guide. New York: Thomas Y. Crowell. Contains detailed maps of mineral localities.

Menard, H. W., 1967. Transitional types of crust under small ocean basins. *J. Geophys. Res.* 72(12):3061–3073.

Pettijohn, F., 1957. *Sedimentary Rocks* (2nd ed.). New York: Harper and Row. [A comprehensive treatise for the advanced student but understandable in itself.]

Ronov, A. B., and A. A. Yaroshevskiy, 1969. Chemical structure of the earth's crust. *Geochem. Int.* 4(6):1041–1066.

Shelton, J. S., 1966. *Geology Illustrated.* San Francisco: W. H. Freeman and Company.

Sinkankas, J., 1955. *Gem Cutting.* Princeton, N.J.: Van Nostrand.

——, 1961. *Gemstones and Minerals.* Princeton, N.J.: Van Nostrand. [An excellent guide for the amateur collector.]

——, 1970. *Prospecting for Gemstones and Minerals.* New York: Van Nostrand Reinhold.

Turner, F., and J. Verhoogen, 1960. *Igneous and Metamorphic Petrology* (2nd ed.). New York: McGraw-Hill. [Guides the advanced student through a complex subject.]

Tuttle, F., 1955. The origin of granite. *Sci. Amer.* 192(4):77–82. (Available as *Sci. Amer.* Offprint 819.) [Experiments indicate that it has an igneous origin.]

DEFORMATION

PETER D'AGOSTINO

5

PLATE TECTONICS

Cecily: That certainly seems a satisfactory explanation, does it not?
Gwendolen: Yes, dear, if you can believe him.
Cecily: I don't. But that doesn't affect the wonderful beauty of his answer.

Oscar Wilde,
THE IMPORTANCE OF BEING EARNEST

Introduction

In this chapter, we will begin a more detailed examination of earth processes and their social effects. We will discuss some of the most exciting new discoveries about the earth and will establish the background for the three chapters that follow. The reader should remember, as he follows the motions of vast crustal plates, that people, buildings, aqueducts, pipelines, and roads drift with them and spread across the boundaries between them.

The concept of moving continents has engrossed earth scientists since it was first elaborated by Alfred Wegener, a German meteorologist and geologist, in 1915. Most of them were very skeptical, however, especially in the United States. A group of geophysicists specializing in paleomagnetism gave increasing support to the notion of continental drift in the 1950s, but general acceptance did not occur until the late 1960s, when the evidence from marine geology and seismology became overwhelming.

Present knowledge of drifting is presented here in a series of simple and logical steps. The reader will appreciate, however, that the actual history of the development of these ideas was quite different. It consisted of many twists in and out of dead ends and through a mass of conflicting data and interpretations. A detailed history of this scientific revolution will one day become part of the common intellectual heritage.

CRUSTAL CONTINUITY

In 1962, the American geologist Harry Hess published what he called an "essay in geopoetry," which contained a hypothesis that crustal plates spread apart at midocean ridges, that new crust fills in the gap, and that the new crust, in turn, splits and spreads apart. This was called sea-floor spreading by the American marine geologist Robert Dietz.

In 1965, the Canadian geophysicist J. Tuzo Wilson focused attention upon the implications of this hypothesis with regard to crustal continuity. If the area of the earth is constant and new crust is created, then an equal amount is destroyed elsewhere. The crust is created at midoceanic ridge crests; it is destroyed under oceanic trenches and island arcs; and the sides of a plate moving from one to the other of these regions are discontinuities that Wilson called transform faults.

EULER'S THEOREM

Hess, Wilson and other earth scientists generally were writing about weak and rather plastic rocks, but only a few years later the crust was shown to be

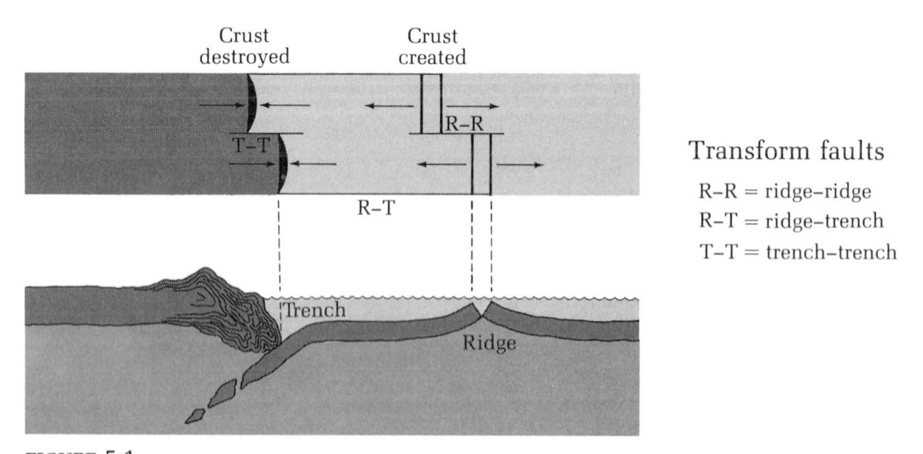

FIGURE 5.1
Crust that is created at spreading centers (usually midoceanic ridges) is destroyed at subduction zones (usually oceanic trenches). The lateral boundaries of moving plates are transform faults that have different names depending on the features they connect.

composed of rigid plates. Geometry provided the unlikely foundation for this new understanding.

A theorem by the Swiss mathematician Leonhard Euler (1707–1783) states that, if one rigid spherical shell moves over another, two diametrically opposed points remain fixed. The motion of any point on the surface of the outer shell may be considered as a rotation about an axis that connects the two points. All other points on the outer shell move in small circles around the fixed points.

The fixed points, which are now referred to as Euler poles, have only geometrical significance. They are quite unrelated, for example, either to the axis of rotation of the earth or to the poles of the magnetic field. The earth, however, is almost a sphere; if the crust drifts rigidly, the motion must accord with Euler's theorem. The crust, in fact, is broken into many large, rigid plates that drift in different directions, but each motion can be described by the theorem. The geometry constrains the motion of each of these fragments of a spherical shell, and they rotate around Euler poles whether they are physically connected to them or not.

The angular displacement around the Euler pole of different parts of a moving, rigid, spherical shell is constant. Consequently, the speed of motion varies with the cosine of the Euler latitude relative to the Euler pole. It is zero at the pole and reaches a maximum 90° away at the "equator." If the velocity at one point is known, it can be calculated for all other points on the same shell or plate.

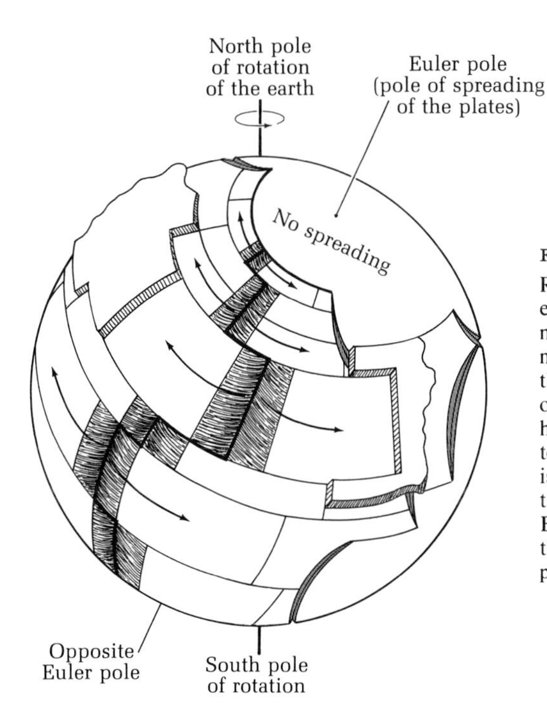

FIGURE 5.2
Rigid lithospheric plates on a spherical earth can move only by rotation. Points near the Euler poles (poles of spreading) move slower and smaller distances than those farther away. The spreading center, offset by several ridge–ridge transforms, has split a continent. The piece drifting to the right has overridden a trench and is moving toward another along trench–trench transforms. Near the northern Euler pole, the crust is not moving, and the boundary between it and the active plates is a pair of ridge–trench transforms.

Modern Plates

A rigid body is defined in the ideal as a connected assemblage of particles that cannot move relative to each other. Evidence that the crust presently consists of large rigid plates comes from the distribution and motion of earthquakes. Real materials only *approach* the ideal of rigidity, and the large crustal plates do not fit the definition as well as do small pieces of pure metal. Small distortions do occur in tectonic plates, but they should not be allowed to obscure the fact that the plates exhibit a remarkable approximation of large-scale rigidity.

EARTHQUAKE DISTRIBUTION

Relative motions in the crust are difficult or impossible to observe by eye, but they generally produce earthquakes that are detected easily even at a distance. Thus, it is reasonable to define a rigid plate as one in which no earthquakes occur. If plates are moving, they jostle each other at the edges, and relative motion and earthquakes should be common there. A plate boundary may be defined, over time, by a relatively continuous line of earthquakes.

In 1969, geophysicists Muawia Barazangi and James Dorman mapped the locations of the 29,553 largest earthquakes that occurred from 1961 to 1967. The distribution is highly suggestive of the existence of crustal plates. Very large regions of the earth, such as the North Pacific, show no more than a score of earthquakes, but such regions are bounded by continuous narrow bands with thousands. It is equally striking that the plates so defined are not the same as continental boundaries. The African plate, for example, encloses Africa but extends out to the center of the Atlantic and Antarctic ocean basins. Likewise, a plate boundary cuts across California, and another approximately separates India from the rest of Asia. Continents are mere chips embedded in the plates like logs embedded in an ice flow.

DIRECTION AND SENSE OF MOTION

Consider two plates that together cover the whole earth. If one moves, it pushes into the other in front, pulls away from it behind, and moves parallel to it on the lateral transform faults. All points on the plate rotate along small circles around an Euler pole; therefore, all motions on fault planes separating the two plates are also such rotations. In 1967, in the first paper published on plate tectonics, the British geophysicists Dan McKenzie and Robert Parker demonstrated that fault motions around the North Pacific plate lie, in fact, on small circles around a single Euler pole. This revolutionary discovery has since been confirmed repeatedly for many plates.

Seismologists can determine not only the direction of motion of distant earthquakes but also whether plates are moving apart, parallel to each other, or together. The sense of motion is consistent along a given edge of a plate. The Pacific plate, for example, is pushing against the Eurasian plate. It is also pulling away from the Antarctic plate, and it is moving parallel to the American plate along the San Andreas fault in California. Once again, the observations of relative motion are readily explained only by motion of enormous rigid plates.

VERTICAL DISTRIBUTION

Earthquakes occur at various depths to 700 kilometers, but the maximum depth is relatively constant at any particular plate boundary. Where oceanic plates are spreading apart, earthquakes may be no deeper than 1 to 3 kilometers. At the sides, the maximum depth is closer to 15 kilometers. Almost

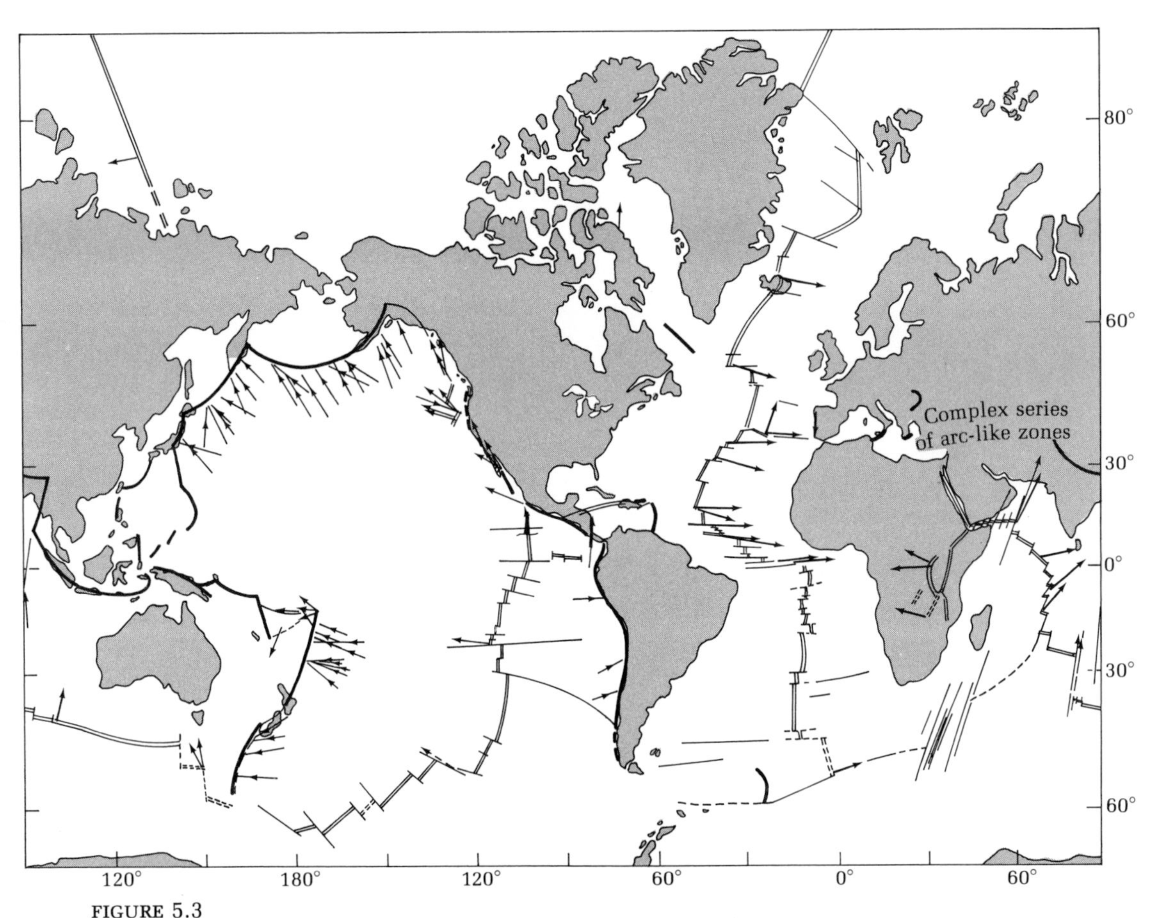

FIGURE 5.3

Slip vectors derived from earthquake-mechanism studies. Each arrow indicates the horizontal component of the motion of the block on which it is drawn relative to the adjoining block. [From Isacks et al., *J. Geophys. Res.* 73:5861 (1968).]

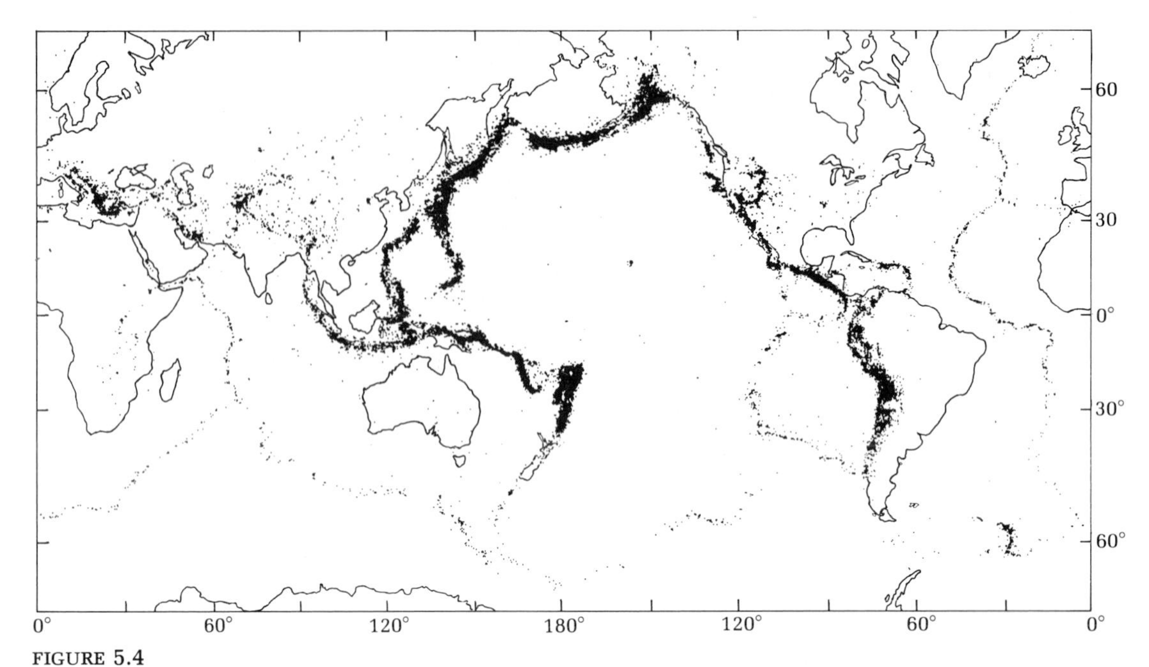

FIGURE 5.4

Epicenters of earthquakes with focal depths of 0–700 kilometers during the period 1961–1967. Continuous lines of earthquakes define plate boundaries. Diffuse quakes in North America, Asia, and Africa indicate significant activity not closely related to plate boundaries. [From Barazangi and Dorman (1968).]

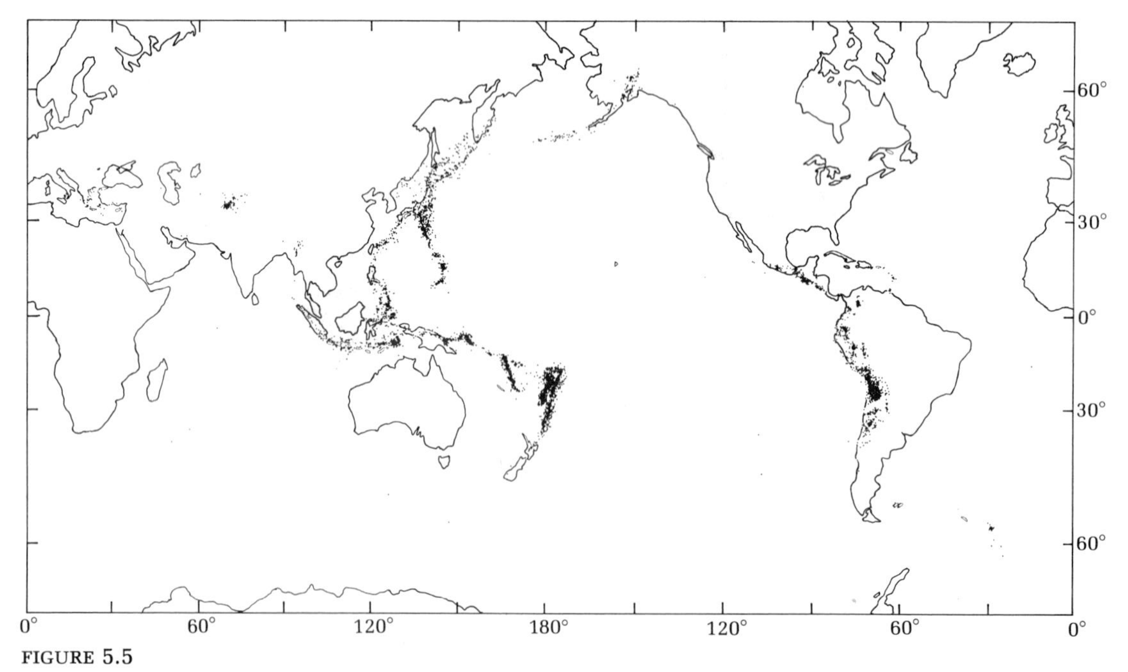

FIGURE 5.5

Epicenters of earthquakes with focal depths of 100–700 kilometers during the period 1961–1967. Deeper quakes occur where plates plunge in subduction zones. [From Barazangi and Dorman (1968).]

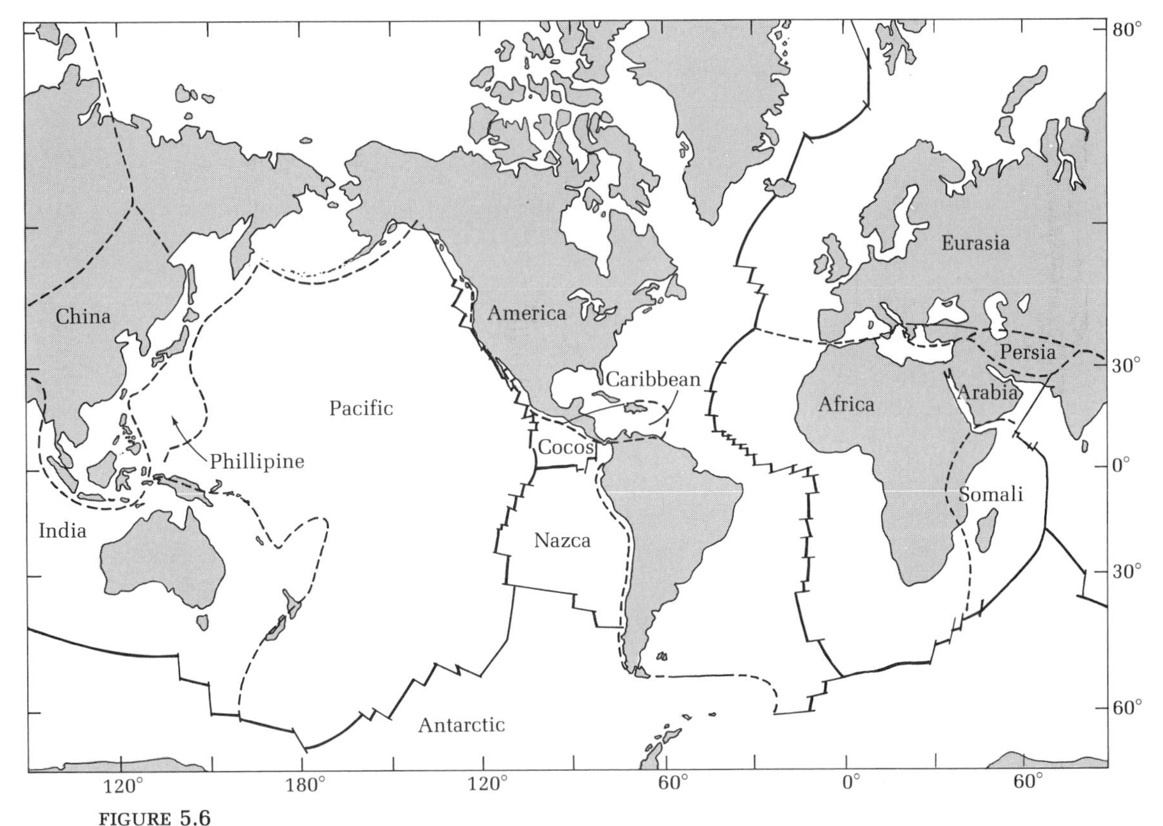

FIGURE 5.6
Names of major crustal plates, according to Morgan (1972).

all the deeper earthquakes occur where plates are pushed together. This distribution is readily explained by plate tectonics. Earthquakes seem to occur only in the rigid lithosphere and not in the weak asthenosphere below. The lithosphere is hot and thin where the new crust forms, so earthquakes there are shallow. The lithosphere thickens as the crust ages, and the earthquakes deepen. The deepest quakes occur where plates overlap and one is pushed or carried down into the mantle until it melts and loses its rigidity.

RATE OF MOTION

The direction and sense of relative motion between modern plates are given by earthquakes and Euler's theorem. The relative motion is not completely described, however, unless the speed is also known, and for this purpose we must turn to other sources.

Geodetic measurements can measure both the speed and direction of relative motion, in some circumstances, on land. Surveyors lay out a network of precisely located points in the area of the plate boundary and measure the slight changes that motion brings. This technique is relatively satisfactory in

California and Iceland, because the plate boundary is identifiable. In most places, however, the boundary is too complex, or the motion takes place too slowly for it to have been determined in the time since the geodetic network was established.

Because most plate boundaries are under the sea, moreover, some other technique is needed. The method now in use depends on the reversals of the earth's magnetic field. Its discovery provides one of those pretty examples of serendipity that scientists prize so highly. The fact of the magnetic reversals was still questionable when Hess wrote his geopoetry. Nevertheless, the British geologists Fred Vine and Drummond Matthews were bold enough to pyramid the conjectures to explain the origin of a distinctive pattern of linear magnetic anomalies in the northeastern Pacific (Figure 5.4).

When tectonic plates move apart, the crack between them fills passively with lava that cools through the Curie temperature and records the magnetic field that is in effect at the time. As the plates continue to move, the cooling lava splits in its turn and new lava wells up into the crack and records the field. The plates move apart so rapidly that crack filling is almost continuous; therefore, the field is recorded as though by a great tape recorder. This process produces a recording of higher fidelity under the sea than it does on the land because of the greater frequency of lava emplacement on the ocean bottom.

There the matter stood until 1965, when Vine and Tuzo Wilson discovered that some of the magnetic anomalies were bilaterally symmetrical around the spreading crack. In other words, the new cracking occurs in the middle of the rock that filled the old crack. The crust, therefore, is a *stereo* tape recorder. By that time, the history of reversals for the last few million years had been established by dating volcanic rocks on land. Given the history, the theory of magnetism makes it possible to calculate the shape of the anomalies that would be produced by spreading at any constant rate. The observed anomalies in the area correspond to those expected by constant spreading at a rate of about 6 centimeters per year.

The history of reversals was extended backward by analysis of sediments from the sea floor, which preserve the record of the magnetic field when they are deposited. Thus, in favorable circumstances, they are vertical magnetic tape recorders. Patterns of bilaterally symmetrical anomalies were also mapped in large areas at sea; these were shown, by theoretical analysis, to be the result of global magnetic reversals plus different rates of spreading.

The American geophysicist W. Jason Morgan published the second paper in plate tectonics using an entirely different approach from the first, which had appeared the month before. He showed that the topographic features that Wilson had identified as plate boundaries conform to the patterns required by Euler's theorem. The fracture zones that mark transform faults on the sides of plates follow arcs of small circles. Moreover, the rate of spreading between two plates varies, just as the theorem requires, with the cosine of the Euler

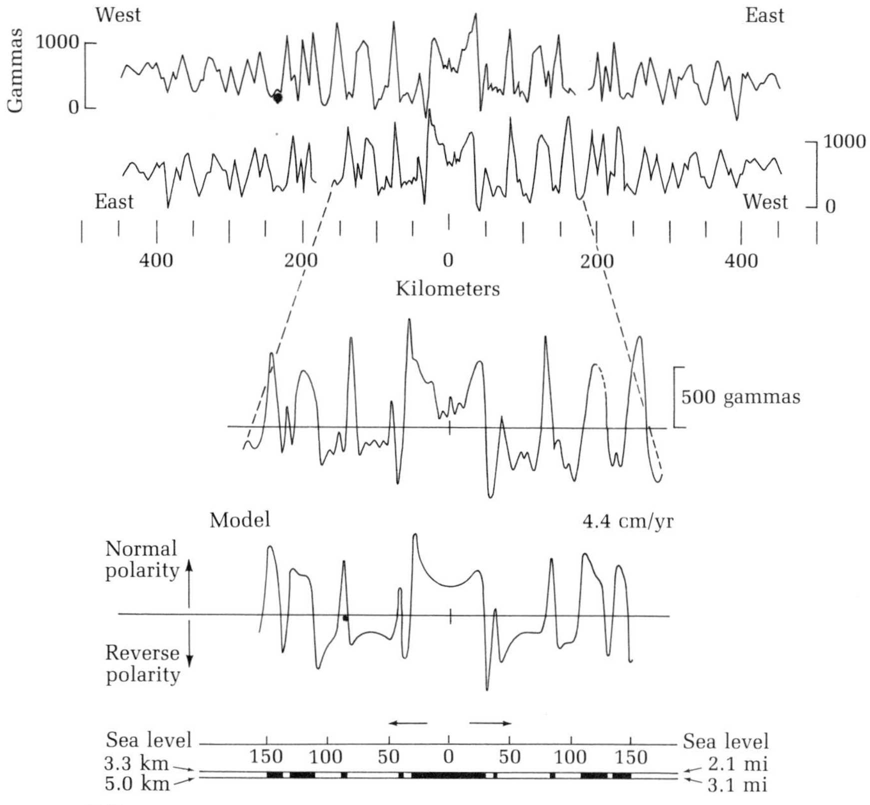

FIGURE 5.7

The upper profile is the observed magnetic anomaly pattern over a midocean ridge. The next profile is merely the observed one reversed. They look the same, showing that the pattern is bilaterally symmetrical. The bottom profile is a theoretical model of the expected anomalies that would be caused by magnetic reversals and sea-floor spreading at 4.4 centimeters per year. It corresponds to the central part of the observed pattern and, thus, confirms the existence of spreading and gives its rate. [After Vine (1966); and Heirtzler, "Sea-Floor Spreading," copyright © 1968 by Scientific American, Inc. (all rights reserved).]

latitude relative to the Euler pole—an elegant demonstration of the simplicity of motions between two plates and of the reality of large-scale rigidity.

Complex Edge Phenomena

MOVING EDGES

The crust consists of eleven large plates and many smaller ones, according to our present state of knowledge. Thus, the geometry of the motions is more complex than we have considered, although it still conforms to Euler's theorem. The most interesting of the new phenomena occur where three plates meet at a triple junction. However, before turning to these, it is fruitful to consider the permanence of the plate shapes and patterns.

Do the edges move as well as the plates, or are the edges fixed while the plate is created, drifts, and is destroyed between them? That spreading centers move can be demonstrated quite simply because, in some places, two of them are not separated by a trench of subduction zone. When new crust forms in both of them and none is destroyed between them, they must be moving apart. This is occurring in the southeastern Pacific and around the southern half of the African plate. Spreading centers may also jump—that is, they may occasionally abandon a plate edge and crack into a nearby plate—but this is rare compared to steady spreading.

The leading edges of plates also move, although this may seem difficult because the plates plunge deep into the mantle. It appears that the leading edge of the Pacific plate, which is subducted in the western Pacific has been abandoned repeatedly there and has plunged anew, each time, into the mantle.

Transform faults merely connect the spreading centers and subduction zones, so they change configuration, or jump, or otherwise move, without difficulty.

Do the lengths of plate edges change? The western and southern edges of the African plate were once against Africa when the Atlantic began to open. The present edge, still in the center of the Atlantic, is much longer than the

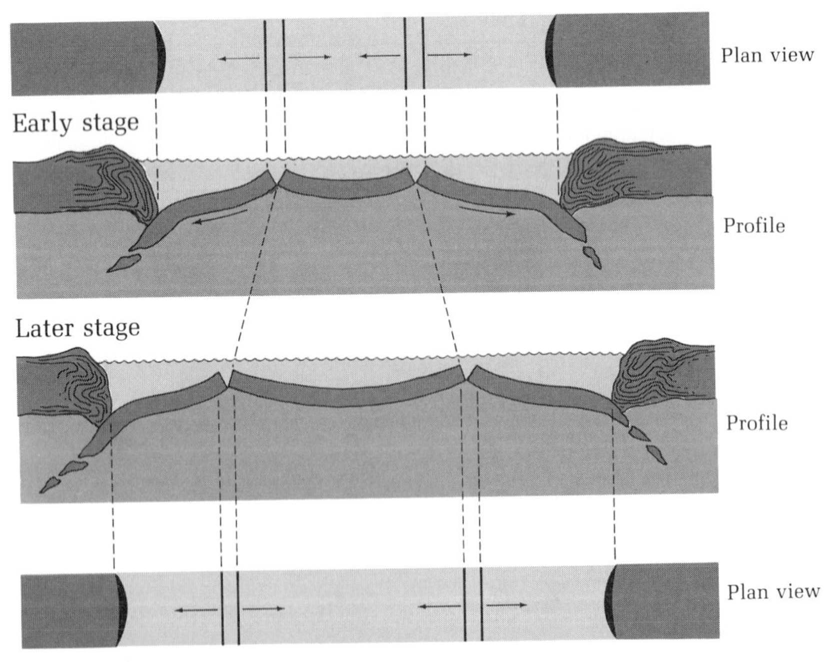

FIGURE 5.8
The fact that spreading centers move is demonstrated by the discovery of pairs that are not separated by subduction zones.

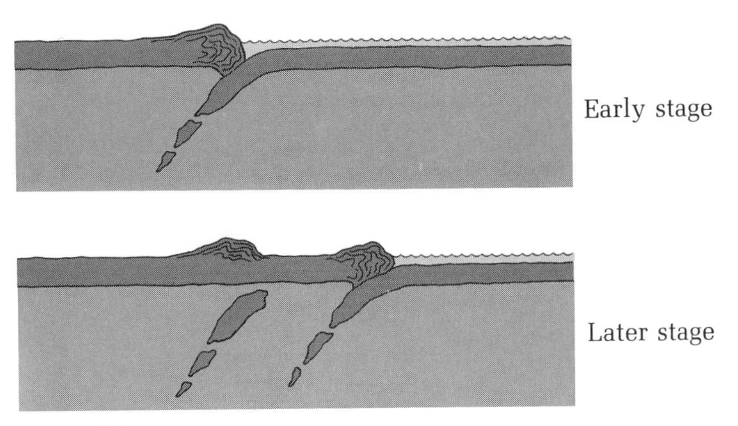

FIGURE 5.9
The fact that subduction zones move is indicated by abandoned trenches and isolated bits of plunging lithosphere.

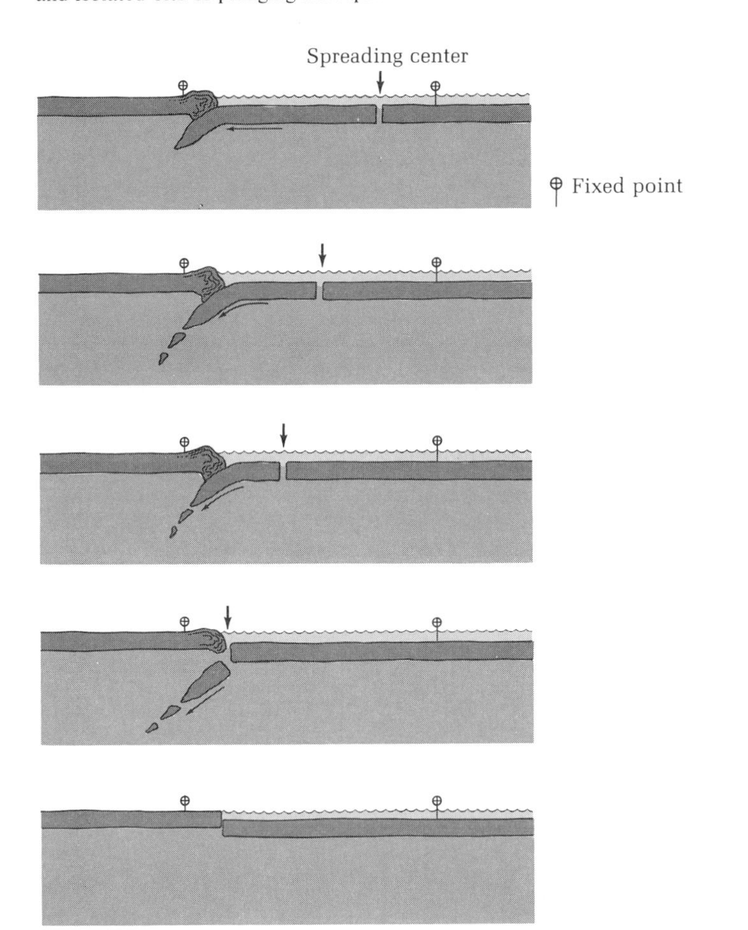

FIGURE 5.10
Migration of a spreading center toward a subduction zone may lead to the destruction of both and the formation of a single plate where once there were three.

margin of Africa, so it has lengthened. Hundreds of other examples of changes in the length of plate boundaries are known.

Are plates permanent? If edges can move, they can intersect or overlap and destroy the plate that separated them. This has occurred many times, and it is the ultimate fate of all oceanic crust.

TRIPLE JUNCTIONS

In 1969, McKenzie and Morgan analyzed all sixteen geometrically possible triple junctions and identified examples of six types that exist at present. These triple junctions consist of various combinations of the three types of plate edges. Most configurations are unstable, regardless of the orientation of the edges, and some others are stable only if the geometry is exactly right. Only the junction of three spreading centers is stable in all orientations. Several of these long-lived triple junctions have been discovered and mapped in the Pacific and Indian Ocean basins. This type of triple junction is possible because the length of the spreading centers increases (Figure 5.11).

Spreading Centers

Spreading centers are characterized by high heat flow, uplift, shallow depths, earthquakes, and linear volcanoes, rifts, fault blocks, and magnetic anomalies. All of these features can be explained more or less quantitatively by the plate-tectonics model.

The crust pulls apart, forming deep, vertical tension cracks that serve as conduits for lava, which rises because it is hot and less dense than solid rock. The heat of the intruding hot rock expands the solid upper mantle and elevates the trailing edges of the spreading plates to form shallow midocean ridges. As the plates spread and age, they cool and sink. This model, in which all the heat-

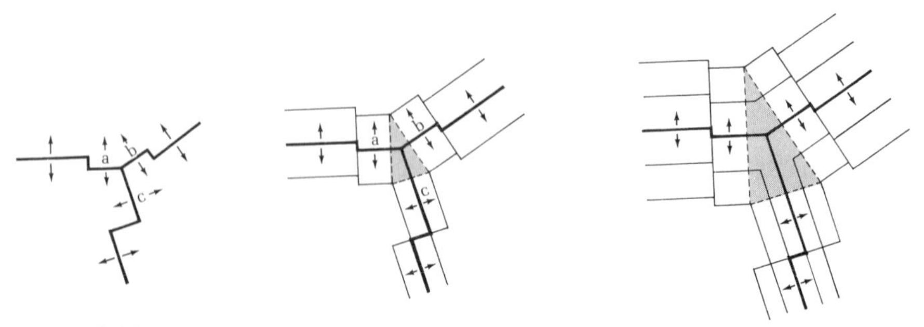

FIGURE 5.11

Evolution of the triple junction of three spreading centers where tectonic plates are moving apart. Shaded triangles show new area in which ridge crest grows longer.

ing is in or near the spreading center, is entirely compatible with several outstanding characteristics of trailing edges of plates—namely, short sections of spreading center alternate with sections of transform fault, and plate edges drift, jump, and can be destroyed. All these seem reasonable if the edges are mere cracks that open passively.

Some of the rising lava solidifies as dikes in the vertical cracks and records the magnetic field. Because the sides of a dike are chilled by the intruded rock,

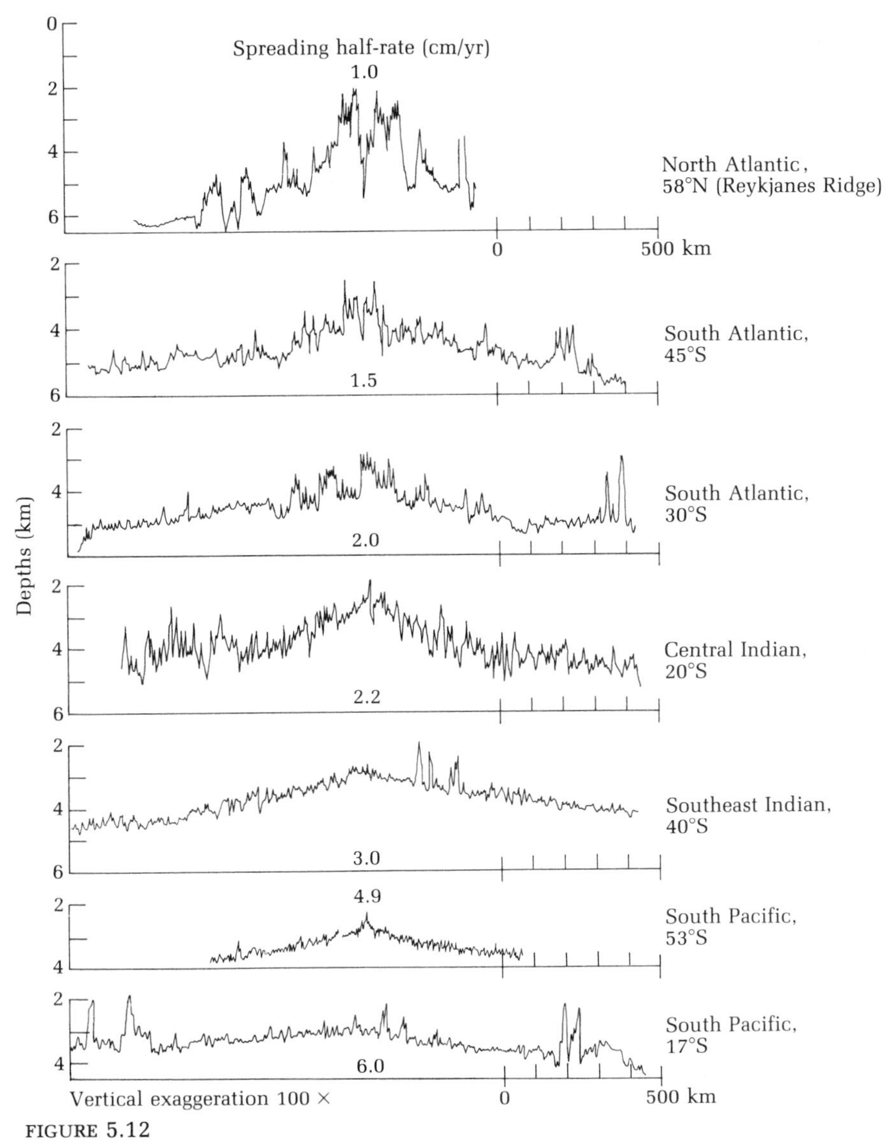

FIGURE 5.12

Slowly spreading ridges have steep side-slopes and local mountains with high relief. Rapidly spreading ones have gentle profiles and low hills.

they solidify first. Because the center is still warm and soft, it is more apt to split in the middle and yield a stereo recording of the earth's magnetic field.

Some of the lava overflows the cracks and pours out on the sea floor to build elongate volcanoes. Normal conical volcanoes rarely occur in spreading centers because there are so many long open cracks. Because Iceland straddles the Mid-Atlantic Ridge, many features characteristic of spreading can be seen there. A narrow band of earthquakes extends along a central volcanic rift zone from north to south across the island. Heat flow is high, and the largest city, Reykjavik, is heated with natural hot water. Long open cracks are so common that they have the special name of *gjá* (rhymes with chow). Geodetic observations show that the cracks are spreading apart. Differential vertical motions occur on some cracks and yield fault-block ridges. Similar ridges and central rift zones with earthquakes occur at many spreading centers elsewhere in the world.

The occurrence of many types of features can be correlated with the spreading rate. If the rate is less than 4 centimeters per year, the relief of the linear volcanoes and fault blocks is generally about 1000 meters. If it is faster, the relief is generally less than 500 meters. Likewise, the layer of dikes and lava flows is thinner if spreading is fast and thicker if it is slow. Further, a central rift is hardly ever observed unless the spreading rate is slower than 4 centimeters per year.

Explanations for these correlations are too new to be firmly established. Norman Sleep, an American geophysicist, reasons that, if the cracks open slowly, the rising lava is chilled by the rocks on each side and becomes viscous. At the surface, it flows only a short distance before it solidifies. Thus, the ridges are steep-sided and high. With rapidly widening cracks, the lava is not so chilled; therefore, it flows farther at the surface and the relief is low.

The relationship between the thickness of the volcanic layer and the spreading rate means that the discharge of lava is constant. It is not controlled by the surface cracking, therefore, but by some more fundamental phenomenon in the mantle that yields molten rock at a constant rate. Because the slowly spreading crust remains longer in the narrow band in which dikes are injected, more lava accumulates.

The central rift is also evidence that a band of intense vulcanism exists in the vicinity of the trailing edge of a plate, rather than just along a mere crack. Sediment on terraces high on the inner sides of the Gorda rift, off California, is of a type deposited by bottom-hugging currents. It is clear, therefore, that the terraces were once the bottom of the rift and that they have been uplifted. The broad midocean ridges that spread rapidly are of the same height as the narrow mountains bordering the rifts of slowly spreading ridges. Thus, the rifts appear to be depressions that are a consequence of lava chilling in cracks that are so narrow that they become choked. Continuing heating, vulcanism, and upward pushing in the central band elevate the crust of the rift as it slowly spreads.

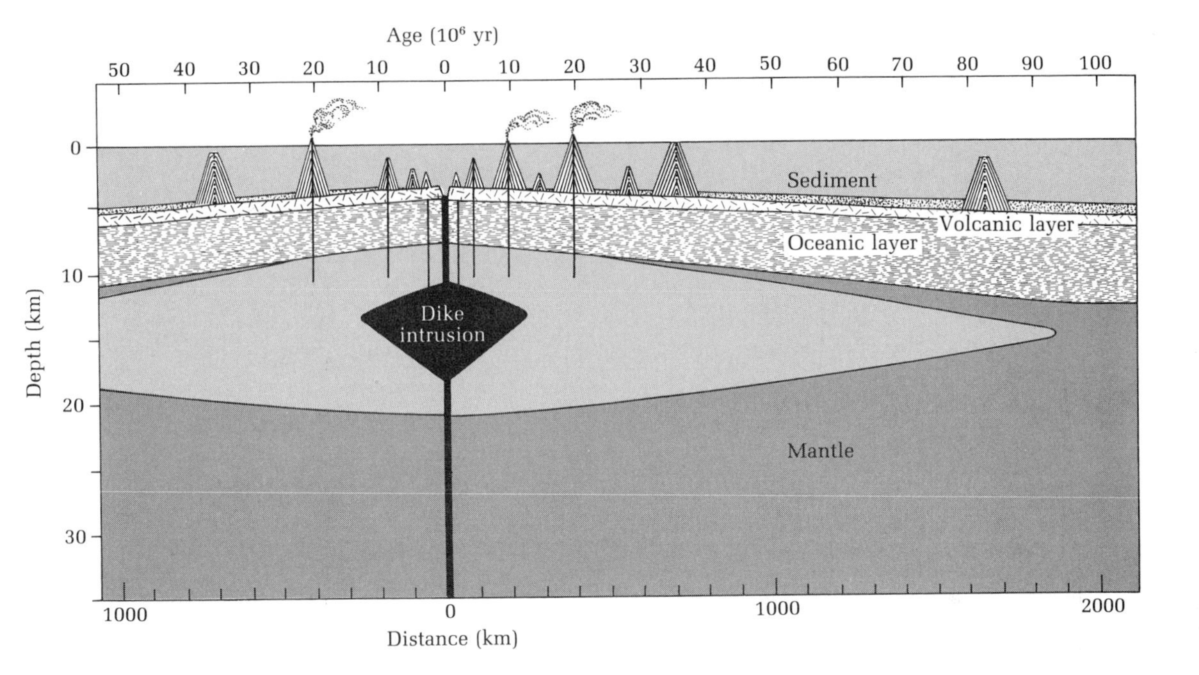

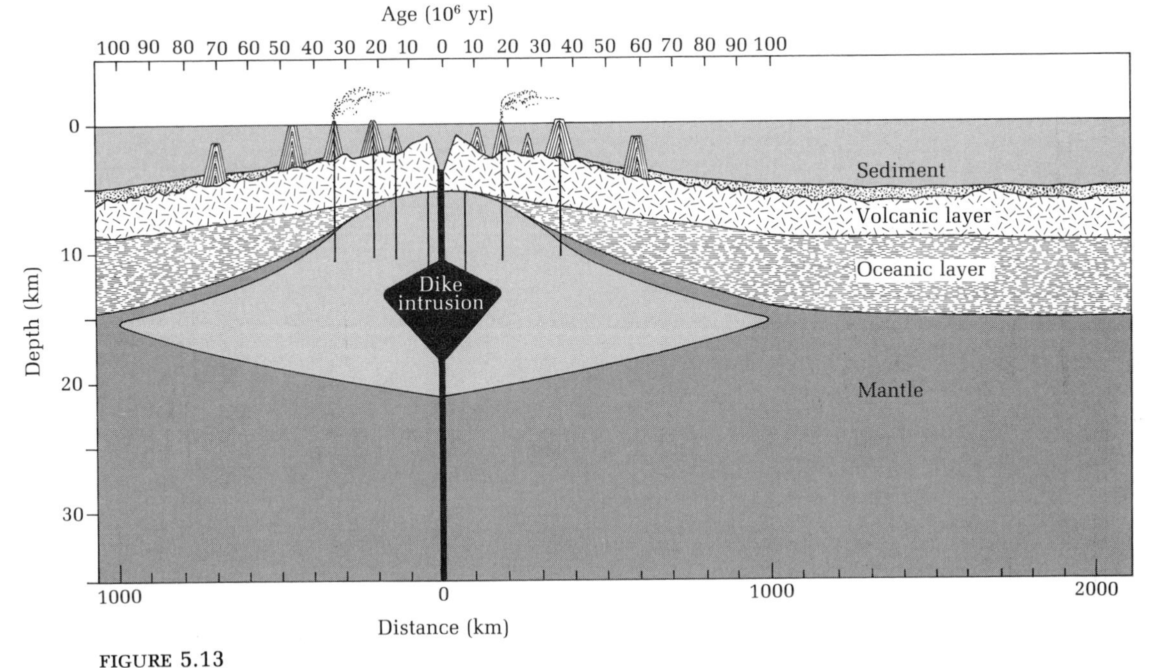

FIGURE 5.13

Rapid spreading (*top*) is accompanied by formation of a thin volcanic layer; growing volcanoes are far from the plate edge before they reach the surface. Slow spreading (*bottom*) produces a thicker volcanic layer; volcanoes growing at the same rates as those in areas of rapid spreading become islands after the same period of time, but they are closer to the plate edge. [From Menard, "The Deep-Ocean Floor," copyright © 1969 by Scientific American, Inc. (all rights reserved).]

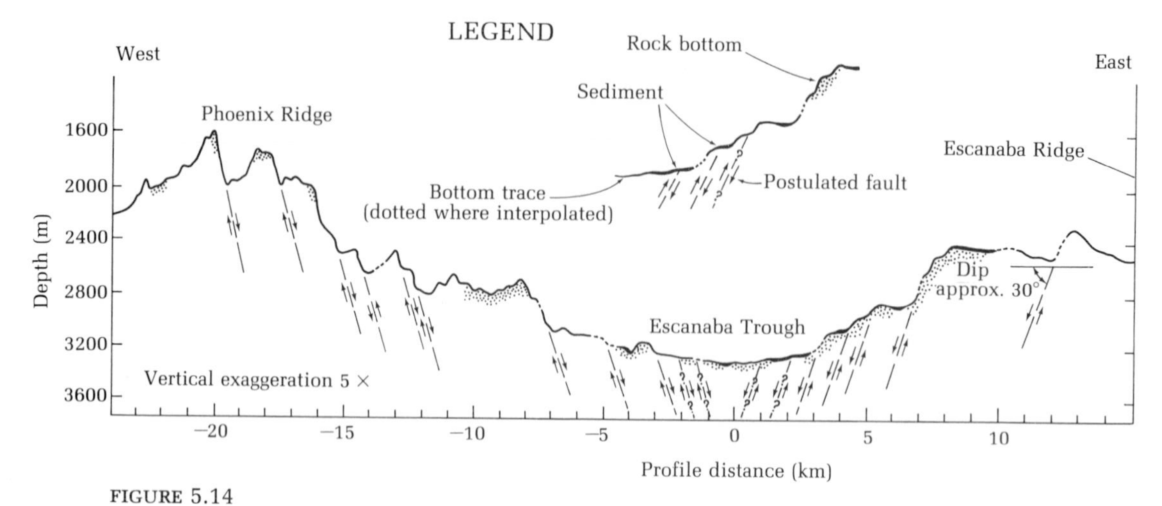

FIGURE 5.14

Topography and sediments of a spreading center, as measured by towing instruments just above the bottom. The inferred structure is a symmetrical system of normal faults that move blocks from the trough upward to form the ridges on each side. The location is in the center of the Gorda Ridge off northern California, and the spreading rate is slow. [From Atwater and Mudie, "Block Faulting on the Gorda Ridge," *Science* 159:729–731, copyright © 1968 by the American Association for the Advancement of Science.]

Fracture Zones

Fracture zones are linear mountain ranges that follow arcs of small circles around Euler poles. They offset, and are roughly perpendicular to, spreading centers, and they include very long ridges and troughs, cliffs separating regions with different depths, and lines of volcanoes. Most of these characteristics can easily be explained by plate tectonics. Virtually all the topography of fracture zones develops along the seismically active centers of the zones that are ridge–ridge transform faults. The great cliffs and regional differences in depth merely reflect the fact that the crust is elevated at spreading centers and subsides as it spreads. Thus, the relief of a cliff is directly related to the ridge offset or, in other terms, the length of the ridge–ridge transform. The relief varies during spreading because of different rates of sinking of the crust, and the slope of the cliff reverses midway along the active transform fault.

The characteristic ridges and troughs of fracture zones are 1–3 kilometers high, 10–20 kilometers wide, and a few hundred kilometers long. They can be accounted for if transform faults sometimes spread very slowly and thereby "leak" a little lava. Thus, the extreme relief is produced by the same phenomena as the rifts and adjacent mountains found in slowly spreading centers.

Beyond the spreading centers, the fracture zones are almost inactive. They are, essentially, the fossil remains of earlier transform faulting. However, the differential sinking on the flanks of midocean ridges produces some vertical motion on faults as the once high cliffs decay. This accounts for a

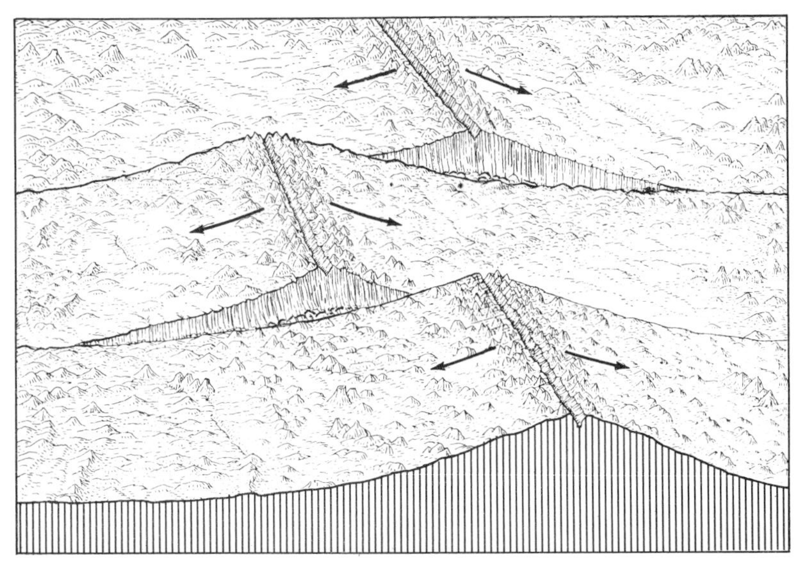

FIGURE 5.15

A ridge-ridge transform fault appears between two segments of a ridge that are displaced from each other. Mountains are built, earthquakes shake the plate edges, and volcanoes erupt in such an area because of the forces generated as the plates, formed at the spreading centers under the ridges, slide past each other in opposite directions. On outer slopes of the midocean ridges, however, this intense seismic activity appears to subside. [From Menard, "The Deep-Ocean Floor," copyright © 1969 by Scientific American, Inc. (all rights reserved).]

few earthquakes and for vertical offsets of the thickening sediment that accumulates as the crust ages on the ridge flanks. Flank fracture zones can become active transform faults again at any time; if they do, however, they become the sides of smaller new plates rather than the trailing edges of old ones. This is, at most, a rare phenomenon.

If the Euler pole remains relatively fixed, the fracture zones extending into the plate follow the same small circles as the active ridge–ridge transform at the plate edge. This occurs for some distance into the flanks along most fracture zones. However, at greater distances, and in crust more than about 10 million years old, the trends frequently differ from the active ones. A change can be dated by the age of the unfaulted magnetic anomalies next to it. In the northeastern Pacific, where they are best known, the trend changes were simultaneous on many fracture zones, and the magnetic anomalies of the same age also changed trend and remained perpendicular to the zones.

Intraplate Vertical Motion

The lithosphere created in a spreading center gradually moves into the interior of a plate as newer material is added to the trailing edge and older material is destroyed at the leading edge. The new crust takes the form of, or mani-

expansion. It is not yet certain that heating accounts for all of the elevation of a midocean ridge, but theory and observation are moving toward agreement on this possibility. As the lithosphere ages, it moves out to become a flank of a ridge, and it cools and contracts. According to this model, the relative elevation of any place on the ridge is merely an indication of the age of the lithosphere at that point.

The approximate age of any part of the ridge can be determined if magnetic anomalies can be identified. Consequently, the rate of sinking can be measured, provided the flanks in a given region were once at the same depth as the ridge crest is now. The fact that the ridge crest is about the same depth in most places and that its height above the deep sea basins can be accounted for theoretically by thermal expansion makes this priviso reasonable. Moreover, the results are pleasingly consistent (Figure 5.16). On the average, the crust sinks about 90 meters per million years for the first 10 million years after it forms, about 30 meters per million years for the next 30 million years, and 20 meters per million years thereafter. In a few places, however, the crust has remained level—or has even risen—for as long as 20 million years.

The rate at which the trailing edges of plates move away from spreading centers may be anything from less than 5 kilometers per million years to more than 50 kilometers per million years, but the rate is relatively constant in a given region. The plates sink at relatively constant rates that depend on their age. Consequently, the width and side slopes of a midocean ridge are approximately proportional to the spreading rate. Fast spreading yields a ridge with

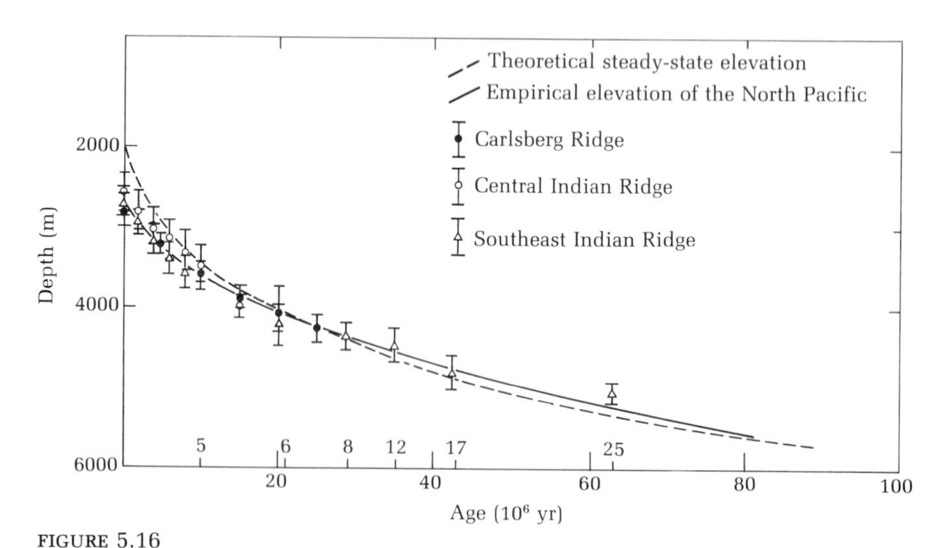

FIGURE 5.16

The sea floor sinks generally at a rate that depends on its age but is independent of the speed at which it drifts. The sinking agrees closely with a theoretical curve computed from observed heat flow and explained by assuming that the lithosphere cools as it ages. Numerals above the age scale are magnetic-anomaly numbers. [From McKenzie and Sclater (1971).]

a broad elevation with gentle slopes, such as the East Pacific Rise. Slow spreading yields a ridge with steep slopes and a narrow elevation, such as the Mid-Atlantic Ridge. The decreasing rate of sinking makes all ridge profiles concave upward.

Intraplate Vulcanism

Vulcanism is not confined to the margins of plates although it is concentrated there. The linear mode of vulcanism decreases rapidly in intensity away from spreading centers, but active, roughly circular volcanoes occur even in the middle of tectonic plates. Hawaii is a conspicuous example. Small conical volcanoes exist near spreading centers on crust that is only a few hundred thousand years old. Thus, it appears that the great cracks that are the sites of dikes and linear vulcanism are soon sealed as a plate ages and spreads. Vulcanism is then concentrated in central vents, which are created at different times and places.

There appear to be two modes of intraplate vulcanism in ocean basins. In the first, growth is slow and the volcano moves relatively far while it is active. In the other mode, vulcanism is very rapid and terminates before much movement occurs.

In the first mode, many central vents remain open for tens of millions of years. This is suggested by dated volcanic extrusions and by the distribution of different sizes and classes of marine volcanoes. Both big and little volcanoes occur in many areas, and the big ones are bigger at greater distances from a spreading center. Volcanic centers with a lava discharge less than about 100 cubic kilometers per million years never become islands because the sea floor sinks too fast for them to reach the sea surface. Drifting volcanoes may remain active, or become active, on crust 100 million years old, just as the Canary Islands have done. However, in many places, they become inactive by the time the crust is 20–30 million years old. This is demonstrated by the distribution of atolls and guyots (drowned ancient islands), which become submerged by gradual sinking of the aging crust. They occur almost entirely on crust more than 30 million years old.

The second mode of vulcanism, although probably sparsely distributed within the plates, can be very intense. The localities in which this mode of vulcanism occurs are called hot spots. The best data available on hot spots come from the island of Hawaii, where about 4 cubic kilometers of lava has flowed out during the last 100 years. The whole enormous volume of the subaerial and submarine portions of the volcanoes and their roots could have accumulated in only 5–6 million years at the present rate of discharge. It appears that some hot spots remain fixed in the mantle for long periods as the lithosphere drifts above them. This produces chains of islands (see Figure 5.18 and the first section of Chapter 6).

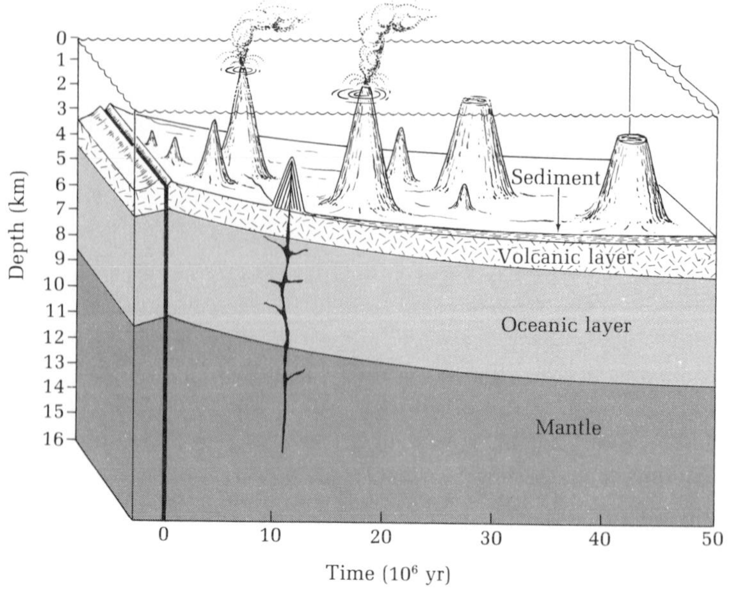

FIGURE 5.17

Many undersea volcanoes normally begin to rise near spreading centers. They may quickly become inactive, but, if not, they ride along on the moving plate as they grow. If a volcano rises fast enough to surmount the original depth of the water and the sinking of the ocean floor, it emerges as an island such as St. Helena in the South Atlantic. In order to rise above the water, an undersea volcano must grow to a height of about 4 kilometers in 10 million years. Island volcanoes sink after 20 to 30 million years and become the sediment-capped seamounts called guyots. [After Menard, "The Deep-Ocean Floor," and Wilson, "Continental Drift," copyright © 1969 and 1963, respectively, by Scientific American, Inc. (all rights reserved).]

Continental Plate Effects

With regard to plate tectonics, the continents are relatively dull places, because they are usually in the middle of plates and, therefore, away from the action. Normal spreading centers, for example, are always oceanic because, were they to penetrate a continental plate, they would split it into two continents separated by an ocean basin. This is happening now in the Red Sea and Gulf of California, and it happened as Africa was split away from South America, and Australia from Antarctica, and so on. It is difficult to tell whether continents are elevated before they are split, because much of the evidence is eroded away. If they are, the effect apparently is localized. Once the splitting begins, the continental margins rise as vertical cliffs above the inrushing sea. This is too much for the rock to bear, and tensional cracking occurs parallel to the main, opening crack. Great blocks collapse into the ocean until the slope is stabilized. Vulcanism has occurred, too, along the secondary cracks in the Red Sea and Gulf of California. Nevertheless, the effects of vulcanism and collapsing margins are small: even after they have occurred, the continental margins generally can be matched almost exactly.

Almost all transform faults, active or dead, are in ocean basins. Notable exceptions occur, at present, in California, New Zealand, and West Pakistan, but it does not appear that they will endure for long as active intra-continental features. Consider the San Andreas fault, which is a ridge–ridge transform fault (or a system of closely related faults) 1600 kilometers long. During the last 4 million years, California west of the fault has been moving at the rate of 6 centimeters per year relative to the continent to the east. In about 15 million years, at present rates, the separation will be almost complete and coastal California will be an island or a peninsula off Oregon and Washington. Moreover, the San Andreas system of faults, as a relatively simple transform, is uniquely long. Most continental blocks would move past each other and separate in only a few million years of transform faulting.

Once continents are separated, either by spreading or transform faulting, they may become passive blocks that merely shed sediment because they are high. They are eroded and rise isostatically, while the surrounding ocean basins, to which they are attached, collect sediment and sink. However, the continents may also be subject to intraplate vulcanism and broad arching as they drift.

Eventually, continents reach the leading edges of plates after any oceanic crust in front of them is consumed. They themselves are not consumed. Small continental plates are being pushed together in the Mediterranean at present without plunging. Likewise, it appears that, when India drifted into Asia, it merely became part of Asia rather than plunging under it. This point is not entirely certain, but, under the Himalayas and Tibet, where a great overlap might have occurred, the crust, although somewhat unusual, is by no means the thickness of two continents. It appears likely that the material of the continents is conserved at the surface. Continents can be split apart and rejoined in random ways, but not consumed from below. Because erosion eats away the tops and sides, the material of the continents is gradually recycled by erosional, tectonic, and igneous processes.

Subduction Zones

Trenches and young mountain belts form where plates push together. The French geophysicist Xavier Le Pichon concludes that the occurrence of one or the other is a function of the rate at which this pushing together occurs. If the rate is less than about 5 centimeters per year, the crust can absorb the compression by shortening, and it forms great mountain ranges by folding and overthrusting. If the rate is faster, the lithosphere breaks free and sinks into the mantle, apparently at a constant rate. It is spreading horizontally, however, at different rates, and the proportion between horizontal and vertical components determines the angle at which the lithosphere plunges.

Before plunging, the lithosphere bends, cracks, and forms fault blocks parallel to the trench. This is confirmed not only by the fact that the sediment in the region is faulted, but by the motions of earthquakes.

The simple plate formed in a spreading center is usually encrusted with volcanoes and coated with sediment by the time it reaches a trench and turns down. The plate in the western Pacific that is plunging under the Marianas arc is an example. Because the volcanoes stick up like barnacles and the sediment is mere muck compared to solid rock, they both should be scraped off when the plate plunges. Direct evidence of this scraping is not yet available from the sea floor, but it is abundant in continental margins or geosynclines that were once above subduction zones. California, for example, was once bounded by a subduction zone, and the Coast Ranges now contain distinctive sedimentary and igneous rocks of types typical of ocean basins. It appears that these rocks were scraped off at the continental edge, or just under it, whence they were squeezed up along faults.

Once it starts to plunge, the relatively cold lithosphere begins to heat up, both because of conduction from the hot mantle and because of friction with it. The various minerals in the lithosphere melt at different temperatures and pressures, and various new minerals replace them. The lower-melting components bleed out at a depth of about 100–160 kilometers and float upward to form a line of relatively silicic volcanoes, which are typical of island arcs even though they are surrounded by oceanic crust that is more basic.

The heat generated by friction also penetrates the overlying mantle; it and the physical disturbance of the moving lithosphere cause some minor motion above the plunging plate. This, in turn, splits the overlying plate a little and produces interarc basins floored by youthful lava flows like those of major spreading centers. Drilling in the deep sea by JOIDES has revealed sedimentary rock west of the Marianas arc that is younger than the old Pacific plate to the east. The American geologist Daniel Karig explains this by suggesting that the island arc and subduction zone migrated eastward, leaving behind a series of fossil arcs and interarc basins.

Origin of Plate Motions

The cause of plate motions is still a matter of conjecture. There are only two major energy sources, the sun and heat flow from the interior, and only the latter can have a powerful and very direct effect upon the plates. Thus, the question is reduced to how the energy of the heat is applied, and whether the plate is given potential energy, which drives the motion, or is in some way more directly driven by the heat.

The plate obtains potential energy in two ways: (1) the materials in it are elevated through the mantle by heating and may be denser than the mantle when they cool, and (2) a whole lithospheric plate is elevated at one edge by thermal expansion of the mantle. Physicist Egon Orowan has suggested that the potential energy of the elevation of the edge may provide the driving force. However, the slope is a small fraction of a degree, and a surface without friction is required to separate the lithosphere and the asthenosphere. Moreover, the fastest plate motion is associated with the gentlest slopes, instead of the implied reverse.

Physicist Walter Elsasser has suggested that the denser lithosphere pulls itself into the mantle once chance or a plate collision causes it to start sinking. This basic instability is reinforced because the lithosphere is compressed as it sinks and its minerals are transformed into denser phases. Thus, the lithosphere moves over an almost passive mantle which moves only at depth and in the opposite direction in order to maintain continuity.

The lithosphere can also be moved directly by the drag of thermally driven convection in the mantle. Two modes of this convection have been suggested. The "classic" convection pattern of earth scientists since the 1930s has been

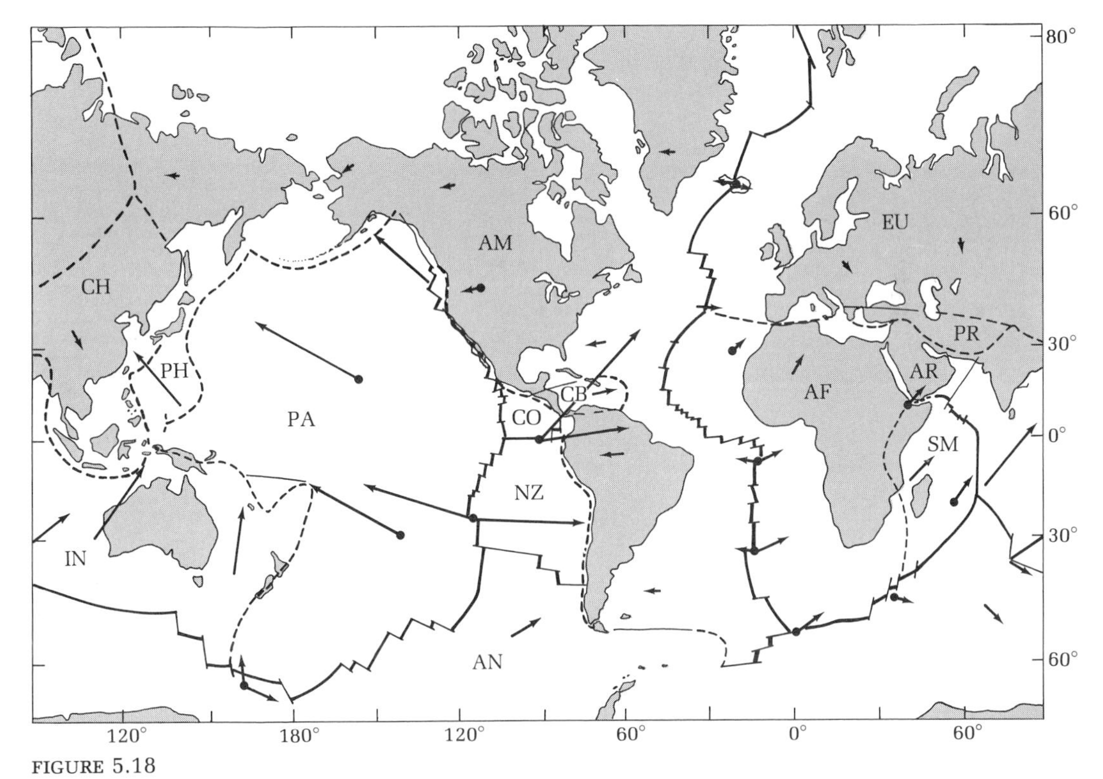

FIGURE 5.18

Present motion of plates over hot spots in the mantle, as deduced by Morgan (1972). The lengths of the arrows are proportional to plate speed. The letters are abbreviations of the names of the plates (see Figure 5.6).

an overturn of great cells extending deep into the mantle. This was the pattern visualized by Harry Hess when he proposed sea-floor spreading: the mantle rises beneath the trailing edge of a plate and sinks beneath the leading edge. Its plastic motions are directly linked to the rigid motions of the lithosphere—which is difficult to understand. Moreover, its configuration is the same as that of the trailing edge of the plate—it is offset by transform faults, and so on. Likewise, it changes direction when the relative motion between plates does. The newer facts make this pattern of convection highly unlikely.

The second style of convection that has been visualized involves small rising cells or plumes, like the rising thunderheads of the advective circulation of the atmosphere. The existence of such plumes is required because some intraplate vulcanism originates during the drifting of plates over relatively fixed hot spots in the mantle. The small cells or plumes can flow in different directions. The rigid plate merely moves off in the direction of the (vector) sum of all the stresses of the various cells. If the many stresses are too opposed to be summed, the plate breaks and spreading begins.

The origin of plate motion is still an intriguing question for further study. Even without an answer, enough is known about plate motion that, in the next few chapters, we can examine both its past history and its future social effects with some confidence.

Summary

1. The crust is continuous; therefore, if a plate moves, it interacts with adjacent plates. The interactions cause lines of earthquakes at the boundaries. The interiors of plates are much less prone to earthquakes.
2. Rigid plates on a sphere move over the asthenosphere in accordance with Euler's theorem: if one rigid spherical shell moves over another, two diametrically opposed points remain fixed. The points, which may be called "Euler poles," have no reality except in a geometrical sense.
3. The rigidity of large plates is indicated by the horizontal and vertical distribution of earthquakes, by first motions of the earthquakes, by the distribution of spreading rates, and by the pattern of fracture zones.
4. The rates of relative motion can be measured geodetically on land and by analysis of patterns of magnetic anomalies at sea.
5. The edges of plates move and jump about, and plates can be completely destroyed.
6. Spreading centers form where plates move apart. They produce oceanic crust and are characterized by high heat flow, uplift, shallow depths, linear volcanoes, magnetic anomalies, faults, and earthquakes. All these can be explained by plate tectonics.

7. Fracture zones form where plates move beside each other along transform faults. They consist of very long mountain ranges and troughs that follow enormous arcs of circles. They separate regions of different depth along great cliffs.
8. The hot oceanic crust that forms at spreading centers generally sinks as it drifts away and cools. Thus, the age of a certain piece of oceanic crust and depth are related, and the shape of midocean ridges can be explained by differences in spreading rates.
9. The interiors of large tectonic plates are not wholly inactive. Large volcanoes may form, local uplift occurs, and there is some minor faulting.
10. Continents are usually within plate boundaries because plate motions split them apart if they overlie the boundaries.
11. Subduction zones usually occur where plates push together. The lithosphere may be crushed into mountains, but generally one plate plunges into the mantle and a trench, volcanic arc, and zone of deep earthquakes is produced.
12. The origin of plate motions is uncertain at present.

Discussion Questions

1. What is the main evidence that indicates that the crust is made of large, rigid, moving plates?
2. How can the rate of relative motion between plates be measured? What are normal rates? Can you measure rates of relative motion in subduction zones?
3. Do all natural things that are bilaterally symmetrical become so by splitting down the middle? If not, can the magnetic anomalies of the sea floor be caused by something other than spreading?
4. If there are three plates that are separated by spreading centers, as in Figure 5.8, is the middle plate necessarily immobile?
5. How does the theory of plate tectonics explain the geologic features observed at spreading centers?
6. If fracture zones are curved along arcs of circles, how does the plate on the concave side avoid pushing over the one on the convex side?
7. What factors determine that the Mid-Atlantic Ridge is narrower and has steeper sides than the East Pacific Rise? Both are midocean ridges that have formed at spreading centers.
8. Why was plate tectonics not discovered by the study of continental rocks?
9. Why are deep earthquakes produced in a subduction zone?
10. If we do not yet know why plates move, are we justified in questioning whether they do move?

References

Atwater, T. M., 1970. Implications of plate tectonics for the Cenozoic tectonic evolution of western North America. *Bull. Geol. Soc. Amer.* 81:3513–3536. [Migrating triple junctions and their effects.]

——, and J. D. Mudie, 1968. Block-faulting on the Gorda ridge. *Science* 159:729–731. [The bottom of the rift splits and migrates up the sides.]

Barazangi, M., and J. Dorman, 1969. World seismicity maps compiled from ESSA, Coast and Geodetic Survey, epicenter data 1961–1967. *Bull. Seismol. Soc. Amer.* 59(1):369–380. [Precise plots.]

Bullard, E., 1969. The origin of the oceans. *Sci. Amer.* 221(3):66–75. (Available as *Sci. Amer.* Offprint 880.) [The ocean basins, not the water.]

Elsasser, W. M., 1971. A two-layer model of upper-mantle circulation. *J. Geophys. Res.* 76:4744–4753. [The plunging lithosphere pulls itself down.]

Heezen, B. C., 1960. The rift in the ocean floor. *Sci. Amer.* 203(4):98–110. [Photos of Iceland splitting apart.]

Heirtzler, J. R., 1968. Sea-floor spreading. *Sci. Amer.* 219(6):60–70. (Available as *Sci. Amer.* Offprint 875.)

Hess, H. H., 1962. History of ocean basins. *In* A. E. J. Engel, et al., eds., *Petrologic Studies: A Volume in Honor of A. F. Buddington,* pp. 599–620. New York: Geological Society of America. [The "essay on geopoetry" that revitalized the idea of continental drift.]

Hurley, P. M., 1968. The confirmation of continental drift. *Sci. Amer.* 218(4):52–64. (Available as *Sci. Amer.* Offprint 874.) [A detailed match of geology in South America and Africa.]

Isacks, B. L., J. Oliver, and L. R. Sykes, 1968. Seismology and the new global tectonics. *J. Geophys. Res.* 73:5855–5899. [Different types of earthquakes correspond to different parts of interacting plates.]

Karig, D. E., 1971. Structural history of the Mariana Island Arc system. *Bull. Geol. Soc. Amer.* 82:323–344. [Interarc basins develop behind trenches.]

Luyendyk, B. P., 1970. Dip of the downgoing lithospheric plate beneath island arcs. *Contrib. Woods Hole Oceanogr. Inst.* (2463). [Dip depends on rate of subduction.]

McKenzie, D. P., 1969. Speculation on the consequences and causes of plate motions. *Geophys. J.* 18:1–32. [A comprehensive analysis.]

——, and W. J. Morgan, 1969. Evolution of triple junction. *Nature* 224:125–133. [Essential information for the professional.]

——, and R. L. Parker, 1967. North Pacific: an example of tectonics on a sphere. *Nature* 216:1276–1280. [The first paper in plate tectonics; based on seismology.]

——, and J. G. Sclater, 1971. The evolution of the Indian Ocean since the late Cretaceous. *Geophys J.* 24:437–528.

Menard, H. W., 1969a. The deep-ocean floor. *Sci. Amer.* 221(3):126–142. (Available as *Sci. Amer.* Offprint 883.)

——, 1969b. Elevation and subsidence of oceanic crust. *Earth Planetary Sci. Letters* 6:275–284.

——, 1969c. Growth of drifting volcanoes. *J. Geophys. Res.* 74(20):4827–4837.

——, and T. E. Chase, 1971. Fracture zones. *In* A. E. Maxwell, ed., *The Sea,* vol. 4, pt. I, pp. 421–443. New York: John Wiley & Sons.

Morgan, W. J., 1968. Rises, trenches, great faults, and crustal blocks. *J. Geophys. Res.* 73(6):1959–1982. [The second paper in plate tectonics; based on geology.]

——, 1972. Deep mantle convection plumes and plate motions. *Bull. Amer. Ass. Petrol. Geol.* 56(2):203–213. [Mantle plumes, resembling thunderheads, may move plates.]

——, 1972. Plate motions and deep mantle convection. *In* R. Shagam, ed., *Studies in Earth and Space Sciences (Geol. Soc. Amer. Mem.* 132), pp. 7–23. New York: Geological Society of America. [Moving volcanoes that form over fixed mantle hot spots show the course of plate motions.]

Orowan, E., 1969. The origin of the oceanic ridges. *Sci. Amer.* 221(5):103–119. [Influence of gravity and effects of brittle fracture.]

Sleep, N. H., 1969. Sensitivity of heat flow and gravity to the mechanism of sea-floor spreading. *J. Geophys. Res.* 74(2):542–549.

Vine, F. J., 1966. Spreading of the ocean floor: new evidence. *Science* 154 (3755):1405–1415.

——, and D. H. Matthews, 1963. Magnetic anomalies over oceanic ridges. *Nature* 199:947–949. [Magnetic anomalies are caused by field reversals recorded by sea-floor spreading.]

——, and J. T. Wilson, 1965. Magnetic anomalies over a young oceanic ridge off Vancouver Island. *Science* 150(3695):485–489. [Anomalies are symmetrical.]

Wegener, A. L., 1915. *Die Entstehung der Kontinente und Ozeane.* Brunswick: Vieveg und Sohn. (English translation by John Biron, 1966. *The Origin of Continents and Oceans.* New York: Dover.) [The book that generated the debate that lasted fifty years. Like most original syntheses, it is simple and easy to understand.]

Wilson, J. T., 1963. Continental drift. *Sci. Amer.* 208(4):86–100. (Available as *Sci. Amer.* Offprint 868.) [Interesting to compare with Hurley's article in the same journal in 1968.]

——, 1965. A new class of faults and their bearing on continental drift. *Nature* 207:343–347. [The concept of transform faults expounded.]

6

CONTINENTAL DRIFT

While I was writing this chapter (April 1830), I attended a meeting of the Geological Society of London, at which the President, in his address used the expression "a geological logician." Smiles appeared on the faces of all who were present, and some, like Cicero's augurs, could not suppress a laugh; so amusing appears the combination of geology and logic.

Charles Lyell, PRINCIPLES OF GEOLOGY

Motion Relative to the Pole

Many consistent types of observations linked by theory have been used to establish that large crustal plates are moving over the surface of the earth. The history of their *relative* motions for more than 100 million years can be deciphered from the sea-floor magnetic anomalies. This chapter, however, is not confined to discussion of such relative motions; it is also concerned with motions relative to a geographic frame of reference. For convenience, the axis of rotation and the north pole can serve as the geographic reference points, and motions relative to the north pole can be termed latitudinal.

Latitudinal motions can be derived from any phenomenon related to latitude that is preserved in the geological record. The most useful information comes from the studies of paleomagnetism and paleoclimatology, because the climate is always tied to the axis of rotation, and the magnetic pole has been so for a minimum of about 100 million years.

The latitudinal motion of a tectonic plate can be determined by observing the changing attitude of the paleomagnetic field as it has been recorded in rocks of different ages. Data can be collected by hand on land or by magnetic mapping and computer analysis at sea. Consider a hypothetical experiment in which we assume, for convenience, that the magnetic pole coincides with the pole of rotation: We go to a place in the northern hemisphere, where the magnetic field trends north and dips 45° downward, and take a suitable sample of volcanic rock. In the laboratory, we determine that the rock is 500 million years old and that it cooled through the Curie point in a horizontal magnetic field that now has an orientation of due north. From the dip, we can be sure that the rock cooled at the magnetic equator, and from its present position, we know it drifted 45° north (although we do not know where it may have been in the meantime—it might have been to the south pole, for all we know).

The ancient magnetic field had the same declination as the present magnetic field, which might seem to suggest that the plate has merely drifted up a line of longitude. Consider, however, a plate initially on the equator but on the opposite side of the earth: were it to drift first halfway around the world on the equator and then to drift north, it would record the same magnetic field as if it had remained fixed in longitude. *Paleomagnetism determines latitude but not longitude.*

The complete motion of tectonic plates relative to each other can be determined for the last hundred million years from magnetic anomalies on the sea-floor. That is, if we knew that some place was fixed in longitude but drifting in latitude, we could describe the motion of all the other plates in terms of latitude and longitude both. Unfortunately, we have no certain way of knowing this at present. Perhaps the reference point drifts east, perhaps west; we know only relative longitudinal motion. Jason Morgan has suggested that hot plumes remain fixed in the mantle and, by building volcanoes, brand their locations on the plates that drift above them. If so, both latitude and longitude can be established. The plumes may not last long, however, relative to geologic time; moreover, the mantle may also drift.

PERMANENCE OF THE DIPOLE

With two poles—that is to say, with two points each crossed by all lines of longitude—geography is simple. But what if we tried to navigate using maps that sometimes include lines of longitude connecting three or four poles? But surely, you say, no mapmaker would be so irrational. But what if the maps were made for some purpose—decoration, perhaps—other than navigation? With paleomagnetism, we chart the drift of tectonic plates, but the magnetic field was not designed for that purpose. Indeed, the magnetic field may have had more than two poles at some time in the past. Although the magnetic charts appear to give reliable and consistent results, even in the remote past,

BOX 6.1 DIRECTIONS OF DRIFTING

The motion of a plate can be arbitrarily related to another plate or to the geographical framework of the earth; in either case, however, it is a rotation around an Euler pole. For this reason, it is generally inaccurate, although convenient, to say, for example, that a continent drifts "southeast." To describe the motion accurately, there is no substitute for identifying the Euler pole and the angular rate of rotation.

The only correspondence between plate motions and geographical directions arises if the Euler pole is a geographic pole. Then a plate can drift "east" or "west" because the directions are along lines of latitude, which are small circles around the pole.

A whole plate cannot drift "north" because *all* points on a plate rotate in small circles around an Euler pole. The most favorable case is when an Euler pole is on the equator. Then the points along the Euler "equator" of a plate can drift "north" but, even so, no others on the same plate can do so.

it is necessary to assume that the surface field was a dipole—that is, there were only two poles. If not, large apparent motions of very ancient plates might merely reflect variations in a complex field. A relatively stationary plate might have been near a weak pole that vanished, which would make it appear once to have been remote from another, stronger pole from which it had never strayed.

The sun has a secondary (toroidal) component of its magnetic field that breaks the surface in sunspots, and the dynamo theory suggests the existence of a similar field in the interior of the earth. Can it emerge? Has it done so? These questions are unanswered as yet. Apparently, we should not be uncritical in examining the paleomagnetic evidence for very ancient continental drift. We need not worry about any evidence for drift of ancient oceanic plates, because no such plates exist that are more than a mere few hundred million years old.

PALEOCLIMATES

Global atmospheric circulation is inevitably related to the tilt of the earth's axis and the speed of its rotation, which, apparently, have not changed very much during the last several hundred million years. Thus, the moraines and striations of an ancient continental glaciation are evidence of polar conditions, rather than equatorial ones. A global girdle of contemporaneous desert sand dunes is an indicator of the subtropical divergence, which occurs at a certain latitude with a wind pattern that may be preserved in the orientation of ripples in sand that has since been consolidated into sandstone. Climate, like magnetism, is independent of longitude (except secondarily because of the distribution of continents). It indicates only latitudinal drift.

ANCIENT CONTINENTAL DRIFT

If the magnetic stripes produced by sea-floor spreading are successively removed, the continents rotate back together. They join at the continental slope, which is the real margin of continental crust. (The shoreline varies with sea level, and is not as useful for matching.) Africa is rejoined to the Americas, Antarctica, and India. Australia is rejoined by Antarctica, and North America to Europe. The remarkable fit was not appreciated until Euler's theorem and a computer were used by Sir Edward Bullard and others in 1965 to obtain a fit of the two sides of the Atlantic. This technique is now in common use. Robert Dietz and John Holden have made a complete reconstruction of continental positions 200 million years ago in which the gaps and overlaps at the rejoined margins are remarkably few. The great Amazon and Niger rivers have built deltas on the continental margins since drift began; naturally, these overlap. Spain is awkward—but it appears to have rotated, so it can be accounted for. The biggest overlap is between the Bahamas Islands and Africa, but this is interpreted as evidence that the Bahamas are coral islands built up since drift began rather than continental crust.

Many slightly different reconstructions exist, and much research remains to be done. The general picture is becoming accepted, however, and the Dietz–Holden reconstruction will be used here with the understanding that it is a tentative model.

Two separate continents may have existed 200 million years ago—a northern one, Laurasia, and a southern one, Gondwanaland. In this model, however, they are joined into one, called Pangaea ("all lands"), covering 40% of the earth. Does this mean that continental drift occurred only once in the history of the earth, or was Pangaea assembled from still older fragments? On this point there is no agreement. Many geologists interpret ancient rocks as evidence of intermittent, persistent continental drift. Tuzo Wilson, for example, observes that certain anomalous rocks along the opposite margins of Europe and North America can be explained if the North Atlantic opened more than once.

Geophysicists Patrick Hurley and John Rand take an opposite view. The continental nuclei are made up of rocks largely more than 800 million years old, and with large areas more than 1700 million years old. Hurley and Rand observed that, in a Pangaea reconstruction, all the continental nuclei are restricted to a relatively limited region. They suggest that it is unlikely that this is due to chance, and they propose that no drifting occurred between 1700 million and 200 million years ago. Drifting before 1700 million years ago is neither eliminated nor supported by this line of argument.

It is apparent that the early history of drifting continents has many gaps and uncertainties. The various views remain to be reconciled. If the present drifting of continents is unique, or even a rare phenomenon, a fascinating question arises—namely, what triggered it? If it is commonplace, why did

FIGURE 6.1
The continents restored to their positions before the breakup of Pangaea. The ancient shield areas (*darker shading*) and occurrences of deep-seated rock facies (*black dots*), also of ancient age, are closely clustered. This suggests that they did not move between the time they formed and the time Pangaea began to drift apart. However, other lines of evidence appear to contradict this. [From Hurley and Rand (1969), copyright 1969 by the American Association for the Advancement of Science.]

the drifting continents join to form Pangaea? This question gives less difficulty. Many random patterns of plate motion are conceivable that would sweep the continental fragments into one place, and continents seem to be able to fuse readily.

Pangaea and Panthalassa

What was the world like 200 million years ago? The enormous continent Pangaea was large enough to reach from pole to pole, but was it in the right position? The overwhelming evidence of Late Paleozoic continental glaciation in the center of southern Pangaea seems to provide an answer. Fossil moraines occur in southern Africa, South America, India, and Australia, and the orientation of glacially cut grooves indicates that Antarctica was the center of the glaciers.

The latitudinal positions of the continents can be derived from paleomagnetism. These positions are not entirely consistent, but a "best fit" can be derived by computer that correlates with the reconstructed map of Pangaea—

BOX 6.2 THE PLATE-TECTONICS GAME

It may be useful to think of a children's game while considering continental drift. The game is called Paper-Scissors-Stone, and two children announce one of the three objects simultaneously. Paper covers stone; scissors cut paper; stone breaks scissors. The one who picks the dominant member of a pair wins the game. Ties don't count.

Plate tectonics lends itself to the same game:

1. Rift and transform cut continent.
2. Continent covers trench.
3. Trench eats rift and transform.

Oceanic crust doesn't count.

that is, the south pole was near the center of the supposed glaciers at the right time, and so on.

In earlier times, the geography was quite different. Even if Pangaea had been unbroken for 1500 million years, it could have drifted as a unit. Roughly 450 million years ago, for example, the north magnetic pole was at the point where Africa, South America, and North America joined.

By Triassic time, 200 million years ago, glaciation had ceased, and paleomagnetism indicates a generally northerly drift that brought South Africa from within the Antarctic circle to about 50°S latitude. The equator was then near a line extending through Mexico, Florida, western Africa, northern Egypt, and what is now central Asia. The north pole was near the Bering Sea.

This reconstruction, which probably approaches reality, permits some consideration of the world at the time. The enormous continent had profound effects on climate and drainage, which were totally different from what they are at the present. The whole interior, apparently, was high and arid—rather like central Asia. Rivers arising in the interior traversed the whole width of what is now Africa and South America before reaching an ocean.

The oceans formed one great Panthalassa ("all oceans"), or proto-Pacific, covering 60% of the earth from pole to pole. Storm winds blew unimpeded, and the global patterns of surface water circulation were ideally simple. Of the floor of that enormous sea we know nothing, except that it existed. There may have been moving plates and volcanoes, but whatever existed has disappeared long since into subduction zones.

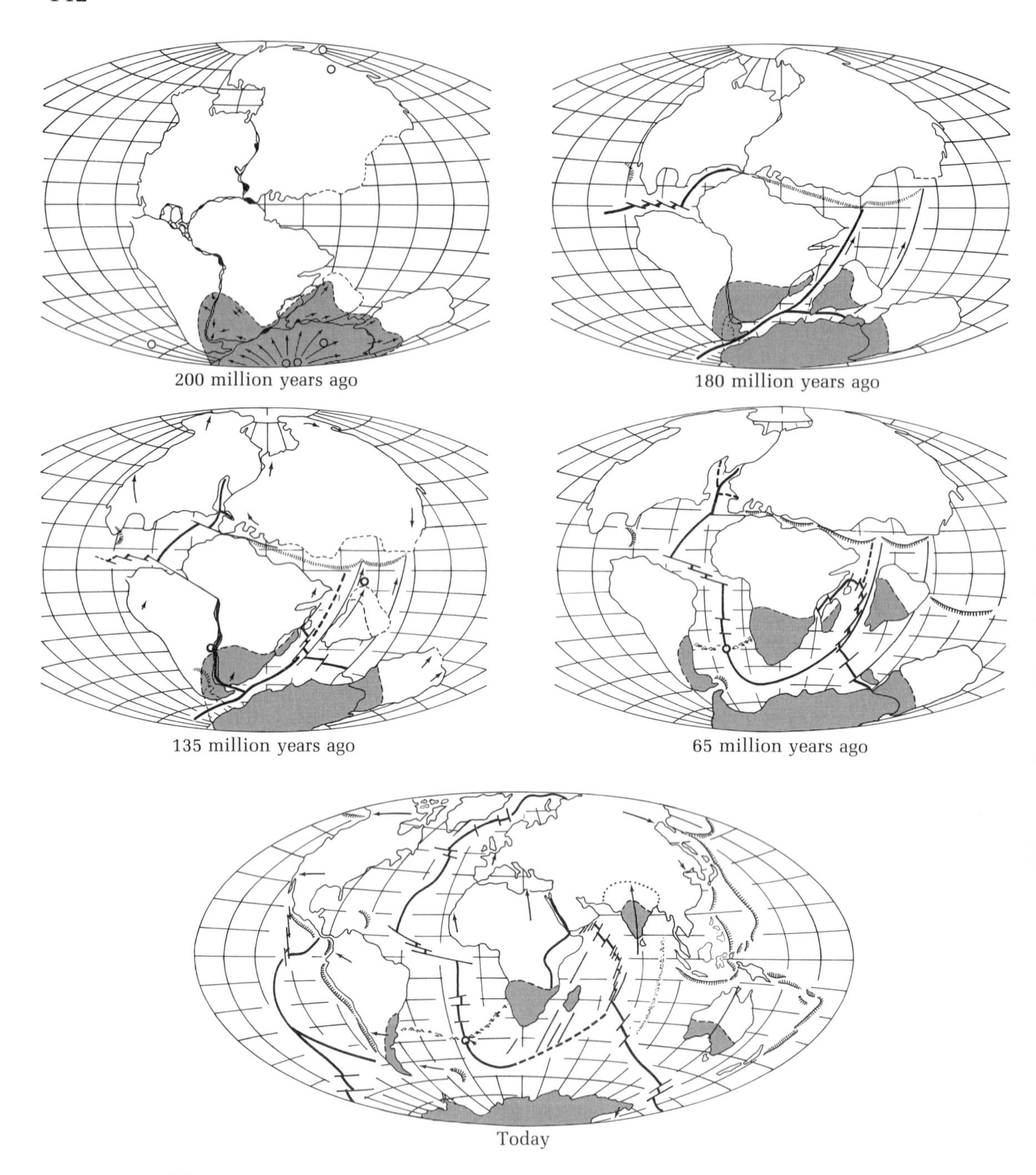

FIGURE 6.2

The breakup of Pangaea, according to Dietz and Holden (1970). Certain ancient glacial deposits that were all together 200 million years ago are widely spread today. The North and South Atlantic opened gradually and bordering seas formed. India, originally joined to Antarctica, moved through the whole Indian Ocean to collide with Asia. The Mediterranean formed as Africa moved past Europe. Australia split off from Antarctica last of all. [After Dietz and Holden, "The Breakup of Pangaea," copyright © 1970 by Scientific American, Inc. (all rights reserved).]

Expansion of Pangaea

For some reason, the world began to come apart about 200 million years ago. Basaltic lavas welled out along opening rifts in such places as New Jersey when the Atlantic began to spread. We may visualize a likely series of subsequent events: As a rift opens in a continent, it is flooded by the sea, and the first marine organisms and salt are deposited in a region where deserts and savannas existed before. This sequence of deposits occurs in many places around the margins of Africa.

180 MILLION YEARS AGO

The history of the breakup of Pangaea is still a matter of conjecture, but it seems probable that it went in steps. By 180 million years ago, three continents may have come into existence. North America rotated away from South America around an Euler pole near Spain. This created a northern continent, Laurasia. The Caribbean Sea and Gulf of Mexico were the locus of a spreading center that reached from the Pacific into a gradually opening North Atlantic. The rotation around Spain began to close a vast embayment between Africa and Asia and to form a long, narrow, equatorial sea—the Tethys Sea—in which sediment accumulated that now forms mountain ranges from the Himalayas to the Alps.

A second rift, not necessarily contemporaneous, began to open to the south of Africa and South America. Antarctica, Australia, India, and Madagascar formed a third continent.

180–135 MILLION YEARS AGO

Three continents probably still existed during the time from 180 million to 135 million years ago, although the earlier rifts were widening and branching. In the North Atlantic, the rift probably branched by successive changes from transform faults to spreading centers. The Bay of Biscay opened as Spain rotated toward the east. Then the Labrador Sea opened between Canada and Greenland.

The main branch from the southern rift began to open between Africa and South America. The South Atlantic resembled the modern Red Sea, and a ridge-ridge transform connected it to the North Atlantic. Along this fault the moving continents remained in near contact, not unlike Baja California and mainland Mexico along the present Gulf of California.

135–65 MILLION YEARS AGO

The marine magnetic anomalies are of use in interpreting the history of continental drift for the interval from 135 million to 65 million years ago. The

Atlantic continued to rotate open, and the South Atlantic had more than half its present width. At the northern end, a new branch was opening between Greenland and Europe, and the older branches were dead. The rotation of Africa moved it east of Spain, which jutted into the Atlantic, and began to crush the Tethys Sea sediments against Asia. Madagascar separated from the east side of Africa.

India split off from Antarctica–Australia during this interval, or perhaps before. Dan McKenzie and the British geophysicist John Sclater put the separation at 3000 kilometers 75 million years ago. The leading edge of the Indian plate was a subduction zone thrusting oceanic crust under Asia. Thus, oceanic crust was destroyed at the north and created at the south, and India migrated through the plate without breaking it in any way.

65 MILLION YEARS AGO TO THE PRESENT

The Atlantic basin continued to open and penetrate northward—ultimately reaching Asia after crossing the Arctic Basin. The differential opening in the north moved Europe east relative to Africa and created the Mediterranean as we now know it.

India put on a remarkable spurt between 75 million and 55 million years ago: it moved away from Antarctica at an average rate of 18 centimeters per year. Then, motion apparently ceased until about 36 million years ago. Australia finally separated from Antarctica during this time. The present episode of spreading began about 36 million years ago, and the Gulf of Aden, the Red Sea, and most of the African Rift valleys began to open, or continued to open, while India squeezed up the Himalayas when it reached Asia.

Contraction and Drift of Panthalassa

CONTRACTION

Panthalassa covered 60% of the earth, but its remnant, the Pacific Ocean, covers only half of that. The losses have occurred in two forms, by pinch-offs to form other oceans, and by crustal loss in subduction zones.

The northward motion and break-up of Australia and Melanesia effectively cut off the western arm of Panthalassa and created the Indian Ocean. It is noteworthy that this creation will not endure forever. The drift of India illustrates what probably will happen. India sailed right through the Indian Ocean until it became attached to Asia. Likewise, Australia and New Guinea are drifting north as a unit, and there is nothing to stop them short of China and Siberia. If present trends continue, the Pacific and Indian Oceans will again be united.

The Arctic Ocean is also a pinch-off of Panthalassa that formed when Alaska rotated into Siberia. It is now being incorporated into the intruding Atlantic.

A quarter of Panthalassa was pinched off; meanwhile, the opening of the Atlantic was balanced by the loss of yet another quarter. The configuration of the Pacific prior to 80 million years ago is still obscure. The American oceanographers Clement Chase and Roger Larson have identified a pattern of triple junctions in the central Pacific that indicates the existence of many plates without intervening trenches in Mesozoic time. Thus, midocean ridges were migrating in various directions, but where they were, relative to the continents, is unclear.

DRIFT

A consistent drift of the Pacific plate can be deduced from paleomagnetism, paleoclimatology, and the geography of presumed hot spots in the mantle. The surface waters of the sea in the equatorial zone are highly productive, and the microscopic shells of various organisms rain to the bottom at much faster rates than in most other places. Thus, the equator is "drawn" on the sea floor as a narrow ridge of sediment, which has been identified by subbottom profiling on oceanographic ships. Deep-sea drilling by JOIDES shows that the ridge consists of overlapping ridges that are successively younger to the south. Apparently, the Pacific plate has drifted latitudinally at the rate of about 1 centimeter per year for the last 37 million years.

The paleomagnetic poles of small older seamounts in the Pacific indicate that they were formed far to the south of their present positions at times ranging from 25 million to 90 million years ago. The rate of latitudinal drift between 37 million and 90 million years ago was apparently as great as 5 centimeters per year.

The linear island groups of the Pacific typically show a progression from an active volcano at one end, through dead volcanoes and volcanoes truncated by wave erosion, to drowned volcanoes capped by coral atolls at the other end. Tuzo Wilson proposed that this sequence was caused by the oceanic crust drifting over hot spots in the mantle. Jason Morgan later showed that the concept was compatible with plate tectonics and thereby it could define the rates of drift of plates and demonstrate that some of the hot spots remained relatively fixed in the mantle for 100 million years. This refinement has added a new precision to the geography of plate motions because it gives longitude as well as latitude.

The most useful hot spot, for this purpose, is now under Hawaii. The motion of the Pacific plate over it can be defined by the geology of the Hawaiian chain and the Emperor seamount chain that connects to its northern end. From 100 million to 60 million years ago, the plate moved almost due north at a few centimeters per year. Roughly 40 million years ago, it changed direction to the northwest and accelerated to about 5 centimeters per year (although

BOX 6.3 A GEOLOGIST, JOHN HOLDEN, DRAWS GEOLOGISTS LOOKING AT CONTINENTAL DRIFT

Reams of paper are being expended and gallons of ink spilled . . .
but the argument is only just warming up.

Good fits are easy to achieve on a ping-pong-ball-sized globe.

In Australia, the belief is popular that the continents have been dispersed by an expanding earth.

The continental-drift jigsaw game is not without constraints. It is not cricket, for example, to play continental leapfrog, although one South African geologist claims that some jump fits are as good as any of those usually promoted by drifters.

Source: Dietz (1967)

it may have accelerated to 10 centimeters per year about 5 million years ago). The fact that large volcanoes have time to form, but not a continuous ridge, requires that the hot spot generates an extraordinary but intermittent flood of lava through the plate. Historical eruptions in Hawaii indicate the fastest known outpouring of lava in the world.

Complexities in California

The geologic history of the west coast of North America has been exceedingly complex for at least the last 60 million years because of the interaction of various mobile plate boundaries. In some ways, it seems almost *too* complex to examine, except in the greatest possible detail. It is worthwhile, however, to attempt to generalize about some aspects of the history of this region in order to emphasize two important matters: first, that the geography we now see has changed rapidly and dramatically; and second, that changes at similar rates and on similar scales are occurring now, and presumably, will continue to occur.

About 80 million years ago, the spreading centers in the Pacific basin were pushing lithosphere against western North America to form a subduction zone. This is evident from the sedimentary and igneous rocks that are preserved along the continental margin, which include equivalents of the typical sediments of the deep-sea floor (Chapter 16) that have been cemented into rock or metamorphosed as they plunged into the mantle. Likewise, the coastal zone contains volcanic rocks characteristic of island arcs—which are curved lines of volcanoes that form along subduction zones (Figure 6.3). At present, the different types of rocks are broken by faults and do not display an obvious pattern like an island arc (Figure 6.4). However, the American geologist Warren Hamilton has reconstructed the geography before faulting, and the paleogeography does resemble a long chain of arcs parallel to the coast (Figure 6.5).

These island arcs and the related subduction zones were destroyed piecemeal as the Pacific spreading center—the crest of the East Pacific Rise—gradually intersected the continent. When that happened, it became evident that the Pacific plate (Figures 6.6 and 6.7) had been moving northwesterly relative to the American plate. The evidence is that a new ridge–ridge transform fault developed between them. It still exists as the San Andreas fault, although it has lengthened and changed in other ways since it began to develop 38 million years ago, according to the reconstruction of the American geologist Tanya Atwater.

By 10 million years ago, the San Andreas fault cut across California from Los Angeles to San Francisco, but the boundary between the Pacific and American plates included a series of long transforms and short spreading

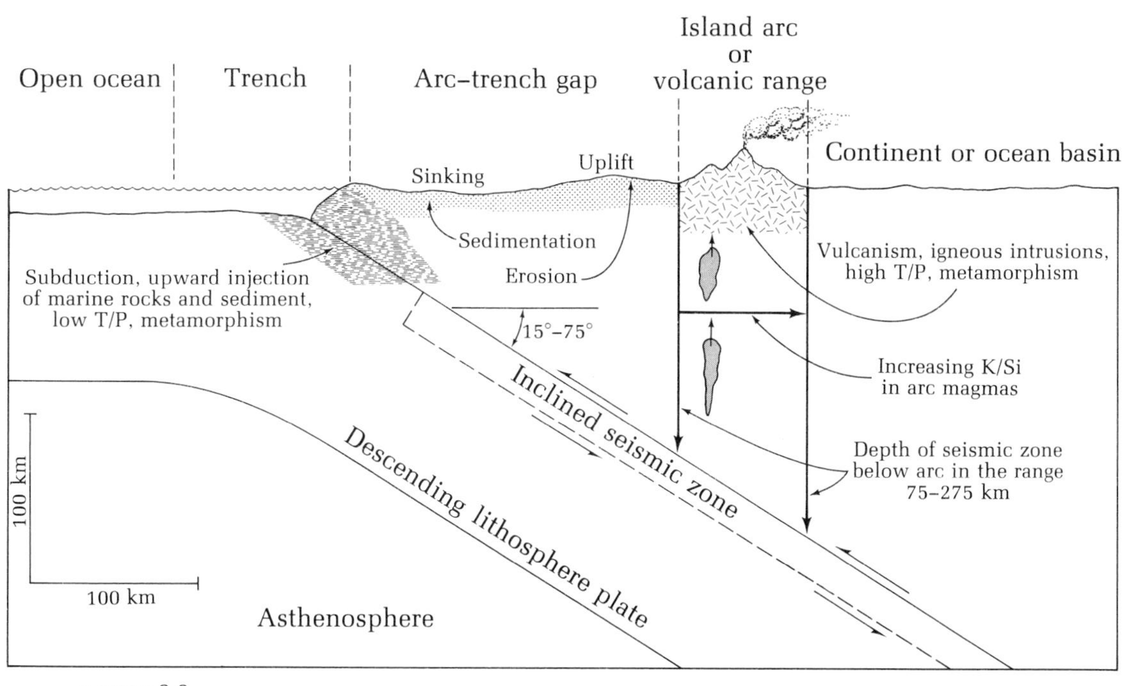

FIGURE 6.3

Generalized transverse section of an arc–trench system showing characteristic features of the three main parallel tectonic elements: oceanic trench; arc–trench gap; and island arc, which is either insular or marginal to a continent. *T/P*, temperature/pressure ratio; *K/Si*, potassium/silicon ratio. [After Dickinson (1971), copyright 1971 by the American Association for the Advancement of Science.]

centers fringing the coast of Baja California as far south as the present mouth of the Gulf of California. About this time, Catalina, San Clemente, and the other continental islands off Southern California drifted away from the mainland.

Five million years ago, the northern end of the San Andreas fault grew to Cape Mendocino in California and the southern end became complicated by the initial opening of the gulf. By 4 million years ago, the junction jumped into the Gulf of California, and the long and complex interaction between the Pacific and American plates at last reached its current state of development. There seems little reason to believe that it will remain as it is.

Two widely separated remnants of the once continuous subduction zone still exist. The Cascades are a line of still active or recently active andesitic volcanoes extending from northern California to British Columbia. Although there is no trench offshore, there is convergence between plates, and these volcanoes show that the lithosphere was plunging there until very recently, and it probably still is. An active subduction zone off the east coast of Mexico is marked by deep earthquakes, a trench, and andesitic volcanoes. None of

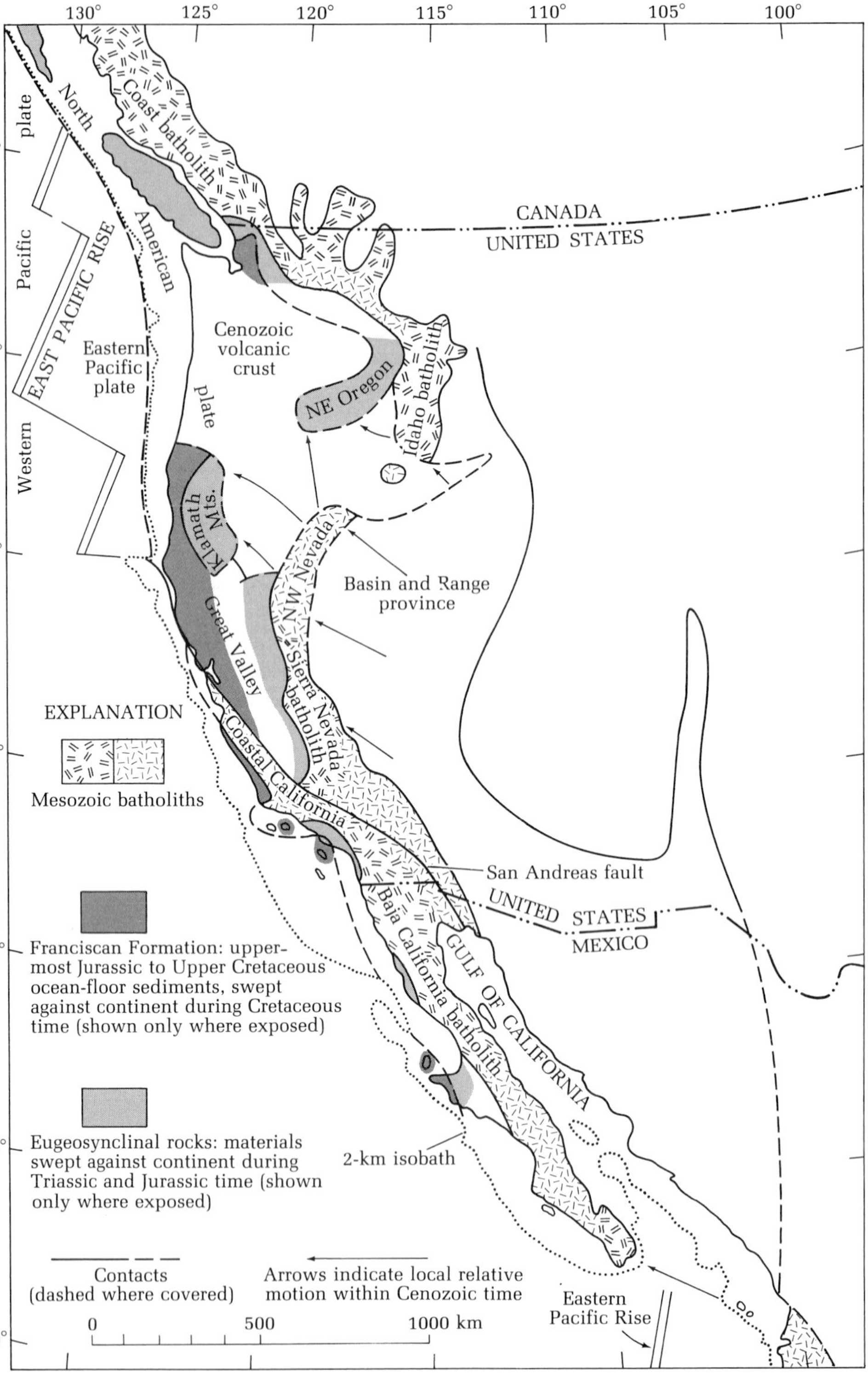

FIGURE 6.4

Upper Mesozoic complexes of west-central North America, showing inferred motion relative to the continental interior during Cenozoic time. [From Hamilton (1969), courtesy the Geological Society of America.]

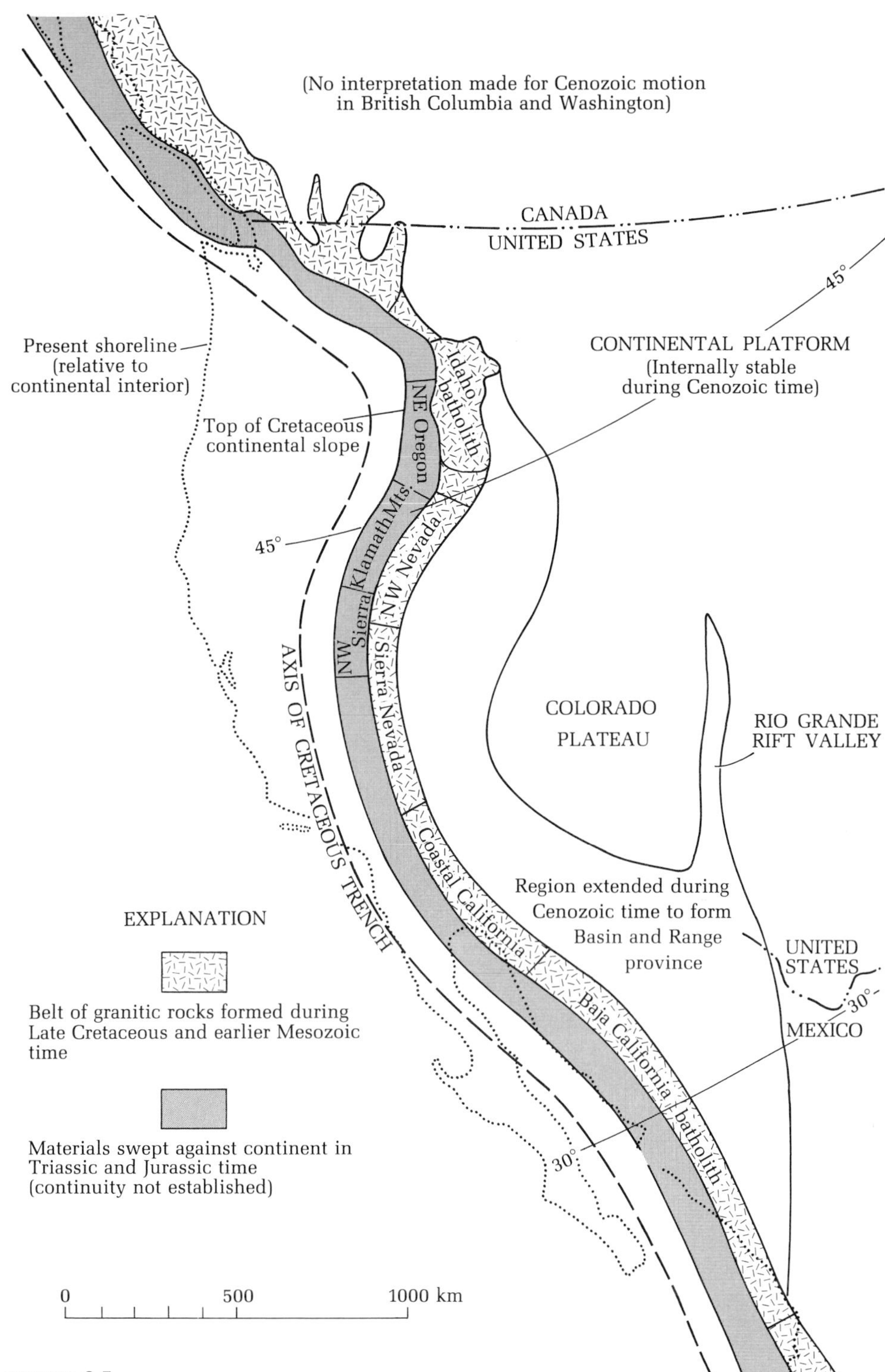

FIGURE 6.5

Paleogeographic map of the continental-margin complexes of west-central North America in middle Late Cretaceous time (80 million years ago). Cenozoic extension has been reversed to derive this map from that of present geography. The paleolatitudes correspond to a pole at 70°N, 175°E, as suggested by paleomagnetic orientations. [From Hamilton (1969), courtesy the Geological Society of America.]

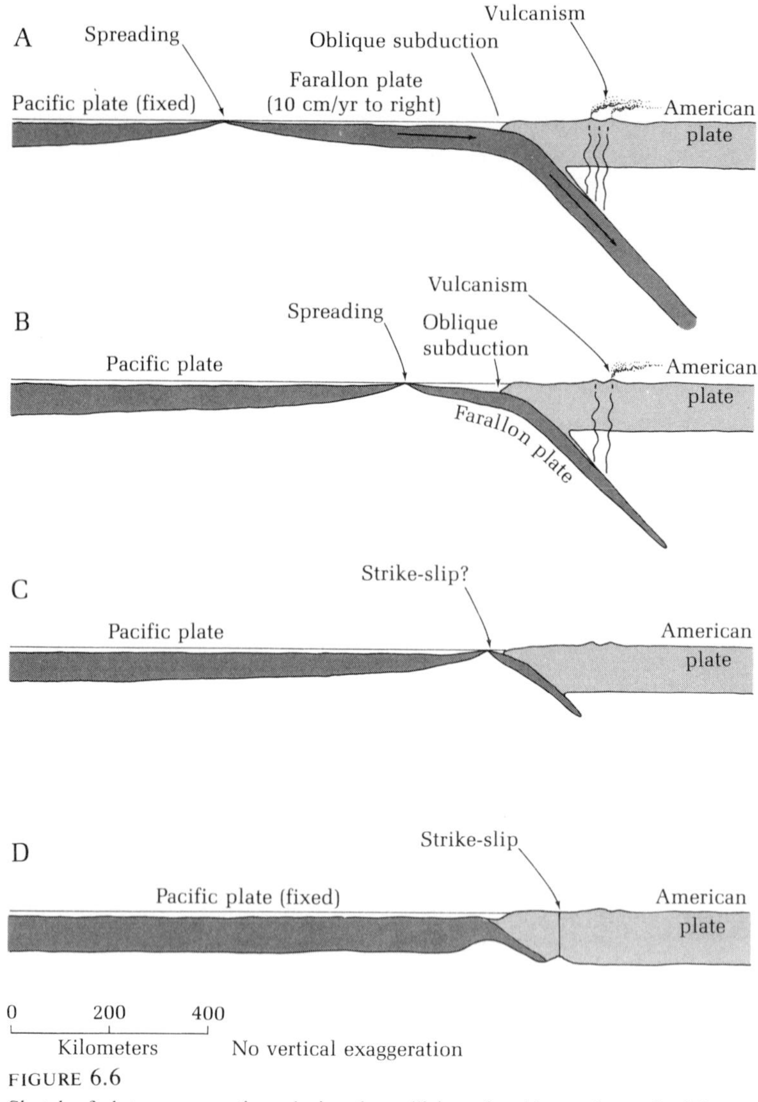

FIGURE 6.6

Sketch of plate cross-sections during the collision of a ridge and trench. (The Pacific plate is held fixed for the purposes of this illustration.) The spreading center accretes material onto both plates and moves to the right, while the Farallon plate moves to the right even faster. The American plate moves out of the page. Consumption at the trench is oblique. As the Farallon plate moves away from the ridge, it thickens by cooling; as it plunges, it is thinned by being heated. *A*, the plates in early Tertiary time; *B*, the spreading center as it migrated toward the continent; *C*, the situation when the ridge and trench collided; *D*, the plate configuration in central California at the present time. [After Atwater (1970), courtesy the Geological Society of America.]

these features occur between central Mexico and northern California because the subduction zone there began to disappear as the Farallon plate narrowed. Andesitic vulcanism generally ceased when the transform junction between the American and Pacific plates was established.

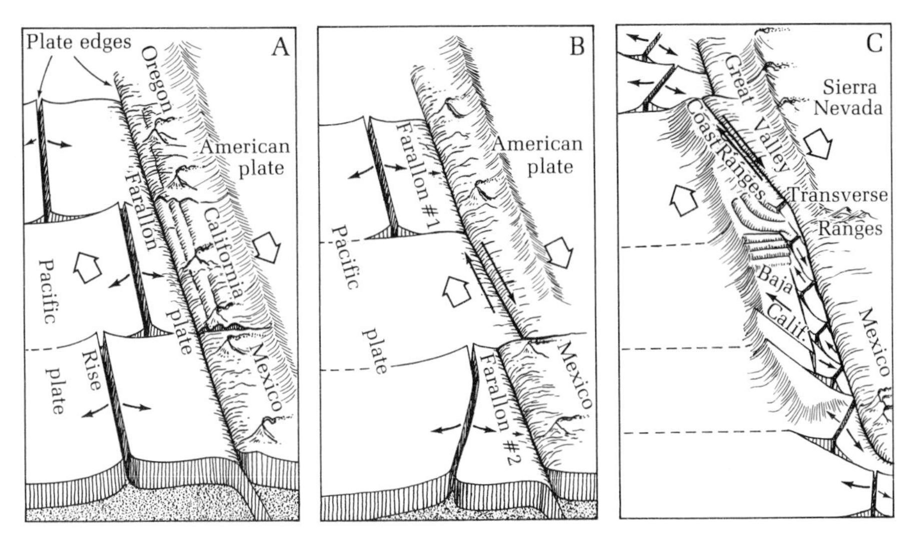

FIGURE 6.7
The evolution of the west coast of North America is shown here in a highly diagramatic form. About 50 million years ago (*A*), three plates existed: the Pacific and American plates separated by the Farallon plate. Mountains, trenches and volcanoes formed at the subduction zone. Perhaps 38 million years ago (*B*), the irregular crest of the East Pacific Rise intersected the continent, and the Farallon plate broke in two. The new boundary between the American and Pacific plates was a ridge–ridge transform and, by 10 million years ago, had evolved into the San Andreas fault, reaching the length of central California. About 4 million years ago (*C*), the crest of the East Pacific Rise jumped into Mexico and the Gulf of California began to open. The Cascade volcanoes in Oregon and the volcanoes of central Mexico are the only remnants of a once continuous line.

Future Drift

If present plate motions continue, it is a matter of little more than geometry to predict what will occur: In a mere 50 million years, coastal California and Baja California will form an arrow with the point beginning to pierce Alaska and the feathers still touching northern California; the Mediterranean will be largely crushed between Africa and Europe; East Africa will be splitting off from the rest of the continent; and Australia and New Guinea will have traversed from the Antarctic to southeast Asia and the Indian and Pacific oceans will be reunited.

The Atlantic may continue to spread until eventually it becomes a new Panthalassa, as the other oceanic plates are eaten and the continents recongregate into a new Pangaea. And, perhaps, that is the way of the world.

Summary

1. Latitudinal motion of drifting plates can be determined from the present distribution of ancient indicators of climate and the magnetic field.
2. Longitudinal motion cannot now be traced unless there are fixed hot spots that generate trains of volcanoes.
3. Opinion is divided concerning the existence of continental drift prior to the breakup of Pangaea.
4. About 200 million years ago, the world had one continent (possibly two) and one ocean basin. The southern part of the continent had been glaciated while it was near the south pole.
5. The great continent, Pangaea, slowly fragmented, and ocean basins were created between the modern continents as they drifted away. Other ocean basins were isolated from the original Panthalassa by drifting continents.
6. The geology of California has been complicated because different plate boundaries have intersected the continental margin at different times and places.
7. The San Andreas fault is a ridge–ridge transform between the American and Pacific plates. It formed about 40 million years ago when the plates came in contact.
8. The Gulf of California began to open only a few million years ago when the American–Pacific plate boundary jumped into the continent.

Discussion Questions

1. Can sea-floor magnetic anomalies ordinarily indicate the direction of plate motion relative to the north pole?
2. Why can't a whole plate drift north?
3. Place the ocean basins in order of their formation: Antarctic, Arctic, North Atlantic, South Atlantic, Indian, North Pacific, South Pacific.
4. How does the past drift of India suggest the future fate of Australia?
5. Why did the San Andreas fault form in central California?
6. Why is the Gulf of California opening while California is not?

References

Anderson, D. L., 1971. The San Andreas fault. *Sci. Amer.* 225(5):52–68. (Available as *Sci. Amer.* Offprint 896.)

Atwater, T., 1970. Implications of plate tectonics for the Cenozoic tectonic evolution of western North America. *Geol. Soc. Amer. Bull.* 81:3513–3536.

[Unravels the interactions of the Pacific and American plates.]
Bullard, E., J. E. Everett, and A. G. Smith, 1965. The fit of the continents around the Atlantic. *Phil. Trans. Roy. Soc. London Ser. A Math. Phys. Sci.* 258:27–41. [The first application of computer fitting of continents.]
Dickinson, W. R., 1971. Plate tectonics in geologic history. *Science* 174:107–113. [A reinterpretation of some classical concepts of geology.]
Dietz, R. S., 1967. More about continental drift. *Sea Frontier* 13(2):66–82. (Illustrated by John Holden.)
——, and J. Holden, 1970. The breakup of Pangaea. *Sci. Amer.* 223(4):30–41. (Available as *Sci. Amer.* Offprint 892.) [The article around which much of the chapter is constructed.]
Francheteau, J., 1970. *Paleomagnetism and Plate Tectonics.* Ph.D. thesis, University of California, San Diego. [Used as a check on various stages in the breakup of Pangaea.]
Grow, J., and T. Atwater, 1970. Mid-Tertiary tectonic transition in the Aleutian arc. *Geol. Soc. Amer. Bull.* 81:3715–3722. [The demise of the Kula plate.]
Hamilton, W., 1969. Mesozoic California and the underflow of Pacific mantle. *Geol. Soc. Amer. Bull.* 80:2409–2430. [Origin of the Sierra Nevada and related features.]
Hurley, P. M., and J. R. Rand, 1969. Pre-drift continental nuclei. *Science* 164:1229–1242. [Pangaea may have existed from 1700 to 200 million years ago.]
King, L. C., 1967. *Morphology of the Earth* (2nd ed.). New York: Hafner.
Laughton, A. S., 1971. South Labrador Sea and the evolution of the North Atlantic. *Nature* 232:612–617.
Le Pichon, X., 1968. Sea-floor spreading and continental drift. *J. Geophys. Res.* 73:3661–3697. [The first quantitative global synthesis of plate tectonics.]
McKenzie, D. P., 1970. Plate tectonics of the Mediterranean region. *Nature* 226:239–243. [Small plates move laterally to fill in gaps as Africa and Europe fuse together.]
——, and J. Sclater, 1971. The evolution of the Indian Ocean since the Late Cretaceous. *Geophys. J.* 24:437–528. [The first quantitative unraveling of the complex history of the Indian Ocean basin.]
Menard, H. W., 1969. Growth of drifting volcanoes. *J. Geophys. Res.* 74:4827–4837. [Presents evidence for the growth.]
——, 1972. History of the ocean basins. *In* E. Robertson, ed., *The Nature of the Solid Earth,* pp. 440–460. New York: McGraw-Hill. [Latitudinal motion of the Pacific plate based on paleomagnetism and paleoclimatology.]
Morgan, W. J., 1972. Deep mantle convection plumes and plate motions. *Bull. Amer. Ass. Petrol. Geol.* 56:203–213. [Develops plate motions from patterns of marine volcanoes.]
Vacquier, V., and S. Uyeda, 1967. Paleomagnetism of nine seamounts in the western Pacific and of three volcanoes in Japan. *Bull. Earthquake Res. Inst. Tokyo Univ.* 45:815–848. [The northward drift is calculated according to techniques described in the paper.]
Wilson, J. T., 1963. Continental drift. *Sci. Amer.* 208(4):86–100. (Available as *Sci. Amer.* Offprint 868.) [The qualitative consequences of drift and crustal continuity before plate tectonics made them quantitative.]
——, 1966. Did the Atlantic close and then re-open? *Nature* 211:676–681. [Between 600 million and 200 million years ago.]

7

EARTHQUAKES AND DEFORMATION

It is one of the advantages of geology that it is truly a recreative science. If most branches require precision of knowledge in biology and physics, in chemistry and mineralogy, and much work has to be done in the museum and laboratory, to say nothing of the arm-chair, yet the fulness of the science can never be attained without the vivifying influence of mountain and moor, of valley and sea-coast.

Horace B. Woodward

THE HISTORY OF THE GEOLOGICAL SOCIETY OF LONDON

Inevitable Deformation

Dissipation of the energy of the sun in the atmosphere results in erosion and deposition of sediment, which are accompanied by isostatic elevation and depression of the land. Dissipation of the internal energy of the earth causes crustal plates to drift and to move up and down. Thus, it is inevitable that the surface of the earth will be deformed as long as these sources of energy are available.

We have already examined the great deformation that occurs at the boundaries of moving plates; in this chapter, we turn to smaller features that may occur anywhere. Near plate edges, the rocks in the upper few miles of the crust may be intensely deformed into tight folds and broken by faults. In the interior of plates, the deformation is less apt to be intense: there are great gentle warps, the folds are broad, and the faults widely spaced or offset only small amounts.

These are generalities, however; deformation may be intense anywhere. Moreover, what are now the interiors of plates may once have been the edges; thus, ancient rocks anywhere tend to be highly deformed.

Notably intense deformation is associated with various phenomena that are energized by gravity. The circumstances favorable for gravity tectonics depend on the type and thickness of sediment, among other things. Thus, masses of salt may move upward through layers of denser rock and vast sheets of sedimentary rock slide downward. These phenomena occur within tectonic plates as well as at the edges.

These rather localized folds and faults, with dimensions ranging from meters to kilometers, are quite important to society. Folds and faults may mark the locations of mineral resources, particularly ground water and oil. Moreover, distinct plate boundaries are quite rare on land. Most large faults, and most earthquakes that break the ground, are at plate edges, but earthquakes that demolish structures may be anywhere.

Intensity, Magnitude, and Nuclear Explosions

The intensity of an earthquake is a measure of its local effects upon people and construction. Its magnitude is a measure of the energy released at the source, as determined by the motion of a seismometer. An intensity scale was devised as early as 1883 by Rossi and Forel, but the scale currently in use is based on the work of Guiseppe Mercalli in 1902, as modified in 1931 by H. O. Wood and Frank Neumann. The intensities on the modified Mercalli scale are assigned Roman numerals from I to XII, and due regard is given to different types of construction:

I. Not felt except by a very few people under specially favorable circumstances.

II. Felt by only a few persons at rest, especially on upper floors of buildings. Delicately suspended objects may swing.

III. Felt quite noticeably indoors, especially on upper floors of buildings, but many people do not recognize it as an earthquake. Standing motorcars may rock slightly. Vibration like the passing of a truck. Duration can be estimated.

IV. During the day, felt indoors by many, outdoors by few. At night, some awakened. Dishes, windows, and doors rattle; walls make creaking sounds. Sensation like a heavy truck striking building. Standing motorcars rocked noticeably.

V. Felt by nearly everyone; many awakened. Some dishes and windows are broken; a few instances of cracked plaster; unstable objects are overturned. Disturbances of trees, poles, and other tall objects are sometimes noticed. Pendulum clocks may stop.

VI. Felt by all; many are frightened and run outdoors. Some heavy furniture is moved. Some instances of fallen plaster or damaged chimneys, but structural damage is otherwise slight.

VII. Nearly everyone runs outdoors. Damage is negligible in buildings of good design and construction, slight to moderate in well-built ordinary structures, and considerable in poorly built or badly designed structures; some chimneys are broken. Noticed by persons in moving motorcars.

VIII. Damage is slight in specially designed structures; it is considerable in ordinary, substantial buildings, which may partially collapse, and great in poorly built structures. Panel walls are thrown out of frame structures. Chimneys, factory stacks, columns, monuments, and walls may fall. Heavy furniture is overturned. Sand and mud are ejected from the ground in small amounts, and there are noticeable changes in well water. Persons driving motorcars are disturbed.

IX. Damage is considerable in specially designed structures, and well-designed frame structures are thrown out of plumb; damage is great in substantial buildings, with partial collapse. Buildings are shifted off their foundations. The ground is cracked conspicuously and underground pipes are broken.

X. Some well-built wooden structures are destroyed, and most masonry and frame structures are destroyed with their foundations. The ground is badly cracked and rails are bent. Landslides are considerable from river banks and steep slopes, sand and mud are shifted, and water is splashed (slopped) over banks.

XI. Few, if any, masonry structures remain standing. Bridges are destroyed. There are broad fissures in the ground and underground pipelines are completely out of service. The earth slumps and land slips in soft ground. Rails are bent greatly.

XII. Damage is virtually total. Waves are seen on ground surfaces and objects are thrown upward into the air.

After an earthquake in a populated region, interviews and observations can establish the intensity in various places, and the locus of the quake and trend of the fault can be established. However, because intensity is an assessment of the local social effects of an earthquake, the same earthquake in an unpopulated region, or at sea, would have an intensity of zero.

Magnitude is simply an instrumental measure of the energy of an earthquake that is released in the form of waves, and the computed value takes into account the distance from the source to the point at which the measurement is made.

The magnitude scale most commonly used is the one developed by Beno Gutenberg and Charles Richter. This "Richter scale" is logarithmic—that is, the ground motion of larger quakes increases tenfold for an increase of only one scale unit.

The magnitude and intensity scales can be compared in populated regions: an earthquake of magnitude 2.5 can be felt nearby, one of 4.5 causes local damage, one of 6.0 causes considerable damage, and one of 7.0 or more is a major earthquake. By one method of calculation, the largest measured quakes have magnitudes of 8.6; by others, magnitudes of up to 8.9. Such earthquakes can cause enormous damage. We can be happy that only a small fraction of the thermal energy of the earth is converted into earthquakes. Otherwise, intense quakes would be so frequent and widespread that our lives would be constantly affected.

The total energy of earthquakes, including waves and heat, may be compared with that of underground tests of nuclear weapons. Magnitude 5.0 equals a small atomic bomb, equivalent to 20,000 tons of TNT. Magnitude 6.0 equals a one-megaton thermonuclear bomb. Magnitude 8.6 equals 60,000 bombs with one-megaton yield, which is more than the total nuclear armament of the nations of the world. It is evident that underground nuclear explosions like those that have occurred in Nevada, the Aleutians, and Siberia cannot generate major earthquakes. It has been observed in Nevada, however, that they appear to be able to *trigger* natural quakes. This possibility, which might be desirable were it to prevent an even bigger quake later on, is discussed in the next chapter.

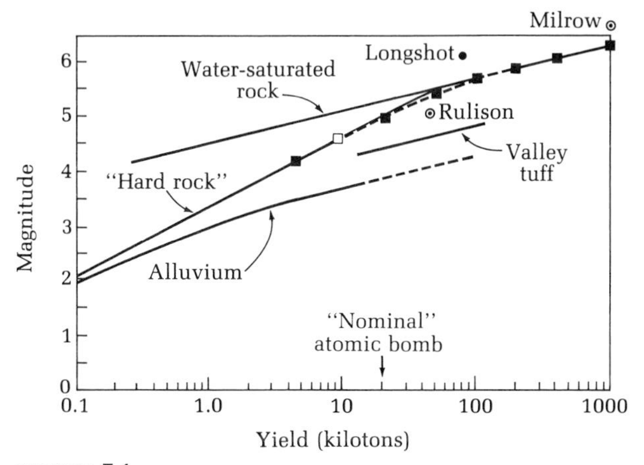

FIGURE 7.1

Magnitude versus yield in kilotons of explosions in different materials. Lines are for data at the Nevada Test Site. Points are other explosions and data calibrated against observations in the eastern United States. Earthquakes of magnitude 6, which are quite common, are equivalent to a 1000-kiloton hydrogen bomb. Fortunately, earthquakes are not aimed at targets. [After Evernden and Filson (1971).]

Faults

Faults are dislocation surfaces in the earth along which relative motion has occurred by shearing. Faults are classified according to the direction of the relative motion and the inclination or slope of the surface. The types of faults are readily visualized in terms of a few features: the hanging wall, which is above the fault; the footwall, which is below it; and the dip, which is the inclination of the fault from the horizontal. A normal fault is one whose hanging wall moves *down* relative to the footwall; the motion extends the crust, and faults of this type are the result of tension. A reverse fault is one whose hanging wall moves *up* relative to the footwall; motion of this type of fault shortens the surface as a consequence of compression. Lateral faults have a vertical dip, and the relative motion is horizontal. If you face toward the fault and the block on the other side has moved to the left, it is a left-lateral fault. A fault with the opposite motion is termed right-lateral. Various combinations of vertical and horizontal motion also occur.

Some of these faults, namely the lateral ones, are mainly plate-edge phe-

BOX 7.1 SEISMOMETERS

The most common instrument for measuring earth movements is a seismometer which records the waves generated by earthquakes. The everyday acoustical tape recorder also records waves, but they have much higher frequencies. The seismometer has a special problem, as a recorder, because it is attached to the earth whose vibrations it seeks to record. The two must be disconnected as much as possible. This can be achieved by suspending a large mass by a small wire. Inertia holds the mass in place as the earth moves beneath it, and all that is needed is a device to record the relative motion between the earth and the inertial mass.

Modern instruments are triumphs of ingenuity. They record motion in three dimensions and are so sensitive that they can detect the passage of a truck on a highway as much as a kilometer away. On the other hand, almost every city has one or more amateur seismologists with simple, homemade instruments suitable for detecting and timing earthquakes. Indeed, a calm pond or swimming pool is an excellent detector.

The earth moves not only in response to earthquakes but in response to tides and long-term or secular geologic phenomena. Tides are detectable with very sensitive gravimeters and tiltmeters. Strain is a change in the distance between points in a body, and laser strain gauges can measure such changes in the earth.

The recordings made by seismometers and similar instruments include different frequencies; for the purposes of modern science, these frequencies are filtered, as treble or bass is suppressed on a hi-fi set. Increasingly, they are recorded in a form suitable for computer processing, and signal enhancement is employed to retrieve otherwise undetectable signals from noise. The same technique has been used to change blurred television photos of the moon and Mars, transmitted to earth from spaceships or satellites, into understandable pictures.

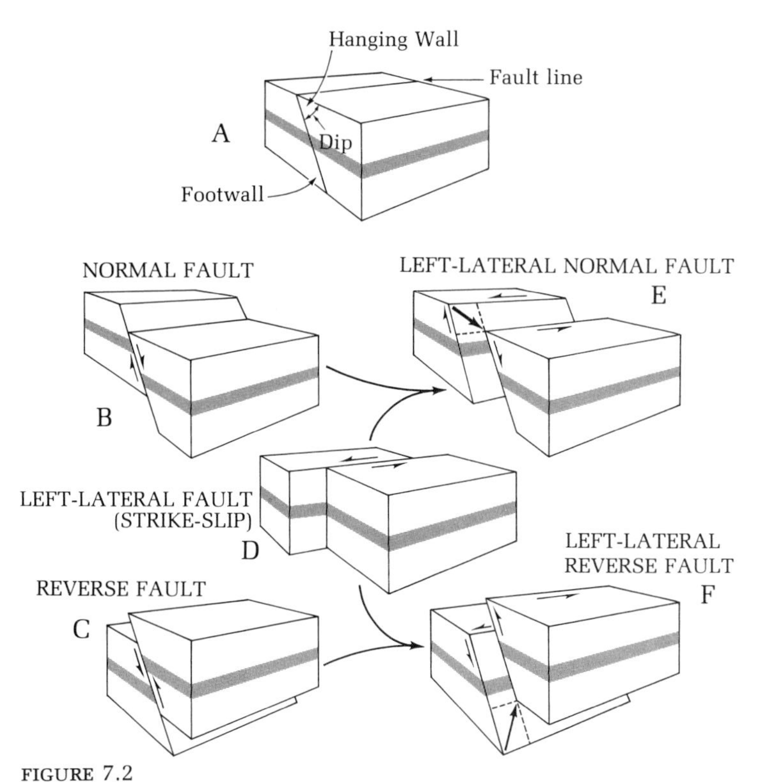

FIGURE 7.2

Types of fault movement: *A*, names of some of the components of faults; *B*, a normal fault, in which the hanging wall has moved down relative to the foot wall; *C*, a reverse fault, in which the hanging wall has moved up relative to the foot wall; *D*, a lateral fault, sometimes called a strike-slip fault, in which the rocks on either side of the fault have moved sideways past each other (the fault shown is termed left-lateral because the rocks on the other side of the fault have moved to the left, as observed while facing the fault; had the other side of the fault moved to the right, it would be termed right-lateral); *E*, a left-lateral normal fault, in which the movement is a combination of normal faulting and left-lateral faulting; *F*, a left-lateral reverse fault, in which the movement is a combination of left-lateral faulting and reverse faulting. [After Clark and Hauge (1971).]

nomena. The San Andreas fault, for example, is genetically a ridge–ridge transform and descriptively a right-lateral fault. Most faults of all classes, however, are within plates and have offsets of a few meters or a few kilometers—trivial motions compared to those between the plates. These faults are highly important, however, because they are so abundant.

Large regions of surface rocks are more or less chopped by cracks and faults. All of the oceanic crust, which is generated as a filling of tensional cracks, is probably sliced into almost vertical planes. These cracks or joints, it should be noted, are not faults in the geologic sense if they just open and separate the two blocks. Shearing, or relative motion along the crack, is required in the definition of a fault.

BOX 7.2 STRESS AND STRAIN

The strength of rocks has been tested experimentally in laboratories, and a number of concepts have emerged about the relation between the stress (what is done to the rock) and the strain (how it responds to the stress). These relations were summarized by John Handin in 1966.

Elastic strain is deformation that is restored when the stress is removed.

Strength is qualitatively defined as resistance to failure.

Flow is any deformation, not instantly recoverable, with permanent loss of cohesion.

Fracture is complete loss of cohesion and resistance to stress, which results in separation into two or more parts. It is accompanied by release of stored elastic energy.

Visualize various rock samples subjected to increasing stresses on the sides and ends: The absolute amount of the stresses is not as important as the differences between them. For example, if the stresses are large but all the same, the samples are compressed but will not fracture. At low-stress differences, the strain is elastic and reversible. At the yield stress, permanent strain is initiated. The ultimate strength of a rock sample is the greatest stress difference that the sample can stand under given conditions. This is difficult to determine if what is known as work hardening occurs in the sample as it is being tested, because the sample then requires additional stress for additional permanent strain. Breaking strength is the stress difference at fracture, and crushing strength is the compressive stress required to break the sample at atmospheric pressure. Some materials are brittle—that is, they break easily. Other materials are ductile, which means that they can undergo large permanent deformation without fracture. Most rocks are brittle at low temperatures, low pressures, and high strain rates but are ductile at high pressures, high temperatures, and low strain rates. Thus, a rock will break if hit by a sledge hammer on a mountain, but, if it is buried deep in the hot crust under a burden of 10 kilometers of rock, it will flow in response to the slightest persistent stress.

Rocks deform in a variety of ways. Very brittle rocks break by extension fracture in a direction perpendicular to a tension or parallel to a compression. This is a type of break familiar to all of us because most of the common materials we can break are very brittle. It is a very important mode of deformation on the earth because it occurs where the crust is split and tectonic plates move apart. Faulting, which is localized displacement in a plane inclined to the direction of the principal stress at angles of 45° or less, produces most earthquakes. Faulting can occur in brittle or moderately ductile rocks; in the latter, however, the fault zone tends to broaden.

Uniform flow, which is the macroscopically homogeneous deformation of ductile materials, is the normal mode of deformation at depth. It is accomplished by three microscopic mechanisms: cataclasis, or crushing of individual grains and localized regions; intracrystalline gliding between molecular slip planes; and recrystallization on a molecular scale by solid diffusion, local melting, or solution and deposition.

The continental crust is faulted in different ways and on a scale ranging from millimeters to kilometers, although some essentially unfaulted regions exist. This generally pervasive faulting influences almost every aspect of geology. Roads may be broken and hospitals leveled by movements on very small faults. The faulting reduces the strength of crustal rocks, and it directs and focuses the energy of earthquakes. It also determines the location of springs and oil fields, because fault planes tend to impede or facilitate the movement of underground fluids. Likewise, it exerts a strong influence on the deposition of metallic ores, because ore-bearing solutions tend to flow along the faults. Later faulting may offset the ore bodies, however, and thereby complicate mining.

At any given time, most faults are inactive—that is, neither motion nor earthquakes occur along them. Even though they may never become active again, they may exhibit all of the effects mentioned in the preceding paragraph. However, if a new plate boundary forms in an ancient stable crust, or if intraplate warping occurs, a stable region returns to an environment of higher stress. The old fault planes are zones of weakness in which new faulting tends to occur. Thus, the pattern of faults may reflect ancient stresses as well as active ones. It appears, for example, that Africa and South America split apart along ancient lines of weakness.

UPLIFT OF THE SIERRA NEVADA

The Sierra Nevada of California provides an interesting example of normal faulting on a very large scale. Topographically, these mountains form an asymmetrical ridge, gently sloping on the west and steeply cliffed on the east. The peak of Mt. Whitney, the highest mountain in the contiguous United States, is an easy climb because of the gentle slopes. Structurally, the ridge is a tilted fault block with a normal fault to the east. Because of their great height, the mountains have a distinctive fauna and flora. Moreover, this same height localizes heavy rain and casts an arid rain shadow over the state of Nevada, thereby affecting *its* fauna and flora as well.

Have conditions always been thus? Are the mountains ageless? These questions can be answered by studying fossil organisms and the ancient stream valleys (which are well known because they contained most of the gold discovered in 1849). It appears that the whole environment of the Sierra Nevada is quite new. The tilting movement on the normal fault has elevated the crest 1200–2100 meters in the last several million years; locally, parts of the crest have been elevated at least 300 meters in the last 700,000 years. A meter every 230 years may not seem like much, but look at the mighty Sierra!

Strain, Earthquakes, and Creep

ELASTIC STRAIN

After the great earthquake that destroyed San Francisco in 1906, the American geologist H. F. Reid proposed that the strain, so suddenly released, had accumulated gradually, and that it would accumulate again until relieved by another quake. If so, the strain, which is a kind of relative motion, should have been detectable by repeated geodetic surveys of points crossing the fault. For several decades, none was detected; but this was because of the conventional method of adjusting errors uniformly in all parts of a geodetic survey. This was the logical procedure to obtain the best adjusted location of all the survey points. Because it assumed the absence of local geologic effects, such as faulting, it could detect none. In 1949, C. A. Whitten of the U.S. Coast and Geodetic Survey adopted a different philosophy of data reduction and confirmed evidence of the predicted strain. This is a nice example of the pitfalls of scientific research on the environment, and the tendency toward futility in research that is unguided by critical hypotheses.

Repeated geodetic surveys of active faulting in California, Japan, and New Zealand all show accumulating strain. The surface of the land is distorted both horizontally and vertically at rates as great as several centimeters per year. Ultimately, this strain will be relieved by movement on faults. Whether it will also produce earthquakes depends on the frequency and size of the movements.

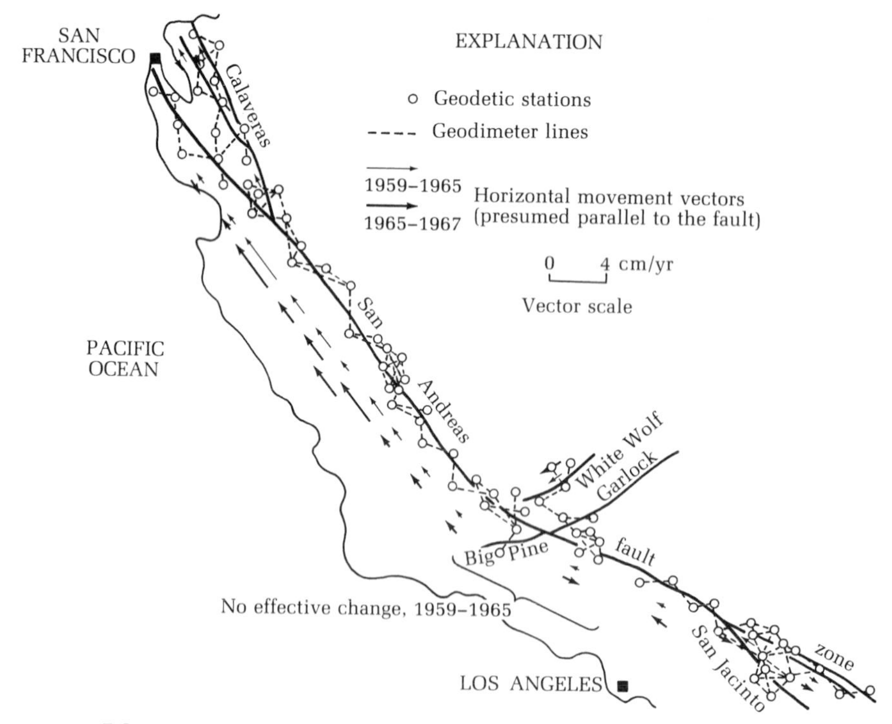

FIGURE 7.3

Average annual California fault movement for the periods 1959–1965 and 1965–1967. [From Hofman (1968).]

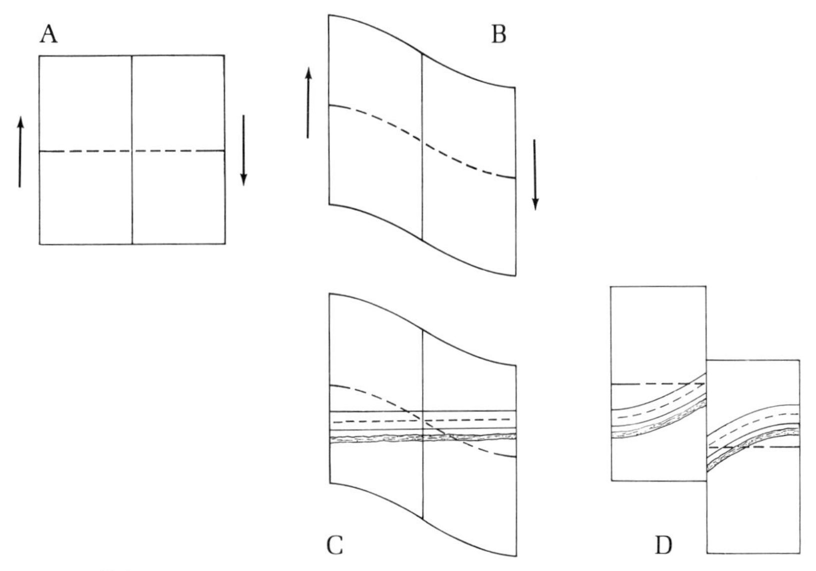

FIGURE 7.4

Strain accumulation and release: *A*, an area in an unstrained condition; *B*, the same area strained prior to earthquake; *C*, a canal and a road built straight across the strained region; *D*, faulting has returned the region to a condition of no strain, the canal and the road are broken, and their fragments are deformed.

MOVEMENT ON FAULTS

Movement on fault planes, apparently, is episodic rather than continuous, and the amount observed so far varies from 1 millimeter to 14 meters per episode. This is as might be expected from the behavior of all types of materials. For example, if two smooth, flat pieces of metal are rubbed together, they become worn by sticking for an instant and then slipping, sticking and slipping, and so on. If two rough, flat pieces are rubbed together in the same way, they tend to hang up on projections for a bit and then slip some distance until they hang up again. It appears that faults generally behave in a similar fashion. When they slip, they generate earthquakes. The magnitude of an earthquake is related to the length of the slipping surface, which, apparently, is also related to the time between slippages.

California, split by the boundary between the Pacific and American plates, is an admirable natural laboratory for the study of these phenomena. The continuity of plate boundaries requires the same amount of movement along all parts of the boundary, but it does not specify how this movement should occur. Among other things, the boundary is not a simple transform fault plane, as it appears to be at sea. Instead, it consists of a zone of variable width that contains one or more parallel lateral faults. The movement at the boundary can be distributed among the faults like a sheared deck of cards, rather than concentrated on a single plane, and some faults move more than others.

The style of motion can also vary from place to place, even though the amount may be the same. A few big slippages on one part of a single fault can be equaled by many small ones elsewhere. The greatest historical quakes along the San Andreas system occurred in 1857 north of Los Angeles and in 1838 and 1906 near San Francisco. The one in 1906 had a magnitude of 8.3; since that time, more than twenty quakes of magnitudes from 6.0 to 8.0, and countless smaller ones, have occurred on this fault system. Earthquakes are so frequent that they can be analyzed statistically. If the past is any indication, two magnitude 6.0 quakes should occur every five years, plus one magnitude 7.0 every 15 years. The estimated average interval between quakes with a magnitude greater than 8.0 is 100 years—and the last one was almost 70 years ago.

Where will the earthquakes occur along the 1100-kilometer length of the San Andreas fault system? Less than a decade ago it was believed that faulting did not break the surface of the ground unless there was an earthquake of magnitude 6.5 or greater. Ground is broken along almost the whole length of the San Andreas, and this was interpreted as evidence of a uniform distribution of earthquakes over a long period of time. Because large quakes have occurred in historical times only along some parts of the fault, the other parts appeared to be in increasing peril. This view was based on insufficient information about the nature of faulting. It now appears that movement is occurring along all parts of the fault—which is why the ground is broken. In some places, however, it occurs as small and frequent jumps, or creep; elsewhere, it occurs as the big jumps we experience as earthquakes. This suggests that future earthquakes will occur along the same sections of the fault where they have in the past, rather than where they have not.

CREEP

Creep is so subtle that it tends to escape attention as a natural phenomenon. In the vicinity of Hollister, California, sidewalks, curbs, and walls have been bent right-laterally along major faults (Figure 7.8) and walks and curbs on other faults have been buckled and overthrust. Landowners and city officials have generally viewed this as the consequence of poor workmanship or sinking ground, but the little breaks line up when plotted on aerial photographs.

Creep appears to be just like larger faulting in every respect but scale, although we still know too little to be absolutely sure. Creep, like faulting, occurs during brief intervals along significant lengths of faults. The best documented creep event occurred along a 6-kilometer length of the Calaveras fault at Hollister on July 17, 1971. The maximum offset was 0.9 centimeters. Probably related to this was a series of creep events observed along a 60-kilometer section of the main San Andreas fault to the west: between July

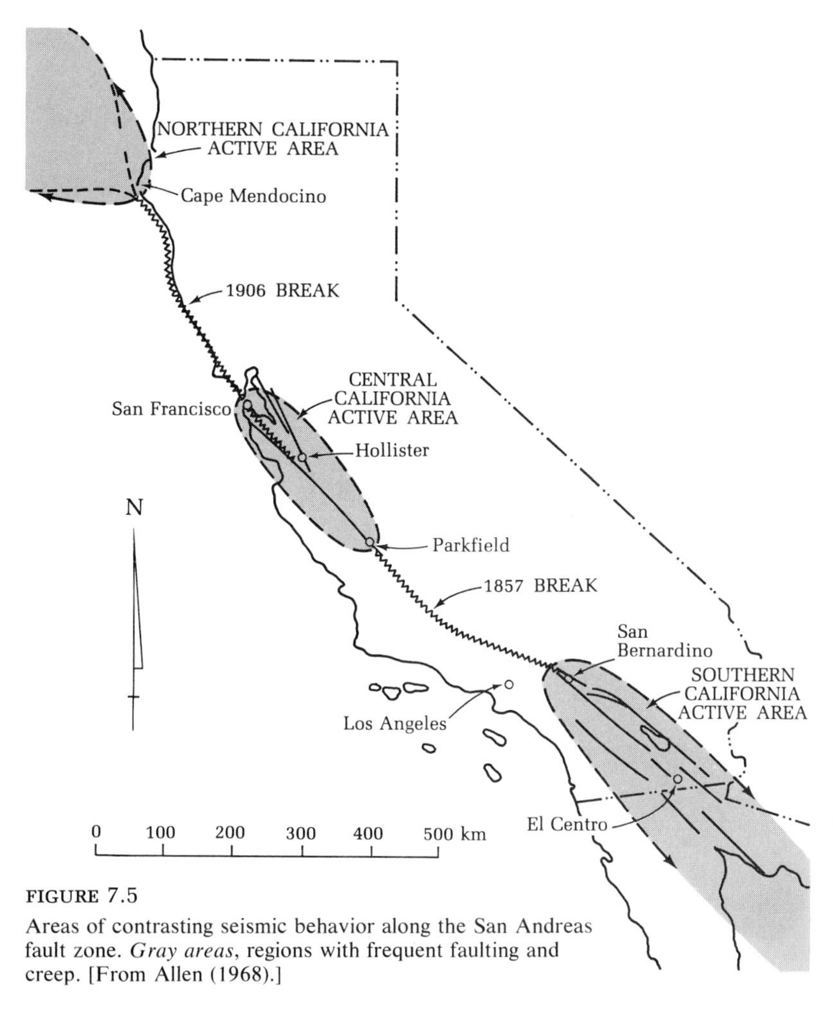

FIGURE 7.5

Areas of contrasting seismic behavior along the San Andreas fault zone. *Gray areas*, regions with frequent faulting and creep. [From Allen (1968).]

FIGURE 7.6

A fence offset by a fault. The view is to the northwest. The camera was alined with the straight part of fence beyond the disturbance, in order to show the flexure as well as the offset of the fence. [Photo by G. K. Gilbert, courtesy of the Carnegie Institution of Washington, from Lawson (1908).]

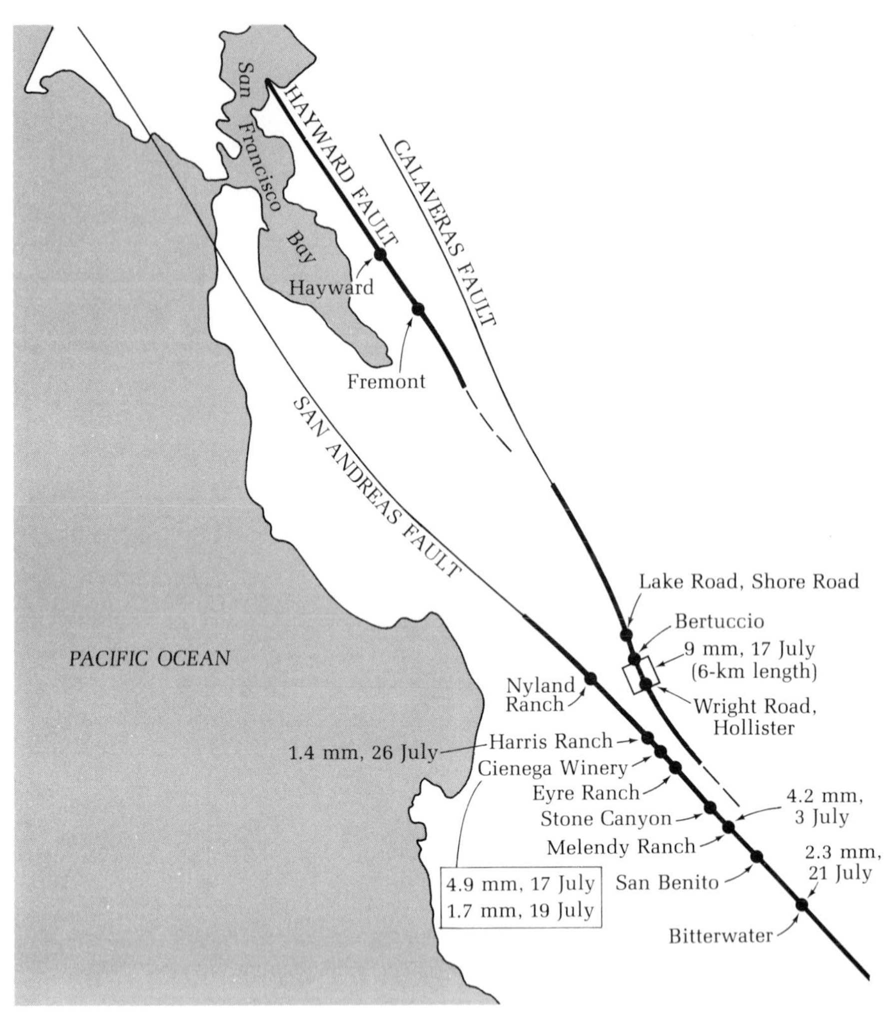

FIGURE 7.7
Creep events in the vicinity of Hollister, California. [After *Geotimes*, November 1971, p. 33.]

3 and July 26, creep events were recorded at four different stations, although none resulted in an offset larger than 0.5 centimeters.

Creep, like faulting, is episodic. This has been determined in the vicinity of Hollister, where concrete sidewalks and walls laid from 1910 to 1934 are offset about the same distance, about 25 centimeters. This indicates no offset from 1910 to 1934 and an average of about 0.8 centimeters per year since then. Likewise, in Fremont, California, structures as much as 50 years old were offset right-laterally along the Hayward fault by a uniform 15–18 cm during the period 1950–1957, but not at all before.

FIGURE 7.8
Curb and street offset by fault creep, Hollister, California. [Photo by Robert D. Nason.]

Folds

A fold is a bend in layered rock or any other tabular or slablike body. Geologists distinguish various types according to the shape, direction, and intensity of the folding. Elongate up-folds are anticlines (Figure 7.9) and elongate down-folds are synclines. Equidimensional up-folds are domes and equidimensional down-folds are basins. These simple folds can be overturned until they lie recumbent, or they may be refolded, in turn, into the most intricate structures.

Folds are rather spectacular evidence of crustal deformation because erosion commonly exposes them to even casual inspection. The hills covered with oil derricks in the Los Angeles area are anticlines that are still growing upward from the flat plain, but they are already partially eroded. Some of the great folds in the Rocky Mountains are also only partially eroded; the relatively resistant sandstone beds sketch the structure of the remainder. Folding, however, is a low-intensity phenomenon, and it cannot compete for long with erosion and deposition. Thus, folds tend to be worn away until only long, gently curving ridges remain to mark the position of resistant layers. These may occur in belts more than 1000 kilometers long, like the Appalachian Mountains of the eastern United States.

Consider a rigid tabular body such as a sheet of plywood. If it is bent into an anticline, the top of the sheet is stretched and the bottom is compressed. If several sheets are glued together to form a thick slab of plywood, it is difficult to bend, because the amount of stretching at the top and the amount of compression at the bottom increase accordingly. On the other hand, if the plywood sheets are merely stacked without gluing, it is easy to bend them, because one sheet is able to slip past another (Figure 7.10). Each individual

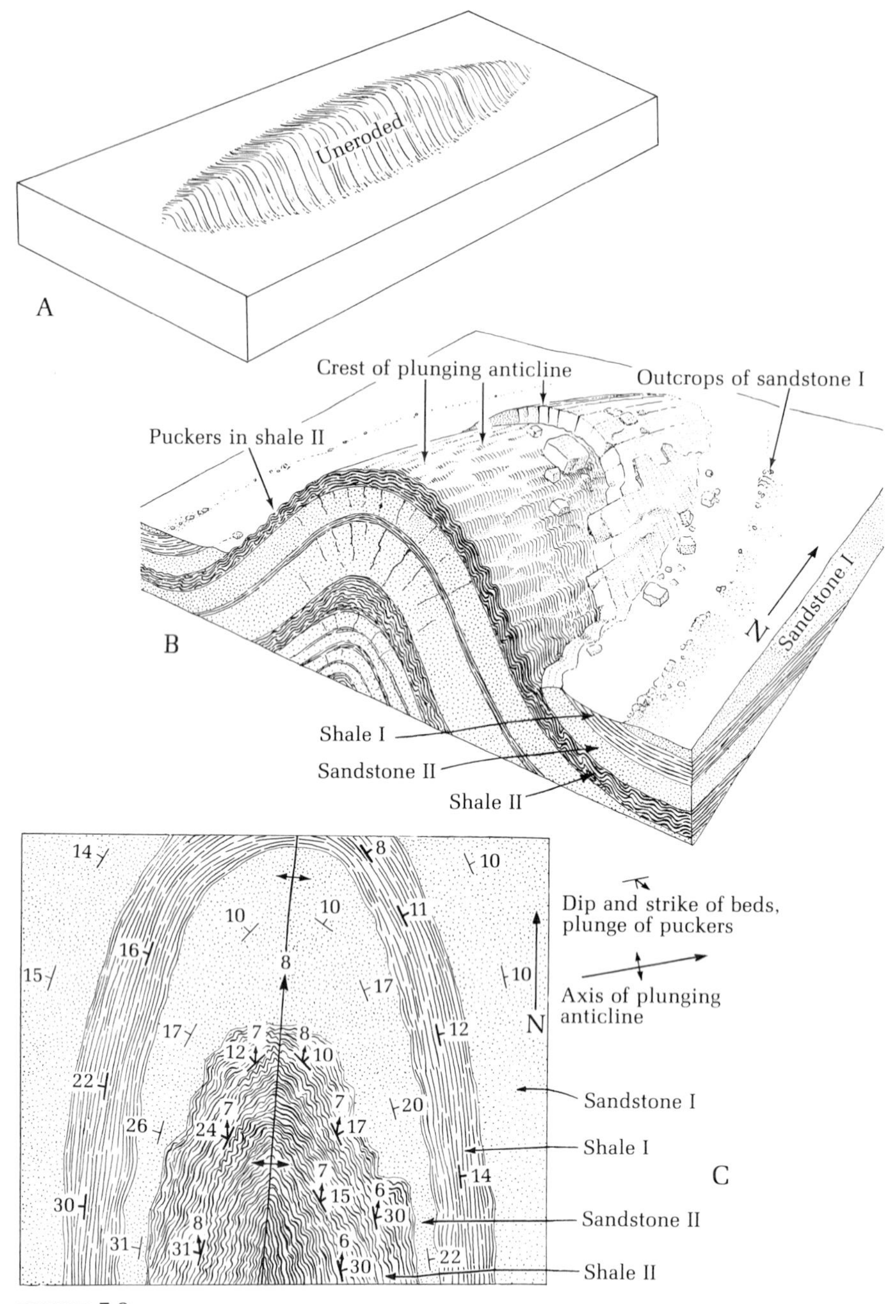

FIGURE 7.9

Aspects of an anticlinal fold: *A*, an anticlinal fold as it would appear if uneroded; *B*, a block diagram of one end of an eroded anticline in Arkansas, showing cracks and puckers produced by bending and slippage between layers; *C*, a geologic map of the same anticline, showing the attitudes of the layers and the boundaries between them (such a map would reveal the fold structure even if it were wholly leveled by erosion). [After Gilluly, Waters and Woodford, *Principles of Geology*, W. H. Freeman and Company, copyright © 1968.]

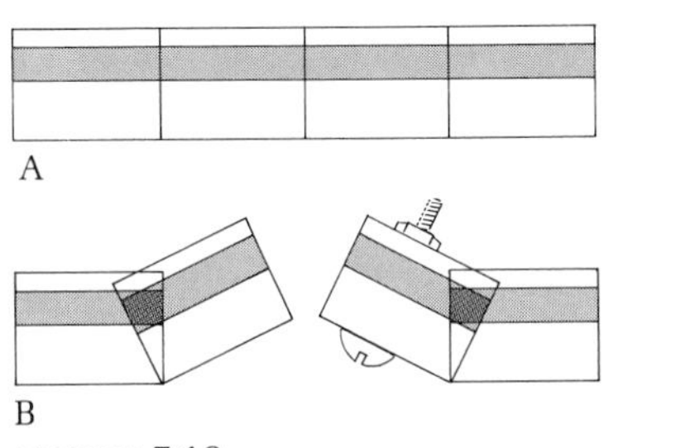

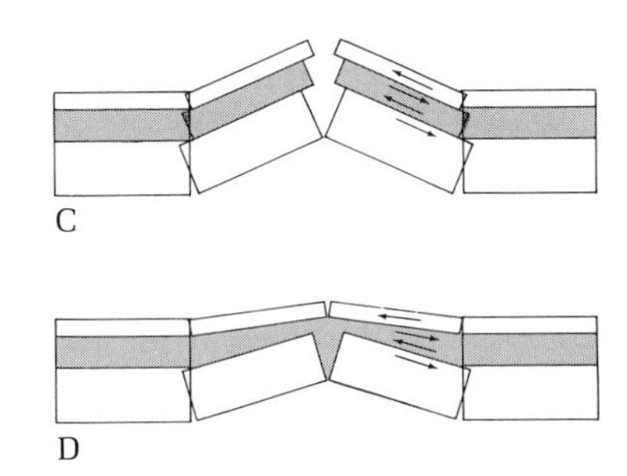

FIGURE 7.10
Simplified diagrams illustrating differential movement required in folding: *A*, unfolded layers; *B*, if layers do not shift when folded, a large tensional gap and compressional overlaps result; *C*, sliding between layers minimizes gaps and overlaps; *D*, sliding combined with the flow of an incompetent layer further reduces the necessity for gaps.

sheet is stretched at the top and compressed at the bottom. If the bending becomes more acute, however, the sheets on the top and bottom of the stack are bent by different amounts and it becomes impossible to make them follow even an approximation of the same curves. It is possible, however, if we sandwich layers of some more plastic material, such as lard, between the plywood sheets. Then the sheets have room to bend, and the lard flows away from the places where the plywood moves together and into the places where it moves apart.

Sedimentary rocks behave in these same ways when they are folded. Beds of sandstone and limestone are called competent because they tend to be stiff like the plywood. Beds of shale are called incompetent because shale, like the lard, will flow readily into the intricate little gaps between the competent beds. Thus, the sequence of rock layers has a major influence upon the nature of the folds produced by a given stress. Layer on layer of thick, massive sandstones can only bend into gentle folds without breaking in tension at the crests of the anticlines. Thinner sandstones can be folded more without breaking, but folding is easiest in sequences of alternating competent and incompetent beds. At the other extreme, normal folding is rare in a sequence of incompetent beds, such as thick shales. Instead, the beds tend to become contorted like scrambled eggs.

Gravity Sliding

The folding of some continental mountain belts has been shown, by erosion and drilling, to be a surficial phenomenon. The Applachian folds, for example, are in a sheet of rock with an area of many thousands of square kilometers but with a thickness of only a few kilometers (Figure 7.11). This sheet has slid over the underlying, unfolded rocks for a distance of perhaps tens of

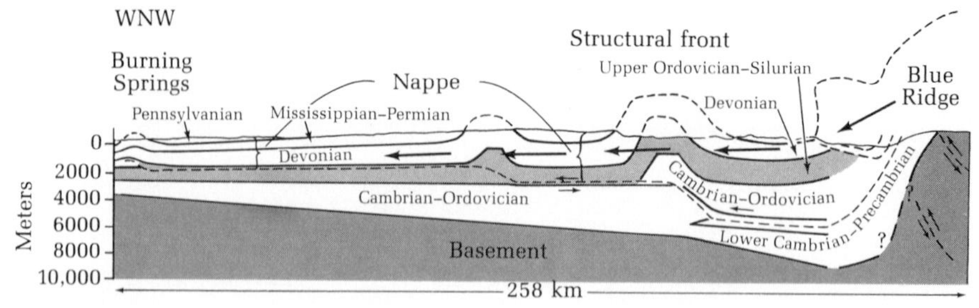

FIGURE 7.11

A schematic cross-section of the Central Appalachians, from Virginia to West Virginia, demonstrating the near-horizontal present-day attitudes of a thrust plane. Folds in the nappe are associated with warps in the thrust plane (*dashed line*), which lies largely along bedding planes. [After Gwinn (1964), courtesy the Geological Society of America.]

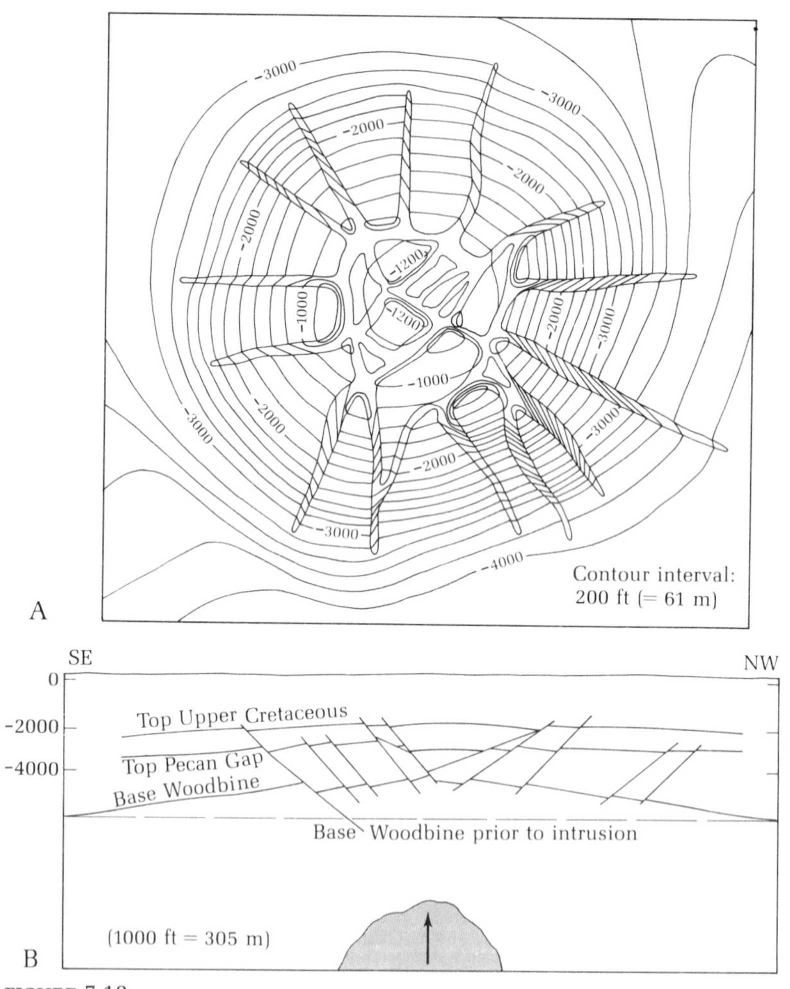

FIGURE 7.12

The structure of domes: *A*, radial cracking above a salt dome shown by a structure map of Clay Creek dome, Texas; *B*, a cross section of the Hawkins dome, illustrating normal faulting. [From Gussow (1968).]

kilometers. The proportions of the dimensions of the sheet are not unlike those of a piece of typing paper. Apparently, the rock is deformed into a fold just as the paper would be were it, while lying on a flat surface, to have its two ends pushed somewhat toward each other.

Thin, broad, moving sheets of rock, called nappes, also exist. The discontinuity between a nappe and the rock below is a thrust fault, which is generally somewhat folded, although almost horizontal on the average. Several nappes may be superimposed. Apparently, the folding is rather incidental to the motion of these sheets, and it may be triggered by a sheet passing over a bump in the rock below the thrust fault.

How are these sheets formed and how do they move? Thrust faults are so called because, when they were discovered, geologists believed that they were the result of crustal shortening and the thrusting of one sheet over another. In 1959, the American geologists M. King Hubbert and William Rubey rejected this hypothesis because the crushing strength of the sheets is too low. Instead of moving, they would merely crumble.

Somehow, nappes must move along the thrust faults with hardly any friction with the rock below. Hubbert and Rubey showed that this can occur if ground water is under pressure along the fault. This condition is favored in some parts of the Appalachian Mountains where the thrust plane follows bedding planes. Thus, the possibility exists that there are impermeable rocks above the fault that can entrap ground water.

If the thrust plane is lubricated enough, it will move under the gentle urging of gravity. It appears that, when the core of a mountain range is elevated, some sheets of sedimentary rock on the flanks gain enough potential energy to break free and glide slowly downhill.

Diapirs

If they can be mobilized enough to flow, low-density rocks will move upward through the rocks above like oil through water. The most common types of geologic materials that behave in this way are salt and molten lava, but gypsum, shale, and serpentine also form piercement structures or diapirs. They move upward in circular blobs, generally 1–10 kilometers in diameter. As they approach the surface, these blobs exert enough force to buoy the overlying rocks into domes and break them with distinctive patterns of faults (Figure 7.12). Mud and peat may produce the same phenomena on a small scale in unconsolidated sediments.

Molten rock loses its mobility as it cools, and it may stop rising short of the surface. Salt, however, often penetrates the surface, where it may be dissolved to form a shallow basin ringed with ridges of eroded rock. This occurs in humid regions, such as those that rim the Gulf of Mexico. In Iran, where it is very arid, the salt forms hills and flows like lava into the nearby valleys.

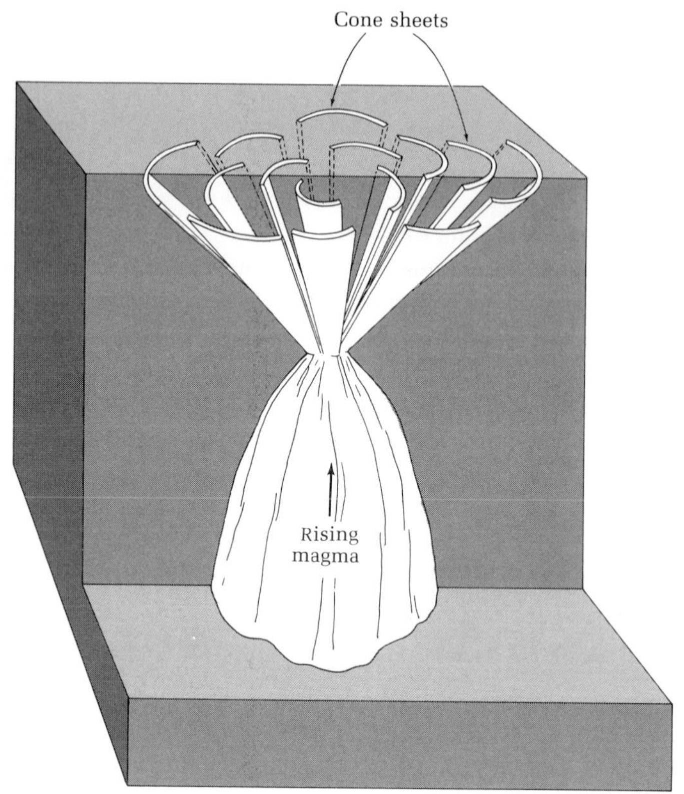

FIGURE 7.13
Cone sheets formed by the injection of magma along conical normal faults over an intruding domical mass of magma.

The faults associated with a diapir typically form along vertical radial lines or about concentric cones above the top of the rising blob. This is exactly what would be expected from a tensional fracture above a point. Thick glass breaks in a circular cone above a point where it is struck, for example, and dry dirt breaks in radial lines over an emerging mushroom (or an emerging mushroom-shaped cloud).

Summary

1. Earthquakes and deformation of the earth are inevitable as long as plates continue to move.
2. Intensity is a measure of the local social effects of an earthquake. Magnitude is an instrumental measurement that takes distance into account. An earthquake of intensity XI causes considerable damage; an earthquake of magnitude 6.0

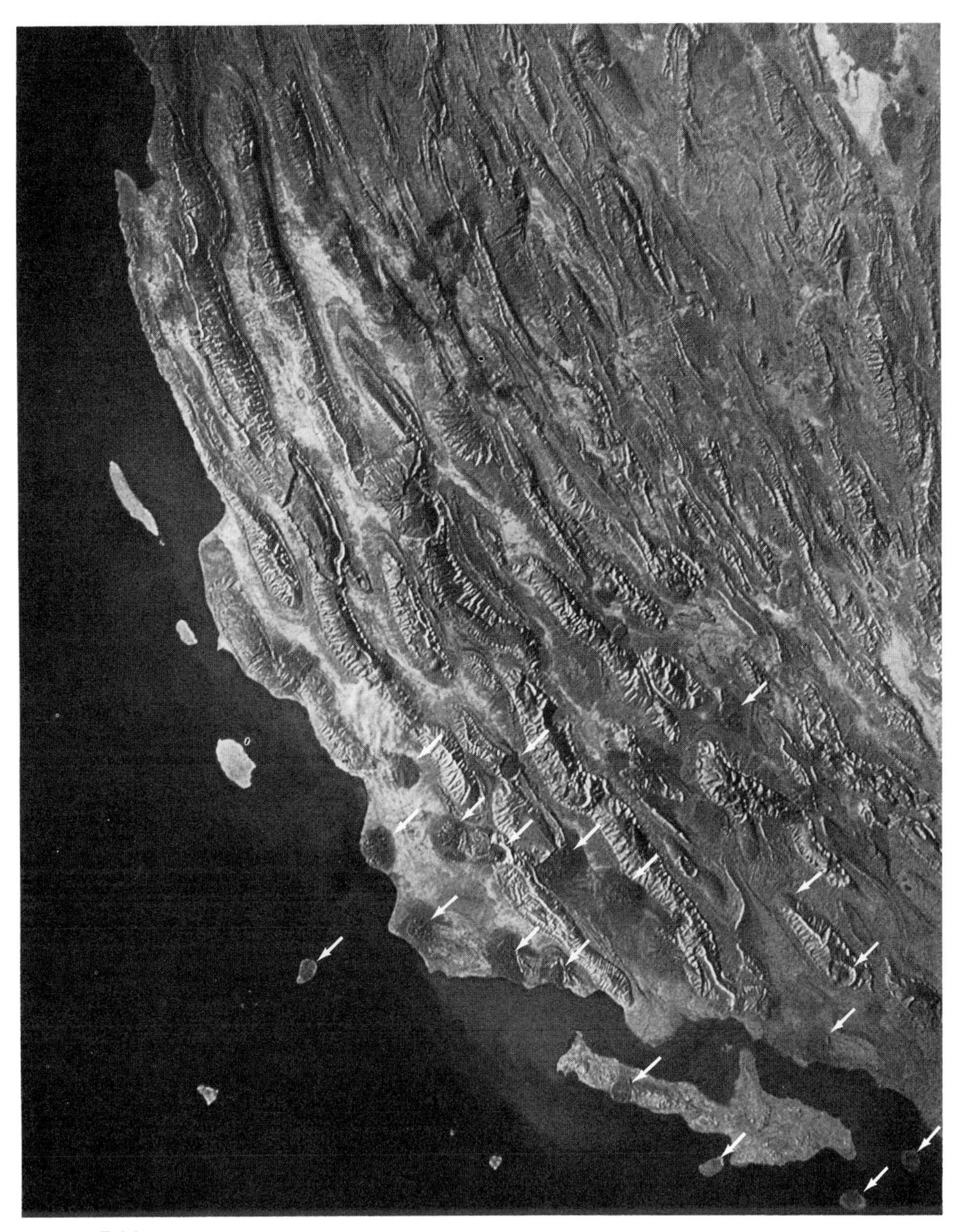

FIGURE 7.14

Elongate folds of the Zagros Mountains along the coast of the Persian Gulf, photographed by Gemini 12. The anticlines are in different stages of erosion. The dark circles resembling smallpox scars are salt domes (indicated by *arrows*). Note how the coastline and islands are shaped by the structures. [NASA photo.]

may have an intensity of XI if it occurs near anything that can be damaged. The magnitude 6.0 quake has the energy of a one-megaton thermonuclear bomb.

3. Faults, or dislocation surfaces in the earth, are of various types, depending on the relative motion of the material on each side. They are extremely common features.
4. Elastic strain builds up in the earth, and it is relieved by creep or larger movements on faults.
5. It appears that some parts of the San Andreas fault are prone to large earthquakes and that others are prone to creep.
6. A fold is a bend in layered rock, and various kinds can be differentiated by shape. Folding is generally a low-intensity phenomenon. The shape of a fold depends on the character of the layers that are bent.
7. If rocks are elevated enough, gravity may induce them to fracture and slide along a thrust fault. If the fault is lubricated by water under pressure, gravity sliding may extend for tens of kilometers.
8. Thick layers of low-density rocks, such as salt, may penetrate upward through overlying rocks in diapirs. They may deform the surface into domes marked by radial or conical faults.

Discussion Questions

1. What is the source of energy for earthquakes?
2. How are we able to compare the magnitude of an earthquake with that of an underground nuclear explosion? How can we calculate the size of a "secret" underground nuclear bomb test in Siberia?
3. Why does a fault form? Why does it become inactive?
4. In Figure 7.4, why is the road curved after it is broken by faulting?
5. How often will a quake with an energy release equivalent to that of a one-megaton (1000-kiloton) bomb occur in California?
6. Describe the movement between layers that occurs when rocks are folded.
7. How does a thrust fault resemble a landslide? How does it resemble a skier?
8. At which type of plate boundary would you expect the following: thrust faults, folds, normal faults, lateral faults, diapirs?

References

Allen, C., 1968. The tectonic environments of seismically active and inactive areas along the San Andreas fault system. *In* W. R. Dickinson and A. Grantz, eds., *Proceedings of Conference on Geologic Problems of San*

Andreas Fault System (Stanford Univ. Publ. Geol. Sci., v. 11), pp. 70–83. Stanford, Calif.: School of Earth Sciences, Stanford University.

Christensen, M. N., 1966. Late Cenozoic crustal movements in the Sierra Nevada of California. *Geol. Soc. Amer. Bull.* 77:163–182. [Uplift of 4000–7000 feet.]

Clark, W. B., and C. J. Hauge, 1971. The earth quakes . . . you can reduce the danger. *Calif. Geol.* 24(11):203–219. [Social effects of California earthquakes.]

Cluff, L. S., 1968. Urban development within the San Andreas fault system. *In* W. R. Dickinson and A. Grantz, eds., *Proceedings of Conference on Geologic Problems of San Andreas Fault System* (Stanford Univ. Publ. Geol. Sci., v. 11), pp. 55–69. Stanford, Calif.: School of Earth Sciences, Stanford University.

Evernden, J., and J. Filson, 1971. Regional dependence of surface-wave versus body-wave magnitudes. *J. Geophys. Res.* 76:3303–3308. [Yield of underground nuclear explosions compared to earthquake magnitudes.]

Geotimes, 1971. Newsnotes: San Andreas creep. *Geotimes* 16:11, 33.

Gilluly, J., A. C. Waters, and A. O. Woodford, 1968. *Principles of Geology* (3rd ed.). San Francisco: W. H. Freeman and Company.

Gussow, W. C., 1968. Salt diapirism: importance of temperature, and energy source of emplacement. *In* J. Braunstein and G. D. O'Brien, eds., *Diapirism and diapirs* (Amer. Ass. Petrol. Geol. Mem. 8), pp. 16–52. Tulsa: American Association of Petroleum Geologists.

Gwinn, V. E., 1964. Thin-skinned tectonics in the plateau and northwest valley and ridge provinces of the central Appalachians. *Geol. Soc. Amer. Bull.* 75:863–900.

Handin, J., 1966. Strength and ductility. *In* S. P. Clark, Jr., ed., *Handbook of Physical Constants* (Geol. Soc. Amer. Mem. 97), pp. 223–229. New York: Geological Society of America.

Hofman, R. B., 1968. Recent changes in California fault movement. *In* W. R. Dickinson and A. Grantz, eds., *Proceedings of the Conference on Geologic Problems of the San Andreas Fault System* (Stanford Univ. Publ. Geol. Sci., v. 11), pp. 89–93. Stanford, Calif.: School of Earth Sciences, Stanford University.

Hubbert, M. K., and W. W. Rubey, 1959. Role of fluid pressure in mechanics of overthrust faulting. *Geol. Soc. Amer. Bull.* 70:115–206. [Nappes can float on a cushion of water.]

Iacopi, R., 1964. *Earthquake Country*. Menlo Park, Calif.: Lane. [An intimate account of earthquakes in California.]

Lawson, A. C., chairman, 1908. *The California Earthquake of April 18, 1906* (Report of the State Earthquake Investigation Commission; Carnegie Inst. Wash. Publ. 87). Washington, D.C.: Carnegie Institution of Washington.

Richter, C. F., 1958. *Elementary Seismology*. San Francisco: W. H. Freeman and Company. [Interesting descriptions.]

Rogers, T. H., and R. D. Nason, 1968. Active faulting in the Hollister area. *In* W. R. Dickinson and A. Grantz, eds., *Proceedings of Conference on Geologic Problems of San Andreas Fault System* (Stanford Univ. Publ. Geol. Sci., v. 11), pp. 42–43. Stanford, Calif.: School of Earth Sciences, Stanford University.

8

EARTHQUAKES AND SOCIETY

That was the year when Lisbon-town
Saw the earth open and gulp her down.

Oliver Wendell Holmes,
THE DEACON'S MASTERPIECE

Physical Effects and Costs of Earthquakes

Earthquakes and their side effects are the greatest natural destroyers of property, ranking in this respect only below the human activities of war and accidental burning. They are not such high-ranking killers, being easily surpassed by war, pestilence, and starvation, but they have, on occasion, taken a terrible toll. It seems that the worst killer among earthquakes was the one that struck Shensi, China, in 1556: 830,000 are reported to have died. Among the other disastrous earthquakes are the one that struck Kansu, China, in 1920 (with 180,000 dead) and the one that hit Kwanto, Japan, in 1923 (with a loss of 100,000). Most of the great killers have been in the ancient and heavily populated countries of the Far East and the Mediterranean region. The only other place to suffer a loss of 10,000 or more is Chile, where 25,000 were killed in 1939, and 10,000 in 1960.

The magnitudes of the earliest killer quakes are unknown, but all the more recent ones with fatalities exceeding 10,000 have been of magnitude 7.9 or greater—that is, with an energy release equivalent to that of a large hydrogen bomb. It may be worthwhile, at this point, to reflect upon the effects of the population explosion. Fatalities may be expected to increase for various reasons, in addition to the obvious one that earthquakes can kill people in large numbers only where the population is concentrated. That is why so many are killed in the Far East and around the Mediterranean. A more subtle and more important reason why fatalities may increase is that the expanding population tends to be concentrated in weaker buildings on less stable land. Destruction of buildings by earthquakes is largely determined by foundation materials, as is shown in the following sections. The death toll, in turn, is largely determined by buildings collapsing upon their inhabitants. Thus, the time of day is very important: If people are in the streets and away from buildings when an earthquake strikes, there are few fatalities. The difference between a major human tragedy and a minor one may depend on whether the children are inside school buildings or are playing in the yards when a quake strikes. This point emphasizes the importance of the possibility of earthquake predictions and warnings.

A warning can also be of benefit in reducing destruction to property. It cannot compensate for foolish construction on active faults or for shoddy building on unstable ground, but it can provide time to turn off fires and pilot lights and to take steps to combat the accidental fires that inevitably occur. It was fire that destroyed much of San Francisco after the earthquake of 1906, and fire was also highly destructive after the earthquakes in Lisbon in 1755 and Tokyo in 1923. Earthquakes cause damage in many ways: (1) by faulting that breaks the ground; (2) by vertical movements that drop land below sea level with catastrophic flooding; (3) by shaking buildings, bridges, and other structures; (4) by disturbing earth into slumps, landslides, and other mass movements; (5) by disturbing water into large moving and standing waves; and (6) by triggering such secondary phenomena as fire storms and pestilence.

The damage caused by earthquakes is largely a function of the concentration of vulnerable structures, and these increase rapidly along with the growth of the population and the advance of technology. The effect is readily shown by a comparison of the damage from the Long Beach earthquake of 1933 and that of the San Fernando quake of 1971. The two were in the same general region and the magnitudes were rather similar—6.3 and 6.6, respectively. During the interval, however, the population had expanded and a vast complex of buildings and public facilities had been built. Estimated damage in the earlier quake was $50 million; in the latter, it was $1000 million. Even allowing for inflation, the damage was much more costly simply because there was more to damage.

The damage to a modern city that would result from a nearby major quake is difficult to imagine. The loss in San Francisco in 1906 was $400 million, which is roughly equivalent to $1600 million in modern debased dollars. Judging by the San Fernando quake, damage from a great shock in a major industrial city could readily cost several billion dollars.

BOX 8.1 GREAT EARTHQUAKES AND THEIR EFFECTS

The following table lists chronologically some of the great earthquakes of the last six centuries and gives data, where known, on their magnitudes and the deaths and destruction they caused.

Location of earthquake	*Date*	*Magnitude*	*Deaths*	*Notes on destruction*
Naples, Italy	1456		30,000	
Shensi, China	1556		830,000	
India	1668			30,000 houses
Algiers, Algeria	1716		20,000	
Lisbon, Portugal	1755		60,000	17,000 of 20,000 houses; fire burned six days
Baalbek, Lebanon	1759		20,000	
Calabria, Italy	1783–1786		50,000	33 villages eliminated, 148 villages destroyed but rebuilt; no quakes for preceding six centuries
Kutch, India	1819		1,800	10,000 houses
Ischia, Italy	1883		2,300	
Mino-Owaii, Japan	1891		7,300	
Sanriku, Japan	1896		28,000	6,000 houses, from accompanying tsunami only
Assam	1897	8.7	1,542	6 major towns and all villages in 30,000 square miles leveled
Alaska	1899	8.6		
Tibet	1902	8.6		
Kangra, India	1905	8.7	19,000	
San Francisco, California	1906	8.3	700	
Colombia	1906	8.9		
Formosa	1906		1,250	
Valparaíso, Chile	1906	8.6		
Messina, Italy	1908		82,000	98% of houses ruined
Avezzano, Italy	1915		30,000	

Prevalence of Earthquakes

Many residents of the United States have never experienced a major earthquake and may feel little personal concern about the possibility. The reader, at this point—knowing most earthquakes occur at plate boundaries, and that

(continued)

Location of earthquake	*Date*	*Magnitude*	*Deaths*	*Notes on destruction*
Kansu, China	1920	8.6	180,000	Enormous landslides
Kwanto, Japan	1923	8.2	100,000	Resulting fire storm in Tokyo killed 38,000
Tango, Japan	1927	7.9	3,000	25,000 houses, some destroyed by fire
Sanriku, Japan	1933	8.9	3,000	8,800 houses, from accompanying tsunami only
Concepción, Chile	1939	8.3	25,000	
Turkey	1939	7.9	25,000	30,000 dwellings
Tottori, Japan	1943	7.4	1,400	
San Juan, Argentina	1944		5,000	
Fukui, Japan	1948	7.3	5,300	
Ambalo, Equador	1949	6.75	6,000	
Assam	1950	8.7	1,526	
San Salvador	1951		4,000	
Kurdistan	1957	7.25	2,000	
Agadir, Morroco	1960		10,000	
Chile	1960	8.9	10,000	58,600 houses

The short table below lists, according to magnitude, six California earthquakes of the twentieth century, giving the number of deaths and the dollar cost of the damage caused by each.

Location of earthquake	*Date*	*Magnitude*	*Deaths*	*Cost of damage ($)*
San Francisco	1906	8.3	700	400 million
Bakersfield	1952	7.7	0	48 million
Imperial Valley	1940	7.1	9	6 million
San Fernando	1971	6.6	65	1,000 million
Long Beach	1933	6.3	120	50 million

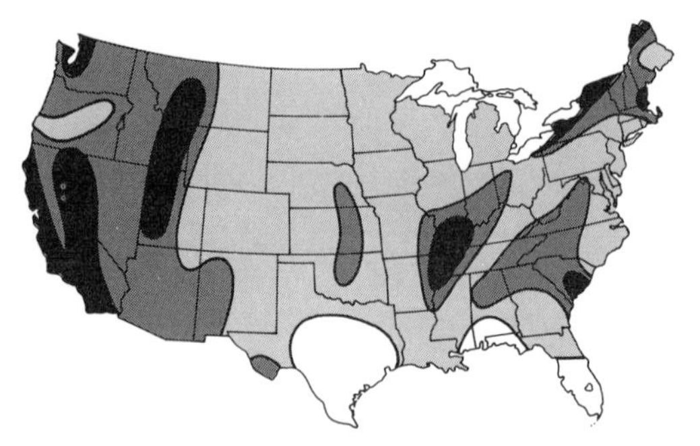

Zone 0 No reasonable expectance of earthquake damage
Zone 1 Minor earthquake damage can be expected
Zone 2 Moderate earthquake damage can be expected
Zone 3 Major destructive earthquakes may occur

FIGURE 8.1

Earthquake hazards in the United States, assuming that more will occur where they have in the past. [From National Earthquake Information Center (May–June 1971), courtesy NOAA Environmental Research Laboratories.]

all of the territory of the United States, except for California and Alaska, is within plates—may feel reinforced in this view. It is, however, far from correct. During historical times, major destructive earthquakes have been experienced in Nevada, Utah, Idaho, Washington, Montana, Wyoming, Illinois, Indiana, Kentucky, Missouri, Tennessee, Arkansas, Mississippi, South Carolina, New York, New Hampshire, Maine, Vermont, and Massachusetts. Earthquakes are everyone's concern.

Judging by the extent of intense effects, the three quakes of 1811 and 1812 in the New Madrid area in the southeastern tip of Missouri were the greatest in historical times in the United States. The area was sparsely populated; neither life nor property was much affected, even though chimneys were damaged in a very large area. The shock of the largest earthquake of the three was felt from Boston to New Orleans. The most spectacular effects of that quake were the formation of Lake St. Francis in Missouri and Reelfoot Lake in Tennessee, the former being 70 kilometers long and 1 kilometer wide. Near the lakes, land was heaved upward, fault scarps caused waterfalls on the mighty Mississippi, and fissures and water spouting from the ground were widely reported. Whole forests were destroyed by movement of the soil or by submergence in the new lakes.

Another surprising series of earthquakes struck Charleston, South Carolina, in 1886. Damage was extensive in a wide area; 14,000 chimneys were destroyed in Charleston, and the streets were almost blocked by debris from buildings. Log cabins in the surrounding country, however, were undamaged because of their construction. Large earthquakes were also experienced in southeastern Canada and New England in 1663 and 1755. A less intense quake in the St. Lawrence River area in 1925, however, may have been felt by more people in the United States than any other, simply because of the growth of the population of that region since the seventeenth and eighteenth centuries. The intensity was 4.0 or more in Boston and New York.

By mapping and cataloguing the distribution of these and many smaller earthquakes, regions of equal earthquake frequency and intensity have been identified. These are equated to earthquake risk by making the assumption that future quakes will be distributed as they have been in the past. In most places, this assumption can hardly be justified by any theoretical understanding of what is happening. At the plate boundary in California, of course, the assumption seems justified by plate-tectonic theory; all parts of the plate boundary move in accordance with Euler's theorem, and earthquakes will continue to occur. However, Charleston, New Madrid, and New England are *within* plates, and the origin of earthquakes in such places is a mystery. Are the earthquakes relieving accumulated strain in the whole drifting plate? If not, what is happening in the regions without earthquakes? Perhaps earthquakes are less apt to occur now where they have in the past. This is the argument once advanced for predicting earthquake risks along the San Andreas fault in California. We know little of these matters; until we know more, we will have no basis for predicting earthquakes within tectonic plates.

Tsunami—an Example of Useful Warning of Disaster

One of the most terrifying sights known to man is the advancing wall of water colloquially called a "tidal wave." This name is inappropriate because the phenomena has nothing to do with the tides (although "tidal bores," which are rushing walls of water in estuaries and rivers, are superficially similar). Scientists usually adopt the Japanese term *tsunami* instead.

Tsunamis are mainly generated by sudden subsidence of large areas of the sea floor; this, apparently, is a rare occurrence, except in subduction zones, where the lithosphere always tends to curve downward. When the sea floor drops, it creates a hollow in the overlying water, which is then very rapidly filled by the surrounding water. A train of waves spreads outward from this disturbance like ripples from a dropped pebble, and the front of the train is a low trough rather than a high crest. An earthquake usually accompanies the subsidence, and seismic waves move through the solid earth at speeds of 16,000–32,000 kilometers per hour. These waves are recorded on many seismographs within a few minutes of the event. From the known characteristics of such waves, seismic observations from several suitably located stations can be used to determine the location of the earthquake.

FIGURE 8.2
Generalized diagram of sources of tsunamis detected in the Hawaiian Islands. Almost all damaging and distant ones originate in subduction zones at plate boundaries.

Meanwhile, the sea waves move outward from the source at about 800 kilometers per hour, although the speed decreases with the depth of the water. The tsunami of April 1, 1946, is a particularly well-studied example, because it had the effrontery to rout the marine geologist Francis Shepard out of his bed in Hawaii. The source was 130 kilometers south of Unimak Island, Alaska, and a few minutes after the earthquake, a lighthouse on the island was struck by waves 18 meters high. The waves spread southward in deep water, but there they were imperceptible. Their height was only a meter or so, and the distance between their crests was about 160 kilometers. The shock of an earthquake may be felt by a ship far out to sea, but a tsunami cannot be.

The long, low waves reached the Hawaiian Islands about $4\frac{1}{2}$ hours after they started. There, they were slowed by the drag of the bottom in shallow depths, and the water piled up locally into waves 11 meters high at Oahu and almost 17 meters high at the big island of Hawaii. They killed 159 people and caused $25,000,000 worth of property damage. The disaster was not uniform, because some places were more sheltered than others. Nor was it as great as it might have been, had it not been early morning. This is because the initial pulse of a tsunami is downward upon arrival, just as it is at the source—that is, a trough precedes the first crest. Inexperienced people, seeing the sea recede, may walk out on the glistening sea bottom. There, the first crest overwhelms them. Experienced people, whether Fijians or Chilean fishermen, shout: "The water recedes! Take to the hills!"

The tsunami of April, 1946, spread throughout the Pacific, and the waves were still 1 meter high at Valparaíso after traveling 18 hours and more than 12,800 kilometers. The waves were reflected from the shores of South America and other continents and continued to slosh back and forth in the Pacific for days (although, by then, they were detectable only with instruments).

Tsunamis are quite frequent along most Pacific shorelines. This is because they are generated in most subduction zones, particularly those near Japan and Chile, and each wave can be damaging on the other side of the ocean. Hawaii, in the center of the Pacific, has experienced 40 large ones in the last 150 years, and many more small ones may have passed unobserved. In Hawaii, deaths from tsunamis have always been relatively few, not only because there are few people, but also because the land rises steeply, the islands tend to protect each other from exposure, and most of the people live in Honolulu, which is high and protected. Deaths and damage are more common in Hilo, the second city of the islands, because it is relatively low and exposed.

TSUNAMI WARNING SYSTEM

It was apparent, as soon as the velocities of waves in the solid earth and in the sea were calculated, that it is possible to predict tsunamis at points distant from earthquakes. By 1923, the seismologists of the Hawaiian Volcano

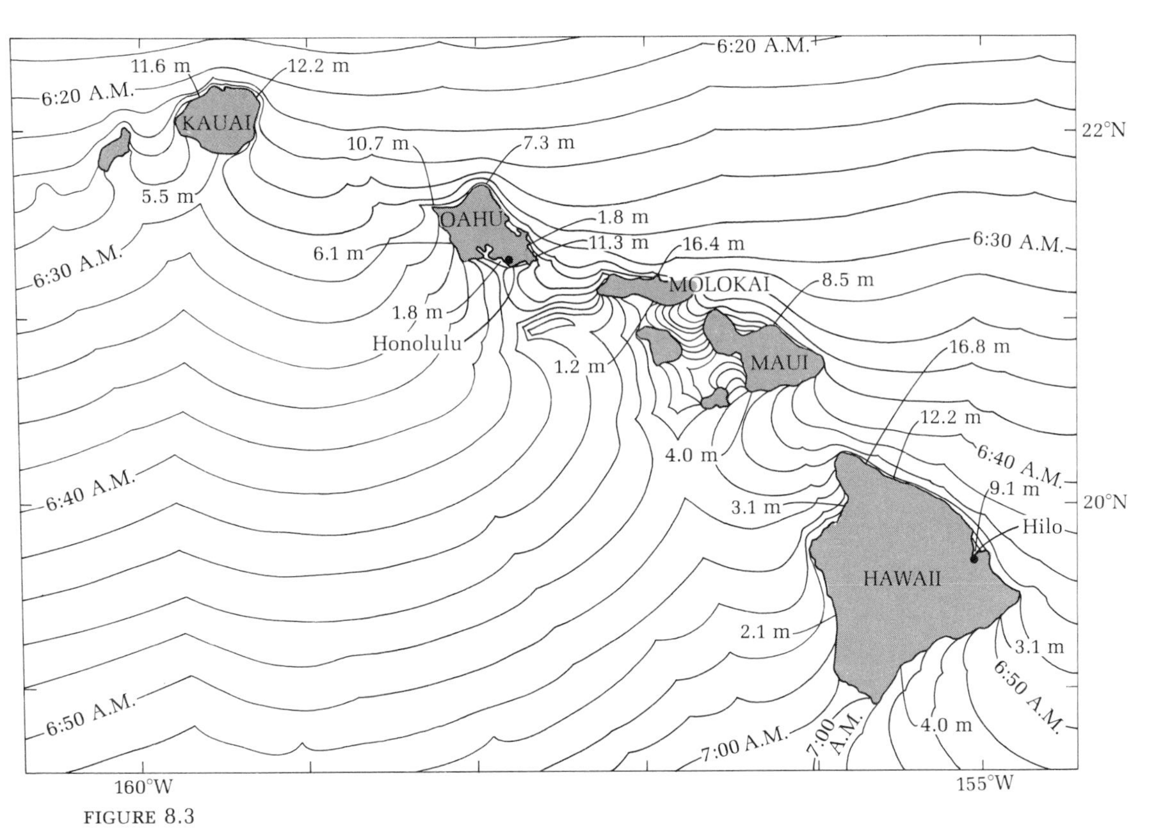

FIGURE 8.3

Wave fronts from the Alaskan tsunami of 1946. Note refraction around the Hawaiian Islands. Minor changes may be associated with water depth. Maximum water heights, in meters, are shown for particular points on the islands; times refer to the passage of the first wave. [Data from Shepard et al. (1950).]

Observatory were successful in predicting arrival times of tsunamis. However, a high-speed communications network and a counterdisaster organization are necessary to issue warnings and make them effective. These did not exist until the tsunami of 1946 illustrated the folly of doing without them. A warning system was created in Honolulu in 1948 by what was then the U.S. Coast and Geodetic Survey (now a part of the National Oceanic and Atmospheric Agency). Whenever a strong quake is recorded there, seismological observatories in the western United States and Alaska are queried by telephone. An accurate location and an approximate magnitude can be determined within minutes, and the probability of a tsunami can thus be evaluated. After considering past tsunamis from the same locality, a warning of the arrival time of possible waves is issued to endangered coastal areas. The system alerts not only Hawaii, but the whole Pacific rim as well. Once an active tsunami is detected, it can be followed from point to point and the warnings can be made more precise. The tsunami generated by the Alaskan earthquake of 1964 was minor, but its progress down the coast of the United States was followed by television news bulletins that helped to alert the maximum audience and to reduce damage to a minimum.

WEAKNESSES OF THE WARNING SYSTEM

The warning system has two weaknesses: (1) it works only at a distance from the earthquakes, and destruction is greatest near them; and (2) usually, it can only predict the time of arrival of a tsunami, not whether it will be a damaging one. The second of these is more important, because it is related to a problem that arises in all environmental warnings, but let us dispose of the first one first. The greatest potential danger area for tsunamis is Japan, because it is a land of low elevation and dense population, and because it is very near a subduction zone in which tsunamis are generated. In the Sanriku district, 28,000 people drowned in 1896, and 3000 in 1933. Since then, the population has grown much denser along the coast. Chile is similarly exposed to tsunamis, and there, too, the population along the coast is expanding, although the numbers are still relatively small. In Chile, the danger in the great quake of 1960 was augmented by subsidence of the coastal region. A region 480 kilometers long and 20–30 kilometers wide sank as much as 2 meters in a few seconds, and then the tsunami rolled in. The damage from quakes and tsunamis cannot be clearly separated, but roughly 10,000 people died and 58,000 houses were destroyed along that still sparsely populated stretch of the Pacific coast of South America.

Submarine volcanic explosions and collapse may also generate tsunamis that strike before the warning system can act. The explosion of Krakatoa in Indonesia in 1883 is a famous example. It is less well known that Hawaii is subject to some tsunamis of local origin that may be of this type. Such a tsunami drowned at least 81 people on the south coast of Hawaii in 1868. A smaller local wave occurred in 1872 at Hilo.

The greater flaw in the warning system is a sociological one: people pay no heed if you cry "Wolf!" too often. The tsunami warning system duly alerted Hawaii on the occasion of the 1960 Chilean earthquake. Nonetheless, 61 people were drowned and 300 were injured when the waves hit Hilo, about as predicted, at 1:00 A.M. Subsequent investigation indicated that many people had responded to alerts in 1958 and 1959, when nothing had happened. This time, 15% of the survivors said that they knew of the warnings but ignored them, and 53% said they were awaiting more precise information. This same tsunami caused far more damage in distant Japan, where 180 died and damage amounted to $500,000,000. The reason for the deaths in Japan was partially that no alert was issued there. No one dreamed that waves could cross the whole Pacific and still be so destructive. Herein we can see an important generality. An ideal warning system, or even a merely successful one, must have the following characteristics:

1. It is able to determine when the danger is certain.
2. It alerts only when the danger is certain. Thus, when it does alert, people respond.
3. It specifies the time of the predicted event with enough accuracy and sufficient warning to permit people to plan appropriate action.

Foundation and Ground Effects

AMPLIFICATION IN SEDIMENT

In 1952, a quake shook Kern County, California, and damaged the city of Bakersfield. Naturally, those who experienced the surface shaking were anxious about a party of tourists walking through nearby Crystal Cave. When the tourists emerged, however, not only were they safe, they were unaware that a quake had occurred. This phenomenon has been observed repeatedly for earthquakes in mining districts. The intensity of earthquakes is less on hard rock than on sediment, and mines are mostly in rock.

Instruments record the scale of this effect. One is located on bedrock near Pasadena, and another on sedimentary fill 300 meters thick only 5 kilometers away. Nearby quakes have four times as great an amplitude at the recorder on the sediment.

CAUSE OF AMPLIFICATION

The cause of the amplification is not a simple physical effect; if anything, the sediment might be expected to provide a cushion. However, we can gain some insight about the amplification of earthquake intensity by sediment if we consider the seismic behavior of sediment that is not covered by buildings.

Sediment may exhibit various stages of compaction and various degrees of saturation with groundwater. When a quake vibrates the sediment, it tends to become more compacted; when this occurs, the sediment must void the water that fills the spaces among the grains. Thus, on flat ground, "earthquake fountains" may pour out water and lift sand. The effect is most pronounced where the ground is saturated almost to the surface, as it is near rivers and springs. The spouts may emit enough material to cover fields and roads, and leave craters that are several meters across.

The general upward motion of water may also eject posts from the ground. It might seem that this would occur only in soil deposited since the last earthquake—that is, that the ground in a seismically active region might become completely compacted after a few earthquakes. This, however, is not correct. The great Tokyo earthquake of 1923 ejected posts from the ground of a river flat where resident farmers had not known they existed. Historical research showed that these posts were the foundations for a bridge built in 1182 that had long since vanished.

Sediment may lose all coherence while it is consolidating and water is flowing out. For a moment, the grains are barely in contact and the sediment has no more strength than quicksand. Thus, as water and wooden posts float out, heavy rocks and buildings sink. Spectacular effects have been noted in India, where buildings have sunk half a story or more without much tilting or other disturbance.

So much for saturated sediment on the level. If it is on a slope, gravity may put masses of sediment in motion. Thus, slumps and landslides cascade down from hillsides to the plains below. In mountainous areas, such as the Andes of Peru and Chile, the chief direct cause of fatalities and destruction may be landslides that bury whole villages. Such landslides also hurl dams across rivers and impound lakes. These dams have little strength, however: as soon as the water is deep enough, they collapse or are catastrophically eroded, and a flood is released down the valley. Little can be done about these natural dams, except to blast them away before such water collects. There is no excuse for the collapse of man-made dams, which is another effect of some earthquakes. The lower Van Norman dam was built by the hydraulic earthfill method in the San Fernando Valley north of Los Angeles. This mode of construction is cheap but it may leave super-saturated sediment in the dam, which, in any event, is not at maximum consolidation. The San Fernando Valley earthquake of February, 1971, caused a partial collapse of the upstream face of the dam. A catastrophic flood would have poured down upon the suburbs below, had not the engineers already lowered the water level as an unrelated safety measure a few days before. As it was, the danger was so acute that a large area had to be evacuated for a four-day period.

A natural hillside is not necessary for slumps. The banks of rivers and canals are places where the strength of the surrounding earth prevents it from flowing into the low channels. When the strength decreases even momentarily, slumps are widespread, and deep cracks form parallel to the banks. The common mode of freeway construction on flat ground is to cut and fill alternatively in order to bridge over or tunnel under the intersecting older streets. The sides of the cuts and fills are prone to the same earth movements as river banks, except that the ground is not so saturated.

Because sediment on underwater slopes is completely saturated, it is highly subject to disturbance. This is one reason that the continental slopes are relatively bare of sediment in seismically active regions. The effects may be greatest, however, in regions where large earthquakes are rare, because there is more time for sediment to collect on the slope. The 1929 quake on the Grand Banks off Newfoundland caused a vast slump on the continental slope. This slump produced a turbid bottom current that flowed for hundreds of kilometers, at more than 16 kilometers per hour, and broke most of the deep-sea telegraph cables between Europe and the United States.

All of the things that happen to bare sediment can also happen to sediment that is covered by buildings. This seems to be the explanation of the amplifying effect of unconsolidated foundations.

DEGRADING QUALITY OF FOUNDATIONS

The Bible, which was written in earthquake country, recommends building one's house on solid rock, and people often do so when they can. Unfortunately, many cities are sited along river mouths or banks or in flat valleys along natural trade routes, and the sediment in such places tends to be thick and saturated. Thus, the French geologist Haroun Tazieff observed that cities were leveled by the 1960 quake in Chile, whereas the surrounding villages were not. The villages had good foundations on old, thin soil, whereas the cities were on recently deposited alluvium or artificial fill. This is a universal phenomenon, because people tend to move into cities (particularly coastal cities) even faster than they breed. More and more buildings are constructed on worse and worse land. Estuaries are filled, bay margins advance, and houses are not far behind.

The margin of San Francisco Bay is a prime example of an area that has been extensively filled. In a magnificent study of the 1906 earthquake (Lawson, 1908), it was demonstrated that, if the intensity of damage on solid rock was rated as 1.0 in the city, the intensity on natural sand was 2.4–4.4 and that on marsh and man-made land was 4.4–12.0. The maximum effect is not inevitable if the artificial fill is carefully compacted, but apparently it rarely is. In any event, the amplification of damage increases with thickness, even on natural sediment.

Since 1906, encroachment on the bay margins has continued. By this time, there are countless freeway overpasses, several airports, naval yards, commercial docks, and whole subdivisions of houses built on newly filled land, most of it emplaced hydraulically.

What can be done about foundation problems? It will help to educate the contractors and buyers to potential dangers, to strengthen foundation codes, and to increase the number of inspectors responsible for enforcing them. Even so, filled tidelands are always saturated with water, which usually amplifies earthquake damage. Thus, special construction methods may be necessary to enhance margins of safety. Of course, if the population does not grow, the need for building on ground that provides inferior foundations will vanish.

Physical Breaks

Nothing can resist the movement on a fault, but it is quite possible to minimize the total damage that results from a fault breaking the ground. For example, no buildings should be constructed on active faults. It is sad to relate, but this seemingly obvious rule is regularly violated in California as new subdivisions mushroom. Professors of geology leading students along familiar traces of major active faults often find, to their dismay, that someone's living room now covers a bit of it. That such a building is doomed is so close to inevitable that one can only wonder why so many people—contractor, carpenter, zoning commissioner, building inspector, insurance agent, and the hapless purchaser—were ignorant of it.

There are subtler dangers. Less vigorous faults that may not have been active in historical times cannot be separated from the endless network of older faults that must be disregarded lest construction cease entirely. Previously unmapped faults became known when they caused some of the damaging California earthquakes of 1892, 1940, 1952, 1954, and 1971. This suggests that construction is covering faults faster than geologists can map them, now that the obvious ones have been discovered.

In a properly informed world, buildings need not be constructed on active faults, but there is no way to avoid building highways, canals, aqueducts, and other linear structures across them. The best that can be done is to minimize such crossings. It is ironic, with this thought in mind, that Los Angeles has built the greatest aqueduct system in the world across the boundary between the Pacific and American plates. Likewise, California's system of freeways, pipelines, and canals is without parallel in the world, because Los Angeles is almost ringed by a wide desert through which all resources must be imported and all products exported.

Many steps have been taken to minimize the risk. Water and fuel are stored in large reservoirs on the Pacific plate, and electrical lines should not be broken by earthquakes. Thus, there are reserves of major utilities. As to the

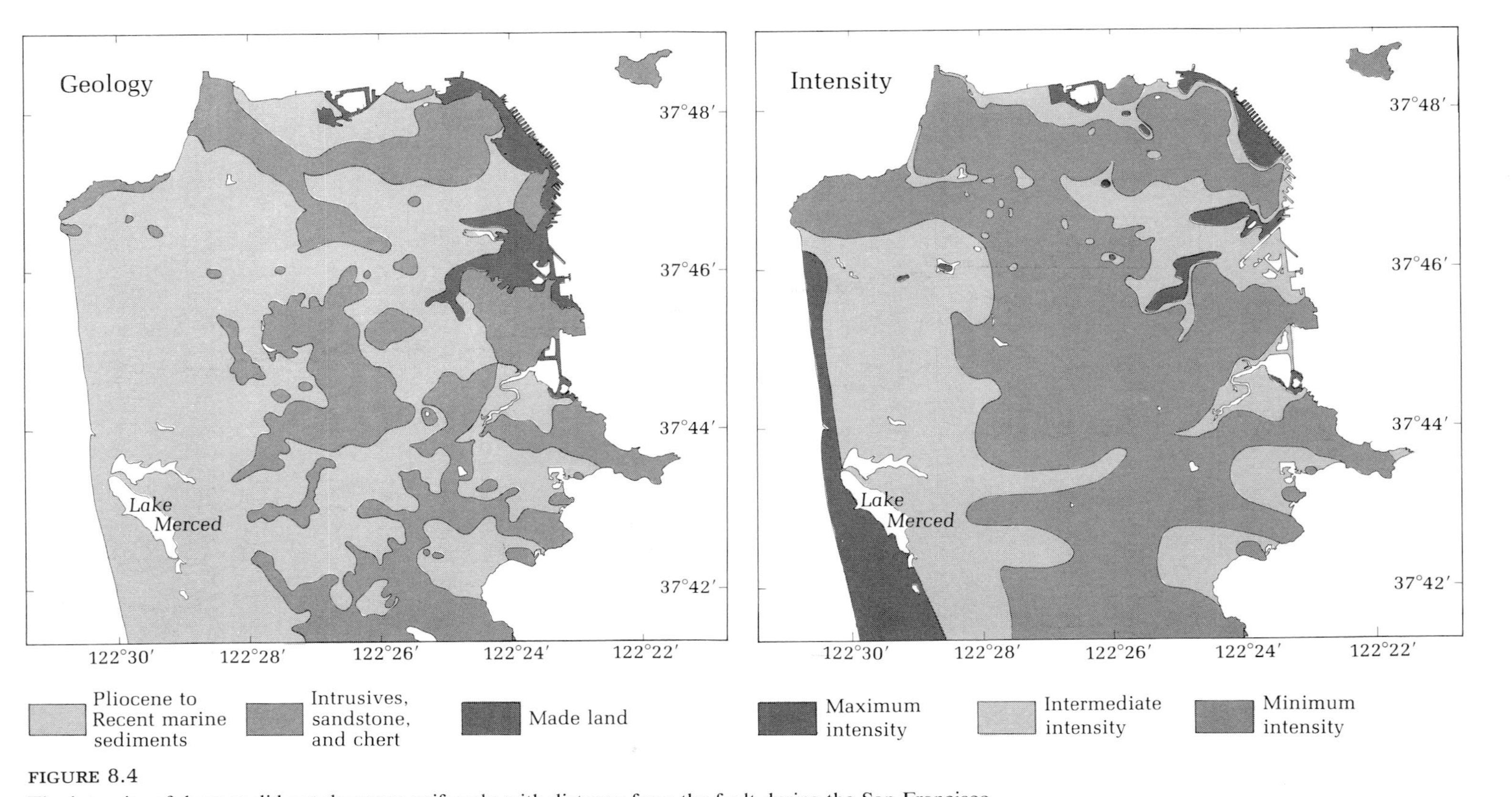

FIGURE 8.4

The intensity of damage did not decrease uniformly with distance from the fault during the San Francisco earthquake of 1906. Foundation effects were more important in San Francisco, and filled land suffered most of all. [After Lawson (1908), courtesy of the Carnegie Institution of Washington.]

roads and canals, they can be repaired much more rapidly at the fault lines than at many other places, such as overpasses and pumping stations, where damage may also occur. Fault-line damage should be endurable outside of the cities, although it would help to keep pipelines above ground where they cross active faults. In a city, however, the creation of a two-meter scarp or three-meter horizontal offset could break a large number of surface and subsurface structures and cause a crisis that could only be countered by a disaster-control organization with adequate resources and response time.

Shaking

Cymbals, bells, and gongs vibrate at different frequencies, depending on their composition, structure, shape, and size. So do buildings, bridges, and towers, when the energy of an earthquake hits them. Resonance is a phenomena in which a solid vibrates in response to, and with the same frequency as, waves (such as sound waves) that impinge upon it. The vibration is *reinforced* because each wave imparts a tiny push to the solid at exactly the right time. A glass tumbler can be destroyed by resonance when a soprano sings exactly the right high note. So can buildings, bridges, and towers, when struck by the various waves of an earthquake.

RESISTANT DESIGN

The resistance of structures to earthquake damage depends on their design and construction. An ideal structure is flexible, tied together as a unit, low, and with a light roof, and it has a natural frequency unlike that of most earthquakes. A log cabin is highly resistant. Few were damaged in the Charleston earthquake of 1886, which shattered great houses and government buildings.

A Charleston family in a one-story wooden house slept right through the earthquake. The residents of such houses in Southern California are not likely to do so, because the floors of their houses are elevated a few feet to keep the termites out. This violates the rule that the weight should be near the ground. If the foundation posts are not an integral bridge from ground to house, the structure may collapse.

In many parts of the world, men have long since eliminated wood as a construction material by burning all the trees, or they have turned to masonry, stone, or concrete because they are more permanent or more fire resistant. Unfortunately, such materials are generally much less resistant to shaking. Unless bonded together with steel rods, walls of these materials crack and shatter with relative ease. The little wooden houses of old Los Angeles are tough, but their brick chimneys topple to the ground.

Because modern buildings with steel frames are both flexible and integrated, they are generally very resistant to quake damage. No modern highrise buildings were damaged structurally by the San Fernando earthquake of 1971. However, that relatively small quake generated little energy in the frequency range, roughly 1–5 seconds, at which such buildings resonate. The buildings did sway, but this had little effect on them, except to incapacitate their elevators. The effects on the terrified occupants of those elevators are easy to imagine.

HIGHWAYS

Limited-access highways span, and are in turn spanned by, countless reinforced-concrete bridges. The cloverleaf intersection of two American freeways is a dazzling display of curving masses of concrete supported by slender pillars. The interstate highway system and most state thruways have been constructed since the last major earthquake. How will they fare when one hits?

The moderate San Fernando earthquake, which had a magnitude of 6.4–6.6, suggests an answer. In a matter of seconds, sixty-two bridges on five freeways were damaged, six of them beyond repair, as sections simply collapsed. Only two people were killed, and the freeways were reopened along detours within days. The results would have been far different had the quake not occurred

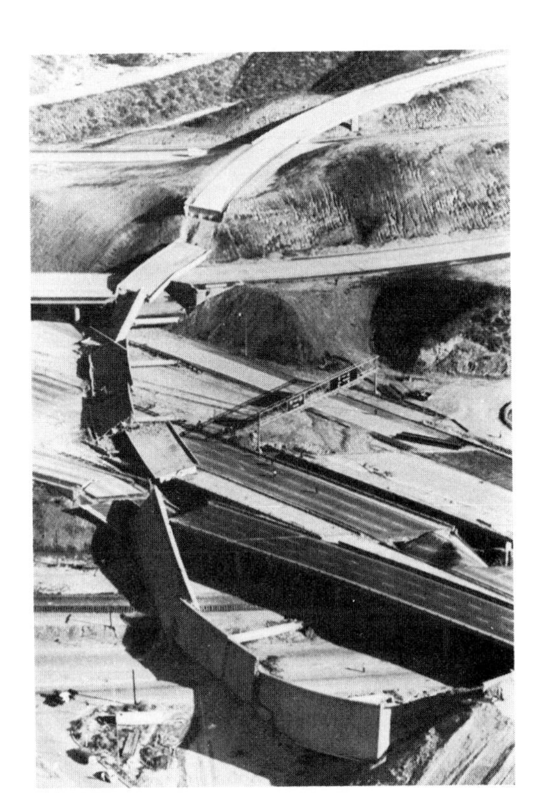

FIGURE 8.5
Collapsed overpasses crossing the main inland highway from Los Angeles to San Francisco after the moderate San Fernando earthquake of 1971. Each collapse stopped traffic on at least two highways. [UPI photo from the *Newhall Signal* (February 26, 1971).]

an hour before sunrise—one of the few times when the freeways are not congested with cars. It is impossible to make these structures invulnerable to earthquakes, but the Division of Highways aims to design them so they will not collapse. Judging by this example, the freeway system may become inoperable in the vicinity of any major earthquake, and collapsing bridges may cut cross-traffic as well.

CONSTRUCTION

Even the mildest seismologist waxes bitter when he discovers shoddy or fraudulent construction in the debris of an earthquake. Thus, Tazieff writes of "poor concrete, improperly or insufficiently reinforced with cut-price steel" and "open scandal" when new government buildings became unusable after the Chilean earthquake of 1960.

In the same vein, the official *Earthquake History of the United States* (Eppley, 1965) refers to the fact that old brick buildings stood up when younger ones were damaged because the "type of bond used in earlier buildings, though no stronger than that used later, was better adapted to prevent concealing a lack of sufficient mortar."

Richter remarked that, in 1948 at Fukui, Japan, and in 1957 in Mexico City, "defects of design, materials, and execution were so grave as to add little to our data for sound construction." He has also pointed out that, in California, "mortar has often been so bad that after several destructive earthquakes it has been profitable to collect bricks from damaged buildings, wash off the remains of mortar with a hose, and sell secondhand bricks practically as good as new."

BUILDING CODES AND INSPECTION

After the Long Beach earthquake of 1933, the junior high school I was attending was closed for several days for examination of cracks. After the San Fernando quake of 1971, the high school from which I had long since graduated was condemned. The latter quake, which was a moderate one, caused the immediate closing of 75 school buildings, and 28 more will soon be vacated. The State of California has the most comprehensive building codes regarding earthquakes in the nation, but its children are better protected by a providence that permits quakes only when the schools are empty. Somehow, it is the government buildings that suffer most.

An engineering study of the 1952 earthquake in Kern County, California, discusses the Cummings Valley School, nominally of modern reinforced concrete. It collapsed because it was "classic in poor design, poor material and poor workmanship." The Olive View Hospital, newly constructed of reinforced concrete, responded to the 1971 San Fernando quake by "pancaking" part of its second floor into its first. Four five-story stairwell wings separated from the main building and three of them toppled over.

These things are not new; they happen in most great quakes. They are noteworthy only as a clear demonstration that building codes may be inadequate and insufficiently enforced even where they take potential earthquake damage into consideration. What of regions in which great quakes have occurred in the past but have been ignored by those who draft and revise the building codes?

Fire Storms and Aggregate Effects

The San Fernando quake was moderate, but it was centered in a heavily populated area. It knocked out part of the freeway system and 50% of the capacity of the main power substation; it cut the water lines in 1400 places; it seriously damaged or destroyed four hospitals; and it barely missed causing the collapse of a large earth-fill dam. A year later, all of these facilities were still being repaired or were abandoned. The area is served by the two Los Angeles aqueducts that cross the San Andreas fault on the way from Owens Valley. If these had been broken, the area would have had only a four-hour reserve water supply.

Many of the greatest disasters associated with earthquakes result from the degrading of public services, such as occurred in the San Fernando Valley. Both damage and loss of life in San Francisco in 1906, as well as in Tokyo in 1923, were largely attributable to fire storms like those caused by massive bombing. Urban organization is increasingly complex and correspondingly vulnerable. It is not unlikely that a major quake will simultaneously knock out most utilities, including limited-access highways, while causing widespread fires from broken gas pipes and electric lines.

Much can be achieved by improvisation, but improvisation is no substitute for planning. San Francisco leads the way. The fire in 1906 was uncontrollable because broken mains drained the whole water system. Now there are valves throughout the system that serve to isolate each neighborhood's water supply in case of broken mains, and each neighborhood has its own reserve supply in cisterns under the streets. Likewise, a disaster plan is ready for implementation, should it be needed. Other cities would be wise to take similar precautions.

Sequences of Quakes

Earthquakes on a given fault usually come in groups. These quakes may include both foreshocks and aftershocks which are defined in relation to one or more main shocks. On occasion, quakes occur in swarms of more or less

BOX 8.2 THE LOS ANGELES COUNTY EARTHQUAKE COMMISSION: CONCLUSIONS AND RECOMMENDATIONS

1. Hazardous Old Buildings

Thousands of pre-1933 buildings in Southern California constitute the most serious threat to public safety because of the probability of their collapse during strong earthquakes in the future. The San Fernando Veterans Administration Hospital buildings that collapsed are an example. Such buildings should be brought up to modern standards of seismic resistance, or they should be demolished. Because of the economic and human consequences of requiring repair or demolition, a phased program is recommended with those buildings that present the greatest hazard—relative to use, location and nature of construction—receiving the first and most urgent attention. This program also might include incentives to help ease the burden. The Commission believes that full compliance by 1980 is an achievable goal.

2. Unsafe Dams

The severe damage sustained by the old earthen dams, which retained the water of the lower and upper Van Norman Reservoirs, very nearly caused a catastrophe. All existing dams in California should be brought up to modern standards of safety or their use restricted.

3. Highway Structures

Present standards of earthquake design for highway bridges and other roadway structures should be revised and improved to conform with the current state of knowledge of earthquake engineering and should provide sufficient resistance to survive very strong shaking.

4. Code Revisions

Building-code provisions that require earthquake-resistant design have been in effect since the 1933 Long Beach earthquake. The greater survivability and increased protection of human life and limb provided in buildings constructed according to these newer code provisions are testimony to their efficacy. Nevertheless, the results of the February 9, 1971, earthquake indicate that further revision of building codes is needed to insure that the degree of damage will not be so great as to be hazardous to life and limb.

5. Facilities Vital in Emergencies

Certain types of structures and facilities, which are particularly important in post-disaster operations, such as hospitals, emergency power installations, emergency operating centers, public safety facilities and essential elements of key communications systems, should be designed and constructed to withstand strong earthquake shaking and yet be able to continue to function. Building codes should be amended to accomplish this.

6. Federal Construction

Federal action—which may include the enactment of legislation—must be taken to require that all construction by federal agencies, whether of new structures or the remodeling of older buildings, comply with local building-code provisions for earthquake resistance where the federal-agency building regulations are silent on the matter, or where such regulations establish a standard that is below the minimum standard of the applicable local code.

7. Schools

While most modern school buildings performed very well during the earthquake, some potentially hazardous damage was sustained by a few of them. Such damage should be studied and the code requirements for earthquake resistance should be revised to eliminate these hazards. The use of old school buildings, which were not designed to resist earthquakes, should be prohibited until such buildings are brought up to modern standards of safety. If they cannot be made safe, they should be vacated immediately, even if classes must be held in tents.

8. Houses

Most typical, modern, one-story, wood-frame houses performed well during the earthquake ground shaking in that no severe hazard was created nor were the major economic losses widespread. Some modern, non-typical houses were severely damaged by earth shaking. Studies should be made of the applicable building-code specifications to work out practical revisions so as to improve the earthquake resistance of such houses.

9. Earthquake Insurance

Earthquake insurance should be made readily available to the homeowner by its inclusion in extended-coverage riders on "standard homeowners insurance" and on "standard fire" policies. It is recommended that all insurance underwriters use this form and that some method for reinsurance against catastrophic loss be provided, perhaps through Federal reinsurance and/or through a change in allowed reserves. Similar earthquake insurance should be made available to owners of small commercial buildings as well. Lending institutions should establish the principle that earthquake insurance—in California—is as essential as fire insurance.

10. Non-Structural Damage

Damage to non-structural building elements, such as partitions, ceilings and windows, and to electrical and mechanical equipment, was sustained in buildings that had adequate strength to resist the earthquake forces. Much of this damage was costly and potentially hazardous. Earthquake-resistant design and construction of such items must be improved.

11. Utilities

The results of the February 9 earthquake indicate that standards for designing and constructing utility systems—electric, water, gas, telephone, waste and sewers—should be reviewed and revised so that future earthquake damage will be within acceptable limits.

12. Instrumentation of Major Structures

The recordings of the strong-motion accelerographs, located in the general Los Angeles area, provided valuable engineering data about the seismic motions of the ground and of structures. These data will lead to improved methods of designing earthquake-resistant structures.

Many of these accelerographs were installed in major structures in the City of Los Angeles as the result of legislation by the Los Angeles City Council. Jurisdictions, which do not require such instrumentation, should make it mandatory within their boundaries. Satisfactory maintenance and logical expansion of the strong-motion accelerograph network in Southern California must be encouraged.

13. Research

Research on earthquakes and their effects should be continued and encouraged, for only through such research can the understanding of earthquakes be increased and the ability to minimize their hazards be improved. All public authorities should cooperate with private and public agencies in collecting and publishing information concerning future earthquake probabilities. These authorities should support field work and studies in geology, geophysics, soil dynamics and physical response of structures to earthquakes.

BOX 8.2 *(continued)*

14. Strong Ground Shaking and Faulting

Strong shaking during an earthquake typically extends over many square miles, while permanent fault dislocation is a very localized phenomenon. Therefore, a vastly greater number of structures will be affected by shaking than by quake damage, and the locations of faults are often unknown or poorly defined. Nevertheless, the building of structures directly across known active fault traces should be avoided whenever possible. Communications, transportation and utility lines should not be run along major active faults for long distances if feasible alternatives exist.

15. Emergency Operations for Earthquakes

Federal, State and local government agencies, the American Red Cross, and non-governmental groups and associations took effective measures to minimize the disastrous effects of the earthquake and recovery was rapid. However, weaknesses were noted in emergency operations that would have been magnified had this been a great earthquake. Governmental agencies performed independently at a time when coordination and team effort would have been mutually helpful. A need clearly was shown for local governments to provide emergency operating centers where information could be pooled and coordination achieved from a single, central location. Since such disasters usually affect many local governments, provision should be made within Los Angeles County for interjurisdictional coordination and exchange of information in the event of an emergency.

Radio communcations systems must be constructed to survive strong ground shaking without loss of function.

On the basis of the February 9, 1971, experience, critical analyses and updating of plans, procedures and measures for coping with the effects of destructive earthquakes should be made. Responsible local officials should apply this experience to preparing for the possibility of an earthquake of very large magnitude.

equal intensity without any particular main shock. It should be emphasized immediately that the term foreshock does not imply *foreknowledge*. It is only in retrospect that a small shock can be identified as an isolated event, or the first of a swarm.

The mere fact that earthquakes tend to occur in groups is useful to know. A single earthquake generates different types of waves, which, because they travel at different speeds, arrive one after another, usually with increasing intensity. In areas near the source, they all arrive within a few seconds. Other quakes in the same group may occur minutes, hours, or even days later. Even though of lesser magnitude than the main quake, they may cause intense damage in weakened structures. They also may intensify the terror of people who may justifiably wonder if the *main* quake has yet occurred. No matter how strong a particular shock may be, there is no certainty that it will be the strongest of its series.

Prediction

USEFUL PREDICTION

It would be very useful if earthquakes could be predicted. Is it possible? Apparently, some people believe so. A group of people recently fled California to escape from a great quake predicted by their leader. The quake, however, didn't occur on schedule. Herein we see the problem. It is evident that plate motions, which generate earthquakes, cannot be stopped. Thus, unless or until we can disconnect the quakes from the motions of tectonic plates, it is safe to predict that they will occur, but it is still impossible to say when.

At present, most predictions merely say that a certain number of quakes of a certain intensity will occur in a specified time in a given area. This is useful for preparing building codes and disaster plans, but for little else. It is likely that meaningful predictions will result from current research. Meanwhile, let us imagine that we can predict exactly when an earthquake will occur but not how big it will be. Experience with the tsunami warning system indicates that, after full alerts for a few minor quakes, few people will pay attention to such earthquake warnings.

Alternatively, let us imagine that we can predict the magnitude of a quake exactly but not predict the time more accurately than to within a week or a month. This would be more useful, but the life of a great city does not cease under the threat of quakes any more than under the threat of bombing. Children might be evacuated, for example, but industry probably would go on. The prediction might be certain; more likely it would only offer odds of two or three to one. Thus, the chance of loss of life by a possible quake would have to be balanced against the certain disruption and the cost of evacuation. Only a perfect prediction will be of much help. Fortunately, the first glimmerings that may lead to prediction have appeared in recent years.

PREDICTORS

The surest way to predict an earthquake is to trigger it yourself when you want it to occur. We shall consider this possibility in the next section. If nature is to continue on its own, we must seek relationships that are useful as predictors, such as the build-up of strain on a fault or the triggering of quakes at times when other phenomena occur.

Moonquakes occur on the largely quiescent moon when it undergoes maximum tidal stresses because of the attraction of the earth and sun. Thus, it is easy to predict them. Presumably, tidal stresses also trigger some earthquakes, but the effect is not important because the earth is too dynamic a body. If the moon and sun have no significant effect, the lack of influence of the planets and stars is evident. Lacking the regularity of astronomical agents, there seems little possibility of ever discovering any rhythms with which quakes naturally occur.

Various means exist for measuring the accumulation of strain along faults for the purpose of assessing the probable maximum intensity of any quake that may occur. The most useful procedure is to make repeated observations in a geodetic net. This determines strain by measuring the distortion of angles and distances. Unfortunately, because the relief of strain may occur either as several small quakes or as one large one, the intensity remains unpredictable. Nonetheless, this monitoring of strain by the Geological Survey and the National Oceanic and Atmospheric Agency offers promise for the future.

Accumulated strain is ultimately released by some trigger, the proverbial "last straw." The region in which Boulder Dam was constructed was one in which the earth was already strained. The load of the deepening waters of Lake Mead triggered many small quakes. Subsequently, the frequency of shocks was correlated with changes in water depth. This does not occur at all reservoirs, only where the region is already strained.

This correlation opens interesting possibilities. If a fault is highly strained, then the occurrence of even a small natural quake may trigger a large shock. In such circumstances, it may be possible to identify foreshocks before a main quake occurs and to issue a useful warning. The other possibility is even more intriguing—namely, the deliberate triggering of earthquakes, to which we now turn.

Amelioration

In the early 1960s, no experienced seismologist had any conviction that man could ever predict or control earthquakes. Then it was shown that man was already inadvertently triggering them, and the possibility opened that he could do it deliberately. The U.S. army Corps of Engineers began to dispose of poisonous waste fluids in 1962 by injecting them into old wells more than two miles deep near Denver, Colorado. Later analysis showed that small earthquakes were triggered by this injection. Likewise, small quakes tend to occur in the vicinity of the Nevada Proving Ground after an underground atomic blast.

What can we hope to do? The natural forces involved are enormous, but the triggering forces are small and within human capability. We may at least hope that we can frequently relieve the accumulating strain. Thus, ideally, the San Andreas fault can be induced to creep steadily instead of making jumps that generate earthquakes. This may be impossible, however, if the jumps result from irregularities in the geometry of the fault. Perhaps large quakes are the normal mode of motion in some parts of the fault. If so, it still would be very useful to be able to trigger the big quakes at desired times rather than to wait for the inevitable. Loss of life could be virtually eliminated and property damage could be minimized.

Just how the lubrication or triggering might be accomplished is unknown. An obvious possibility is to pump water down wells drilled along the fault. Small underground nuclear explosions might provide triggers. Experiments with active faults will be essential to resolve the matter, and the legal and insurance problems involved may be formidable. Nevertheless, the decade 1962–1972 has seen the rise of serious study of earthquake control. Our ignorance is great, but so are the opportunities for important advances.

Summary

1. Earthquakes are the greatest natural destroyers of property and sometimes cause many deaths as well. The cost per quake is increasing in regions where building continues to expand.
2. Earthquakes capable of destroying property have occurred almost everywhere in the United States.
3. Tsunamis are giant waves that are generated mainly by sudden sinking of large areas of the sea floor in subduction zones. The sinking also produces earthquakes, which travel much faster than tsunamis and, thus, provide the basis for a warning system.
4. Destruction by earthquakes is most intense on soft, water-saturated ground of the sort produced by improper artificial fill.
5. The only way to avoid destruction of property resting on active faults is not to put it there.
6. Earthquakes shake structures in ways that are influenced both by the nature of the structures and their resonance with the waves. Poor design and construction have been the cause of much damage.
7. The aggregate effect of a major earthquake in a large city may be largely the result of degrading of the public services, notably fire control, by which a city normally defends itself.
8. Earthquakes occur in groups; the most intense member of a group may be anyone from the first to the last.
9. Few quakes can now be predicted in any useful way, but earth scientists are exploring the possibility of common predictions.
10. Men cause earthquakes without meaning to; thus, it follows that they can cause some of them when and if they want. It is possible that earthquakes can be partially controlled in the future.

Discussion Questions

1. How do earthquakes cause damage?
2. Where in the United States did the largest known earthquake occur?

3. In the Hawaiian Islands, why is Hilo more subject to damage by tsunamis than Honolulu?
4. Why would anyone fail to heed a warning that a tsunami or an earthquake will occur?
5. Why may fence posts pop out of the ground during earthquakes?
6. How can a soprano shatter a glass without touching it? Does the same effect influence the safety of buildings with regard to shaking by earthquakes?
7. Does your city building code take potential earthquake damage into account?
8. If there are foreshocks, why can we not use them to predict earthquakes?
9. Why does the ability to create earthquakes open the possibility of partially controlling them?

References

California Geology, April–May, and November, 1971. [A publication of the California Division of Mines and Geology, Sacramento, CA 95814. This admirable monthly series of publications is highly informative on environmental geology, natural resources, and environmental issues in the state that needs the information most. Subscriptions $2.00.]

Davison, C., 1936. *Great Earthquakes.* London: Thomas Murby. [Written by an established scientist, but open to question where based on newspaper accounts.]

Eppley, R. A., 1965. *Earthquake History of the United States,* Parts 1 and 2. Washington, D.C.: U.S. Government Printing Office. [Part of an authoritative continuing series.]

Iacopi, R., 1964. *Earthquake Country.* Menlo Park, Calif.: Lane. [Very well illustrated tour of the major faults of California and a history of what has happened along them. Realistic and unemotional discussion of social problems.]

Lawson, A. C., chairman, 1908. *The California Earthquake of April 18, 1906* (Report of the State Earthquake Investigation Commission; Carnegie Inst. Wash. Publ. 87). Washington D.C.: Carnegie Institution of Washington.

Los Angeles County Earthquake Commission, 1972. An official report on the San Fernando earthquake. *Eng. Sci.* 35:3, 21.

National Earthquake Information Center, May–June, 1971. Computerized earthquake surveys. *Earthq. Inform. Bull.* 3(3).

Newhall Signal, February 26, 1971. Earthquake (Special Edition).

Richter, C. F., 1958. *Elementary seismology.* San Francisco: W. H. Freeman and Company. [Almost a labor of love for the world-famous creator of the magnitude scale. Conservative and reliable.]

Shepard, F. P., G. A. Macdonald, and D. C. Cox, 1950. The tsunami of April 1, 1946. *Bull. Scripps Inst. Oceanogr.* 5:391–528. [A very detailed study by scientists who were in it.]

Tazieff, H., 1964. *When the Earth Trembles* (translated from the French edition of 1962). New York: Harcourt, Brace and World. [The author studied the effects of the Chilean earthquake of 1960. A very readable and thoughtful analysis of earthquake problems, but wild errors appear here and there.]

VULCANISM

JOHN S. SHELTON

9

VULCANISM AND VOLCANIC HAZARDS

To yawning gulfe of deep Avernus hole.
By that same hole an entrance, dark and bace,
With smoake and sulphur hiding all the place,
Descends to hell.

Edmund Spenser, THE FAERIE QUEEN

When Spartacus encamped his army of ten thousand gladiators in the old extinct crater of Vesuvius, the volcano was more justly a subject of terror to Campania, than it has ever been since the rekindling of its fires.

Charles Lyell, PRINCIPLES OF GEOLOGY

The internal heat of the earth radiates through the crust more or less uniformly or is expended in moving the crustal plates. A tiny fraction, however, moves upward in local plumes or thermal cells, which focus the heat and metamorphose or melt rock. Ice floats because it is less dense than water; but solid silicates, unlike ice, are denser than their liquid forms. Thus, liquid silicates tend to rise through the solid rocks above them and, thereby, to transfer heat rapidly toward the surface. The manifestations of focused internal heat, their social effects, and their hazards are the subject matter of this chapter.

Magma

The liquid source material for igneous rocks is called magma. It is a very complex solution of silicate liquids and gases, often containing suspensions of silicate crystals of various sizes. The common magmas range widely in chemical composition, but they fall roughly into two principal types: The first, basalt, is a dark, dense material with a relatively low viscosity; it is relatively low in silica and rich in calcium, iron, and magnesium. Andesite is a typical member of the second group of rocks, which are rich in silica, lower in iron and magnesium, higher in sodium and potassium, and relatively more viscous.

The gas content of magma generally amounts to only a few percent, but it largely determines the behavior of the emerging lava, the mode of vulcanism, the shape of volcanoes, and whether explosions occur. Magma in the interior of the earth is very hot and under enormous pressure, and gas, as such, does not exist. As the magma rises, however, the pressure decreases until the gases begin to come out of solution and form bubbles. The gas bubbles expand with decreasing pressure. Lava at the surface is at a temperature of about 1000°C, but under only one atmosphere of pressure; and, if it is very fluid, the bubbles may simply escape as they do from boiling water. This has been occurring in the incandescent lava lake in the crater of Nyiragongo in Africa for more than fifty years. If the lava is viscous, however, the gas may be unable to escape passively. The gas may be trapped in spherical or deformed bubbles in the rock if the lava cools rapidly, or it may cause intense explosions, bursting the lava into tiny fragments called ash or cinders that are blown violently into the atmosphere.

The gas emerging at the surface is composed chiefly of hydrogen, oxygen, carbon, and sulphur in compounds whose composition depends on varying chemical equilibria. Water vapor and either sulphur dioxide or hydrogen sulphide are almost always present. Some of these materials come from the deep interior of the earth, some from interaction of the magma or its gases with the sedimentary rocks near the surface, and some by vaporization of ground water. The proportions from each source are difficult to determine, and they may vary with each volcanic eruption. However, isotopic analysis of the water from volcanoes and hot springs shows that all of it is merely recycled ground water which has fallen as rain. The measurements allow for a small error, so perhaps as much as 1% of juvenile water may be hidden within the meteoric water, but no more.

Other minor elements in the hot gas come mainly from the magma itself. They include argon, boron, nitrogen, chlorine, and fluorine. These last two elements appear, respectively, as hydrochloric acid and the extremely corrosive hydrofluoric acid, which etches glass. These gases are detected as they emerge from volcanic vents and pipes drilled to tap volcanic steam; as the gases cool, such metals as copper, iron, mercury, and zinc may be deposited. In addition, concentrations of barium, iron, and manganese occur in the sediments at the crests of midocean ridges. Presumably, the opening cracks of the spreading centers emit gases rich in these materials.

An enormous quantity of gas emerges in a volcanic eruption, but it is difficult to determine precisely how much. A suggestion of the magnitudes involved can be derived from study of the ash and other materials deposited in the Valley

FIGURE 9.1
Mauna Loa in Hawaii is the vast low mountain in the distance. It is shaped like an inverted saucer because it has been built by fluid lava that flows far from the center of eruption. The little cones in the foreground are built of ash that erupts in a later stage of volcanic development. They are on the flank of the older volcano, Mauna Kea. [U.S. Air Force photo.]

of Ten Thousand Smokes in Alaska by the eruption of Katmai Volcano in 1912: For some time, this volcano emitted water at the rate of 23,000,000 liters per second, and it had an annual output of more than a million tons of hydrochloric acid and roughly 200,000 tons of hydrofluoric acid.

Types of Thermal Activity

Under "thermal activity" we may group all phenomena generated by the local surficial concentration of the internal heat of the earth. The dominant phenomenon by far is the filling of the opening cracks of spreading centers by basaltic dikes and lava flows. This results in the emplacement of about 4 cubic kilometers of new rock every year, which is roughly four times as great as the known output of all other types of igneous rock.

Next in importance with regard to volume are the intrusive rocks that cooled at depth but are exposed to view by deep erosion. The largest known masses of such intrusive rocks, which are called batholiths, are roughly 80 kilometers by 650 kilometers and are found in the western United States—which is a typical region for such features. However, if Warren Hamilton is correct, these are merely the fragments produced by faulting in a zone of intrusion, once parallel to the subduction zone between the American and Farallon plates, that was more than 1600 kilometers long. This vast zone was

FIGURE 9.2

Vulcanism was widespread in the western United States in relatively recent times. This illustration shows Humphreys Peak in the distance. Just beyond it is Flagstaff, Arizona. Ash and lava have spread over a desert surface of older sedimentary rocks. [From Shelton, *Geology Illustrated*, W. H. Freeman and Company, copyright © 1965.]

not created in a single event; instead, hundreds of individual intrusions were emplaced over a period of perhaps 20 million years, beginning roughly 100 million years ago. This timing illustrates a fundamental difference between batholith intrusion and the essentially continuous vulcanism of spreading centers. Even during the relatively brief period of its emplacement, the volume of intrusive igneous rock per unit of time, assuming a thickness of 8 kilometers, was only 1.5% of that at spreading centers.

Small intrusions are unimportant with regard to the gross upward movement of magma and internal heat. They may be the dominant igneous phenomenon in many localities, however, and may be of particular social importance because they may be feeders for volcanoes. Except in Iceland, people are little influenced by the vulcanism at spreading centers. Likewise, the intrusion of a batholith at the depth should give no cause for alarm. Local continental vulcanism above small intrusions is another matter, however, because people may be directly affected.

If magma reaches the surface it constructs features of various sizes and shapes. Most common are the elongate abyssal hills, created at spreading centers. Next most voluminous are the great basalt plateaus that occur within plates or at very slowly opening spreading centers, as in Africa. These have been few in number during geological time, but they are mighty in their effects. The Columbia River Plateau of the northwestern United States is an example. Very fluid basalt flowed out of long cracks and completely buried pre-existing hills and valleys during the last 30 million years. In places, the lava flows have a total thickness of more than 2 kilometers. Accumulations of this sort form plateaus, because they are relatively flat on top and are resistant to erosion.

Perhaps equal in importance but more difficult to recognize, are vast sheets of volcanic ash deposits that were so hot when deposited that their constituent fragments were welded together. These are the silicic (or rhyolitic) equivalent, in many ways, of the plateau basalts. Both accumulate in vast floods that fill valleys and bury hills. The difference is that fluid basalt flows as a normal liquid from which most gas has escaped; rhyolite flows as a solid suspension in a hot gas, which is escaping from the viscous incandescent particles as they flow.

Individual volcanoes have forms that reflect the viscosity and gas content of lava. Fluid basalt flowing in random directions from a single conduit tends to construct a shield volcano like an inverted saucer. Andesite and rhyolite, when they explode into individual particles of ash, tend to build steep-sided cinder cones whose shape may be affected by the prevailing winds. Welded ash sheets, however, form plains. The beautiful stratovolcanoes, such as Fuji in Japan, which have graceful curving slopes and are almost circular, are constructed of interbedded lavas and ash.

Hot water emerges from different types of vents in thermal regions. These include hot springs, geysers, and pools of boiling mud. The water, apparently, is all from the surface, but the heat is from the depths. The thermal region in Yellowstone National Park is an example of the release of heat from the earth without modern vulcanism.

Distribution of Active Volcanoes

By the restricted definition, an active volcano is one that has been observed to erupt, but this excludes many with thermal activity and recent lava flows that, by chance, were not observed. For most purposes, it is useful to call such volcanoes active on the grounds that they are probably dormant rather than dead, or else so recently dead as to be good indicators of present activity. By either definition, most active volcanoes are associated with plate boundaries. The type of vulcanism is even determined by the type of plate boundary: fluid basalt pours out at spreading centers, and rhyolitic and andesitic flows and granitic intrusions are associated with the plunging lithosphere of subduction zones.

Intraplate igneous activity, although quantitatively trivial, may be important as a manifestation of the rising mantle plumes that Jason Morgan proposed as the driving force for plate motions. This activity may occur wherever there are plumes—which, as yet, do not appear to be in an organized pattern. Vulcanism may also occur within plates for reasons that are still unknown. The character of intraplate igneous activity seems to be dominated by the crust in which it occurs. If it occurs in a continent, the rocks are mainly silicic; if in an ocean basin, they are basalts, although with a slightly different chemical composition than those at spreading centers.

The Hawaiian Volcanoes

The Hawaiian volcanoes in Hawaii National Park are the site of a volcano observatory, and they have been intensively studied by numerous techniques, both on land and at sea. These volcanoes are admirable sources of information, both about oceanic vulcanism and about fluid-basalt vulcanism (which need not occur together). It should be remembered that vulcanism in the Hawaiian Islands is very intense and that it may be occurring over a hot spot in the mantle. This point, however, is not essential to an understanding of the following paragraphs. For some reason, volcanic rock pours copiously from an initial opening in the sea floor at a depth of 5000 meters, and at that time this story begins.

The lava flows forward in the form of pillows that bud outward from a crust that forms a protective thermal insulator from cold water. Even so, chilling prevents these pillows from spreading very far, and they pile up to form a

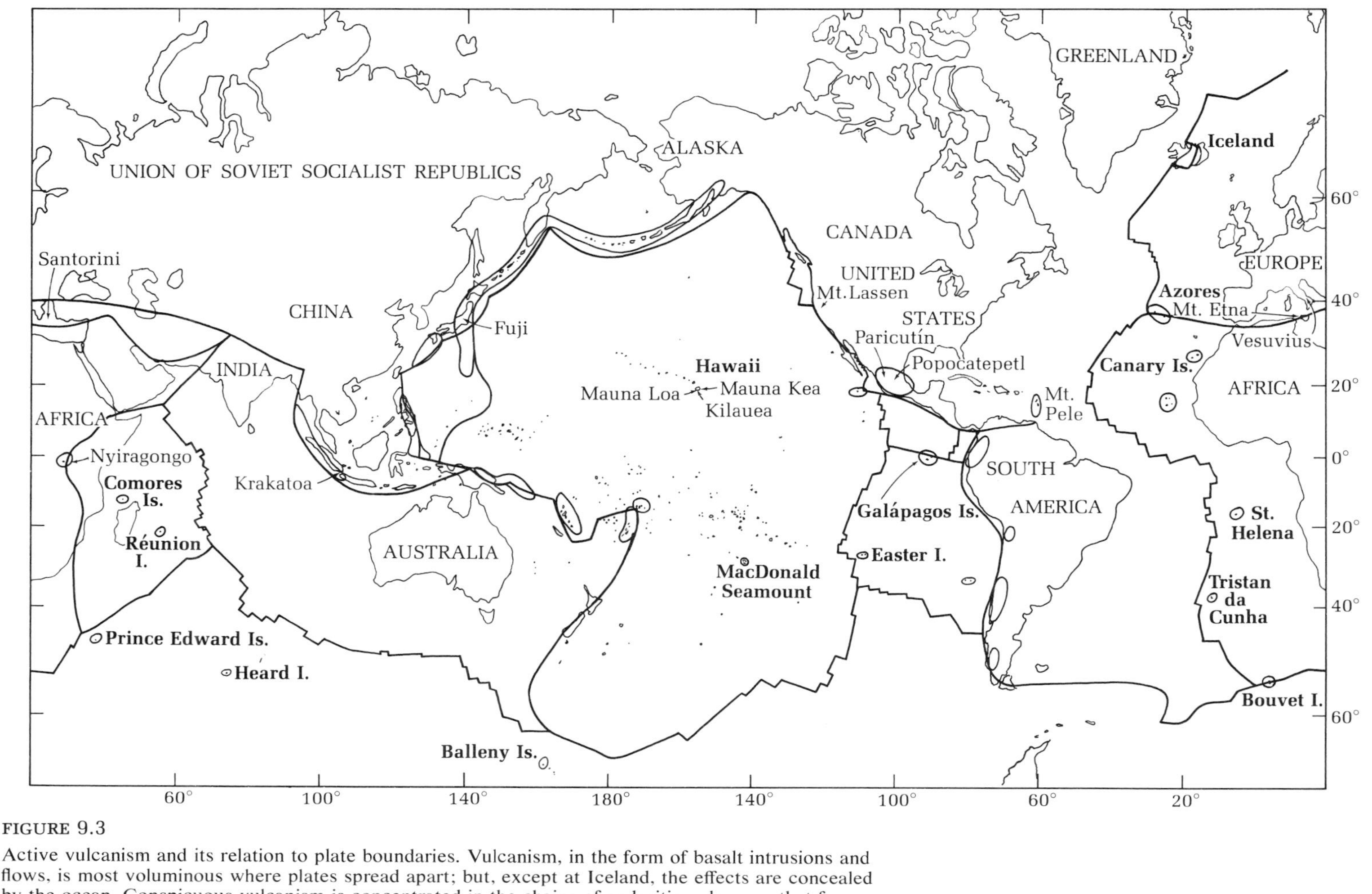

FIGURE 9.3

Active vulcanism and its relation to plate boundaries. Vulcanism, in the form of basalt intrusions and flows, is most voluminous where plates spread apart; but, except at Iceland, the effects are concealed by the ocean. Conspicuous vulcanism is concentrated in the chains of andesitic volcanoes that form over subduction zones where plates come together. Plates move over hot spots in the mantle, which produces volcanoes such as the Hawaiian Islands. The hot spots (according to Morgan, 1972) are labeled in boldface type.

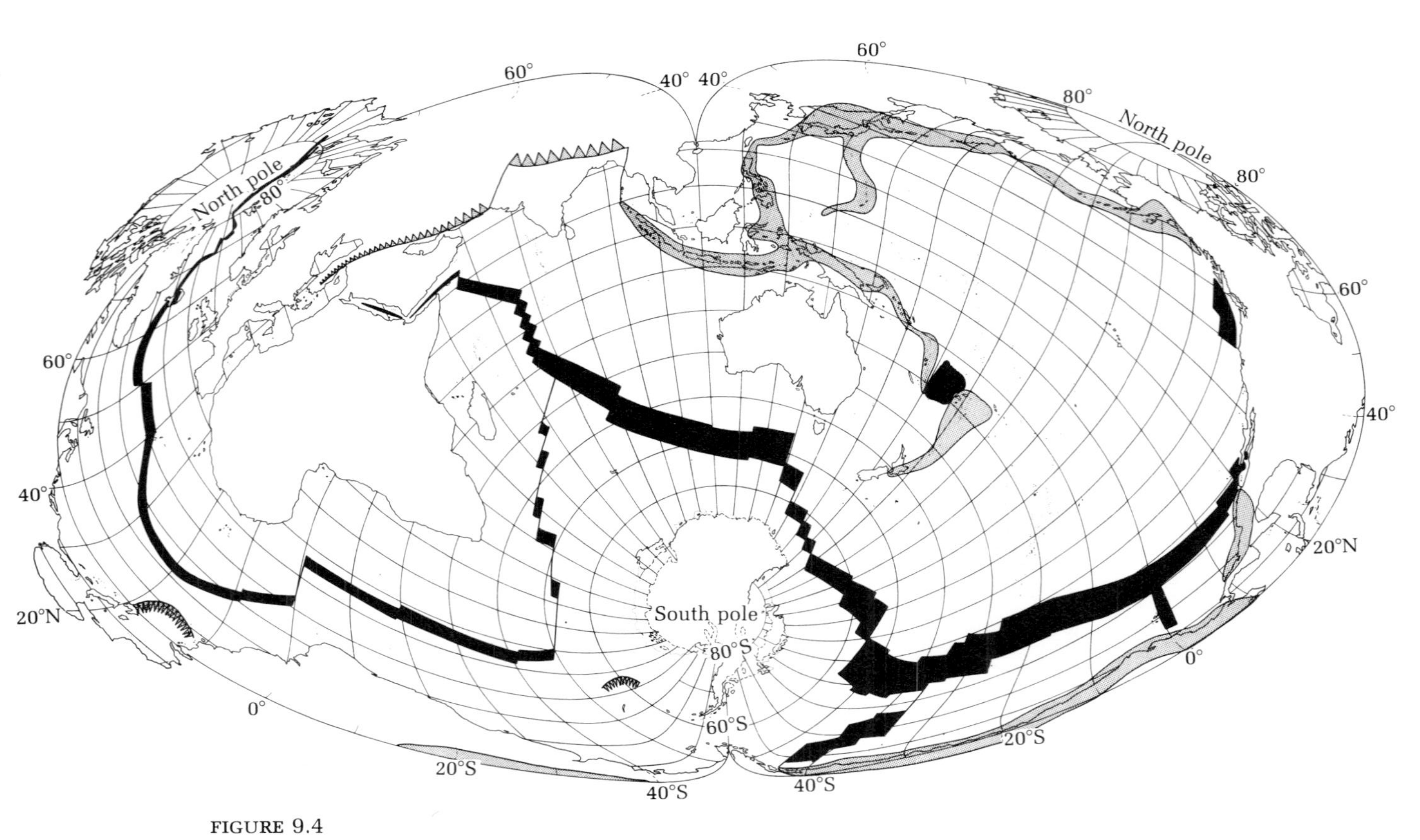

FIGURE 9.4

Crust created during the last 10 million years by the spreading sea floor being filled with basaltic lava. This projection is such that each area on the surface is proportional to an area on the globe. Thus, it is immediately apparent that most vulcanism and crust formation have occurred in the South Pacific and Indian Ocean basins. The lightly shaded area shows the distribution of deep earthquakes and, presumably, shows where the crust is plunging. The area of plunging crust is about equal to that of the newly created crust, which suggests that the earth is not expanding. [From Menard (1971).]

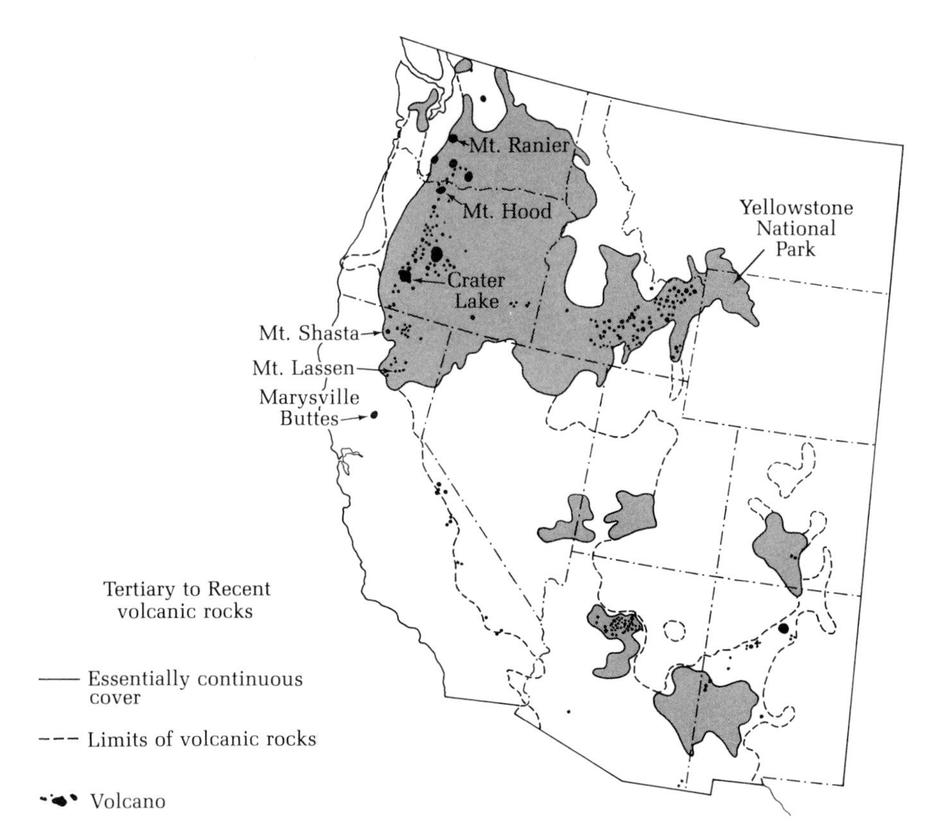

FIGURE 9.5

Areas of Tertiary and Recent vulcanism in the western United States. Vulcanism has not occurred in some areas for millions of years, but other areas have been active in historical times. Eruptions in the immediate future probably will occur near where they have in the past, but the time and place cannot now be predicted. [After *Tectonic Map of U.S.A.*, 1960, U.S. Geological Survey and the American Association of Petroleum Geologists.]

steep-sided volcano. Gas tends to remain in the lava because of the great confining pressure. It is for this reason that deep submarine vulcanism is seldom detectable at the surface.

The volcano exerts a load on the crust; consequently, it sinks somewhat as it grows upward, The effect would be greater, were it not for the buoying effect of the sea water. This may account for the fact that individual submarine volcanoes are rarely surrounded by moats, which might be expected as a consequence of sinking.

When the peak grows so high that the top is only about 300 meters below the water's surface, the style of eruption begins to change. The gas pressure begins to prevail over the hydrostatic pressure, and the lava tends to be fragmented into ash. This spreads as a mantle over the fundamental pile of pillow lavas, making it broader and reducing the slope at the bottom. When the volcano reaches the surface, waves and streams tend to distribute the ash

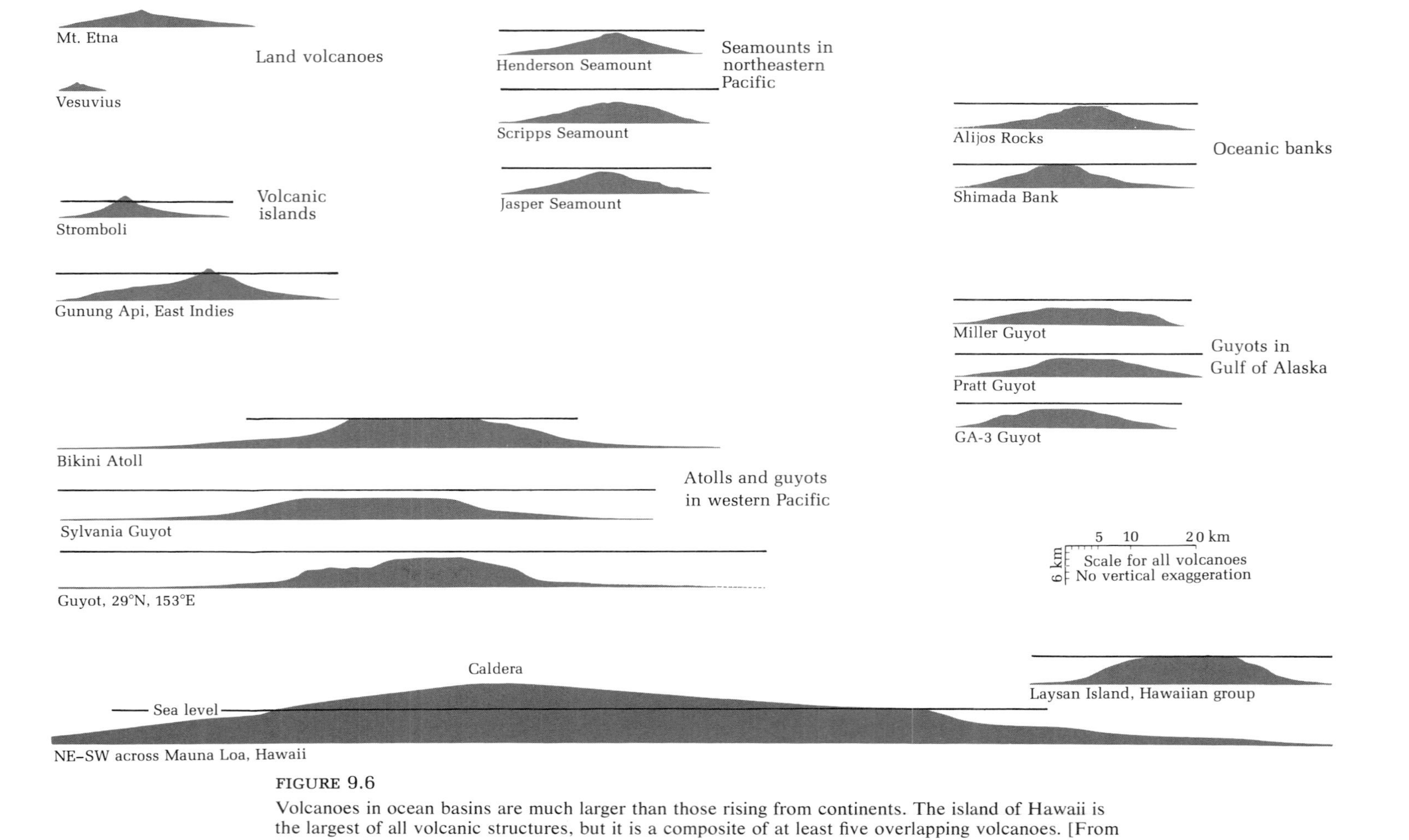

FIGURE 9.6

Volcanoes in ocean basins are much larger than those rising from continents. The island of Hawaii is the largest of all volcanic structures, but it is a composite of at least five overlapping volcanoes. [From Menard (1964).]

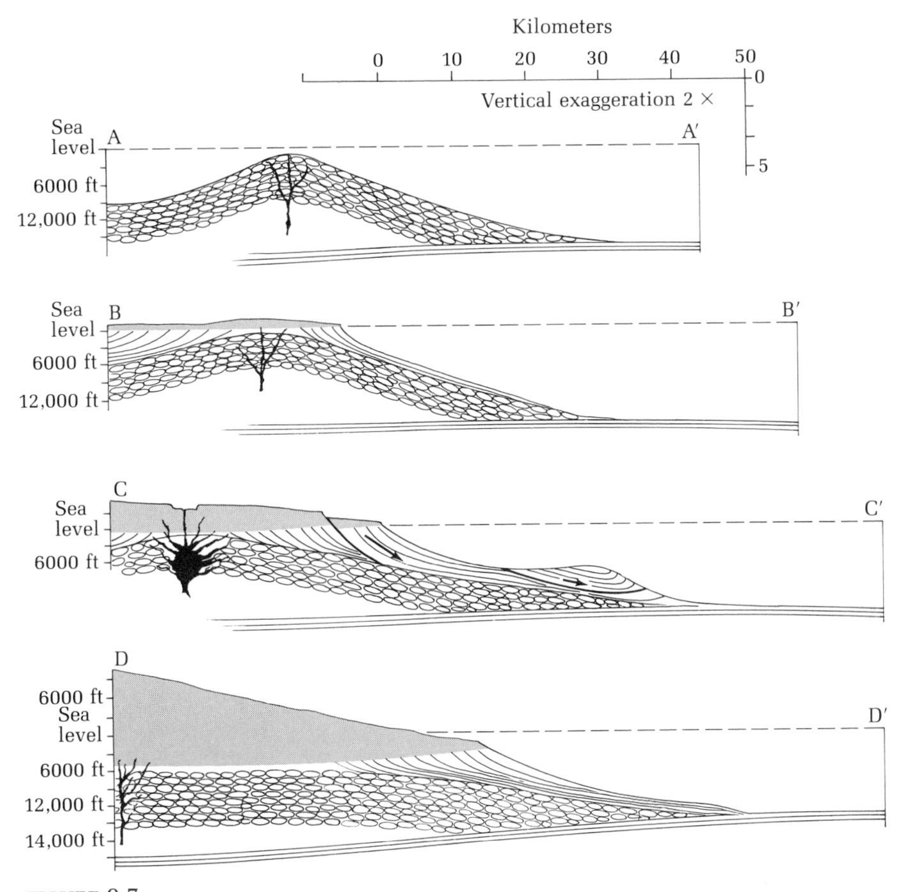

FIGURE 9.7
Cross sections of the island of Hawaii based on submarine dredging and photography: *ovals*, submarine pillow lavas; *white layers*, ash formed near sea level; *shaded areas*, a capping consisting of the thin flows of basalt visible to the surface observer. Presumably, the different sections can be taken as examples of a historic sequence of development for a volcanic island. [From Moore and Fiske (1969), courtesy the Geological Society of America.]

into a broad, shallow submarine bank. After repeatedly rising above sea level and being eroded below, the volcanic pile becomes wide enough to reduce wave action, and fluid lava flows begin to form a protective carapace over the ash. This structure may be undermined periodically by slumping of the island margins, particularly along bedding planes in the steeply dipping ash. Finally, the fluid lavas accumulate to build great shield volcanoes. In this way, the island of Hawaii was formed.

Part of the growth of the volcanic edifice is accomplished by intrusion of dikes and sills. The amount of swelling of Kilauea, one of the active Hawaiian volcanoes, indicates the average flow of magma into an underlying storage chamber, and only 60% of it appears as flows at the surface. Presumably, much of the other 40% is permanently emplaced as intrusions. Alexander McBirney

FIGURE 9.8
Pillow lava. This photograph, which was taken at a depth of 1900 meters, shows the sacklike pillows that form when lava flows underwater. [Photo by Fred Dixon, 1964, courtesy Scripps Institution of Oceanography.]

of the University of Oregon has observed that molten basalt is denser than pillow lavas and ash and, thus, tends to cool as deep intrusions rather than to rise to the surface. Consequently, many submarine volcanoes may contain a dense core of interlaced intrusions through which subsequent ones may flow.

A volcano has a circular summit pit like the depression at the top of a large candle. This pit is the source of the lava that builds it higher. A volcano is a complicated structure, however, that is constantly weakened by swelling as magma is intruded from below. Consequently, the molten rock, seeking an outlet, may be diverted into a lateral crack from which it escapes without ever reaching the top. This produces a flank eruption, which may quickly drain a large magma-storage chamber. Among the Hawaiian volcanoes, Kilauea has more summit eruptions than flank eruptions, and Mauna Loa, which is much higher, has most of its eruptions on the flanks. Both volcanoes, however, tend to have their *largest* eruptions from the flanks.

Calderas are enormous circular or oval pits that occur on the summits of volcanoes. In the Hawaiian Islands, they are 3–20 kilometers across, have vertical walls, and tend to form in lines. They form by collapse when, as a consequence of the withdrawal of lava in flank eruptions, the summit loses support. Not all Hawaiian volcanoes have calderas, but they commonly form in the latter stages of growth as flank flows become more common.

Calderas are hundreds of meters deep. Once they form, only the most voluminous summit eruptions can fill the vast bowl and spread over the side slopes. The caldera, thus, tends to pond thick massive lava beds, which may remain as massive plateaus when the volcano dies and erosion wears away the less resistant flanks.

In the final stage of growth, the copious basaltic vulcanism ceases, and relatively minor volumes of more silicic (and, thus, more viscous) lavas and ash are extruded. These cap the knobby summit of Mauna Kea, which appears quite different from the smooth profile of Mauna Loa, a volcano that is still in the basalt stage.

Ash formed by internal gas explosions rarely accompanies fluid basaltic vulcanism. However, Hawaiian eruptions often feature firefountains, in which droplets of molten lava are hurled 300 meters into the air. The chilled droplets may accumulate in cinder cones around the orifice or they may be blown mainly to one side by the trade winds. The lava that flows from the volcano generally has a temperature of about 1000°C, and it moves at speeds of several kilometers per hour. Hawaiians have a special name, pahoehoe, for the rocks, resembling chilled fudge or tar, that form from these flows. However, if lava continues down slope, cooling and losing gases, its character changes to a blocky clinker that they call aa. When a pahoehoe flow reaches the sea, it may enter passively and begin to extrude pillows underwater.

Alternatively, a pahoehoe flow may heat the water sufficiently to produce steam explosions; so may an aa flow. This phenomenon builds littoral ash cones, such as the one that forms the foundation for a hotel near Lahaina on the island of Maui.

GROUPS OF VOLCANOES

The island of Hawaii comprises five volcanoes, of which two are still active, and most of the other islands are made up of several. This tendency to cluster is normal, and it is the cause of many characteristic features of large oceanic volcanoes. The islands, for example, rise from vast submarine ridges that result from ponding of overlapping lava flows from individual volcanoes.

Groups of oceanic volcanoes go through a typical cycle of development, whether they form over hot spots in the mantle or carry their magma source with them as the crust drifts. As they grow very large (much larger than any continental volcano) and overlap in ridges, they form unsupportable loads on the crust and begin to sink. This depresses the surrounding crust, and groups of volcanoes commonly are encircled by moats as a consequence. The crust does not wholly lack strength, and some of the stress is distributed to a wider region that is bowed into a broad arch encircling the moat. These structures extend more than 200 kilometers from the volcanic islands that produce them.

The moat and arch are mappable around the actively growing Hawaiian

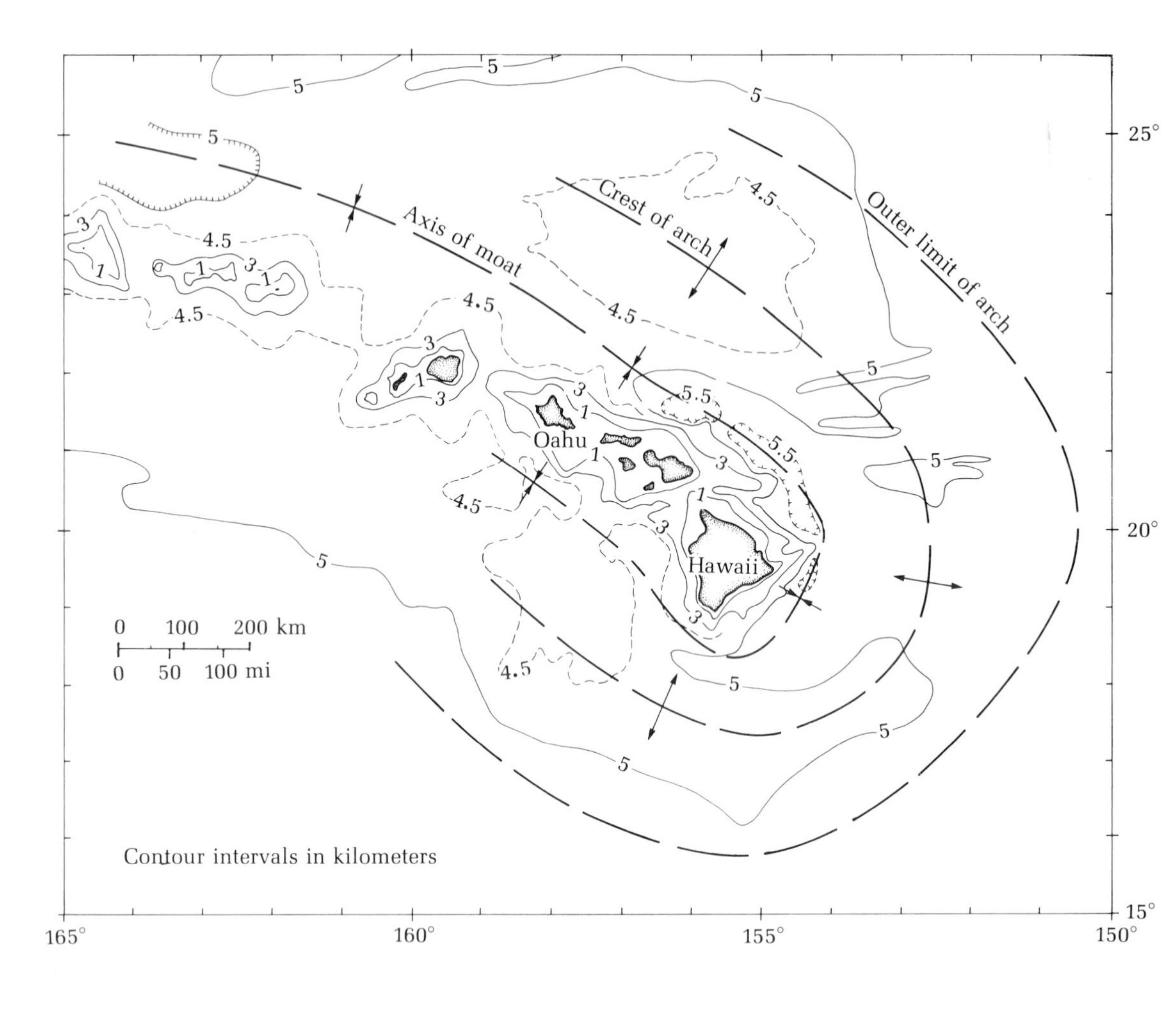

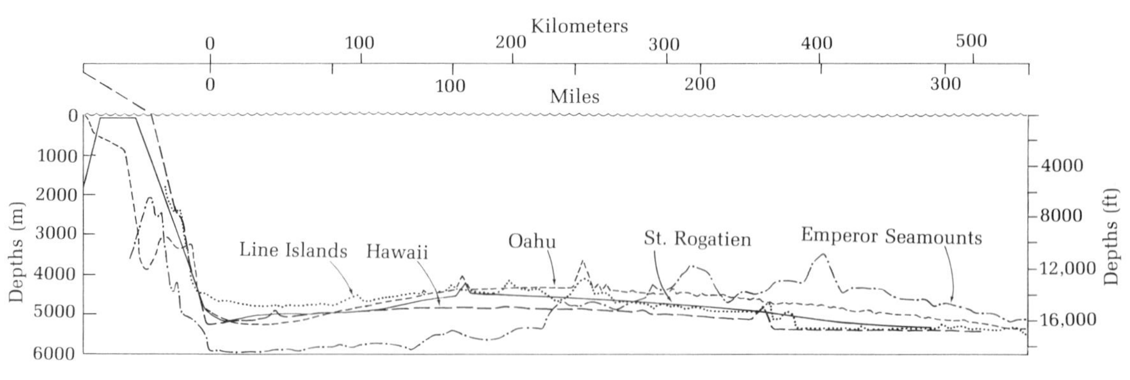

FIGURE 9.9

Above, a chart of the Hawaiian Islands showing the moat that partially encircles the group and the arch that partially encircles the moat. *Below,* profiles of the sea floor near various Pacific island groups. The moats and arches are rather similar around many such island groups. [After Menard (1964).]

Islands. Similar features appear to exist around the Samoan and Marquesan islands, but they are buried under archipelagic aprons, which are thought to be intermixtures of abyssal lava flows and erosion products from the islands, with the former predominating. With the formation of the aprons, the development of the group seems to end. It drifts passively with the crust (unless it encounters another hot spot) until, in the course of time, it is engulfed in an oceanic trench and rejoins the mantle.

Ash Emitters

The foundations of many ash-emitting volcanoes are solid, low, basaltic shield volcanoes, but the upper parts are made of layers of ash or interbedded ash and lava flows. These are relatively weak and subject to faulting and collapse when the volcano swells as magma intrudes it. As a consequence, flank flows are very common and summit flows are relatively rare. The flank flows are loaded with gas, however, and, as they emerge, they tend to explode and build small secondary cinder cones on the flanks of the main peak. Swelling produces radial cracks that divide the volcano into pie-shaped sectors. In the grandest form of collapse, a whole sector may slump and open up the core of the volcano.

Ash emitters occur either near sea level or higher on continents in a low-pressure environment that favors explosive expansion of the gas in the magma. However, the viscosity of the magma is so great that bubbles of gas have difficulty in escaping. Some magmas are so viscous that they hardly flow at all. Instead, they are intruded as massive plugs that rise above ground level as bulbous domes or even as spires 30–100 meters high. Such material tends to clog the vents of a volcano, thus trapping the gas until sufficient pressure builds up to power a great explosion that blows out the plug. Any crack in the structure will trigger the release of the gas, which boils out carrying the liquid with it, just as gas bubbles may carry the liquid out of a bottle of warm beer when it is opened.

CALDERAS

Calderas form on ash emitters as well as on shield volcanoes, but they originate on the former by a combination of explosions and collapse, instead of just by collapse. Indeed, all the factors—weak structure, formation of viscous plugs, prevalence of flank flows, tendency toward explosions—favor caldera formation at some stage in the development of large ash emitters.

Crater Lake in Oregon, a widely known example of a caldera, has been studied in great detail. Some aspects of its history are still obscure, but Howell Williams, an American leader in vulcanology, has reconstructed the main cataclysm. A 3700-meter volcanic peak, Mt. Mazama, existed at the site 6500 years ago. The magma chamber was full, but it emptied during the emission of enormous volumes of pumice and ash in glowing avalanches. The eruption of 40 cubic kilometers of material was too rapid for the magma to be replaced from below, and the peak, unsupported, collapsed along conical cracks into the magma chamber. After the eruption stopped, the caldera filled with water, and subsequent minor eruptions built Wizard Island, a small cone rising above the surface of the lake.

VOLCANIC ASH

The feared but familiar mushroom-shaped cloud of a nuclear explosion is an excellent copy of the cloud of ash and steam that rises from a major volcanic eruption, except that the latter lasts much longer. The reason for the similarity is simply that sufficient energy is available, in either case, to punch a mass upward 30,000 meters, or to a level where it spreads laterally in a mushroom. The work expended by the eruption of Krakatoa has been estimated to be roughly equivalent to a hundred nominal atomic bombs, or 2,000,000 tons of TNT. The principal differences in appearance derive from the fact that the

FIGURE 9.10

Crater Lake, Oregon, with Wizard Island rising from the lake. The caldera has the very steep sides and regular form characteristic of collapse features. [From Shelton, *Geology Illustrated,* W. H. Freeman and Company, copyright © 1966.]

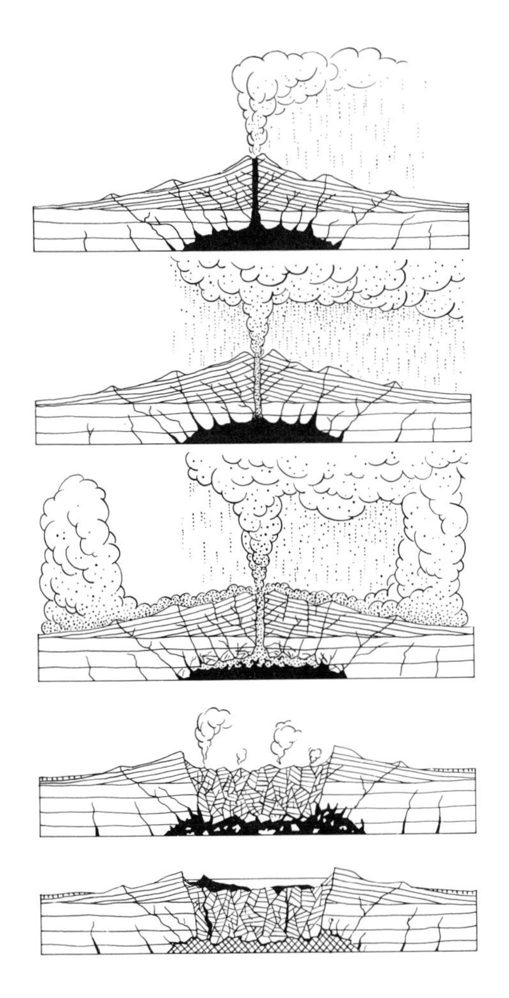

FIGURE 9.11
The presumed history of Crater Lake: An eruption begins and the magma chamber is full; when the chamber is partially emptied by explosions, the volcano collapses; a lake fills the caldera, and the cinder cone, Wizard Island, is formed by minor eruptions. [From Williams, "Volcanoes," copyright © 1951 by Scientific American, Inc. (All rights reserved).]

nuclear explosion is a single event that follows the almost instantaneous generation of a temperature of more than 1,000,000°C, whereas the volcanic one is a series of superimposed events generated at 1000–2000°C. Secondary differences derive from the fact that the solids in the nuclear blast are relatively inert (except for radioactivity), while the volcanic ones contain large quantities of expanding gas that generates constant small explosions as the cloud rises.

In high-speed photographs of nuclear tests we can see phenomena that are obscured in the overlapping eruptions. The ash hurled into the air consists of particles of all sizes from blocks that are meters in diameter down to dust as fine as powder. These are sorted out by gravity, with the coarsest falling near the crater. In nuclear explosions, a similar fall of solids generates a base surge, which spreads radially from the outer edge of the base of the mushroom cloud. Eruptions also generate base surges that spread coarse material down the slopes of volcanoes. The finer ash and dust are carried much more widely in the mushroom top of the cloud and may fall out as layers centimeters deep scores of kilometers from the volcano.

The finest dust may circulate in the upper atmosphere for years and travel around the earth repeatedly before it settles. Dust in the upper atmosphere scatters sunlight and produces sunrises and sunsets of surpassing beauty, so it is relatively easy to map the movement of dust clouds just from local newspaper reports. The Royal Society of London made a thorough study of the 1883 eruption of Krakatoa, and thereby mapped the global circulation of volcanic dust for the first time. At one time, the skies were so bright in Poughkeepsie, New York, that the fire department was called out to quench the celestial fires. At the opposite extreme, the dust sufficiently obscured the sun at the observatory at Montpellier, France, to reduce the average intensity of solar radiation by 10% for four years. A similar effect is observed after most large eruptions, and the suggestion that it is a factor in climatic change will be discussed in Chapter 11.

GLOWING AVALANCHES

> There was a tremendous explosion about 7:45 soon after we got in. The mountain was blown to pieces. There was no warning. The side of the volcano was ripped out, and there hurled straight toward us a solid wall of flame. It sounded like a thousand cannon. The wave of fire was on us and over us like a lightning flash. It was like a hurricane of fire, which rolled in mass straight down on St. Pierre and the shipping. The town vanished before our eyes, and then the air grew stifling hot and we were in the thick of it. Wherever the mass of fire struck the sea, the water boiled and sent up great clouds of steam. I saved my life by running to my stateroom and burying myself in the bedding. The blast of fire from the volcano lasted only for a few minutes.

Assistant Purser Thompson, the author of the preceding paragraph (quoted in Leet, 1948), was fortunate that his ship was in the harbor and that he was somewhat protected, therefore, from the glowing volcanic avalanche that had just killed all but two of the 30,000 people who lived in St. Pierre, Martinique, called "the Paris of the West Indies," before nearby Mt. Pelée erupted in 1902. The temperature of the avalanche was 600–650°C, sufficient to soften stacks of glass plates and to fuse metal cutlery into grotesque masses.

This calamity introduced the glowing avalanche or glowing cloud (*nuée ardente*, in French) to the world. Subsequent study has shown that these avalanches are very common phenomena, but that they had escaped notice because of their transience. They flow down the sides of volcanoes at speeds of 80–180 kilometers per hour or more, carrying boulders as big as a house like wood chips in a torrent and leveling all that they encounter.

How do glowing avalanches work? Because they flow down valleys and are deflected by hills, they are clearly a form of bottom-seeking density current. Hot as they are, why do they not rise like most mushroom-shaped clouds from volcanic eruptions? Careful observation has shown that incandescent, viscous lava spills out of a crater while still highly charged with gas. The weight of the liquid and solid ash carries the mass downward while the gas continues to expand, explode, and give it mobility. The glowing avalanche is a solid–liquid suspension that flows freely on and in a cushion of self-contained gas and compressed air. During its flow—and after it comes to a halt, its potential energy expended—the hot gas rises from the avalanche to form a great curtain of smoke and steam, which tends to obscure the flow itself. Thus, the nature of glowing avalanches went undiscovered for years.

LAHARS AND GLACIER BURSTS

Lahars are mudflows made of volcanic ash, and they are particularly apt to form during an eruption in a crater lake. This happened in Java in 1919, when the volcano Kelut exploded its caldera after 18 quiet years. About 30,000,000 cubic meters of water mixed with ash, suddenly released, flowed down the radial valleys of the volcano to engulf villages and kill 5500 people. Less catastrophic lahars may be generated when the voluminous steam of a volcanic cloud condenses in an intense local rain that mixes with loose and unstable ash.

Somewhat related phenomena are glacier bursts (a translation of the Icelandic *jökulhlaup*), which are torrents of unbelievable violence that result when a volcano erupts under a glacier (or when an ice-dammed lake is drained). A lake of hot water melts the glacier until it reaches the edge of the ice and collapses it. The maximum discharge may be twenty times that of the Amazon River as an irresistible wall of water loaded with sediment and debris moves to the sea. Similar but smaller glacier bursts occur in the high Andes, and probably occurred much more commonly during the Pleistocene glacial ages.

Initiation and Persistence

What determines when an episode of vulcanism begins or ends? How long does a volcano lie dormant between eruptions? Are eruptions cyclical? These are questions of social as well as scientific interest, because they are related to the potential hazards of volcanic eruptions.

INITIATION

We may distinguish between regional and local vulcanism. The initiation of the former is largely the result of the creation or migration of a plate boundary or the drifting of the interior of a plate over a hot spot. Thus, copious basalts erupted in a new episode of vulcanism when Gondwanaland began to split apart. Likewise, lines of silicic intrusions and volcanoes presumably begin to develop when a lithospheric plate begins to plunge in a new subduction zone. In addition to these plate-related phenomena, vulcanism of unknown origin may occur, but it is not very common. Many areas of ancient continental crust within plates have been free of vulcanism for more than 1000 million years.

Locally, somewhere within the earth's several volcanic provinces, new centers of eruption develop perhaps once a year. The number depends on whether new cinder cones on the outer flanks of established volcanoes are counted as new volcanoes. Most of the new centers of eruption are mere ash cones, 30–100 meters high and active for a few years. Examples are the two volcanoes that have been born in historical times in Mexico. These arose in a volcanic province and amid hundreds of inactive cinder cones but far from major volcanoes. Perhaps, in due course, they will become large volcanoes, although they seem dead at present. One was Jorullo volcano, which rose from a farm on September 29, 1759, following three months of subterranean noises and minor earthquakes. By October 10, the ash cone was 250 meters high and the region was devastated, but there were no deaths and no lava flows. Activity ceased after 15 years, by which time the peak had grown to 300 meters and four lava flows had occurred.

Paricutín arose from a farm about 80 kilometers southeast of the dead Jorullo on February 20,1943. After a week of emitting ash, it was 140 meters high, and after a year, 330 meters. Thereafter, it changed its style: for the next eight years, it extruded mainly lava. After nine years, it became inactive, and it has remained so for 20 years.

PERSISTENCE AND CONTINUITY

Paricutín may rise again, for a volcano inactive for only two decades can hardly be counted as dead forever. Provinces presumably stay active as long as the plate-boundary conditions persist, and the subduction zone on the west coast of Mexico is still active. In general, provinces persist for tens of millions of years, individual large volcanoes for a few million years, and cinder cones for a decade or two. The problem is that large volcanoes may begin as cinder cones, and no one can tell which will continue to grow.

The continuity of eruptions, or duration of dormancy between eruptions, is highly variable. Kilauea, for example, was active most of every year for decades and then dormant for a number of years. More interest attaches to the dormancy of silicic volcanoes that are prone to explosions. Some of these

---

FIGURE 9.12

Periods of activity and dormancy of two Hawaiian volcanoes. These records show little evidence of periodic cycles. [After Stearns (1946).]

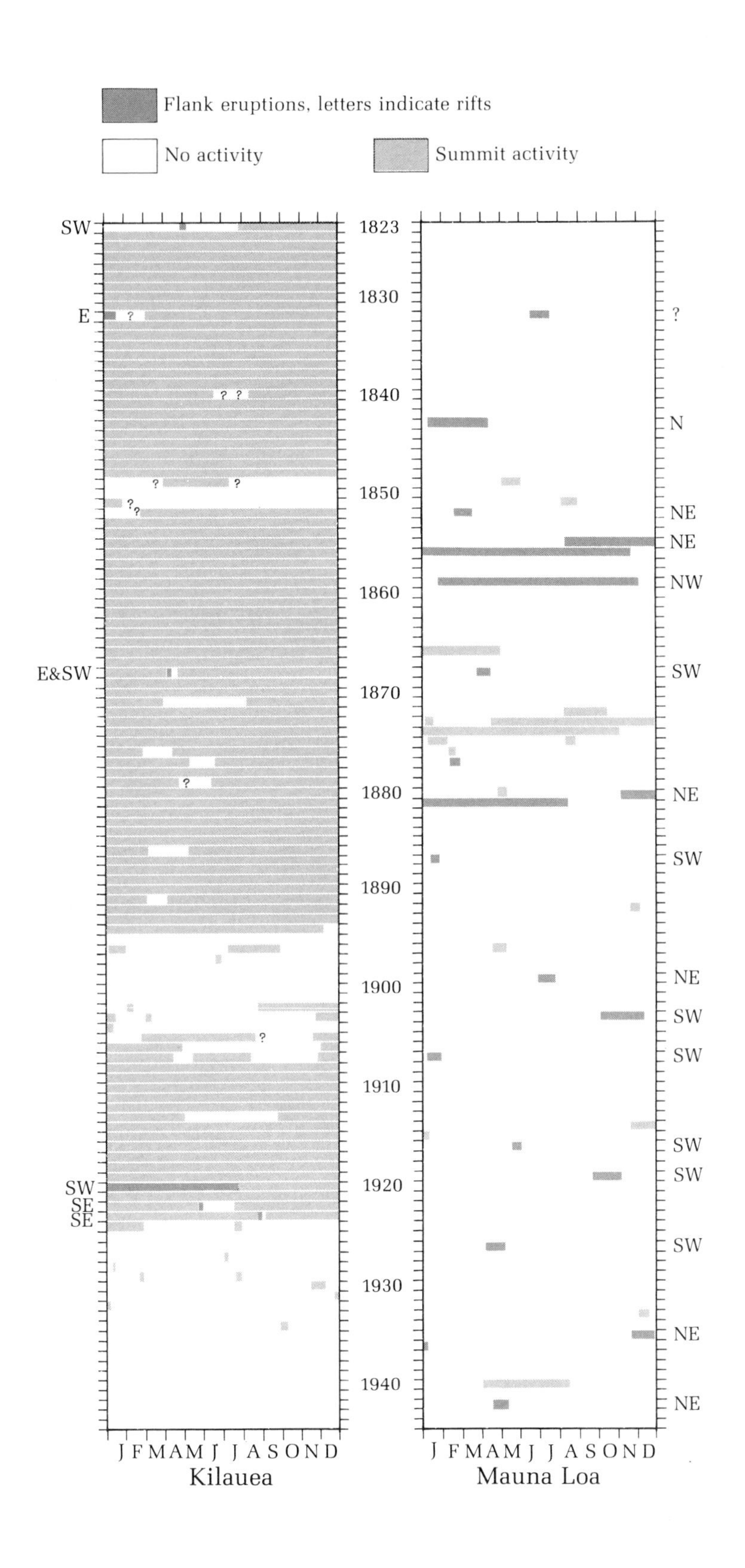

Flank eruptions, letters indicate rifts
No activity
Summit activity
SW
E
E&SW
SW
SE
SE
1823
1830
1840
1850
1860
1870
1880
1890
1900
1910
1920
1930
1940
?
N
NE
NE
NW
SW
NE
SW
NE
SW
SW
SW
SW
SW
NE
NE
J F M A M J J A S O N D
Kilauea
J F M A M J J A S O N D
Mauna Loa

erupt in small puffs almost constantly. The great explosions take more time to build up a head of steam. Mt. Pelée first entered historical records in 1635; and, although the area was a busy commercial center, only two small eruptions were observed, in 1792 and 1851, before the great glowing avalanches of 1902. Likewise, although it is in a very busy strait, Krakatoa was not seen to be active from the time it first appeared in European records, in 1680, until it destroyed itself in 1883. Vesuvius appears to be an even more extreme example of long dormancy. It was the site of Greek and Roman occupation for about 800 years, and it was cultivated to the peak when it erupted in 79 A.D. and buried Herculaneum and Pompeii. The conquerors of the Mediterranean were very familiar with active volcanoes: Mt. Etna is only a day's sailing to the south. Still, the extensive historical record makes no mention of activity at Vesuvius, and the location of the farms argues strongly against any.

The persistence of vulcanism extends far beyond the historical record, and the duration between eruptions, in most places, approaches the length of historical observations. At present, there is no way of setting a maximum period after which a volcano is safely dead. Even if there were, a new volcano might erupt very near a dead one. New molten crust has to fill in spreading centers, and the heat generated in subduction zones must escape somewhere.

CYCLES

Some of the most careful observers of volcanoes have been convinced that they erupt in cycles; others have held a contrary opinion. This is a matter of some importance, because cycles would form the basis for accurate prediction of eruptions. Two issues are involved—as are two uses of the term "cycle," which may have led to some confusion. There can be little question but that a regular sequence of events occurs when many volcanoes erupt. In the long term, the character and chemistry of the eruptives of a large volcano may change, for example, from fluid basalt to more ash-generating rhyolite. Likewise, ash formation regularly precedes the appearance of fluid lavas in the development of small cinder cones. In Hawaii, it has also been observed that a sequence of eruptions may occur at ever higher elevations along one of the great flank rifts. These sequences of events are called "cycles," and little dispute seems to exist about their reality.

Their utility for prediction is quite another matter. These cycles, like most encountered in the environmental sciences, are far from regular. Stages may be omitted or duplicated, and so on. Moreover, they are quite variable in duration and, hence, almost useless for predictions. Finally, they are of no use at all in predicting catastrophic explosions of volcanoes thought to be dormant—and this volcanic activity is the sort most important to society.

Extent and Prevalence of Volcanic Hazards

The general attitude of people toward active volcanoes is suggested by the fact that Java, where there is an eruption every 3 years, has 460 people per square kilometer and nearby Borneo, with no volcanoes, has 2 per square kilometer. The additional fertility that results from the ash falls is well worth the risk. The vulcanologist A. Rittmann (1962, p. 53) expresses the same view:

> Only the big explosive eruptions of volcanoes with viscous magmas are really dangerous. But even these appear relatively harmless in comparison with earthquakes, storm tides and hurricanes—to say nothing of wars, revolutions and pestilence. Nowadays, traffic fatalities are much more numerous than those due to volcanic eruptions.

That is not to say that we should not try to do something about them.

The main reason that volcanoes are not much of a social problem is that almost all the effects they produce are localized to the actual slopes. If these are avoided, the risk is minimal. In any event, lava moves too slowly to endanger many lives, and some flows can even be diverted away from habitations. Against glowing avalanches, lahars, glacier bursts, and their sort there is no defense—except simply to stay away from areas of volcanic activity.

Volcanoes sometimes radiate concentrations of matter and energy for some distance from an eruption site; when this happens, the remote social effects may be significantly greater than the local ones. Volcanoes in shallow water can generate a type of giant wave like a tsunami. Waves from the explosion of Krakatoa in 1883 spread throughout Indonesia, killing 36,000 people along those populous shores and destroying 295 towns. Ash can also be a killer. After the eruption of Tambora in 1815, crops were smothered in the surrounding area of Indonesia and 82,000 people died of starvation. The eruption of Mount Skaptar in Iceland in 1783 was even more of a national disaster: Crops were damaged as far away as Scotland and Norway, and ash and floods destroyed many farms near the volcano. The greatest toll, however, was from a bluish haze, probably containing sulphur dioxide, that plagued the country during the summer of the eruption. An intense natural smog, it blighted the grass, and famine followed. Iceland lost 10,000 people—about a fifth of the population—plus about three-fourths of its sheep and horses and half of its cattle—230,000 head. Presumably, in these days of rapid communication and transportation, the famines in Indonesia and Iceland would have been relieved. Deaths from smog, however, are not yet within our control, except by mass evacuation.

The German vulcanologist Karl Sapper estimated that thousands of eruptions by 500 volcanoes killed 190,000 people between 1500 and 1914. The preceding paragraph accounts for nearly 130,000 of the victims—and these were killed away from the site of the eruption and largely by starvation, which is preventable. This leaves a total of 60,000, or about 120 per volcano, or 150 per year. These are still dolorous figures; but, if we further reduce the total by the 40,000 killed in the cities of St. Pierre and Pompeii, we come to 40 per volcano, or 50 per year, or only a few per eruption. Many of these may have been spectators, such as the 22 people trapped by lava flows when they went onto Vesuvius to observe the eruption of 1872.

In spite of their violence and the minor property damage that they cause, it appears that volcanoes are relatively unimportant environmental hazards to life, and that, despite the expansion of the population, they will remain so. They might, however, have been a greater potential danger, had we remained as ignorant of volcanic processes as we were when the glowing avalanche struck St. Pierre, or had vulcanologists not shown that most eruptions can be predicted and their effects partially ameliorated. They remain a hazard to farms and to any construction near them.

PREVALENCE OF RISK

Active volcanoes are well known in Hawaii and Alaska. Some of the Cascade volcanoes of the northwestern United States have been active in historical times, and the others are suspect as long as plate motions remain as they have been. The extent of vulcanism in the last few million years in the western states may come as more of a surprise. Large areas of many states have been buried by lava flows and are dotted with young cinder cones. The risk in most of these areas may be minimal, but it is far from nonexistent. An explosion of Crater Lake or one of the larger Cascade volcanoes, for example, would have widespread effects.

Volcanic hazards in California have recently been evaluated by the California Division of Mines and Geology, which has identified more than twenty areas of Pleistocene and recent volcanic rock spread mainly along the sparsely populated eastern margin of the state. Twelve volcanic events have occurred in California in historical times, almost all of them in the northeastern corner. For example, ash, pumice, and lava erupted from Mt. Lassen in 1915, and there were both lahars and glowing avalanches. Mount Shasta, another of the great Cascade volcanoes, apparently erupted in 1786, and there are several lava flows along its flanks that are probably only 300–500 years old.

The analysis of potential hazards from volcanoes in California has led geologist Charles Chesterman to some interesting conclusions:

1. Since the last eruption in 1915, the population of California has increased from 2,800,000 to about 22,000,000. Although most of the newcomers live in coastal cities, a certain number "have settled in small communities

BOX 9.1 PHYSICAL AND SOCIAL EFFECTS OF SOME MAJOR VOLCANIC ERUPTIONS

Year	*Volcano*	*Location*	*Discharge (km^3 of rock)*	*Deaths*	*Property destroyed*	*Notes*
1919	Kelut	Java		5,000		Deaths by lahars
1912	Katmai	Alaska	21			
1902	Mt. Pelée	Martinique		30,000	City of St. Pierre	
1883	Krakatoa	Indonesia	21	36,000	295 towns	Deaths by tsunami
1835	Coseguina	Nicaragua	21	0		Ash
1815	Tambora	Indonesia	150	92,000		82,000 died of starvation
1783	Laki	Iceland	50	10,000		Deaths largely from famine
1772	Papandayang	Java		3,000		
1631	Vesuvius	Italy		4,000		Deaths in Naples by ash
1006	Merapi	Java				End of Mataram culture
79	Vesuvius	Italy		10,000?	City of Pompeii	Death of Pliny the Elder
79	Vesuvius	Italy		few	City of Herculaneum	
1400 B.C.	Thera	Greece		few		End of Minoan civilization

within and near the several active or dormant volcanic areas" (Chesterman, 1971, p.141). With the rapid expansion of recreational communities and "second homes," this settlement must be multiplying.

2. The expansion of the population has brought highways, power sources, water supply systems, and agriculture into areas where volcanic activities are likely to occur in the future. Widespread ash falls would cause severe crop losses.

BOX 9.2 HISTORIC VOLCANIC EVENTS IN CALIFORNIA

Year	*Locality*	*Activity and products*
1951	Lake City, Modoc County	Mud volcanoes: low cones of mud formed and hot water discharged
1930	Mt. Lassen, Shasta and Tehama counties	Strong earthquake shocks, presumably originating near Mt. Lassen, were recorded on the seismograph of the Mt. Lassen Volcano Observatory
1917	Mt. Lassen, Shasta and Tehama counties	Emission of steam; volcano became dormant in 1917
1915	Mt. Lassen, Shasta and Tehama counties	Great blast: ash and pumice eruptions, new lava formed, mud flows and glowing avalanches
1914	Mt. Lassen, Shasta and Tehama counties	Explosive action: new lava emplaced and fall of ash
1890	Mono Lake, Mono County	Sublacustral eruption: emission of steam and sulfurous fumes in puffs; boiling water and hot mud from formerly cold springs
1857	Mt. Lassen or Mt. Shasta	Ash eruption
1851–1852	Cinder Cone, Lassen County	Surface eruption: development of cinder cone and flows of basalt
1786	Mt. Lassen or Mr. Shasta	Steam and ash eruption (observations made by La Perouse while voyaging along the California coast in 1786)
1450–1500	Burnt Lava Flow, Siskiyou County	Surface eruption: development of cinder cones and flows of basalt
1100–1500	Inyo Crater Lakes, Inyo County	Explosions: development of crater whose walls are composed of interlayered lava flows and pyroclastic deposits
900	Big Glass Mountain, Siskiyou County	Extensive pumice eruptions followed by emplacement of domes: pumice deposits, and flows and domes of obsidian

Source: After Chesterman (1971).

3. "It may even be foolhardy to believe that another eruption will not occur within the coming 50 or 100 years" (Chesterman, 1971, p.147).

Society grows more vulnerable as it grows more complex, unless it takes corresponding steps to protect itself.

Prediction

Minor explosions may result from the penetration of surface water down to a magma pool perched within a volcano. Most eruptions, however, follow the injection of new magma, which rises by forcing its way upward along old cracks or by opening new cracks in the overlying rock. This generates small earthquakes, initially at depth but gradually approaching the surface. The intrusion increases the volume of the volcano and changes its shape, even before an eruption occurs. The changes in shape are readily measured with tiltmeters, and changes in volume are detectable with a geodetic network.

Several other phenomena facilitate the prediction of eruptions. The temperature of hot springs may increase and their chemistry may change as new gases stream out of the interior. Likewise, the magnetic and electrical fields of the volcano may change; the intruding molten magma is too hot to register a magnetic signature.

Volcano observatories have existed for decades in most places in the world where there has been sufficient reason to issue warnings about possible eruptions. As the population expands, and with it the recreational use of volcanic areas, it will be prudent to expand the monitoring of volcanoes.

It is not necessary to detect the initial stage of an eruption to prevent most major disasters. The 1902 eruption of Mt. Pelée began on April 2, but St. Pierre was virtually untouched until it was annihilated on May 8. The alert for the eruption of Vesuvius in 79 A.D. was given by 16 years of small earthquakes in the area where none were noted before. Krakatoa also issued a warning by erupting for 14 weeks prior to the final paroxysm that destroyed it on August 26, 1883.

Apparently, scientists can predict the initiation of an eruption, and the volcano itself usually gives fair warning that greater explosions may follow. Wherein is the problem? It is the same one we encountered in the discussion of earthquakes—namely, that there is no way, as yet, to predict the intensity of an eruption, and people who have experienced minor eruptions at a distance refuse to be alarmed by yet another warning of possible disaster. On May 7, the day before the end, *Les Colonies*, the newspaper in St. Pierre, declared dutifully that alarm was unwarranted: "Mount Pelée is no more to be feared by St. Pierre than Vesuvius is feared by Naples. We confess we cannot understand this panic. Where could one be better off than at St. Pierre? (quoted in translation in Bullard, 1962). One can only wonder if any newspaper in Naples has ever inverted the statement and questioned the safety of the city.

Amelioration

People have been living with volcanoes for a long time, and apparently rather successfully. They have developed various ways of minimizing risk, deflecting lava flows, and even controlling some types of eruptions. The possibility of further amelioration of volcanic hazards is opening for the future.

ZONING AND SITING

The safest approach to volcanoes is to fence them off from encroachment by habitations but to leave them open for observation, study, and enjoyment. The United States has been remarkably successful in achieving this end by putting most known active volcanoes, and the Yellowstone thermal area, into National Parks or National Monuments. The only active volcanoes that are excluded are those in the Aleutian Islands, which are almost uninhabited. So far, so good. Unfortunately, other volcanoes—especially those in the high Cascades—may be dormant instead of dead, and it might serve the public interest to make them National Parks as well.

With regard to essential buildings on the flanks of volcanoes, much can be done to reduce volcanic hazards merely by siting them on high ground. The really dangerous phenomena, glowing avalanches and lahars, flow down valleys. So do the less dangerous but still destructive lava flows. The danger of destructive ash falls can be minimized by building steeply pitched roofs. These are simple steps, but ones that require zoning restrictions before construction—not after the area is full of recreational retreats.

MONITORING AND WARNING

The U.S. Geological Survey already has a program of monitoring changes in volcanoes. It is relatively inexpensive, and it would be foolhardy not to expand it to include any volcanoes that may merely be dormant. However, the knowledge that an eruption is imminent will not be enough to provide the basis for a successful warning system. If Mt. Lassen has minor eruptions for 16 years, will the National Park be evacuated and be kept that way? Pompeii and Herculaneum were not. Much research remains to be done to refine methods, particularly those for predicting the intensity of eruptions.

COUNTERACTIONS

When lava from Mt. Etna penetrated the town of Catania in 1669, Diego Pappalardo conceived a counterattack. He organized fifty men, protected with wet hides, to hack a breach through the chilled side of the flow above the town. This induced a lateral flow, which, had it occurred earlier, might have saved the town. But perhaps not: because the diverted lava was thought to threaten the town of Paterno, 500 armed men from there put Pappalardo and his peaceful environmentalists to flight, and the lava again flowed toward Catania.

The citizens of Hilo, Hawaii, have been more successful: they have had both science and the army on their side. T. A. Jaggar, the resident vulcanologist, arranged to have aerial bombs dropped on lava streams approaching Hilo in 1935, and his successor, R. H. Finch, did so in 1942. Both times the attempt was partially successful. Gordon Macdonald, a later resident vulcanologist, attributed this success to the favorable nature of the Hawaiian flows. Pahoehoe at first flows in open streams; but after a few hours or days, it crusts over and flows through a self-made tube that may be 15 meters in diameter. The famous Thurston lava tube in Hawaii National Park is an inactive example through which tourists may walk. Bombs dropped on an active tube may break it open, clog it with debris, and cause a diversion, which may cut off a threatening flow. Aa flows remain open, but they build natural levees on each side as lava spills over and cools. Thus, these flows commonly rise above the surrounding ground, and bombs can divert them by smashing the levees. The lava pool in a cinder cone also rises above the ground—which is why it flows out—and bombing the cone may cause a diversion at the source.

Despite the favorable results of bombing, Macdonald believes that the only long-term solution to Hilo's problem is to build a permanent barrier to deflect the flows. Experience has shown that such barriers can be surprisingly effective, and many have been constructed. The Japanese village of Nomashi lies directly below a low gap in the caldera wall of Mihara volcano. When lava rose near the level of the gap in 1951, the villagers built a masonry wall, a few meters high and thick, with the hope that the lava would be deflected to another, lower gap and would flow away from the village. It was not put to the test because the lava stopped rising, but this shows how an adequate warning system and a vigorous (albeit small) response may hope to protect a city. The justification for believing that the wall would have succeeded is that lava that flows into villages commonly does not knock down the ordinary masonry walls of houses, although these walls have very little lateral strength. A Hawaiian pahoehoe flow in 1920, ponded by a wall of loose stones three feet high and a foot and a half thick, eventually spilled over the wall without damaging it. Bulldozers can construct effective earth barriers almost as fast as lava can flow on low slopes, so the barrier solution looks quite promising.

Circumstances favorable to the diversion of lava flows occur much less frequently on composite volcanoes because of the steep slopes and the presence of deep radial valleys. It may be possible, however, to breach a crater to deflect welling lava toward uninhabited ground. Moreover, low transverse walls may pond and thicken the flows and thereby restrict them to the upper slopes of the volcano.

An imaginative approach to the problem of lahars from the volcano Kelut in Java shows what is possible when the effort seems worth the cost. The lahars originated in the crater lake, so the lake was simply drained through a tunnel.

CONTROL

Countermeasures are a step toward control. Are further steps possible? It appears that the volcanoes in the United States have not been considered hazardous enough to merit evaluation of this question by government agencies. The attitude of the California Division of Mines and Geology, like that of the U.S. Geological Survey, reflects some governmental reassessment of the importance of volcanoes. Even so, if volcanoes are merely liabilities, the logical approach is to minimize contact, and control may be unwarranted even if it is possible.

A different conclusion emerges if volcanoes are regarded as assets. Human resources cannot control the localization of heat emerging from the interior of the earth, but, in favorable circumstances, they may be able to divert it from vulcanism and put it to work. Cheap, pollution-free energy is now developed in many places by utilizing natural steam. This is discussed in Chapter 18. Is it beyond possibility that volcanoes near cities, such as Vesuvius, can be controlled or moderated by dissipating some of their heat in the recycling of steam to generate power?

Summary

1. Magma, the liquid source material of igneous rock, commonly contains gases in solution, fragments of rock, and solid crystals in suspension.
2. Most of the characteristics of volcanic eruptions depend on the chemistry, viscosity, and gas content of the magma.
3. Steam is common in volcanic eruptions, but most of it, if not all of it is recycled rain water.
4. Magma may reach the surface and flow as lava, or it may solidify below the surface in intrusions that are later exposed by erosion.
5. Volcanoes occur at spreading centers and subduction zones because of tectonic plate motions. They also occur within plates over hot spots in the mantle or elsewhere for unknown reasons.
6. A group of volcanoes is a load on the lithosphere, which bends down around it to form a moat. The bending also causes an arch to form around the moat.
7. Volcanic explosions occur when the gas in magma is confined. Not infrequently, the whole top of a volcano is blown away. The coarsest debris is deposited near the volcano, but the finest may float all around the earth in the upper atmosphere.

8. Glowing avalanches sometimes spread downslope at great speed and with disastrous effect. Enormously destructive floods may occur if a crater lake is breached or a volcano melts a hole in a glacier.

9. New centers of eruption form somewhere in the world every year, but they rarely last very long. New volcanic regions develop much less frequently. They are commonly associated with changes in tectonic plate boundaries, and usually persist for tens of millions of years.

10. Volcanoes are generally only a local hazard to society.

11. Because rising magma causes earthquakes and other conspicuous phenomena before it emerges, many volcanic eruptions can be predicted.

12. Volcanoes pose only small hazards while people stay at a distance, but the risk increases as the population closes in upon them. Even so, careful siting of buildings and monitoring of possible eruptions can reduce the danger.

Discussion Questions

1. How do the different gases that come out of a volcano get into the lava in the first place?
2. Why is vulcanism more intense at spreading centers than along transform faults?
3. Why does the magma that produces pillow lava in deep water produce ash explosions at sea level?
4. What are the reasons for believing that Crater Lake, Oregon, was once the site of a volcanic peak?
5. Compare a volcanic explosion with the explosion of an atomic bomb.
6. Arrange volcanic phenomena in order of their danger to life and property.
7. How long must a volcano be inactive before it is safe to predict that it is dead?
8. How can a volcano kill people beyond its slopes? Is this common?
9. How can volcanic eruptions be predicted?
10. In what ways do thermal phenomena benefit society?

References

Bullard, F. M., 1961. *Volcanoes.* Austin, Texas: University of Texas Press. [An interesting, informative, nontechnical summary in the classical style.]

Chesterman, C. W., 1971. Volcanism in California. *Calif. Geol.* 24:139–147. [Active and recently active areas identified and catalogued.]

Leet, L. D., 1948. *Causes of Catastrophe.* New York: Whittlesey House.

Macdonald, G. A., 1958. Barriers to protect Hilo from lava flows. *Pac. Sci.* 12:258–277. [Experiments in counterattacking lava flows.]

——, 1965. Hawaiian calderas. *Pac. Sci.* 19:320–334.

McBirney, A. R., 1971. Oceanic vulcanism; a review. *Rev. Geophys. Space Phys.* 9:523–556. [A comprehensive, scholarly summary.]

Menard, H. W., 1964. *Marine Geology of the Pacific.* New York: McGraw-Hill.

———, 1971. The Late Cenozoic history of the Pacific and Indian Ocean basins. *In* K. K. Turekian, ed., *Late Cenozoic Glacial Ages,* pp. 1–14. New Haven: Yale University Press.

Moore, J. G., and R. S. Fiske, 1969. Volcanic substructure inferred from dredge samples and ocean-bottom photographs, Hawaii. *Bull. Geol. Soc. Amer.* 80:1191–1202.

Morgan, W. J., 1972. Deep mantle convection plumes and plate motions. *Bull. Amer. Ass. Petrol. Geol.* 56:203–213.

Rittman, A., 1962. *Volcanoes and Their Activity* (translated from the German by E. A. Vincent). New York: John Wiley & Sons. [A book for the professional, but still full of fascinating description by one of the world's leading authorities.]

Sapper, K., 1927. *Vulkankunde.* Stuttgart: J. Engelhorn, Nachfolger. Translated by A. L. Day.

Shelton, J. S., 1966. *Geology Illustrated.* San Francisco: W. H. Freeman and Company.

Stearns, H. T., 1946. Geology of the Hawaiian Islands. *Terr. Hawaii Div. Hydrogr. Bull.* (8). [The very model of a well-illustrated geological guidebook readable by the layman.]

Williams, H., 1951. Volcanoes. *Sci. Amer.* 185(5):45–53. (Available as *Sci. Amer.* Offprint 822.) [A summary by the dean of American vulcanologists.]

ATMOSPHERE AND HYDROSPHERE

PETER D'AGOSTINO

10

AIR AND WATER

Here may we sit and view their toil
That travail in the deep. . . .

Samuel Daniel, ULYSSES AND THE SIREN

And Noah he often said to his wife when he sat down to dine,
"I don't care where the water goes if it doesn't get into the wine."

G. K. Chesterton, THE FLYING INN

This chapter, which is the first of three dealing with air and water, is essentially a description and analysis of their properties, motions, and interactions. Chapter 11 deals with the geologic history of changes in their properties and motions, particularly with the effects of glacial periods, when large areas were covered with ice and the sea level was lower. Chapter 12 is about climatic changes that have occurred in historical times—and that may be expected to recur in the immediate future and, thus, to affect society—and the problems and opportunities of weather prediction, weather modification, and the control of weather by man.

This closely knit group of chapters does not include all the material concerned with air and water. Indeed, the chapters on the sedimentary cycle, weathering, and mineral resources are all related to the interactions of the solid earth with the thin layers of liquid and gas that surround it.

Water is an uncommon compound everywhere in the solar system but on the surface of the earth, and it is water that makes the earth habitable for the life forms found on it. It is the protecting shield, the moderator of the extremes of space, the transfer agent for solar energy, and the molder of the surface land forms. Its role is determined not only by its prevalence but also by its chemical and physical properties. It is a powerful solvent that, given enough time, is capable of dissolving the solid rock of the land or converting it into soil and the salt of the sea. Physically, it occurs as a solid, liquid, or gas within the temperature range commonly found on the earth, although it would not do so if the earth were farther from the sun or the sun were hotter or colder. Moreover, the energy required to change the state of water from liquid to gas without change of temperature at 20°C – 585 calories/gram – is exceptionally large compared to that required to so change other substances. The conversion from a solid (ice) to a liquid (water) requires about 80 calories/gram, which is also larger than for most compounds. In contrast, it takes only 100 calories/gram to elevate the temperature of water from the temperature of solidification to that of vaporization. Thus, much of the energy circulation at the surface of the earth is the result of changes in the state of water.

Atmospheric and oceanic interactions are highly complex. Winds drive ocean currents and masses of water move into regions with different temperatures, a condition, that, in turn, generates winds. Likewise, the winds make waves by friction with the water surface, the waves break and enclose bubbles, the bubbles rise to the surface and inject salt into the air, the salt droplets in the air act as the nuclei for rain drops, and on and on and on. Similarly, hurricanes derive energy from the heat of the ocean; when they move ashore and are cut off from the heat source, they rapidly decay. Thus, it is clear that the earth's air and water are intimately related. The separation of the atmosphere and ocean in the following sections is only for simplicity's sake.

Global Atmospheric Circulation

We are now in a position to follow the energy of the sun as it is distributed over the surface of the earth by air and water, transformed, and dissipated. The process of distribution is highly effective. For example, the solar energy received per unit area varies with latitude, with a maximum at the equator twice a year when the sun is directly overhead. At these times, hardly any radiation is received at the poles, but the latitudinal temperature gradient is minor because of the rapid circulation of heat. Most solar energy readily penetrates clear air and is converted to heat at the earth's surface.

Heating generally causes expansion, but the effects are totally different depending on the location of the material being heated. The *top* of the ocean is heated by the sun; thus, the low-density material is on top and the density structure of the ocean is stable. The atmosphere heats differently, however: because it is the bottom of the atmosphere that is heated, the low-density

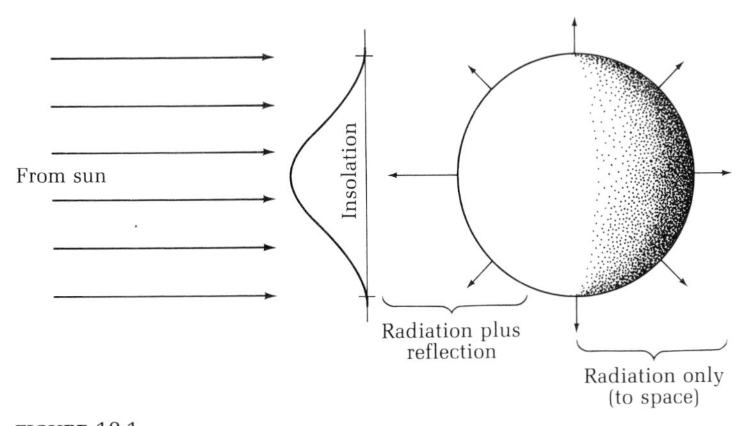

FIGURE 10.1
The solar energy received per unit area, insolation, varies greatly with latitude because of the angle between the incoming radiation and the spherical earth. Earth radiation is radial and, thus, much less variable, although direct reflection occurs only during the day.

material is on the bottom and the configuration is unstable. The depth of heating is also important. Continents are warmed to a depth of a few meters at most, and they cool and reheat rapidly in daily and seasonal cycles. Wind-induced stirring of the ocean mixes warm surface water to a depth of a hundred meters or so and produces an enormous reservoir of heat. Thus, the daily and seasonal cycles are far less intense and the sea is a stabilizing climatic factor.

The vertical transfer of heat by radiation alone can be computed theoretically. The surface would heat rapidly and the air would cool even faster by radiation to space. The speed of radiation outward is readily observed when frost forms on exposed metal on dry, clear nights, even though the air temperature is above freezing.

In the real world, the circulation of air overwhelms the effect of radiation both by direct transfer of heat and by transfer of latent heat, which is released when rising water vapor condenses into rain. The mechanism of mixing is by turbulent eddies, which grow as they rise and are most obviously manifest in great thunderheads and cumulus clouds.

The ratio of incoming and outgoing radiation varies with latitude, because solar radiation is essentially parallel and, thus, is incident at low angles at the poles, whereas earth radiation is radial and, thus, is more equal everywhere. Inasmuch as incoming and outgoing radiation for the whole earth are equal, it follows that there is an excess of radiant heating in the summer hemisphere and a deficit in the winter one. This is partially balanced by withdrawals or additions to the heat stored in the ocean. Even so, the temperature gradient would be very large, were it not for the fact that it drives a global circulation and is, in turn, moderated by it.

BOX 10.1 THE GREENHOUSE EFFECT

Black-body radiation is the type of radiation that would be emitted at a certain temperature by a theoretical body that absorbs all of the radiation that falls upon it. The sun emits like a black body at 6300°C with maximum radiation in the visible wavelengths of 0.4–0.7 microns. Much of the small amount of shorter-wavelength radiation, in the ultraviolet, is absorbed in the upper atmosphere by oxygen and ozone and drives a weak circulation. Most of the energy readily penetrates clear air and is converted to heat at the earth's surface.

The geologic record demonstrates that the average temperature of the earth is relatively constant. Thus, the earth radiates as much energy to space as it receives. However, the properties of the energy are changed because the earth emits energy like a black body at a temperature of only −18°C and, thus, with a broad spectrum with a low maximum at a wavelength of 12 microns (see figure at left). The atmosphere is not transparent to such wavelengths largely because of the presence of water vapor, carbon dioxide, and ozone. This transparency to incoming short radiation but not to outgoing long radiation is the famous greenhouse effect, which may be warming the earth as the burning of fossil fuels pours carbon dioxide into the atmosphere.

The efficiency of the atmospheric transfer of heat can also be demonstrated by comparing the extreme values of the black body temperatures of the earth with the mild variations observed (see figure above).

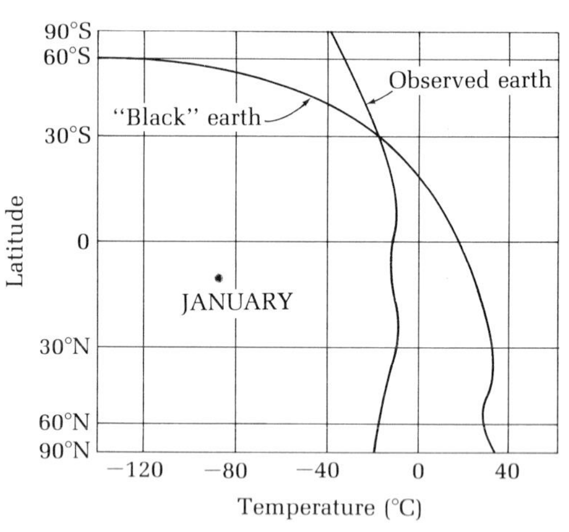

The importance of atmospheric dynamics in moderating the earth's climate is demonstrated by this graph, which compares the calculated radiative-equilibrium temperature for a "black" earth, with the observed vertical mean temperature as a function of latitude, during January. At this time, no sunshine reaches the earth north of the Arctic Circle; neglecting any lag effects due to the storage of heat, the radiative-equilibrium temperature in the polar cap would go down to absolute zero (−273.2°C), while the summer hemisphere would tend to become extremely hot. [After Oort, "The Energy Cycle of the Earth," copyright © 1970 by Scientific American, Inc. (all rights reserved).]

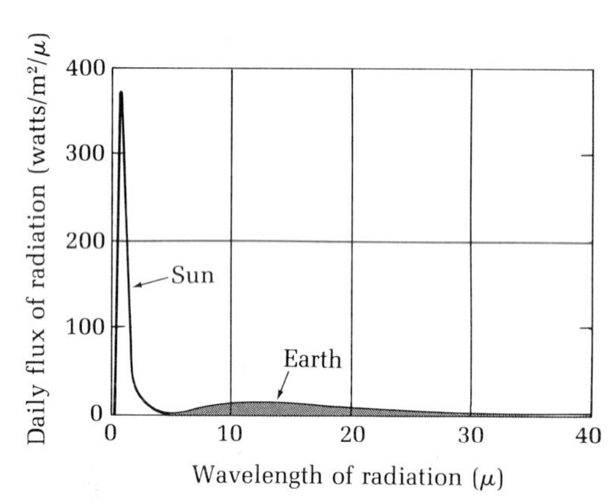

Approximate emission spectra of the sun and the earth under the assumption that they radiate as "black bodies" with temperatures of 6300°C and −18°C, respectively. The areas under the two curves are equal; in other words, the earth emits as much radiation as it absorbs. The important change in the character of this radiation from the short-wave to the long-wave part of the spectrum is evident. [After Oort, "The Energy Cycle of the Earth," copyright © 1970 by Scientific American, Inc. (all rights reserved).]

HADLEY CELLS AND THE LATITUDINAL CIRCULATION

The simplest circulation conceivable includes hot air rising at the equator, cold air sinking at the poles, and horizontal flow aloft and at the surface to connect the vertical motions into a continuous cell. Such a circulation was proposed by George Hadley in 1735, and cells directly driven by such rising and sinking are named after him by meteorologists. Because the directions of the motions are controlled by the Coriolis effect as well as by the temperature gradient, they are at various angles instead of just north and south.

The actual climatic belts of the earth indicate a more complex circulation with three cells from equator to pole, as proposed by William Ferrel and others in the nineteenth century. It is important, however, to realize that only the equatorial and polar cells are Hadley cells, driven by rising hot air and sinking cold air, respectively. The middle cell is driven against the temperature gradient, in that cold air rises and warm air sinks. The driving energy for the middle cells is derived from a longitudinal circulation, by way of great atmospheric waves, which we shall touch on shortly.

The evidence for the third latitudinal cell can be seen by following the circulation from the equator to the pole. The hot air that rises at the equator

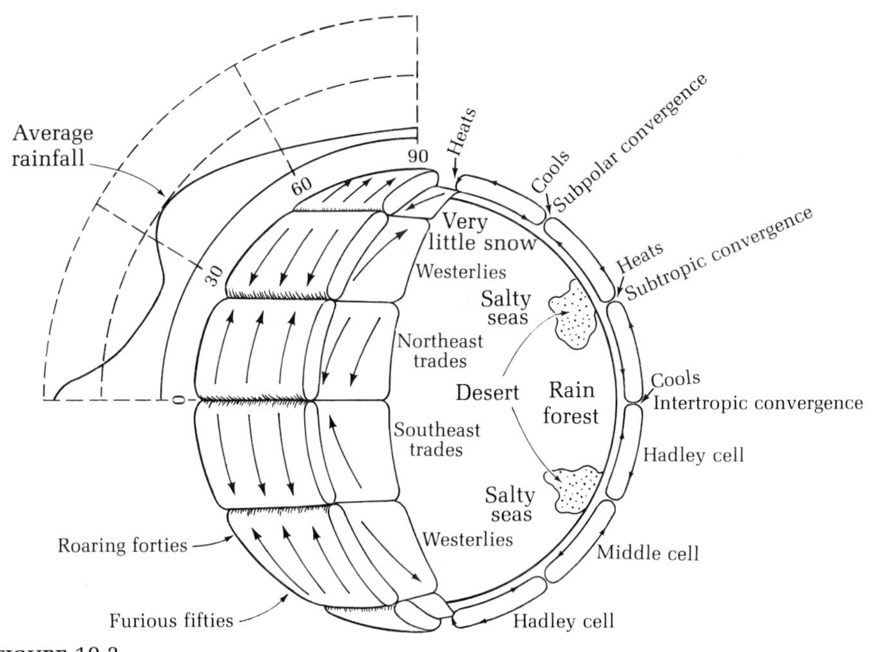

FIGURE 10.2
A cellular model of atmospheric circulation and its correlation with the climatic belts on the earth. Maximum rainfall is in zones of rising air. The winds are deflected from polar paths by the Coriolis effect.

moves over the oceans and absorbs moisture and heat. The surface winds are the balmy northeast trades of the northern hemisphere and the southeast trades of the southern one. The deviation of these winds from north and south is one of the most obvious effects of the Coriolis acceleration.

The hot, moist air rises in a narrow intertropical convergence, cools, and dumps rain back into the equatorial belt, which has a higher average rainfall than any other latitude. The locus of the intertropical convergence varies with the seasons in response to changes in tropical heating. Thus, there are "rainy" seasons and "dry" seasons in the tropics, rather than seasons based on temperature. Rain and sun combine to produce the wide belt of lush rain forest that we associate with the valleys of the Amazon and the Congo and the

BOX 10.2 ROTATION AND MOTION

The rotation of the earth affects not only its shape but also all motion, both on the surface and in the interior. The effect of rotation upon motion is called the Coriolis effect. Consider a cold-hearted warlord floating with a ballistic missile on an iceberg far to the north of England: Filled with a hatred of science, he decides to destroy the tomb of Sir Isaac Newton in Westminster Abbey. The spirit of Sir Isaac, at peace among the celestial spheres, is at first filled with horror, but then notes with satisfaction that the missile is aimed straight toward his dusty bones. The rocket is launched, and Sir Isaac observes that the warhead obeys his first law (naturally) and travels in a straight line. Meanwhile, the earth turns beneath it, and, instead of hitting Westminster Abbey, it falls harmlessly into the Irish Sea. The ignorant warlord merely sees the missile curve to the right in its path. This is fanciful, of course; but the Germans in World War I actually aimed their great gun called Big Bertha to the left of its intended target in Paris in order to correct for the rotation of the earth.

Consider the rocket on the iceberg: Although still on the pad, it has a velocity appropriate to its latitude on the spinning earth. At the pole its velocity would be zero; at the equator, it would be about 40,000 kilometers in 24 hours; and in between its velocity would vary as the cosine of the latitude. When the rocket fires toward the south, it retains its velocity component due to rotation, and, to a celestial observer, it goes in a straight line. It soon moves into lower latitudes, however, where points on the surface of the earth are moving east with a greater velocity due to rotation. Thus, the earth appears to move the target away from the approaching rocket. Meanwhile, the warlord, sitting in a fixed position relative to the surface of the earth, sees the rocket curve futilely off to the right.

Inasmuch as warlords are apt to have more than one rocket, let us assume that this one steers his iceberg to the south of England and fires another. Being to the south, it moves from the outset at a higher velocity due to rotation than the prospective target. Sir Isaac observes with pleasure that, although it goes in a straight line, England does not have time to rotate under it, so it falls harmlessly into the English Channel. The warlord, to his amazement, sees his rocket apparently veer to the right again.

He has one last mad hope. Perhaps he can hit the target from the west. But no, when he fires it, it still goes to the right. He is frustrated, just as he would have been had he fired islands of Ceylon, Borneo, and New Guinea. The rising air cools and gains potential energy at the expense of the latent heat of the water vapor as it condenses into rain. Aloft, the rising air turns to flow parallel to the earth's surface toward the poles.

The limit of the northern motion of the air that rose at the intertropical convergence is indicated at 30° latitude by a belt in which the ocean tends to evaporate at a relatively rapid rate (and, thus, to have a relatively high salinity) and in which continental deserts abound. Among these are the deserts of the Sahara and the Middle East, the Atacama desert of Chile, the Kalahari of South Africa, the Sonoran Desert of Mexico and the United States, and much of the desert of Australia and Asia. The air that originated in the tropics,

it from the east, by the fact that the rocket flies in a straight line, as seen from space, and a straight line on a sphere is a great circle (that is, a circle that is the intersection of the surface of a sphere with a plane passing through the center of the sphere). Meridians of longitude and the equator are all great circles, but any other parallel of latitude is not. Thus, it is impossible to shoot exactly east or west along a line of latitude; and if you shoot in any other direction, the effect of rotation is felt. We can also visualize the Coriolis effect on eastward motion in terms of the necessity for a body to remain on the earth. Moving east, it has a higher velocity than the earth, so it moves right to a lower latitude where its velocity matches that of the earth.

This Coriolis effect (or Coriolis acceleration) always acts perpendicular to a force and, therefore, by definition, does no work. For the reasons discussed above, it acts in the opposite sense in the Southern Hemisphere. That is, all motion is deflected to the left instead of to the right. The acceleration is largest at the poles and decreases to zero at the equator. It is never very great, but neither are many of the other accelerations that affect the motion of the atmosphere and ocean. When a low-pressure area develops in the air of the Northern Hemisphere, the surrounding dense air does not move directly into it. Instead, the winds curve to the right to form a clockwise pattern. Under the same circumstances, winds in the Southern Hemisphere curve to the left. We shall return to the Coriolis effect again and again when discussing motion on the earth.

We have considered the effect of the earth's rotation upon a body in linear motion. It also has an effect upon bodies in oscillatory motion. Consider a heavy ball suspended from a long, thin wire at the north pole: If we push the ball to one side and then release it to swing as a pendulum, no force but gravity acts upon it, so it continues to move back and forth on the same line (as seen from space) while the earth turns beneath it. This experiment was done in Paris in 1851 by Leon Foucault, who attached a thin rod to the ball so as to mark its motion in a bed of fine sand. The rod traced out the area of a circle, just as it would have done at the pole. Tides produce oscillatory motion of particles suspended in the ocean, so the Coriolis effect causes them to move in circles whose size varies with the latitude and the velocity of the particle.

now cool, begins to sink at this latitude. As it does, the pressure increases, which causes it to heat up again. Thus, part of the potential energy imparted by wet heat in the tropics is converted back into dry heat.

From this subtropical convergence, the surface air flow is toward the equator in the gentle trade winds or poleward in the westerlies of the middle-latitude circulation cell. The westerlies converge, owing to the shape of the cells, and tend to be very intense. They storm through the latitudes of the "roaring forties" and the "furious fifties," famed in sea stories and notorious to seasick ocean voyagers. At roughly 60° latitude, at the subpolar convergence, the winds become calm but rain and snow increase. The change of weather with latitude is quite spectacular in the southern oceans, where continents do not distort the basic circulation. The subpolar convergence is generally like the intertropical one, except that the temperature is lower. Precipitation is more abundant here than at any latitude except the equator. Because precipitation exceeds evaporation at sea, the water has a relatively low salinity. The air pressure is low as the air rises; rain forms from cooling water vapor; and the winds aloft split in a divergence that moves one stream of air north in a Hadley cell and the other south in the return flow of the middle cell. At the poles, the air on the Hadley cell sinks and diverges to move weakly back toward the subpolar convergence over the dry polar wastes.

ZONAL OR LONGITUDINAL CIRCULATION

The circulation in latitudinal cells does not transport enough energy poleward to counteract the temperature gradient caused by radiation from the sun. Thus, an additional longitudinal (or zonal) circulation is required. Consider only the latitudinal cells: The temperature at the equator rises, and that at the poles

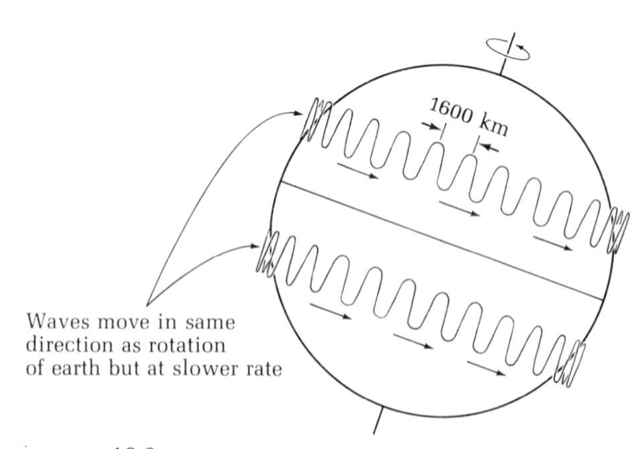

FIGURE 10.3

Zonal circulation occurs as waves that break and mix polar and equatorial air horizontally, thereby reducing the latitudinal temperature gradient.

drops, until the gradient reaches an intolerable critical value. Then zonal or longitudinal protrusions of warm and cold air begin to move north and south in the form of waves about 1600 kilometers long, which drift east a few days to a week apart.

These waves are the familiar cyclonic highs and lows that bring storms to the middle latitudes. The poleward heat transfer is predominantly by horizontal mixing as the waves break. This reinforces the vertical mixing of the latitudinal cells.

CONTINENTAL EFFECTS

With mention of cyclonic storms, we approach the subject of weather—but first the effects of land masses should be discussed. The continents become warmer in the summer and cooler in the winter than the oceans, which are heat reservoirs. In winter, the air over the cold continents contracts and more air flows in to fill the void aloft. This produces high pressure over continents. A compensating air flow outward occurs at ground level.

The circulation is reversed in the summer, when the pressure is low over continents and the surface winds are toward them. In India, this circulation is called the monsoons. The winter monsoon is dry, because it moves out from the land. The summer monsoon brings almost all the annual rain, because it moves moist air in from the oceans. The advent of the monsoons is reasonably predictable and its abruptness, after a brief period of stagnant weather, is famed in the literature of India.

GEOLOGIC FACTORS AFFECTING THE WEATHER OF THE PAST

Finally, before leaving the global circulation, we should consider possible changes that might be expected during geologic time. The major factor driving the circulation is the latitudinal gradient in incident radiation, and there is little reason to believe it has changed. On the other hand, the outgoing radiation, which affects the latitudinal temperature gradient, can be altered relatively easily by changes in cloud cover or by increased reflection from snow (increased relative to that from the soil that it covers). These effects would tend to intensify the circulation because they would increase the temperature gradient by increasing cooling in high latitudes.

The pattern of circulation is also affected by the gradual slowing of the earth's spin and consequent reduction of the Coriolis acceleration, but this change is difficult to detect in the geologic record.

Continental drift could produce more obvious effects on climate. For example, if a ring of continent covered the equator, the rising hot air of the Hadley cells adjacent to it would be dry and the transfer of heat by latitudinal cells would be much less effective. Because such a transfer must occur somehow, the zonal or longitudinal circulation would presumably be much more intense.

Likewise, continental agglomeration would have a major effect on circulation. At present, the Southern Hemisphere is largely water, but it once contained the great continent of Gondwanaland, which has since broken into Africa, South America, Australia, India, and Antarctica. Gondwanaland once exerted an influence over atmospheric circulation even more powerful than that of Asia today.

Weather and Storms

We may define as weather all such transient and more or less intense atmospheric events as rain, snow, hail, fog, sudden changes in temperature and pressure, winds, and other phenomena that may occur alone or grouped into storms. The key to understanding these diverse phenomena lies in correlating them with the global circulation, and in an understanding both of the factors that control winds and of the effects of vertical motions of air masses.

GEOSTROPHIC WINDS

Consider a high-pressure zone generated by a breaking wave in the zonal circulation of the atmosphere. Surfaces of equal pressure have the form of an invisible mountain, or, to put it in other terms, the pressure is greater at a given elevation in the high than in the surrounding normal air. The surface of equal pressure has a slope, and this generates a current in a fluid. If the earth did not rotate, a wind would blow down the slope just as water runs down a hill. However, the Coriolis acceleration turns the air so that it does not flow directly downslope. If there were no friction, it would be deflected until the Coriolis acceleration exactly balanced the acceleration of gravity, and the wind would blow at right angles to the slope or parallel to the contours. There is friction between moving air and the underlying surface, however, and the acceleration of gravity is balanced by the sum of the frictional drag and the Coriolis acceleration. This means that the wind is diverted by less than a right angle and flows somewhat downslope, therefore, instead of parallel to the contours. The direction of the wind depends on the hemisphere in which the high occurs. For a low-pressure zone, all the arguments are the same but the sense of motion is reversed: at the ground, friction causes winds that flow toward the center rather than away from it as in a high.

These patterns of flow can be summarized in what is called "Buys-Ballot's law" after a nineteenth-century meteorologist: *With regard to the Northern Hemisphere, if an observer stands with his back to the wind, the pressure is lower on his left than on his right.* An identical rule exists in oceanography,

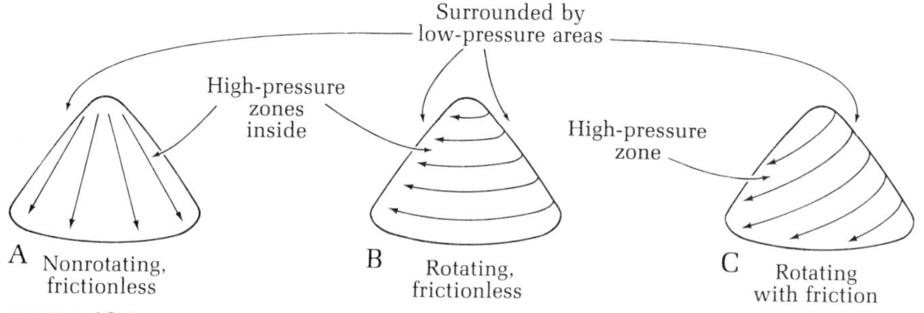

FIGURE 10.4
Wind patterns in an anticyclone (high pressure zone) in the Northern Hemisphere: *A*, assuming that the earth is nonrotating and there is no friction; *B*, assuming rotation but no friction; and *C*, on the real earth, with both rotation and friction.

although it is couched in different terms: *With regard to the Northern Hemisphere, if an observer has his back to the current, the density is higher on his left than on his right.* These statements may seem contradictory, but they merely reflect the fact that air is easily compressible and water is not.

CYCLONIC WINDS

Midlatitude zonal mixing of the air commonly occurs in the form of cyclonic storms, which are great masses of rotating and moving air characterized by abnormal pressures. A cyclone is a low-pressure air mass that, in the Northern Hemisphere, has a counterclockwise circulation that is deflected inward by friction. The air piles up in the center and, to maintain continuity of flow, rises, cools, and produces clouds and precipitation. Low-pressure areas, therefore, are regions in which the weather is generally considered to be "poor," except by city dwellers. The flow is clockwise in a cyclone in the Southern Hemisphere, but the air is still deflected inward and the weather is the same.

An anticyclone is a high-pressure air mass that, in the Northern Hemisphere, has a clockwise flow that is deflected outward by friction. Thus, cool air from aloft is sucked down in the center and, as it warms, is accompanied by what used to be called "good," clear weather. In the cities where people are now concentrated, however, smog tends to accumulate in clear, calm weather and to dissipate in cloudy, stormy weather. Thus, city dwellers are apt to call stormy weather "good" because they can breathe the air.

The most intense cyclonic storms are the lows that arise in the tropics and curve upward into middle latitudes. These are called hurricanes in the Atlantic and typhoons in the Pacific, and they are highly concentrated at the western margins of those oceans. In these great storms, wind velocities may surpass 160 kilometers per hour in the circular zone around the renowned "eye," which is calm. However, the eye, too, has its perils. The lowered air pressure and the dynamic stress of air moving toward the eye cause the level of the sea

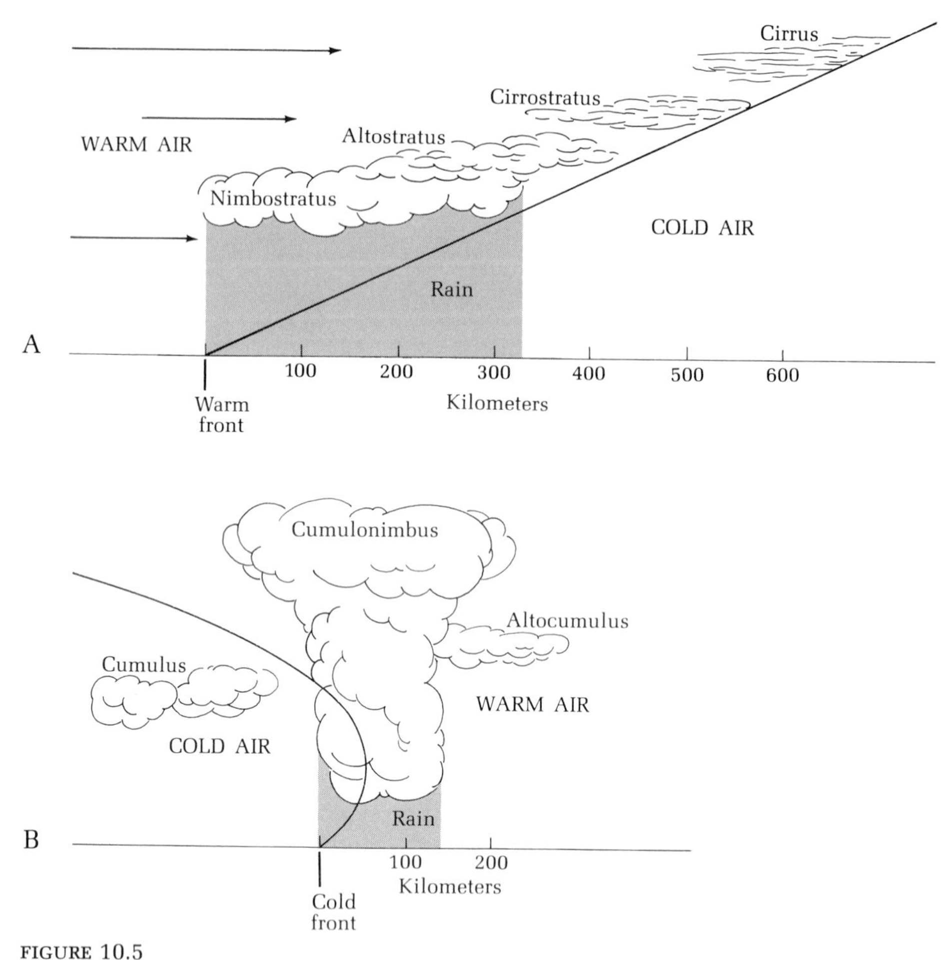

FIGURE 10.5
The interaction of air masses: *A*, a mass of warm air overriding cold air along a gently sloping surface, with high clouds heralding the approach of the front; *B*, a steep-faced cold-air mass displacing warm air, which produces a narrow front with relatively intense weather. [After Krauskopf and Beiser (1966).]

to rise as much as several meters in a storm surge. Enormous waves are generated by these storms and are superimposed on the storm surge. Inasmuch as the surge persists until the whole storm drifts away, unmatched flooding and destruction can occur when a hurricane or typhoon stalls at a shoreline. The devastation in the upper Bay of Bengal in 1971, which killed hundreds of thousands of people, was the consequence of a storm surge and waves.

As the warm air masses push north and the cold ones push south in the zonal mixing, they come together in the highs and lows along boundaries called fronts. A cold front is one in which the steep front of a moving cold mass of air pushes warm air aloft. The passage of such a front is accompanied by a regular sequence of weather and wind patterns that depends both upon the type of cyclone and upon the hemisphere. Warm fronts are those in which a

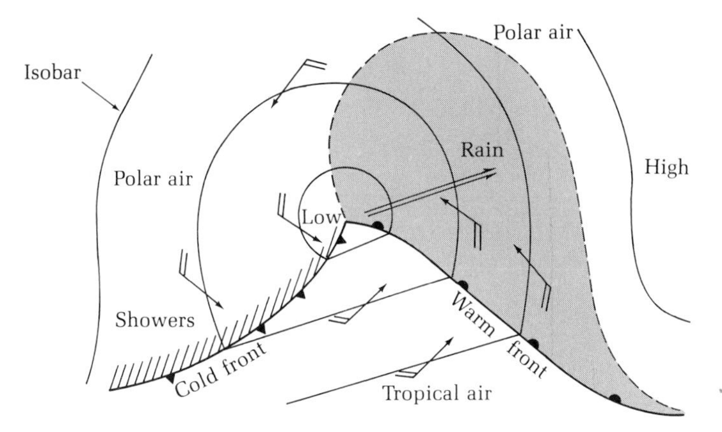

FIGURE 10.6

A typical weather map of a middle-latitude cyclone in the Northern Hemisphere. Warm, moist, tropical air is pushing northeast and, in turn, is being displaced by cold, dry, polar air moving southeast. The isobars are lines of equal pressure. The intensity of the wind is indicated by the number of lines in the flag of a wind arrow, which shows wind direction. Winds are curving into the center of the low-pressure area, which, in turn, is moving northeast (*double arrow*). [After Spar (1965).]

warm mass of air rides over a cold air mass. Because this type of front is gently sloping, the zone of rain is broad and the weather is not as intense as when a cold front passes.

OROGRAPHIC EFFECTS AND RAIN SHADOWS

An air mass contains a certain amount of water, and it can obtain only a little more by evaporation if it is over land. Thus, rain is not quickly replaced once it falls, and coastal regions tend to have higher rainfall than the interiors of continents. Precipitation may occur, however, if air is moved aloft during the passage of weather fronts, so rain may fall almost anywhere as the fronts shift from place to place. An air mass may also be elevated by the necessity of flowing over a mountain range, such as the Sierra Nevada in California, which is transverse to the prevailing air currents and strips almost all the moisture out of the air masses that move west from the Pacific Ocean. Inasmuch as the mountains are fixed, the area of Nevada to the east lies in a "rain shadow" and is a desert. These circumstances have not always existed. In the recent geologic past, the biota of Nevada was characteristic of a much more humid climate, which suggests that the Sierra Nevada was not very high at the time.

The effects that result from the passage of an air mass over a mountain are called orographic. They are particularly spectacular on oceanic islands. Because islands, including atolls (which are rings of low land around lagoons),

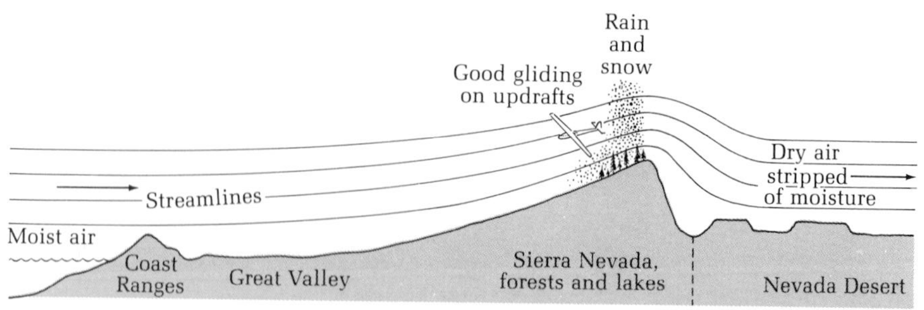

FIGURE 10.7
The moist Pacific air masses are elevated over the Sierra Nevada, cooled, and stripped of moisture. The air reaching Nevada is very dry and a desert results.

have different thermal characteristics than the surrounding sea, they tend to be the locus of formation of clouds, which help the mariner to locate them. Low islands without lagoons, however, produce no orographic effect, and they tend to be the arid, barren homes of sea birds, sea mammals, and turtles. The high Hawaiian Islands present a contrasting picture of lush greenery, but have marked climatic zones at different elevations. The rainfall at Honolulu, for example, is only a tenth as great as at the Pali, which is only 10 kilometers northeast and about 900 meters higher. A rain shadow exists on the leeward sides of high islands, and even in Hawaii such shadows tend to be deserts. Occasionally, the arid conditions may be interrupted by torrential downpours, if storms change the normal, persistent wind patterns.

A rain shadow may also exist downwind from a coast off which there is an upwelling of cold, deep-ocean water, a phenomenon that will shortly be discussed. The air above this water contains little moisture because it is cold. Thus, if it is heated, it does not yield rain; and if it rises and cools in a front or over a mountain range, it still yields little water as precipitation.

Water Budget

The distribution and movement of water are considered in this section in the broadest terms. In addition to geography, we shall draw upon the concept of residence time for insights into this subject. Residence time can be defined as the volume of material in a container divided by the rate at which it moves in or out of that container. If the volume in the container is constant, the flow in and the flow out are equal. Residence time gives an average value that indicates nothing about extreme variations. Some fraction of a material may remain for a long time in the container while another fraction moves in and out quickly. Nevertheless, the average values themselves are revealing and useful to consider.

TABLE 10.1
Distribution and residence time of various components of the global water budget. The world water budget is a tiny though important portion of the world water supply.

Water item	*Volume (km^3)*	*Percentage of total*	*Residence time (yr)*
SUPPLY			
World ocean	1,370,000,000	97.2	40,000
Icecaps and glaciers	29,200,000	2.13	10,000
Ground water to depth of 4000 meters	8,350,000	0.59	5,000
Fresh-water lakes	125,000	0.0089	100
Saline lakes and inland seas	104,000	0.0074	100
Soil moisture and vadose water	67,000	0.00475	1
Atmosphere	13,000	0.00092	0.1
Rivers (average instantaneous volume)	1,250	0.00009	1
Total, all items	1,410,000,000	100	
BUDGET			
Annual evaporation from world ocean	350,000	0.026	
Annual evaporation from land areas	70,000	0.005	
Total annual evaporation	420,000	0.031	
Annual precipitation on world ocean	320,000	0.024	
Annual precipitation on land areas	100,000	0.007	
Total annual precipitation	420,000	0.031	
Annual runoff to oceans from rivers and icecaps	32,000	0.003	
Annual ground-water outflow to oceans	1,600	0.0001	
Total annual runoff and outflow	33,600	0.0031	

Source: Data, in part, after Nace (1967), modified to agree with ocean volume in Menard and Smith (1966).

DISTRIBUTION

Over 98% of the surface water of the earth is in the oceans, which is merely another way of saying that water runs downhill and takes little time to do so. It is elevated by solar energy as water vapor, but only 0.001% of all the earth's water is in the air. The fraction in rivers is even smaller, another indication of rapid flux. The remaining water, perched temporarily between sky and sea, is surprisingly small. Lakes, fresh or salt, are few and shallow and contain little more than 0.01% of the water. Soil, more widespread but shallower, contains a similar fraction. The pore spaces in near-surface rocks, plus cracks, channels, and caves, contain about 0.5% of the world's water because they may extend for as much as 4000 meters below the surface.

Ice caps and glaciers store about 2% of the earth's surface water at present; thus, they are by far the most important loci of perched water. This is a particularly significant fact, because of the relative ease with which this store can vary. Even if lakes and swamps covered the land, they would be too shallow to contain much water. Likewise, variations in ground water are not usually very significant. Glaciers, however, could cover the land to a depth of a few kilometers and capture most of the water in the sea. The shallow sea bed itself could be covered with glaciers and the deep sea with floating ice. This has never happened, but glaciers have been much more extensive in the past than they are now.

BUDGET

Evaporation and precipitation are in approximate balance over the whole world, but evaporation exceeds precipitation over the oceans and precipitation exceeds evaporation over the continents. Equilibrium is restored by outflow of rivers and ground water. The global circulation of water is geologically very fast, as can be seen from its residence time in various places. Evaporation is so great compared to the volume of water in the air at any one time that the residence time is only about 12 days. In short, much atmospheric water falls back as rain soon after it is evaporated.

The average evaporation from the ocean is 1 meter per year, although it is much greater in some places than in others. Thus, an ocean 5000 meters deep would be completely evaporated in 5000 years, were there no replacement. Even with recycling, the residence time is short: about 40,000 years is required to transport the oceans to the continents and let their entire volume flow back at normal rates through river channels. Most salt is left behind by evaporation in this water cycle, and the rains and rivers bring more of it to the sea. However, salt has its own cycle, and it is removed from the world ocean by the complete evaporation of small seas.

The residence times for water in fresh lakes is trivial because of the small volume, the rapid through-flow, and the speed of evaporation. Because wastes remain behind, pollution may build to undesirable levels, but the water comes and goes quickly.

The average residence time for ground water is only 5000 years; but because water, like oil or gas, can be trapped underground for tens of millions of years, the *range* of ages is great compared to the range for water in lakes.

Glaciers also store water for at least thousands of years by piling one fall of snow upon another. By the time they are a few kilometers thick, however, they cannot support their own weight, and the old central ice moves to the edges where it melts. The residence time for water in glaciers is probably a few tens of thousands of years, which is longer than for any other reservoir—except, perhaps, the ocean.

EFFECTS OF CHANGES IN BUDGET

About 20,000 years ago, large areas of North America and Eurasia were covered by thick continental glaciers and the ocean was reduced enough to expose the continental shelf. Melting then began, and it reached its present stage about 5000 years ago. Similar glacial periods had occurred before. Did they require enormous changes in the water budget? Available data suggest that they did not.

At their maximum extent, the glaciers had a volume of roughly 100 million cubic kilometers, which is perhaps four times as great as their present volume. Consider an ice-free world: If evaporation from the sea increased by only 1% above the present, and if the additional water vapor collected as ice, glaciers equivalent to the great glaciers of the ice ages could be built in only 30,000 years. If evaporation did not change at all, but all precipitation on the continents was in the form of snow with no runoff, such glaciers could accumulate in only 3000 years. Likewise, a small increase of continental reflectivity would occur if permanent snow covered the ground, and this would decrease evaporation, enhance cooling, and favor the formation of more snow and ice. The general conclusion to be drawn from these calculations is that it is quite reasonable to suppose that small variations in the water budget could cause enormous variations in the extent of continental glaciers. This may account for the number of glacial periods in the recent geologic past. Their relative rarity in earlier times may indicate that average climatic conditions have been significantly different from those of the present during most of the history of the earth.

Oceanic Circulation

The density of the sea varies because of heating in the tropics, cooling at the poles, differences in salinity and evaporation, and the introduction of fresh water by rivers and ice. These differences generate persistent and large currents. In addition, the continuing stresses of the global wind pattern generate currents and surface waves. The tides also create currents that may be very intense in shallow water. Other sources of motion are much less important, but they include storm surges, splashes, and tsunamis, which are long waves generated by sea-floor motion. Of the various motions of the sea, we will consider currents first.

Sea water has a low viscosity, currents generally move at speeds of 1 kilometer or so per hour, and the dimensions of the currents and the ocean are very large. Hence, the motion of the sea is highly turbulent. Nevertheless, the area of the sea is enormously greater than its depth, and the dominant motion of currents is horizontal. Water flows downhill; it also flows across a level floor from a place where the water is deeper (has a higher surface) to where it is shallower (has a lower surface). The sloping surface between the two places is called a hydraulic slope. In the ocean, the actual slope of the sea surface is still almost impossible to measure for lack of a frame of reference. Thus, it is useful to define a level of surface as an imaginary one on which no component of gravity acts—the surface of a fluid at rest is a real example. An isobaric surface is an imaginary one along which the pressure remains constant—any level in a homogeneous, isothermal liquid at rest is a real example. However, if the isobaric surface and the level surface are not parallel, a pressure gradient exists and a force is exerted on the fluid in the direction of the gradient and toward the lower pressure.

The force does not generate a current that flows down the gradient, however, because the Coriolis effect acts as it does with winds, and bends a current moving down a pressure gradient until the component of gravity down the gradient just balances the Coriolis acceleration up the gradient. The result is rather curious but important: currents generated by density differences in the ocean move parallel to the contours of the sloping isobaric surface. This means that, in the Northern Hemisphere, an observer looking in the direction of the current finds lighter water on his right hand. The reverse effect occurs in the Southern Hemisphere—the lighter water is to the left. This effect is of geologic importance. For example, rivers introduce low-density water at the margins of the oceans. Off the northeastern United States, the nearshore currents generally set toward the south, thus keeping the low-density water to their right. The opposite effect occurs in the Southern Hemisphere. In general, river waters hug the shore in high latitudes. The effect does not exist at the equator: as it happens, the only large river exactly on the equator is the Amazon, and its water moves far out to sea.

The Coriolis effect also influences the motion of currents generated by winds. Ekman calculated theoretically, and it has been found by observation, that a surface current or wind drift is at an angle of 45° to the generating wind stress in the Northern Hemisphere. The surface current likewise stresses the water directly below, and it also moves in a direction more to the right and at a diminished velocity. This continues downward until a weak current moves in the opposite direction to the surface current. Projected to the surface, the velocity vectors form a logarithmic spiral called the Ekman spiral. The depth at which the current is reversed, which can be regarded as the one to which mixing by the wind extends, varies with latitude.

FIGURE 10.8

Surface currents in the ocean basins in an unusual projection to emphasize the central role of the Antarctic Ocean. The Equatorial Countercurrent, which is present in all the oceans, lies five degrees north of the geographic equator. [After Munk, "The Circulation of the Oceans," copyright © 1955 by Scientific American, Inc. (all rights reserved).]

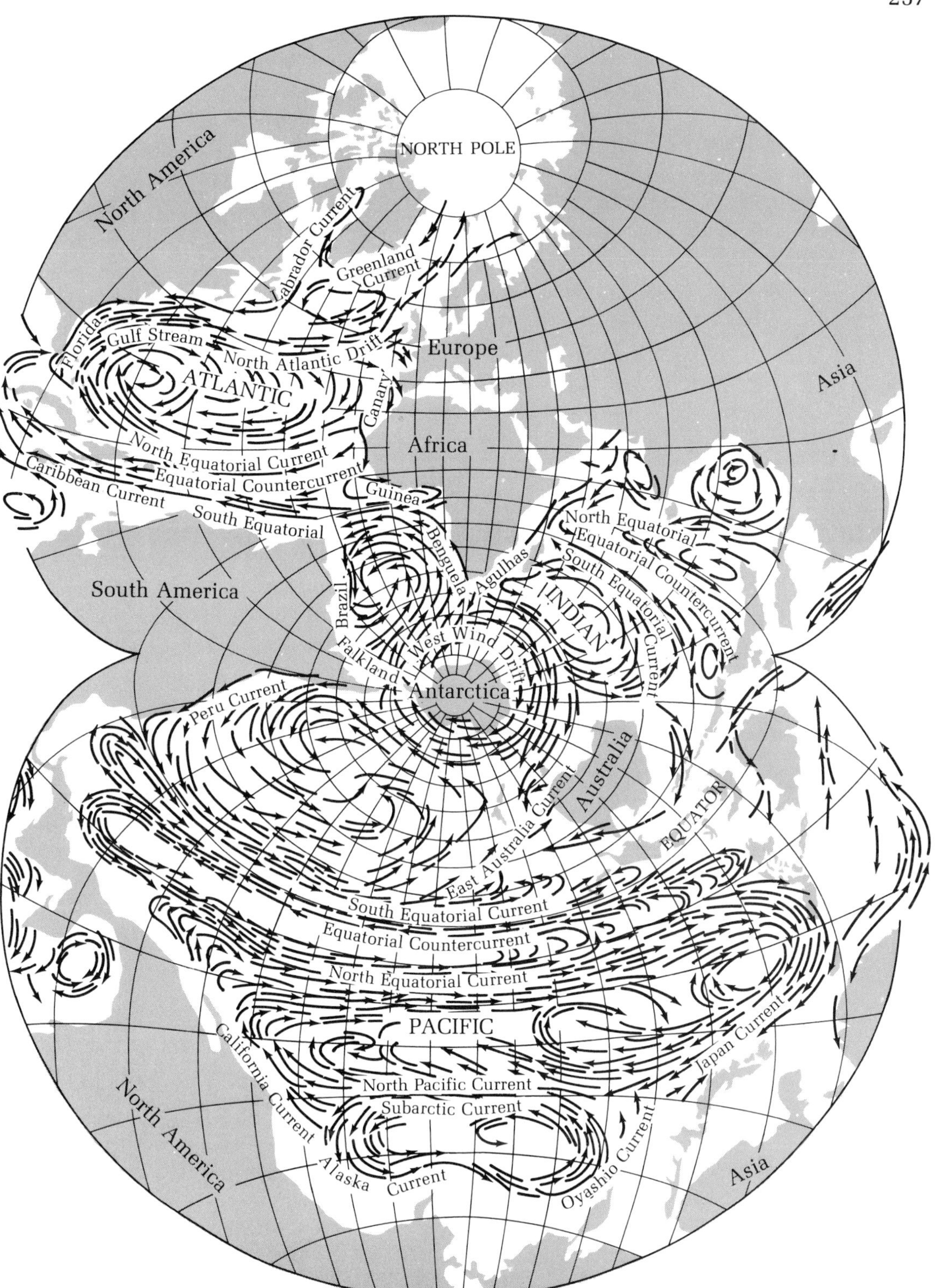
NORTH POLE
North America
Labrador Current
Greenland Current
Florida
Gulf Stream
North Atlantic Drift
Europe
Asia
ATLANTIC
Canary
Africa
North Equatorial Current
Equatorial Countercurrent
Caribbean Current
Guinea
South Equatorial
Benguela
North Equatorial
Equatorial Countercurrent
South Equatorial Current
Agulhas
INDIAN
South America
Brazil
West Wind Drift
Falkland
Antarctica
Peru Current
Australia
West Australia Current
EQUATOR
East Australia
South Equatorial Current
Equatorial Countercurrent
North Equatorial Current
PACIFIC
California Current
Japan Current
North Pacific Current
Subarctic Current
North America
Alaska Current
Oyashio Current
Asia

At the equator, the Coriolis effect is zero and there is no Ekman spiral. Elsewhere, the effects are striking and important. Consider a wind moving south in the Northern Hemisphere along the western coast of a continent: It generates a current moving offshore at the surface with a return of water from depth to the shore. The upwelling water is relatively cold and loaded with the nutrients characteristic of deep water. Thus, it discourages swimmers, generates fog, and creates prosperous fisheries based on the proliferation of organisms. This is the description of coastal southern California, coastal Chile (reversing the geography), and so on.

Tidal currents are also affected by the Coriolis effect, but it is inconsequential in the narrow bays and shallow waters in which tidal currents reach a velocity of several kilometers per hour and have their most important geological effects. They scour deep holes at the narrow mouths of bays like San Francisco Bay and deposit submarine bars on opposite ends of the hole. They move vast drifts of sand around on the shallow floor of the North Sea and the Strait of Dover. The northwest and southwest winds of high latitudes drive easterly ocean currents, and the northeast and southeast trade winds drive equatorial currents to the west. The equatorial currents are separated by a countercurrent, which forms where the wind stress is at a minimum. Except around Antarctica, these latitudinal circulations are interrupted by the Americas, Eurasia-Africa, and Eurasia-Australia. In order to maintain continuity, the West Wind Drift of high latitudes and the equatorial currents must be connected by currents toward the equator on the east sides of oceans and by currents away from the equator on the west sides. The equatorward currents are relatively weak, but those away from it include the Gulf Stream and Kuroshio Current, which are the most intense in the oceans. The American oceanographer Henry Stommel has shown that the westward intensification of these currents is caused by the variation of Coriolis effect with latitude. The western boundary current in the South Atlantic is less intense than the Gulf Stream, apparently because the intensification of Coriolis effect is approximately balanced by a deepening of the current. The discharge increases rather than the velocity.

These surface currents have various effects of geologic significance. By shifting isotherms north and south, they affect the distribution of coral reefs and, thus, the erosion of volcanic islands and continental margins. The Gulf Stream impinges on the continental margin directly south and east of Florida and Georgia and inhibits deposition in the Straits of Florida and on the nearby Blake Plateau. The West Wind Drift and the equatorial currents are the loci of intense shear, which favors nutrient flux and high productivity with the resultant deposition of siliceous and calcareous oozes. Cold currents favor the equatorward movement of icebergs and organisms; thus, penguins live in the Galápagos Islands on the equator.

MIDWATER CURRENTS

Various major currents also exist in midwater, but, touching neither the surface nor the bottom, they have little geologic effect. They are of two types—those generated by differences in density and those that result from counterflows to surface currents. Counterflows exist under many western boundary currents, but most are of low velocity. However, the extraordinary Cromwell Current lies exactly on the equator below the Equatorial Current, and it has a velocity of more than 1 kilometer per hour toward the east. The water that flows from the Mediterranean to the Atlantic is an example of the other type. The dense saline water of the Mediterranean flows west across the floor of the Strait of Gibraltar and plunges to a depth of about 1000 meters. There it encounters even denser water and spreads horizontally for more than a thousand kilometers. Perhaps it intersects the submarine slopes of an occasional volcano and slightly inhibits sedimentation, but otherwise it hardly affects marine geology.

DEEP CIRCULATION

The movement of deep currents, in contrast, has profound geologic effects. The surface currents extend down only a kilometer or so and intermediate currents also are largely horizontal. Thus, the deep water, if it has a circulation, moves quite independently of the overlying ocean. The only way to drive the circulation is to plunge dense water from the surface in enormous quantities. Density increases as the salinity rises, which occurs generally in low and middle latitudes, and as the temperature drops, which occurs in high latitudes. Thus, water dense enough to cascade to the bottom is rare and forms at present in only two circumstances: (1) high-salinity water moves into high latitudes and is cooled; and (2) Relatively high-salinity water freezes. The only important places where these phenomena occur at present are in the North Atlantic and the Antarctic, respectively. The Gulf Stream carries saline warm water to the Irminger, Labrador, and Norwegian seas where it is cooled, plunges to the bottom, and moves slowly south, hugging the bottom, to the southernmost Atlantic. Most Antarctic bottom water originates on the continental shelf of the Weddell Sea. As shelf ice forms, it removes water and concentrates a residual brine that is very cold and cascades down the continental slope. Thence it spreads north along the bottom of the Atlantic and also mixes with the North Atlantic bottom water. Together, they flow into the Indian and Pacific oceans and spread slowly to their northern boundaries. The residual flow returns to the Atlantic through Drake Passage, between South America and the Antarctic Peninsula.

Throughout their route, the bottom currents are guided by the bottom topography. They are confined to the deepest parts of basins, and cannot pass

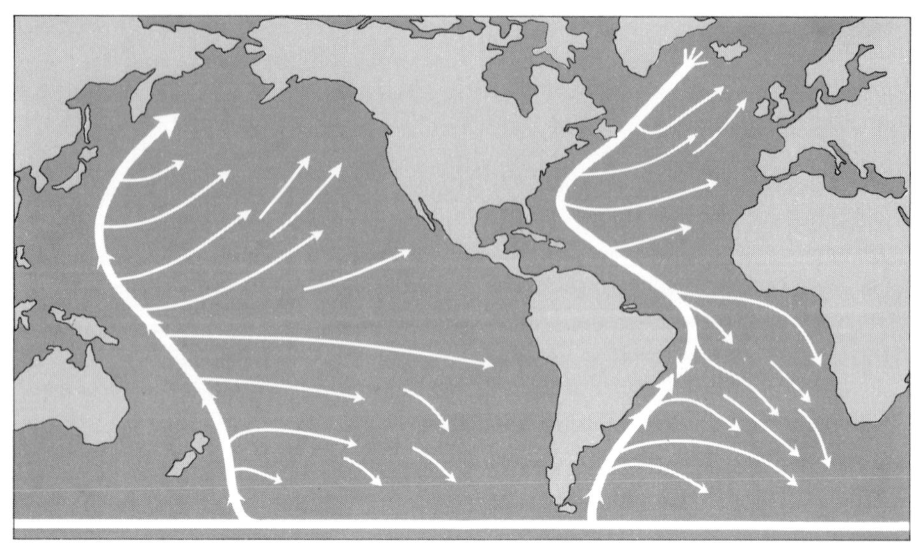

FIGURE 10.9
Cold deep water originates in the Antarctic and North Atlantic, flows in western boundary currents along the sea floor, and spreads laterally. This illustration is highly diagrammatic. [After Stewart, "The Atmosphere and the Ocean," copyright © 1969 by Scientific American, Inc. (all rights reserved).]

over midocean ridges or the great volcanic ridges of island archipelagoes, island arcs, and fracture zones. On the other hand, they can flow *through* midocean ridges where they are cut by the deep troughs of fracture zones, and this is observed in many places in the Atlantic. Likewise, they can flow through the passes in volcanic groups, as they do in the central Pacific, and around island arcs and into isolated basins, as they do in Melanesia. In most places, the currents have velocities of only a few centimeters per second, but they accelerate where they go through narrow passes. There they scour oozes, expose bedrock, and concentrate lag gravels of manganese nodules. Where the passes are broader, they produce current ripple marks in oozes and prevent the deposition of sediment carried in suspension. The bottom currents also are geologically important where they are concentrated by the Coriolis effect against the western boundaries of basins. This produces contour currents, which are powerful enough to shape the morphology of the continental rise and even the basin floor. For this reason, the turbidites on the east side of the Pacific remain deposited in topographic fans, whereas those on the west side of the Atlantic are reworked and the fans are reshaped into a continental rise.

Before leaving ocean currents, we should consider the fact that the present pattern of circulation, shallow and deep, is not the only possible one. The principles that govern the circulation do not alter, but the geography and climate that influence the importance of various factors are mutable. For example, if the Mediterranean water had a slightly higher salinity, and if no dense bottom water existed, the Mediterranean midwater would plunge to the bottom of the Atlantic and the circulation would be totally different. Likewise, when there were no polar continents, because of continental drift, water may not have been chilled enough to plunge and the deep ocean may have been much quieter. Eventually, heat flow from the interior may warm the bottom water until it rises, or evaporation may increase surface salinity to the point that the surface water is dense enough to plunge. JOIDES drilling indicates that deep ocean circulation was quite different in the past.

Waves and Tides

Waves are a delight to the surfer, a distress to the voyager, a danger to the beach dweller, and a dilemma to the coastal engineer. We are concerned here only with phenomena of interest in environmental geology—namely, tides, wind waves, tsunamis, storm surges, and various sorts of rare but startling waves. We have already considered the body waves that move through the earth, triggered by earthquakes or explosions. Now we are involved with surface waves that are also generated in the earth by earthquakes and explosions but that have characteristics different from those of body waves. Surface waves are, for example, influenced by the depth or thickness of the medium through which they travel.

As with other waves, we distinguish between the motion of a particle of water in a wave and of the wave form itself. However, even the particle motion depends on water depth, and so we should begin with a consideration of the wave itself.

The phase velocity of a wave is the velocity at which the wave form moves forward. The equations of motion vary with depth; for geologically important phenomena, however, the water may be considered shallow, and the equation is the same as that for a gravity wave in which the velocity is equal to the square root of gravity times the water depth. Shallow water may be defined as water having a depth of less than one-twentieth of the wavelength of a wave—the length from crest to crest. The average depth of the deep sea is about 5 kilometers, and the wavelength of the tides is about half the circumference of the earth while that of tsunamis is 150–250 kilometers. Thus, tides and tsunamis both are shallow waves, even in the deep sea. In fact, the first reasonable estimate of the average depth of the sea (4 kilometers, made by A. D. Bache in 1856) was based on the phase velocity of tsunamis. Shallow waves produce oscillatory or orbital motion from the surface to the bottom of the sea. Whether the bottom motion of such a wave is capable of moving sediment depends on the sediment and on the configuration of the bottom. Such waves offer a reasonable mechanism for the wafting of sediment from

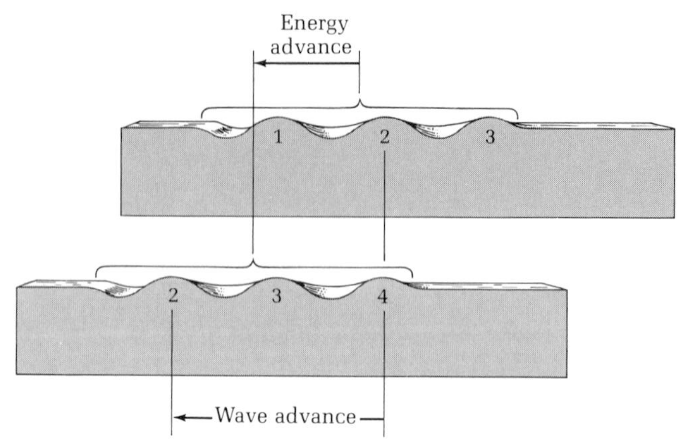

FIGURE 10.10

A moving train of waves advances at only half the speed of its individual waves. At top is a wave train in its first position. At the bottom, the train and its energy have moved only half as far as wave *2* has. Meanwhile wave *1* has died, but wave *4* has formed at the rear of the train to replace it. Waves arriving at shore are thus remote descendants of waves originally generated. [From Bascom, "Ocean Waves," copyright © 1959 by Scientific American, Inc. (all rights reserved).]

abyssal hills to the deep-sea floor in regions above the influence of bottom currents.

Waves erode vigorously, which explains the existence of sea cliffs, but how deep under the sea do they erode? Consider the expenditure of wave energy on the continental shelf: This varies with the stress exerted on the bottom, which, in turn, is a function of the orbital velocity. A particle of water in a wave moves in a circle or an ellipse as the wave form passes, with little or no translation in the direction of the wave motion except when the wave breaks. The velocity at which the particle moves in this manner is its orbital velocity. The equations describing changes in orbital velocity with depth vary according to the type of wave, but the orbital velocity is always a function of the phase velocity and the ratio of wave height to water depth. For a small wave with a height of only 1 meter in water 3 meters deep, the maximum horizontal orbital velocity is more than 1.5 meters per second, and it is half that at the bottom. The velocity at the bottom for even this small wave is about ten times as large as that required to transport sediment. Consequently, common waves lose some energy even on the outer parts of the continental shelf, but the amount they lose is relatively unimportant.

Indeed, almost all wave energy is expended in the surf zone or above the surf depth. The percentage of the energy available where the wave breaks varies inversely with the period of the wave and directly with the slope of the beach. Even small waves have 30% of their energy remaining when they break after crossing an extremely flat continental slope. On a steep slope, they have

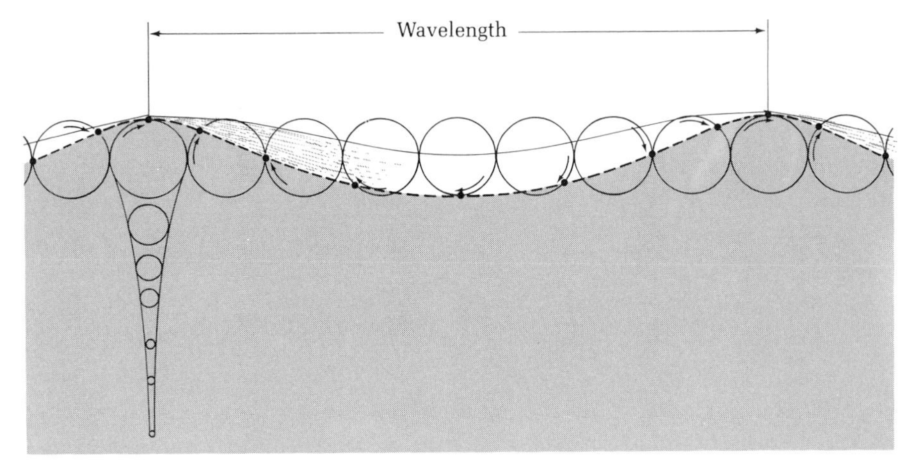

FIGURE 10.11
A cross section of an ocean wave traveling from left to right, showing the wavelength as the distance between successive crests. The time between the passage of two crests is the wave period. The circles are the orbits of water particles in the wave. At the surface, their diameter equals the wave height. At a depth of half the wavelength (*left*), the orbital diameter is only 4% of that at the surface. [From Bascom, "Ocean Waves," copyright © 1959 by Scientific American, Inc. (all rights reserved).]

almost all their energy when they break. Larger waves retain more than 60% of their energy when they break, regardless of the bottom slope. With regard to the expenditure of energy above surf depth, it should be noted that the smaller waves hardly affect the bottom in depths greater than those in which the larger, longer-period waves break. In short, a highly significant discontinuity in energy expenditure occurs at surf depth, and it is the only one in the ocean. Evidence of erosion of a rock platform at greater depths is apt to be an indicator of a relative change in sea level.

The second geologic phenomenon that is influenced in an important way by waves is the movement of sediment below the surf depth. At continental-shelf depths, orbital velocities for very fast waves, such as the tides and tsunamis, are capable of stirring sediment. At greater depths, the issue is not so clear, because the height of the long-period waves is so small. A tide 1 meter high would be quite exceptional in the open ocean, and its orbital velocity in deep water is only about the minimum required to move sediment, although it could produce scour if concentrated by an obstruction. Orbital velocities likely to erode all sediment seldom occur below 1800–2100 meters—that is, only on the continental slope and on such isolated highs as volcanoes.

Yet another type of long-period wave may affect sediment at all depths in the ocean—namely, the rare waves generated by meteorite impact. R. J. Seymour of Scripps Institution of Oceanography recently analyzed the effect of meteorite-generated waves on islands, and such waves also affect the sea floor. Assuming the same rate and size of cratering on earth as on the moon, about seven cratering events occur in the oceans every million years. Of these,

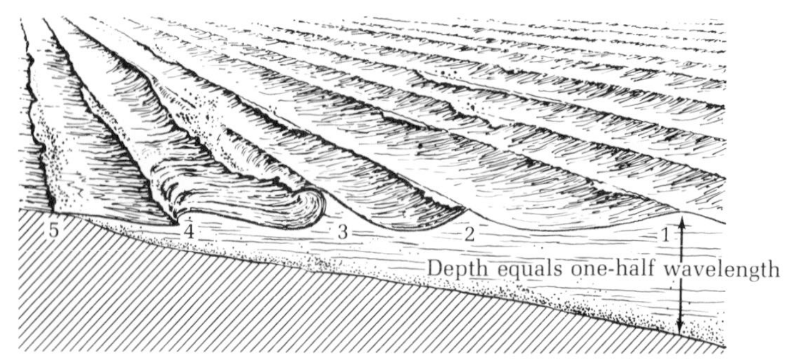

FIGURE 10.12

A wave breaks up at the beach when the swell moves into water shallower than half the wavelength (*1*). The shallow bottom raises wave height and decreases length (*2*). At a water depth 1.3 times the wave height, water supply is reduced and the particles of water in the crest have no room to complete their cycles; the wave form breaks (*3*). A foam line forms and water particles, instead of just the wave form, move forward (*4*). The low remaining wave runs up the face of the beach (*5*). [From Bascom, "Ocean Waves," copyright © 1959 by Scientific American, Inc. (all rights reserved).]

one results from the impact of a meteorite about 400 meters in diameter. On the average, a meteorite nearly 1100 meters in diameter hits the ocean every 5 million years.

The effects of meteoritic impact on the ocean are very similar to those of an underwater nuclear explosion. The meteorite mainly vaporizes and forms a large cavity in the ocean's surface that collapses and projects a great column of water into the air. This, in turn, collapses and generates a train of waves that dissipate as they spread. These are very long waves; hence, they behave according to the equation for shallow-water waves, even in the deep sea. Seymour calculates that a meteorite 400 meters in diameter would generate waves that would still be 12 meters high after traveling 1600 kilometers.

Many uncertainties remain, but it appears highly probable that sediment at all depths is occasionally put in suspension by the passage of a train of waves generated by a meteorite. What if the meteorite that made Barringer Crater had hit only a thousand kilometers to the west in the Pacific? Although such events are rare, their effects may appear in pelagic sediment because it accumulates extremely slowly. Red clay, the most common type of sediment, is deposited at the rate of about 1 meter every million years. Consequently, the sedimentation of 30 meters of red clay is accompanied by the impact of scores of large meteorites in the global sea.

Consider now a third geologic problem—the influence of wave action upon shoreline erosion and coastal morphology. Ocean waves are refracted like other waves, if the velocity of part of the wave form changes. This occurs, if parts of a shallow wave are in water of different depths, because of variations in bottom friction. The geologic effect resulting from refraction is that wave

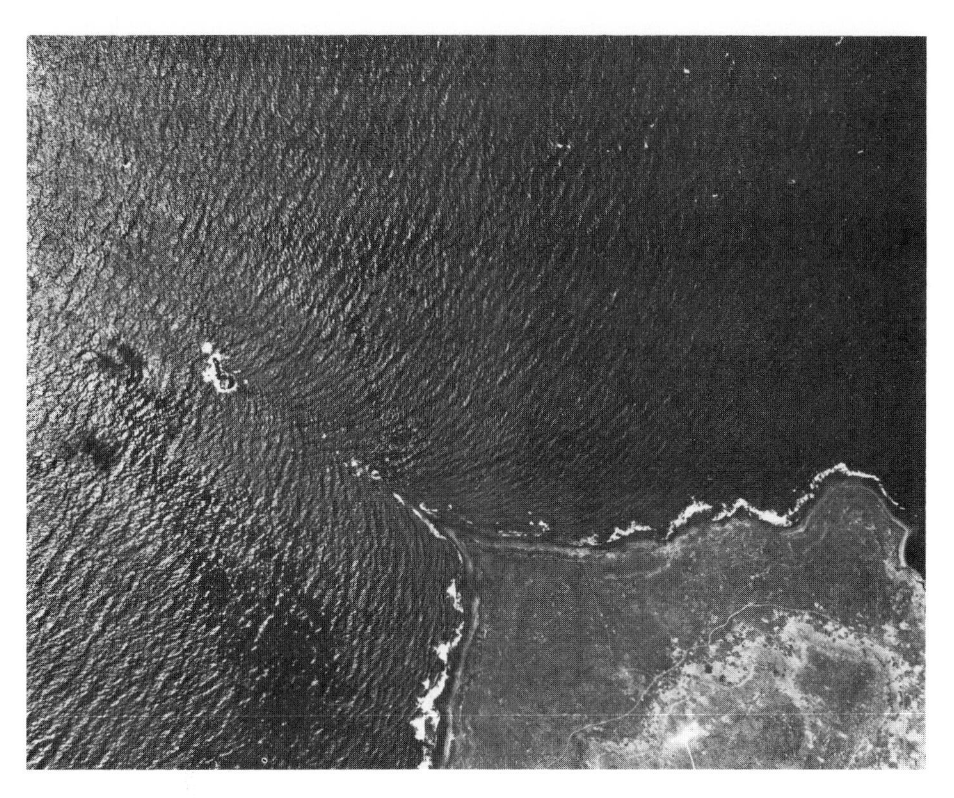

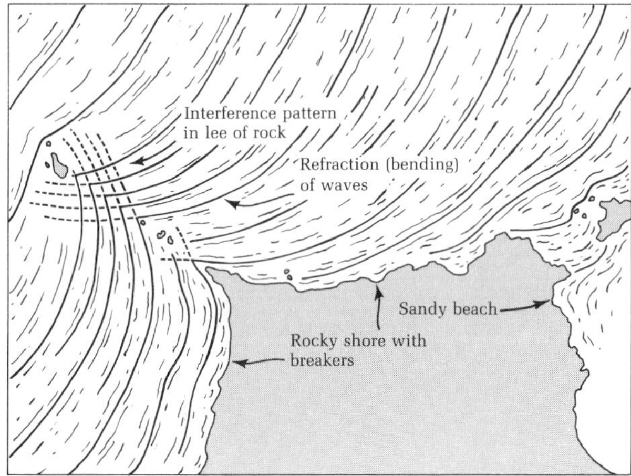

FIGURE 10.13
Vertical air photo and diagram showing bending of the waves (wave refraction) around a point of land. The photograph was taken at the north end of San Clemente Island, California, on June 20, 1944. [Official U.S. Navy photo.]

energy is concentrated on headlands and away from coves and, thus, tends to straighten the shoreline. The refraction of the waves also generates longshore currents, which have a strong influence on sediment transport.

The shape of coasts is likewise affected by various extreme types of waves, such as storm surges and tsunamis, the rare and local landslide splashes, and the very rare but far-reaching waves generated by the impact of large meteorites. These waves are refracted like wind waves; thus, the important question is whether they are so infrequent that they have little effect on coastal geology, but as to that we have no answer. The situation with regard to coastal structures is different. All types of large waves undermine, batter, and demolish structures that are within their reach. The bigger waves will reach structures that are safe from more common ones, but a building may collapse of old age before a truly big wave rolls in. These and other effects of shore processes are discussed in Chapter 16.

Summary

1. The energy of the sun drives the atmospheric circulation, which distributes heat relatively uniformly over the globe.
2. The circulation of the atmosphere consists primarily in hot air rising at the equator and cold air sinking at the poles, but other effects cause a more complicated latitudinal circulation. The excess solar energy is distributed by the longitudinal circulation, which produces most of the weather.
3. The midlatitude zoning mixing commonly occurs as cyclonic storms, which vary according to the movements of cold and warm air masses.
4. Air masses that ride over mountains are cooled and stripped of their moisture. This leaves a dry area, or rain shadow, downwind.
5. Globally, evaporation balances precipitation, but evaporation dominates at sea and precipitation dominates on land. Evaporation could remove all the ocean's water in 5000 years, if there were some place to put it. Glacial ice on land is a place in which large quantities of water are (and have been) stored.
6. The density of the sea varies with salinity and temperature, and gravity causes dense water to flow downward, displacing water that is less dense. The direction of the resulting current is controlled by the Coriolis effect of the rotation of the earth.
7. The bottom water in the ocean originates at the surface in polar regions where it reaches a maximum density. It cascades to the bottom and spreads.
8. Many kinds and sizes of waves exist in the sea. They derive energy from winds, gravity, volcanoes, earthquakes, and other sources, and they expend most of it in a shallow zone at the shoreline.

Discussion Questions

1. Why is water so important on earth? Why not elsewhere in the solar system?
2. Why do the trade winds north and south of the equator move in different directions?
3. Why is a cold front steep and a warm front gently sloping?

4. How do we know that global evaporation approximately balances global precipitation?
5. Why is the residence time of ground water more variable than that of ocean water?
6. What feature of oceanic circulation dominates the coastal climate of southern California?
7. Why are ocean waves mainly parallel to the shoreline?

References

Bascom, W., 1959. Ocean Waves. *Sci. Amer.* 201(2):74–84. (Available as *Sci. Amer.* Offprint 828.)

Forchhammer, G., 1865. On the composition of sea water in the different parts of the ocean. *Phil. Trans. Roy. Soc. London* 155:203–262.

Goldberg, E. D., 1963. The oceans as a chemical system. *In* M. N. Hill, ed., *The Sea,* vol. 2, pp. 3–25. New York: John Wiley & Sons.

——, and M. Koide, 1962. Geochronological studies of deep-sea sediments. *Geochim. Cosmochim. Acta* 13:417–450.

Ion, D. C., and F. W. Shotten, 1965. *Salt Basins around Africa.* Amsterdam: Elsevier. [Salt basins occur on the continental shelf, indicating the Atlantic was once a very narrow sea as it began to open.]

Krauskopf, K., and A. Beiser, 1966. *Fundamentals of Physical Science.* New York: McGraw-Hill.

Lowman, P. D., Jr., 1968. *Space Panorama.* Zurich: Weltflugbild. [A collection of 69 annotated photographs from American space craft. Many weather phenomena are self-evident.]

Malkus, J. S., 1957. The origin of hurricanes. *Sci. Amer.* 197(2):33–39. (Available as *Sci. Amer.* Offprint 847.)

Mason, B., 1966. *Principles of Geochemistry.* New York: John Wiley & Sons. [A broad but technical summary.]

McDonald, J. E., 1952. The Coriolis effect. *Sci. Amer.* 186(5):72–78. (Available as *Sci. Amer.* Offprint 839.)

Menard, H. W., and S. M. Smith, 1966. Hypsometry of ocean basin provinces. *J. Geophys. Res.* 71(18):4305–4325.

Munk, W., 1955. The circulation of the oceans. *Sci. Amer.* 193(3):96–104. (Available as *Sci. Amer.* Offprint 813.)

Nace, R. L., 1967. Water resources: a global problem with local roots. *Environ. Sci. Technol.* 1:550–560. [Describes the nature and purpose of the global cooperative study called the International Hydrological Decade.]

Oort, A., 1970. The energy cycle of the earth. *Sci. Amer.* 223(3):54–63. (Available as *Sci. Amer.* Offprint 1189.)

Peterson, M. N. A., 1966. Calcite: ratios and dissolution in a vertical profile in the central Pacific. *Science* 154:1542–1544. [Experiments show that calcium carbonate is dissolved more rapidly at depth.]

Shepard, F. P., 1963. *Submarine Geology* (2nd ed.). New York: Harper & Row. [Chapter 3, by D. L. Inman, contains a comprehensive analysis of ocean waves.]

Spar, J., 1965. *Earth, Sea, and Air*. Reading, Mass.: Addison-Wesley.
Stewart, R. W., 1969. The atmosphere and the ocean. *Sci. Amer.* 221(3):76–86. (Available as *Sci. Amer.* Offprint 881.)
Stommel, H., 1965. *The Gulf Stream* (2nd ed.). Berkeley: University of California Press.
Sverdrup, H. U., M. W. Johnson, and R. H. Fleming, 1942. *The Oceans*. New York: Prentice-Hall. [The first modern summary—and, in some respects the only one. Intended for the specialist.]
Turekian, K. K., 1968. *Oceans*. Englewood Cliffs, N. J.: Prentice-Hall.

11

CLIMATIC CHANGE

For the moving of large masses of rock, the most powerful engines without doubt which nature employs are the glaciers. . . .

John Playfair

ILLUSTRATIONS OF THE HUTTONIAN THEORY OF THE EARTH

Say when, and whence, and how, huge Mister Boulder,
And by what wond'rous force hast thou been rolled here?
Has some strong torrent driven thee from afar,
Or hast thou ridden on an icy car?

P. Duncan

(about 1840, quoted in Embleton and King, 1968)

The climate has changed at various times in the geologic past, and, at one extreme, glaciers have spread over large areas of the continents. Such changes are of particular interest because we are in the midst of a period of extremely variable climate that began several million years ago. Ice now covers the Antarctic continent, Greenland, and the peaks of many mountains. Even so, the weather is mild compared to what it was a mere 12,000 years ago when Scandinavia and much of Canada were covered with glaciers. The ice age has not ended, and we are left to wonder what will happen next. Is it coming to an end? Will the remaining ice melt and the climate become even milder, or will glaciers once again advance?

It is quite possible that the stresses of staying alive during past climatic changes have had much to do with the development of modern man, the most versatile and adaptable of all living creatures. At present, it appears that man has lost some of the flexibility that enabled him to flourish during the glacial ages of the past. All he had to do then was move with the herds. Now there are too many people to move without total modification of the world's social structure. If the glaciers advance, half a billion people will be displaced. If the glaciers melt, all the shore dwellers, far more than a billion people, will need to move inland. These are extreme possibilities, but less extreme and more probable climatic variations may cause major upheavals in our densely populated world.

Ancient Climatic Change

The nature of a climate affects an enormous range of interrelated phenomena. A polar region, for example, has bare rock, mechanical weathering, few species of organisms (but often abundant life), no trees (or else evergreens with growth rings indicating seasons), and animals with lots of white hair. A tropical region, in contrast, has thick soil, chemical weathering, many species (but often less abundant life), a dense rain forest of broad-leaved trees often without annual growth rings, and animals with bare, dark skins. Many of these phenomena are preserved in the geologic record, and it is a common observation that the ancient climatic indicators in many places are not those of the present climate. Until recently, most geologists have tended to interpret this as indicating that the climate has changed at a particular place. That is, if warm climate indicators are found in Antarctica, it means that the climate of the south polar region was once warm. An alternative explanation is that it is always cold at the south pole but that Antarctica has drifted there from a warmer region. This is more commonly accepted now that continental drift has been confirmed in so many ways.

We know that climates vary and that continents drift, and it is inherently difficult to separate the two effects unless the climatic change is extreme or unless the positions of all the continents can be resolved for the time in question. There is widespread evidence of continental glaciation several times in the very ancient past, again about 250–300 million years ago in Pangaea, and again at present. It is obvious that, during such periods, the global climate changes regardless of the location of the glaciated continents. The mere existence of an enormous expanse of ice is adequate to alter the global atmospheric circulation, shift climatic zones toward the equator, alter rainfall, and so on.

With regard to the positions of continents, we know where they have been for the last 10 million years, and there is not too much uncertainty about where they have been for the last 200 million years. Before that, we hardly know enough to have a bearing on this question, because climate can be highly sensitive to small variations in the positions of continents. For example, the opening of a strait by sea-floor spreading can permit warm water to enter a previously isolated polar ocean and, thereby, ameliorate the whole global climate. The closing of a strait has the opposite effect. Likewise, the orientation or location of a coastal region determines whether there is upwelling of cool water, coastal fog, and other local climatic effects.

As more and more is learned of paleogeography, it will be increasingly possible to detect global climatic changes by analysis of ancient temperature and wind patterns. Even so, the interpretations will be difficult, because the paleontological indicators of temperature are so difficult to calibrate. A promising development is the discovery of a paleothermometer by the imaginative Harold Urey. Many marine organisms deposit calcium carbonate shells by extracting the elements from sea water. The sea water contains the isotopes oxygen-16 and oxygen-18; Urey discovered that the ratio of the isotopes in some shells depends on the temperature of the water from which they were extracted. All that was necessary was to make the reasonable assumption that the ratio of isotopes in sea water in the past was the same as it is at the present. This thermometer proved its utility by showing seasonal temperature variations in the layering of the skeleton of a belemnite, a squidlike animal. This particular belemnite was born in the summer 150 million years ago, and it died in the spring four years later!

The seasons could be determined from the variations in temperature, but the absolute value proved more elusive because it was found that the isotopic ratio varies from place to place in the ocean. This difficulty still remains, and it becomes particularly vexing in glacial periods just when it would be most

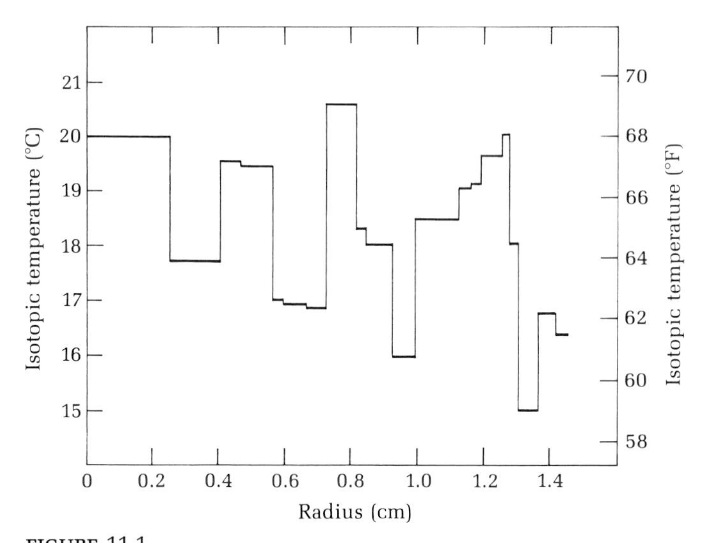

FIGURE 11.1
Ancient temperatures measured in the skeletal-growth layers of a belemnite 150 million years old. Seasonal changes indicate that the animal was born in the summer and died in the spring four years later. [After Emiliani, "Ancient Temperatures," copyright © 1958 by Scientific American, Inc. (all rights reserved).]

informative. This is because oxygen-16, being lighter, tends to evaporate more readily from sea water. Thus, when enormous quantities of water are stored in glaciers, the sea becomes richer in oxygen-18, the ice richer in oxygen-16, and interpretation more difficult.

Because of these various problems, much of our understanding of climatic change and its consequences is derived from study of the current ice age, which has the ideal characteristics of extreme variability, known geography, and a multitude of impressive effects. The onset of the glaciation was gradual. Icebergs, presumably from glaciers in coastal ranges, rafted gravel into the deep sea off Antarctica as much as 40 million years ago. By 7–10 million years ago, continental glaciation was widespread in Antarctica, as was mountain glaciation in southern Alaska. The chronology of succeeding events is surprisingly unsatisfactory, because different methods of measurement do not agree very well. It appears that continental glaciation began rather abruptly in northern North America and Europe one or two million years ago.

Everyone agrees that there have been several episodes of ice advance and retreat in most places during the current ice age, although Antarctica may

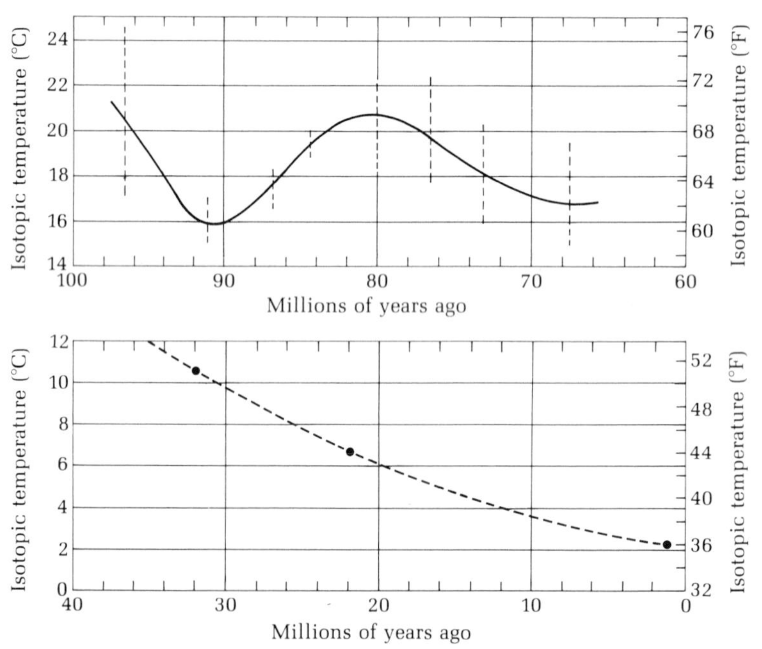

FIGURE 11.2

Marine temperature changes indicated by the abundance of oxygen-18 and oxygen-16 in the shells of marine organisms: *solid line*, surface waters; *dashed line*, bottom waters in the Pacific derived from the south polar seas. Trends are more certain than absolute values because the isotope ratios depend on some factors in addition to temperature, and surface and bottom waters are at different temperatures. [After Emiliani, "Ancient Temperatures," copyright © 1958 by Scientific American, Inc. (all rights reserved).]

have been glaciated continuously for the last 5 million years. Elsewhere the glacial deposits alternate with more normal sediments, such as fossil soils, which can only be explained by episodic glaciation. The question is how many episodes, and on this matter (as well as that of the onset of glaciation), there is a marked disagreement, which may reflect the importance of regional variations in intensity. In a general way, glaciologists working on land have tended to fit their observations into a historical framework derived from glacial events in Europe. This indicates four major glacial advances and a few minor ones.

Marine geologists, however, tend to find many layers of alternating glacial and nonglacial sediments, or oozes with faunas indicating alternating cold and warm water. This suggests a longer sequence of glacial events than the continental interpretation. Where is the correct answer to be found? Geophysicist Maurice Ewing, director of the Lamont-Doherty Geological Observatory, contrasts the problem of estimating the number of glaciations from continental evidence with estimating it from marine evidence by saying, "The first may be compared in complexity to estimating the number of times the blackboard has been erased; while the second may be compared to finding the number of times the wall has been painted." The advantages of the marine sequences include the fact that they can be dated by paleomagnetism, whereas the land ones generally cannot.

Sequences of glacial marine sediment, dated by the magnetic-reversal time scale, indicate six periods of ice rafting in the last 700,000 years, at least eleven periods during the last 1.2 million years, and four more in the North Pacific from 1.2 to 2.5 million years ago. There are four periods of rather more intense ice rafting beginning 0.8 million years ago, and these may be equivalent to the four standard continental glaciations.

Cause of Glaciation

The cause of glaciation is not agreed upon at present. This is not to say that it is unknown, because it appears that every logical possibility has been suggested. We just do not know which one to believe. It is agreed that the important facts to explain are: (1) that brief glacial periods have alternated with long warmer periods through much of the history of the earth; and (2) that very brief, intense episodes of cooling and warming have occurred within the glacial periods.

The only possible basic cause of the cooling is a change in the net heating from the sun—that is, the incoming solar radiation, minus the outgoing earth radiation, must have decreased. Otherwise, part of the earth would have heated as other parts cooled, and the global circulation is too efficient to permit this. The radiation may be affected in many ways—each of which has been proposed by one or more scientists as the origin of glaciation.

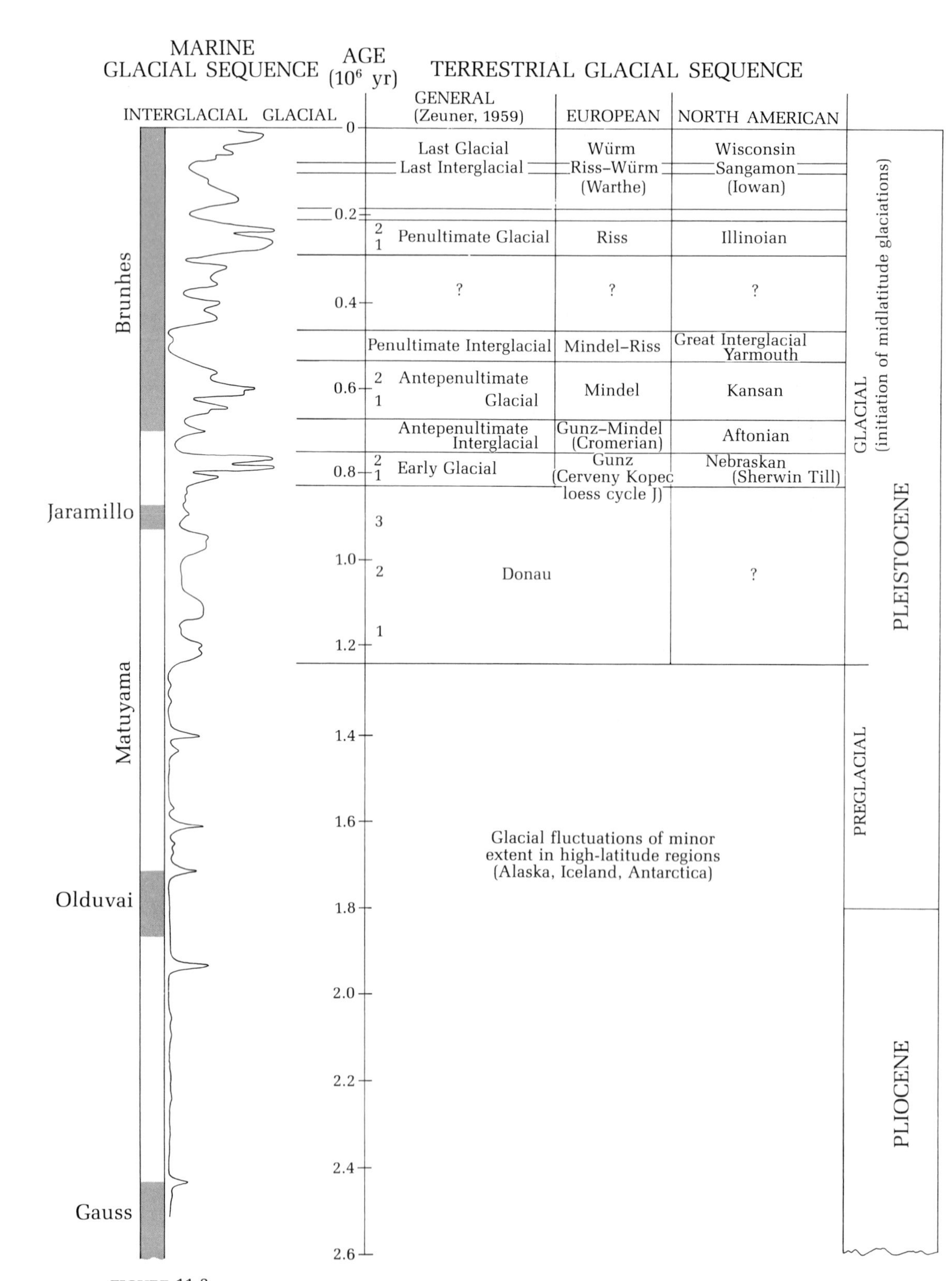

FIGURE 11.3

Correlation between glacial marine sediments in the North Pacific and continental glaciation sequences, with dating by paleomagnetism. Glaciation began even earlier in the Antarctic. Other scientists suggest quite different correlations between land and marine glacial evidence. [From Kent et al. (1971), courtesy the Geological Society of America.]

1. The radiative output of the sun may vary.
2. The distance to the sun may vary.
3. Passing dust or ice clouds in space may reduce the transmission of radiation through space to the earth.
4. Owing to such things as volcanic eruptions or changes in carbon dioxide content of the atmosphere, transmission of radiation through the atmosphere may vary.
5. The radiation reflected back from the earth may change, perhaps because of the greenhouse effect in the atmosphere or variations in surface reflectivity.

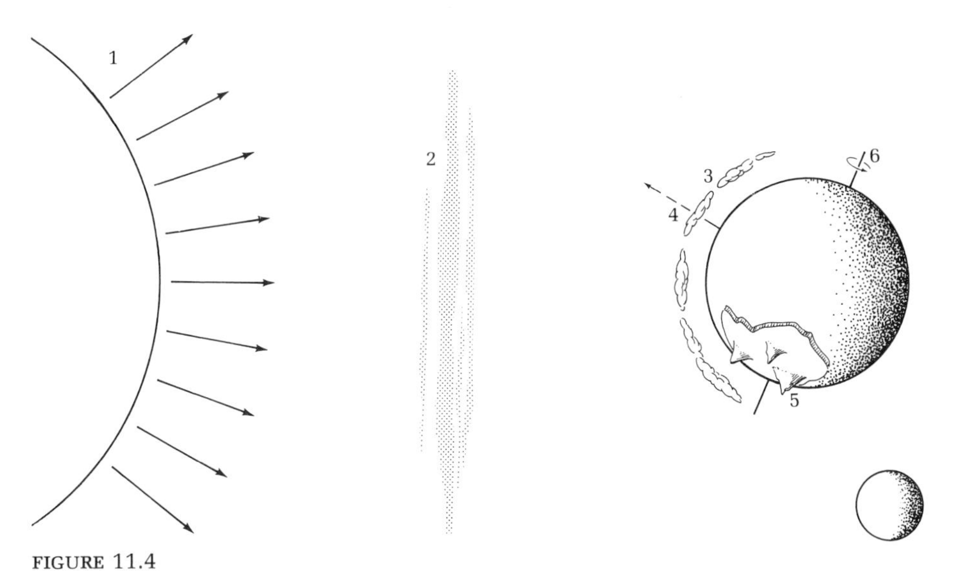

FIGURE 11.4
The origin of glaciation—but which one? Variations in solar output of radiation (*1*), or earth input (*2*, *3*), or earth output (*4*) are possible. The position and height of continents also affects back radiation (*5*). Wobbling of the axis of rotation (*6*) and changes in the distance to the sun appear the most likely because of brief fluctuations. The moon, which is subject to only *1* and *2*, may provide the answer.

These are the major possibilities, although there are many important subproposals remaining to be considered. The important point here is that the moon has no atmosphere, and, thus, by studying the moon, it is theoretically possible to separate effects 1 and 3 from 4 and 5. This has now been done by analysis of the effects of solar particle bombardment on two moon samples collected by Apollo 11. R. C. Finkel and colleagues believe "that long-term variations in solar thermal flux, on a scale of more than [one] percent, must now be considered unlikely."

That leaves the earth. It is noteworthy that the reflectivity of the surface increases and cooling occurs if the ground is snow-covered during the year.

Thus, self-refrigeration can be induced by rather small changes that inhibit summer melting. Geologically, conditions are favorable if the continents drift into polar regions or if they are particularly mountainous. There were continents in polar latitudes during the glaciation of Pangaea, as there are now, so this effect may be all-important.

None of the continental effects are capable of explaining the short-term glacial episodes in any simple way. For this purpose the slow, regular motions of the axis of rotation, and the similar changes in the shape of the earth's orbit, are without peer. The insolation at a given latitude depends on these motions. In 1920, Milutin Milankovitch presented calculations of the temperature changes that would result from the combination of the various motions. Since then, many attempts have been made to correlate the glacial sequence with this precise astronomical calendar. These attempts have met with a mixed response. Some experts reject them, some embrace them with enthusiasm.

In sum, major glacial periods appear to be the result of variations in the reflectivity of the earth induced by the fortuitous drift of continents into high latitudes. The shorter glacial episodes may be a consequence of regular motions of earth's orbit and axis of rotation, but this is also conjectural.

Movement of Ice

When more snow falls than melts during the year, it accumulates layer on fluffy layer, compacting as it accumulates until the bottom layers are transformed to solid, crystalline ice. If this ice reaches a critical thickness that depends on the bed slope and other factors, it begins to flow outward as a glacier from the area of accumulation. On a steep-sided peak, a glacier need be only a few scores of meters thick; on a flat plain, it must be at least several hundreds of meters thick. In Antarctica, the glacial ice is as much as 4000 meters thick where it is partially trapped in a depression.

Although glaciers flow like other crystalline solids, they have some peculiar properties connected with the fact that they are near their melting temperature. If the melting temperature is exceeded at the base, they flow partially on a very thin layer of water, although most projections of the underlying rock reach into the solid ice. If the base is solid, it deforms plastically or by pressure melting and so flows around projections. Because the bottom layer is loaded with sand and gravel, the glacier erodes the rock like a vast sheet of sandpaper pressed down by an enormously powerful hand. Glaciers erode solid rock far more rapidly than rivers, winds, or waves. They produce rock flour just as a grist mill grinds grain into flour, and where this flour rubs against the bedrocks, it commonly polishes it to a glistening smoothness.

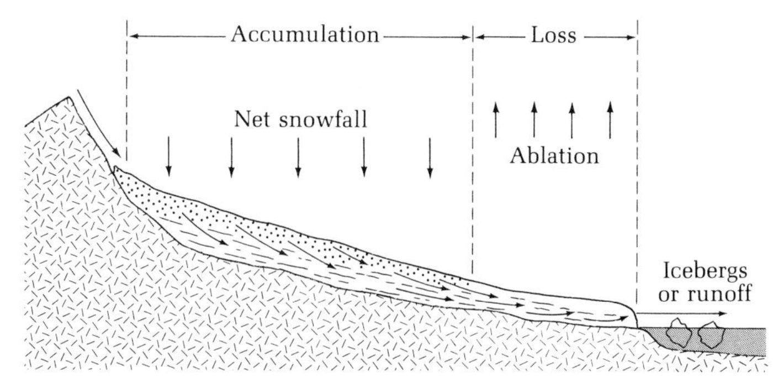

FIGURE 11.5

Accumulation, loss, and motion of a glacier in a mountain valley. Average particle motion is indicated by flow arrows. The width of the zones varies with climate and geography.

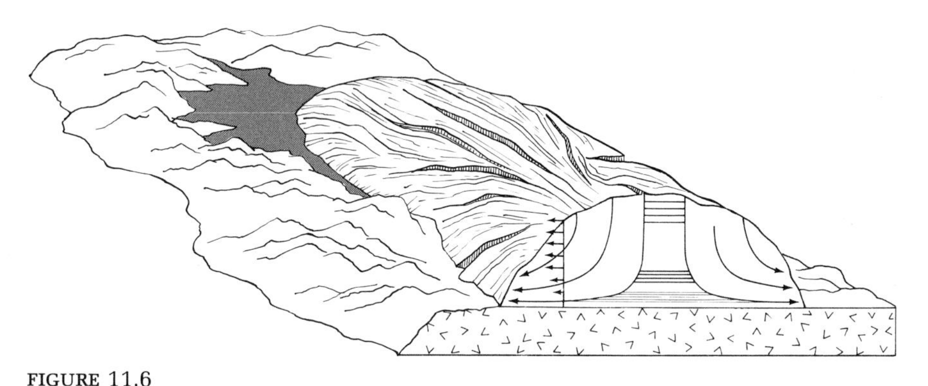

FIGURE 11.6

Idealized flow in an ice sheet or continental glacier: *long arrows*, the path followed by an average snowflake that falls on a given part of the surface; *small arrows*, the velocity profile from top to bottom. Velocity is relatively constant except near the bottom. The central mass of ice is plastically deformed, as indicated by the spacing between and length of the horizontal lines, which represent surfaces of deposition at a given time. This deformation is observed in ice drilled from ice sheets. [Data from Dansgaard et al. (1971).]

We may follow the ideal path of snow that falls near the head of a mountain glacier or near the center of a continental one. It moves down and out as it is gradually buried, and then up and out as it passes the limit of the area of accumulation and enters the area of loss of mass by ablation or melting. Before it reaches the upper surface, it commonly reaches the edge, which may be scores of meters high. There it melts, if on land, or calves into icebergs, if it reaches the sea. As it melts, it deposits the gravel it carries in various types of moraines. If it calves, the icebergs drop their gravel on the sea floor while they drift away and gradually melt. The velocity of flow varies, but it is usually in the range of scores to hundreds of meters per year.

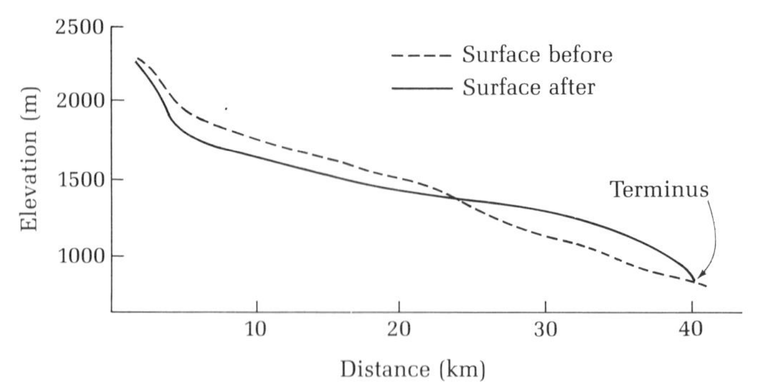

FIGURE 11.7

Changes in surface elevation resulting from an ice surge in Muldrow Glacier, Alaska. [Reprinted, with permission, from Paterson, *The Physics of Glaciers,* copyright 1969 by Pergamon Press Ltd.]

The flow is not always so simple. Undulations called kinematic waves sometimes move down glaciers at about four times the velocity of the ice. They are not like ocean waves, but they are similar to the waves that move through the traffic on a freeway when an accident causes temporary bunching of cars. The back of the bunch accumulates as the front speeds away, and so a kinematic wave moves back through the traffic even after the wreck is removed. The wave in the glacier outspeeds the ice because thicker ice moves faster than thin.

Kinematic waves are understood; ice surges remain puzzling. In a surge, the whole accumulation area may drop and the ablation zone rise as though in a giant kinematic wave.

The advance and recession of the glacier front reflects a combination of the velocity of the glacial ice, the rate of accumulation and ablation, and the occurrence of kinematic waves and surges. Thus, the front may retreat as the ideal snowflake in the glacier advances. During a surge, the front may spurt ahead. In the Karakoram mountains of Asia, one glacier front advanced 13 kilometers in 3 months in 1953.

Pleistocene Glaciers

Western Europe and northern North America, the centers of industrial civilization, were glaciated in Pleistocene time. The last major glaciers began to retreat only 12,000 years ago, and a great many profound aftereffects of these glaciers are all around us.

FIGURE 11.8

The extent of continental glaciers at the time of the last major glacial maximum. The inferred limit of pack ice is shown at sea. Icebergs drifted into much lower latitudes. [From Lisitzin (1972) and Lobeck (1939).]

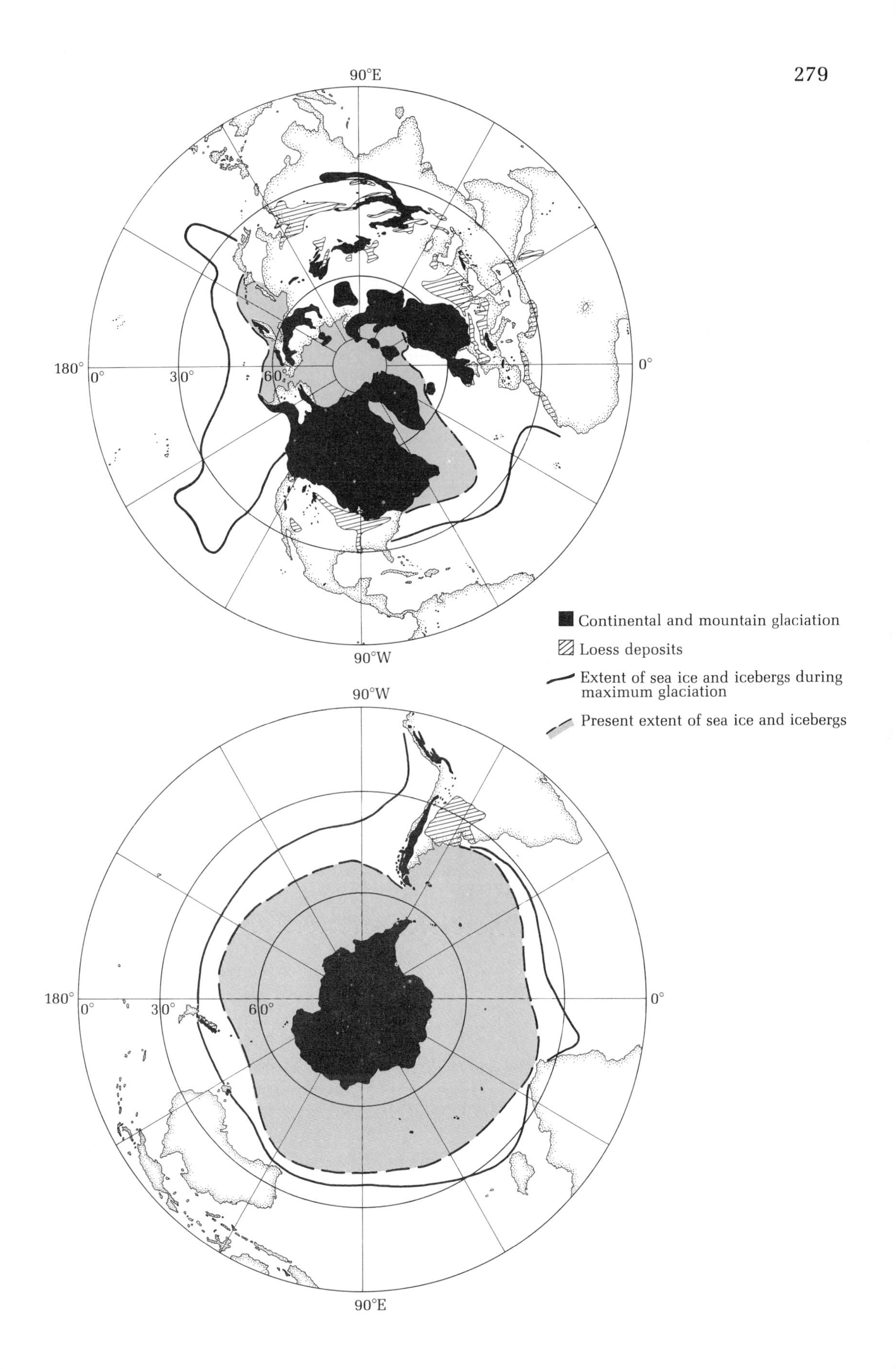

90°E
180°
0°
30°
60°
0°
90°W
Continental and mountain glaciation
Loess deposits
Extent of sea ice and icebergs during maximum glaciation
Present extent of sea ice and icebergs
90°W
180°
0°
30°
60°
0°
90°E

The major known effects can be visualized as logical consequences of the wax and wane of a vast, thick, mobile sheet of ice: The cold of the ice shifts climatic zones toward the south and with them the fauna and flora. The load of the ice depresses the crust beneath it and bulges the surrounding area upward. The extraction of water from the oceans to form the ice depresses sea level and exposes the continental shelves. The abrasive glacier strips the soil and erodes the rock beneath it. It transports the sediment to its periphery and deposits it in mounds as it melts. Most things reverse as the glacier melts: the climatic zones, fauna, and flora move north; sea level rises and the continental shelves are flooded; and the crust returns to equilibrium. But the glacier *melts,* it does not flow backward. Thus, the central erosion and peripheral deposition are not restored. Moreover, the response times of the various phenomena are different. The ice is now gone, but the crust is still striving for its equilibrium levels. The sea is refilled, but the crust is still adjusting to the shifting of its load away from and back onto the continental shelf. Many life forms adjust quickly to change but not all have been able to survive the extreme stresses of the Pleistocene. Thus, the biota of today is not entirely the same as that of the relatively recent past.

DISTRIBUTION

The catalogue of changes is vast because the glaciation was on a scale that is difficult to comprehend now that the ice has melted. In the Northern Hemisphere, glaciers up to 3 kilometers thick spread from several centers and coalesced to cover much of the area north of 50° latitude. The major exception was in Siberia, where precipitation presumably was too slight to exceed ablation. Large mountainous regions were glaciated in central Asia down to 30° latitude, and, even on the equator, the tops of individual peaks supported glaciers.

In the Southern Hemisphere, glaciers covered the Antarctic but, because of the extent of seas, were otherwise mainly confined to mountains.

EROSION

The depth of erosion under the center of the glaciers is impossible to reconstruct, but the American geologist William White observes that Precambrian rocks are rarely exposed except where glaciation has occurred. Moreover, it is only in these ancient rocks that there are widespread exposures of rocks that were once deep in the crust. Thus, glacial erosion may have reached down many kilometers. Near the periphery of the glaciers, the erosion was shorter in duration and not so deep, but it may have scooped out the basins of the Great Lakes as well as those of the marine gulfs among the Arctic islands.

DEPOSITION

Active glaciers deposited peripheral moraines and medial moraines where different ice sheets coalesced. In addition, there were distinctive deposits from subglacial streams, and also streamlined hills, called drumlins, that were molded by the ice. Some drumlins are all glacial debris, or till, and others are part bedrock. They are very numerous in some regions, as indicated by the following rough census: western New York, 10,000; Wisconsin, 5000; south-central New England, 3000. They commonly dominate the topography in the otherwise flat areas in which they occur.

When glaciers stagnate, they dump their load of suspended material in place. If they move forward over their own debris, they produce a ground moraine, which may be very extensive.

LOADING AND UNLOADING

The sea eroded and deposited around parts of the edges of the Canadian and Scandinavian glaciers. These ancient strandlines are now well above sea level, and the amount of uplift is greater toward the glacial centers. Maximum uplift in both regions has been about 300 meters. Thus, the crust was depressed at least that much by the load of the ice and has rebounded that much since the load was removed. The actual maximum was certainly greater for two reasons. First, the strandline did not reach the center until the glacier was thinned by melting and part of the rebound had already occurred. Second, the rebound is still continuing. This is demonstrated by tide gauges in the Baltic Sea, which indicate that the northern end of the Gulf of Bothnia is

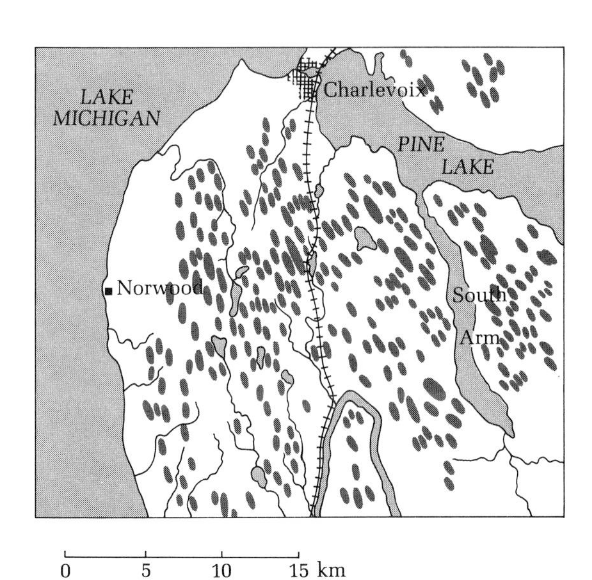

FIGURE 11.9
Some drumlins of Michigan, showing their abundance and their oval forms. They are elongate parallel to the direction of flow of the overlying glaciers that shape them from clay and rock debris. [From Daly (1934).]

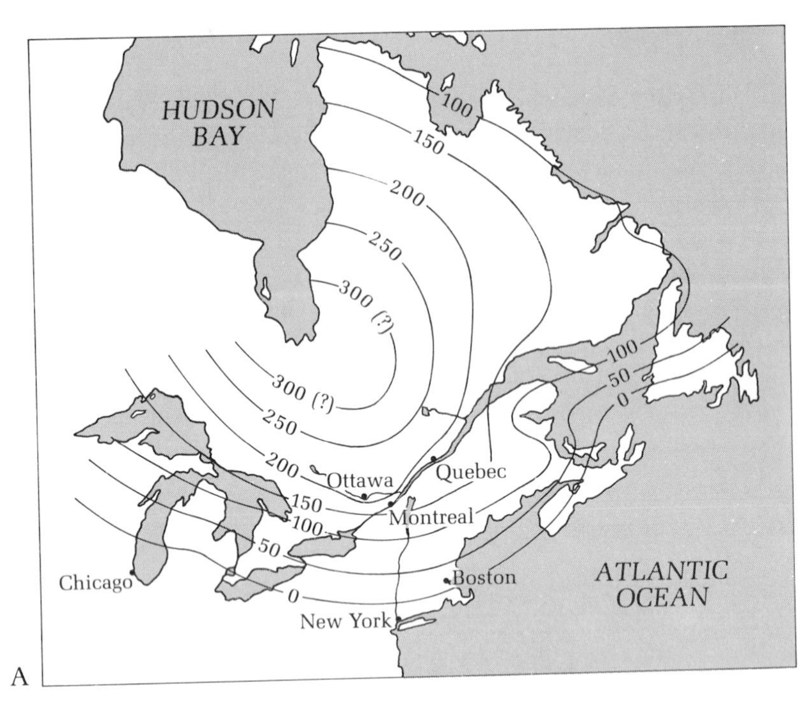

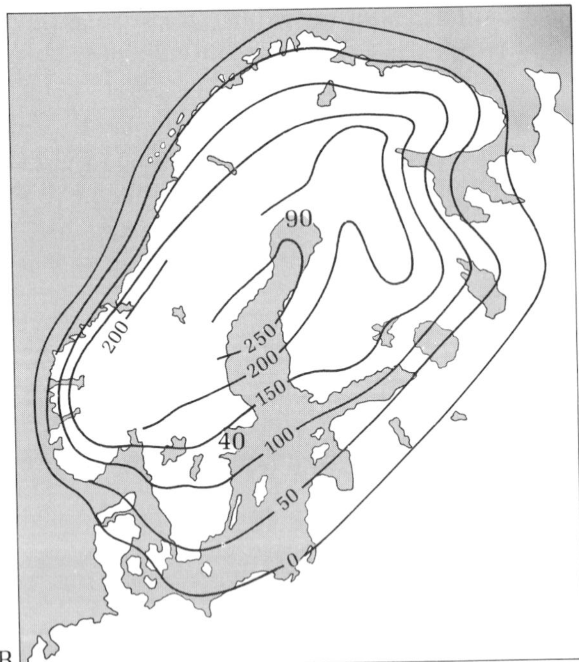

FIGURE 11.10

Rebound of the crust of glaciated regions of eastern North America (*A*) and Scandinavia (*B*), in meters, after the retreat of the ice. Continuing uplift is demonstrated by changes in the tide gauges: *large numbers,* uplift in centimeters per century; *small numbers,* total uplift in meters. [After Daly (1934).]

rising at the rate of 89 centimeters per century, that Stockholm is rising at the rate of 40 centimeters per century, and that Copenhagen, further south, is stable. (Similarly, the town of Churchill by Hudson Bay is rising at the rate of 91 centimeters per century.) These very fast changes have had significant social effects around the populated Baltic. Harbors have become unusable and fishing grounds have become farms.

PLUVIAL AND PERIGLACIAL EFFECTS

The existence of broad continental ice sheets compressed the general atmospheric circulation toward the equator, more so in the Northern Hemisphere, because the ice was more widespread there. This intensified circulation changed weather patterns with widespread results—notably, the formation of lakes in what are now deserts, and the intensification of frost and wind action in the belt adjacent to the ice.

LAKES

Lakes form only where there are suitable basins, commonly in faulted or folded regions lacking integrated river drainage. Only deserts normally lack river drainage, and in this we see why a change in climate tends to put lakes in deserts more than elsewhere. Clearly, the lake phase is temporary. If it persists, the lakes overflow the basins, erode the natural dams, and establish an integrated river pattern without lakes. Continental glaciers can also act as dams, and great temporary lakes form as a result; the Columbia River was dammed in this way during the Pleistocene.

The number and extent of the lakes varied rapidly and is still changing. The maximum effects in the western United States were striking. Much of western Utah was covered by Lake Bonneville, which had an area of 52,000 square kilometers and a maximum depth of 300 meters. During part of its history, it drained north to the Columbia River through what are now great empty valleys. Nevada, southeastern Oregon, and eastern California were dotted with a hundred lakes that waxed and waned. Some were occasionally linked by rivers, but most were in isolated valleys.

Dry lake beds and wave terraces on the flanks of the desert mountains mark the sites of these lakes. They are most conspicuous around Great Salt Lake, which is salty because the greater Lake Bonneville, of which it is a remnant, has largely evaporated, leaving behind an increasing concentration of salt as it has done so. The former lakes of the Mojave Desert evaporated completely, leaving deposits rich in potassium. Such deposits in the dry bed of Searles Lake are now mined commercially.

Prehistoric man trudged across the Bering Sea when it was exposed by glacial lowering of sea level. He arrived in the southwest to find a land of

BOX 11.1 DISCOVERY OF THE ICE AGE—A SCIENTIFIC REVOLUTION

Louis Agassiz's *Discourse of Neuchâtel**

Louis Agassiz (1807–1873), the brilliant young Swiss specialist on fossil fish, regarded the curious hypothesis that glaciers had once been widespread as neither probable nor tenable. However, while spending one summer vacation exercising in the Alps, he chanced to join some of his friends who were espousing these wild and unsupported ideas. In a few weeks, after he had sorted out all the data they had collected in seven years, he realized that their general concept was correct. In subsequent years, Agassiz encouraged them to publish their findings, but they did little.

A year later, Agassiz was due to present his presidential address about fossil fish, but he could contain himself no longer. The night of July 23, he dashed off the *Discours de Neuchâtel,* as it came to be called, and launched a scientific revolution. We know that he was excited as he wrote because, for the only time in his long career, he italicized words for emphasis. This is a not unusual example of how scientists break out of their normal mode when they launch revolutionary ideas.

Agassiz told his astounded audience about the evidence he had seen and its interpretation in terms of widespread glaciation. So far, so good. However, he continued on to explain the glaciation not as spreading from the Alps but as originating in the north as a vast sheet of ice that spread over the area and was disrupted by the rise of the Alps underneath them. (It is difficult not to use italics here to emphasize the number of outlandish mistakes he was making.) He continued to explain the ice sheet as a result of general cooling of the earth with a superimposed temporary cooling at the end of the last geologic period. He compared this latter to the cooling that follows the death of an animal.

"The general reaction of the audience was astonishment and incredulity, both during the talk and the lively discussion following the delivery of the address," says Albert Carozzi, who edited a recent edition of Agassiz's address. The famous geologist von Buch was in the audience. He raised his hands to the sky, bowed his head toward the Bernese Alps, and called upon the spirit of a pioneering alpine geologist to pray for them. There was so much commotion that the next speaker did not dare to present his own theory on the existence of facies in rocks—a theory that also changed many ideas when it finally appeared.

Agassiz was a brilliant man and a fighter, and although, in the following years, he mercifully said little more about his explanation of glaciation, he forced acceptance of the fact that continental glaciation had indeed taken place. The furor did not subside for twenty years.

So the savants argued. One of Agassiz's friends, de Charpentier, had met a woodcutter while hiking through the Alps several years before Agassiz made his controversial talk. The woodcutter knew that the large boulders they were passing were from a distant region, and de Charpentier asked him how he thought they had been transported. He answered, "The Grimsel glacier transported and deposited them on both sides of the valley, because that glacier extended in the past as far as the town of Bern, indeed water could not have deposited them at such an elevation above the valley bottom without filling the lakes." One man's revolution is another man's common sense.

*Presidential address delivered at the opening session of the meeting of the Societé Helvetique des Sciences Naturelles, at Neuchâtel, July 24, 1837, by L. Agassiz.

lakes quite unlike the terrible deserts that now exist. Lakes were also characteristic of now drier basins in Africa, where the most primitive men evolved. The changes in these lakes may have influenced human development in many ways. The cradle of civilization in the near east was likewise affected. Glacial meltwater poured into the Caspian Sea, which expanded enormously to link with the Aral Sea to the east and the Black Sea, then a deep lake, to the west. The whole system eventually overspilled into the Mediterranean Sea, which

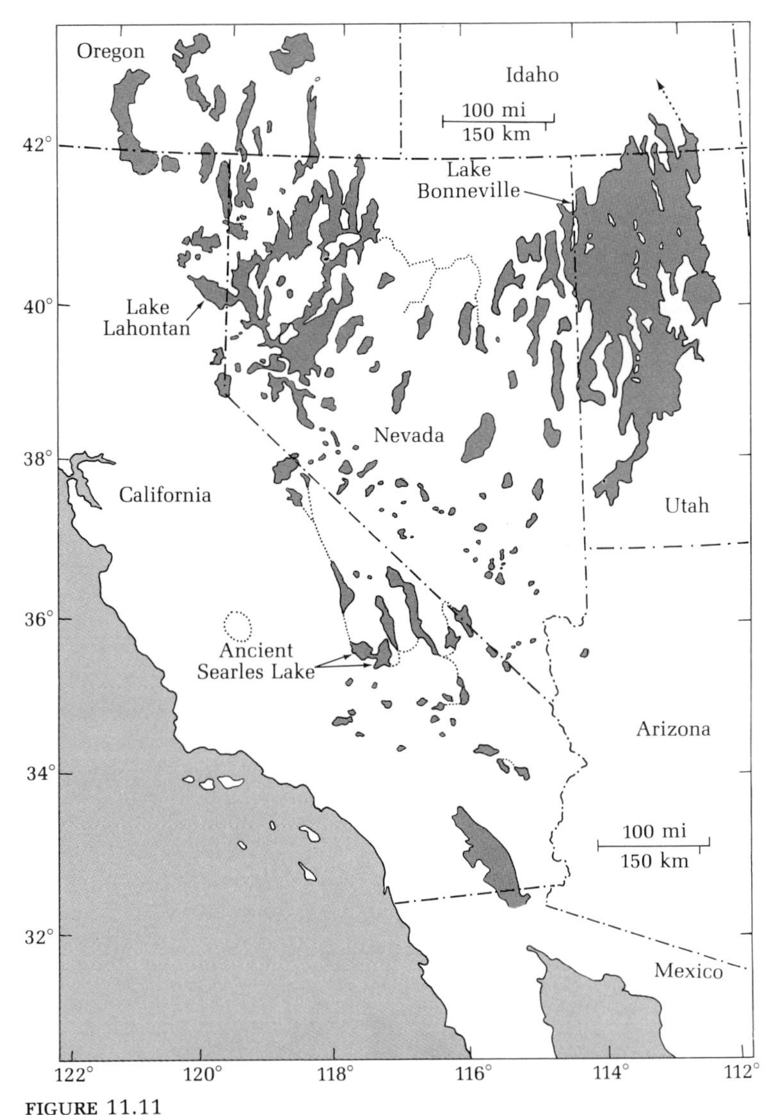

FIGURE 11.11

Ice-age lakes in what are now the western deserts. Transient rivers (*dotted lines*) connected some lakes. The arrow in the upper right corner indicates Lake Bonneville draining toward the Columbia River. [After Flint (1971).]

was much lower at the time. The Dead Sea was also much larger, about 460 meters deeper, and fresher when man first occupied its shores. Many of the arid basins of the Middle East, like those of the western United States, were lakes at that time, so the present desert was no barrier to human migration.

The periglacial regions did not stop man either, but they were hardly hospitable. The ground was generally frozen, except for a thin surface layer, and a polar desert existed. Soil flowed intermittently, as it does in polar regions now, and this and frost heaving exposed the subsoil to intense weathering. (These effects are discussed at greater length in Chapter 14.) In addition, the rivers draining the glaciers were milky-white with rock flour and silt, which was spread over their flood plains.

Conditions were ideal for wind erosion and the transport and deposition of the airborne dust that forms the sediment called loess. It is extremely widespread in the central United States, central Europe, and central Asia, where it forms a blanket tens to hundreds of meters thick. It yields a relatively new and immature fine-grained soil that is ideal for grasses; thus, such soils support some of the principal grain harvests of the world.

SWINGING SEA LEVEL

The major marine effect of glaciation was a lowering of sea level as water was transformed into ice. The maximum lowering of about 120 meters occurred more than once, and lesser fluctuations were frequent, judging by glacial advances and retreats. The cutting edge of oceanic waves crossed the continental shelves repeatedly and planed them smoother than before. Rivers, such as the Hudson and Mississippi, flowed across the exposed shelf and cut channels through the soft sediment. The Hudson River channel still exists, although it is now deeply submerged. A whole river system developed on the broad shelves of Indonesia, connecting rivers that are now isolated on separate islands. A number of these abandoned channels are mined for "sea tin," which is cassiterite, a dense mineral transported from the islands by once active rivers.

These river channels on the shelf are not to be confused with the great submarine canyons that extend down the continental slope to abyssal depths. The exposure of the shelves and intense glacial erosion sent a pulse of sediment into the ocean, and much of it traveled down the submarine canyons in the form of turbidity currents, which, like the glowing clouds of volcanoes, are driven by their excess density. The sediment spread across the deep-sea floor and was deposited as fans and plains. Many of these are still sites of sedimentation, but others were cut off from the continental source of sediment when sea level once again rose to flood the shelf. As it did so, the sea left behind beaches and lagoons and offshore bars, which now litter the shelf at depths of tens or scores of meters.

The fossil biota of the drowned shorelines can be dated by the carbon-14 method and used to construct the detailed history of the rise of sea level. The chief problems are identifying the exact depth in which the organisms lived and assessing the effects of local vertical motions of the land. There is also a danger that the remains being dated—oyster shells, for example—were from a kitchen midden instead of the site of their natural habitat. However, the best carbon-14 results give a reasonably consistent history. The sea level was 110 meters lower 20,000 years ago, and then it rose very rapidly until it was only about 6 meters lower 7000 years ago. Events since then are uncertain, and may reflect local movement of the land.

The American geologist Arthur Bloom has observed that the ice sheets melted by about 6000 years ago, that water was therefore restored to the continents, but that, on the average, sea level appears to continue to rise. He

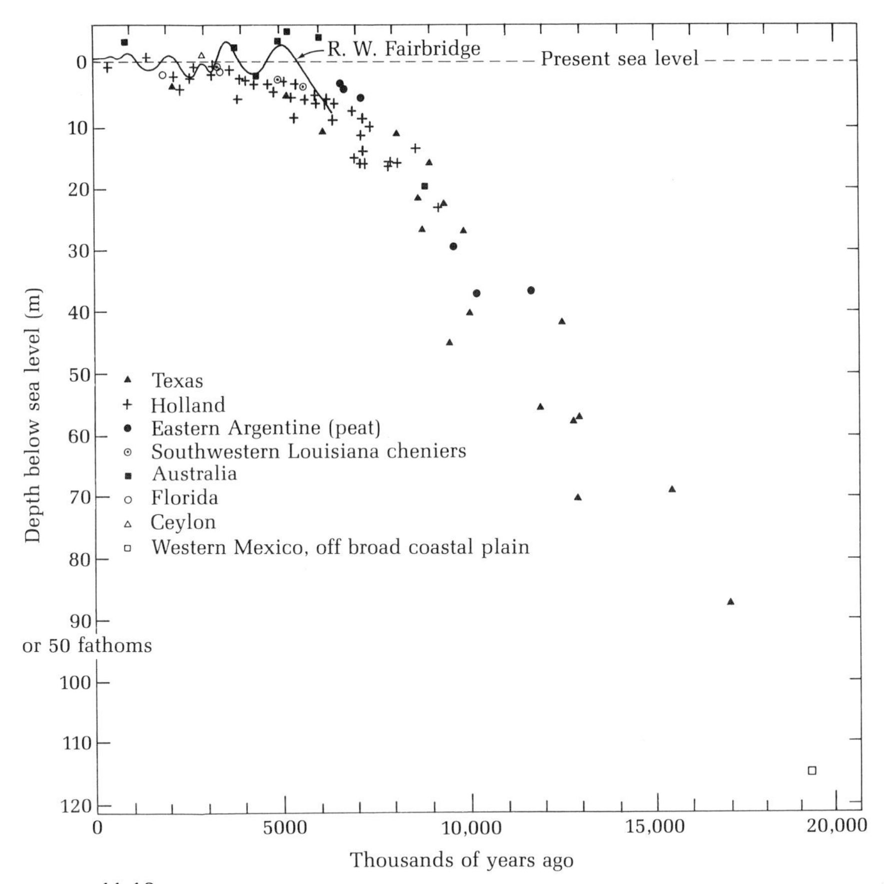

FIGURE 11.12
The rise of sea level as ice sheets melted during the last 20,000 years, based on carbon-14 dates of fossils from drowned shorelines on the continental shelves. Sea level may have been higher than at present. [After Shepard (1963).]

ascribes this to isostatic loading of the continental shelves by the rising sea. Sea level appears to rise in most places because the continental margin, where we make our observations, is slowly sinking in response to the loading.

In the tropics, there are extensive living coral reefs just at sea level. The drop in sea level during glacial periods exposed all such reefs and killed many of them, although they were largely repopulated at a later time. The exposed dead reefs were eroded by waves and streams and were dissolved by rain. When sea level again rose so abruptly, most reefs grew rapidly upward from their eroded platforms. Others were flooded too fast or were not repopulated in time. Many such reefs are still 30 meters or more below sea level, although reef-building corals once again live on them.

Changes in Life

Plant and animal life have changed in many ways since the beginning of the ice age in Antarctica and the later spread of northern ice sheets. The descent of man is the event of this period that is of most interest to us—indeed, all other significant events of the period may be related to it.

MAN

Two million years ago, primitive manlike primates, the australopithecine hominids, were using tools in southern and eastern Africa and Java. The region of Olduvai Gorge, and other parts of east Africa, were warm, fertile, and dotted with volcanoes and lakes.

The genus *Homo* appeared about 800,000 years ago, and the first species, *Homo erectus,* endured and prospered for 400,000 years. During that time, he spread from the warm tropics to northern China and central Europe, and there encountered the edge of the ice sheet.

Early forms of the species to which modern man belongs, *Homo sapiens,* apparently evolved from *Homo erectus* about 300,000 years ago and spread over the Old World. The form known as Neanderthal man developed 100,000 years ago, and Cro-Magnon man, who differed from us only in culture, appeared perhaps 30,000 years later. Man did not complete his occupation of the earth until he crossed into the Americas, which may have been only 12,000 years ago.

For the last 2 million years, the family of man has been subject to the stresses of climatic change. The lakes of Africa fluctuated; and in southern Europe and the Near East, lake levels, rainfall, and temperature were extremely variable. Continental shelves were exposed, land bridges opened and closed, rivers shifted, biting winds carried dust clouds over the land around the ice sheets, and the ice itself waxed and waned. So much for man. The rest of the biota was exposed not only to the hazards of nature but to those of man as well.

LARGE ANIMALS

Alfred Wallace, co-discoverer of evolution, wrote in 1876 that the remarkable extinction of large mammals in late Pleistocene time was a consequence of climatic change. Darwin, however, found that the last giant ground sloth, among other such mammals, lived long after the ice sheets began to melt. By 1911, Wallace wrote (quoted in Martin, 1967):

> What we are seeking for is a cause which has been in action over the whole earth during the period in question, and which was adequate to produce the observed result. When the problem is stated in this way, the answer is very obvious.

The answer, he believed, was that man exterminated those large herbivorous mammals that were suitable for food, and that the carnivores that had preyed on them died of starvation. This question is unresolved. *Pleistocene Extinctions* (see Martin, 1967) is a whole book that presents differing viewpoints on the subject as of 1967.

The general arguments in support of climatic change as the cause of the extinctions are that it occurred at about the right time and that mass extinctions are recorded in the geologic past long before the appearance of man.

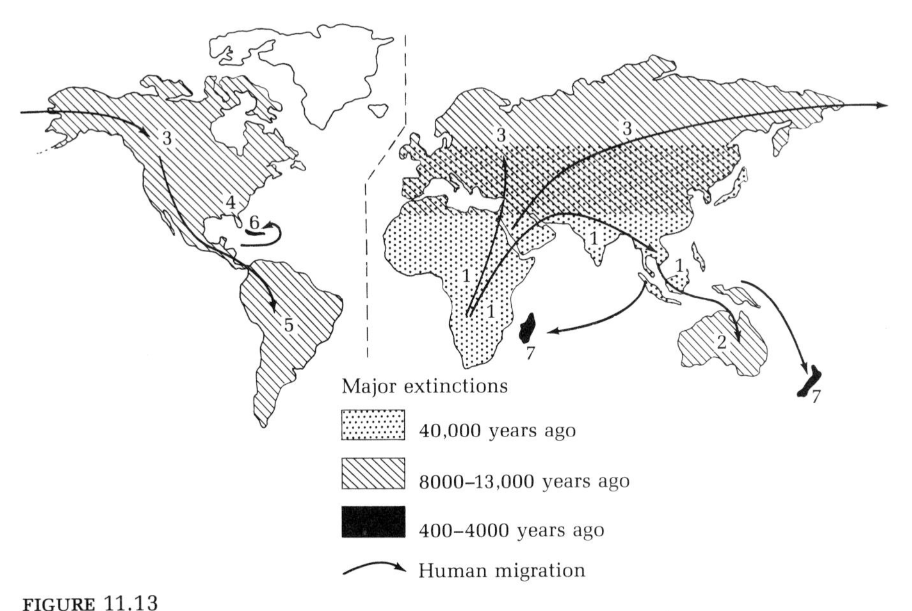

FIGURE 11.13

Direction and sequence of human migration compared with the global pattern of accelerated extinction of large mammals. [Data from Martin (1967).]

Paul Martin argues for prehistoric overkill by man on the grounds that *accelerated* extinction closely follows but never precedes man's arrival on a continent or island. In Africa, and probably southeast Asia, many extinctions occurred roughly 50,000 years ago. In northern Eurasia, extinctions were common 13,000–11,000 years ago; in Australia, 13,000 years ago; in North America, 11,000 years ago; and in South America, 10,000 years ago. This corresponds to the human migration schedule for outlying regions. Man did not reach the remote islands of New Zealand until the Polynesian canoes arrived from Tahiti about 900 years ago. The major extinctions in that area followed shortly thereafter. The pattern was repeated in Madagascar about 100 years later.

PLANTS

Bogs, lakes, and swamps receive pollen from plants that grow nearby, and much of it is preserved in bottom sediments in a stratigraphic sequence. The fossil pollens are distinctive enough that the plants that produced each kind can easily be identified; and the identity of the plants that produced them can reveal the climate that prevailed at the time the pollens were deposited. Thus, fir pollen indicates a fairly cold climate, whereas the pollens of oaks and beeches indicate a mild one. At present, a botanical profile from the Arctic through the Alps to the Mediterranean is zoned from tundra through coniferous forests to deciduous forests. In times of glacial maximum, the zones were both displaced and modified. A polar desert bordered the ice; tundra and steppe reached to the southern sea. Forests vanished, except on the southern slopes of the Alps.

The forests returned as the glaciers melted away, but man has since cut them in large areas. The role of primitive man in this is uncertain. He cleared forest with flint axes in England, but elsewhere he drove game by setting fires, and he may have burned large forests in the process.

Some Major Effects of Climatic Change in the Past 25,000 Years

Only 25,000 years ago, Cro-Magnon man co-existed with Neanderthal man in the inhospitable reaches of western Europe. Glaciers covered the northern lands and icy winds swept great dust clouds across the polar desert. In the Dordogne Valley of southwestern France, the Cro-Magnon hunters had been living off herds of reindeer and horses since the decline of the aurochs and wooly rhinoceros. It is lovely country now, but then it was bitterly cold, and men wore furs and spent much time painting on cave walls.

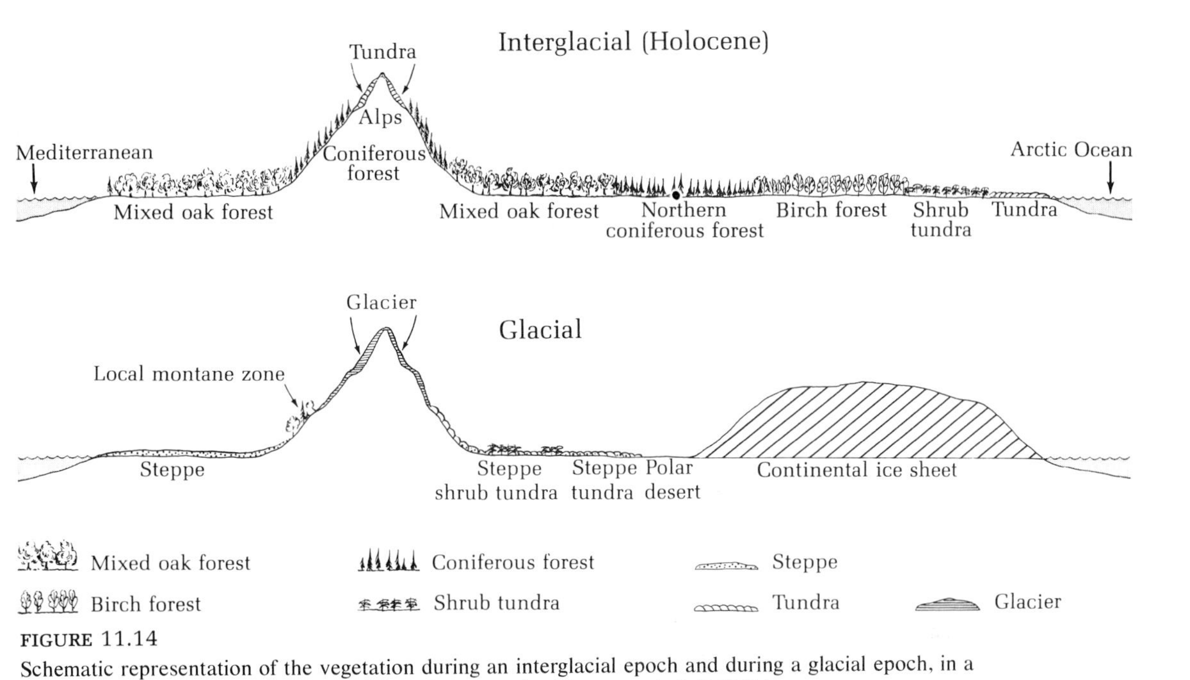

FIGURE 11.14

Schematic representation of the vegetation during an interglacial epoch and during a glacial epoch, in a south–north section through Europe. [Data from van der Hammen et al. (1971).]

The continental shelves were exposed, and great rivers ran through gorges to the sea. Because the former shoreline is now under 110 meters of water, records of what went on there are difficult to observe. However, the shoreline has always been attractive for its abundant shellfish, and it seems reasonable to assume that people lived near the shore. If so, they would have noticed that, for generation after generation, the sea encroached upon the land as the ice melted and water flowed back to the sea. This flooding must have been most noticeable in such regions as Indonesia, where the shelves are extremely broad.

By 15,000 years ago, according to the American anthropologist Wilhelm Solheim, the highlands of southeast Asia were occupied by people of the Hoabinhian culture, who made pottery and had domesticated plants and animals. The flooding of the shelf must have been obvious to these people. From 13,000 to 7000 years ago, sea level rose about a third of a meter in a 30-year lifetime; in Indonesia, that was enough to move the shoreline one kilometer inland—which could hardly have been missed.

Glacial retreat, northward spread of the fauna and flora, and flooding of the shelves continued for millennia. The effects nearest the ice front were the most spectacular, in part because a peripheral crustal depression persisted after the ice edge melted. Thus, about 9800 years ago, a great Baltic lake existed around the south and east sides of the Scandinavian ice sheet. It was separated from the sea by a land bridge connecting Denmark to southern

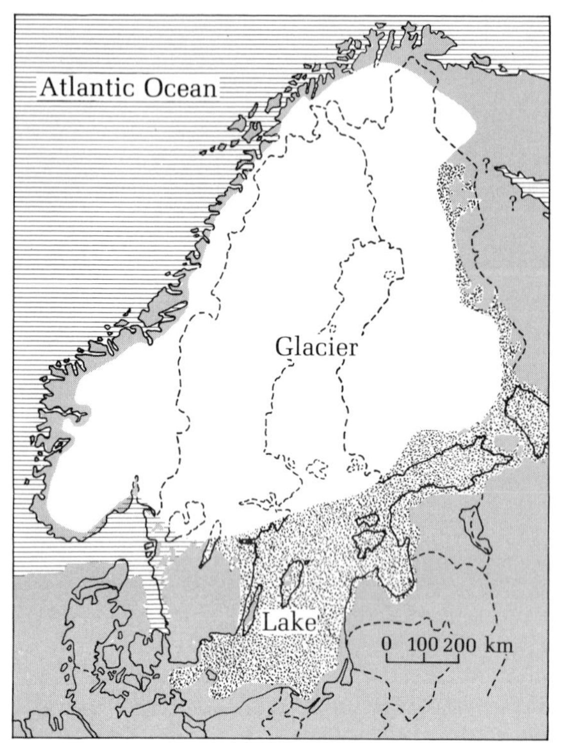

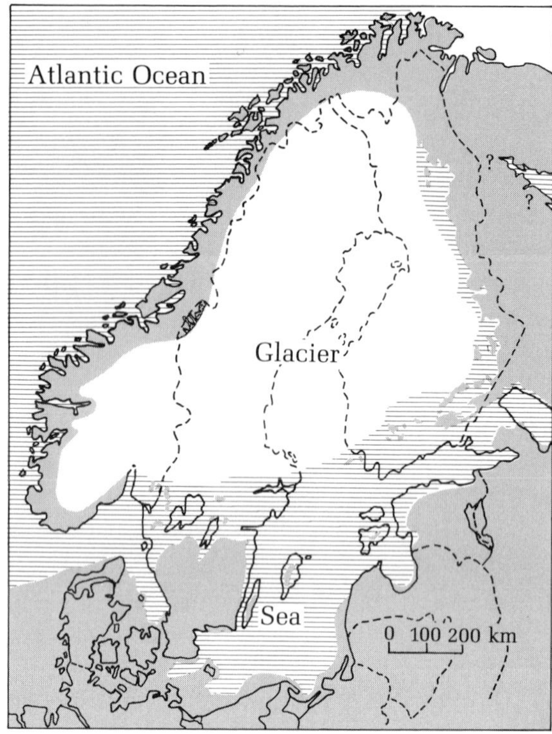

FIGURE 11.15
The Baltic lake about 9800 years ago (*left*) and the invasion of the sea a century later through the crustal depression in central Sweden (*right*). [Data from Granlund (1936), cited in Magnusson et al. (1957).]

Sweden. A century later, the ice retreated enough to expose a depression across what is now central Sweden, and a brackish sea invaded the Baltic region as far as eastern Finland.

A flood of glacial meltwater collected in the basins of the Great Lakes, as the ice front retreated, and intricate changes occurred: About 12,000–11,000 years ago, the area of the lakes was covered by ice most of the time, although it was exposed at least once. Various large lakes appeared, drained in different directions from time to time, and vanished. By about 9000 years ago, all but northern Lake Superior was free of glacial ice. The southern "Lake Duluth" drained to the Mississippi, as did Lake Michigan. The eastern lakes drained past Rome, New York, into the Hudson River, because the St. Lawrence valley was then under the ice. By about 6000 years ago, the ice was far to the north, the sea invaded Lake Ontario, and the western lakes drained into the sea through Canada rather than Lake Erie.

The ice retreat can be followed by carbon-14 dating of associated plants and by the location of shorelines and moraines. The ice lingered over most of Canada until about 8000 years ago, when it broke into two lobes separated by Hudson Bay. The western lobe vanished about 6000 years ago and the

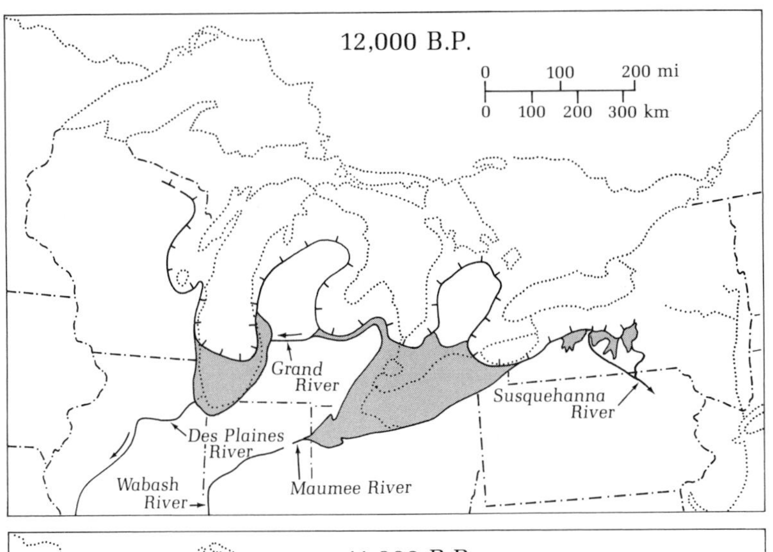

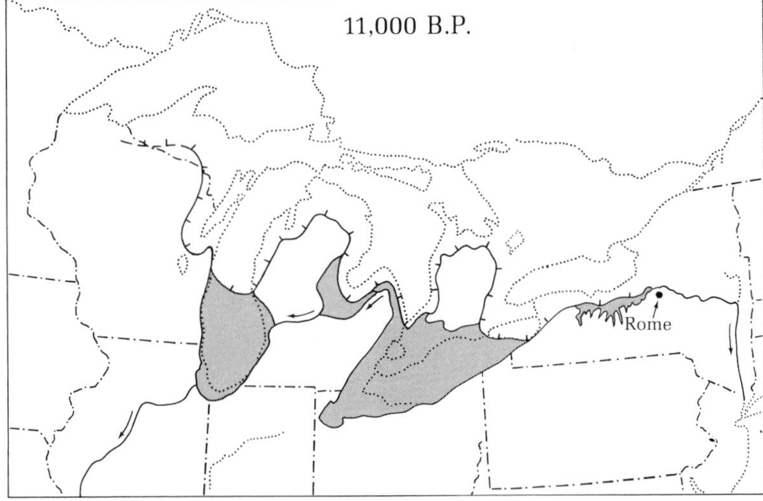

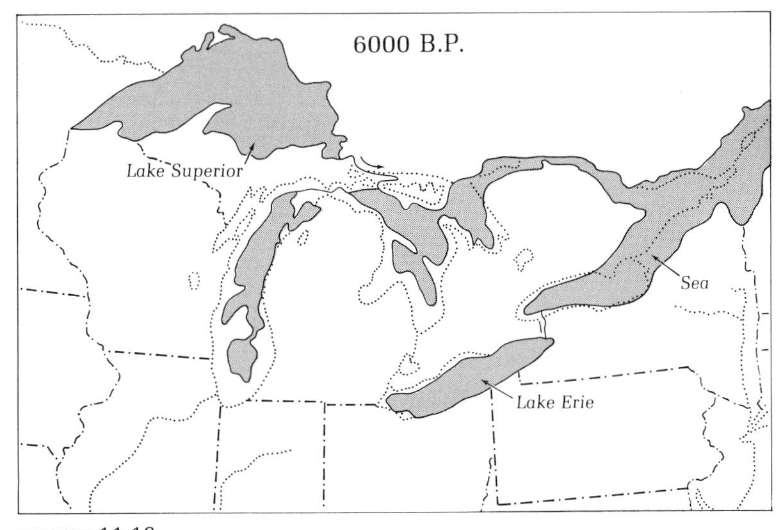

FIGURE 11.16
The history of the Great Lakes.
[After Flint (1971); dates from Bryson et al. (1969).]

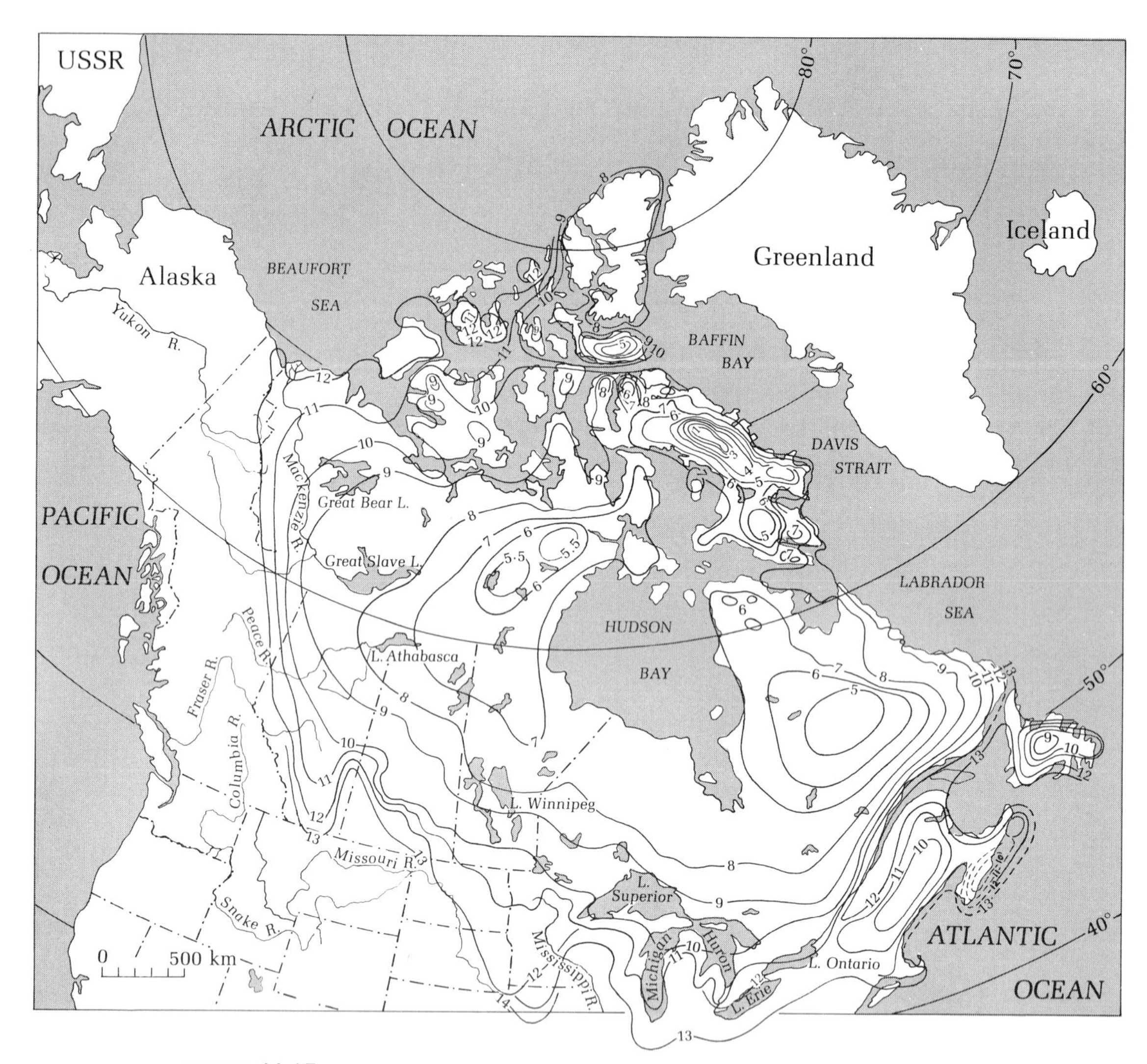

FIGURE 11.17

Successive positions of the continental ice front after the last glaciation. Contours give the age of the ice front in thousands of years B.P. [After Bryson et al. (1969).]

eastern one 1000 years later. No wonder that the crust is still adjusting to the unloading and that lakes and rivers continue to change.

The Postglacial Optimum or Hypsithermal

It was much colder 10,000 years ago, and the reader may entertain the supposition that it has been warming ever since. This is not correct. Very roughly 5000 years ago, it was generally warmer than now. In high latitudes, a little heat is an improvement, and geologists there called this episode the Postglacial or Climatic Optimum. Farther south, the attractiveness of more heat is in question, and geologists there called it hypsithermal or megathermal time.

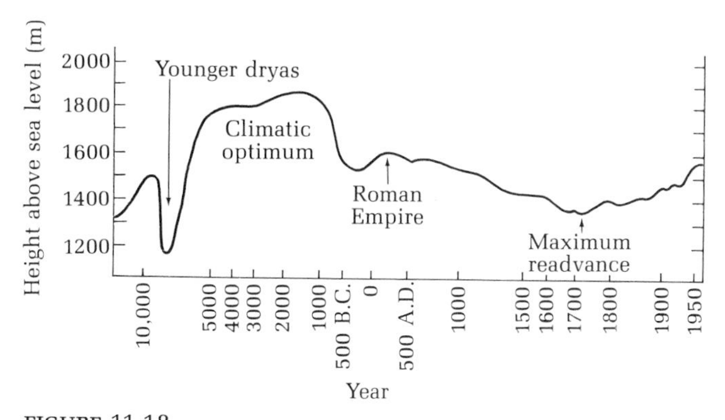

FIGURE 11.18
Snow line in Norway during the last 12,000 years. The time scale is logarithmic. [After Schwarzbach (1963).]

Warming suggests glacial melting and a rise in sea level, but scientists are divided regarding the validity of evidence for such a rise. Except for this uncertainty, the global picture of warming seems quite consistent.

Although confusing during major glaciations, the oxygen-isotope thermometer is less affected during periods of relatively constant oceanic volume, such as the last 6000 years. An ice core from the Greenland glacier indicates that the temperature of sea water, from which the snow came, was definitely warmer than now from 8000 to 5000 years ago and somewhat warmer from 2400 to 1000 years ago.

Similar effects in Scandinavia can be detected in the changes in the kinds of pollen that were preserved in bogs, as well as by means of other biological indicators. Between 9000 and 3000 years ago, for example, the snow line in Norway was at about 1860–1890 meters. By 2000 years ago, the climate had cooled enough to drop the snow line to 1590 meters, where it is today. About 1700 A.D., however, it dropped to 1370 meters during the "Little Ice Age." During the Climatic Optimum between 7000 and 5000 years ago, the hazel, water chestnut, and marsh turtle were more widespread than today, and treeless tundra almost disappeared from Eurasia.

Periods warmer than the present have also occurred in North America during the last 10,000 years, but the record is complex. In the Middle West, the presence of deciduous forest indicates a moist, cool climate about 8000 years ago and again about 4000 years ago, but, in the time between, a prairie spread over the region. A warm, dry interval occurred in the southeast also, but there it was from 10,000 to 6000 years ago. In the northeast, it was only from 4000 to 1500 years ago.

All this suggests long-term fluctuations in the zonal atmospheric circulation in addition to a global warming. Nonetheless, there are too many indicators in both the Northern and Southern hemispheres, to permit much doubt that the earth has been warmer as well as cooler within the last few thousand years. Presumably, it will be again.

Summary

1. Major changes in climate have occurred from time to time during the entire history of the earth. The most notable of the effects of climatic changes is the spread of large glaciers.
2. We are now in a relatively cold period in the earth's history. Glaciers have advanced and retreated over large areas several times during the last million years.
3. The cause of major climatic changes is unknown, but it is probably not due to a variation in energy received from the sun.
4. A glacier is a mass of ice that flows as a crystalline solid, although its behavior is influenced by the fact it is near its melting temperature.
5. Solid waves and surges move through glaciers and cause rapid advances when they reach the front of the ice.
6. The landscape, soils, and vegetation distribution with which we are familiar in the United States have been changed enormously by widespread glaciation that ended only 12,000 years ago.
7. The effects of cooling are widespread around the margins of continental glaciers. Rainfall is intensified in some places and new lakes form. Winds transport glacial rock flour and deposit it as a top dressing on older soils.
8. Sea level was lowered by about 120 meters when water was stored in continental glaciers. The continental shelf was exposed as a result.
9. The early history of man is probably closely related to the climatic changes of the last 2 million years. Human migration certainly was partially controlled by exposure of the continental shelf.
10. The climate 5000 years ago was generally warmer and wetter than now, and climatic changes have not ceased.

Discussion Questions

1. How can the effects of continental drift be distinguished from those of climatic change?
2. What is the advantage of studying the history of glaciation by use of marine sediments instead of land ones?
3. Could a continental glacier accumulate to a depth of 100 kilometers?
4. What effects do loading and unloading by continental glaciers have on the level of the land?
5. What has happened to the lakes that collected in basins in the southwestern United States during pluvial periods?
6. How do geologists trace the migration of the shoreline during the last 30,000 years? Can they use the same methods for the earlier glacial epochs?
7. What is the evidence that frail man is responsible for the extermination of the saber-tooth cat and other great mammals?

References

Agassiz, L., 1967. *Studies on Glaciers*. New York: Hafner. (Translated from the 1840 edition and edited by Albert Carozzi). [Professor Carozzi presents an admirable introduction to this classical work on continental glaciation.]

Bloom, A., 1971. Glacial-eustatic and isostatic controls of sea level since the last glaciation. *In* K. Turekian, ed., *Late Cenozoic Glacial Ages*, pp. 355–380. New Haven: Yale University Press.

Bryson, R. A., W. M. Wendland, J. D. Ives, and J. T. Andrews, 1969. Radiocarbon isochrones on the disintegration of the Laurentide Ice Sheet. *Arctic Alpine Res.* 1(1):1–14.

Charlesworth, J. K., 1957. *The Quaternary Era*. London: Edward Arnold. [The definitive text for the time.]

Daly, R., 1963. *The Changing World of the Ice Age*. New York: Hafner. (Reprint of 1934 edition.) [An imaginative synthesis of the global effects of glaciation and deglaciation.]

Dansgaard, W., S. J. Johnson, H. B. Clausen, and C. C. Langway, Jr., 1971. Climatic record revealed by the Camp Century ice core. *In* K. Turekian, ed., *Late Cenozoic Glacial Ages*, pp. 38–56. New Haven: Yale University Press.

Denton, G., R. Armstrong, and M. Stuiver, 1971. The late Cenozoic glacial history of Antarctica. *In* K. Turekian, ed., *Late Cenozoic Glacial Ages*, pp. 267–306. New Haven: Yale University Press.

Embleton, C., and C. King, 1968. *Glacial and Periglacial Geomorphology*. New York: St. Martins Press. [A combined analytical and descriptive approach.]

Emiliani, C., 1958. Ancient temperatures. *Sci. Amer.* 198(2):54–63. (Available as *Sci. Amer.* Offprint 815.) [Some conclusions about more recent events are now suspect.]

Ewing, M., 1971. The late Cenozoic history of the Atlantic basin and its bearing on the cause of the ice ages, *In* K. Turekian, ed., *Late Cenozoic Glacial Ages*, pp. 565–574. New Haven: Yale University Press.

Fairbridge, R., 1960. The changing level of the sea. *Sci. Amer.* 202(5):70–79. (Available as *Sci. Amer.* Offprint 805.)

Field, W. O., 1955. Glaciers. *Sci. Amer.* 193(3):84–92. (Available as *Sci. Amer.* Offprint 809.)

Finkel, R. C., et al., 1971. Depth variation of cosmogenic nuclides in a lunar surface rock and lunar soil. *In* A. A. Levinson, ed., *Proceedings of the Second Lunar Science Conference*, vol. 2, pp. 1773–1789. Cambridge, Mass.: The MIT Press.

Flint, R., 1971. *Glacial and Quaternary Geology*. New York: John Wiley & Sons. [Summarizes a staggering amount of material from world-wide localities.]

Howell, F. C., 1965. *Early Man*. New York: Time-Life. [A magnificently illustrated, unpretentious, authoritative account.]

Kent, D., N. D. Opdyke, and M. Ewing, 1971. Climate change in the North Pacific using ice-rafted detritus as a climatic indicator. *Bull. Geol. Soc. Amer.* 82(10):2741–2754.

Lisitzin, A. P., 1972. *Sedimentation in the World Ocean* (Soc. Econ. Paleontol. Mineral. Spec. Publ. 17). Tulsa, Okla.: Society of Economic Paleontologists and Mineralogists.

Lobeck, A. K., 1939. Geomorphology. New York: McGraw-Hill.

Magnusson, N. H., G. Lundqvist, and E. Granlund, 1957. *Sveriges Geologi.* Stockholm: P. A. Norstedt & Söner.

Martin, P., 1967. Prehistoric overkill. *In* P. S. Martin and H. E. Wright, Jr., eds., *Pleistocene extinctions,* pp. 75–120. New Haven: Yale University Press. [Stone-age man may or may not have exterminated the mammoth and starved the saber-tooth cat.]

Milankovitch, M., 1920. *Theorie mathematique des phenomenes thermiques produits par la radiation solair.* Paris: Gauthier-Villars.

Opik, E. J., 1958. Climate and the changing sun. *Sci. Amer.* 198(6):85–92. (Available as *Sci. Amer.* Offprint 835.)

Paterson, W. S. B., 1969. *The Physics of Glaciers.* London: Pergamon Press. [For the advanced undergraduate, but with many insights for those with any background in physics.]

Sass, J. H., A. H. Lachenbruch, and A. M. Jessop, 1971. Uniform heat flow in a deep hole in the Canadian shield and its paleoclimatic implications. *J. Geophys. Res.* 76(35):8586–8596.

Schwarzbach, M., 1963. *Climates of the Past* (translated from the German by Richard Muir). London: Van Nostrand.

Shepard, F., 1963. *Submarine Geology.* New York: Harper & Row. [Documents changes in sea level and their effects.]

Solheim, W. G., II, 1972. An earlier agricultural revolution. *Sci. Amer.* 226(4): 34–41.

van der Hammer, T., T. A. Wijmstra, and W. H. Zagwijn, 1971. The floral record of the late Cenozoic of Europe. *In* K. Turekian, ed., *Late Cenozoic Glacial Ages,* pp. 391–424. New Haven: Yale University Press. [The forests of the northern hemisphere spread north and south with the movement of the glacier front.]

White, W. A., 1972. Deep erosion by continental ice sheets. *Bull. Geol. Soc. Amer.* 83:1037–1056. [Precambrian rocks are exposed mainly where continental glaciers have stripped away younger rocks.]

Wright, H. E., Jr., 1971. Late Quaternary vegetational history of North America, *In* K. Turekian, ed., *Late Cenozoic Glacial Ages,* pp. 425–464. New Haven: Yale University Press.

12

CLIMATE AND WEATHER

The greatest floods in U.S. history devastated the East Coast yesterday from North Carolina to New York. President Nixon declared five states disaster areas as unprecedented flood crests killed scores of persons and caused more than $1 billion in damages.

SAN DIEGO UNION, 24 June 1972

Don't worry the children about the cold, just keep them warm. Burn everything except Shakespeare. [A telegram from Mr. Antrobus upon learning that a new ice age is approaching.]

Thornton Wilder, THE SKIN OF OUR TEETH

I would give part of my life to know the mean barometric reading in Paradise.

G. C. Lichtenberg (eighteenth century)

Weather includes such atmospheric phenomena as precipitation, temperature, pressure, and wind velocity, which vary from day to day. Climate is the summation of weather over a period of years or longer. Both vary on scales that affect our daily lives and those of cultures, cities, and civilizations.

We derive information about past weather from a hundred years of organized instrumental measurements, mostly from a few cities. Such information is collected because, with the invention of the telegraph, it became possible to follow storms as they migrate and, thereby, to make useful weather predictions for periods of 12 to 48 hours. The value of such predictions is universally acknowledged as transcending all political differences short of war. During the cold war, for example, Russia and China never stopped broadcasting weather observations, and the United States continued to produce global predictions.

The historical weather data from any region commonly show "cycles" of hot and cold, or wet and dry, with durations of 10–15 years. An enormous academic effort has gone into attempting to correlate these "cycles" with sunspot cycles. The result has been a statistical morass and the issue is in doubt. Much more success has accompanied the work of meteorologists Jacob Bjerknes and Jerome Namias (of UCLA and of Scripps Institution of Oceanography, respectively), who analyzed these regional weather cycles in terms of variations in zonal atmospheric circulation. Frequently, one region grows drier while an adjacent one becomes wetter. This can be explained by a simple longitudinal shift in the pattern of waves of the zonal circulation.

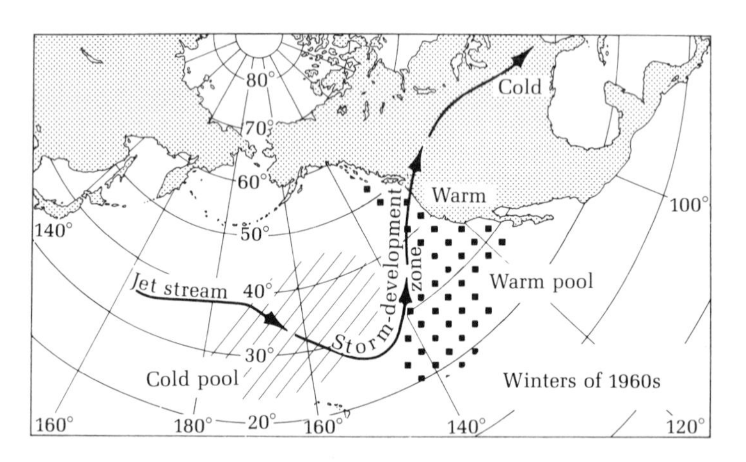

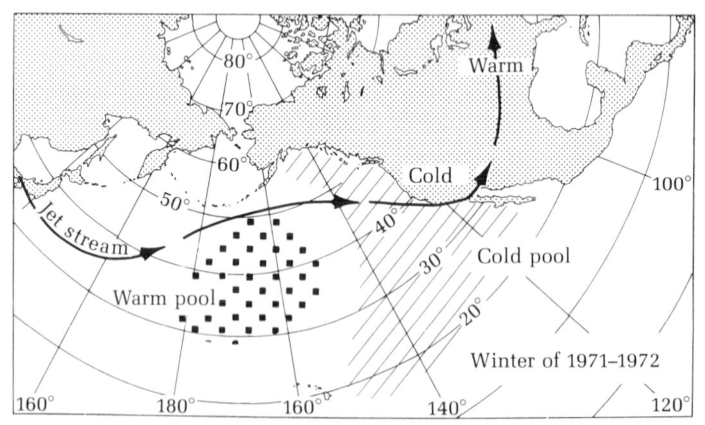

FIGURE 12.1

Zonal atmospheric circulation is partially controlled by decade-long fluctuations in surface-water temperatures in the ocean. This changes local weather on the continents. One region has more rainfall than usual for a decade, another has less. [From Namias (1972).]

The correlation has been strengthened by the discovery that the surface temperature of the ocean varies in large regions over periods of many years. These hot and cold water masses directly determine the zonal circulation and explain the "cycles." We are left to wonder what causes the variations in water temperature.

Most information on weather before about 1800 A.D. comes from daily observations of rain and snow by government or scientific organizations or individual diarists. These can be very useful in identifying extremes of weather. If the extremes persist for decades and centuries, they indicate changes in climate, and these, too, are widespread in historical records. Changes are also perceptible in the natural record of fossil pollen, bones and shells, soils, moraines from glacial advances and retreats, and shoreline terraces from rising and falling lakes. Although the natural record can tell us much about climate, it can tell us nothing about weather, except that it was shifted and modified by climatic change.

We cannot predict climate because we know little of why it changes. It seems entirely reasonable, however, to predict that climate, like the weather, will continue to change. That is enough to present man with a problem and an opportunity.

Alternative Approaches to Some Environmental Problems

The days of man are full of sorrow. Beset with tempest, flood, thirst, and famine, he has fought back with weather predictions, lighthouses, dams, aqueducts, and silos full of grain. For a long time, he was winning.

Consider Indian famines: In 1344–1345, famine was so terrible that the Mogul emperor could not feed the retainers in his own palace; the great famine in Bengal in 1769–1770 killed 10,000,000 people—a third of the population; in 1790–1792 came the Doji Bara, or skull famine, when the dead were too numerous to bury; and so on, and on, with at least a million dead in 1866, in 1869, in 1876, in 1899, and in 1943.

The last of these was in a war zone—otherwise, the British essentially whipped the problem of Bengali famine by building a network of railroads. Famine was a local problem brought on by drought, war, disease, or natural disasters, such as volcanic eruptions. In the days of primitive transportation, it was simply impossible to move enough food into a populous region faced with crop failure. The railroads could move the food, provided it existed anywhere in quantity.

That was a solution for the times, but a new problem has arisen. The population has exploded to the point where there is a very serious question about the availability of food reserves anywhere in the world. If the population continues to expand, local famines will again bring widespread death because the railroad cars will be empty. Just when—or if—this happens is apt to depend on fluctuations of climate and weather. Therein lies the problem: the environment varies, and the population expands to match the temporary optimum (see Figure 1.1).

The limitation on food is well known, but the general applicability of similar limitations is not. Hong Kong, Singapore, and New York have faced water shortages. In Hong Kong in 1963, water was so scarce that naval ships were making water by evaporation in the harbor and barging it ashore. The great cities have survived easily, but, as they continue to grow, the margin of safety diminishes.

Food and water have always presented problems, but, as the population grows and half of it moves into cities, new ones have arisen. The urban dweller survives in moderate comfort only because of the existence of systems that bring energy and supplies into the cities and remove wastes. All of these systems are dependent on weather. The growing urban problems include brown-outs when power is inadequate to heat in winter or cool in summer, clogged highways and paralyzed airports during heavy snows, and the intermittent failure of most public services. Thus, the old problems are replaced by new ones, and the old solutions are no longer adequate.

FIREFIGHTING

The really critical problem of starvation can continue to be solved for a while by "firefighting"—that is, by countering a disaster after it occurs by utilization of standby transportation and reserve supplies. The most recent famine in Bangla Desh, for example, was countered by flying in supplies with giant military transport planes. However, because the food must be grown somewhere, rainfall and the storage of water are increasingly important—and these things cannot be improved by firefighting.

Famine in Bangla Desh was triggered by the drowning of hundreds of thousands of people in a storm surge caused by a hurricane in the Bay of Bengal. Extreme weather, hurricanes, tornadoes, and floods strike frequently around the world, and their aftereffects are often ameliorated by distribution of food and clothing and by gifts or loans to rebuild. This is another form of firefighting that will demand more and more effort as the population expands and concentrates along the vulnerable shoreline.

The less critical problems of water, power, and transportation cannot be dealt with very well by firefighting. Either the problem is intractable or the crisis is over too soon for outside help to be of use. Thus, the residents of a great city gradually become more and more miserable without much prospect of help by firefighting.

PREDICTION

At least a third of the problems caused by weather could be solved by accurate prediction. This is as useful as being able to tell the fire department where and when all the fires will break out. Another third of the problems would vanish if people would take appropriate action, which—as we know from earthquakes and volcanoes—requires very accurate forecasting indeed. Still, the available forecasts are commonly utilized in timing critical military operations, such as the Japanese attack on Pearl Harbor under cover of an advancing storm front, the Allied landing in France in the best weather available when high water occurred at dawn, and the German launching of the Battle of the Bulge at a time when low clouds prevented Allied air reconnaissance. Forecasts also find increasing use during peacetime by farmers, businessmen, and individuals who are planning outside recreation.

Unfortunately, the predictions are, at best, good for only a few days. The possibility of long-range forecasts is discussed in a later section. However, even if we could predict years ahead, we would still have problems. Consider this story of tornado predictions in Kansas: Walter Orr Roberts, an American meteorologist and astrophysicist, relates that, when the Kansas City forecasting center issued a tornado warning and nothing happened, a woman phoned to complain about what a waste of time it had been for her to sit in the shelter under her house. She was even more irate when she phoned some time later after the same thing had happened again. The third time, two years later, she came out of the shelter to find her house demolished by a tornado. She phoned the forecasters yet again and said, "Now, that's more like it." Thus, a perfect prediction, a perfect response, and yet something to be desired in dealing with the weather.

CONCRETE OR CLOUDS

One way to deal with forest fires is to fight them as they occur; another is to build fire breaks in advance to contain and control them whenever they occur. We take comparable measures in dealing with variable and violent weather. Most notably, we have built an enormous system of dams to control floods on major rivers. Many of these are also used to store water for irrigation and to generate power. These uses conflict, because flood control requires minimum water storage before a flood, and irrigation and power generation require maximum storage all the time. Nonetheless, the dams exist and partially serve their purposes, and more of them are proposed.

We also build other enormous public works in order to control wind and water. Aqueducts and canals run the length of California, thousands of kilometers of flood-control channels line our major rivers, and great revetments rise along the eastern seaboard to prevent hurricane-generated storm surges from flooding cities. Thus, we have used a flood of concrete to control the flood of water.

In the past we had no other choice but concrete—except to endure whatever happened. Now we are developing another option: we can partially control the weather. We can make it rain to put out a forest fire or we can prevent it from raining when there is danger of a flood. We can replace concrete with cloud seeding.

INADVERTENT AND DELIBERATE WEATHER MODIFICATION

Man has been inadvertently altering the weather more and more frequently, as population and industry have increased. For example, the burning of coal in growing London provided the smoke particles in the air around which the famous fogs nucleated. The idea that man could do something of this sort deliberately has only come recently. Now the whole of central London is a nonburning zone; even the Queen does not burn wood in the fireplaces of Buckingham Palace. The results have been spectacular. Without the smoke nuclei, the city has emerged from the fog of centuries.

Our power over the environment has grown beyond imagining. The explosion of the first hydrogen bomb at Eniwetok atoll in 1952 blasted a coral island into the upper atmosphere, and the dust produced the same gorgeous sunsets that were remarked when Krakatoa blew up. The power developed, but the demonstration that we could deliberately change the weather was slow in coming.

There have always been men who claim to control the weather, but the results of their "control" have been equivocal. As late as 1965, a committee of the National Academy of Sciences could find no statistically significant evidence that man could make rain. It rained sometimes when people tried to induce it, but it also rained when they didn't, and there was no proof that the attempts had any real effect. The tentative and as yet unpublished conclusions of the committee reached the ears of a group of professional rainmakers who thereupon decided to release all their own private data to the committee. To the general surprise of the scientists, the voluminous new data made a convincing case. Man can make rain—sometimes and in some kinds of clouds. One is reminded of the alpine woodcutter and his knowledge of glaciers before Agassiz (see Box 11.1), but this is different: no one can be expected to accept evidence until it is available for inspection.

Historical Climatic Changes

It is the belief of many archeologists and classical historians that more than one civilization has been overwhelmed by climatic change. Our unique industrial civilization has enormous power and excess capacity that is now expended in war, and presumably, it could withstand any climatic change spread over several centuries. Ideally, it could do so; but our growing numbers may exhaust our capacity, and wars are more apt to be generated by climatic stresses than prevented by them. The starving multitudes of Bangla Desh in 1972 simply walked across the border into India, which was then overtaxed to feed them. No solution was found short of war.

It is clearly important to try to understand the origin of climatic change, even though there is not much known at present. What is certain is that the climate in Europe, where the records are best, was relatively warm from 7000 B.P. to 3400 B.P., cold to 2400 B.P., warm from then to 700 B.P., and then quite cold until 150 B.P. and warmer until perhaps 20 B.P. The sequence of intervals seems peculiar—namely 3600 years, 1000 years, 1700 years, 550 years, and 130 years. Possibly the climatic changes are becoming more frequent, but it seems more likely that we are detecting more as we obtain better records. The "cold" periods of the past may have included an occasional warm decade as the zonal circulation shifted. Perhaps a concentration of such decades would make for a warm century in a cold period. At the opposite extreme, a particularly hot summer or a winter of heavy snow might occur anytime within a wide range of climates. Daily weather variations and cycles were superimposed on the climate of the past just as they are now.

THE END OF THE CLIMATIC OPTIMUM—1400 B.C.

At this point, because we are largely concerned with history, we shall convert to a historical time scale. By about 1400 B.C., the long, warm Climatic Optimum was at last coming to an end. It reached its peak sometime between 5000 B.C. and 3000 B.C., an interval that saw our all-conquering species begin to irrigate and plow the fields of the Nile, write and record a calendar in Egypt, and build cities in Sumeria.

A gradual cooling occurred in Europe between 3000 B.C., when it was warmer than it is now, and 1400 B.C., when it was cooler than now. As the temperature zones shifted, so did the rainfall, and this may have been more important in the Middle East, where momentous events were underway. Early in the interval, urban civilization developed at Harappa and Mohenjo Daro in the Indus Valley, and Elam began to expand in what is now Iran. Later, the Egyptians built the pyramids, Hammurabi dictated the first legal code, and the Babylonians and the Hittites flourished. Near the end of this period, the Minoans peopled Crete and conquered the seas around them.

COLD EPOCH OF THE IRON AGE—1400 B.C. TO 400 B.C.

Beginning about 1400 B.C., the climate of Europe deteriorated markedly, and climates elsewhere in the world changed also. In Europe, dated moraines indicate that the Alpine glaciers advanced far beyond their present fronts. The present glaciers of the Rocky Mountains are not remnants from the ice age that persisted through the Climatic Optimum. It appears, instead, that they formed at about this time. In Russia, forests of cold-climate trees spread to the south. This cooling was not confined to the Northern Hemisphere. By the end of the period, both New Zealand and Tierra del Fuego showed the effects of a northerly shift of zonal circulation.

The most evident effect in Europe was not cold but increased moisture: peat bogs began to develop in Ireland, Germany, and Scandinavia; lake dwellings in central Europe were flooded and abandoned; and ancient tracks across English lowlands were gradually abandoned as they became increasingly marshy.

The climate around the Mediterranean, and probably that of the Near East, may have been colder than now. After this period, around 100 B.C., Roman writers noted that vineyards were spreading northward in Italy into regions that had formerly been too cold. The climate, apparently, was not so dry as it is now (although it was drier then than at the Climatic Optimum), and more rain fell in the summer, when it was most needed.

The Mediterranean and Near East are of particular interest because so many cultural developments of the first magnitude happened there during this time, but they are not alone in this respect. China emerged, in the Shang Dynasty, from a period of legend. By the end of the period, Lao Tzu and Confucius were writing and the emperors were preparing to build the Great Wall. The subcontinent of India was overrun by Aryan peoples at the start of this cool period, and, by the end, the Buddha was alive and the Vedas and the Upanishads were older literature.

Farther west, the beginning of the epoch found Ulysses experiencing, in the lines of Tennyson, ". . . delight of battle with my peers, / Far on the ringing plains of windy Troy." At the same time, the Phoenicians introduced the alphabet; not long afterwards, the Pharaoh Ikhnaton introduced monotheism to Egypt, and Moses led the Exodus. Early in the epoch, the Dorians invaded Greece and replaced the charming culture of Mycenae with more austere standards. By the end, Pericles was building the Parthenon, the Athenians were planning the disastrous attack on Syracuse, and, not much later, Alexander the Great swept the board all the way to India.

Midway in the epoch, the Ethiopian dynasty was founded. Near the end, the first Japanese emperor appeared. Finally, the Mayans began to build cities and to develop a new and superior calendar. The influence of climatic change on all these historical events is obscure because the best climatic data come from northwestern Europe and North America, where little of historical consequence happened. Perhaps that is the point: it was too cold in the north for civilization to develop. Farther south, the Mayans developed a civilization in what is now a tropical jungle; presumably, the climate was more favorable then. Across Eurasia from the Mediterranean to China, there were many cities in what is now desert; presumably, the climate was more favorable then, although overzealous plowing and overgrazing have both been blamed for the barrenness of the land today. At the very least, we can conclude that the climate of this colder and wetter period certainly did no harm to human cultural development.

LITTLE CLIMATIC OPTIMUM— 400 B.C. to 1300 A.D.

From 400 B.C. to 1300 A.D., the climate generally ameliorated in Europe and the Near East, and, perhaps not by coincidence, civilization spread northward. Elsewhere in the world, the record is spotty but generally suggests warming with increased or decreased rain, depending on the latitude and local geography. Apparently, it was wetter in Yucatan and Indochina—both in the tropics. It was drier in southern South America and in the northern Mississippi Valley—both in middle latitudes, although in different hemispheres.

The Mediterranean region at present has a distinctive climate that also occurs in coastal southern California, Chile, South Africa, and Australia. It is dry and mild and generally regarded as ideal both for vineyards and for humans. If it has a flaw, it is its excessive aridity in the summer, for all the rain falls in a few winter months. This flaw was corrected during the Little Climatic Optimum, but at the price of some warming and wetness.

The change is best documented for Alexandria around 150 A.D., when Ptolemy, the distinguished astronomer and naturalist, kept a systematic weather diary. The weather he recorded was quite unlike today's: Now the summer is rainless; then it rained as often in summer as in winter. Now there are no thunderstorms; then they were common. Now the prevalent summer wind is northerly; then it was westerly or southerly. The different weather continued, and the Sahara Desert was moister than now as late as the sixteenth century.

The influence of Mediterranean climatic changes upon civilization is much debated. Rhys Carpenter, for example, attributes the collapse of Byzantium (between 400 A.D. and 750 A.D.) to climatic change. Taking the other side, the warriors of Islam conquered and prospered during the same epoch in a region that now seems too dry to support an advanced culture.

One aspect of the subtleties of interpreting semiarid climates from historical records is illustrated by the famous march of Alexander returning from India in 325 B.C. He took an army with elephants through 290 kilometers of what is now an almost waterless desert. The army suffered, but historians of the time made no comment on the feat. Was rainfall greater then? Ellsworth Huntington of Yale University notes the great variability of rainfall in semiarid climates, and observes that we do not know whether 325 B.C. was a wet or dry year for the times, so the historical fact of the march is not much help in interpreting climate.

Fortunately, the climatic and historical records are easy to interpret in northern Europe and the North Atlantic. One notable example is the Viking exploration and colonization of Iceland and Greenland. The explorers and merchants followed northern routes and sailed near the Greenland coast in regions where passage is now barred by ice. Icelandic glaciers were more restricted than now from 870 A.D. to 1200 A.D., and farms spread over land that has since been overwhelmed by the ice.

The Vikings flourished along the west coast of Greenland. There were 10,000 people on 300 farmsteads in the eleventh century, but the area was almost deserted by 1400 A.D. Eskimo attacks may have played a part, but the climatic effect was important. The early settlers buried their dead in the fertile ground, and tree roots intertwined among the coffins. As time passed, the ground became frozen at depth, and it was too hard to dig new graves. Burials were replaced by sepulchers of piled rocks. Now the colonies are gone and the ground is permanently frozen solid.

The Vikings not only explored the northern regions but also sailed their long ships south along the European coast. They found a very different climate in England from the one that bedevils the residents today. The Domesday Book is the great census that was prepared in 1085–1086 by order of that most successful Northman, William the Conqueror, to provide the usual

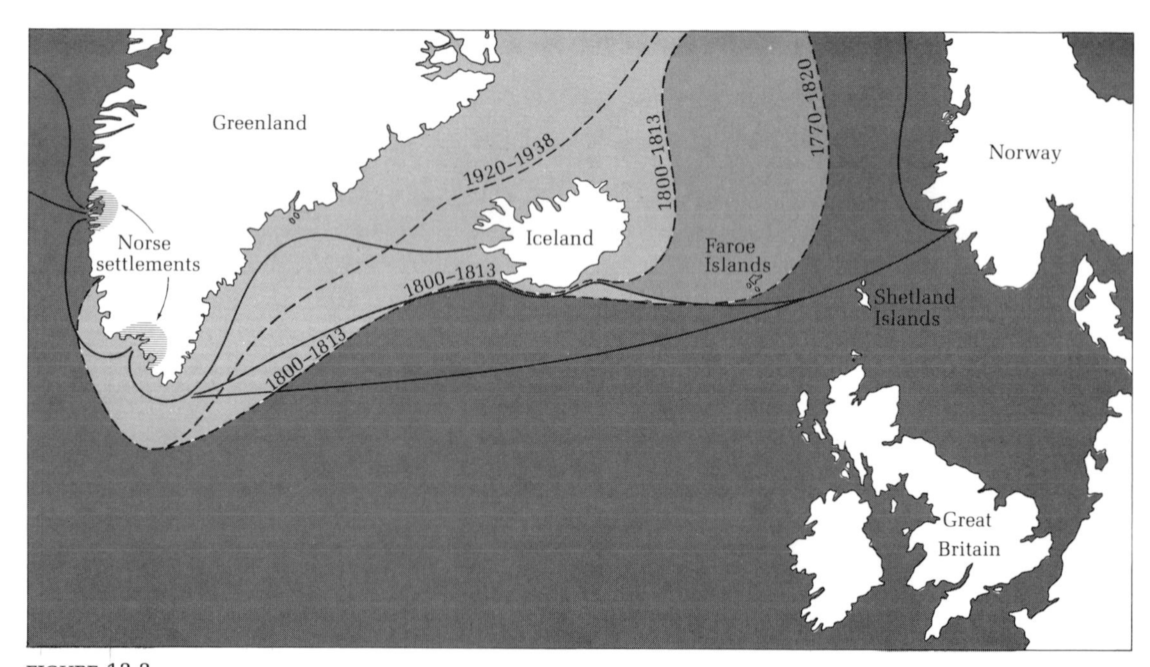

FIGURE 12.2

Norse sailing routes from the tenth to the fourteenth centuries were in regions now commonly blocked by ice. About 1800, the ice boundary was much farther south than it is now. [From Denton and Porter, "Neoglaciation," copyright © 1970 by Scientific American, Inc. (all rights reserved).]

records for taxation. Indirectly, it indicates much about the climate. There were 38 private vineyards in England where today there are none. Not long after, vinedressers were often listed in abbey chronicles as normal members of estate staffs. Grapevines require a mild, warm climate to thrive, and the sugar content and sweetness increase in general with temperature. About 1150 A.D., William of Malmesbury used the wine buff's vocabulary to write of the vale of Gloucester: ". . . nor are the wines made here by any means harsh or ungrateful to the palate, for in the point of sweetness they may almost bear comparison with the growths of France." By the same token, the French wines of that time may have been like those of Morocco at present—which are not highly esteemed because the climate is warmer than optimum.

These were the halcyon days for Europe. How did civilization fare? It began with the Gauls sacking Rome and ended with the Mongols in Europe. In between, the sweet land was overrun by successive waves of Huns, Visigoths, Vandals, and Vikings, and plunged into the Dark Ages. Monasticism and feudalism walled the elect of the populace from the flowering countryside. Perhaps the favorable climate provided the surpluses to fund the castles and cathedrals, but the best climate in several thousand years accompanied the worst decline in civilization.

LITTLE ICE AGE—1300 A.D. TO 1820 A.D.

The following period in Europe was less than optimum for grapes but wonderful for people. The glaciers advanced, but so did the Renaissance and the Reformation. The northern seas were caked with ice, but European sailors circled the globe. Europe was conquered when the climate was optimal, but it conquered the world when the climate was beastly. These relations are complicated.

The evidence for a Little Ice Age from 1300 to about 1820 is wholly convincing. Huntington and Visher devote a whole chapter to climatic stress in the fourteenth century alone. The Baltic was regularly crossed on sleds; and floods, famine, and the Black Death decimated the population and left the farms to lie fallow.

Europe grew generally colder and wetter as the centuries passed, and the alpine glaciers expanded farther than at any time since the great ice sheets vanished 6000 years earlier. The maximum advance occurred at various times during and after the fourteenth century in various places around the world, but the matter is confused by the effect of surges. In Europe, the maximum advance was in the nineteenth century. The Rhone glacier had spread into the Gletsch Valley in 1600 and was still there in 1850, when it was a major attraction for tourists. Now it is a tiny, neglected remnant perched high on the mountain. In Norway, glaciers advanced from 1660 to 1740 or

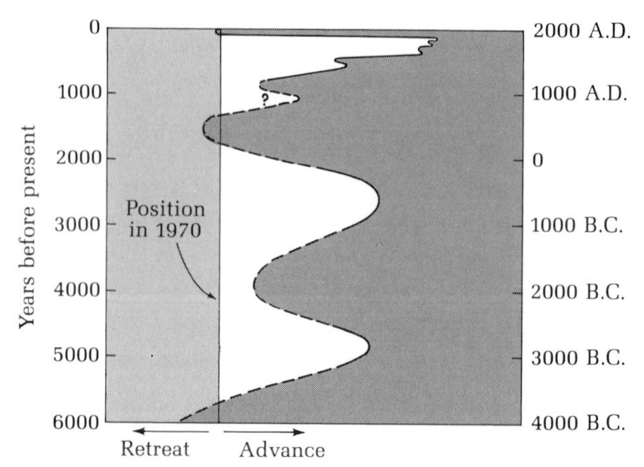

FIGURE 12.3

Estimated average global fluctuations of mountain glaciers, as indicated by the location of dated moraines. The maximum advance was in the middle and late nineteenth century in the Alps, but it occurred earlier elsewhere. [After Denton and Porter, "Neoglaciation," copyright © 1970 by Scientific American, Inc. (all rights reserved).]

so, famine accompanied cooling, and births failed to keep up with deaths. Since then, there have been intermittent advances superimposed on a gradual retreat.

Atlantic ice cut the Viking sea routes by the fourteenth century and reached a maximum at the beginning of the nineteenth. The Faroe Islands, 400 kilometers north of Britain, were apparently surrounded by solid ice reaching to Greenland. At that time, several Eskimos arrived in Scotland by kayak. Not long after, Napoleon invaded Russia at the beginning of one of the worst winters on record and left the Grand Army in the deepening snows.

AFTER (?) THE LITTLE ICE AGE—1820 A.D. TO THE PRESENT

The climate since 1820 is largely a matter of scientific record rather than historical footnote, but we are only somewhat better off than before. Many of the weather records for this period are from cities, and we have learned as much from them about the effects of urban growth upon local weather as we have about changes in global weather. However, evidence for the following is clear: the worldwide average winter temperature rose more than 0.3°C in the 1840s, oscillated to 1880, dropped 0.2°C about 1890, and then rose rather steadily by about 1.0°C (or 1.2°C, depending on how the average is calculated) to 1940. Since 1940, the temperature has dropped by about 0.2°C. Variations in average annual temperatures are smaller than those in average winter temperatures, but the trends are the same.

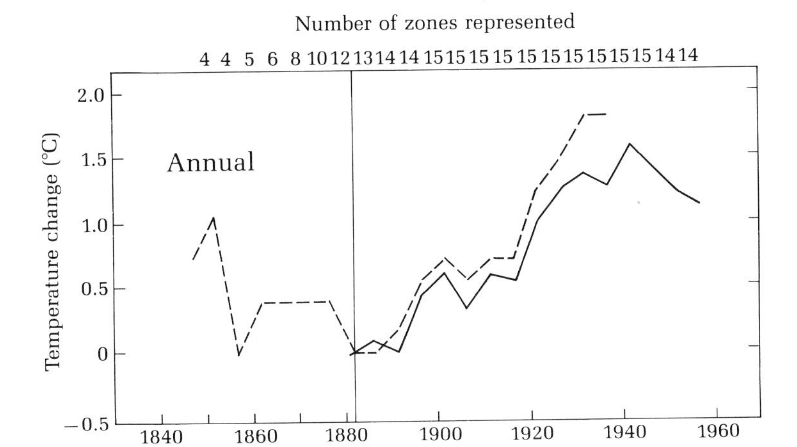

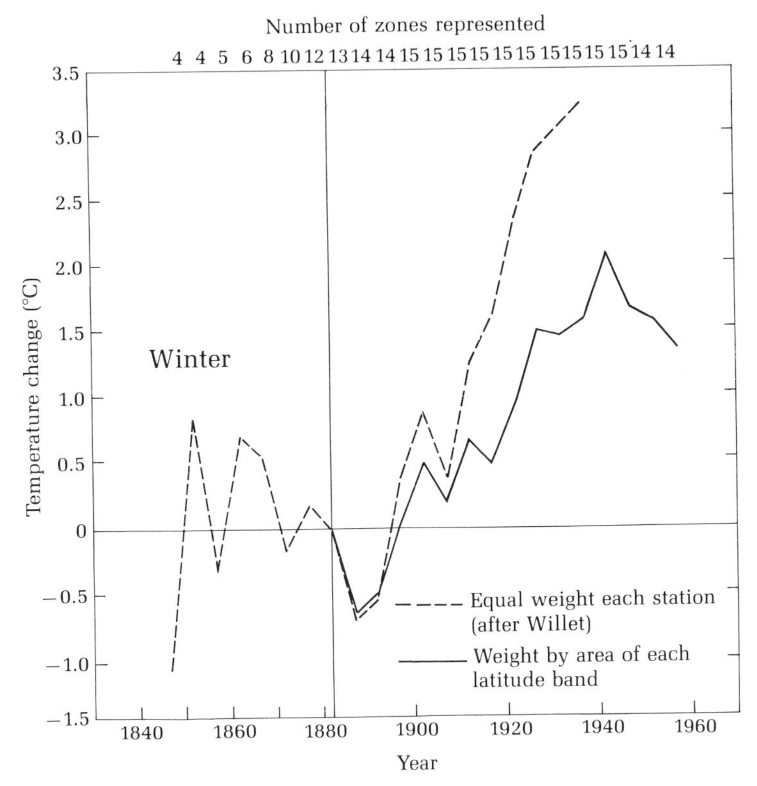

FIGURE 12.4

Trends of annual and winter temperatures averaged over the globe and also averaged over 5-year periods. [After Mitchell (1963), reproduced by permission of UNESCO.]

Because the changes seem very small, it is hard to believe that the climate has been much affected—but it has. The changes have been as dramatic as most in historical times, and this correlation suggests that apparently small factors, such as wobbles of the earth's axis or volcanic ash explosions, can trigger glacial advances or recessions. They even suggest that the inadvertent influence of man may do the same.

The end of the Little Ice Age came abruptly, according to the British climatologist H. H. Lamb. In the 1830s, enormous quantities of arctic ice broke up and drifted past Iceland and Greenland into the Atlantic. In the 1840s, such ice was rarely seen, and it has never since been as extensive as it was before the 1830s. About the same sequence of events occurred in the Antarctic. At the same time, the zonal circulation was generally intensified and many places became notably drier or hotter. There was public concern about changes in the rainfall, and regular measurements of the weather began at that time in many cities.

Because the response time of glaciers is somewhat slower than the breakup of sea ice, those of the Alps did not begin to recede for some decades, but then they all but vanished. Some floating glaciers in bays in Alaska retreated 80 kilometers. The tiny glaciers on the Cascade Range covered about the same area from about 1500 A.D. until the nineteenth century. Since then, they have retreated as much as 5 kilometers.

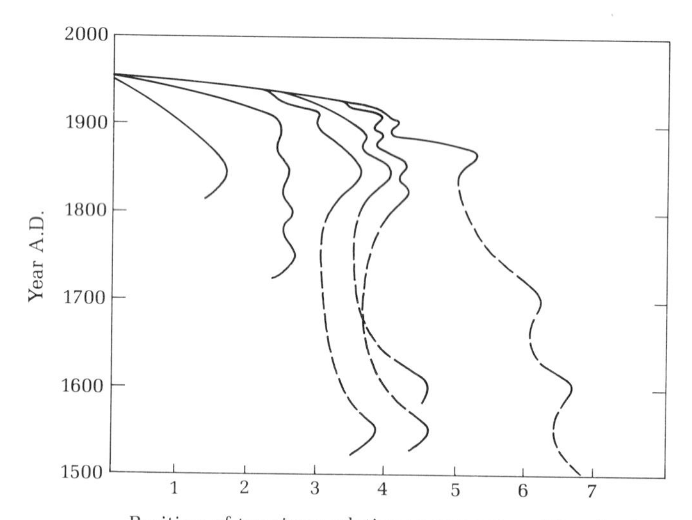

FIGURE 12.5

Small glaciers in the Cascade Range were generally stable from 1500 to 1900 (during and slightly after the Little Ice Age). Since then, they have retreated as the world warmed. They may be advancing again now. [After Denton and Porter, "Neoglaciation," copyright © 1970 by Scientific American, Inc. (all rights reserved).]

Elsewhere, most latitudes have warmed, the deserts have become more arid, and plant and animal habitats have spread poleward. Farming is spreading up the mountains of Norway, and northern coniferous forests are growing faster and migrating into the tundra. The range of the common cod has extended northward and the limit of frozen ground in Siberia has moved far toward the pole. All this is the result of a degree or so of global warming. What of the cooling of a fraction of a degree since 1940? A few glaciers have already begun to advance.

Future Climates

Many scientists believe they can detect regular climatic cycles and, thus, can predict future climates. Perhaps they can; at present, however, there is no clear demonstration of such predictability, nor are the causes of variation understood. What then can be said of future climates? Of the global climate, we can state only that it will fluctuate. If present industrial practices continue, we have some preliminary suggestions about the direction of the global trend; these suggestions merit analysis because we have the power to control the ways in which our industries operate. Moreover, the climate of cities is changing; we know why, and we can predict what will happen. This is not a trivial achievement, considering that half the human race will soon live in cities.

CARBON DIOXIDE

"Hello, carbon monoxide," goes a spirited song from the musical *Hair*, but the relatively inert carbon dioxide is more of a problem. We produce it in enormous quantities, and it may be warming the earth by a mechanism known as the greenhouse effect. The smokestacks of the industrial age have poured out carbon dioxide in ever increasing amounts since the early nineteenth century. The effect in the atmosphere was undetectable at first, because the amount was relatively small and the necessary global sampling and sensitive chemistry did not exist. During the International Geophysical Year, 1957–1958, the sampling network was established from pole to pole, and David Keeling, a chemist at Scripps Institution of Oceanography, started collecting from the top of Mauna Loa in Hawaii—a midocean locality as free from local pollution as can be hoped for. The concentration of carbon dioxide in the air varies from month to month in a regular cycle, but the average trend has been up from 313 parts per million in 1959 to 321 parts per million in 1969.

The well-known industrial output of carbon dioxide may be compared with the observed changes in its concentration in the air. The output greatly exceeds the increased concentration. This means that the residence time of the gas is small, and that about two-thirds of the new carbon dioxide is being captured by solution in the ocean and by increased plant growth, which stores carbon. The total output of carbon dioxide since the dawn of the industrial revolution is known with reasonable accuracy. If a constant fraction has been retained in the air, the total change in concentration has been about 20 parts per million. What is far more important is that both output and concentration are increasing at accelerating rates. Whatever occurred in the past is but the faintest harbinger of the future—unless we do something about it. Extrapolation of present trends indicates 400 parts per million by the end of the century, and as high as 1500 parts per million a century later. Hello, carbon dioxide! It must be emphasized that the factors governing the carbon dioxide cycle in nature, and particularly the residence time of carbon dioxide in air, are not established. Actual concentrations may be either smaller or larger than calculated concentrations.

What has been the effect of the unquestionable increase in carbon dioxide? It occurred just when the earth began to warm; is the warming an effect of the increase? It is tempting to draw such a conclusion, because the greenhouse effect gives a theoretical explanation for the correlation. Atmospheric carbon dioxide has no effect on incoming short-wave radiation from the sun, but it is opaque to some of the longer-wave radiation reflected from the earth. This radiation is captured in the air and heats it. However, water vapor in the air tends to moderate the greenhouse effect—unless it also increases and forms layers of clouds—because the vapor absorbs part of the radiant heat and, thus, limits the temperature change. Only so much heat is available at the pertinent wavelengths, and it can't be captured twice. Rasool and Schneider have computed that, for this reason, even a tenfold increase in carbon dioxide would cause but a small additional temperature change.

AEROSOLS

The global temperature rose only until 1940, even though the carbon dioxide concentration of the atmosphere has continued to increase. If the greenhouse effect causes warming, some more powerful effect is at work to counterbalance it. Rasool and Schneider find it in the outpouring of aerosols, which scatter incoming radiation back into space. The content of aerosols (mainly dust) in the air may have doubled in the last several decades, and the backscattering increases exponentially. Thus, the solid particles in smoke may be far more important than the gaseous carbon dioxide.

The long-term effects of atmospheric pollution on climate may be overwhelming. The warming influence of carbon dioxide, even if it is real, has a low maximum. The cooling influence of the aerosols is nearly unlimited. A fourfold increase in concentration could decrease the surface temperature by more than 3°C. After only a few years, according to Rasool and Schneider (1971), this "could be sufficient to trigger an ice age!"

URBAN CLIMATE

In Washington, D.C., a victim of spring hay fever will suffer, for one month, only during the day at his downtown office and then, the next month, only at night in his suburban home. Cities create their own climate; because such a climate is warmer than normal, spring comes earlier to the city than to the suburbs. For the same reason, rainfall and snowfall are less than normal in cities, and air conditioners are more in use in the summers.

The fact of the heating of cities is indisputable, and it is the main reason for confusion about how much the earth itself has warmed. It amounts to 2–3°C, on the average, and it may be more in certain seasons and at certain times of the day. Smoke merely makes cities dirty; it has not cooled them, since 1940, because the heat in cities is not just from the sun. Smoke and other aerosols retard the back radiation of the heat generated by the city itself.

City heat has many sources. Concrete and other rocklike materials absorb and radiate heat faster than most natural materials in the countryside. The shape and configuration of the buildings tends to trap heat and heated air. Cooling precipitation, whether rain or snow, is speedily removed, and heat-absorbing pollution is retained. Even a dead city, therefore, would tend to be hotter than the country. In a live city, heat is generated by the exhausts of factories, the friction of automobile tires, and the body heat of people. The end effect is a city climate with radial breezes like the onshore–offshore

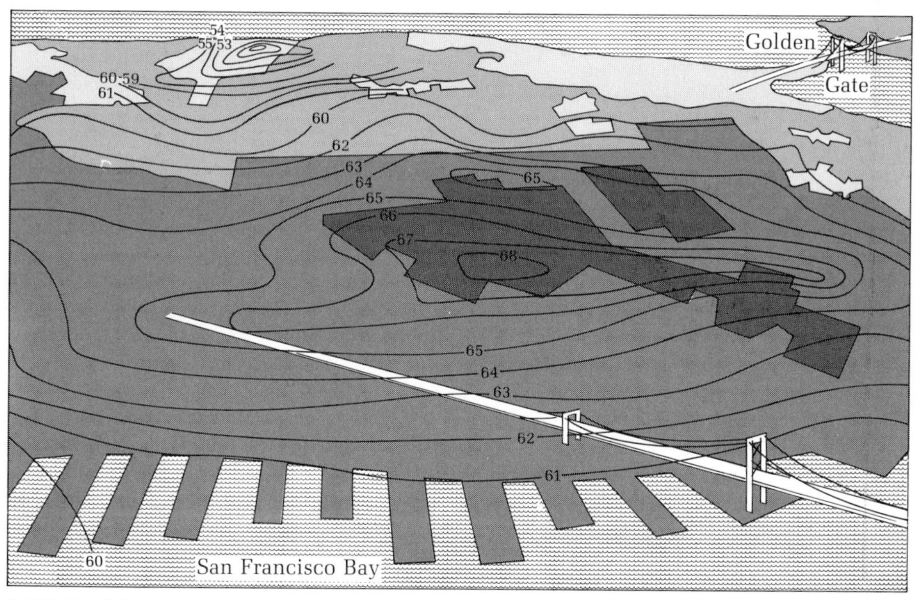

FIGURE 12.6

Temperature distribution (in degrees Fahrenheit) in San Francisco on a spring evening is depicted by lines of equal temperature. The shading ranges from the most densely built-up areas (*dark*), through less dense sections, to open country (*light*). [From Lowry, "The Climate of Cities," copyright © 1967 by Scientific American, Inc. (all rights reserved).]

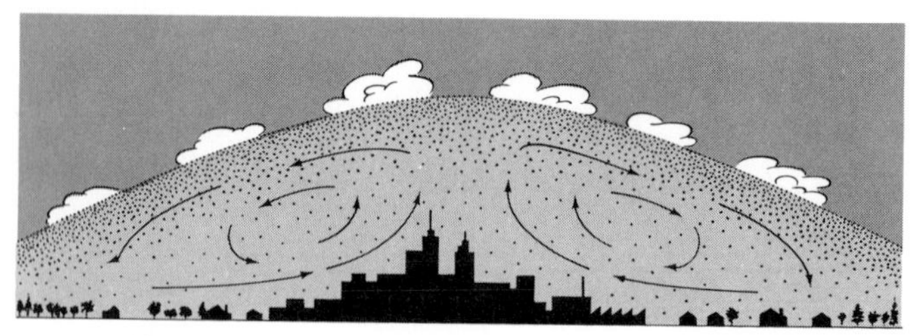

FIGURE 12.7

A dust dome as an example of the climate developed by a city. Air tends to rise over the warmer central part of the city and to settle over the cooler environs, so that a circulatory system develops. This carries the dust and smoke that originate in the city into a dome that is likely to persist, significantly affecting the city's climate, until a strong wind or a heavy rain carries it away. [From Lowry, "The Climate of Cities," copyright © 1967 by Scientific American, Inc. (all rights reserved).]

breezes at the seashore, a rain shadow like a mountain, and other major effects.

The climate-generating influence of cities is a function of their size, although the maximum difference is limited by the mobility of the air. Thus, the maximum temperature difference between a city and its surroundings is 6–8°C, whether it is London or Corvallis, Oregon, which has a population of only a little more than 20,000.

The effects of urbanization will grow, expecially as individual cities merge into great conurbations reaching, for example, from Washington to Boston and from San Diego to Santa Barbara. More city dwellers will be even warmer and drier than in the past, and more power will flow into air conditioners and less into heaters. Will this affect the global climate? That will be a matter for future research.

Long-range Weather Prediction

We turn now to the future. We can't predict the weather very well, but we are learning how. We can't do much about it yet, but we are trying. The new interest came to a focus in the mid-1960s, when scientists began to realize that some things were becoming possible that had not been so before: Computers were fast enough to test theoretical models of atmospheric circulation. Satellites were sampling the weather on a global scale, ships were sampling water, and the computers were plotting the data. The National Academy of Sciences made several studies of the weather at this time.

The congress and the Administration held hearings, committees met, and great plans were made to expand the observation network, build new computers, and begin global experiments to improve theory. Despite continuing interest, these plans have not had a high priority, and progress has been slow. Even so, general agreement has emerged that the potential benefits of weather prediction and control exceed those from any other environmental program, and progress continues.

Anyone can predict the weather an hour ahead with almost complete success by simple extrapolation. With the help of a network of observers, satellites, ships, and computers, the Weather Service can predict the weather for two days with a modest probability of success. For longer intervals, we are almost helpless—although the patterns of zonal circulation can be used to make useful predictions a week or two ahead, as long as details are not essential.

The potential benefits of longer-range forecasts are enormous. Farmers will know when to plant and to harvest, for example, and city managers will know when to conserve power and when to expend it. Achieving a forecasting capability will require inventions of three interdependent sorts, namely: (1) new theories about atmospheric circulation, (2) more capacious computers, and (3) an expanded observation system.

Present theory is inadequate for prediction, but it can be strenghtened by improving the theory of geophysical fluid dynamics, by computer modeling, and by large-scale experiments. Computer modeling involves putting a description of the weather at a given time into a computer, then making small computational changes in accordance with fluid dynamics. After an enormous number of calculations, the new weather pattern that evolves in the computer is compared with what has happened in nature. Then the parameters in the equations are changed slightly, and another computer run begins. Thus, the equations are improved by modeled experiment.

Field experiments offer the opportunity of quick advances in theory and in the determination of the parameters of the equations. The experiments will be global, lengthy, and expensive, but the scientists involved know of no substitute. In a general way, such sensors as satellites, airplanes, and buoys will be concentrated in critical areas for some years in an attempt to correlate changes in the atmosphere with related phenomena. A problem arises, however, because the theory is inadequate to establish the appropriate spacing between sensors.

Bigger computers will be necessary for modeling and, primarily, to handle the enormous flood of data that will be processed in the predictions. At present, numerous data flow into the National Weather Service by telephone links from all the continents and from weather satellites, but this is only a trickle compared to what will come. In 1966, it was estimated that computers would have to be 100 times faster than any then existing in order to keep up with the computations.

The new observation system is mainly needed to fill in the oceanic regions, although there are some continental gaps. Much of the observation can be done by remote sensing from satellites, but sampling the lower air will require airplanes, drifting balloons, anchored buoys, or some other technology that has yet to be invented. A major problem is that the most effective and least expensive way to sample depends on what is to be sampled, how often, and where. These things are uncertain, and they probably will remain so until the Global Atmospheric Research Program illuminates the mysteries of atmospheric circulation some years from now.

Cloud Physics

If we hope to modify the weather to our benefit, we must first understand it. We are already modifying it without understanding it, and among the results are smog and other undesirable effects, possibly including floods. Such successes as now attend our efforts to control the weather are largely based on an understanding of cloud physics and a much more limited knowledge of the formation of rain and ice in clouds.

CLOUD FORMATION

When the air is heated from below, it expands and rises into air that is cooler because it is farther from the source of heat. Consider a mass of warm air at the surface of the land or sea: It is laden with water vapor, but it is not saturated at the warm temperature. Because the degree of saturation varies inversely with the temperature, a rising air mass commonly becomes cool enough, at 1½ kilometers or more, to be saturated, and water condenses to form a cloud. Condensation releases heat, and this keeps the air mass rising and builds the cloud upward. Finally, at a height of 3–4 kilometers, the cloud has the same density as the surrounding air and spreads rather than rises.

This is a minimal effect. If the air mass has more energy, it may punch upward to a level of 8–10 kilometers. Energy is commonly imparted by the zonal circulation and motion of weather fronts—or by the orographic effect, when not just individual clouds but the whole atmosphere is forced up and over mountain ranges.

The heating and circulation of the air move water vapor upward, but it cannot return as rain except in special circumstances—namely, in the brief time that the cloud exists, the vapor must condense into water drops, snowflakes, or ice particles that are big enough to fall through the unsaturated air under the cloud before they evaporate. This suggests why the orographic effect is so efficient in stripping moisture from the air: because the clouds run right into the mountain peaks, the moisture in them does not have to fall through dry air to reach the ground.

NUCLEATION

An average cubic centimeter of air over a continent contains 1200–12,000 solid particles of dust, most of it extremely fine. Water vapor in clouds condenses around these dust particles to form droplets. There is more dust near sources of pollution, and the average amount is increasing; but this increase is of little consequence because, on the average, the air already contains enough nuclei to condense the water vapor in clouds.

The number of solid particles in marine air is only 120–1200 per cubic centimeters, but their average size is relatively large. They form around particles of salt that solidify when droplets of sea water evaporate. The droplets themselves enter the atmosphere when air bubbles, caught in breaking waves, rise to the water surface. There they shatter a thin hemispherical shell of water into roughly a hundred rather uniform droplets, which then move with the air. A few larger droplets are also propelled into the air as the bubble cavity collapses.

DROPLETS IN NONFREEZING CLOUDS

A typical raindrop is about 0.1 centimeter in diameter; there is one per cubic meter of air; and it falls a kilometer in 2.5–3.0 minutes. A typical droplet in a cloud is 0.001 centimeter in diameter; there are a million per cubic meter; and they fall a kilometer in 1870 minutes. Dust nuclei are much smaller still than cloud droplets, they are equally abundant, and they hardly fall at all. The question is, how does a droplet grow so dramatically into a raindrop in a reasonable time?

Small droplets grow by condensation of water vapor; this is relatively rapid when they are small. Condensation slows as they grow, however, and this process is incapable of producing significant rain in a reasonable time. Thus, the only mechanism for forming rain in nonfreezing clouds is the coalescence of small drops.

Droplets that, by chance, are larger than others fall much faster, because the settling velocity increases with the square of the diameter. Thus, coalescence occurs by impact alone, if circumstances are favorable, but often they are not. If all the droplets are about the same size, for example, they fall at the same rate and coalesce only by lateral impact—an unimportant process. At the other extreme, very small droplets cannot readily be captured because they are deflected in the air stream around a bigger falling drop. Because of various effects, it appears that some droplets must grow to about two-thousanths of a centimeter if a cloud is to produce rain by coalescence. Smaller sizes are always available. The minimum size necessary may be less, if the droplets are driven together by violent turbulence or by an electrical field.

The latter effect is obscure, but larger drops are observed to bounce off a water surface, and this effect is eliminated in an electrical field. Rain clouds, commonly, are electrically charged—witness electrical storms—so coalescence of drops may be facilitated in clouds. As it happens, two-thirds of the salt-nucleated droplets in marine clouds are larger than the critical size, whereas less than one in a billion is in dust-nucleated continental clouds. Thus, marine clouds rain comparatively quickly and easily.

FREEZING CLOUDS

Clouds that are below freezing form raindrops in a way that is quite different from that of warmer clouds. Liquid droplets ordinarily exist in all but the coldest clouds—namely, the feathery cirrus clouds at very high elevations. It is the interaction of these drops with ice nuclei that causes rain in freezing clouds.

Ice does not form in pure water vapor if the temperature is lowered below freezing. The vapor can be supercooled to −35.5°C before spontaneous ice nucleation begins. This is reasonable, because it is difficult for the free-moving water molecules to become aligned in the appropriate pattern to form a solid crystal of ice.

Spontaneous nucleation at −35.5°C is probably rare in clouds because of the overabundance of dust and salt particles in the air, which act as nuclei at higher temperatures. The nucleation temperature depends on the material. It appears that a particle with a crystal habit like that of ice is most suitable as an aid to getting water into an icelike form. A clay mineral, kaolinite, generates nucleation at −8°C, and may be the most important natural agent. Silver iodide activates nucleation at −3°C, and is commonly used for that purpose in rainmaking.

The vapor pressure around droplets is much greater than that around ice particles. This is qualitatively explained by the relative ease with which free-moving molecules can escape from a liquid compared to the molecules locked in a crystal. The difference in vapor pressure causes evaporation of the abundant liquid droplets and condensation on the ice. In an hour or so, ice particles can grow enough that they are able to fall to the ground without melting.

The hypothesis of the effect of vapor pressure on rain formation was advanced in 1911 by Alfred Wegener, our old friend of continental-drift fame, who was a meteorologist before he dabbled in geology. It is an interesting sidelight that Wegener did not invent continental drift. He elaborated upon the idea and reinforced the supporting evidence. Still, he gets the credit. On the other hand, he apparently *did* conceive the main mechanism for rain formation, but meteorologists call it the "Bergeron–Findeisen" mechanism, after the scientists who elaborated upon and developed it in the 1930s.

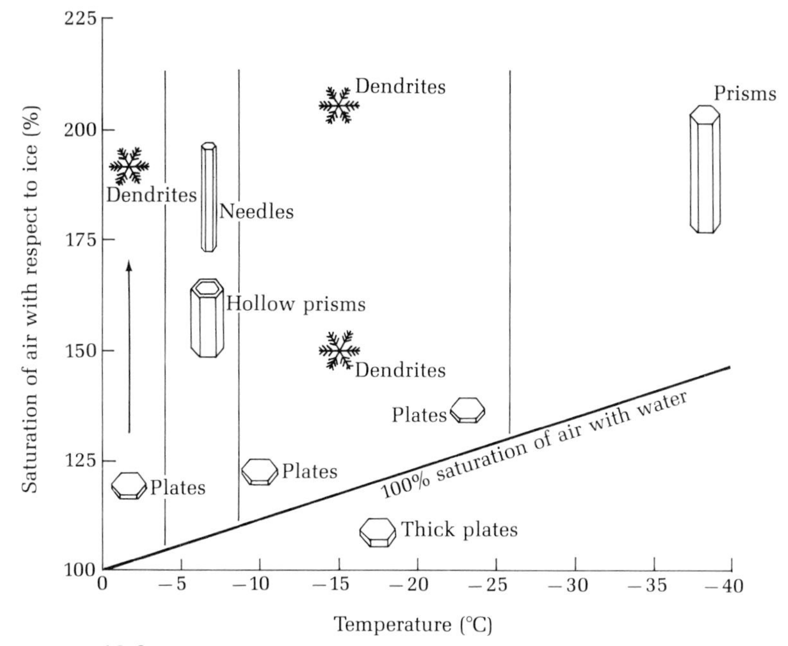

FIGURE 12-8

The form of growth of a snow crystal depends on the temperature and the amount of water vapor in the air in which the crystal forms. High water-vapor content yields more exaggerated forms: dendrites instead of plates, and long, thin needles instead of short, thick ones. Sometimes, crystals are combinations of forms. [From Knight and Knight, "Snow Crystals," copyright © 1973 by Scientific American, Inc. (all rights reserved).]

SNOWFLAKES

Snowflakes take many forms in addition to the lovely and elaborate hexagonal lattices with which we are most familiar. The British physicist Basil Mason grew ice crystals in an elegant experiment that showed that the shape is largely a function of the temperature. Thin hexagonal plates form in three temperature ranges: 0°C to −3°C, −8°C to −12°C, and −16°C to −25°C. Thin needles form between −3°C and −5°C. Hexagonal prisms form from −5°C to −8°C, and begin below −25°C. Dendritic star-shaped flakes occur from −12°C to −16°C.

Weather Control

Weather control has many aspects, ranging from a modest desire to increase local rainfall or to suppress hail to a hope of modifying global weather for peaceful objectives or for war. We produce undesired atmospheric effects, such as pollution, on all scales. As to desired ones, it can be said in a general

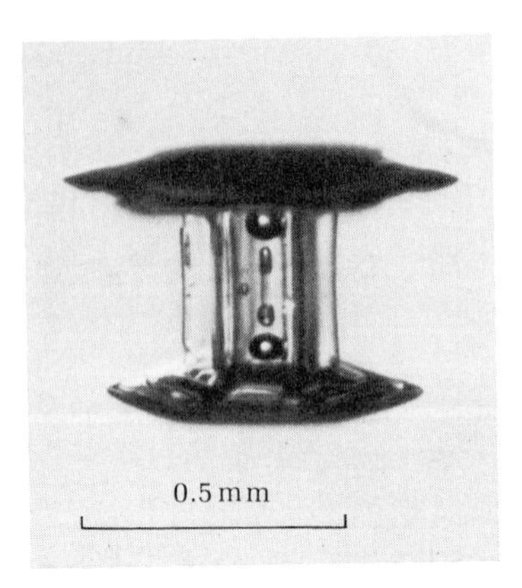

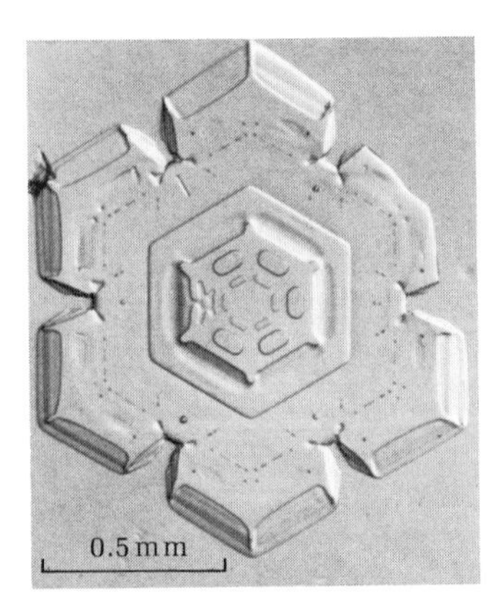

FIGURE 12.9

Tsuzumi crystals, named after a Japanese drum of the same shape, start out as a central hexagonal column that grows a plate at each end. Sometimes, the crystals are clearly of the tsuzumi shape (*left*); at other times, they look almost like an ordinary planar crystal (*right*). Here, the crystal has a very short, thick column with one end plate much larger than the other. The perfect inner hexagon is the smaller end plate, and the markings inside it are the details of the central column. If a central column grows two dendrites instead of two hexagonal plates, one dendrite may develop a few of the six branches and the opposite dendrite the remaining ones. [Photo courtesy Charles Knight.]

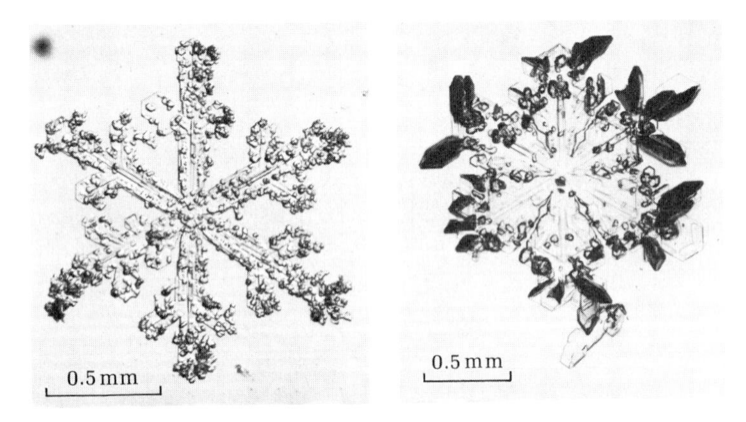

FIGURE 12.10

Rimed snow crystals grow by capturing water droplets as well as vapor molecules. A lightly rimed dendrite (*left*) has a peculair rounded appearance. New orientations of crystals are produced, in many cases, when a rimed crystal grows further from a vapor (*right*). The original crystal was a flat dendrite; the end result is a three-dimensional crystal known as a spatial dendrite. [Photo courtesy Charles Knight.]

way that the more modest the objective, the closer we are to deliberately achieving it. It can also be said that the more we know of the nature of the phenomena, the better we are able to tell whether the net effect of our efforts is desirable or not.

RAIN

From cloud physics, it is evident that rain may be localized by increasing or speeding nucleation in a cloud or by decreasing the evaporation of rain as it falls through the unsaturated air below. All attempts at rainmaking have been directed toward nucleation.

Benvenuto Cellini claimed to have *stopped* a heavy rainstorm by firing artillery shells into the thick of the clouds. As would be expected, he did it for a lady. The Duchess Ottavio wished to make her entry into Rome in sunshine, and he earned her deep gratitude for making it possible. In general, however, cannonades have been used in attempts to trigger rain rather than to suppress it. General Daniel Ruggles obtained a patent for this purpose in 1880, and (self-styled) "General" Robert St. George Dyrenforth set off five barrages in 1891 and 1892 at Washington, D.C., and in Texas. Rain accompanied one of them which could be taken to balance Cellini's account. The effect of bursting shells is, at best, moot. The explosion generates nuclei and turbulence, which are favorable to rain. The heat of the explosion, however, has the opposite effect.

More conventional efforts at making rain usually involve enhancing nucleation by cooling with dry ice or by adding suitable nuclei, such as silver iodide particles. It has been known since the mid-1960s that seeding from an

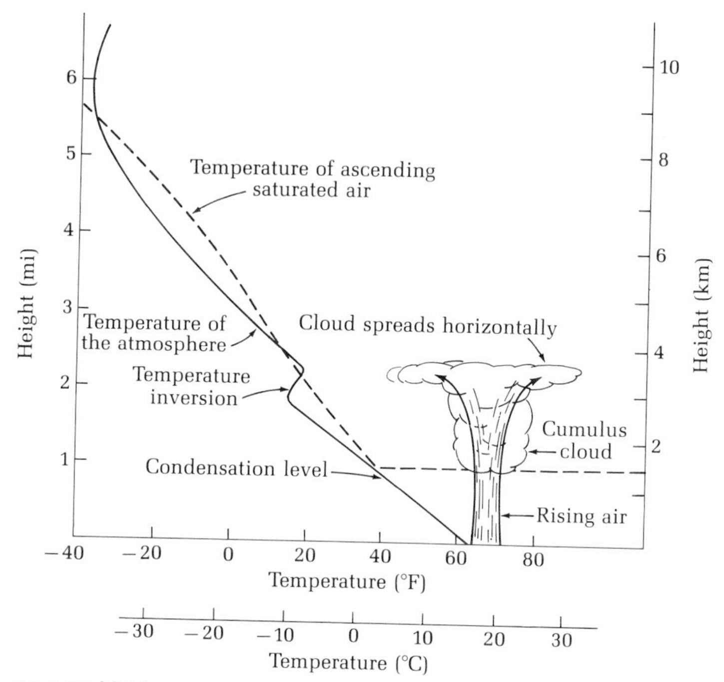

FIGURE 12.11

Formation of a typical warm cumulus cloud. [After Hobbs (1969) and Mason (1960).]

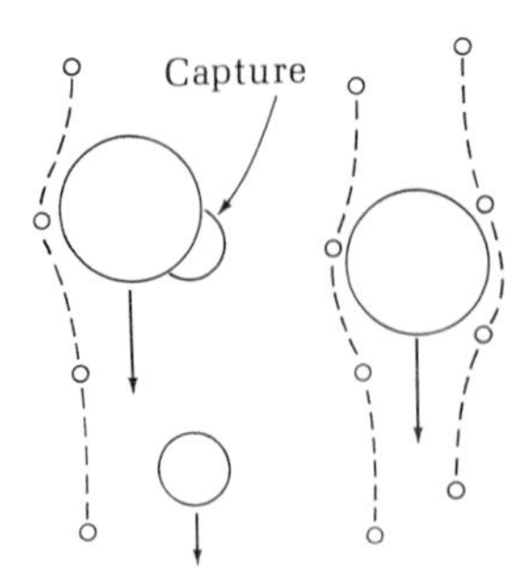

FIGURE 12.12
A large drop pushes aside the air through which it falls so that tiny drops are deflected and do not coalesce. Big drops of similar size fall at the same speed and do not coalesce either.

airplane or a smoke generator on the ground can produce a statistically significant local increase in rain from freezing clouds. The record with warm clouds has been more dubious. It was only in 1971 that the American meteorologist Joanne Simpson demonstrated that such clouds can be seeded successfully. She accomplished this by computer analysis of numerous experiments in which silver iodide "smoke" was released from airplanes over Florida. One experiment triggered 12 centimeters of rain in the Everglades when that area was suffering from a drought. The benefits from the rain exceeded the cost of producing it by 30 to 1—which made it a highly desirable investment.

HAIL

Because hail causes extensive and costly damage, mainly by destroying crops, much effort has been expended over the years in trying to control it. A burgomaster in Austria emplaced 36 "hail cannon" around his district in 1896, and their reported success was widely hailed in Europe. Many copies of these cannon were installed elsewhere before they were discredited by tests in 1902.

The basic idea seems reasonable: Damaging hail stones form in large cumulonimbus clouds, which have strong updrafts that are capable of holding them in suspension. The process may be forestalled if many small ice nuclei are introduced into the cloud to compete for the supercooled droplets that are necessary for accretion. Cannon are a rather economic way to put the nuclei into a cloud. Since 1964, the Russians have regularly fired antiaircraft shells primed with silver iodide into clouds over the Caucasus. They estimate that hail damage has dropped to less than one-third of its former level. Rockets are even cheaper, and the Russians have used these to hoist silver iodide smoke generators into the clouds above Georgia and Moldavia. There, they report that they have lowered hail damage to one-fourth.

The Russian operations are not scientifically controlled, in the sense that they make no random tests to improve the significance of their statistics. Such controlled experimentation has been done in Argentina, however. There, hail damage decreased by 70% when cold-front clouds were seeded. Seeding of other types of clouds, unfortunately, increased hail damage by 100%. We have much to learn.

FOG

Fog, like other clouds, can exist either above or below freezing. Supercooled fogs, like freezing clouds, are relatively easy to nucleate by seeding with dry ice, silver iodide, or liquid propane, all of which are in regular use for defogging airports in France, Russia, and the United States. The success rate is 60–80%.

Unfortunately, if Alaska is excluded, 95% of the airport fog in the United States is warm, and it is more difficult to dissipate. During World War II, British airfields were cleared by heating the air until the fog evaporated or by nucleating it with an artificial dust of materials that attract water, such as calcium chloride. These techniques are generally too expensive for continuing use in peacetime. Some thin ground fogs form in valleys when the earth radiates much of its heat at night. These can sometimes be dispersed by mixing the air by means of the downwash from helicopters or rising hot air from jet exhausts. The dense coastal fogs of California are more tenacious.

TORNADOS

Tornados are terrible, twisting, windy funnels that can form in a moment, move along at 50 kilometers per hour, and spin at literally hundreds of kilometers per hour. A famous one lifted Dorothy and Toto and their house from Kansas, over the rainbow, and into the Land of Oz.

We do not understand much about tornados, cannot predict them, and certainly cannot control them. However, it is worth the effort to try, because they are the principal weather hazard in Kansas, Oklahoma, and the other Great Plains states. More tornados occur in this region than anywhere else in the world—7428 of them from 1960 to 1970. Like volcanoes, they are mainly destroyers of property rather than killers. A typical one hit Fridley, Minnesota, on May 6, 1965. It killed only two people, but it destroyed 425 homes and damaged 1100 more. Advance notices of tornados are already routine. The National Severe Storms Forecast Center in Kansas City identifies conditions that favor storms and broadcasts an alert when one is seen to form. It cannot, however, predict individual tornados, which are only a hundred meters or even less in diameter and last only an hour or so.

Tornados commonly form in association with thunderstorms, and particularly where warm dry air lies above cooler moist air in a temperature inversion. Somehow, the regional energy is focused in the twisting funnel, which descends from a cloud to the ground or to just above it. It literally tears up the ground, and damage to structures borders on the fantastic. Straws are driven through posts by the speed of the wind, and buildings can explode

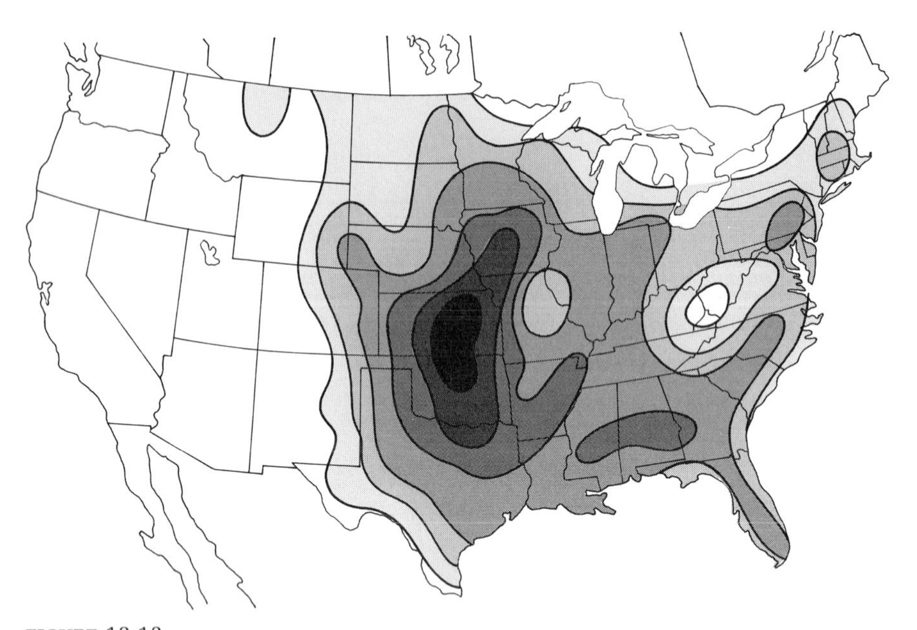

FIGURE 12.13
Relative frequency of tornadoes in the United States from 1916 to 1955. Darker shading indicates greater frequency. [From Tepper, "Tornadoes," copyright © 1958 by Scientific American Inc. (all rights reserved).]

because of the low pressure. It is important to leave the windows open when you leave the house to go to the storm shelter.

Ideas about controlling tornados involve breaking them up individually by means of the usual explosions, or applying heat to destroy the regional inversion that seems to favor their development. The difficulties involved can be perceived in terms of energy: An average tornado generates energy equivalent to 20 tons of TNT exploding per second, or a total energy release roughly the same as a Hiroshima-scale atomic bomb. Any attempt to kill a tornado with explosives would involve the risk of causing more damage than the tornado.

What of the quiet application of heat to break the inversion? This destroys model tornados in the laboratory. However, the Great Plains states are built on a noble scale, and so is a tornado. The energy in a typical twisting funnel is equal to one-third of the total electrical generating power of the whole United States. This indicates the magnitude of the difficulty. The required artificial energy may be as difficult to control as a tornado—or perhaps not. If we know so little, how can we tell that the tornado does not have an Achilles heel, some weak spot where the puny efforts of man can affect it?

HURRICANES

A tornado contains only an infinitesimal fraction of the energy of a hurricane. A single thunderstorm releases energy equivalent to a hydrogen bomb—

1,000,000 tons of TNT—and even a thunderstorm is but a whisper to the roar of a hurricane spinning at more than 160 kilometers per hour and spreading over hundreds of square kilometers. It is particularly encouraging, therefore, to find that we know something about hurricanes, have found a vulnerable point, and appear to be making successful attacks on these monsters.

Hurricanes and their far-eastern brethren, typhoons, draw their energy from the warm tropical seas and move to higher latitudes as extreme examples of zonal circulation. Like Anteas, who lost his strength when Hercules lifted him away from his mother, the Earth, a hurricane dies not long after it moves inland from the warming sea. At the shoreline, however, hurricanes are in an optimum situation to maximize destruction. There are three reasons for this: First, and most important, the hot, rising air is accompanied by extremely low atmospheric pressure, and the inrushing winds drag water toward the storm center. Consequently, the water level rises as much as 3–8 meters in an enormously broad bulge called a storm surge. Second, the wind generates storm waves 6–15 meters high, which are superimposed on the storm surge. Third, because the wind may gust to 240 kilometers per hour, it is highly destructive by itself. These effects combine to make hurricanes the greatest of natural disasters.

The most terrible storm on record occurred on November 12, 1970, when a hurricane brought a storm surge from the Bay of Bengal into the delta of the Ganges in what is now Bangla Desh. Weather satellites photographed the spiral clouds of the storm and warnings were issued, but nothing could be done in such a populous, low region with primitive transportation. Between 250,000 and 500,000 people died.

In the richer and less populous United States, destruction is to property rather than to lives—just as from volcanoes and tornados. Hurricane Betsy, which struck in 1965, killed 75 people but caused more than a billion dollars worth of damage. Hurricane Camille, which smashed into low-lying Louisiana in 1969 with winds of 300 kilometers per hour and a storm surge 7 meters high, caused $1.4 billion in damage. A total of 320 lives were lost, counting 100 in Virginia, where 69 centimeters of rain fell in 8 hours. However, 75,000 people fled the low coast because they had received warnings from the National Hurricane Center well ahead of the arrival of the storm, and transportation was adequate for the emergency. Later analyses indicated that 50,000 people might have drowned without the warnings.

As the population grows and concentrates along the shoreline, the risk of death and damage from hurricanes will increase. What is being done about it? Here in the United States, where the hazard is at a maximum, we can see two contrasting approaches: concrete and clouds. The Army Corp of Engineers has built a concrete wall along the shoreline of the Rockaways on Long Island, New York, where a hurricane caused extensive damage, and has suggested extending it. The National Weather Service, on the other hand, is attempting to master the hurricanes and to dissipate or deflect them while

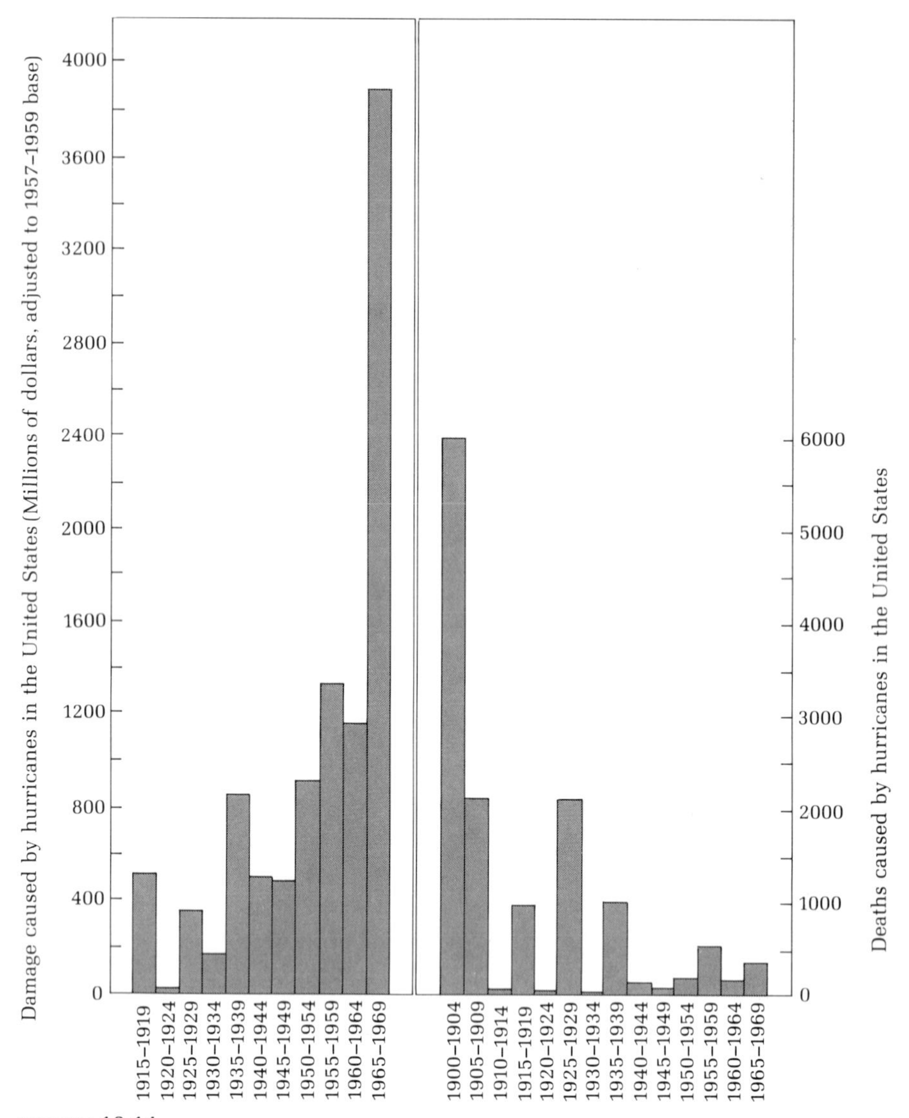

FIGURE 12.14

Trends of losses from hurricanes in the United States. Construction near the shoreline accompanies the population explosion, and, as a consequence, hurricanes caused ten times as much damage in 1965 as in 1930. However, improved warnings have had the opposite effect on loss of life, which has decreased from thousands per year early in the century to a few hundred per year now. [Data from Gentry (1970).]

they are aborning. Project Stormfury is a joint project with the Department of Defense to study the possibility of controlling hurricanes. Robert Simpson of the National Hurricane Center reasoned that, if the high storm clouds near the eye of the storm could be seeded, water vapor and droplets would convert to ice and release heat, and that this might disrupt the hurricane. In 1969, hurricane Debbie was seeded repeatedly by Navy fliers; thereupon, the eye broadened and winds abated by almost one-third. When seeding was halted, the intensity again increased; and when seeding was resumed, the winds dropped by 15%. This sounds like convincing modification, but it could have been mere chance. The experiments continue each year as new hurricanes appear.

There stand our puny efforts to control the weather that causes an estimated $11 billion worth of property damage each year in the United States. We have learned much since Cellini fired his cannon, but the challenge is almost unlimited. In the words of John von Neumann, the famous pioneer of computer use and the inventor of game theory: "The hydrodynamics of meteorology presents without a doubt the most complicated series of interrelated problems not only that we know of but that we can imagine."

Summary

1. The weather and climate fluctuate over periods of years and decades. The fluctuations do not appear to be uniform over the globe. Instead, when one region is especially hot and dry, an adjacent region may be cold and wet. This suggests variations in the pattern of zonal atmospheric circulation.
2. Well-documented climatic changes during historical times appear to be increasingly frequent. Probably, this merely indicates that the records are more complete as time passes.
3. Some aspects of human history and climatic fluctuations are closely correlated, especially exploration and occupation of new regions.
4. The interpretation of scientific records of climatic fluctuations are difficult to interpret, because most instruments were (and are) in cities, and cities themselves change the climate locally.
5. We lack the ability to predict the climate, except to say that it will probably fluctuate as it has in the past.
6. Weather can be predicted with fair accuracy for a maximum of about two days in the future. Accurate long-term predictions will require great advances in theory, observation, and data processing.
7. Most clouds are overheated masses of air that rise until their density matches that of the air around them. As they rise, they cool, and this causes precipitation of rain or snow.
8. Raindrops grow by coalescence of small drops in nonfreezing clouds. In freezing clouds, ice particles grow rapidly at the expense of liquid droplets.
9. Some aspects of the weather can be controlled locally—sometimes. Hail and fog can be suppressed; rain can be coaxed from some kinds of clouds; and hurricanes may be modified.

Discussion Questions

1. What are some of the different approaches to solving environmental problems?
2. Give an example of the effect of climatic variations upon the development of civilization.
3. What would be the effect of adding carbon dioxide to the air by burning fossil fuels? Has this effect been observed?
4. If we can predict the weather a day ahead, why not 100 days ahead?
5. Why does a raindrop have a solid nucleus?
6. Where does the water come from that causes the upward bulge of the sea surface in a storm surge?
7. Which do you believe is the logical approach to the control of hurricane damage at the shoreline, that of concrete or that of clouds?

References

Barzun, J., 1965. *What man has built: Introduction to the Great Ages of Man Series.* New York: Time-Life Books. [Includes valuable correlation chart for major events in all cultures.]

Bjerknes, J., 1963. Climatic change as an ocean–atmosphere problem. *In Changes of Climate.* pp. 297–322. Paris: UNESCO.

Bolin, B., 1970. The carbon cycle. *Sci. Amer.* 223(3):124–132. (Available as *Sci. Amer.* Offprint 1193.) [Traces the intricate path of carbon in nature and assesses the contribution of man.]

Carpenter, R., 1966. *Discontinuity in Greek Civilization.* Cambridge: Cambridge University Press. [Climatic change has had an important effect on the history of civilization.]

Cellini, B., 1960. *Autobiography* (translated by John A. Symonds). New York: Doubleday. [Early attempt at weather modification.]

Denton, G., and S. Porter, 1970. Neoglaciation. *Sci. Amer.* 222(6):101–110. [Mountain glaciers are advancing.]

Fleagle, R., 1970. Background and present status of weather modification. *In* R. Fleagle, ed., *Weather Modification, Science and Public Policy,* pp. 3–17. Seattle: University of Washington Press. [One man's rain may mean the next man's drought.]

———, chairman, 1971. *The Atmospheric Sciences and Man's Needs* (Report of the Committee on Atmospheric Sciences, Nat. Acad. Sci.). Washington, D.C.: National Academy of Sciences.

Gentry, R. C., 1970. Hurricane Debbie modification experiments. *Science* 168:473–475.

Hobbs, P. V., 1969a. The physics of natural precipitation processes. *In* R. Fleagle, ed., *Weather Modification, Science and Public Policy,* pp. 18–29. Seattle: University of Washington Press.

——, 1969b. The scientific basis, techniques, and results of cloud modification. *In* R. Fleagle, ed., *Weather Modification, Science and Public Policy,* pp. 30–42. Seattle: University of Washington Press.

Huntington, E., and S. S. Visher, 1922. *Climatic Changes.* New Haven, Conn.: Yale University Press. [The historical evidence for climatic change is fascinating, even though the chapters on theory are wholly out of date.]

Keeling, C. D., 1970. Is carbon dioxide from fossil fuel changing man's environment? *Proc. Amer. Phil. Soc.* 114:10–17. [Yes.]

Kimble, G., 1950. The changing climate. *Sci. Amer.* 181(4):48–53. [Details on the Little Climatic Optimum and early ideas on weather control.]

Knight, C., and N. Knight, 1973. Snow crystals. *Sci. Amer.* 228(1):100–107.

Lamb, H. H., 1963. On the nature of certain climatic epochs which differed from the modern (1900–39) normal. In *Changes of Climate,* pp. 125–146. Paris: UNESCO. [The climate of the past 5000 years has repeatedly been both colder and warmer than now.]

Lowry, W., 1967. The climate of cities. *Sci. Amer.* 217(2):15–23. (Available as *Sci. Amer.* Offprint 1215.) [They create their own.]

MacDonald, G. J. F., chairman, 1966. *Weather and Climate Modification* (Report of the Panel on Weather and Climate Modification, Nat. Acad. Sci. Publ. 1350). Washington, D.C.: National Academy of Sciences.

Malkus, J. S., 1957. The origin of hurricanes. *Sci. Amer.* 197(2):33–39. (Available as *Sci. Amer.* Offprint 847.)

Mason, B. J., 1960. Cloud physics. In *McGraw-Hill Encyclopedia of Science and Technology,* vol. 3, pp. 214–218. New York: McGraw-Hill.

——, 1971. *The Physics of Clouds.* Oxford: Oxford University Press.

Mitchell, J. M., Jr., 1963. On the world-wide pattern of secular temperature change. In *Changes of Climate,* pp. 161–182. Paris: UNESCO.

Namias, J., 1963. Surface-atmosphere interactions as fundamental causes of drought and other climatic fluctuations. In *Changes of Climate,* pp. 345–361. Paris: UNESCO.

——, 1972. Experiments in objectively predicting some atmospheric and oceanic variables for the winter of 1971–72. *J. Appl. Meteorol.* 11(8): 1164–1174.

Rasool, S. I., and S. H. Schneider, 1971. Atmospheric carbon dioxide and aerosols: effects of large increase on global climate. *Science* 173:138–141.

Roberts, W., 1972. We're doing something about the weather! *Nat. Geogr. Mag.* 141:518–555. [A popular and authoritative account of weather forecasting and control.]

Tepper, M., 1958. Tornadoes. *Sci. Amer.* 198(5):31–37. (Available as *Sci. Amer.* Offprint 848.)

NOT SO SOLID EARTH

JOHN S. SHELTON

13

WEATHERING AND SOIL

This is the land which ye
Shall divide by lot. And neither division nor unity
Matters. This is the land. We have our inheritance.

T. S. Eliot, ASH WEDNESDAY

> Such a picture is somewhat discouraging and might lead one to the conclusion that civilization is running down and that man is gradually but surely destroying the means by which he can exist; but there are means at hand by which this decadence may be indefinitely checked. If we control population increase, adopt means of checking soil erosion, and improve our transportation facilities, it is possible for us to indefinitely postpone the day when the world will no longer support a large human population.

So wrote James Thorp in *Geography of the Soils of China* in 1939. Since that time, not only has the world's population increased from 2.3 billion to 3.7 billion, but the erosion of soil has continued.

The solid rock of the earth's surface, fragmented by gravity, air, water, and organisms, enters the sedimentary cycle, which does not end until millions or billions of years later, when sediment is plunged into the mantle in a subduction zone. This chapter and the next three are concerned with this cycle and its many effects upon our society. Man builds his cities on sediment, floods may bury those cities in sediment, and, most important, man derives his foods from sediment.

We shall begin, in this chapter, with the slow formation of sediment and its alteration into soil, a remarkably delicate material a few centimeters to a few meters thick. We shall continue with the accelerating erosion of soil by man, and its transportation ultimately toward the sea. When reading this material, it is well to remember that mankind will continue to exist only as long as the soil does. At present, there is about one fertile acre per person, but the amount of fertile land is decreasing and the population is increasing.

Five factors influence the development of soil: (1) the parent rock, (2) the topography, (3) the climate, (4) the life forms in the environment, and (5) time. With so many variables, it is hardly surprising that soil is a highly complex material—or rather a highly complex structure, because it consists of layers with different textures and compositions. Many of these factors are closely associated, however, and this reduces the number of types of soil. Examples of important associations are that black soil, a subhumid climate, and grass go together and that deep red soil, a moist climate, and rain forest go together.

The importance of the factors depends on the environment. On steep slopes, gravity and rain remove weathering products and soil does not form. In deserts, because there is no water, there is little life to affect soil formation. Life abounds in tropical jungles, but rainfall is so abundant that all but a top dressing of nutritive material is leached away in the ground water.

As time passes, climate and plant life become the dominant factors in the formation of soils. This is evident from the fact that particular plants and soils are so closely associated in Europe and North America. However, we know that, not long ago, the climatic zones and vegetation advanced and retreated with the glaciers. Thus, the soils that now support oaks and grasses once supported pines. Clearly, the soils have changed. The plants followed the climate regardless of the nature of the soils they encountered along the way. The relative importance of plants compared to climate in soil formation is suggested by the common observation that quite different soils may exist side by side. For example, the soil under kauri trees in New Zealand is totally different from that under nearby broad-leafed trees, and so is the soil under the hardwoods of the northeastern United States compared to that under nearby intermixed pines and spruces.

Plant life dominates the development of mature soils. In youthful soils with only partially developed layering, the effects may be reversed. The coastal hills of California are commonly covered with grass if the underlying rock is shale, whereas sagebrush is the commonest cover on sandstone. A lazy geologist can map the rocks just by examining the plants as seen in an air photo. Moreover, even mature soils differ in minor characteristics that are unrelated to climate, plant cover, or parent rock. Much of Iowa is coated with loess, so the parent material is the same, as are all other variables but slope. Even so, there are numerous types of soil in Iowa that differ enough to deserve recognition for agricultural purposes.

Finally, soils do not persist forever, even if all the variables remain constant. Natural erosion denudes mountains at average rates of 2–10 centimeters per century, and it works on flat land at less than a tenth of that rate. The top layers of most soil are gradually removed, and the weathering profile works down into the subsoil. Not all soil erodes away, however. In the flood plains of many rivers, fertile soil accumulates layer on thin layer every year. Likewise, loess settles out downwind from its source areas. Thus, soils may be buried and preserved in the geological record. They can be very useful. For example, buried soils between ground moraines gave the first indications that continental ice sheets had spread repeatedly over North America.

Weathering

Solid rock is gradually transformed by many processes to become fertile soil, sterile sand on beaches, or limy ooze on the distant sea floor. The processes are logically divided into cracking, physical weathering, and chemical weathering, which are given here in the chronological sequence in which they have their maximum effects.

"Solid rock" is more a poet's fancy than a geologist's observation. It is only solid between parallel cracks called joints, which are spaced a fraction of a centimeter to a few meters apart. Commonly, more than one set of joints exists in a given area, and many sets are at approximately right angles. Thus jointing, which is produced by regional stresses, cracks many rocks into long prismatic fragments with a square crosssection before they reach the surface zone of weathering.

Rocks are fragmented by many other types of cracks or planes of weakness. Faulting, for example, is very common. Sedimentary rocks, in particular, tend to break easily at bedding planes. If joints are at right angles to the bedding planes, as they often are, rocks tend to break into cubes and tabular masses when they reach the weathering zone.

Many thick lava flows cool in a manner that causes hexagonal prisms to develop perpendicular to the surface. The flows also have a flow structure parallel to the surface, so that they fragment into stubby hexagonal prisms.

Quite a different phenomenon, exfoliation, causes great curved slabs of granite and similar rocks to crack away from the surface. This phenomenon is magnificently displayed at Yosemite National Park, where the slabs are in all stages of detachment. The rather spherical top of the granite was originally under pressure from overlying rock. When the load was eroded away, the compressed minerals expanded toward the curved surface, and the curved slabs resulted. Exfoliation also occurs on a much smaller scale in jointed rocks.

In sum, when bedrock is exposed to weathering, it usually already has cracks that make it vulnerable to attack. Physical weathering of bare rock is accomplished through volume changes due to solar heating and cooling of the rock itself and to freezing and melting of water. Solar heating does not seem to be very effective, although it is widespread. Accelerated heating and

A

B

FIGURE 13.1

Jointing: *A*, columnar jointing, produced by cooling in volcanic rock in Devils Post Pile National Monument, California. *B*, vertical jointing produced by regional stresses in the horizontal rocks near where the Colorado and Green rivers join. One set of joints is obvious in the foreground as lines of vegetation in the rock. The cross set in the system can be seen in the rectangular caps of the pinnacles that are being eroded out of the cliff. [Photo A by R. C. Frampton; both photos from Shelton, *Geology Illustrated*, W. H. Freeman and Company, copyright © 1966.]

FIGURE 13.2
Granite dome showing coarse exfoliation, Sierra Nevada. The sheetlike slabs of rock are several meters thick. [Photo by G. K. Gilbert, U.S. Geological Survey.]

cooling in a laboratory have no detectable effect on rock within the temperature range expected on earth. However, the real test is on the airless moon, where rocks have been exposed for millions of years alternately to the direct heat of the sun and the deep cold of space without shattering.

What the moon lacks is water, which is the prime agency in every part of the sedimentary cycle. Water collects in cracks in rock and, if it freezes, opens the cracks a bit by an irresistible expansion. Acting in concert with gravity, this appears to be one of the chief mechanisms for physical fragmentation of bare rock. It is effective, however, only in high latitudes and at high elevations, where freezing occurs, and mainly on bare rock. Thus, the generally invisible but much more widespread fragmentation of rock by plant roots may actually be more important. Tiny rootlets penetrate hairline cracks in soil-covered rock seeking water or nutrition. As the roots grow, they may—perhaps eventually will—expand the cracks.

Physical weathering, which can attack only surfaces, does not compare with chemical weathering, which may penetrate the volume of rock, under soil, and is effective in all latitudes and at all elevations. The most important agent in chemical weathering, as in physical, is water. Water is a very powerful solvent, and could probably dissolve most rocks, if given sufficient opportunity. In fact, it does not have the chance, because it interacts at all times with other natural materials to form weak acids that are even more potent as weathering agents than pure water. Most important of these reactions is the formation of carbonic acid from carbon dioxide, which is abundant in air, soil, or wherever else organic carbon is oxidized. However, carbonic acid is only weakly corrosive, and, in some natural environments, weathering by humic or sulfuric acids is more important. The latter is a potent corrosive that results from the interaction of rainwater and sulfur dioxide, which is a pollutant released by burning most coals and high-sulfur oils. The acidity of rainwater falling on Sweden has increased for decades, and it appears that the principal cause is the burning of low-grade brown coal that contains much sulfur.

Urban pollution accelerates chemical weathering of bare rock, and the effects are particularly spectacular on limestone and marble. In 1880, Archibald Geikie observed that a centimeter of rock had dissolved from some dressed limestone blocks in Edinburgh, and that other blocks had bulged and cracked. The inscription in memory of Joseph Black, the discoverer of carbonic acid, became partially illegible in 80 years as a result of attack by that acid and sulphuric acid in urban rainwater. Likewise, the colleges at industrialized Oxford are noticeably melting away, while those at rural Cambridge fare better. However, weathering is not confined to limestones. Granitic Egyptian obelisks have weathered noticeably in New York, London, and Paris after standing unblemished for thousands of years in the warm, dry, unpolluted air of the desert. One of these, Cleopatra's Needle in New York City, is also exposed to frost wedging, and, within 70 years of its erection, part of the inscription had weathered away. The inscription in stone that Shelley tells us accompanied the shattered statue of Ozymandias—"Look on my works, ye Mighty, and despair!"—would barely have outlasted Ozymandias himself in the climate in which the poem was published.

How do the major rock-forming minerals weather? Quartz is highly resistant to chemical weathering. Thus, when a granite or a similar rock decomposes, the angular crystals and grains of quartz remain as an unaltered residue. These sand grains may be transported thousands of miles in rivers with little effect, but chemical weathering eventually rounds their corners.

Feldspars, the most abundant minerals in igneous rocks, react readily with acidic water to form clay, silica, and soluble ions. Clays, which are crystalline hydrated aluminum silicates, are chemically quite inert, but they interact with many ions in solution because of their layered structure. Clay is slippery because of the layering of the crystals, and its ability to interact with water renders it ideal for pottery making. The three main types of clay—kaolinite, montmorillonite, and illite—contain various additional elements; the different types result from the weathering of different parent materials in different climates. Because clay particles are extremely small when they form, they are quite mobile, and water can move them around in the interstices among the sand grains. As time passes, the relocated crystals grow larger and less mobile.

Chemical weathering of the other major group of rock-forming minerals, the ferromagnesian silicates, also produces clay, silica, and soluble ions. The abundant iron usually combines with oxygen to form various oxides, of which red hematite and yellow limonite (or "rust") give distinctive coloring to many soils. The magnesium, commonly, is loosely incorporated into the surface structure of the clays where it can readily be replaced by similar ions. This occurs through the process of ion exchange: Many small clay particles are negatively charged and, thus, combine with positively charged cations by incorporating them into the crystal lattice. Calcium, magnesium, hydrogen, potassium, and sodium form ions that are readily exchangeable in the clay lattices. Rootlets are also negatively charged, and nutritive cations can move through ground water from the clay lattice to the root. Negatively charged anions can also be absorbed on clays, but this is a much less important phenomenon.

Formation of Soils

Soils consist of several layers that differ from the source material and from each other because of a variety of active processes. Most important are solution, deposition, and chemical changes resulting from the vertical and horizontal movement of water. Countering the layering are the churning activities of man, earthworms, rodents, frost, and the countercirculation of soluble materials within plants from the roots up to the leaves.

The tendency of soils to form layers usually prevails over all counterforces except man. Despite the fact that their nature depends on many factors, three superimposed layers—known as horizons A, B, and C—are almost universally present, and other layers—A_1, A_2, and so on—are not uncommon. The A horizon on top is the familiar one that everyone calls "soil." Much of the rain that falls on it sinks into the ground and passes downward through this zone, dissolving or leaching as it goes. The A horizon is primarily a zone from which soluble material and tiny clay crystals are removed, but it is also a zone of accumulation of organic material in the form of leaves, roots, and so on. The chemical composition of this decaying vegetation is reflected in the leaching solutions, and the concentration of the solutions depends on the proportions of falling vegetation and rain.

The C horizon, which is at the bottom of the soil, grades downward into unweathered rock. It contains fragments of rock, perhaps dislodged by roots and moved upward by frost. Other fragments may be resistant remnants of the bedrock, which is being weathered downward.

The B horizon, which lies between the other two, is the transitional zone of accumulation. The composition of this layer is determined by a balance between the intensity of chemical weathering in the C horizon below and the intensity and volume of leaching in the A horizon above. The weathering commonly leaves a residue of quartz grains and clay, and the descending solutions also deposit clay, in addition to an assortment of other materials that depends

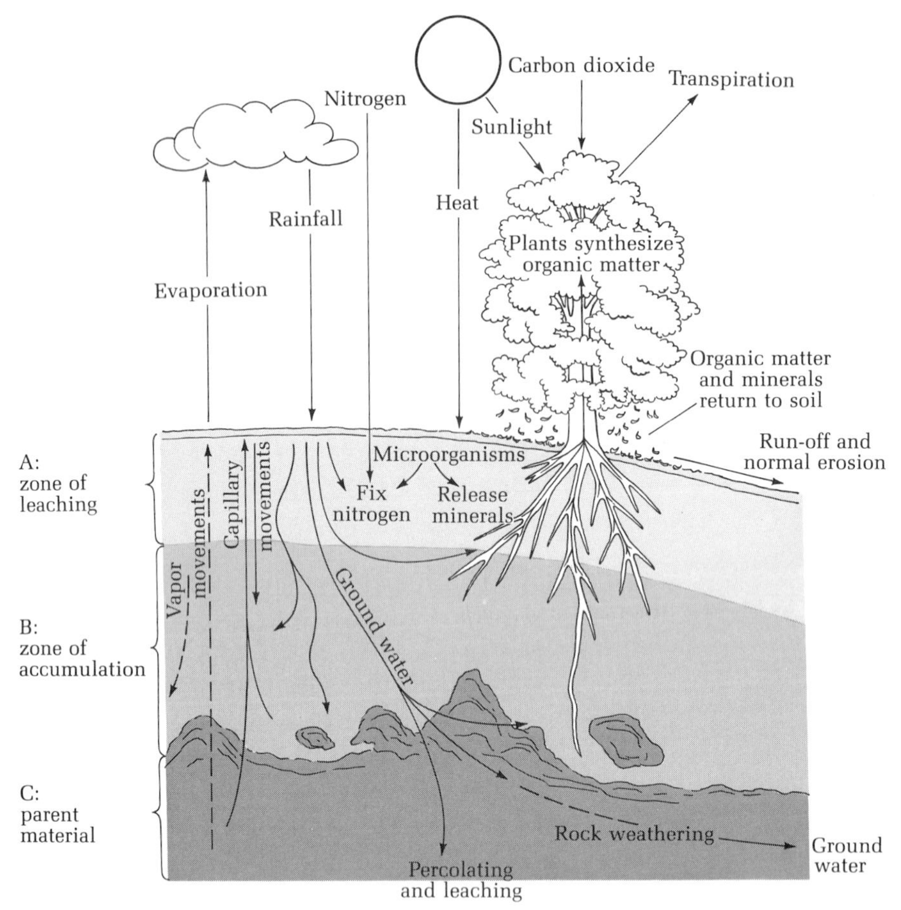

FIGURE 13.3
Physical, chemical, and biological processes participate in the life of the soil. In the biological process, plants remove water and various nutrients from the soil and combine them with carbon from the air. The various plant structures eventually return to the soil and decompose, releasing nutrients for a new cycle of plant growth. The physical processes move water through the soil, and solution and deposition occur to form the layering. [After Kellogg, "Soil," copyright © 1950 by Scientific American, Inc. (all rights reserved).]

upon the climate. In humid climates, iron is deposited in the B horizon; in dry climates, calcium carbonate is deposited. Although the latter is quite soluble in acid solutions, it is precipitated when evaporation exceeds rainfall and the soil becomes dry.

Classification and Distribution of Soils

Farmers have always realized that soils differ, and the critical role of soil type and drainage in agriculture has long been appreciated, especially by those

BOX 13.1 SOIL TEXTURE AND STRUCTURE

Some of the constituents of soil, such as clay, can be defined either by size or by crystal structure. With regard to size alone, soil particles are classified as follows:

Gravel: > 2 mm

Sand: 0.05–2.0 mm

Silt: 0.002–0.05 mm

Clay: < 0.002 mm

The texture of the soil is a measure of the fraction of each of the particle sizes in the soil. Because gravel is rare in soil, the main soil textures can be displayed in a triangular diagram that takes into account only clay, silt, and sand.

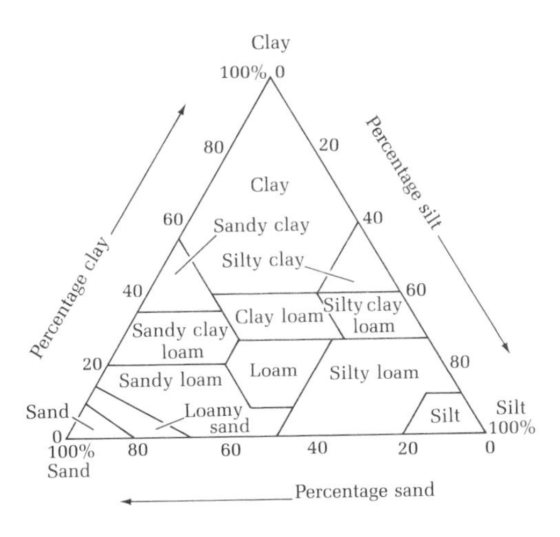

The diagram indicates that loam, for example, may include 25–50% sand, 8–26% clay, and 30–50% silt.

Soil is not a mere mixture of materials of different sizes. The processes of soil formation cause the development of aggregations, in the A and B horizons, that are classified as soil structure. The illustration shows the principal types. They are produced mainly by (1) the physical activity of roots and animals, (2) the decomposition of organic matter and formation of bonding slimes, (3) contraction and expansion caused by wetting and drying, and (4) adsorption of different cations that induce flocculation or dispersion.

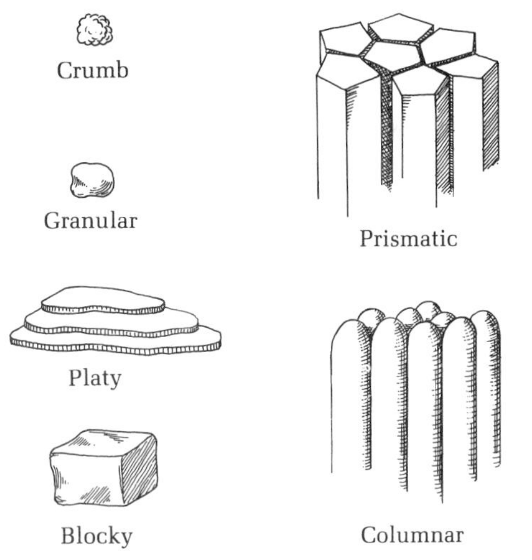

Prisms, columns, blocks, and plates are characteristic of subsoils; granules and crumbs are found nearer the surface. Structure is important in agriculture because it determines the physical behavior of the soil far more than texture.

who grow wine grapes. Early American naturalists took the view that soils differ because they are derived from different materials. This might be expected from people with experience in England and the eastern seaboard of the United States, where rocks were stripped by glaciers and the soils are commonly youthful. The circumstances are different in the vast heartland of Russia: the pioneering Russian soil scientist V. V. Dokuchaev (1846–1903) and others found that the forests there have gray soils and the grassy steppes have black ones, regardless of the bedrock. This is not unreasonable, because much of the bedrock there is covered with a blanket of ice-age loess, and different soils form from it. Russian scientists developed the first correlations with climate and vegetation and used them to classify various types of soils. This is the reason so many major types have such names as chernozem, which means "black earth" in Russian.

The Russian concepts of great soil groups were disseminated in the United States, beginning in the 1920s, by C. F. Marbut, then chief of the Soil Survey (which has since become the Soil Conservation Service in the Department of Agriculture). The utility of a classification varies with the use, and the broad groups proved inadequate for American purposes. Consequently, a vast number of subdivisions appear on the soil maps that are used by farmers to help select crops and to guide conservation. Nonetheless, the concept of the broad groups, identified throughout the world, is useful for understanding the origins of soil characteristics and the influences of those characteristics on society.

TUNDRA

Tundra has covered most of Europe and northern North America during the last 20,000 years; thus, most modern vegetation in these regions has replaced it since that time. Tundra looks like a great pasture in which the "grass" consists of willow trees a few centimeters high and equally tiny shrubs, as well as mosses and grasslike sedges. Tundra forms in areas with a very cold climate and low rainfall comparable to that of a semiarid desert. It is a moist, humid climate, however, because the air is easily saturated at the low temperature. Except where it is well drained, the ground is permanently frozen only a few tens of centimeters beneath the surface. Thus, rain or melted snow cannot migrate downward and leach the soil for any distance. The acidic rain and the humic acids resulting from decomposition of organic matter cannot escape.

At Point Barrow in arctic Alaska, near the great north-slope oil strike, the soil in the summer is 8–10 centimeters of brown loam that is rich in humus and saturated with acidic water. Below for a few centimeters is a gray layer, and at a third of a meter or less is frozen ground that is mottled with white ice.

Tundra supports the grazing of vast herds of reindeer and caribou in Lapland and Canada, but no crops capable of enduring the polar climate have been developed even south of the frozen ground. Tundra is not even able to recover easily if the ground cover is broken, and the scars of exploration and construction remain visible decades after the work is completed.

PODZOLS

Podzols, named from the Russian word for ashes (*zola*), are the principal soils south of the tundra, which they replaced as the glaciers, climatic zones, and forests moved north. Podzol forms in a cool, humid climate under coniferous forests with or without a mixture of broadleaf trees. It presently covers a great band from Scandinavia across Russia and Siberia to the Pacific, and another from Alaska through southern Canada to the Atlantic. Nothing like it—or like tundra, for that matter—occurs in comparable extent in the Southern Hemisphere because the high latitudes of that hemisphere are mainly oceanic or ice covered.

Decomposition is slow in this cool climate, and a mat of leaves and twigs as much as a third of a meter thick lies over the mineral part of the soil. After passing down through this mat, rainwater is unusually acid, which makes the soil below unusually acid, too. Leaching is so effective that, only a centimeter below the organic mat, the soil in the A horizon is nearly ash white—hence the name. Plant roots are largely confined to the decaying mat and the top few centimeters of the white horizon. The B horizon contains abundant iron and other materials leached from above, and it often forms a cemented layer or hardpan. Although resistant, this layer is not impermeable to water, and strong roots can break through.

Podzols are common on glaciated (or otherwise young), ill-drained landscapes and they tend to be immature, full of stones, and only 40–80 centimeters thick over bedrock. Small, shallow lakes and swamps abound in such landscapes, and organic debris and eroded soil gradually convert them to bogs filled with peat. Ireland, for example, has podzols, peat bogs, and also many of the small farms typical of this environment. Potatoes and pasturage for cattle grow well in this soil, but most efforts to farm sandy podzols around the Great Lakes have failed. In sum, podzols are superior to tundra for agriculture, but they grow only a limited range of crops.

GRAY-BROWN PODZOLS

Gray-brown podzols lie south of the podzols in only a few places in the Northern Hemisphere, notably China, Europe, and the northeastern United States. They developed under hardwood forests in a humid, temperate climate that moved north as the glaciers retreated. Most of the area with this type of soil was never covered by ice, and different soil-building processes have had time to make a deep soil with A and B horizons 80–90 centimeters deep.

The leaves of the hardwood forests decay rapidly, so there is no thick blanket of organic matter such as exists on the podzols to the north. Instead, the upper A horizon is a fertile, pale brownish loam, 15 centimeters thick, and teeming with life. Leaching is much less severe than in the podzols, but 13–25 centimeters of lower A horizon is a yellowish gray loam produced by this process. The main B horizon is a third of a meter or more of dark brownish clay with a structure that causes it to break into angular blocks a few centimeters or so on a side. A more massive brown clay lies immediately below in the B horizon. Even deeper, the C horizon is a blocky yellowish clay, regardless of the type of underlying rock.

Both the ancient civilization of China and the modern industrial societies of western Europe and the United States have been fed by the fruits of this type of soil. Most soils have been unable to bear persistent croppings, but these still remain fertile. In China, much of the soil is constantly rejuvenated by flood-deposited silt. The gray-brown podzol of the North Atlantic has no such advantage, but it is an extraordinary soil nonetheless. The oldest agricultural experiment station in the world is at Rothamsted in England. One plot there has grown wheat, a demanding crop, for more than a century without fertilizer or soil conditioning. It still yields 90–100 cubic meters of grain per square kilometer (about 10–12 bushels per acre), which exceeds the world average. Suitably cultured and nourished, the Rothamsted soil yields four times that much wheat, and it will grow a wide variety of crops.

Gray-brown podzol can be hurt by plowing deep and exposing the B horizon to sun and rain, and it can be damaged by neglect; barren and fertile fields may lie side by side. Basically, it is tough because it is rich in humus and clay from which nutrition can be withdrawn slowly by roots. In Europe, moreover, it has had the benefits of manure provided by three thousand years of animal husbandry. Edward Hyams calls it the "perfect artificial soil." It has been saved from excessive cultivation by episodic depopulation of the farms by war and the Black Death. More recently, the population expansion has had little effect on it in Europe because food produced by low-cost labor or machines has been imported from lands where the fertility of the soil is not treasured.

CHERNOZEM

The Russian steppes and the American plains from Canada to Texas are underlain by chernozem, the black soil characteristic of cool, subhumid, temperate grasslands. The thirsty grass sucks moisture and nutrients from below and concentrates all the worth of the soil in the A horizon, leaving deposits of lime in the B horizon. The maximum fertility of these soils exists when they are first plowed, but they remain productive without fertilizer because there is no natural leaching. If the soil is irrigated, salts accumulate unless there is drainage. If drain tiles are installed, there is leaching and fertilizer is needed.

West of the chernozems in America and south of them in Siberia are chestnut soils, which are like chernozems but are typical of an even drier climate and have less humus in the A horizon and more lime below. More irrigation, drainage tiles, and fertilizer are needed; but, even despite such precautions, crops are uncertain because rainfall is unreliable. The American soil scientist Charles Kellogg points out that the farmers on the fertile, durable, gray-brown podzols supported laissez-faire economics, and those from the chernozem and chestnut soils are driven to, and favor, cooperative credit, government control of railroad freight rates, and crop insurance. "Radical" (as opposed to purely capitalist) ideas of economics and government have originated in areas of grassland soils in America during the last century; and in Russia, Lenin was born on chernozem, and Stalin just at the southern edge of it. It was these dry soils that turned into the terrible dustbowl that plagued American farmers during the depression. The top of the dry soil, which commonly was used as a mulch of dust, simply blew away leaving little of value below. Hyams reasons that the farmers, in essence, mined the nutrition in the soil by means of intensive wheat farming, and it took only a generation to mine it out. The fertility of Oklahoma was exported, and the farmers followed on the dusty depression road to California. These black, initially fertile soils also brought disaster to Premier Khrushchev's "virgin land" program to grow grain in Kazakhstan in the southern USSR beginning in 1954. Not even a monolithic socialist program can sustain this soil through drought if it is being cultivated.

DESERT SOILS

In the extreme, deserts have hardly any soil, but many regions that are now arid were humid in glacial times. Soil was formed in such places in the past, and some of it is preserved where the wind has not removed it. The desert blooms in the clear sunlight of Arizona and Israel, but not without effort. Irrigation is essential; thus, drainage tiles may be necessary to permit leaching and to eliminate build-up of salts. Fertilizer is equally important.

TROPICAL SOILS

Tropical soils, like podzols, develop under dense forest, but the speed of decomposition in a hot climate prevents the accumulation of a thick mat of vegetation. The litter is very thin, but leaching is intense as the acids are carried through the ground by heavy rainfall. The A horizon is granular red clay about 0.5 meters thick. Below are roughly 0.6 meters or so of blocky red clay in the B horizon and a meter or so of transitional material above a C horizon of mottled clay and stones.

The very top of this red latosol is full of nutrients, but it is very acid and leached at lower levels. Under shade, its fertility lasts several years, but it is quickly lost in the open. Commonly, tropical soils are merely used until worn out and then abandoned to the encroaching rain forest. With luck, the fertility may be restored in 15 years. A volcanic ashfall is a natural top-dressing loaded with minerals. Thus, the active volcanoes of Java are a magnet, because they permit constant rather than episodic farming.

A soil of particular interest is laterite, named from the Latin word for "brick." Many buildings in modern Thailand are made of lateritic bricks, and the ancient Khmers used them to build structures in Ankor Thom that have stood since the sixteenth century. Such bricks are produced irreversibly by simply exposing and drying lateritic soil. The A horizon in such a soil is leached of all but a very thin layer of humus, below which all of the nutrients are gone, including even the silica. The silica can be traced in tropical ground water: it becomes concentrated in paddy fields, which are typically at the bases of hills, where it "degrades" the soil and damages the rice. What remains after leaching is little more than oxides of iron, aluminum, and a few other minerals. Laterization can readily lead to the formation of a valuable aluminum ore, bauxite, if the iron is leached away, but it also leads to a soil that is almost worthless for agriculture. If the lush forests are cleared, the soil has little to offer to crops because it is so thoroughly leached. Moreover, when the ground is exposed, it can harden to brick while still in place. The Brazilian government attempted to establish an agricultural colony at Iata in the endless jungle of the Amazon basin. Bulldozers cleared the trees efficiently and the planting began. The soil began to disintegrate after a few plantings, and, in five years, the fields became brick pavements. In Dahomey, in tropical west Africa, conditions were not so extreme; but 60 years after the forests were cleared, large areas were transformed to sheets of brick.

Soils are the products of their environments. Some can support farming for long periods, some not at all. Few problems arise with soils of either of these types. On the other hand, some soils can support farming only briefly, and if

---

FIGURE 13.4

Soils of the world are located in a general way on this map. Although each pattern on the map is a rough approximation of the soil type of the region, many of the patterns include thousands of soil types. The symbol *A*, which stands for alluvial soils, denotes only a few of the most important alluvial areas; many small but important alluvial areas are not shown. The pattern on the interior of Greenland is its ice cap. [From Kellogg, "Soils," copyright © 1950 by Scientific American, Inc. (all rights reserved).]

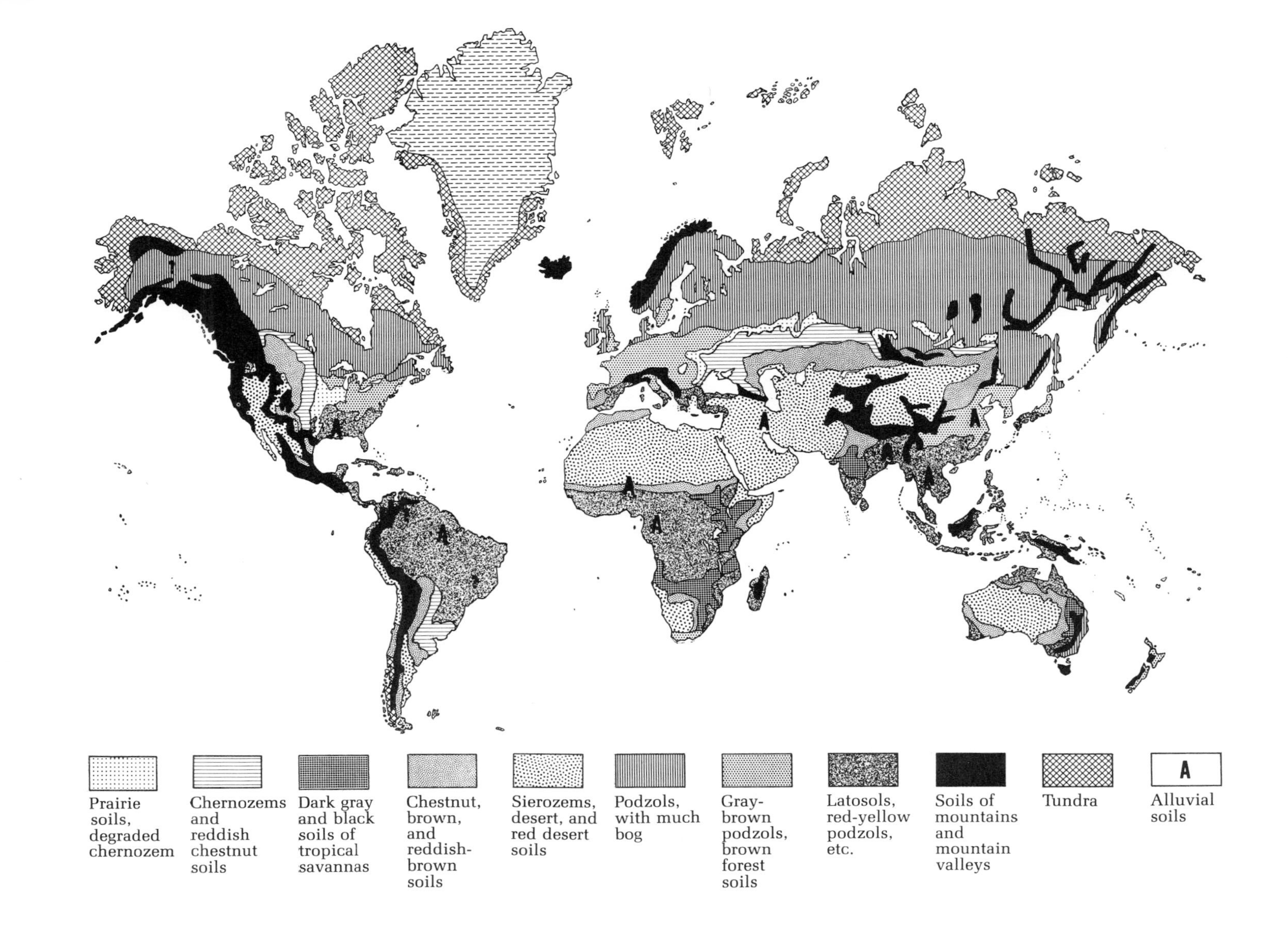
Prairie soils, degraded chernozem
Chernozems and reddish chestnut soils
Dark gray and black soils of tropical savannas
Chestnut, brown, and reddish-brown soils
Sierozems, desert, and red desert soils
Podzols, with much bog
Gray-brown podzols, brown forest soils
Latosols, red-yellow podzols, etc.
Soils of mountains and mountain valleys
Tundra
A
Alluvial soils

BOX 13.2 SOILS OF THE UNITED STATES

ZONAL

Great groups of soils with well-developed characteristics, reflecting the dominating influence of climate and vegetation.

PODZOL SOILS
Light-colored leached soils of cool, humid forested regions.

BROWN PODZOLIC SOILS
Brown leached soils of cool-temperate, humid forested regions

GRAY-BROWN PODZOLIC SOILS
Grayish-brown leached soils of temperate, humid forested regions.

RED AND YELLOW PODZOLIC SOILS
Red or yellow leached soils of warm-temperate, humid forested regions.

PRAIRIE SOILS
Very dark brown soils of cool and temperate, relatively humid grasslands.

REDDISH PRAIRIE SOILS
Dark reddish-brown soils of warm-temperate, relatively humid grasslands.

CHERNOZEM SOILS
Dark-brown to nearly black soils of cool and temperate, subhumid grasslands.

CHESTNUT SOILS
Dark-brown soils of cool and temperate, subhumid to semiarid grasslands.

REDDISH CHESTNUT SOILS
Dark reddish-brown soils of warm-temperate, semiarid regions under mixed shrub and grass vegetation.

REDDISH BROWN SOILS
Reddish-brown soils of warm-temperate to hot, semiarid to arid regions, under mixed shrub and grass vegetation.

BROWN SOILS
Brown soils of cool and temperate, semiarid grasslands.

NONCALCIC BROWN SOILS
Brown or light reddish-brown soils of warm-temperate, wet-dry, semiarid regions, under mixed forest, shrub, and grass vegetation.

SIEROZEM OR GRAY DESERT SOILS
Gray soils of cool to temperate, arid regions, under shrub and grass vegetation.

RED DESERT SOILS
Light reddish-brown soils of warm-temperate to hot, arid regions, under shrub vegetation.

INTRAZONAL

Great groups of soils with more or less well-developed characteristics reflecting the dominating influence of some local factor of relief, parent material or age over the normal effect of climate and vegetation.

PLANOSOLS
Soils with strongly leached surface horizons over claypans on nearly flat land in cool to warm, humid to subhumid regions, under grass or forest vegetation.

RENDZINA SOILS
Dark grayish-brown to black soils developed from soft limy materials in cool to warm, humid to subhumid regions, mostly under grass vegetation.

SOLONCHAK (1) AND SOLONETZ (2) SOILS
(1) Light-colored soils with high concentration of soluble salts, in subhumid to arid regions, under salt-loving plants.
(2) Dark-colored soils with hard prismatic subsoils, usually strongly alkaline, in subhumid or semiarid regions under grass or shrub vegetation.

BOG SOILS
Poorly drained dark peat or muck soils underlain by peat, mostly in humid regions, under swamp or marsh types of vegetation.

WIESENBODEN (1), GROUND WATER PODZOL (2), AND HALF-BOG SOILS (3)
(1) Dark-brown to black soils developed with poor drainage under grasses in humid and subhumid regions.
(2) Gray sandy soils with brown cemented sandy subsoils developed from forests from nearly level imperfectly drained sand in humid regions.
(3) Poorly drained, shallow, dark peaty or mucky soils underlain by gray mineral soil, in humid regions, under swamp-forests.

AZONAL

Soils without well-developed soil characteristics.

LITHOSOLS AND SHALLOW SOILS
(ARID-SUBHUMID)
(HUMID)
Shallow soils consisting largely of an imperfectly weathered mass of rock fragments, largely but not exclusively on steep slopes.

SANDS (DRY)
Very sandy soils.

ALLUVIAL SOILS
Soils developing from recently deposited alluvium that have had little or no modification by processes of soil formation.

This generalized map of the soils of the United States shows the multitude of divisions that are of use in agriculture and soil conservation. Within the areas of the patterns are numerous areas of other soils that are too small to show on a map of this scale. In addition, there are very detailed maps of soil type with regard to potential land use.

Source: From Kellogg, "Soils," copyright © 1950 by Scientific American, Inc. (all rights reserved).

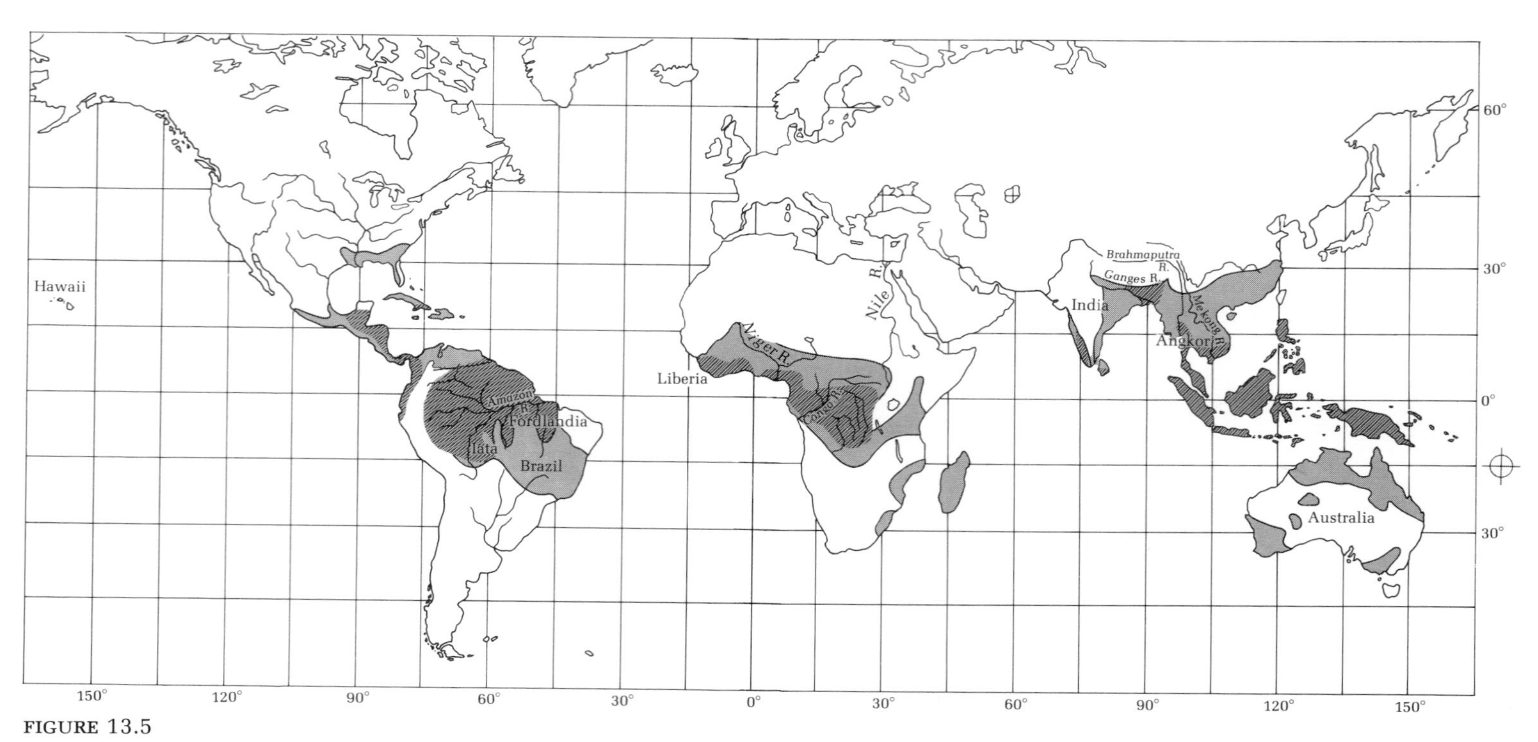

FIGURE 13.5

Lateritic soil (*gray*) is mostly confined to tropical and subtropical regions. Rain forests, shown by hatching, deter hardening to bricklike laterite by insulating the soil somewhat from the effects of tropical climate. Removal of forests, as in efforts to expand agricultural production, tends to quicken laterization, which, in turn, impairs agriculture. Such an evolution has occurred at Iata in Brazil, where the government undertook to establish an agricultural colony. [From McNeil, "Lateritic Soils," copyright © 1964 by Scientific American, Inc. (all rights reserved).]

these are used to feed one more doubling of the population, the resultant problems will be very serious indeed.

Deliberate Destruction of Soil by Man

Homo sapiens is a parasite living off the nutrition stored in a few centimeters of ever changing topsoil. Despite his self-given name of "thinking" man, he has paid little more attention to the health of his host than most parasites. Ignorant or irresponsible, he has inadvertently caused the ruin of millions of square kilometers of topsoil by removing the protective vegetation. Part of the eroded fertility is simply destroyed, but some of it is merely transferred by wind and river to other regions, whence it can be returned by man as exportable food. Some inadvertent loss of fertility in this way is difficult to avoid; but, in addition, man is deliberately destroying soil by mining, construction, and war, and this is less defensible.

MINING

Underground mining destroys soil only to the extent that sterile rock tailings are spread over it. However, many low-grade or shallow deposits are mined at the least initial economic cost merely by stripping the soil away and exposing the ore. This can destroy significant areas of soil, both by the actual mining and by the disposal of the materials that are mined.

Strip mining is a technique in which surface material is mounded to one side to expose the ore—usually soft coal. Such strips wind endlessly through coal country in Ohio and Kentucky and are readily visible from the air. By 1970, 6500 square kilometers had been disturbed by surface coal mining in the United States. The effects are far more widespread because the mining wastes are leached by rainwater and they are extremely acid. Thus, streams and ground water are polluted, and surrounding soils are rendered barren. The damage may be reduced if soil is stockpiled and returned to place after mining is completed.

Hydraulic mining consists of eroding loose soil and the underlying sediment with powerful jets of water, and capturing the ore while letting everything else flow away. This was the principal technique used to acquire the gold that was discovered in California in 1848. The gold was largely mixed in Tertiary stream gravels, and 1200 million cubic meters of soil and gravel were mined hydraulically to separate them from their gold. The mining itself probably destroyed approximately 120–200 square kilometers of the Sierra Nevada. However, the wastes destroyed an area perhaps ten times as great by burying fertile topsoil under sterile sediment. A masterly study of the debris by Grove

Karl Gilbert of the U.S. Geological Survey traced it down the river valleys as a wave that spread over the Great Valley below, partially filled San Francisco Bay, and altered the volume of the tidal flow through the Golden Gate. This, in turn, changed the configuration of the semicircular sand bar deposited on the seaward side of the Golden Gate by tidal erosion. Thus, the effects of surface mining by this method spread far beyond the immediate diggings just as they do in strip mining.

Open-pit mining consists of digging a very large and deep hole to remove low-grade ores, of which the most important by far are sand and gravel. The pit mines in the United States alone had an area of roughly 4050 square kilometers by 1970. However, the materials are used for constructing cities, highways and airports, which for the most part, are thinly spread. Thus, the destruction to soil is mainly by deliberate burial, a subject to which we now turn.

CONSTRUCTION

Cities are the greatest result of construction and, as we all know, they keep growing. Urban and suburban areas are easy to define legally, but a breakdown into land use is more difficult. A golf course is still fertile, a parking lot is not, but both are soil lost to agriculture. It appears that cities in the United States with more than 25,000 inhabitants covered much less than 4000 square kilometers in 1840, about 12,000 square kilometers in 1880, 40,000 square kilometers in 1920, and 85,000 square kilometers in 1960. At present, the urban population is about 150 million. An extrapolation of present trends suggests an urban population of 250 million in the year 2000, and an urban area of 160,000 square kilometers.

Within and connecting the cities are transportation lanes that cover much soil. Foremost are highways, which had an area of 80,000 square kilometers in 1964. A substantial fraction of highways is in cities, and the land they cover is already counted as lost in the tabulation above. However, most of the great Interstate Highway System has been built since 1964, and it has taken at least 4,000 square kilometers of land away from other uses. Even so, the highway system is not growing rapidly compared to the cities. Most roads are used far below capacity and new superhighways are often mere widenings of older roads, which take little land. Other modes of transportation include railroads, which cover a constant area of about 12,000 square kilometers. Airports have an area of 6000 square kilometers and, curiously enough, this area hardly varies. The volume and noise of air traffic have increased much more than the size of airports.

The remaining cover for soil is water. Artificial ponds cover 16,000–25,000 square kilometers of farmland and large reservoirs cover 40,000. Both are increasing rapidly.

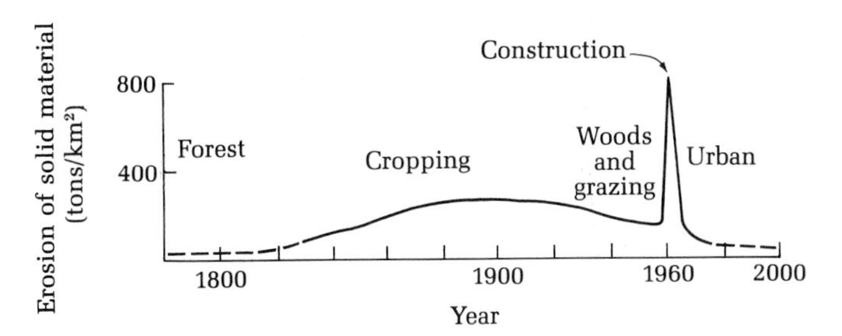

FIGURE 13.6
Variations in intensity of erosion during periods of different land use and urbanization of the Middle Atlantic region of the United States. [After Wolman (1967).]

In sum, mining destroys large quantities of land, but the spreading of mining products over the soil destroys at least ten times as much. Moreover, we dig widely and intensely into the topsoil in the process of urban construction. The clouds of dust arising from highway graders and bulldozers include an abundance of topsoil that is largely lost. This effect has been quantified for the Middle Atlantic region of the United States. Erosion was minimal before 1820 because the area was forested, but it increased 50-fold by 1890, owing to intensive farming. Part of the land was returned to grazing and forest by 1950, and the rate of erosion dropped to only half of the earlier maximum. Then, in the 1960s, urban construction accelerated erosion to an average of three times the farming maximum. Local erosion was ten times as intense as on the high peaks of the Himalaya Mountains, where the natural maximum occurs.

WAR

"War is bad for soil, " said Hyams in 1952. The statement might have been disputed, prior to the invention of gunpowder. Many famous battlefields, in fact, were exceptionally fertile, and the frequent sweep of mobile warfare stopped regular farming and gave the soil a beneficial rest. By World War I, the static situation on the western front combined with the introduction of heavy artillery to destroy the fertile gray-brown podzol after two millennia of tender care. In the Somme salient, two indescribably inept battles churned the moist, fertile soil into a muddy morass into which men, horses, cannon, and fertility disappeared without a trace. At the symbolic bastion of Verdun, the battle lasted two years and the shell craters were still visible half a century later.

Field artillery shells, however, seldom contain even 50 kilograms of explosive, whereas an ordinary bomb contains 100–200 kilograms. Thus, it was not until the invention of massive bombing that warfare moved into the big leagues as a soil destroyer. Europe in the 1940s and Korea in the 1950s were partially spared because of the fluidity of the warfare, the policy of attacking armies and cities, and the modest capacity of the air fleets.

Not until the long stalemate of the Viet Nam War did the true horror emerge. There, the circumstances were the worst possible. One side had complete control of the air over South Viet Nam, enormous capabilities, and no normal strategic targets to attack. A policy of denying the countryside to the enemy was initiated with unbelievable effects. From 1965 through 1971, the United States exploded 12 billion kilograms of munitions in Indochina of which 9.5 billion were in the territory of South Viet Nam. At the reasonable estimate of one crater per 450 kilograms of bombs and shells, that means 26 million craters in a region only slightly larger than Texas.

An average crater from a standard 230-kilogram bomb is roughly 9 meters in diameter and 4.5 meters deep. A typical B-52 raid of seven planes dropped 756 230-kilogram bombs in an area of 4 square kilometers. Bomb fragments, topsoil, and sterile subsoil were spread over an area of 5000 square meters around each crater, although the bombs were closely spaced and fragments usually overlap. South Viet Nam has an area of 173,000 square kilometers, and the craters have removed 140 square kilometers of soil and splattered several thousand more with debris. In much of the area, the soil is lateritic

FIGURE 13.7

Cratered mangrove forest in Gia Dinh province in South Vietnam in August, 1971. Heavy bombing damages trees in three ways: outright destruction, riddling of the timber by missile fragments, and weakening of the trees through subsequent infection. In addition, this particular tract of forest had been subjected to an earlier attack with herbicides. Water-filled craters in both timberlands and croplands have greatly multiplied the available breeding areas for disease-carrying mosquitoes. [From Westing and Pfeiffer, "The Cratering of Indochina," copyright © 1972, by Scientific American, Inc. (all rights reserved).]

FIGURE 13.8
Cratered croplands in Long An province in South Vietnam, photographed in March, 1969. The craters were made by 230-kilogram bombs dropped from B-52s. Such craters, which in this region remain filled with water during much or all of the year, are unsuitable for rice cultivation. For this reason, and because of the dangers of unexploded munitions and sharp bomb fragments buried in the ground, Vietnamese farmers have generally abandoned croplands that have been pocked by craters. [Photo courtesy Gordon H. Orians.]

and, thus, becomes bricklike upon exposure; therefore, many of the craters may become permanent features of the landscape—short of geologic change.

Soil Erosion and Alluvial Soils

Customarily, primitive farmers exhaust the land in a few years and move on. Normal soil offered no other options, until advanced agricultural practices of fertilization and soil renewel were developed. How, then, did fixed agricultural civilizations develop, and, with them, economic surpluses and cities? An interesting and plausible argument can be made that soils had a powerful influence in the matter. Alluvial soils need little or no fertilizer because they frequently receive a top-dressing of fertile soil. The importance of this phenomenon was not confined to ancient times; nearly a third of the global population at present derives a food supply from alluvial soils. Natural top-dressings can come from three sources—river floods, dust storms, and volcanic ash falls. Of the earliest civilizations, those in Egypt, Sumer and Akkad north of the Persian Gulf, and Mohenjo-daro and Harrapa in Pakistan were all founded on alluvial soils. China alone seems a major exception, although its civilization quickly spread to the alluvial soils of the Yellow River. Mayan and Indochinese cultures arose in environments that are questionable in this respect.

EROSION

We shall sketch the history of soils and civilization in an attempt to forecast what may happen as the population expands and agriculture attempts to expand with it. But first it is necessary to examine the process of erosion. We have already considered some of the effects of uncontrolled weather and climate. They vary; thus, land with rainfall that is minimal for farming sooner or later suffers drought and may be abandoned. But how much of the effect should be ascribed to man and how much to climate? The answer has long been debated among different specialists, the agriculturalists emphasizing man and the geologists stressing nature. In part, this was because of conflicting data. Geologists have established that plains are eroded at a very slow rate—roughly 0.75–2.5 centimeters per 1000 years. Soil scientists, in contrast, have found that the topsoil has been partially stripped from more than 1.2 million square kilometers of erstwhile farmland in the United States. If we assume that the fertile A horizon was, on the average, a third of a meter thick, the soil scientists' data show that there has been a 100-fold acceleration in erosion since plowing began. Most of the eroded material moves no farther than the nearest roadside ditch, gully, or valley, but its structure and value as nutritive soil is lost.

Detailed studies of small watersheds have confirmed the extreme acceleration of erosion attendant upon cultivation. In northern Mississippi, erosion increases multifold with increased rainfall regardless of ground cover; even so, the effect is relatively small compared to those accompanying changes in the cover. Consider erosion under a constant annual rainfall of 130 centi-

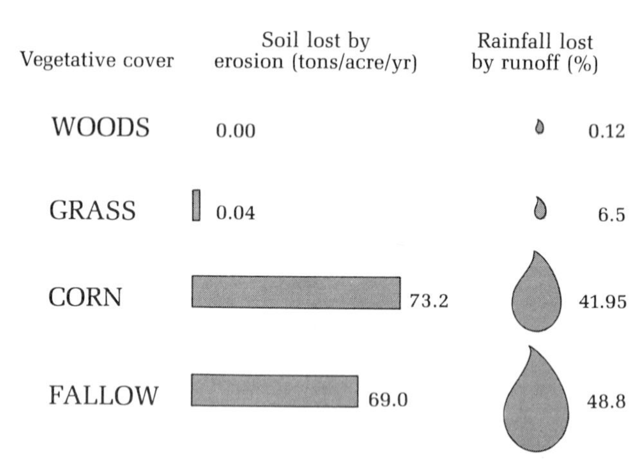

FIGURE 13.9

Results of soil-erosion tests on plots of ground with different vegetative covers. Forests absorb most rain and are hardly eroded at all; grassland is somewhat more vulnerable; but plowing and corn culture are enormously more destructive of soil. [Average values from Gilluly, Waters, and Woodford, *Principles of Geology*, W. H. Freeman and Company, copyright © 1968.]

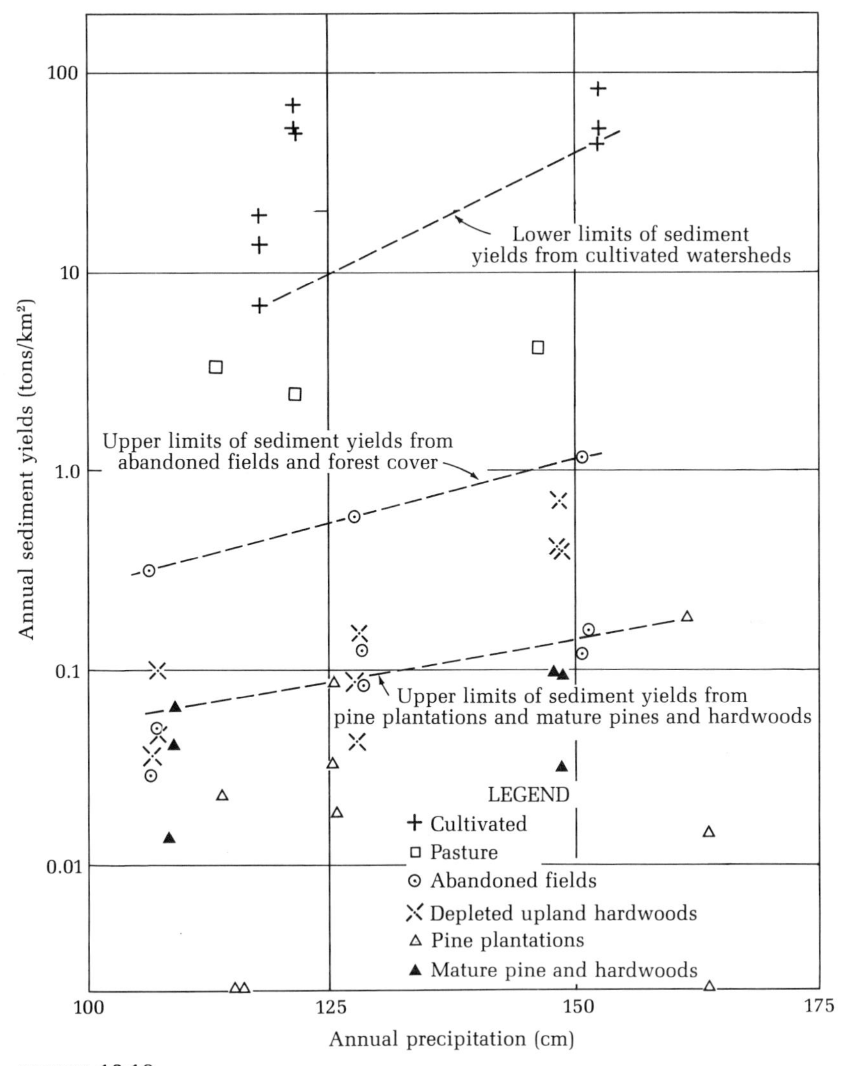

FIGURE 13.10
The slopes of the lines on the graph show the effects of rainfall in detailed studies of individual watersheds in northern Mississippi. The logarithmic scale de-emphasizes the fact that the minimum erosion of cultivated fields is more than 100 times the maximum for forests. [After Ursic and Dendy (1965).]

meters: The maximum annual yield of sediment from forest is about 12.4 tons per square kilometer; the minimum yield from cultivated land is about 1500 tons per square kilometer, or more than a 100-fold increase, and the maximum yield is ten times greater still. With such erosion, some farms have been abandoned and lie fallow partially covered with weeds and grass. These farms have a maximum sediment yield of about 10 times more than the uncut forest. Studies elsewhere give similar relationships. Hardly any rainwater runs off from forest, and erosion is negligible. Grassland loses 9.9 tons per square kilometer, and runoff is small. Corn loses 18,000 tons per square kilometer as more than 40% of the rainwater runs off down the plowed furrows. Loss of soil on fallow land is slightly less, but so is retention of water.

Wind erosion of farmlands can be similarly dramatic in semiarid, grassy plains. Because such land is treeless and flat, there is no impediment to the sweep of the wind except the grass. When the grass is destroyed for any reason, wind erosion strikes. Tilling puts the soil in hazard awaiting the first critical drought. In Oklahoma and the adjacent high plains, a terrible drought occurred in 1933–1934 during the depths of the great economic depression. The soil had been pulverized by constant tillage, which had accompanied the export of fertility in the form of grain. The dust had served as a useful mulch to hold water from evaporation, but, when the water failed completely, it lost all cohesiveness.

A cold front roared across the parched fields of Colorado, Kansas, Oklahoma, and Texas on May 12, 1934, and enormous dust clouds were swept up and toward the east. Some dust settled on the land and enhanced the fertility of the soil where it fell, but the sun was blotted out in New York and

BOX 13.3 THE WIND AND THE PLOW

The last stages of the historical sequence of soil destruction in the American dust bowl are clear: (1) the plains were mainly covered with grass when they were first plowed; (2) a few decades later, in 1933–1934, drought eliminated the cover of crops that replaced the grass; and (3) the wind carried the soil away. The conclusion that the dust bowl was the result of plowing, however, is not only not clear but incorrect. James Malin points out that some sites in the plains areas were occupied repeatedly, but the remains of successive Indian villages on a given site are separated by layers of silt that was blown there by the wind long before anyone undertook to plow the plains. Among other things, the passage of the endless herds of buffalo presumably left strips of bare soil that were eroded by the wind before they were again grassed over.

Plowing in the early twentieth century exposed the soil, but droughts, fire, and (perhaps) natural overgrazing had done so before. There were great dust storms not only in the 1930s but also in the 1830s, which was long before the plains were farmed. Isaac McCoy, a surveyor working in Kansas in 1830, wrote the following at the time (quoted in Malin, 1956): "After we completed our survey, we turned on to a creek, and were looking for an encampment—the day calm and fair—when suddenly the atmosphere became darkened by a cloud of dust and ashes from the recently burnt Prairies occasioned by a sudden wind from the north! It was not three minutes after I had discovered its approach, before the sun was concealed, and the darkness so great, that I could not distinguish objects more than three or four times the length of my horse. The dust, sand, and ashes, were so dense that one appeared in danger of suffocation."

The activities of man merely trigger or accelerate natural processes of erosion. Seen in the opposite context, minor changes in climate may trigger or accelerate erosion even where the activities of man in the past have not destroyed the soil.

clouds of dust swirled over ships 800 kilometers out in the Atlantic. An estimated 300 million tons of topsoil were eroded in this single storm. Wheat farming—and, thus, the rapid exporting of fertility—ceased in whole counties. Farms were abandoned and the wind continued its work.

In time, natural erosion levels the land, unless it is uplifted, but it appears beyond controversy that forest clearing and cultivation vastly accelerate the process. With this background, we can now see what man has done to the soil.

Soil and Civilization

Where some of the first civilizations rose from the fertile alluvial soils of the Tigris, Euphrates, and Indus rivers, there are now barren wastes. Probably, a shift toward a drier climate took its toll, but man may well have brought on the ultimate disaster. That the land of the Middle East was relatively dry when farming developed is suggested by the widespread installation of drainage tiles and the prevalence of water-gathering systems for the cities. Edward Hyams suggests that constant warfare among the Hittites, Assyrians, and Babylonians caused neglect of the drainage, which then caused the land to grow sterile. As to the Indus River valley, it is commonly accepted that the desert there is man-made. The great cities were built of bricks, and whole forests were cut and burned to fire them. With increased evaporation, the soil grew salty and sterile.

CHINA

Chinese civilization originated on the great expanses of loess deposited by the wind at the eastern margin of the deserts of central Asia. We know that people were there before the loess began to settle because their artifacts occur at all depths down to the deeply buried valleys and hills. About 40 centuries ago, they developed a civilization based on farming, but the capital city was moved about, which indicates that there was some instability. The loess is one of the most erodible of materials, and centimeters vanish each year at present. Yet it forms deep soil quickly, and some farming continues on the loess despite the most appalling erosion.

Elsewhere in China, the land has not been so tough. In central southeastern China, it appears that the A horizon, produced under a vanished forest, has been eroded completely away by intermittent farming during periods of population expansion.

Whence comes the food in China? Prior to World War II, the farms were concentrated in the central province of Szechwan, along the lower flood plains of the great rivers, and along the coast. Much of the region of maximum fertility has alluvial soil, annually refreshed by layers of silt derived from the erosion of loess and the topsoil of hills. The silt is captured by the endless

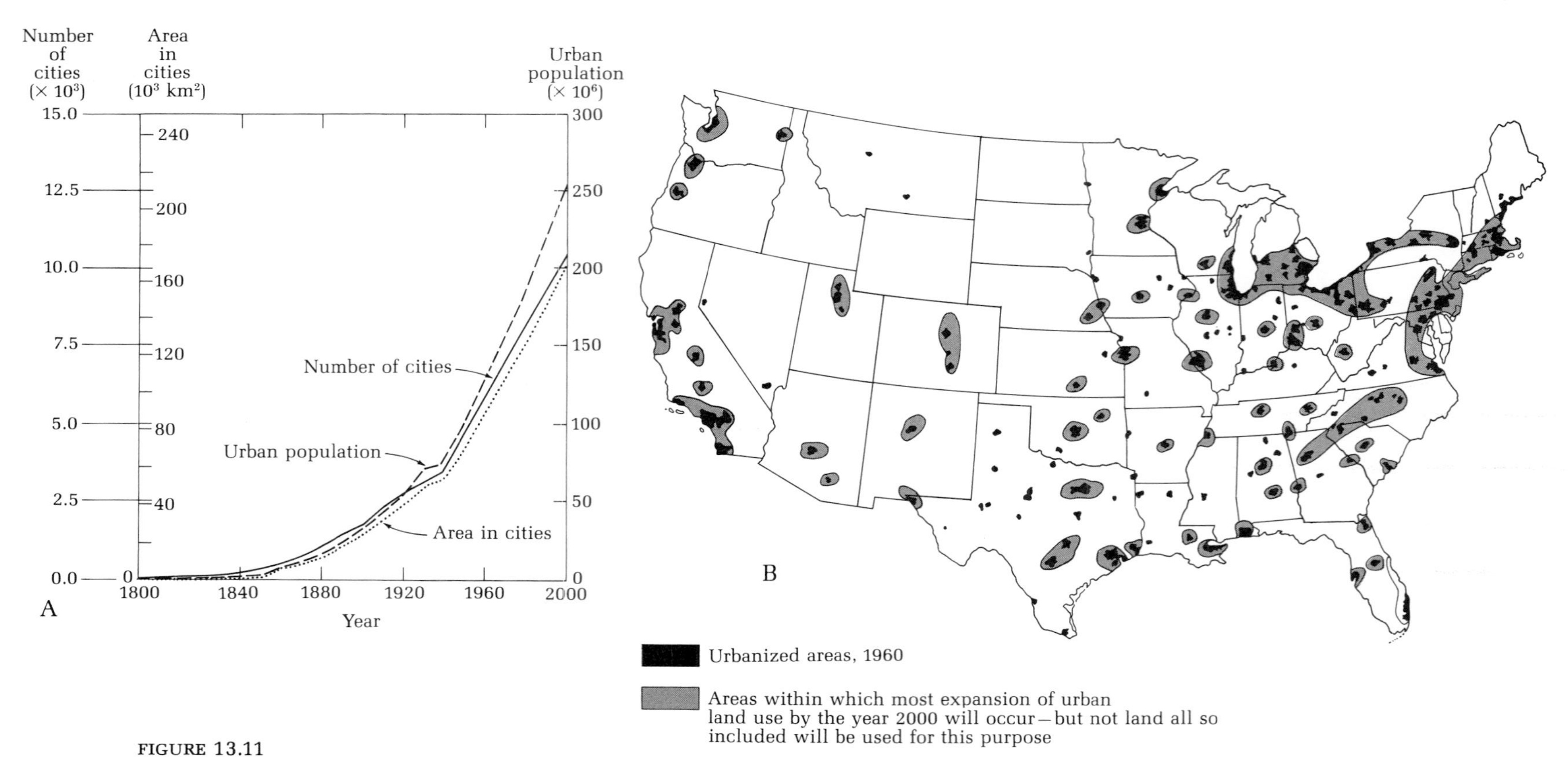

FIGURE 13.11

The history of urban growth in the United States, and predictions based on present trends: *A*, the graph of urban growth shows approximately parallel trends of number of cities and area of cities, but accelerating population; *B*, urban areas in 1960 and predicted expansion by 2000 (all areas are slightly exaggerated, but perhaps not to equal degree, in order to show clearly at this scale). The process of urbanization will be attended by greatly accelerated but temporary erosion. [From Clawson, *America's Land and Its Uses*, copyright 1972 by The Johns Hopkins University Press for Resources for the Future, Inc.]

dams and terraces of paddy fields at the bases of the hills and by enormous canal systems that spread the river water. The still extant canal system of Szechwan was established more than 2000 years ago. Other canals rise high above the lower reaches of the Yellow River. Farmers must scoop muck from the canal bottoms instead of having the river flood over their fields. They now cultivate soil at a level of 3–5 meters above where their ancestors began.

The other salvation of the Chinese soil has been the preservation and use of human waste. Whereas western cities were surrounded by industry, Chinese ones were encircled by agricultural land to which the "night soil" from the cities was transported for use as fertilizer. This land, in turn, provided the vegetable crops consumed by the inhabitants of the cities.

EGYPT

Of all civilizations, the Egyptian was the most blessed by nature. Far in the distant mountains of Ethiopia, the herdsmen cleared the forests to provide pasture, and the monsoon rains eroded new topsoil. Annually, with absolute reliability, a flood of fertile silt flowed down the Nile and spread over the Upper Kingdom on the flood plain in southern Egypt and the Lower Kingdom of the delta. All that was necessary after the flood receded was to raise water from the river for irrigation. During 7000 years, the Nile fed the Egyptians, was the granary of Rome, and sent the finest cotton to the mills of Britain.

What a contrast to the fate of the nonalluvial soils of Greece and Rome, which paid the price of cultivation and overgrazing and generally vanished into the sea. What a difference from the soil of the American dust bowl, which did not survive even two generations of farming.

The matchless fertility of the Nile-borne soil of Egypt faced only one possible peril—that some enemy might dam the river. Ironically, the river has been dammed, not by an enemy but by Egypt herself. The Aswan High Dam was built with the laudable aims of controlling floods, equalizing water flow and generating power. With equalized flow, it is hoped that the Nile flood plain will yield four crops a year instead of one. Unfortunately, while the dam stores water, it stops silt, which settles out in Lake Nasser. Thus, for the first time in 7000 years, the fertility of the delta is not being naturally replenished. Preliminary estimates suggest that, if all the new power generated by the dam were used to fix ammonia from nitrogen in the air, enough fertilizer could be produced to replace the silt to the delta. There is no way to replace the silt and dissolved minerals that the river once poured into the eastern Mediterranean, however, and it appears that the Nile was the principal source of nutrients that fed the now declining fishery in the region. Thus, the dam may well be a disaster. In any event, it did not even address the problem of feeding the Egyptians, because they are doubling their numbers every 24 years. Even if four more equivalent dams were built and the most optimistic estimates of food production were correct, the average level of nutrition would still be about as low as it is now.

People multiply and topsoil subtracts, so rapid construction with bulldozers and concrete may seem an attractive way to solve population problems. It may even *be* attractive, but the risk of calamitous losses is great unless the whole environment is considered. The Vietnamese are among the supposed beneficiaries of the proposed Mekong River plan, but it includes dams that may cut off the supply of fertile silt to the delta as well as open the prospect of laterization in the highlands. Likewise, other tropical soils appear bountiful but readily turn to brick. The prairies seem made for modern industrial farming, but they turn to dust at the first bad drought. Man started with a world covered with topsoil, and he has already sterilized part of it. Many of the remaining areas will be even easier to destroy if the fertility is feverishly mined and exported to feed an exploding population.

Rate of Soil Formation

Land denuded of soil does not remain bare forever. How long does it take to make new soil? It depends on the raw materials and the intensity of the soil-making process. Consider favorable circumstances first. Most rice is grown in warm climates on alluvial soils that are trapped in paddy fields and saturated or flooded throughout the growing period. Flooding commonly induces anaerobic conditions with intense bacterial activity, and mobilization of ions in the soil. Soil profiles may form in 50–100 years under these favorable circumstances. When conditions vary, layers of bleached and unbleached soil occur one above another. In some parts of China, agricultural soils may consist of interlayered river silt, mud from canals, and organic fertilizer. In such soils, ground water may tend to migrate horizontally through the more permeable silt and, thereby, intensify layered leaching.

Conditions are not quite so favorable on the Chengtu Plain of China because of the extremely rapid accumulation of soil: Pottery and other artifacts of the Han dynasty, only 2000 years old, are buried under 3 meters of new soil. However, the silt derived from loess weathers quickly, and definite soil profiles exist—indicating that formation is proceeding faster than accumulation. Loess, in fact, forms soils very rapidly even while being eroded. James Thorp says that, in the loess source area, a "new soil can be made from raw loess in a few years" (Thorp, 1939, p. 446). Thus, the Chinese have been able to cultivate the loess, as it has been eroded from under them for 4000 years.

Soils in humid climates, expecially warm ones, form with surprising speed in a variety of circumstances. A and B horizons indicate weathering of abandoned plowed fields in North Carolina to a depth of 13 centimeters in only 50 years. In the Normandy region of France, a soil profile 40 centimeters deep has formed in a refuse pile at a castle abandoned in 1066. In Indonesia, ash from the eruption of Krakatoa in 1883 developed a soil profile 35 centimeters deep in only 45 years, but Indonesia is warmer than Normandy.

Conditions in the desert result in much slower soil formation. The Russian soil scientist M. G. Konobeeva has analyzed the evolution of soils in ancient oases of central Asia that were occupied intermittently from 700 B.C. until the armies of Genghis Khan in the thirteenth century, and those of Tamerlane in the fourteenth, destroyed the irrigation systems. The evolution of the new desert soil can be traced step by step, and it took 1000–1200 years to form a few centimeters of standard, relatively sterile, desert soil.

It appears that all is not lost if soil is eroded away. In a matter of centuries, in humid climates, it will begin to form again and to show some fertility. This might have been anticipated from what we know about the close correspondence between soils and the shifting climate and vegetation. However, new soil formation will have no effect on the population problem for many generations, and, by then, the problem will have solved itself one way or another. For the next few generations, soil lost or fertility decreased might as well be considered as gone forever.

Summary

1. Physical and chemical processes convert rock to sediment. Some elements go into solution and move through rivers to the ocean. Other elements and minerals are chemically resistant and are transported as solid fragments.
2. Clays are crystalline, hydrated aluminum silicates that grow during many weathering processes and migrate through soil as it is developing.
3. Soil is a thin, structured, layered material with a highly complex origin that determines its character. Climate and plant life dominate soil formation.
4. Soil is in a dynamic balance because erosion tends to remove it and churning tends to destroy its structure. Thus, in order to persist, it must form as fast or faster than it is destroyed.
5. Man is almost completely dependent on the nutrition stored in a few centimeters of ever changing topsoil. Nonetheless, he deliberately destroys soil in many ways.
6. Most soils are rapidly exhausted by farming, but alluvial soils, now and then, receive a natural top dressing of fertile silt, ash, or soil. Most early civilizations developed in regions with alluvial soils. Egyptian farms, for example, were annually covered with fertile soil eroded from the Ethiopian highlands by the Nile River.

7. Rapid erosion by stream and wind may occur when the natural vegetative cover of soil is removed by farming. However, it may also occur without the intervention of man.

8. Soil can by destroyed rapidly, but it can also form within a few years, in the most favorable circumstances.

Discussion Questions

1. What is the evidence that plants play an important role in soil formation?
2. Why do clay minerals become less mobile after they first form?
3. How can soil zones migrate when the climate changes?
4. How does man deliberately destroy soil?
5. Why is erosion as measured by soil scientists much faster than erosion as observed by geologists?
6. What went wrong with the Aswan High Dam project?
7. Why is erosion accelerated during construction of new housing subdivisions?

References

Clawson, M., 1972. *America's Land and Its Uses.* Baltimore: The Johns Hopkins University Press (for Resources for the Future, Inc.). [A broad look at the history of uses of the land.]

Dudal, R., 1968. Genesis and classification of paddy soils. *In* V. A. Kovda and E. V. Lobova, eds., *Geography and Classification of Soils of Asia,* pp. 194–197 (based on a UNESCO seminar in Tashkent in 1962; translated from Russian in 1968). Jersualem: Israel Program for Scientific Translations. [Soils can form in less than a century.]

Ehrlich, P., and A. Ehrlich, 1972. *Population, Resources, Environment: Issues in Human Ecology* (2nd ed.). San Francisco: W. H. Freeman and Company. [An informative overview of biological aspects of the population problem.]

Geikie, A., 1880. Rock weathering as illustrated in Edinburgh churchyards. *Proc. Roy. Soc. Edinburgh* 10:518–532.

Gilbert, G. K., 1917. Hydraulic-mining debris in the Sierra Nevada. *U.S. Geol. Surv. Prof. Pap.* (105). [Traces gold-mining debris from the mountains to the sea.]

Gilluly, J., A. C. Waters, and A. O. Woodford, 1968. *Principles of Geology* (3rd ed.). San Francisco: W. H. Freeman and Company.

Hyams, E., 1952. *Soil and Civilization.* London: Thames and Hudson. [Religion, man, and society. Western Europe managed to preserve its soil.]

Judson, S., 1968. Erosion of the land. *American Scientist* 56:356–374. [Natural and man-induced erosion compared.]

Kellogg, C., 1950. Soil. *Sci. Amer.* 183(1):30–39. (Available as *Sci. Amer.* Offprint 821.) [Full of insights by the then assistant administrator of the Soil Conservation Service.]

Konobeeva, M. G., 1968. Evolution of soils in the ancient oases of the desert zone. *In* V. A. Kovda and E. V. Lobova, eds., *Geography and Classification of Soils of Asia,* pp. 124–131 (based on a UNESCO seminar in Tashkent in 1962; translated from Russian in 1968). Jerusalem: Israel Program for Scientific Translations.

Kubiena, W. L., 1963. Paleosoils as indicators of paleoclimates. In *Changes of Climate,* pp. 207–209. Paris: UNESCO. [Many old soil horizons are preserved.]

Malin, J., 1956. The grassland of North America; its occupance and the challenge of continuous reappraisals. *In* W. Thomas, Jr., ed., *Man's Role in Changing the Face of the Earth,* pp. 350–366. Chicago: University of Chicago Press. [Dust came out of the American plains long before the plow broke the sod, and the Indian may have been facing an ecological crisis.]

McNeil, M., 1964. Lateritic soils. *Sci. Amer.* 211(5):96–102.

Rozanov, B. G., and I. M. Rozanova, 1968. Genesis of "degraded" soils on paddies in the tropics. *In* V. A. Kovda and E. V. Lobova, eds., *Geography and Classification of Soils of Asia,* pp. 243–248 (based on a UNESCO seminar in Tashkent in 1962; translated from Russian in 1968). Jerusalem: Israel Program for Scientific Translation.

Shelton, J. S., 1966. *Geology Illustrated.* San Francisco: W. H. Freeman and Company.

Thorp, J., 1939. *Geography of the Soils of China.* Peking: National Geological Survey of China. [An American expert analyzes the soils of China.]

Ursic, S. J., and F. E. Dendy, 1965. Sediment yields from small watersheds under various land uses and forest covers (Proceedings of the Federal Inter-agency sedimentation conference, 1963). *U.S. Dep. Agr. Misc. Publ.* (970):47–52. [Cultivation causes a 100-fold increase in erosion. Cited by Judson, 1968, a more readily available publication.]

Westing, A. H., and E. W. Pfeiffer, 1972. The cratering of Indochina. *Sci. Amer.* 226(5):21-29. [A horror story.]

Wolman, M. G., 1967. A cycle of sedimentation and erosion in urban river channels. *Geogr. Ann.* 49-A:385–395.

14

UNSTABLE GROUND

During the nine years intervening between 1822 and 1831, the population of Valparaiso was multiplied in an extraordinary manner, and increased from 6,000 to 34,000 inhabitants; in consequence of which every effort was made to preserve the newly acquired sandbanks, and in some places no less than two entire streets have been erected where there was sea before.

Charles Lyell, PRINCIPLES OF GEOLOGY (1834)

AS WATER RISES
SINKING FEELING
ENGULFS HOUSTON

Headline in the SAN DIEGO UNION,
3 October 1972

We have just considered the weathering of surface material less than a meter thick and its social consequences in food production. In the chapters after this one, we shall follow sediment whence it is eroded to its ultimate deposition in the deep sea. In this chapter, we are concerned with the physical phenomena that affect roughly the upper hundred meters of sediment while it is more or less in place. Such phenomena cause the ground surface to rise and sink as much as 5–10 meters without moving horizontally. We shall also consider landslides and the like that occur on slopes and do not move very far.

These phenomena are not primarily the result of plate motions or isostatic loading and unloading, although they may be triggered by earthquakes and tectonic movement. Instead, they are largely the consequences of the interaction of unconsolidated sediment, water, and gravity. The effects may seem minor or highly localized compared to tectonic ones, but the social effects are certainly comparable. This is because any seemingly minor instability of the ground in a city is apt to affect something, whereas a great earthquake in the South Pacific may have no social effect at all. Unstable ground has become increasingly important, like most environmental problems, as the population expands. Construction extends to ever more unsuitable sites: for example, houses are built on marshy land that sinks, pipelines full of hot oil are projected for construction on frozen ground, and carefully graded canals are built across ground that has been sinking for years. Man also faces the long-term consequences of ancient misconceptions about the stability of the ground. Deltas slowly sink, and many cities that are sinking with them, such as Venice and Amsterdam, are menaced by the sea. As the decades and centuries pass, it seems not unlikely that the seemingly minor effects of unstable ground will cause increasing havoc unless we learn to accommodate to them—or even to ameliorate them.

Compaction and Hydrostatic Uplift

If we analyze the transformation of clay into the hard, dense rock called shale, we can gain some insight into the factors that influence many aspects of the stability of the ground. Consider the clay that settles out of suspension when a river enters the ocean or a deep lake: It spreads out to form a thin layer that is saturated with water and may range in consistency from thick soup to fluffy mud. The water is all connected between flakes of clay, and so it exerts a hydrostatic uplift, or buoyancy, just as it does on any solid surrounded by a liquid. As time passes, layers of fresh clay are deposited one above another, and the lower ones are squeezed and compressed. Part of the water is driven upward and out, but the sediment is still saturated and hydrostatic uplift continues.

Burial and compression gradually increase the density of an average clay from 1.40 grams per cubic centimeter at the surface to 1.80 grams per cubic centimeter at a depth of 300 meters. This first occurs by mechanical rearrangement of grains as the water is squeezed out from the pores among the grains. Then the water absorbed on the clay is driven out and the grains are compressed together. Most of the compression is irreversible, but roughly 5% is elastic and is recovered by expansion if the load is removed. At approximately 300 meters depth, the water is effectively discontinuous and hydrostatic uplift ceases. The closely packed solids bear all the weight of the overlying sediment. These effects are not instantaneous; there is some lag between an increase in loading and the final adjustment of the lower layers.

The depth at which hydrostatic uplift ceases depends on the type of sediment, and so does the compaction under loading. A sand, for example, does not compact as much as a clay because the grains of quartz are much more difficult to deform. Nevertheless, compaction affects all sediment, and hydrostatic uplift affects all that are saturated. Several important generalizations

BOX 14.1 POROSITY AND PERMEABILITY

It is important to distinguish between the amount of space among sediment grains and the continuity of the spaces, because these properties affect many important phenomena, including the stability of the flow of ground water and the extraction of oil.

Porosity is a measure of the space among grains of a sediment. It is the ratio of pore volume to total volume and is expressed as a percentage. It commonly ranges from 12–45% in sediments, but it may reach 90%, depending initially on the shape, sorting, and packing of the grains and later on how much of the pore space is filled by the natural cementation that makes sedimentary rocks solid. In addition to pores, among grains, solid rocks may be porous because of fracturing. Plates and flakes of freshly deposited clay form a loose lattice that has a very high porosity but is also easy to compress. The porosity of sand is not so high, but it reaches a maximum if the grains are perfect spheres of uniform size and are stacked one above another. If uniform grains are closely packed, the pore space decreases. If the grains are somewhat angular, the porosity may be less, because the flat surfaces of the grains are not necessarily separated by gaps. If the grains are of different sizes, the smaller ones fill the pore spaces among the bigger ones. The degree of uniformity of particle sizes in a sediment is called sorting, and it is termed "good" by geologists if it is uniform and "good" by civil engineers if it is not. It depends on your viewpoint: uniform sorting yields weak concrete.

Permeability is a measure of the capacity of a porous substance to transmit a fluid. If the pores are isolated, there is no permeability, but a single open crack may permit voluminous flow. Permeability is measured in a unit called a darcy, which takes into account both the fluid viscosity and the pressure gradient.

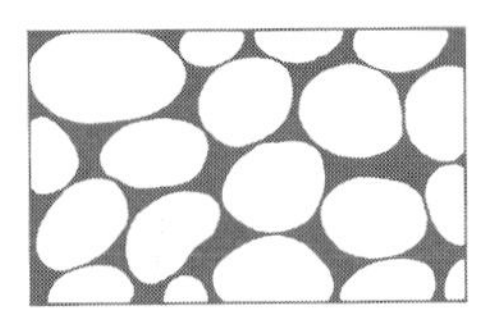

Well-sorted sediment is highly porous.

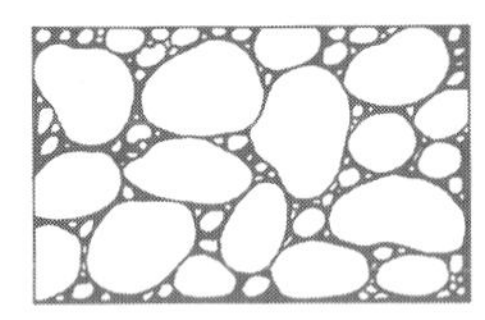

Poorly sorted sediment is less porous.

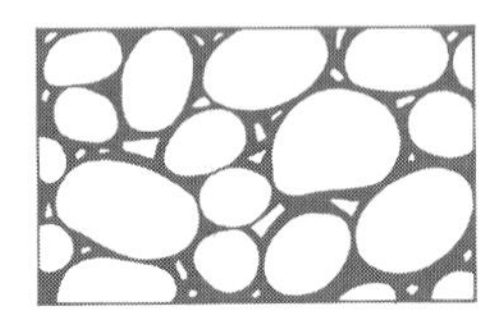

Mineral-filled sediment is slightly porous.

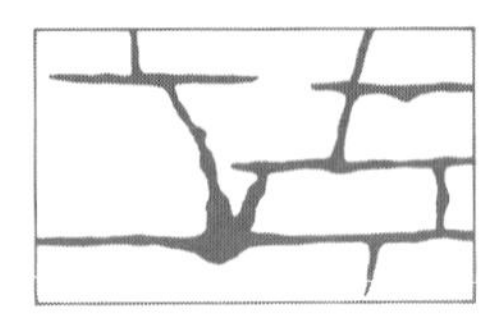

Soluble rock is often porous because water dissolves cavities in it.

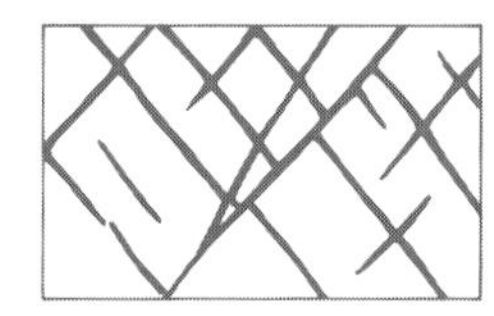

Fractured rock has porous structure similar to that of soluble rock.

Clay-rich zones in or around the strata that yield oil, gas, or water can be the source of considerable subsidence. Before the withdrawal of fluid (*left*), clay particles are kept separated by the fluid and by the pressure applied to it by the overlying land. When fluid is withdrawn (*right*), a claylike soil is more likely than most others to undergo compaction.

Poorly consolidated rocks also can lead to subsidence when fluid is withdrawn, although the subsidence will be less because the rock is stronger than clay. In such rock, the individual grains are weakly held together and will be compressed by additional pressure.

Well-consolidated rocks, in which individual grains are held together strongly, retain their structure even if fluid or gas is removed and give rise to little or no subsidence.

follow from these and related facts: First, if liquid is removed from the pores of a sediment, the hydrostatic uplift vanishes and the sediment sinks under the new load. Second, if a sediment is flooded, its surface is elevated but only by the small amount of elastic compression. Third, a natural or artificial load compresses a sediment. Fourth, the compression occurs gradually after the load is applied.

Withdrawal of Fluids

The fact that fluid withdrawal causes subsidence was obscured by several factors until the middle of this century. First, the effects were initially subtle, and detection required the establishment of a precise geodetic network and the passage of enough time so that resurveying could detect something. Second, subsidence didn't amount to much until the rate of pumping of ground water for agriculture and oil for industry expanded to keep pace with the population. Third, the subsidence caused damage, which resulted in lawsuits and a spirited defense by the pumpers to the effect that they were not responsible.

The geologic evidence for subsidence due to pumping was presented for the Wilmington oil field in California by geologists James Gilluly and Ulysses Grant in 1949. By 1969, the correlation was so widely accepted that UNESCO held a symposium on the subject in Tokyo. Carefully documented examples of major subsidence due to ground-water extraction were available from Osaka, Tokyo, Mexico City, London, Taipei, and, in the United States, from Arizona, California, Georgia, Louisiana, Nevada, and Texas. Sinking ground had been identified in more than 21 oil fields in California and in 7 more in Texas. The American geologist J. F. Poland, in a paper authorized by the Director of the U.S. Geological Survey, wrote: "We can anticipate that in the next few decades areas of land subsidence will multiply and hence that problems will become more widespread. The problems caused to date by land subsidence due to the withdrawal of fluids are a principal reason for this International Symposium at Tokyo."

So much for the scientists. What of the courts? Could anyone who suffered damages collect compensation from the pumpers? From the discussions in Tokyo in 1969, it appears that the matter is moot:

> DISCUSSION [following Mayuga and Allen, 1970]
> *Intervention* of Dr. J. F. Enslin (Republic of South Africa)
> *Question:*
> What machinery, if any, exists in the U.S.A. whereby property owners may legally be entitled to claim compensation in the event of damage to their properties as a result of surface subsidence due to extraction of oil?
>
> *Answer* of Dr. Mayuga:
> As you may have heard, we did have a lawsuit in Long Beach. The U.S. Navy, with their naval shipyard in the area and who has no interest at all in the oil,

TABLE 14.1

Notable examples of major areas of land subsidence caused primarily by ground-water extraction. All the regions are underlain by thick layers of recently deposited sediment that may also be compacting under its own weight.

Location	*Depositional environment and age*	*Depth range of compacting beds below land surface (m)*	*Maximum subsidence (m)*	*Area of subsidence (km^2)*	*Time of principal occurrence*
Japan, Osaka and Tokyo	Alluvial (?); Quaternary (?)	10–200 (?)	3–4	?	1928–1943 1948–1965+
Mexico, Mexico City	Alluvial and lacustrine; late Cenozoic	Chiefly 10–50	8	25+	1938–1968+
Taiwan, Taipei Basin	Alluvial; late Cenozoic	30–200 (?)	1	100±	?–1966+
Arizona, central	Alluvial and lacustrine (?); late Cenozoic	100–300+	2.3	?	1952–1967+
California, Santa Clara Valley	Alluvial; late Cenozoic	50–300	4	600	1920–1967+
California, San Joaquin Valley (three areas)	Alluvial and lacustrine; late Cenozoic	90–900	8	9,000	1935–1966+
Nevada, Las Vegas	Alluvial; late Cenozoic	60–300 (?)	1	500	1935–1963+
Texas, Houston–Galveston area	Fluviatile and shallow marine; late Cenozoic	50–600+	1–2	10,000	1943–1964+
Louisiana, Baton Rouge	Fluviatile and shallow marine; Miocene to Holocene	40–900 (?)	0.3	500	1934–1965+

Source: Poland (1970), reproduced by permission of UNESCO.

TABLE 14.2
Documented surface deformation over U.S. oil and gas fields

Field	Discovery year	Maximum producing area (km^2)	Median depth of production (m)	Differential subsidence*		
				Maximum measured (m)	Area of subsidence (km^2)	Period of measurement
CALIFORNIA						
Buena Vista	1910	48	1130	0.27		1957–1964
				2.3		1942(?)–1964‡
Dominguez	1923	7	1430	>0.07		1945–1960
Edison§	1928	6	1100	>0.09		1926–1965
Fruitvale§	1928	14	1370	>0.04		1953–1965
Greeley§	1936	9	3235	>0.01		1953–1965
Huntington Beach	1920	16	930	1.22	37	1933–1965
Inglewood	1924	5	900	1.73	11	1911–1963
Kern Front	1912	19	745	>0.34		1903–1968
Long Beach	1921	7	1690	>0.61	31	1925–1967
McKittrick	1898	6	360	(?)		
Midway-Sunset§						
Central area	1901	65	555	>0.49		1935–1965
Globe anticline	1912	15	1020	>0.43		1935–1965
Sunset area	1900?	20	590	>0.18		1935–1965
Paloma§	1939	23	3800	>0.07		1957–1965
Playa del Rey	1929	2	1520	>0.29		1925–1937
Rio Vista (gas)	1936	98	1300	>0.30		1939–1964
River Island (gas)	1950	18	1250	>0.23		1935–1964
San Emidio Nose§	1958	4	3900	>0.06		1935–1965
Santa Fe Springs	1919	6	1300	0.66	16	1927–1963
Tejon, North§	1957	10	2800	>0.09		1935–1965
Torrance	1922	27	1230	>0.10		1953–1960
Unnamed						
(Orange Co.)	1909	5	1480	>0.05		1959–1964
Wilmington	1936	29	1000	>8.84	>75	1928–1966
TEXAS						
Clinton	1937	12	820			
Eureka Heights	1935	6	2540			
Goose Creek	1916	6	600	>1	11	1917–1925
Mykawa	1929	6	900	>0.1		
Saxet	1930	25	1800	>0.93		1942–1959
South Houston	1935	7	1200	0.09		
Webster	1937	11	1670			

Source: Yerkes and Castle (1970), reproduced by permission of UNESCO

*Determined with respect to local control near margins of fields; any regional subsidence gradient removed. In many cases, determined only for a point within the field and maximum subsidence not known.

†*H*, high angle; *L*, low angle; *n*, normal; *r*, reverse.

‡Data courtesy R. D. Nason (1969).

§Data on differential subsidence courtesy B. E. Lofgren (1969).

Horizontal displacement		Surface faulting			
Maximum measured (m)	Period of measurement	Type†	Displacement (m)	Length (m)	Year first observed
0.39	1932–1959	*Lr*	0.74	2600	1932
0.76	1934–1963	*Hn*	>0.15	700	1957
		Hn	>0.34	5000	1943?
		Hr	0.030	>8	1932
		Hr	0.05	305	1968
3.66	1937–1966				
		Hn	>0.64	>900	1962
		Hn	0.24	>1000	1962
		Hn	0.41	>700	1925
		Hn	Undermined	1000	1962
		H	>0.61	2500	1950 (?)
		Hn	0.46	>1000	1962
		Hn	Undermined	>1000	1962

sued the City of Long Beach, the State of California, who is our partner in this oil production, and all the oil operators in the area for damage to their installations allegedly due to subsidence. That case was never actually adjudicated, because a compromise of financial settlement was made. So the responsibility was not actually established by the courts.

Most people who have been damaged in the Long Beach area are themselves involved in oil operation and received benefits from the oil production. I do not know yet what the responsibility would be or who would be liable, if someone files a claim. I think this is a good case for lawyers. This will be argued for some time. We thought the U.S. Navy's case against the oil operators, including the City of Long Beach, would establish liability but it did not. The case was not adjudicated.

DISCUSSION [following Castle et al., 1970]
Intervention of Dr. Manuel N. Mayuga (U.S.A.)
Question:
Yesterday I was asked about the legal aspects of subsidence. I wonder if you would be free to comment on the legal aspects of the Inglewood oil-field subsidence with respect to structural damage generated in this area?

Answer of Dr. Castle:
As Dr. Mayuga has indicated, serious structural damage has been associated with the Inglewood oil-field subsidence. A reservoir located on the edge of the subsidence dish failed in 1943. The California Department of Water Resources concluded that the failure was attributable to faulting through the floor of the reservoir. We are not engineers and cannot make judgments on the failure of engineered structures. We have, however, attributed the indicated faulting to oil-field operations. Suits have been brought against the operators of the oil field and it is my understanding that they will be tried this fall.

Comment of Dr. Mayuga:
What I was trying to point out is that subsidence has led to a number of suits now pending in U.S. courts. Accordingly, there exists a problem of legal responsibility for any damage on the surface associated with subsidence.

Answer of Dr. Castle:
I agree. I might make one observation that could be considered an indirect response to your comment: this relates to litigation over the subsidence centering on the Goose Creek oil field in Texas.

The state of Texas retains title to tidelands oil. When the Goose Creek field sank beneath the waters of Galveston Bay the state claimed that title automatically transferred to them. The operators, according to Pratt and Johnson, contended that they were responsible for the subsidence and that it could not be considered "an act of God." A decision was rendered in favor of the defendant; that is, the operators retained title.

These discussions highlight an important aspect of the social effects of environmental hazards. Just because a cause–effect relationship is scientifically established does not mean that anyone is legally liable for the effect. The law is not established until a test case arises. Ironical situations, such as that of the Goose Creek oil field in Texas, are rare, and zoning restrictions, building codes, and the law may lag far behind scientific knowledge. As a consequence, there are few places where restrictions would prevent the construction of yet another dam that might fail because of the subsidence of yet another oil field or ground-water reservoir.

SINKING TOKYO

Tokyo, one of the largest cities in the world, is among the many that are sinking and in danger of being flooded by the sea. Some 3.3 million inhabitants and 150,000 businesses are in the heart of the city where it meets Tokyo Bay. As a result of excessive withdrawal of ground water, primarily for industry, parts of the district have subsided more than 4 meters and are now more than 1 meter below sea level. The area has been repeatedly flooded by the storm surges and high waves of typhoons.

The center of Tokyo lies on roughly 30 meters of uncompacted sediment deposited as a delta by small rivers in recent times. Below is a buried topography and hundreds of meters of deltaic sediment deposited earlier by the rivers. The fact that the very center, the Koto district, was sinking was in the records after the resurvey of 1915 showed that the levels of markers fixed to the ground had dropped since the original survey of 1892. However, the amount was small and unremarked.

After the 1923 earthquake that struck the Kwanto Plain region around Tokyo, a program of frequent resurveys began, and it was discovered that the Koto district was sinking at the rate of about 10 centimeters per year. It continued to do so for two decades, and the survey and research programs were expanded until the district was destroyed by fire-bombing in World War II.

In 1947 and 1949, the district was flooded by typhoons, and concern mounted. Meanwhile, industrialization continued, ground water dropped, and sinking resumed at the rate of 10 centimeters per year. The correlation between water withdrawal and subsidence was established and, in 1961, the withdrawal of ground water was partially regulated. Control of pumping increased; by 1966, the level of ground water began to rise. Ground subsidence continued, however, at about 2 centimeters per year. The *rate* of subsidence decreased, but that was all.

It appears that consolidation of the recently deposited delta will continue. An interesting correlation by Shigeru Aihara and his colleagues shows that the annual industrial output of the Koto district has increased by almost 1200 billion yen since 1950. The production was accompanied by a drop in the ground-water level of about 20 meters. Subsidence of the ground amounted to about 2 meters, and, in 1966, a barrier was constructed to prevent floods. If present trends continue, the barrier will grow ever higher until flooding ultimately restores equilibrium. Meanwhile, the subways and highways require constant attention as the levels change.

TABLE 14.3
Historical review of land subsidence and related explorations in Tokyo

Year	*Phenomena, disasters, countermeasures*	*Explorations, surveys*
1923	(Kwanto Earthquake occurred)	Precise levelling started
1924	Abnormal subsidence of bench marks in Koto Delta discovered	
1929		Geological maps of Tokyo and Yokohama published
1932	The degree of land subsidence became remarkable	
1933		Observation well with compaction recorder constructed, and soil tests undertaken
1938	Land subsidence advanced	Repeated precise leveling in Koto Delta
1940		Fukugawa observation well of 35 m depth constructed
1941	(World War II broke out)	Tokyo Metropolitan Conferences for Land Subsidence Problems made plan for land subsidence exploration; six observation wells of 32–35 m depth constructed
1942		Azuma observation well of 19 m depth constructed
1944	Land subsidence had not advanced	
1945	(Koto region attacked and devastated by bombers, World War II ended)	Observations well destroyed; precise levelling for all bench marks stopped; local levelling continued
1947	Subsided area invaded by high water influx caused by typhoon	Relevelling started
1949	Subsided area invaded by high water influx caused by typhoon, land subsidence revived	
1951		A series of land subsidence exploration for five years started
1952	Iron Tube of observation wells sinking	Kameido observation well of 60.5 m constructed; amount of industrial ground water withdrawal surveyed; chemical components of ground water studied

Year	Phenomena, disasters, countermeasures	Explorations, surveys
1953	The degree of land subsidence in the Johoku district became remarkable	Two observation wells of ground water table constructed
1954		Regional hydrogeological survey in and around Tokyo started by Geological Survey of Japan; amount of industrial ground water withdrawal resurveyed
1955	Land subsidence in the Koto Delta remarkably advanced	Azuma-B observation well of 115 m depth constructed; future amount of land subsidence for five years estimated
1957	Local land subsidence occurred along river valley in upland area of Tokyo	Adachi observation well of 115 m depth constructed; survey on amount of ground water withdrawal in whole lowland area started
1959	(Ise Bay Typhoon struck Nagoya City)	Geological map of Tokyo published
1960		Repeated precise levelling for embankment of subsidence started; future amount of land subsidence reestimated
1961	Control of industrial ground water withdrawal in the Koto Delta decided	Group of observation wells of different depth constructed at several localities, i.e., Nanagochi No. 1 (50.5 m), No. 1 (27 m depth), No. 4 (290 m depth), etc.
1962	Control of ground water withdrawal for building use decided	Hydrogeological map published
1963	Control of industrial ground water withdrawal in Johoku district decided	Three observation wells constructed
1966	Rate of land subsidence decreasing; ground water table rising; shrinkage of deeper soil layers advanced; construction of outer embankment of Koto Delta completed	Observation well of 450 m depth constructed
1968		Four observation wells constructed in upland area; geological map of Tokyo published

Source: Inaba et al. (1970), reproduced by permission of UNESCO.

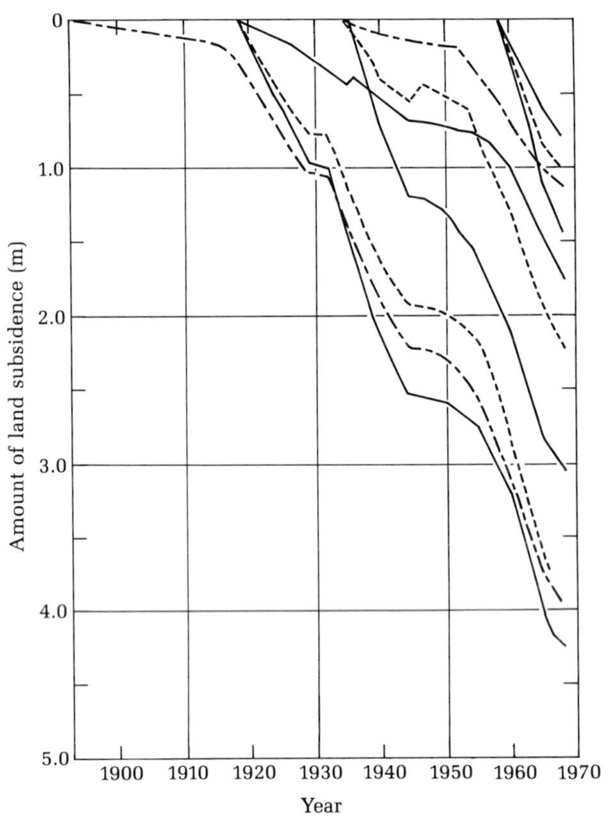

FIGURE 14.1

Total subsidence of several bench marks emplaced at different times in the lowlands of Tokyo. [After Inaba et al. (1970), by permission of UNESCO.]

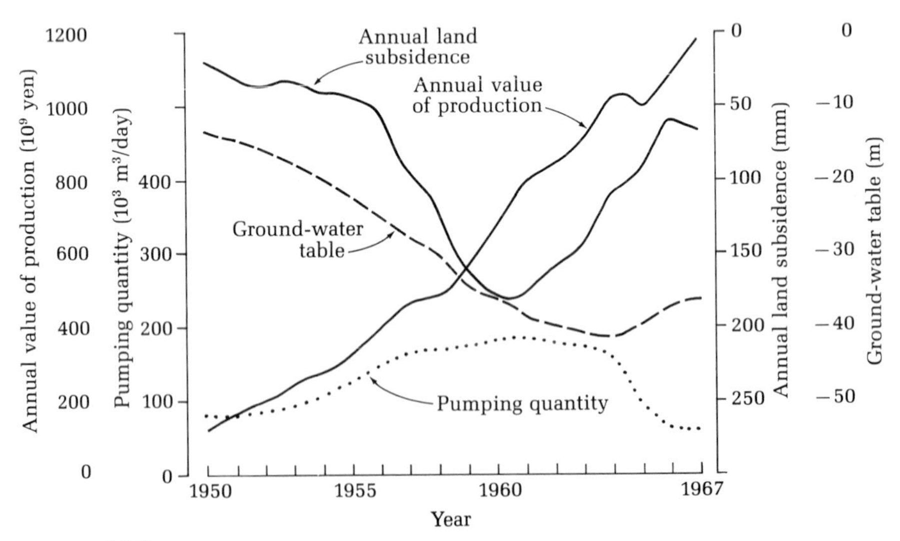

FIGURE 14.2

Correlation between annual industrial output, pumped ground water, depth to water, and annual rate of land subsidence in the Koto district of Tokyo. Control of ground water withdrawal began in 1961. The land continues to sink, but at a slower rate. [After Aihara et al. (1970), by permission of UNESCO.]

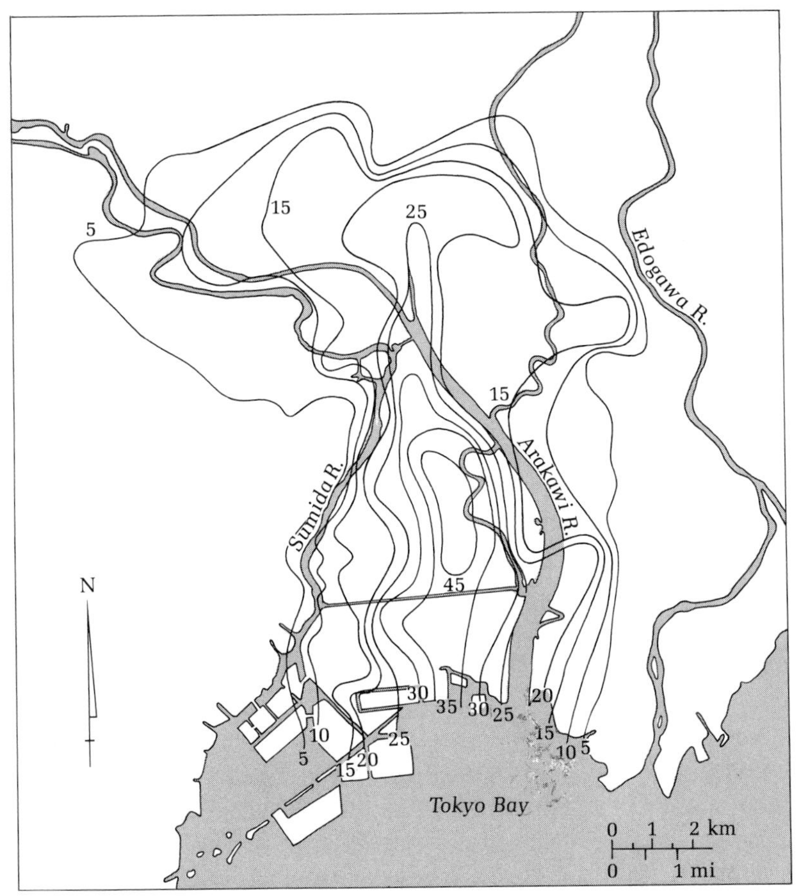

FIGURE 14.3
Land subsidence (in millimeters) in central Tokyo during 1951–1952 alone. Maximum subsidence is in the area of maximum thickness of recently deposited sediment and maximum withdrawal of ground water. [From Inaba et al. (1970), reproduced by permission of UNESCO.]

The experience of Tokyo illustrates once again the interplay between scientific knowledge and social response. A social response to the terrible earthquake of 1923 established the scientific surveys that discovered the sinking in 1924, but it was not until the 1950s that the cause was established. Within a decade, controls were established; within another five years, the sinking slowed. This was a remarkably fast response, probably because of the increasing possibility of an unprecedented disaster. It serves to illustrate the point that many geologic phenomena that affect society are undetected for lack of research, and that the problems may grow acute before they can be controlled.

SINKING FARMLANDS

Excessive pumping of ground water for agriculture has caused 9000 square kilometers of the fertile San Joaquin Valley in California to sink as much as 8.5 meters and at rates exceeding 0.4 meters per year. This is the most extensive subsidence on such a scale that has been caused by man. It still continues. Through the area passes the California aqueduct, which is the largest (and, at a total cost of $2 billion, the most costly) water distribution system ever built. The flow of water is by gravity and is critically dependent on stable ground. Thus, this is an ideal area to study the social effects of subsidence on a grand scale.

The western part of the valley, where sinking is at a maximum, is underlain by hundreds of meters of unconsolidated sediment derived from the Coast Ranges and the Sierra Nevada in the last 3 million years. About 600,000 years ago, a great lake spread over this region, and the clays deposited in it serve as a useful level marker. This Corcoran clay lies at depths of 60–250 meters. Thus, the average rate of subsidence has been roughly 0.03 centimeters per year.

There is no dispute, at present, that pumping causes sinking. Depending on the area, a decline of 3–8 meters in the level of ground water causes a subsidence of a third of a meter. Indeed, the correlation improves as more sensitive instruments are used for measurement. There is dispute among scientists about when the subsidence was recognized, in part because the relationship became evident at different times in various places. Ben Lofgren of the U.S. Geological Survey states that subsidence was "first recognized in the valley in 1935" and correlated with pumping. Certainly, sinking is evident on graphs of changes in level south of Tulare and west of Fresno. At the latter site, one bench mark sank about 2 meters between 1940 and 1955.

A lack of communication among government agencies is suggested by the fact that, after the Bureau of Reclamation built the very large Delta–Mendota Canal through the area in 1951, Nikola Prokopovich of that agency said that the "existence of subsidence in the area was not known during preconstruction and construction studies." As a consequence of this lack of information and, thus, of a lack of adequate planning, 55–65 kilometers of the canal had sunk more than 2 meters by 1966, and extensive damage resulted.

The situation was rectified by the time the San Luis Canal was constructed in the years 1963–1968. Subsidence was mapped, predicted, and taken into account as much as possible. Parts of the area had subsided as much as 6 meters when construction began, and some of this was not due to simple compaction by loading but to compaction caused by rearrangement of grains. Some sediments in the area are very loosely packed, and it is surmised that they may have been above ground-water level since they were deposited. When the ground was flooded in an attempt to replace the water lost to pumping, these sediments essentially collapsed, and sinking of about 4–5 meters occurred along the site of the Delta-Mendota Canal. To avoid this, the bed of the San Luis Canal was flooded in sections and up to 2.5 meters of settling occurred. The ground was regraded after subsidence and before the concrete liner was emplaced.

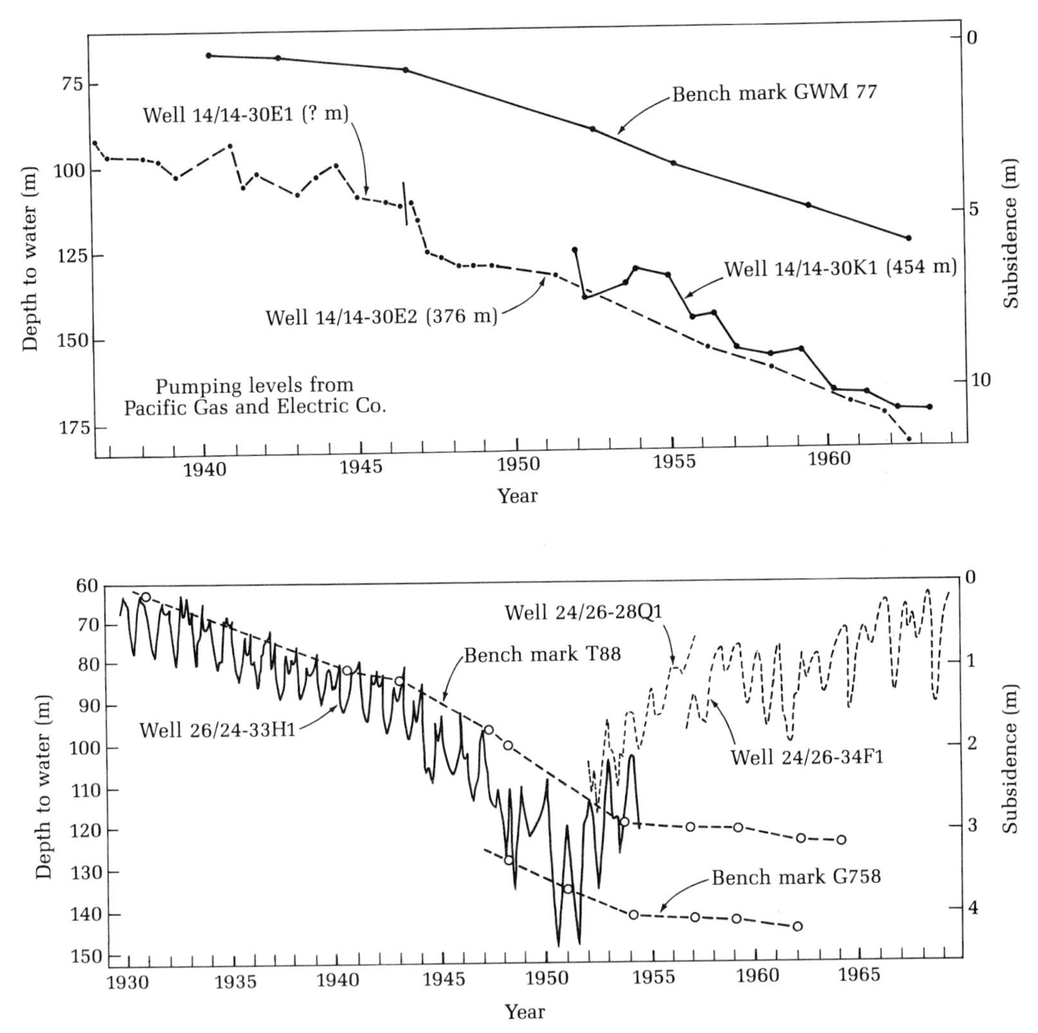

FIGURE 14.4
Correlation of water-level fluctuations and subsidence near centers of subsidence west of Fresno, California (*above*), and south of Tulare, California (*below*). The water level fluctuated but gradually sank, and the ground south of Tulare sank about one-third meter for each 12-meter drop in the water. In the early 1950s, measures were taken to raise the water level and keep it at about 70 meters. The ground continued to sink, but at the much slower rate of about 2.5 centimeters per year. [After Lofgren (1970), by permission of UNESCO.]

The Bureau of Reclamation has observed that the rate of subsidence due to pumping water decreases rapidly after withdrawal ceases and hydrocompaction is completed. Thus, it is able to predict that only 0.3–0.6 meters of further sinking will occur, and that it will be virtually complete in 20–25 years. However, the region has been sinking for at least 600,000 years, and stabilization of water level near Tulare and in Tokyo has not yet stopped subsidence in those places.

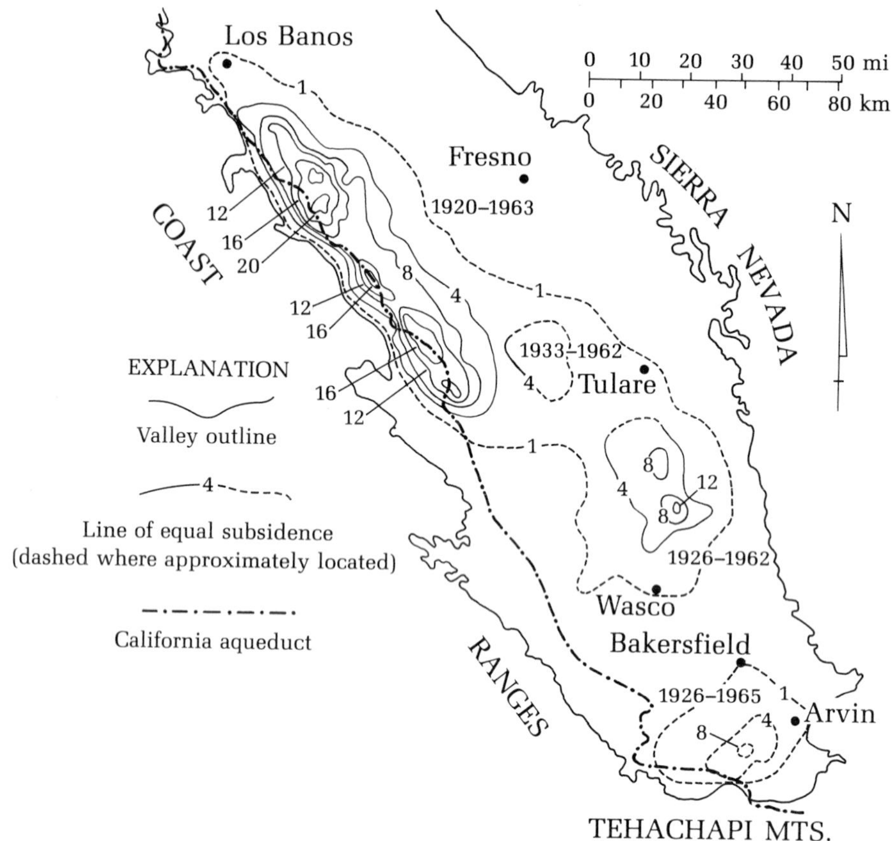

FIGURE 14.5

Extensive areas in the San Joaquin Valley have subsided because of excessive pumping of ground water and, to a lesser extent, other causes. Sinking continues at a slow rate and threatens the stability of the California aqueduct, which was built through the area. The numerals indicate subsidence in meters during the time intervals indicated. [After Lofgren (1970), by permission of UNESCO.]

SINKING OIL FIELDS

Presumably all oil fields sink as fluids are removed, but this affects few structures except those owned by beneficiaries of the removal, and they are prepared to bear the cost. However, both the physical and the social effects of this phenomenon are spectacular at Wilmington oil field in California, because it is a highly productive field at the shoreline and subject to flooding, and it is the site of a large power plant, a Navy shipyard, and the Port of Long Beach. It will serve well to illustrate what happens in similar situations elsewhere and what can be done about it.

The Wilmington field was discovered in 1936, and seven major producing zones were opened by drilling into late Tertiary sands and shales. Minor subsidence was noted but ignored prior to 1939, when major production began. By 1945, the center of the field had sunk 1.2 meters and both the rate and the area of sinking increased in the following years. The ground was only a few meters above sea level to begin with, and eventually the tides began to flood the area by backing up through the drains. The area was diked at the

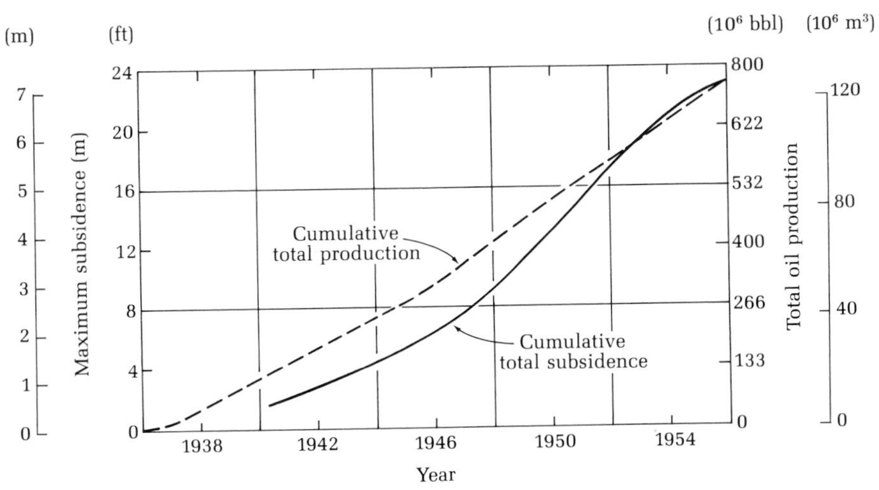

FIGURE 14.6
Relation of oil production to subsidence at Wilmington Oil Field in California. Note that subsidence is inverted to show the approximately parallel trends. [From Mayuga and Allen (1970), reproduced by permission of UNESCO.]

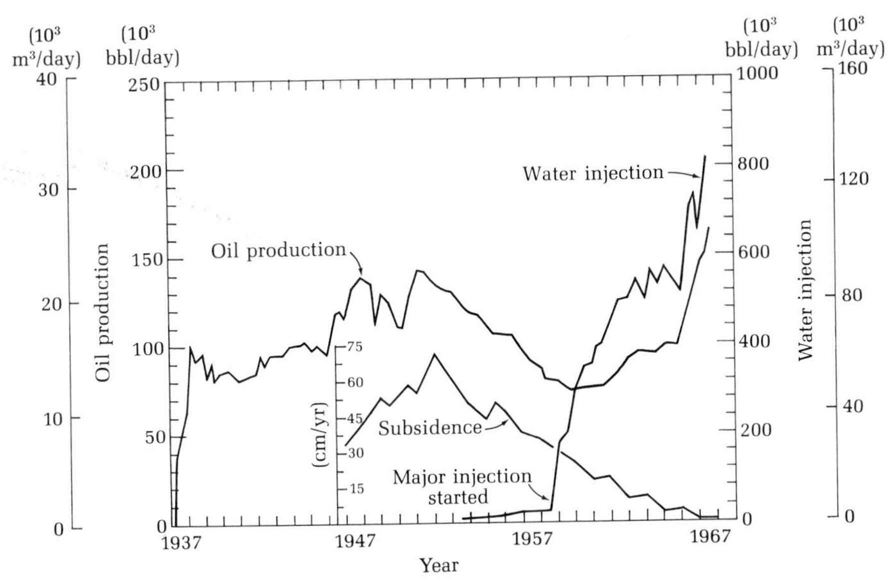

FIGURE 14.7
History of *rates* of oil production, subsidence, and water injection to balance subsidence of Wilmington Oil Field. [From Mayuga and Allen (1970), reproduced by permission of UNESCO.]

margins and filled in the middle, and structures were elevated to the new level. The center of the field sank at a maximum rate of 74 centimeters per year in 1952 to a total of 9 meters by 1968. The vertical motion was accompanied by horizontal shifting and by the opening of tension cracks—a common phenomenon elsewhere as well.

The rate of oil production was reduced until, by 1960, it was less than half of its peak in 1952. The maximum rate of subsidence dropped to only a quarter during this period, so the procedure for alleviation was successful. The oil output was down, however, and the ground was still sinking. A program of injecting water to balance the oil removed began in 1953. By 1957, the City of Long Beach had spent $90 million to repair damage caused by subsidence, and, in 1958, the rate of injection was accelerated. Shortly thereafter, the removal of oil began to increase, but the rate of subsidence continued to decrease and it is now only a fraction of a centimeter per year. Thus, it appears that the technology to minimize subsidence caused by pumping is available. This does not mean that oil fields on thick young sediment, particularly on deltas, will not continue to subside, but perhaps the subsidence of such an area will be no faster than that of its surroundings.

ORGANIC SOILS

Organic soils form near water level in bogs, marshes, and swamps. They are doubly valuable: they are fertile, and therefore desirable for farming, and they are at the shoreline, and therefore desirable for housing and industry. It is relatively easy to dig canals in organic soils and, thereby, to drain them and open them for use. This has occurred in many countries—notably England, the Netherlands, Russia, and the United States—and is increasingly common. These soils are quite peculiar, and serious social consequences often accompany their development.

The Everglades of southern Florida is a well-documented example, although it is difficult to follow historically because of the repeated and massive human intervention related to flood control, assurance of water supplies, farming, and road and airport construction. The organic soils of the region were formed in a broad, shallow, limestone trough leading from Lake Okeechobee south to the sea. The lake usually overflowed during the rainy season in a "river" 160 kilometers long, 65 kilometers wide and 0.3–0.6 meters deep. Soils gradually developed by anaerobic decay of the soggy vegetation. In 1906, artificial drainage was introduced to permit the development of the excellent soil for farming, and, by 1926, a large area south of Lake Okeechobee was in use.

In 1926, a hurricane caused extensive damage to the area and killed 300 people. Only two years later, another hurricane hit Lake Okeechobee like

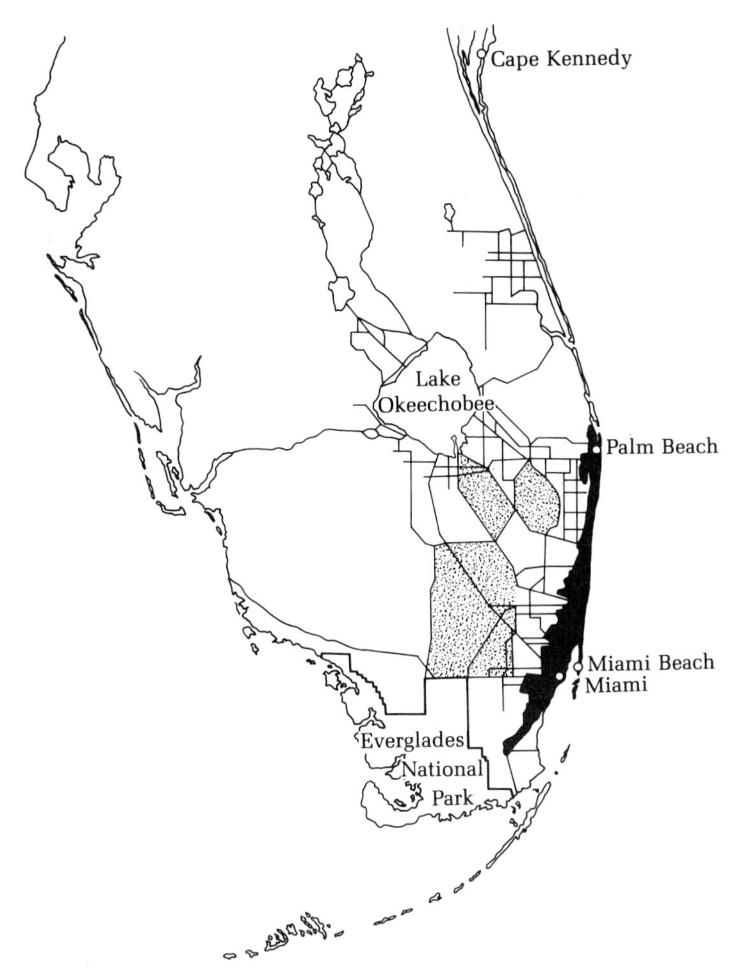

FIGURE 14.8
An intricate system of canals and water controls has been constructed between Lake Okeechobee and the Everglades National Park. The urban expansion (*black*) has increased demands for the same water that maintains the park. Organic soils between canals are sinking. Shaded areas are water control districts.

a giant hand and splashed it to the south over the farmland. More than 2000 people died in one of the worst natural disasters in American history. The Army Corps of Engineers, directed to prevent future floods, began to build a dike around the lake and a system of dikes and canals to speed floods to the sea.

Congress authorized a study of the Everglades in 1928, and, in 1934, established the Everglades National Park at the southern end of the region. Miami and Miami Beach expanded rapidly, and, when the east coast of Florida was flooded in 1947–1948, the Corps of Engineers was directed to take even more extensive measures to control floods. Water-conservation districts were established as open land in which to store water from Lake Okeechobee in the hurricane season but also to supply ground water for the growing needs of Miami.

The national park was on the end of the flow, but no serious problems arose until one of those minor, but inevitable, climatic fluctuations occurred. From 1961 through 1965, south Florida suffered a severe drought. The floodgates on the canals were closed to protect the Miami water supply, and the park began a period of extreme hardship. This is a classic example of the population expanding to the temporary limit of a fluctuating water supply. The problem would have worsened had a proposed Florida jetport been constructed just north of the park, because a city gradually develops around any major airport.

So, step by step, the environment of the region has changed. We return to the subject of organic soils to see other consequences of the logical program to control floods. The organic soils drained by early canals in 1914 began to shrink as they dried, and the surface of the ground dropped at the rate of 10 centimeters per year. When the total drop was about 1 meter at one site, the surface of the ground approached the water table and shrinkage slowed. However, the soil was of much less use for agriculture, so pumps were installed in the fields to keep the water down. Another site, at a different distance from a canal, sank a total of 1.5 meters before pumps were required. Sinking has continued at both sites and now totals nearly 2.5 meters.

The trends have been extrapolated by Stephens and Speir. In 1970, some of the fields were almost as low as the canals, and the danger of flooding was greatly increased. By the year 2000, both the farms and the canals will have dropped about 0.6–1.0 meter and the organic soil will be only a third of a meter thick on the average. Some 88% of the original soil will be gone. Thus, farming in its present style has a foreseeable end, and the crops, vegetables for the cities of the east coast, will have to come from somewhere else.

In cooler climates, organic soils may drop at the slower rate of 2.5 centimeters per year while they are drying. Later, the sinking continues as the once anaerobic soils are oxidized until a large proportion of the soil above the water table vanishes. This will be of great importance in Byelorussia in the Soviet Union, where enormous areas of organic soil are to be reclaimed by 1975. Sinking must be expected wherever organic soils are drained for farming or construction.

Collapse Structures

Features like the great, circular, vertical-sided pits that vulcanologists call calderas occur on a much smaller scale in surface materials in many environments when vertical tension is induced by withdrawal of support. This may occur because of tunneling, ditching, or mining, as well as natural undermining and cavern collapse. The social consequences vary with the intensity, frequency, and location of the pits relative to human activities.

In general, the greatest social consequences result from digging in or under cities, because some structure is almost certain to be damaged or destroyed. Tunneling for highways and rapid transit systems has been widespread and probably will become more so, but it is absolutely essential that such tunnels be self-supporting. They are designed not to collapse and they do not, so no pits form above them. This is enough to show that collapse pits over other diggings are the result of economic rather than technological constraints.

The most widespread (and very likely the most annoying) type of surface collapse results from poor consolidation of foundation materials. Filled land that is not consolidated to maximum density is apt to subside, producing minor cracks and putting doors and windows out of plumb. On thick fill, damage may be more extensive. The common practice of ditching streets to bury additional utility lines and pipes is rarely followed by careful consolidation of the dirt that refills the trench before the road surface is restored. Consequently, shallow creases form as the road surface sinks into the gap left by consolidation. If water enters such refilled trenches, the dirt may be repacked into a solid mass and drop several centimeters. Above such gaps, the asphalt may crack and drop down to form small circular pits. These forms of collapse could be prevented by proper application of the well-known laws and practices of soil mechanics.

Mining differs from subway tunneling mainly in the fact that there has been little incentive to build mines in such a way that they will stay open after the ore is removed. As a consequence, mining districts all over the world are marked by collapse pits and subsidence zones. Several phenomena short of outright pit formation are known to occur. The ground sags in the center, the surrounding ground tilts and is displaced toward the center, and cracks form. When the cracks form a circle, a pit may develop gradually, but it may also drop almost without warning. Collapse is not uncommon in Pennsylvania coal districts; in 1963, at Wilkes-Barre, one occurred so suddenly under Route 309 that a car was trapped, and the driver barely scrambled out before it disappeared down a pit 9 meters in diameter. The State of Pennsylvania passed *The Bituminous Mine Subsidence and Land Conservation Act of 1966* to regulate the mining practices that caused surface collapse. More coal will be left in the mines to support the ground above, although back filling with mine tailings serves the same purpose. Unfortunately, this does not control fires underground, and some fires cause extensive collapse as the coal is consumed. Measures for control are being developed.

Natural processes also make tunnels and cause collapse at the surface. This occurs everywhere that a hard layer lies over one that is easy to erode.

BOX 14.2 QUICKSAND

Quicksand is a deposit of sand that is supersaturated with water that is flowing upward and, thus, exerting a pressure from below that separates the grains and makes the sand deposit swell or increase in volume. When the upward pressure is equal to the weight of the sand, the grains are suspended and frictionless and will not support a load. However, the surface of the sand still looks solid, and an unwary walker may be trapped and sink beneath the ground.

Gerard Matthes observed in 1953 that "quicksand" was seldom heard of anymore because people did little walking, but that "the malevolent phenomenon figured prominently in the landscapes of 19th-century romances (where it was frequently a convenience to authors needing to dispose of unwanted characters) and in the pioneering of the U.S. West." Now that walking is more popular, it may be helpful to repeat some of Matthes' lore.

Quicksand can form in any type of sand but only quite locally over springs, which may be under water, near river banks, or wherever a small amount of water emerges under pressure. It cannot exist in large areas or under high pressure because the sand would be washed away. Quicksand is uncommon in mountains and plains except near rivers. It is more common in hilly ground, especially in karst topography where springs are abundant.

Quicksand cannot be identified by eye, but a light pole or stick—longer than a cane—can be used to probe in front of a walker. The bed of a swiftly running stream is safe because the sand would be eroded away if the grains were lifted by a spring. A quicksand under a pond more than a few feet deep is no problem, because the water itself is a support and the walker will not sink into the sand more than a few inches.

What if a walker is suddenly mired? He should immediately lie on his back and stretch out his arms, just as though he were trying to float on water. Quicksand does not suck downward; it is, in effect, a dense fluid and, consequently, is easier to float on than water. If a walker has companions, they can build a broad, lightweight platform—a raft, as it were—and use this as a base to pull him out. If alone and too far away to call for help, his problem is greater. If he has the long pole, he should slowly work it down from his shoulders until his hips rest upon it. Then his feet can be lifted out and he can *roll* back to solid ground. Without the pole, he can only hope that, when he sinks slightly until he is floating, he may be able to raise his feet until he is entirely on the surface and can roll away.

Fortunately, quicksands are small and the stroller will begin to sink as soon as he steps on one. If he merely walks indolently instead of energetically, he will be on solid ground when he falls back.

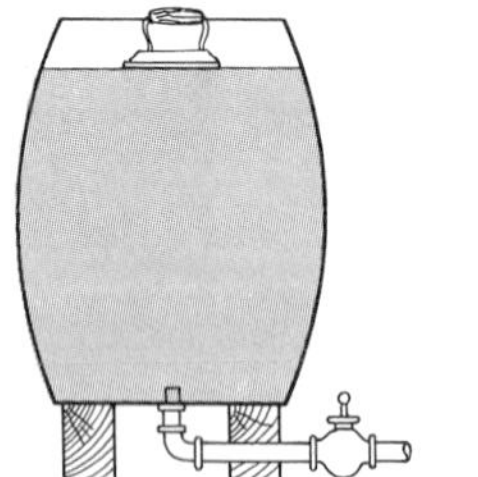
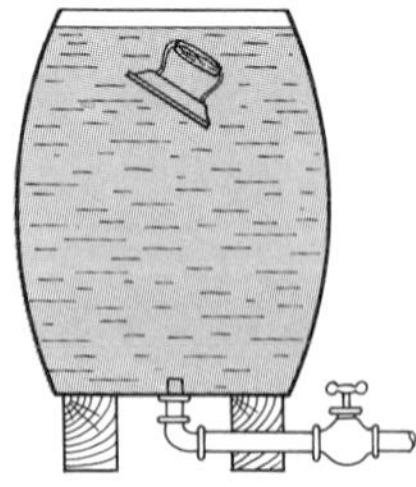
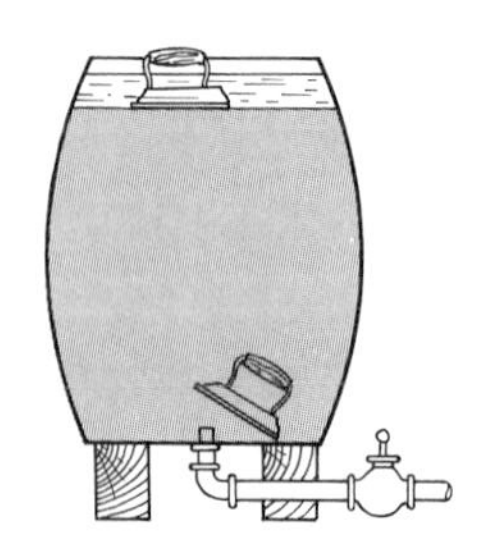

An experiment showing how ordinary sand becomes quicksand: A barrel full of sand is fitted at the bottom with a water inlet. When the water is off, the flatiron rests securely atop the sand (*left*). When the water is turned on and a weak pressure maintained, the sand swells and its grains separate; the flatiron sinks (*center*). When the water is turned off again, the sand settles and squeezes the water to the surface; though it is wet and slightly larger in volume than before, it readily supports a second flatiron (*right*). [From Matthes, "Quicksand," copyright © 1953 by Scientific American, Inc. (all rights reserved).]

The naturally cemented soils of the semiarid California coast lie on a relatively weak subsoil, and both waves and streams undercut them deeply. Collapse under tension follows.

Limestone is particularly subject to solution by acid-bearing ground water, and most of the great caverns of the world form in this way. A topography pitted with sinkholes is characteristic of karst regions, which are named after the type example in the Karst district in Yugoslavia. Karst regions occur widely in the United States. In Florida, houses and highways have been known to drop into newly formed sinkholes as large as 160 square meters in area. In Kentucky, a sinkhole formed under a cement truck as it was pouring a foundation for a house. This illustrates that loading of the thin surface rock may trigger collapse. Caverns that may collapse can be detected in karst regions by borings or various geophysical techniques, but the expense of the survey must be borne before construction, not after. An extreme case of collapse has occurred in the Far West Rand district of South Africa, which is the source of 20% of the world's gold. The mines lie under a thick layer of water-saturated limy rock and are subject to flooding. Consequently, the water has been pumped out, and all the caverns and subsurface sinkholes have been weakened. Great collapse pits have opened under the mine workings and the town above. Both borings and geophysical surveys have been used to identify areas where the risk of future collapse is great, and future construction will be planned to avoid them.

Loading

We have seen that the surface of the ground sinks if fluids or solids are withdrawn from below; it also sinks if loaded from above. This is by no means a widespread phenomenon, because sediment has a high bearing strength if it is sandy and the grains are in contact. Some foundations driven into soft sediment have supported large buildings for more than a thousand years. Thus, sinking under loading is confined mainly to unusual sedimentary environments—namely, active deltas and filled lake basins.

Deltas are among the most favorable locations for the development of civilizations because they are extremely fertile, have access to both fresh water for farming and marine fish for protein, and are at the junction of trade routes from the sea to the interior. The problem with deltas is that they sink. Farmers deal with this automatically by piling annual layers of sediment on their land to increase soil fertility. In the delta of the Hwang Ho, for example, some of the land is just above the rivers, but pottery and other human artifacts uncovered in the area show that the original surface that was farmed is now 5 meters below. Thus, the problem is really confined to houses, roads, and cities, which cannot be elevated a fraction of a centimeter per year.

Deltaic subsidence appears to be the consequence of several factors. Large deltas form where large rivers dump sediment in basins that sink. Thus, the bottom of the delta goes down—partially in response to the load of the delta and perhaps also because of vertical motions of crustal plates. In addition, each new layer compresses the ones below, and this causes the top surface to sink and make a space for deposition of the next layer.

The rate and amount of sinking are easy to document because, among other attractive characteristics, deltas are commonly oil-bearing and, thus, have been drilled and analyzed. The Mississippi delta at the Louisiana shoreline is underlain by 6000 meters of sediment deposited during the last 20 million years. Lagoonal and beach facies occur throughout the section, and so it has sunk at an average rate of 0.03 meters per century. Not a very disturbing rate, but it is only an average.

The surface of the delta 20,000 years ago, when sea level began to rise, can be identified underground (Figure 14.9). It slopes toward a point offshore from the delta. A coastal highway constructed at that time would now be buried, in places, under 150 meters of delta mud and 110 meters of water. However, the amount of sinking varied with local deposition. At New Orleans, about 65 kilometers inland, the sinking has been only 0.073 meters per century.

The delta of the Po River in Italy is of exceptional interest because it has been the site of cities for 2000 years and, thus, the effects of geologic factors have had time to become evident. The shoreline has advanced an average of about 8 kilometers in the last 3000 years, and the Roman ports of the first century A.D., identified by Pliny the Elder, are well inland (Figure 14.10). Ferrara also has ceased to be a port, but for a different reason—namely, that the Po could not be prevented from shifting its channel in 1150 A.D.

The channel shifted from the Po di Ferrara to the Po di Venezia, and the approaches to the island republic of Venice were threatened with silt. The Venetians dug a diversionary channel in 1604 to solve this problem. However, they have had other problems, and some of them now threaten the existence of the city. Venice was founded in 451 A.D. on an island in a lagoon—a wretched site, but a good one to withstand barbarian marauders when the Roman Empire fell. The city was never assaulted, although it finally capitulated to Napoleon after almost 1400 years of unparalleled prosperity and grandeur. It is only at the periphery of a delta, but it illustrates what can happen as sinking continues.

The river mouth is sinking at the rate of 13 centimeters per year, or 13 meters per century. It is no place to build anything. However, the load of the newly deposited sediment is localized by the elaborate dikes built to control the river. Only 10–13 kilometers to the north or to the south of the river mouth the sinking decreases to only 8 centimeters per year, or 7.6 meters per century, but these are still not good sites for a city. Halfway to Venice, the rate of sinking is 2.5 meters per century; at Venice, however, it is only 20 centi-

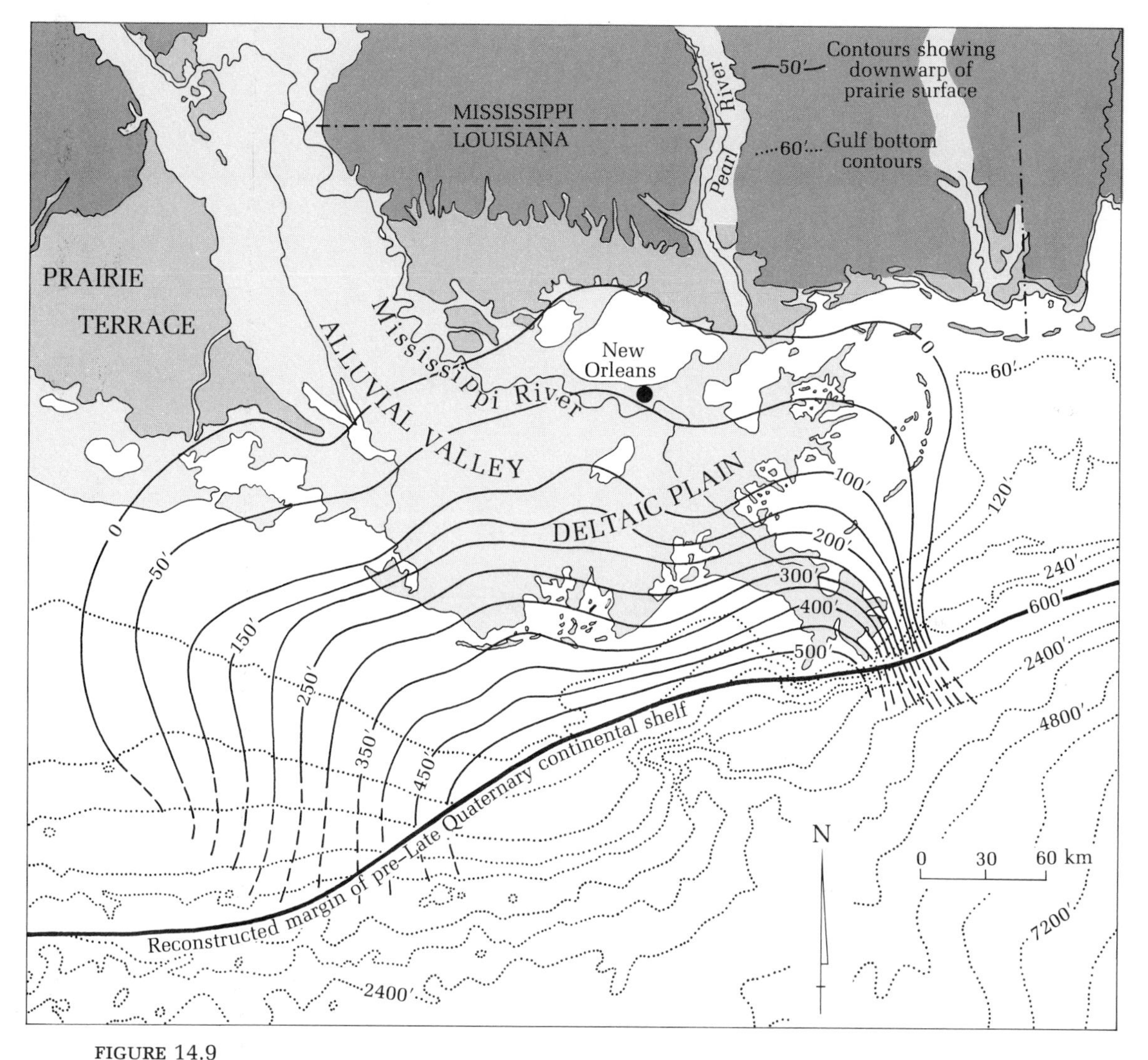

FIGURE 14.9
Subsidence (in feet) of the "Prairie surface," the ground level of 20,000 years ago, under the load of self-compacting sediment in the Mississippi delta. Data are corrected for changes in sea level. [After Gould (1970).]

meters per century, and a city can endure that much for a while. Part of the apparent modern sinking may be the consequence of a minor fluctuation of sea level as well as subsidence. This is an unresolved uncertainty for minor changes in level. One way or another, the city has dropped relative to sea level. At 20 centimeters per century, the drop is 3 meters since Venice was founded on a mud bank that was a small but unknown amount above sea level. As a consequence, the city has been flooded by storm surges more and

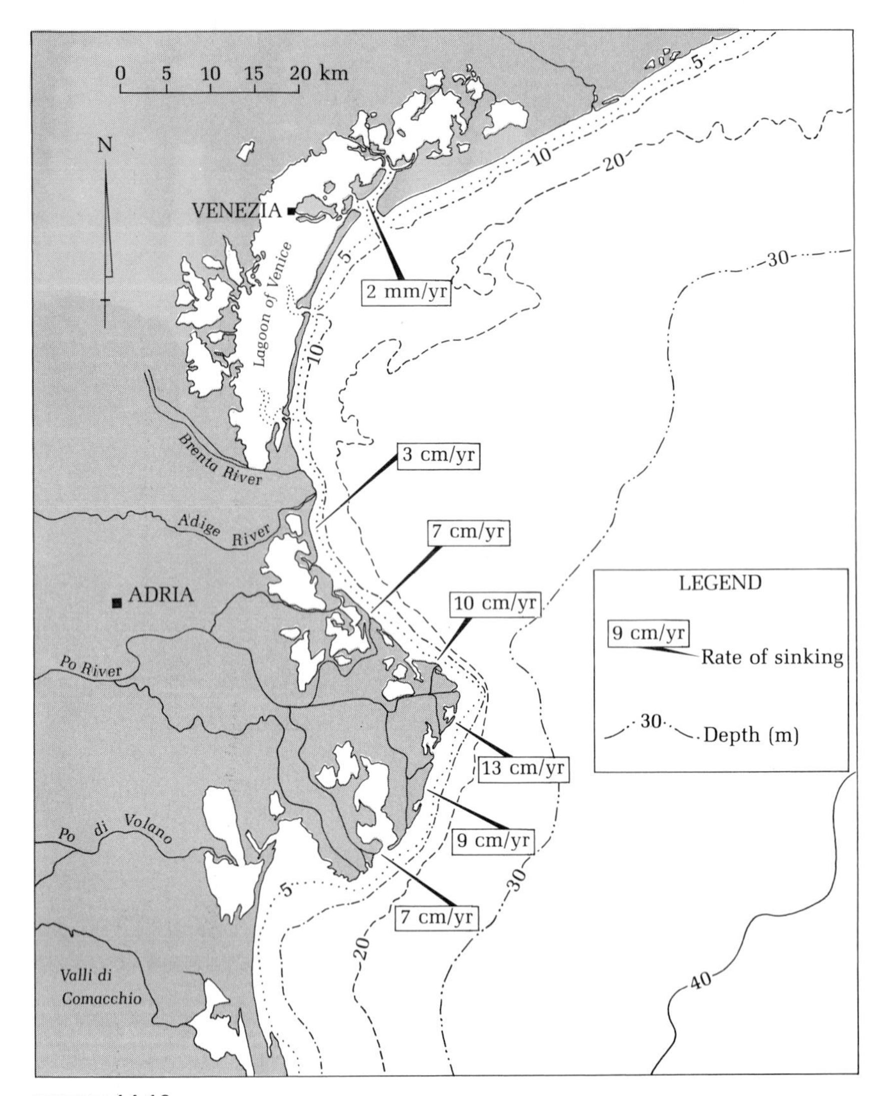

FIGURE 14.10

Sinking along the Adriatic coast is closely correlated with loading caused by deposition at the mouth of the Po River. [After Nelson (1970).]

more frequently, as the centuries have passed, until now the central Piazza San Marco is awash several times a year. The Venetians have not been passive. Walter Munk of Scripps Institution of Oceanography has established that the level of the Piazza has been raised several times and that the pillars of the facade of the Palace of the Doges are partially buried as a result. John Ruskin, in his book *St. Mark's Rest,* had remarked upon the brilliant functional design of the pillars. They are cylindrical rather than widening at the base, and he noted that this increased the ease of passage between them. It seems that the pillars do taper, like most others, to give them strength, but the tapered part is buried.

So rich city dwellers make essentially the same response to delta subsidence as poor farmers when they pile one layer of roads and plazas on another. The Venetian problem is buildings that cannot be extended upward. The problem is even more acute for two reasons—pumping of ground water at a nearby industrial port is accelerating subsidence, and some of the most important buildings are so heavy that they are intense local loads in themselves. The twelfth-century mosaic floor of St. Mark's, the cathedral of Venice, has been rebuilt, but it is still undulating because of differential subsidence, and it curves down into moats around the great pillars. Fortunately, the Italian government has embarked on a program to shield Venice from the sea and to protect these treasures—even though sinking will continue, just as it will in Tokyo and other cities built on deltas.

Mexico City is more than a kilometer and a half high, and so it escapes the problem that Venice has with the sea, but its foundations are even more troublesome. It rests on an ancient lake bed, where the Aztecs built a city that, like Venice, was relatively immune to assault. The population, now 6,500,000, is growing extremely rapidly, and the demands for ground water are correspondingly acute. The whole area is sinking and roads and underground utilities are damaged. Moreover, the clays compact easily, and older buildings are as much as half a floor below the adjacent ground level. The Palacio de Bellas Artes was not built until 1904, but it is so massive, like St. Mark's, that it has sunk 3 meters since then and rests in a definite dimple in the asphalted surface. It was still sinking at the rate of about 3.8 centimeters per year in 1962, or almost as fast as the average rate since it was built. The Mexican government has now outlawed pumping under the center of the city, which is a major step in the right direction. However, the ban on pumping will not control differential subsidence from loading. Moreover, with the population growing—or even if it is stabilized—it is difficult to see how general subsidence of the area can be stopped.

Permafrost and Pipelines

Near Prudhoe Bay, on the Arctic coast of Alaska, there are several gigantic, although undeveloped, oil fields that hold a confirmed 10 billion barrels of oil and 730 billion cubic meters of gas. The *estimated* reserves are about twice as great, and the total resource probably has a value of more than $100 billion.

The central problem in developing this colossal resource is to bring the oil and gas out of the Arctic to the consumers of the United States, Canada, and, possibly, Japan. There are three major possibilities: tankers breaking through the thick sea ice, a pipeline through Canada to connect into existing pipelines, or a pipeline through Alaska to tankers at the port of Valdez. The tanker route has been tested in a dramatic experiment, but it appears risky. Most everyone seems agreed that the gas should go up the Mackenzie River valley through Canada. The proposal for an oil pipeline through Alaska, however, has roused an intense and prolonged controversy, both because of the possibility of degrading the environment during construction and because of the possibility of oil spills afterward.

Both the socioeconomic and the environmental challenges are extraordinary. Robert Zelnick has called the development of this oil resource the largest undertaking in the history of private enterprise. In September of 1969, oil companies paid $900 million to the state of Alaska for permits to explore for oil on the north slope. The companies have spent an estimated $100 million to survey the proposed pipeline and to purchase a quantity of 122-centimeter pipe for the 1270 kilometers of the Alaska route. Much of this pipe is stacked and waiting for Federal authorization to start building. The cost of the pipeline is uncertain, but the three proposed gas pipelines through Canada have an estimated cost of $2.0–2.5 billion.

As to the environment, the Alaska pipeline would cross hundreds of kilometers of frozen ground, 350 streams, and several ranges of mountains, and would approach the seismically active boundary between the America and Pacific plates. The oil in the line will be hot as it comes out of the ground, and it will flow the whole distance at about 63°C. The oil needs to be transported hot, because the viscosity increases at lower temperatures and makes it too difficult to pump. The problems attendant upon putting hot oil and frozen ground together are suggested by the words—rather dramatic words, for a committee report—of the Environmental Impact Statement prepared for the Department of the Interior (Morton, 1972, p. 93):

> A large hot oil pipeline buried for hundreds of miles in permafrost has no precedent. Unique effects, unfamiliar in pipeline experience, would arise primarily from the loss of strength and change in volume of ice-rich soil when it is thawed by heat from the pipe. Unless these processes are anticipated, the consequences could affect the mechanical integrity of the pipe and the stability of the adjacent terrain.

Environmental problems of unprecedented scale will also attend the actual construction of the pipeline, which will last for several years. A road will parallel the pipeline, and several airfields and a communication system will be required for maintenance. The pipe will be buried, where possible, to

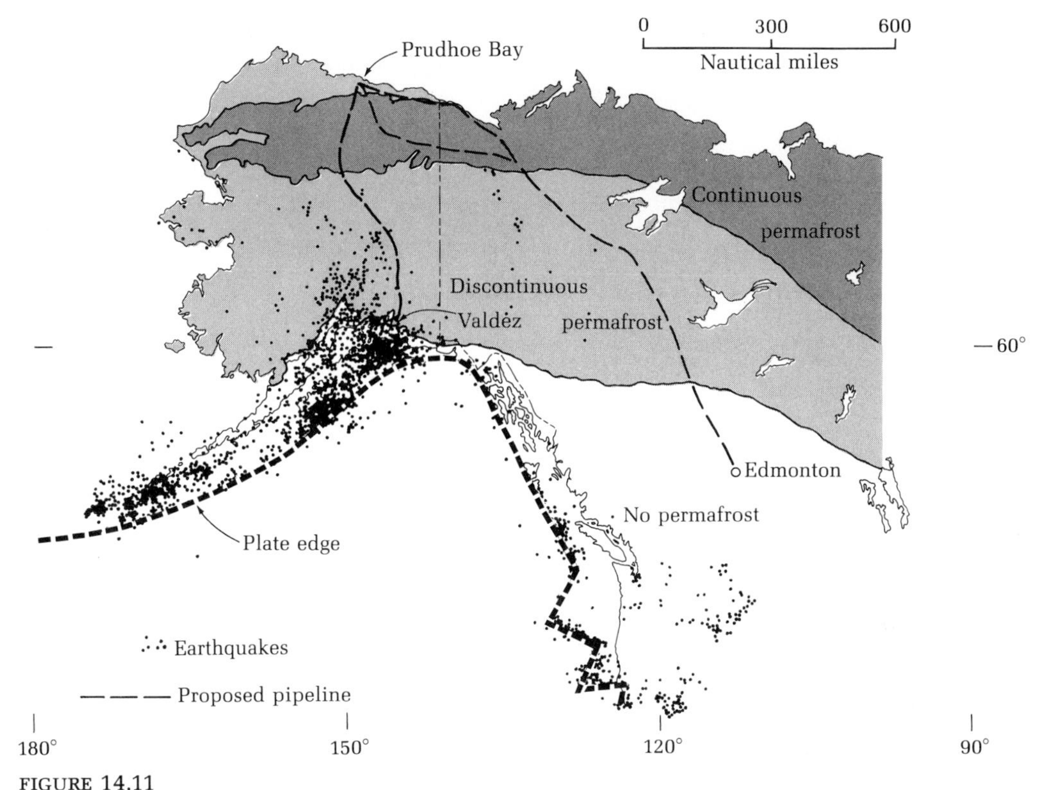

FIGURE 14.11

Proposed transportation systems for oil and gas from the north slope of Alaska, and the environmental hazards that confront them.

conserve the heat and minimize sabotage. Experiments in Canada suggest that burial is feasible in frozen ground but these experiments have been too short-term to show what may ultimately happen. The buried pipe will be insulated by a bed of gravel up to 1 meter thick. The construction of the pipeline and roads will displace 52 million cubic meters of material, and another 12 million cubic meters will be moved to build the oil field. Construction of the gas pipelines may involve similar volumes of sediment. On this score, the Environmental Impact Statement says (Morton, 1972, p. 95): "The excavation of construction materials in areas of ice-rich permafrost would cause the permafrost to thaw. The materials could become unstable and flow or slide, especially on slopes."

PERMAFROST AND PINGOS

The ground is frozen to a depth of tens to hundreds of meters in northern Alaska. An "active layer" a few centimeters to a meter thick thaws in the summer and refreezes in the fall. Permafrost is forming in some places in Alaska and thawing in others, and its distribution varies with the particle size and permeability of soils, the extent of insulating ground cover, and other effects, as well as temperature. Some frozen ground merely has ice in the pores between mineral grains that are in contact, and thawing and freezing cause only minor vertical movement.

Volume and surface changes may become much larger over buried masses of clear ice. Segregated ice occurs in layers within fine-grained sediment (sediment with a maximum particle size of less than 0.01 millimeter) when water migrates into a zone in which freezing has begun. The water is derived largely from ground water below the zone of ice formation, because the permafrost above is an effective seal against surface water percolating downward. The water is drawn in by surface tension, and this is most effective in the small spaces between fine grains. Ice lenses parallel to the ground reach a thickness of 4 meters in the silts of Alaska. Segregated ice generally causes a broad, slow heaving of the surface, but may arch it into one of the hundreds of large mounds called pingos that rise from the Arctic plains. The effect of melting is rapid collapse for the thickness of the lens of ice. This occurs naturally and generates a landscape of pits and crevasses, which resembles karst topography in limestone.

Vertical ice veins and wedges form in thick organic-rich silts. Such materials contract when they freeze, and the ground cracks with such speed that it may be audible. The cracks fill with water, which later freezes to form a thin vein. Then, year after year, like the spreading sea floor, the crack continues to open and be filled with widening ice. Cracks may open as much as 10 centimeters in a year. Ice wedges may become extinct, be buried, and cause ground collapse when they melt. Some are more than 9 meters wide, and they form in polygonal patterns that affect enormous areas of the north slope.

The phenomena discussed above occur whether the ground is flat or sloping, but, if it is sloping, there are additional effects, because the active layer moves readily over the permafrost below. This occurs in two modes that may interact. When fine-grained sediment thaws, it retains moisture and loses much of its strength, so that it flows slowly but easily. Coarse material is more affected by the combined action of freezing and thawing. Freezing of pore water causes a radial expansion of the sediment, with part of it moving upslope and part downslope. Melting, however, is accompanied by a vertical drop that is all downslope. Thus, sand and gravel move gradually downslope. The ground cover usually prevents or inhibits downslope motion on gentle slopes, but it is easily initiated if construction has removed the cover. The movement generally amounts to only a few centimeters per year on slopes of 15–20°, but, on steeper slopes, it may trigger much faster phenomena, such as landslides.

These are some of the geologic perils that will confront the oil and gas pipelines. Considering the sums involved, an elaborate survey for buried ice and other hazards seems likely, and it could prevent much of the possible damage. Experiments in a loop of 122-centimeter pipe 609 meters long on the frozen ground of the Mackenzie River valley suggest that hot oil can be insulated from the ground. Moreover, the Russians are already operating an oil pipeline across frozen ground. Thus, the range of technological responses to the environmental challenge is just beginning to be explored. However, the Arctic is little known and variable, and it may be that we are only beginning to understand the environmental challenges that may arise during the decades that the pipelines will operate.

Mass Movement Downslope

We have discussed many phenomena that cause surface materials to lose strength, such as earthquake shocks, the less dramatic upward flow of water in quicksands, and freezing and thawing. Gravity acts on such materials, and, if they are weak enough, or if the slope is steep enough, they will move downslope in a more or less fragmented mass. The importance of such motion is illustrated by the fact that young valleys generally have gentle side slopes but the rivers that run through them cut mainly downward. The great width of the Grand Canyon is largely a result of mass movement, rather than cutting by the Colorado River.

Mass movement occurs in a wide variety of modes, scales, and intensities, all of which may affect society. Even the very slow sheet motion called creep eventually will topple walls and gravestones on sloping ground. Likewise, slowly moving, coherent masses of soil (or debris streams) moving by freezing and thawing can affect upland farms and roads, but they are not difficult to control if control is necessary. Thus, it is the rapid mass movements, plus the slow ones directly associated with construction, that are of the most social concern. These occur generally on steep slopes of high mountains and canyons, at the waters edge, and wherever man creates steep slopes and saturates the ground.

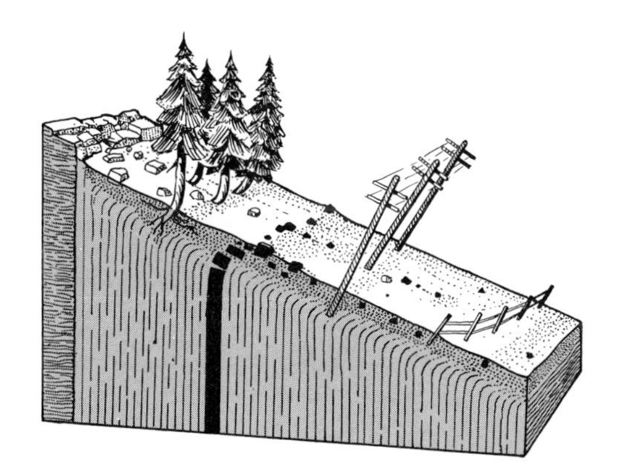

FIGURE 14.12

Common effects of creep. [From Gilluly, Waters, and Woodford, *Principles of Geology*, W. H. Freeman and Company, copyright © 1968.]

EXTREME WEAKNESS

Quick-clay is a material that can be so weakened by minor shocks that it will flow with hardly any slope at all. The undisturbed clay apparently has an open structure, with the platy clay crystals acting like a house of cards. A shock collapses the structure, and the material becomes essentially liquid until the water is squeezed out. Slides of quick-clay occur mainly in Norway, Sweden, and Canada in regions where glacial rock flour was deposited gently in the sea. The salt of the sea usually helps stabilize and strengthen clays, but, after crustal rebound, these areas were uplifted, the ground was leached of salt, and the quick-clay formed.

The slides are fast and can be quite damaging, especially in those regions where steep topography concentrates people in the flat-bottomed valleys in which quick-clay occurs. Much of the town of Surte on the Göta River in Sweden was destroyed by a slide in 1950. Apparently, it was triggered by a pile driver in use for new construction. Suddenly, a huge mass, 2.96 million cubic meters of soil and gravel bearing 31 houses, began to move; within 3 minutes, it plunged into the river and almost blocked it. Although the slide was more than 600 meters long, individual houses and a highway moved only a fraction of that distance. Only one person died, but other such slides have killed more than a hundred.

EXTREME SLOPES

Hard rock will stand on vertical slopes long enough to symbolize solidity in the trademark of an insurance company, but eventually it will fall in small or large pieces. The results can be truly spectacular. In 1958, an earthquake in southern Alaska dislodged a mass of rock weighing 90 million tons, which fell into Lituya Bay from a maximum height of 900 meters. It made a gigantic splash that scoured away a mature forest and left bare rock as high as 490 meters. An analysis of tree ages and distributions around the bay shows that this has happened there before.

The residents of Elm, Switzerland, were victims of an extraordinary fall in 1881. A quarry for roofing slate had undercut a mountain slope near the

---

FIGURE 14.13

The effects of quick clay: A great slide took place in 1950 at Surte on the Göta River in Sweden. Cross sections along broken line *A–A* in the plan view (*bottom*) show low slope and other conditions before the slide (*top*) and after the slide (*middle*). The slide involved both liquefaction and rotation of the soil. Houses were rafted, as shown by arrows in bottom diagram, and the highway and railroad also moved. Occupants of houses were made aware of rocking, rising, and falling motions, but not of sliding. As the structures came to rest at crazy angles, one house split in half and another overturned. The entire slide took less than three minutes. Water from the clay formed several large ponds. [From Kerr, "Quick Clay," copyright © 1963 by Scientific American, Inc. (all rights reserved).]

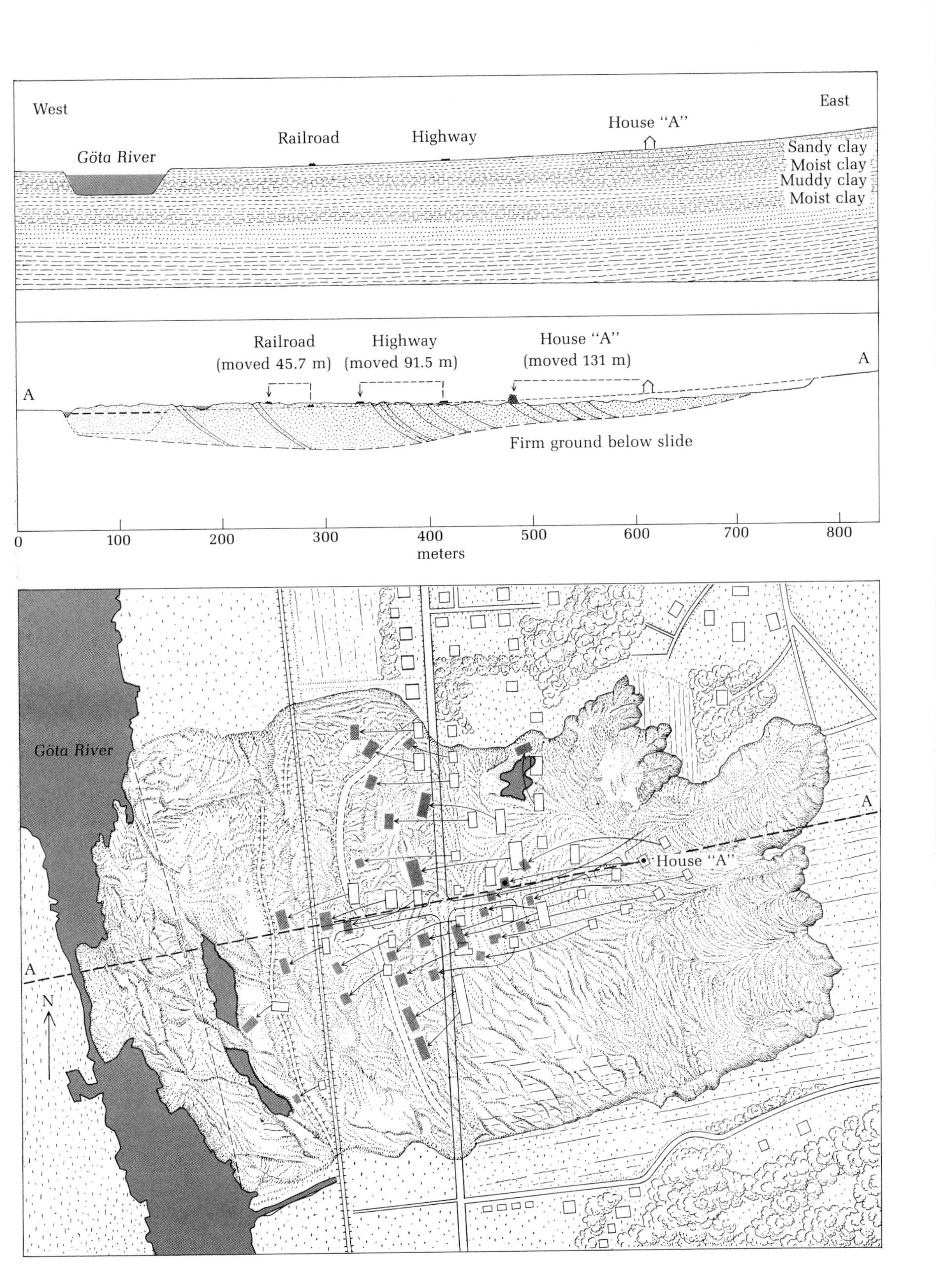
West
East
Göta River
Railroad
Highway
House "A"
Sandy clay
Moist clay
Muddy clay
Moist clay
Railroad
(moved 45.7 m)
Highway
(moved 91.5 m)
House "A"
(moved 131 m)
A
A
Firm ground below slide
0
100
200
300
400
500
600
700
800
meters
Göta River
A
A
House "A"
N

village, and a tensional fissure opened in the rock about 304 meters above the quarry. The fissure filled with water, and the slate became saturated until it eventually gave way and fell in a mass into the quarry. There it shattered and bounced outward in free fall. The villagers looked out across the valley *under* the cascade. The energized rock struck the valley floor, rolled and bounced up the other side to a height of 91 meters, and then turned almost 90° and rolled and bounced back down again. Many houses were destroyed and 115 people were killed by this cannonade of stones, some of which were moving at about 160 kilometers per hour across the valley.

It appears that this fall was cushioned on a layer of compressed air that was trapped by the downrush of fragmented rock; indeed, this is a common phenomenon in large falls. The great earthquake of 1964 triggered a rock fall of about 25 million tons into the Sherman Glacier in the Chugach Range, which lies east of Valdez in southern Alaska. This moved across the glacier and up the other side of the valley, but it did not even scrape off the wet snow on the glacier's surface.

FLOODS

Mass movements can be direct killers; one rock fall killed 3500 people and destroyed seven villages in Peru in 1962. However, a greater danger may occur when falling rock generates floods by splashing water out of a lake, as in Lituya Bay, or by damming a river temporarily until it cuts through and releases a torrent. Disastrous splashes are more common than might be supposed. Waves up to 90 meters high caused heavy mortality in Norwegian fjords in 1905, 1934, and 1936, and the Vaiont Reservoir disaster killed nearly 3000 people in northern Italy in 1963, even though the concrete dam held firm when the splash went over it.

As for temporary natural dams formed by mass movements, their failure is a common enough cause of floods. Even the great Indus River is sometimes dammed by rock falls where it cuts through a gorge in the Himalayas. In 1840, a dam 300 meters high backed up a great lake, which, when released by erosion, overwhelmed an army encampment on the flood plain of the river below.

CUT AND FILL

Few lives are lost by nature's attempts to restore natural slopes after man's cutting, filling, and other construction, but the cumulative damage is enormous, and it is growing as less desirable land is occupied. The examples are endless: The harbor of Valdez, Alaska, the proposed southern terminus of the Alaska pipeline, was built out over the edge of a delta. When the 1964 earthquake struck, the harbor slumped into the fjord. At the other extreme of climate, in Los Angeles, both the coast highway north of Santa Monica and

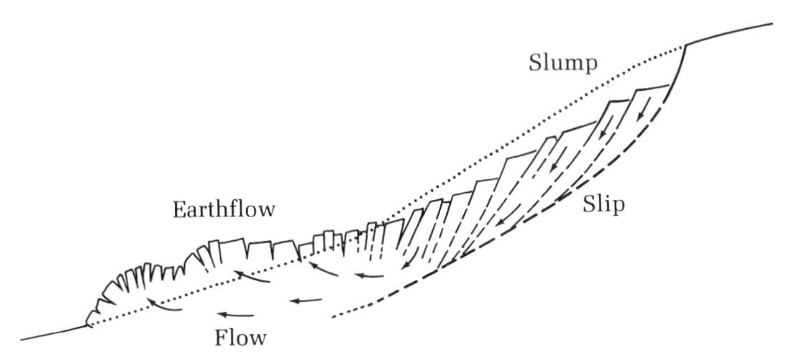

FIGURE 14.14
Longitudinal section through a slump, showing the slip surface and earthflow below the slump. Soil formation smooths the surface, and level areas like the one in the middle of this slump are then popular sites for houses. [From Sharpe (1938).]

the older road beside the Los Angeles River channel have repeatedly been buried under slumps from oversteepened road cuts.

Im Southern California, there is a related problem in some hilly areas because the most attractive sites for expensive houses appear to be the relatively level floors of amphitheaterlike hollows in the slopes. Unfortunately, these are the scars of natural landslides. Even so, they may be stabilized. They are not unsuitable for housing, provided the occupants simply do not plant lawns and water them. If they do so, they are apt to lubricate and reactivate the landslide slip surface. People are so desperate for space, or so uninformed, that they build houses on the coastline at Point Fermin, where earlier houses have slid away. These new houses slide away, too; and nearby, at the exclusive Portuguese Bend area, enthusiastic gardeners caused whole blocks of houses to slide toward the Pacific below.

Mountain road cuts and recreational-housing pads bulldozed on slopes may be subject to slumping. The problem is particularly acute around newly formed lakes, because old slumps and landslides may be reactivated when the surrounding area becomes saturated. The Eel River Reservoir in northern California has been partially filled by slumps that flowed in from its margins.

MINE TAILINGS

We earlier considered the consequences of the collapses that may occur if mines are not backfilled with the tailings removed from them. Another social cost may arise because some tailings are left at the surface in inhabited areas. They are uniformly fine-grained, loosely packed, nearly saturated sediment piles with steep side slopes. In short, they are ripe for liquefaction and flow. Exactly such a disaster occurred in 1965 at Aberfan, Wales, when rain saturated a tailings dump from a coal mine. The muck flowed downslope, overwhelmed the miners' school, and killed more than 100 children. The residents of other British coal mining towns were still expressing fears of similar flows on television interviews in 1971. The situation in the United States proved even more remarkable: The Corps of Engineers states that the Bureau of Mines did not enforce its own regulations, and unconsolidated mine tailings were used to build a dam across a small stream in Pennsylvania. When it ultimately gave way in 1972, it swept down a whole small valley, killing 118 people and smashing towns as it went.

Summary

1. The interactions of unconsolidated sediment and water under the attraction of gravity are minor and local but can cause enormous damage.
2. Clay is generally deposited as a fluffy mud, but continuing deposition loads the lower clay, squeezes out water, and increases the density and solidity of the sediment.
3. Fluids in the pores of a sediment help to support it. Withdrawal of fluid, water or oil, causes subsidence, which may damage structures in the area.
4. Much of the subsidence caused by the withdrawal of oil can be prevented by injecting water as a replacement.
5. Collapse structures are vertical-sided pits that form in surface materials when support is withdrawn from underneath.
6. Unconsolidated sediment sinks under loading, but this has social consequences only where there is construction on active deltas or filled lake basins, where subsidence is relatively large and rapid.
7. In polar regions, the expansion and contraction that accompany repeated freezing and thawing of ground water may cause cracking, heaving, sinking, collapse, flowing, and slumping. These affect all types of construction.
8. Mass movement occurs where slopes are steep enough and materials are weak enough so that gravity overcomes any resistance. There are many modes of mass movement, ranging from rock falls on steep mountains to flowage of saturated clays on nearly level ground at the water's edge.

Discussion Questions

1. Why does a layer of clay compact more than a layer of sand if they are equally loaded?
2. Why was the relationship between fluid withdrawal and subsidence obscured for so long?
3. What happens to an organic soil when it is drained to permit farming or construction?

4. Why does subsidence of a delta bother city dwellers more than farmers? What if a dam cuts off the annual supply of new sediment?
5. Why does the landscape in some polar regions resemble karst topography?
6. How can a rock avalanche flow over snow without scraping it?

References

Aihara, S., H. Ugata, K. Miyazawa, and Y. Tanaka, 1970. Problems on the groundwater control in Tokyo. In *Land Subsidence*, vol. 2, pp. 635–644. Paris: UNESCO.

Budel, J., 1966. Deltas—a basis of culture and civilization. In *Scientific Problems of the Humid Tropical Zone Deltas and Their Implications*, pp. 295–300. Paris: UNESCO. [The proceedings of an international symposium in Dacca in 1964.]

Castle, R., R. Yerkes, and F. Riley, 1970. A linear relationship between liquid production and oil-field subsidence. In *Land Subsidence*, vol. 1, pp. 162–173. Paris: UNESCO.

Flawn, P., 1970. *Environmental Geology*. New York: Harper & Row.

Gilluly, J., and U. Grant, 1949. Subsidence in the Long Beach Harbor area, California. *Geol. Soc. Amer. Bull.* 60:461–529. [Pumping fluids out of the ground causes the surface to sink.]

——, A. C. Waters, and A. O. Woodford, 1968. *Principles of Geology* (3rd ed.). San Francisco: W. H. Freeman and Company.

Goldberger, M., and G. J. F. MacDonald, co-chairmen, 1970. *Environmental Problems in South Florida, Part II*. Washington, D.C.: National Academy of Sciences. [The problem was too many people trying to do too many things with the same resources.]

Gould, H., 1970. The Mississippi delta complex, *In* J. Morgan, ed., *Deltaic Sedimentation, Modern and Ancient* (Soc. Econ. Paleontol. Mineral. Spec. Publ. 15), pp. 3–30. Tulsa: Society of Economic Paleontologists and Mineralogists.

Hamilton, E., 1959. Thickness and consolidation of deep-sea sediments. *Geol. Soc. Amer. Bull.* 70:1399–1424.

Hedberg, H., 1936. The gravitational compaction of clays and shales. *Amer. J. Sci.* 231:241–287.

Inaba, Y., S. Aoki, T. Endo, and R. Kaido, 1970. Reviews of land subsidence researches in Tokyo. In *Land Subsidence*, vol. 1, pp. 87–97. Paris: UNESCO.

Kerr, P., 1963. Quick clay. *Sci. Amer.* 209(5):132–142. [Takes Scandinavian houses for a ride.]

Lofgren, B., 1970. Field measurement of aquifer-system compaction, San Joaquin Valley, California. In *Land Subsidence*, vol. 1, pp. 272–284. Paris: UNESCO. (Quotation from p. 272.)

Marsden, S. S., Jr., and S. N. Davis, 1967. Geological subsidence. *Sci. Amer.* 216(6):93–100.

Matthes, G., 1953. Quicksand. *Sci. Amer.* 188(6):97–102. [And how to avoid it.]

Mayuga, M., and D. Allen, 1970. Subsidence in the Wilmington oil field, Long Beach, California, U.S.A. In *Land Subsidence,* vol. 1, pp. 66–79. Paris: UNESCO.

Morton, R., responsible official, 1972. *Final Environmental Impact Statement, Proposed Trans-Alaska Pipeline.* A (mimeographed) report required by the National Environmental Policy Act of 1969 and prepared by a special interagency task force for the federal task force on Alaskan oil development.

Munk, J., and W. Munk, 1972. Venice hologram. *Proc. Amer. Phil. Soc.* 116(5): 415–442. [Science and art in happy harmony.]

Nelson, B., 1970. Hydrography, sediment dispersal, and recent historical development of the Po River Delta, Italy. *In* J. Morgan, ed., *Deltaic Sedimentation, Modern and Ancient* (Soc. Econ. Paleontol. Mineral. Spec. Publ. 15), pp. 152–184. Tulsa: Society of Economic Paleontologists and Mineralogists.

Oakeshott, G., 1971. *California's Changing Landscapes.* New York: McGraw-Hill.

Poland, J., 1970. Status of present knowledge and needs for additional research on compaction of aquifer systems. In *Land Subsidence,* vol. 1, pp. 11–20. Paris: UNESCO. [Gives major occurrences of land subsidence.]

Prokopovich, N., 1970. Prediction of future subsidence along San Luis and Delta-Mendota canals, San Joaquin Valley, California. In *Land Subsidence.* vol. 2, pp. 600–610. Paris: UNESCO. (Quoted on p. 603.)

Ruskin, J., 1877–1884. *St. Mark's Rest: The History of Venice.* Orpington, Kent: George Allen. [Recent research indicates that some of his ideas were based on misconceptions.]

Sayre, A. N., 1950. Ground water. *Sci. Amer.* 183(5):14–19. (Available as *Sci. Amer.* Offprint 818.)

Sharpe, S., 1938. *Landslides and Related Phenomena.* New York: Columbia University Press. [The standard authoritative work.]

Stephens, J., and W. Speir, 1970. Subsidence of organic soils in the U.S.A. In *Land Subsidence,* vol. 2, pp. 529–534. Paris: UNESCO

Wohlrab, B., 1970. Effects of land subsidence caused by mining to the groundwater and remedial measures. In *Land Subsidence,* vol. 2, pp. 502–511. Paris: UNESCO.

Yerkes, R., and R. Castle, 1970. Surface deformation associated with oil and gas field operations in the United States. In *Land Subsidence,* vol. 1, pp. 55–65. Paris: UNESCO.

SEDIMENT IN MOTION

JOHN S. SHELTON

15

THE SEDIMENTARY CYCLE ON LAND

Our solid earth is every where wasted, where exposed to the day. The summits of the mountains are necessarily degraded. The solid weighty materials of those mountains are every where urged through the valleys, by the force of running water.

James Hutton, THEORY OF THE EARTH

The last two chapters have dealt with the weathering of a surface of solid rock to soil, solutions, and loose sediment, and with the initial, localized movement of these materials. In this chapter, we begin to follow the sediment on its long journey to the deep sea and its ultimate return to the continents in subduction zones. The chapter ends with the sediment at the shoreline, but this is not indicative of any break in the continuous flux from mountains to ocean basins and back. The processes of transport and the involvement of society in the processes both change at the shoreline—but, even so, it is merely an arbitrary place to divide two chapters.

The sedimentary cycle on land begins with the generation of sediment by erosion and weathering. Sediment is transported almost entirely by the integrated network of rills, streams, and rivers for the simple reason that it *is* integrated and generally leads to the sea. Infrequently, a river may not flow directly to the sea. Instead, it drains into a lake or a desert and deposits the sediment it is carrying. There the sediment remains for a while, but not forever.

Eventually, the lake fills, the dam is destroyed, and the river, or some river, continues on its way. Eventually, the rainfall increases, or a great river cuts across the desert and it, too, begins to yield sediment to the sea. Sediment resides longest on the continents if it falls into a structural trap, accumulates to depths of many kilometers and becomes rock. Rock erodes at an average rate of only a few centimeters in a thousand years—when it is being eroded at all. There is no loss when it is covered by shallow seas. Thus, some of it may remain for at least a billion years before it rejoins the active cycle of sedimentation.

Seen with the geologically long view, whole continents are being transported by rivers to the sea floor. If the flux is interrupted by deposition in a structural basin, then responses are triggered all through the cycle. Solution of other sediment may occur on the distant floor of the deep sea. Responses may occur for years and centuries—the time scale that most concerns society—because the course of rivers is also the course of a continuous stream of moving sediment. This "river of sand" is essentially continuous and in equilibrium with the flow of water. If it is disturbed in any way at any point, it responds in ways that may affect the level of the sand surface for hundreds of kilometers up or down stream. We shall study the effects of such natural and man-made disturbances after first examining the flow of rivers and sand in the equilibrium state.

The Equilibrium Stream

Water at any level above the sea possesses potential energy, which enables it to flow, provided it is not trapped in a lake. The potential energy is converted to kinetic energy and, gradually, is dissipated as heat. This occurs because of external friction with the sides of channels and internal viscosity during turbulent motion. The rate at which potential energy is made available depends on the slope, and the speed of flow varies as the square root of the slope. The rate at which potential energy is used up depends on the side friction and on several minor factors, including density and temperature, that influence viscous internal turbulence. The important factor in friction is the ratio of the cross-sectional area, which is not in contact with the sides, to the perimeter, which is. This ratio, called the hydraulic depth, approximates the actual depth in normal-shaped river channels. Thus, a very broad and shallow river has a much larger frictional loss than a deep one with a semicircular cross section. The speed of a river varies with the square root of the hydraulic depth.

The discharge of a river is the amount of water that flows past a given point in a unit of time. It is the product of the speed (distance/time) and the cross-sectional area. An essential point to remember is that the discharge increases

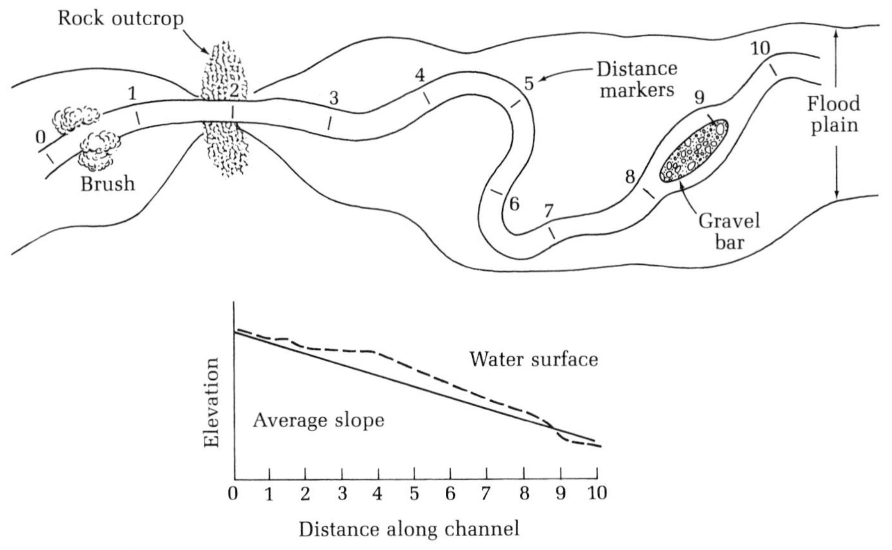

FIGURE 15.1
Relation of channel features to profile of water surface. Note effects of obstruction, curves, and central island. [After Leopold, Wolman, and Miller, *Fluvial Processes in Geomorphology*, W. H. Freeman and Company, copyright © 1964.]

downstream in a river because of the addition of water from tributaries. The speed of flow is relatively constant, however, and this is because the slope upstream is steep where the river is small, and the hydraulic depth downstream is large because of the tributaries. Thus, the broad picture is that a river has a longitudinal profile that is steeper at the source and flatter near the sea; it is concave upward. In detail, the profile is much more complex. Barriers, banks, and natural dams break it into some lengths that are relatively fast-flowing, steep, and shallow, and others that are slow-flowing, gently sloping, and deep. This is necessary to maintain the continuity of flow.

FLOODS

Rain falls episodically in amounts that vary with the weather and climate, and, if the discharge from the sky is excessive, the rivers below become flooded and the flood propagates downstream as a wave. Rivers, like glaciers, flow faster when they become deeper; that is, they flow faster and deeper where before they were fast and shallow and also faster and deeper where before they were slow and deep. Thus, the flood wave moves faster than the river, just as such waves outspeed the glaciers in which they occur. The flood water does not all escape quickly, however, because a natural river overflows its banks and spreads a thin sheet over the adjacent valley floor, or flood plain.

Floods occur in most rivers every year following the spring melting of mountain snow, but the volume varies. The river system readily accommodates the smaller floods, and society expects them. Great floods are another matter. When these occur, the rivers are overloaded and disasters may follow, especially if dams or levees are destroyed. No one, as yet, can predict when a great flood will occur, or control the rain in a way that would prevent it. The common method of control is to build a dam to contain the flood and discharge it slowly. Although the time of a flood cannot be predicted, the average frequency of floods of a given intensity can be determined from the historical record of past floods, and this is very important information for the proper design of flood-control structures.

A study of two river basins in Pennsylvania indicates the general relationship between the frequency and intensity of floods (see Figure 15.2). The "annual," which is an *average* flood, occurs about every two years. Floods half as great as average occur almost every year. Floods twice as great as the annual one appear every 18 years, and ones three times as great occur less than once in a century. The records do not extend for more than a century, but the trend can be extrapolated to imply that, even in 1000 years, no flood would exceed 3.5 times the annual one. Such extrapolations are capable of providing some design criteria for flood control, but, all too frequently, population pressure over a twenty-year period will cover a flood plain with

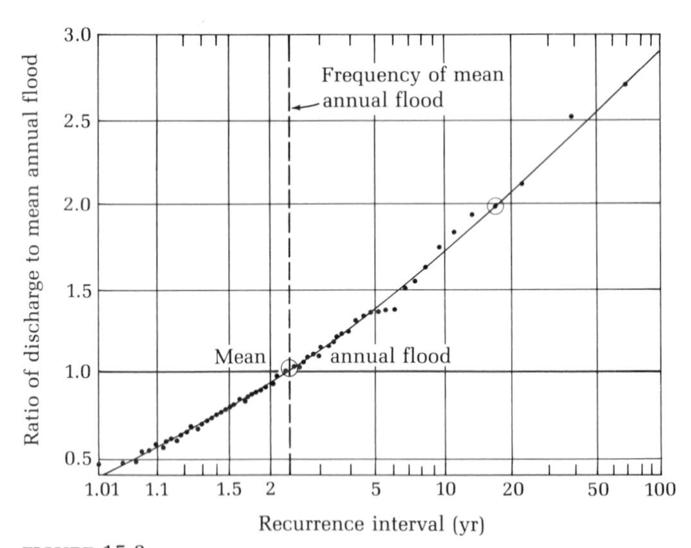

FIGURE 15.2

Regional flood-frequency curve for Youghiogheny and Kiskiminetas river basins, Pennsylvania. For example, a flood twice as great as the mean annual (or average) flood occurs only once in 18 years (*see large circles*). [After Leopold, Wolman, and Miller, *Fluvial Processes in Geomorphology,* W. H. Freeman and Company, copyright © 1964.]

houses, as though the concept of the "20-year flood" or the "100-year flood" were unknown.

The relation between flood intensity and frequency varies with the size of the source area of a river and also with climatic and man-made changes. Climatic changes are reflected in consistent trends in the annual flow, such as the average decrease of river discharge in the whole United States from about 1905 to 1935 and then the average rise. The records for individual rivers are quite striking, particularly the large ones that integrate vast drainage basins. For example, the maximum annual flow of the Columbia River from 1920 to 1945 was never as high as the minimum from 1880 to 1920. The social effects of these variations should now be familiar. The population expands until the use of water matches the available supply. If the supply then drops, as it frequently does, a more or less acute problem arises.

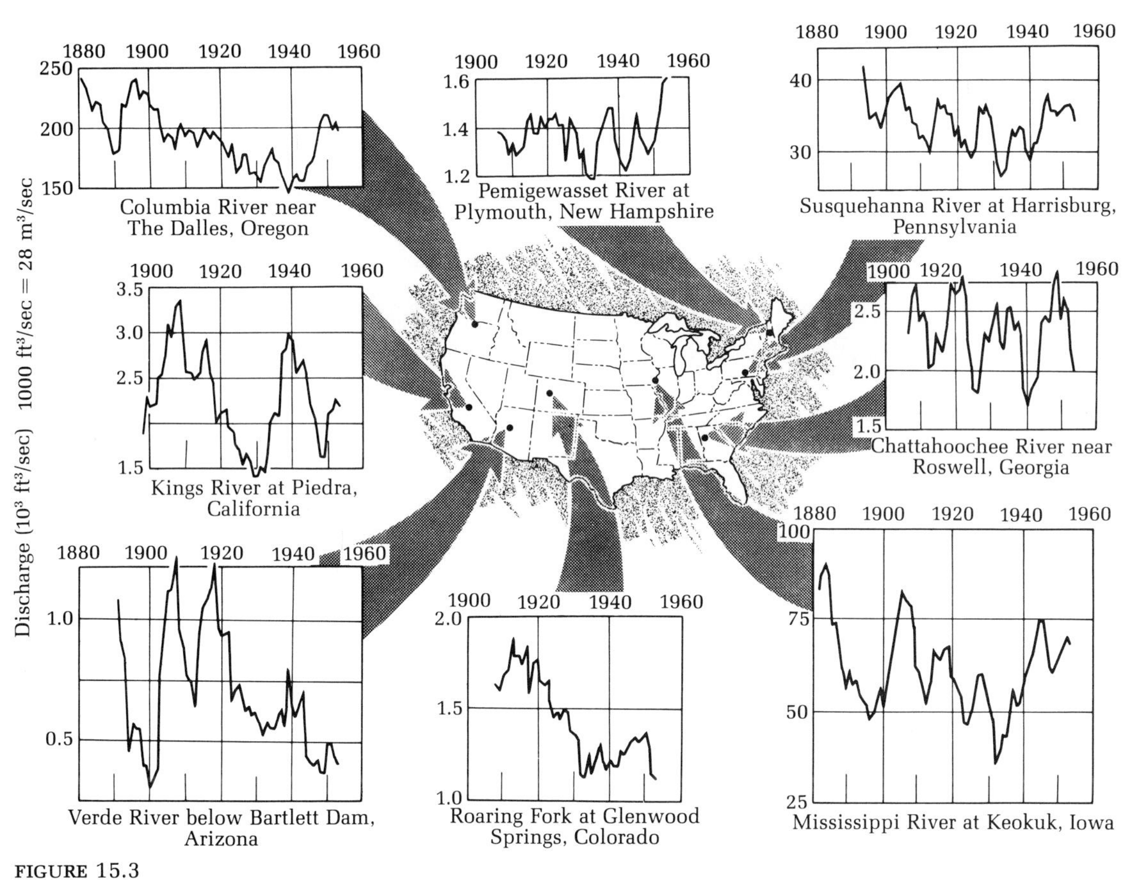

FIGURE 15.3

Historical records of annual flow in some American rivers. Consistent trends indicate cyclical weather and climatic changes. [From Leopold, Wolman, and Miller, *Fluvial Processes in Geomorphology*, W. H. Freeman and Company, copyright © 1964.]

SEDIMENT DISCHARGE

Sediment moves in a river either as bed load along the bottom or as suspended load within the water, depending on the speed and discharge of the river and the size of the sediment particles. In a general way, it takes a higher

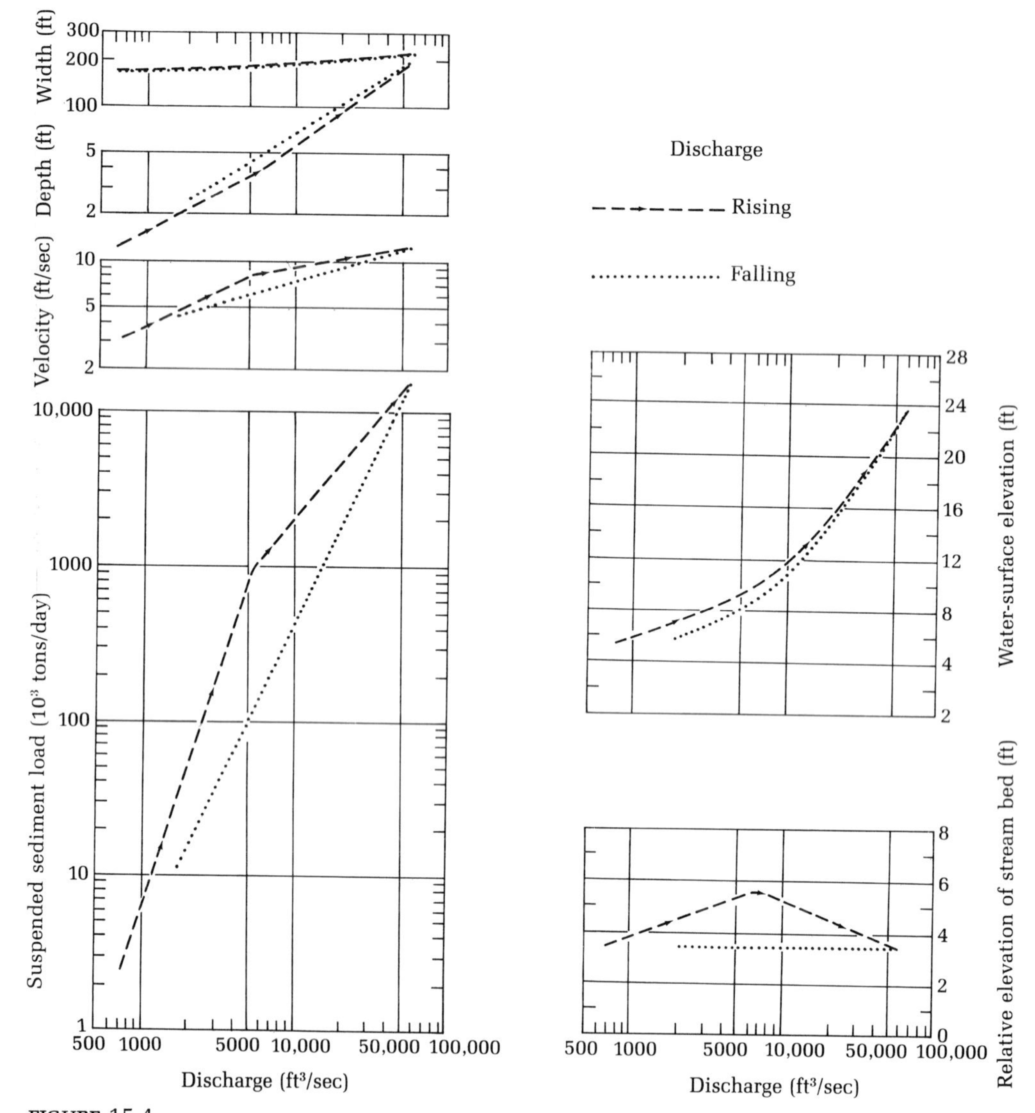

FIGURE 15.4

Changes in width, depth, velocity, water-surface elevation, and stream-bed elevation with discharge during the rising and falling stages of the flood of the San Juan River (near Bluff, Utah) between September and December, 1941. Note the very large increase in suspended sediment compared to velocity, and the very small change in width compared to depth. [After Leopold, Wolman, and Miller, *Fluvial Processes in Geomorphology*, W. H. Freeman and Company, copyright © 1964.]

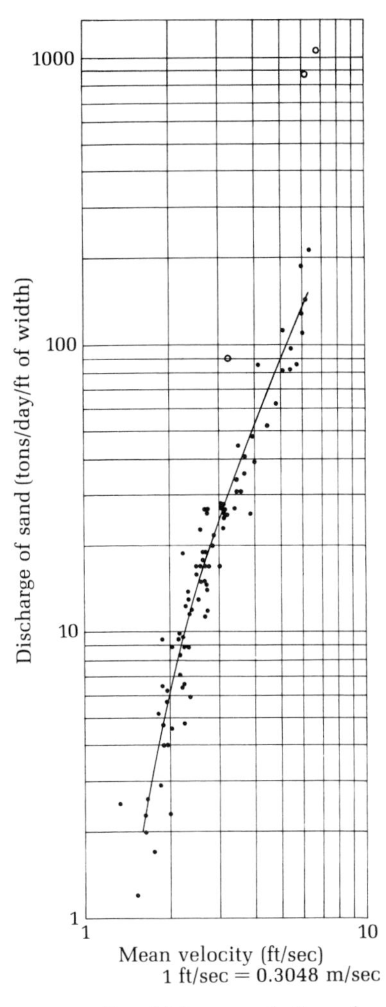

FIGURE 15.5
Discharge of sands plotted against mean velocity for the Rio Grande in New Mexico. [From Leopold, Wolman, and Miller, *Fluvial Processes in Geomorphology,* W. H. Freeman and Company, copyright © 1964.]

velocity to move bigger grains of sand and gravel, and these flow along the bottom in ripples, dunes, and bars—which are different scales of sand waves. Silt and clay, however, have peculiar characteristics relative to velocity. They settle slowly and, thus, are readily transported in suspension. However, if they settle to the bottom, they tend to form a very smooth layer. The speed of the river water is slowed by friction in a boundary layer next to the channel walls, even though it is fast in midriver. This matters little for sand and gravel, because they protrude through the boundary layer. Silt and clay do not, so, on the average, they are influenced only by slowly moving water. Thus, a fast-moving river may be able to transport clay and sand, and to erode sand but not erode clay.

Taking these different factors into account, we find that the discharge of sediment in a river increases very rapidly with river speed at a given depth, and generally increases with depth at flood speeds. The Rio Grande in New Mexico illustrates the velocity–discharge relation. At a water speed of 0.5 meters per second, the sand discharge is 6 tons per day per meter of width. The sand flux is 100 times greater when the water speed is 2 meters per second.

The significance of the relation is that most rivers transport almost all their load when they are in flood. They run on a bottom of sand, which is ordinarily moving only at the surface of the bed. When the water level rises in a flood, the bottom is simultaneously eroded downward. In one annual flood of the Colorado River in 1956, as the water level went up and back down about 2.5 meters, the bottom went down and back up about 2 meters at the same time. At the peak of the flood, the river was three times its depth at the beginning, but, by the end, everything was about the same as at the start, and equilibrium was restored.

SAND WAVES

Sand and gravel, like highway traffic, glaciers, and floods, can be seen to move in waves under the influence of moving water. The sand in the bottom of an experimental flume collects in "ripples," which are small sand waves. The grains are eroded from the upstream side of a ripple and are deposited on the downstream side. The ripple itself migrates downstream, and each grain that was deposited in front reappears in back and goes through the wave again. This is what takes place at slow water speeds. At higher speeds, the sand becomes smooth; at even higher ones, ripples again form, but they have a different shape and they move upstream.

Similar, but larger, sand waves called bars occur in rivers, and they move downstream when the water flows slowly and there is an abundance of sand. If the sand supply is insufficient to maintain the river bed, erosion occurs as a local scour. This moves upstream along the bed of the stream, even though the sand itself moves downstream. The scour behaves as a "negative" sand wave.

The social implication of these observations is rather important. If sand is artificially piled in a river, the bed forms a wave; if sand is dredged from a river, the bed forms a scour or negative wave, and sand waves move to places where they may not be expected or wanted.

MEANDERS

An unconfined stream that is long relative to its width undulates randomly and develops orderly horizontal waves called meanders. Meanders are observed in such diverse features as rivers, the Gulf Stream, channels on the

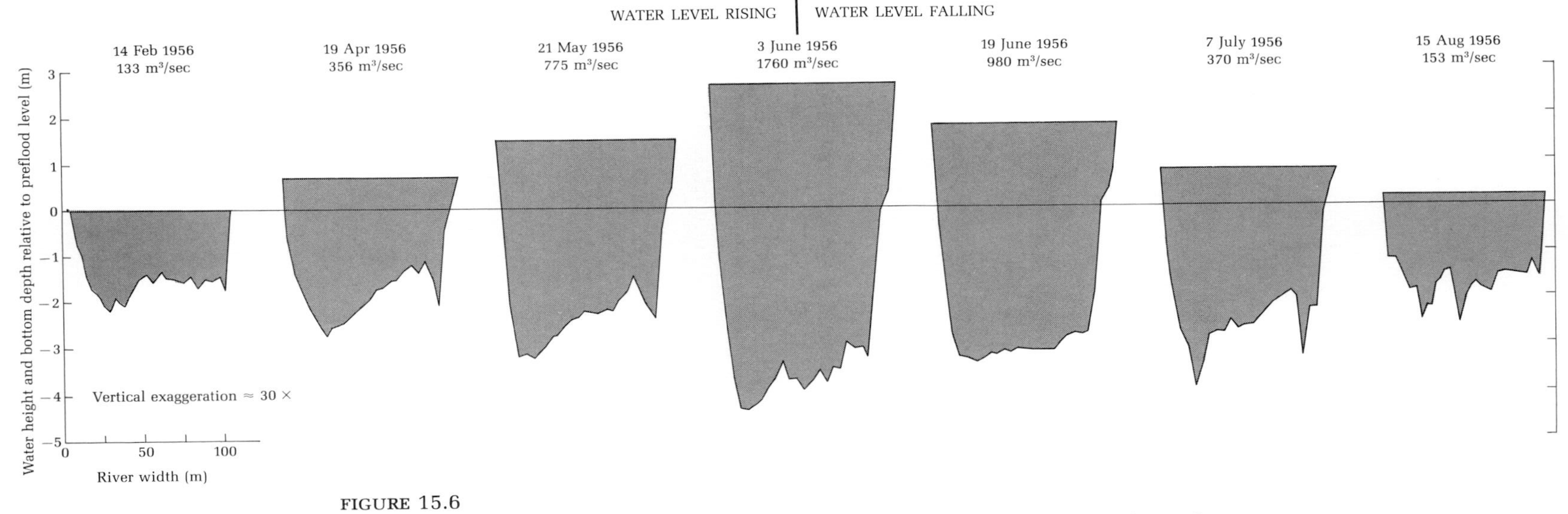

FIGURE 15.6

Cross sections showing scour and subsequent fill during flood passage of the Colorado River at Lees Ferry, Arizona, during water year 1956. [After Leopold, Wolman, and Miller, *Fluvial Processes in Geomorphology*, W. H. Freeman and Company, copyright © 1964.]

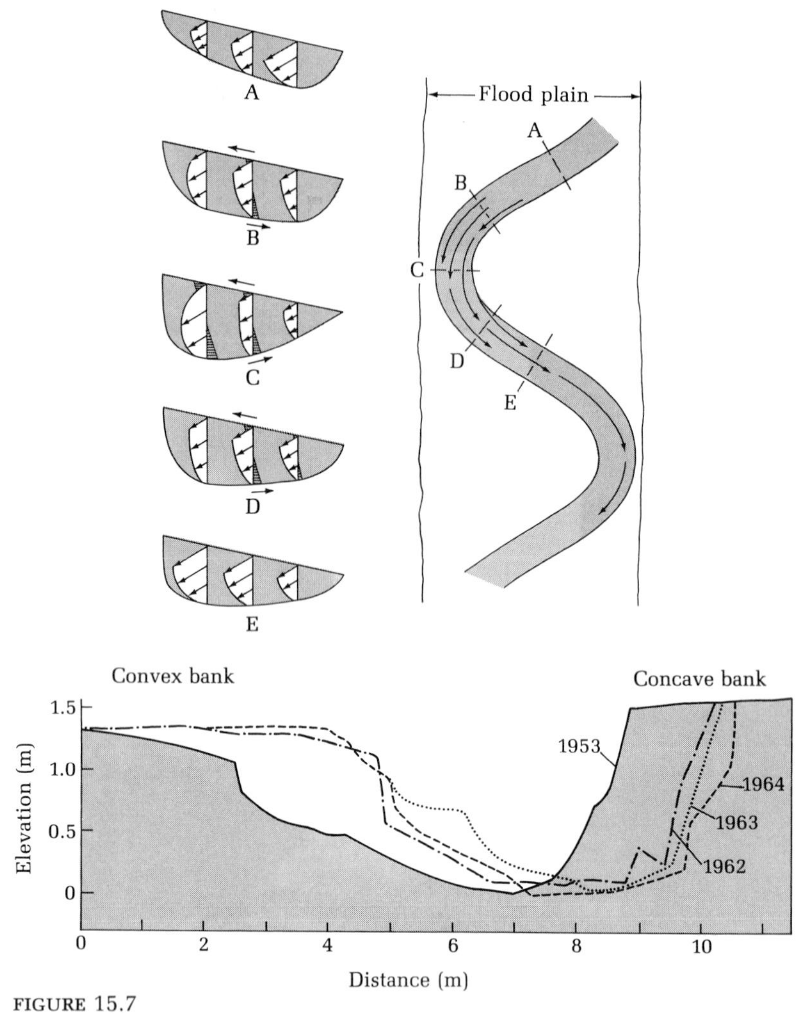

FIGURE 15.7

River meanders: *Above*, an idealized flow pattern of a typical meander. Indicated are the velocity vectors, in a downstream direction, for five cross sections across the curve (*above left*), with the lateral component of the velocity shown by the triangular hatched areas, and the streamlines at the surface of the meander (*above right*). *Below*, the lateral migration of a typical meander is demonstrated in four successive cross sections, surveyed between 1953 and 1964, of Watts Branch (a small tributary of the Potomac River near Washington), which show that the lateral migration of meanders, by the erosion of the concave banks and deposition on the convex banks over many years, results in a river channel's occupying every possible position between the valley walls. [After Leopold and Langbein, "River Meanders," copyright © 1966 by Scientific American, Inc. (all rights reserved).]

sea floor, and jets from garden hoses. Rivers are only rarely straight for as far as ten times their width, and, once they begin to curve, they gradually develop a regular pattern of meanders. This is because the meander curvature enables the river to turn while doing the least work.

In turning, a river tends to move from side to side in the channel, eroding the upstream side of the meander and depositing on the downstream side.

Thus, meanders are large, horizontal waves of sediment that slowly drift downstream at a rate of perhaps a third of a meter a year. From their very slow motion, it appears that they are relatively efficient at transporting large concentrations of sediment. In the course of time, a meandering river moves through every part of its flood plain and reworks the top layer of sediment.

GRADED STREAMS

The capacity of a wild river to do work is generally in equilibrium with the work it does. If not, it speeds up, and, thus, automatically does more work eroding its bed and transporting more sediment. Likewise, if it is overworked, it deposits sediment until it again reaches equilibrium. Deposition and erosion change the bed slope, which, in itself, changes the energy available locally. Thus, a normal river, flowing on an unconfined sedimentary bed, is in a delicate equilibrium that geologists call graded. Like any equilibrium system, its response to a change in any of the controlling factors is to displace the equilibrium in a direction that will tend to absorb the effect of the change. We shall now consider some natural and man-made changes and the effects that inevitably absorb them.

Natural Changes in Graded Streams

For anyone trained to read the signs, a casual inspection of many river valleys is enough to show that deposition and erosion have alternated and that the bed slope has changed repeatedly. The most obvious indicators are those terraces that are fragments of abandoned flood plains into which the modern stream channel has been cut. If the substructure of the valley can be established—by drilling wells, for example—similar but buried terraces may be found. In some regions, the terraces can be correlated in some way, and it is often found that each contemporaneous group formed an integrated flood plain for a stream with quite a different grade from the existing one. What causes these changes, and how will they affect cities and farms that are concentrated near rivers?

Fundamentally, most long-term changes in the flow characteristics of rivers are in response to changes in the available potential energy. This can occur either by a fluctuation of the quantity of water (mass) or the difference of elevation (head) from the headwaters to the mouth of a river. The quantity can change with climate or vegetative cover or lake formation; the elevation difference varies with sea level, tectonic elevation or subsidence, and the erosion of mountains. The quantity of water in a river and the relief of the terrain through which it flows can both be changed by the process of stream capture. This occurs at drainage divides, such as the one along the Rocky

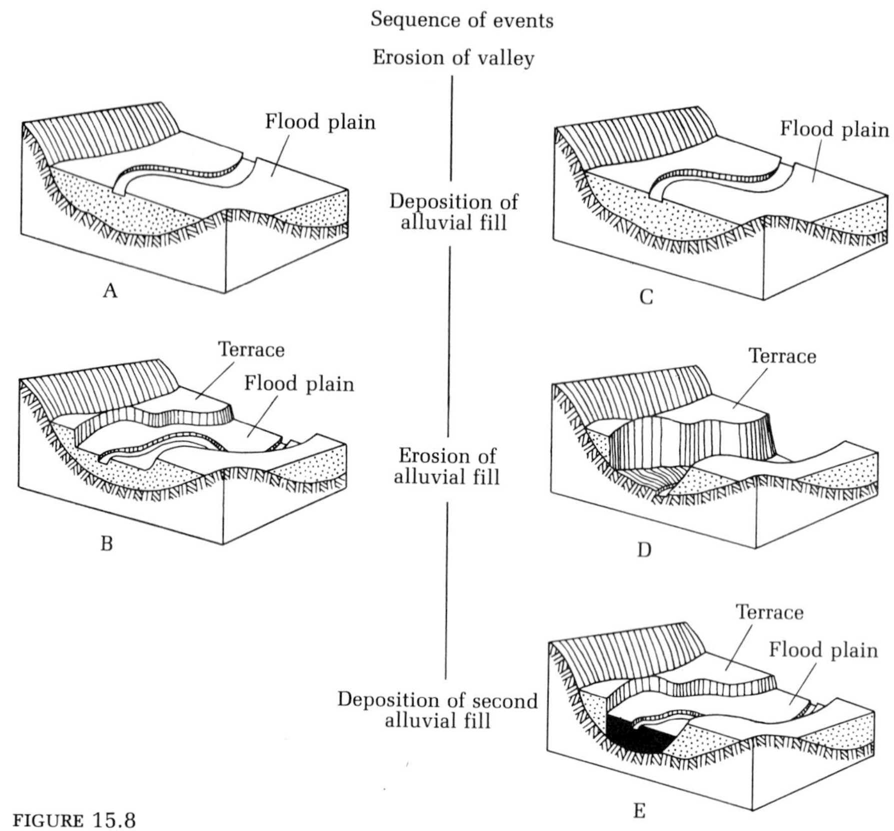

FIGURE 15.8

Stages in the development of a terrace. Two sequences of events leading to the same surface geometry are shown, one in diagrams *A* and *B*, the other in diagrams *C*, *D*, and *E*. [From Leopold, Wolman, and Miller, *Fluvial Processes in Geomorphology*, W. H. Freeman and Company, copyright © 1964.]

Mountains, which separates the area draining to the Atlantic from that draining to the Pacific. If a Pacific stream erodes headward rapidly, it may intersect a stream that is in the Atlantic drainage basin. The waters of the second stream, above the point of intersection, are captured by the steeper, more vigorously eroding one. The quantity of water in the shortened Atlantic-basin river is reduced, while that in the other is increased.

In the United States and Europe, many of the terraces, buried and elevated, can be readily explained in terms of the waxing and waning of glaciers and the accompanying climatic changes. When the glaciers spread, they carried an enormously increased load of sediment to nearby rivers. The overloaded rivers filled the upper reaches of their valleys with silt as they established a steeper grade, and the effect of the loading was felt for hundreds of kilometers downstream. Meanwhile, sea level dropped as the glaciers grew larger. Thus, the bed slope increased at the mouths of the rivers and narrow gorges were eroded.

When the glaciers melted, the headwaters increased, but so did the sediment load. However, as the ice front retreated, the coarse sediment was deposited near it, and, at the old ice front, the rivers had an excess of energy and eroded deep channels. Meanwhile, sea level rose at the mouths of the rivers, the available energy decreased, and the narrow gorges were filled with sediment.

These great events shaped most existing river valleys; even in the tropics, the fluctuations in sea level had their effects. However, such great events are not required to alter the shape of a river. Quite substantial changes in river grade have occurred in historical times when the possible range of causative factors was quite minimal. Within a period of 60 years, the Cheyenne River in Nebraska, for example, first elevated its bed 2.5 meters and then cut a channel down below the original level. Nothing very noteworthy has happened to explain this, although presumably something has influenced the runoff in the source area. A brief cycle of rainfall might have been the cause, and the rise and fall may have marked the passage of a wave of sediment.

The mere evolution of a river drainage basin is enough to change a river valley. In the course of time, the mountains of the interior are eroded, transported grain by grain to the river mouth, and deposited there, unless the sea carries the sediment away. Ideally, at some intermediate point, the elevation of the surface does not change. At increasing distances upstream and downstream from this point, the lower course of the river gradually rises while the upper reaches are worn down. An integrated river basin may be segmented, at least temporarily, into several smaller ones by a variety of natural events. In the mountainous reaches, a landslide may dam a stream, and a tributary stream pouring out coarse gravel may partially dam even the lower reaches of a great river. The segments evolve, as long as they last, just as if they were independent basins. Their "headwaters" are eroded downward and their lower courses are sedimented until the main basin is again integrated and the processes are reversed.

It might seem that folding and faulting of the underlying rock would commonly dam rivers, but this occurs only very rarely. If a fold grows upward under a river, the river normally has no difficulty cutting down through it and maintaining its grade. Rivers are far more potent than they appear, as well as far more changeable. If mountains cannot conquer rivers, how have men fared in their attempts to control them?

Human Modifications of Rivers

It is simple to obtain agreement among scientists that certain events transpire if *nature* erodes a channel or dams a river, but, as we observed in our study of land subsidence, there is more diversity of opinion if the changes are made by man. An analysis of the events that followed the construction of Elephant Butte Dam will illuminate how opposite views may honestly arise.

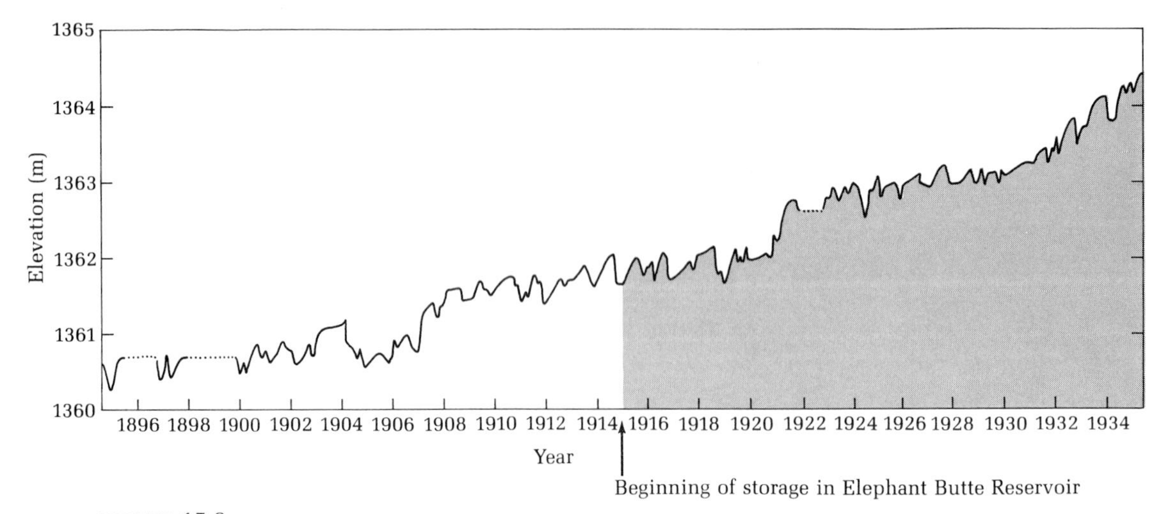

FIGURE 15.9

History of the elevation of the low-water surface of the Rio Grande River opposite San Marcial, New Mexico. The rise after 1915 was blamed on the storage until it was realized that a similar rise had occurred beforehand. [After Leopold, Wolman, and Miller, *Fluvial Processes in Geomorphology,* W. H. Freeman and Company, copyright © 1964.]

The dam was built in New Mexico in 1915 to control the Rio Grande. Deposition subsequently occurred not only in the new reservoir but also for tens of kilometers upstream, where it raised the river bed by 4 meters at San Marcial. This was hailed as a classic example of a river seeking to re-establish grade and of the dangers that may accompany human intervention. Geologist Hoover Mackin, a master of the concept of graded streams, pointed out that, when the rise in level propagated upstream, "a few tens of feet of upbuilding to the latitude of Albuquerque would destroy highways and railways, towns, irrigation works, and farmlands with an aggregate value that may exceed the cost of the dam."

However, the case has not proved to be so simple, and the influence of the dam cannot be demonstrated with clarity. Records of bed level of the Rio Grande at San Marcial exist from 1896, and, when they were all analyzed, it was discovered that the bed was rising almost as fast for the 20 years before the dam was constructed as for the 15 years after. Was the river just naturally aggrading rather than at grade? Was man not involved at all? Here the problem grows more complicated.

The level of the Rio Grande was being measured for a reason. From 1880 to 1915, so much water was diverted from the river for irrigation that the flow at San Marcial was halved. That meant that the slope of the tributary streams became steeper, they flowed faster, and they began to erode more vigorously. One of them, the Rio Puerco, developed into an enormous gully 190 kilometers long, 8.5 meters deep, and 87 meters wide during the period from 1885 to 1928. All this material was dumped into the Rio Grande, which, meanwhile, had lost much of its ability to transport sediment, and, thus, it aggraded its bed. It appears that human intervention may have caused much of the aggradation before the dam was built, but how can we be sure? The measurements of river level and the intervention began at the same time. Some natural aggradation may have been occurring just before the farmers arrived.

The confusion about the cause of changes in the Rio Grande serves to illustrate an important point. A river responds in the same way to man-made changes as to natural ones. If a city suddenly draws off half a river for drinking water, the river will dump part of its load, aggrade its bed, and propagate a wave of sediment downstream. However, without a complete understanding of the recent geologic history of the river, it may be impossible to prove that the same or similar events would not have occurred anyway. This is especially true of "proof" in the legal sense. Still, rivers do behave in predictable ways. By 1931, the rate of aggradation at San Marcial began to increase, and, within a decade, it was reasonably clear that the dam was causing aggradation that was being superimposed on what had been going on before.

With regard to the threat to Albuquerque, Mackin's warning was an extrapolation of a known trend. It applied regardless of the cause of the aggradation.

INDUCED EROSION

It is difficult to tell that a river is aggrading its channel, because it buries its history under layers of mud. Degradation, on the other hand, is easy to recognize, because the river cuts its channel deeper. It may be for this reason that induced erosion of previously graded, or aggrading, riverbeds is less controversial. The Dennison Dam on the Red River in Texas provides one of many undisputed examples. About 390 million tons of sediment has been deposited by the river in the reservoir created by the dam. Of this, 77 million tons is sand. The water that pours over the dam has no load of sediment, so it has eroded the riverbed 1.5–2.0 meters deeper for almost 30 kilometers downstream. The locus of erosion does not move upstream, because it is trapped by the dam. The total amount eroded is about 67 million tons, or about the same as the sand deposited behind the dam. Considering that some water evaporated or was diverted from the reservoir, the transport of sand downstream in the river is made up immediately by erosion. However, the finer material deposited behind the dam is not so obviously replaced.

LEVEES

The Po River drains much of northern Italy, including the southern slopes of the Alps. Man has been trying to bring it under control since the Dark Ages, so it provides an excellent example of how a river attempts to restore equilibrium. The river began to wreak havoc locally in Roman times by meandering

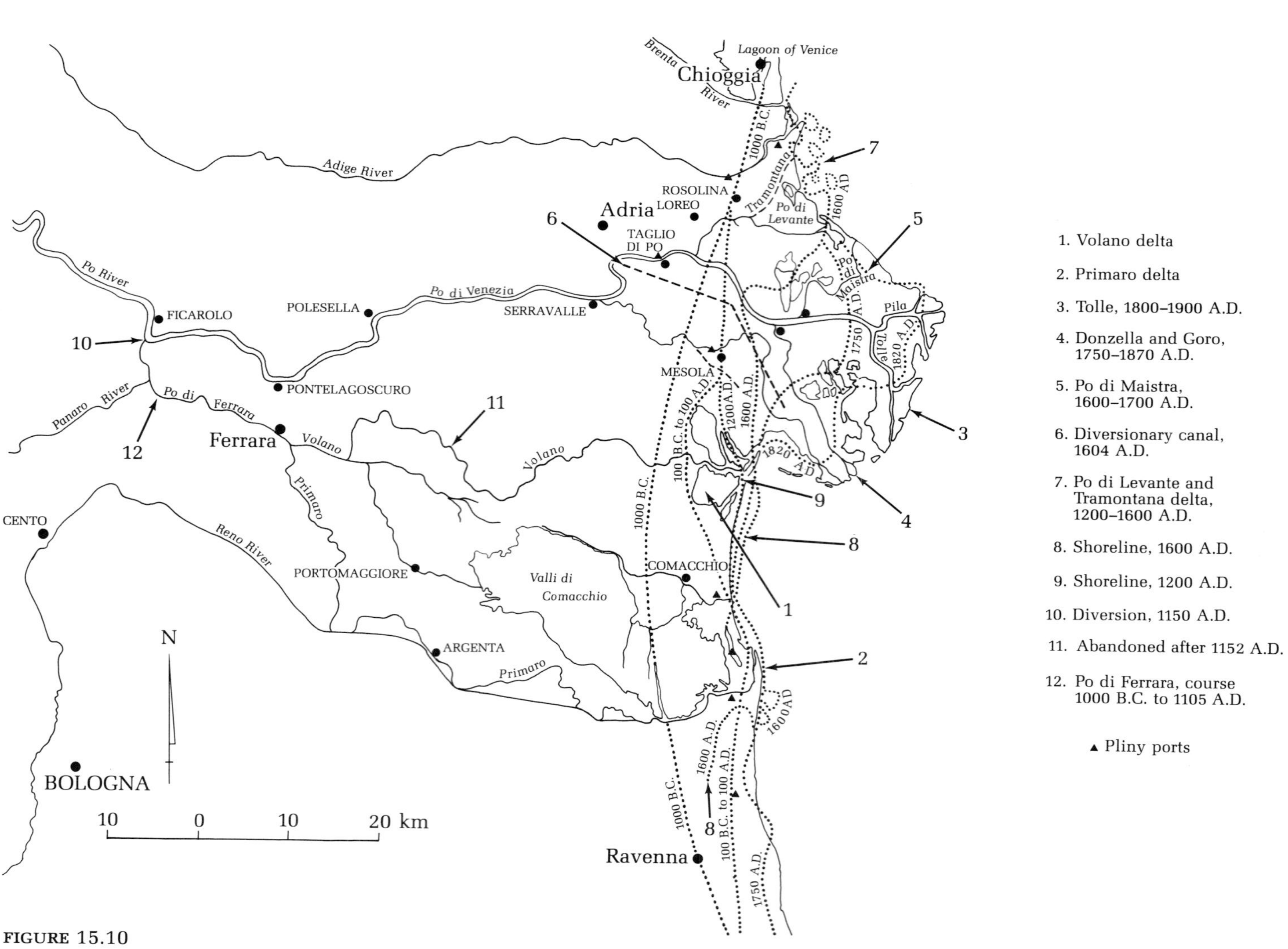

FIGURE 15.10
Historical growth and development of the Po channel and delta. [After Nelson (1970).]

and shifting its bed, but the problem did not become acute until the valley became filled with towns and cities divided by the river into provinces. As the river meandered, a town was first on one bank of the river and then on the other, and its political allegiance shifted accordingly. Town after town had to tear down the parish church and rebuild it farther from the river, and a whole monastery was moved in 1471.

To counter these destructive meanders, the Italians began, in the thirteenth century, to line the channels with dikes. Since a bit of dike merely diverts a flood elsewhere, soon almost the whole river system was diked all the way to the sea.

The early importance of such dikes or levees was suggested by Lyell, who observed that Dante, writing in the fourteenth century, described in the "seventh circle of hell, a rivulet of tears separated from a burning sandy desert by embankments 'like those which, between Ghent and Bruges, were raised against the ocean, or those which the Paduans had erected along the Brenta to defend their villas on the melting of the Alpine snows."

The Po River was diked, but it contained the same amount of water and sediment, and it had no option but to carry it straight to the sea. The delta front, which had advanced only slowly, began to spurt forward at the rate of 27 meters per year. This threatened Venice, and the northern distributary channel was deflected toward the main channel in 1604. Being even more concentrated by dikes, the delta front advanced at the rate of 87 meters per year for 150 years. Since that time, it has continued to advance at differing, but still rapid, rates as more and more dikes have been constructed or old ones have been elevated.

By building forward, the river grew longer, so it had to raise its whole bed in order to restore equilibrium. By 1880, in the vicinity of Ferrara, more than 50 kilometers inland, the Po was almost as high as the roof tops. The whole river system became elevated like the veins on an aging hand. In some years, it was necessary to pile a third of a meter of mud on the levees to prevent flooding. In the subdued phraseology of modern science, the American geologist Bruce Nelson (1970) states: "Thus, the present delta is poised in metastable equilibrium with the regional landforms." We might say the same of the river. After 600 years, men are still winning, but the Po has hardly begun to fight.

DREDGING CHANNELS

If building the levees up leads to undesired and expensive consequences, what about dredging the channel down? This is a very controversial subject indeed in the United States, if for no other reason than because powerful government agencies are arrayed on opposite sides of the question. Two quotations collected by Robert Gillette illustrate this point: "Stream channel alteration under the banner of 'improvement' is undoubtedly one of the most destructive water management practices . . . the aquatic version of the dust-bowl disaster" (Nathaniel P. Reed, Assistant Secretary of the Interior for Fish, Wildlife, and Parks). "American agriculture couldn't survive without it" (Eugene C. Buie, Assistant Deputy Administrator, Soil Conservation Service).

The Corps of Engineers of the U.S. Army began "improving" great American rivers at the direction of the Congress in the 1870s, and it has done a magnificent job of moving sand economically. This was in aid of navigation. Alteration of small rivers in support of agriculture was relatively unimportant until the Watershed Protection and Flood Prevention Act of 1954 was passed. Less than 20 years later, the Soil Conservation Service of the Department of Agriculture had widened, deepened, and straightened more than 13,000 kilometers of streams draining 40,000 square kilometers. The Corps of Engineers dredged only 2400 kilometers during that period.

As the quotations above indicate, there is no consensus about the effects of these changes in stream channels. We may visualize thousands of streams trying to restore equilibrium and tens of thousands of sand waves moving upstream and downstream, but there is no historic precedent for evaluation. No country has ever before had both the urge and the technological resources to alter so many rivers at once. Perhaps in other lands they remember another American experience. Navigation of the Mississippi was modified by cutting straight channels where the river made great curved meanders. Between 1722 and 1884, the cutoffs shortened the river by about 400 kilometers, but the other meanders continued their leisurely growth and, by 1929, the Mississippi was as long as ever.

Lakes

Lakes are transient, trapped bodies of water that are in relatively delicate equilibrium compared to the rivers that feed them. Geologically, they have been rare, and they are unusually abundant now simply because river drainage has had little time to readjust since it was disrupted by the last glaciation. People are accustomed to lakes, however, and are concerned when they are changed; the fact that the lakes will inevitably be drained, filled, evaporated, or poisoned in the natural course of events does not provide a reason to lessen mankind's concern for their immediate future. Nevertheless, it is useful to understand that lakes change naturally, and to understand why they do and how fast. If we attempt to prevent natural change, there may be side effects that are less desirable than the change itself.

WATER LEVEL

A lake is a river to the sky: on the average, a layer of water about 100 centimeters deep evaporates from the whole surface each year. Most lakes are shallow—Lake Erie averages only 18 meters in depth—and, thus, they would not last long if evaporation were not balanced by an influx of replacement water. Consider, for simplicity, a lake without an outlet. If the influx of rivers is reduced by withdrawal of river water for agriculture, cooling water for a power plant, or by climatic change, the lake level will fall. Likewise, if rainfall increases or the lake basin is partially filled with sediment, the level will rise.

Lake Erie illustrates the natural variability of lake levels and the difficulties they pose for society. The lake is typically low in the winter and about a half a meter higher in the summer. Longer cycles of changes in level are more interesting. Rainfall in the Great Lakes drainage basin was above normal from 1875 to 1886 and significantly below normal from 1917 to 1923. Lake Erie is a pool along a water system that connects all the Great Lakes with the St. Lawrence River and the sea. Thus, presumably, its level fluctuates only if that of the whole system does. Nonetheless, the drought of 1923 apparently was accompanied by a pronounced drop in the level of Lake Erie. The drop elsewhere in the lakes caused alarm, and, by 1930, a series of lawsuits successfully limited the withdrawal of water from Lake Michigan at Chicago on the grounds that it was necessary "*to prevent diversions* which were alleged to be lowering the levels of the upper lakes," according to Thomas Langlois. Rainfall in the basin was above normal from 1941 through 1952, and, by the latter year, the level of Lake Erie was 1.5 meters above the recorded minimum of 1926. Langlois notes that "resolutions were introduced in Congress in 1952 *to authorize greater diversions* at Chicago as a means of reducing high lake levels." The idea that the level of the whole Great Lake system can be controlled by turning a vast spigot at Chicago is attractively simple, but the problem is more complex. The Army Corps of Engineers, for example, has diverted more water into the lakes from the natural drainage to Hudson Bay than Chicago withdraws from them. This is an example of artificial stream capture.

Population increases and industrial expansion have intensified the concerns of people in earlier decades with changes in lake levels, but attention has now shifted to the more acute problem of pollution in Lake Erie.

Fluctuating lake levels are not merely a concern of Americans. The Russians also live in a vast, flat land that was glaciated, and there the cumulative effects of climatic change and persistent human tampering have become spectacular. The Volga River once flowed in full volume into the Caspian Sea. Now part of it evaporates behind two great dams, and much more is withdrawn for agriculture. The level of the salty Caspian has dropped at the same time, and its shoreline has migrated 100 kilometers across the gently sloping shelf of the northern part of its bed between 1930 and 1970. One of the main habitats of the sturgeon was the Volga, and, as a result of the degradations that river has suffered, the Russians have had to invent synthetic caviar. Presumably, the lakeshore tourist hotels of the 1930s are in as much trouble as the fisheries.

Man has changed the Volga and, thus, the Caspian. Would there have been stability without his intervention? If tampering with the river stops, will the lake level be stabilized? It seems unlikely. The Caspian may be showing the effects of a major climatic change instead of a mere cycle reinforced by man. The lake level was 100 meters higher at the end of the last ice age, and it was relatively high as late as the end of the Little Ice Age of 1810–1820 A.D. Then it declined very slowly to 1930–1940, when the present rapid drop began. Thus, the trend toward evaporation has persisted for a very long time, and the importance of human influence is difficult to evaluate.

EUTROPHICATION

The health of a lake is a complex function of its biology, chemistry, physics, geology, and age, and of the climate in which it exists. Take first its geology and age (or stage of development). Other things being equal, a lake begins deep and with an irregular bed, but it gradually fills with sediment and becomes a smooth, shallow saucer. Meanwhile, debris from a succession of plants around the edge of the lake helps to convert it to marsh that is invaded by mosses and bog-adapted plants, and these help to convert the lake silt to fertile soil.

Before this happens, however, the life within the lake also goes through various cycles, which depend on the other factors. Algae and other microscopic plants live in the surface waters because they require light for photosynthesis. If a lake is shallow—roughly, less than 15 meters deep—light may penetrate the whole depth and plants can live throughout. Plants cannot live at greater depths, but dead ones rain down from the surface water. The fish and bacteria in the food chain live on the plants, but they can exist at any depth provided food and oxygen are adequate. Oxygen is manufactured in water by the plants and is dissolved from the air at the water's surface. Thus, it forms at or near the surface and must be carried to the depths by plunging of surface waters, which, in turn, depends largely on the climate.

In the tropics and subtropics, lakes are warmest at the top; thus, the water is least dense there, and will not plunge unless driven by winds. Large quantities of organic waste rain down from the warm, fertile waters, and bottom-dwelling bacteria consume it. Commonly, the thriving bacteria exhaust the meager oxygen and the lake becomes poisonous with hydrogen sulphide. A lake in which the oxygen is exhausted in this way by excessive fertility is called eutrophic.

Water has the peculiar property that it reaches its maximum density at a temperature a few degrees above freezing. In polar regions, where the surface rock is always cold, dense water collects on lake bottoms and it is not warmed. When the winter freezing cycle cools the surface waters, they plunge but do not displace the bottom water, which easily becomes stagnant. The organic productivity of the surface waters is low, but, because of the stagnancy, deep polar lakes are commonly eutrophic.

In intermediate latitudes (where industrial civilization is concentrated), the mud of lake bottoms is not as cold as near the poles, but a freezing cycle creates dense water at the surface. Thus, the water of lakes overturns every fall and spring, and oxygen is carried to the bottom. Even so, temperate-zone lakes may temporarily become naturally eutrophic if, for any reason, bacteria exhaust the oxygen before the next overturn. This occurs annually in many lakes.

Lake Erie is a shallow, temperate lake. Why, then, is there so much concern about it becoming eutrophic if this is the natural state of affairs at some stage in the history of a lake? Why is the Federal Water Pollution Control Administration advocating an anti-pollution program that will cost almost $3 billion? The reason is that increased erosion and intense pollution have enormously accelerated the aging. The FWPCA believes that it would have taken Lake Erie another 15,000 years to reach the eutrophic stage naturally and that the lake can be restored by pollution control. Pollution-induced aging of lakes has been observed in many places, and the reality of the acceleration is unquestioned. Lake Zürich in Switzerland was deep and clear only a century ago, and now the end near the city of Zürich is thoroughly polluted and eutrophic. There are many similar examples. What may be questioned is whether the pollution controls, which are worthwhile for many other reasons, can successfully reverse the process. The population living and working around the lake, which was 13 million in 1968, is expected to rise to around 26 million by 2010, and occasions for degrading the lake will be more frequent. Meanwhile, climatic cycles will continue, and use of the lake may expand to fit the maximum that it will bear under the most favorable conditions.

Although the prognosis for temperate lakes is not very favorable as people concentrate around them, it is well to remember that it is better than for either tropical or polar lakes. If population pressures cause increased use of these naturally eutrophic lakes, it may take only a few decades to induce massive pollution.

Estuaries and Deltas

When a river reaches sea level, all its potential energy is gone, although it may have enough momentum to push a jet of fresh water into the sea. The river dumps its load of sediment in a delta. However, the sea is full of energy—in the form of currents, waves, tides, and fluctuations in sea level—that acts upon the sediment. Thus, the sediment is ordinarily eroded and transported away from the delta and distributed along the shoreline and in the deep sea. There are only two circumstances in which delta sediments are likely to stay in place: first, if the environment is such as to protect them from the energy of the sea; and, second, if a great river deposits them so fast that the sea cannot carry them away.

ESTUARIES

Sheltered near-shore environments arise in many ways. Among other things, the tidal range is small in many regions, and wind-driven waves are relatively ineffective in small seas and enclosed gulfs. The most widespread protective environments, however, are estuaries, or drowned river mouths, and lagoons between long islands of sand and the mainland. Estuaries and lagoons form some 80–90% of the Atlantic and Gulf coasts of the United States and 10% of the mountainous west coast.

Estuaries and lagoons grade into one another because they are produced by the same forces acting in the same sequence in more or less different geography. When sea level was last lowered during the ice ages, rivers cut their beds downward to reach an equilibrium grade. When sea level rose, the deep river valleys became drowned. Wider drowning has occurred in regions that have subsided because of glacial loading or tectonic deformation. The drop in sea level also exposed the sediments of the continental shelf to wave action and created long straight beaches—some of which migrated inland when sea level rose. Lagoons migrated inland behind the beaches.

The coastal geography we see is mainly an expression of the local imbalance between coastal deposition and marine erosion during the last 4000–6000 years since sea level became relatively stabilized. The estuaries and lagoons are filling, like lakes, with sediment and organic debris, and, in the geologic twinkling of an eye, most will vanish. Meanwhile, they are of great social importance, at least in the United States, because their protected waters are the focus of recreation, commerce, commercial fishing, and housing. With such multiple usage, the main problem is pollution, as it is in lakes. However, in California, where coastal flatland is rare and valuable, there is a tendency to destroy small estuaries completely by filling them with sediment to provide foundations for buildings.

Human earth movers can easily conquer almost any estuary or lagoon, but they only accelerate natural processes which can be surprisingly fast. The Amazon River valley was an estuary 2400-kilometers long when sea level stabilized. Since then, the main river has filled the valley with a narrow delta from the interior to the sea. Meanwhile, the smaller tributaries, unable to keep up, have become partially dammed. They form long, narrow lakes,

TABLE 15.1
Estuarine habitat areas lost to dredging and filling operations in the United States. Most of the loss in California is around San Francisco Bay. (One acre = 4050 square meters.)

	Area of estuaries (10^3 acres)			
State	*Total area*	*Basic area of important habitat*	*Area of basic habitat lost by dredging and filling*	*Habitat lost (%)*
Alabama	530	133	2	1.5
Alaska	11,023	574	1	0.2
California	552	382	256	67.0
Connecticut	32	20	2	10.3
Delaware	396	152	9	5.6
Florida	1,051	796	60	7.5
Georgia	171	125	1	0.6
Louisiana	3,545	2,077	65	3.1
Maine	39	15	1	6.5
Maryland	1,046	376	1	0.3
Massachusetts	207	31	2	6.5
Michigan*	152	152	4	2.3
Mississippi	251	76	2	2.2
New Hampshire	12	10	1	10.0
New Jersey	778	411	54	13.1
New York (Atlantic)	377	133	20	15.0
New York (Great Lakes)	49	49	1	1.2
North Carolina	2,207	794	8	1.0
Ohio*	37	37	<0.5	0.3
Oregon	58	20	1	3.5
Pennsylvania*	5	5	<0.5	2.0
Rhode Island	95	15	1	6.1
South Carolina	428	269	4	1.6
Texas	1,344	828	68	8.2
Virginia	1,670	428	2	0.6
Washington	194	96	4	4.5
Wisconsin*	11	11	<0.5	<0.1
Total	26,618†	7,988†	569†	7.1‡

Source: Stratton (1969).
*In Great Lakes, only shoals (areas less than 6 feet deep) were considered as estuaries.
†Discrepancy caused by rounding.
‡Average.

some of them as much as 30 meters deep, where they join the Amazon.

Estuaries may be filled with mud eroded from the sea floor, in favorably situated bays. Cambridge, England, for example, was very near the sea in historic times, and the farmland to the north is called the "Isle of Ely" because it was once an island. Marine muds are filling in the remainder of the ancient

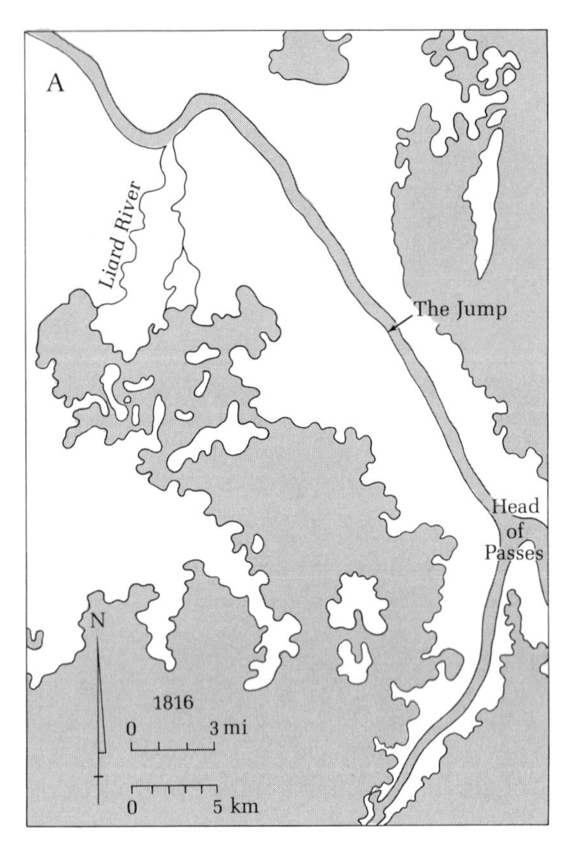

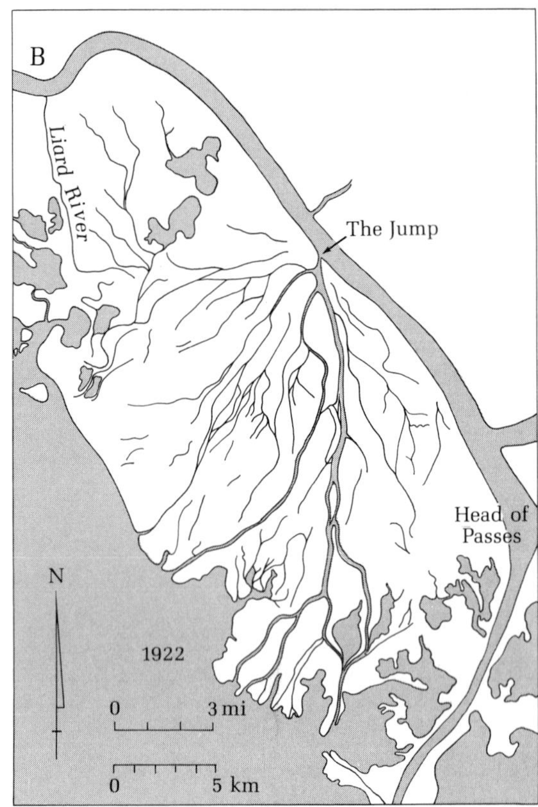

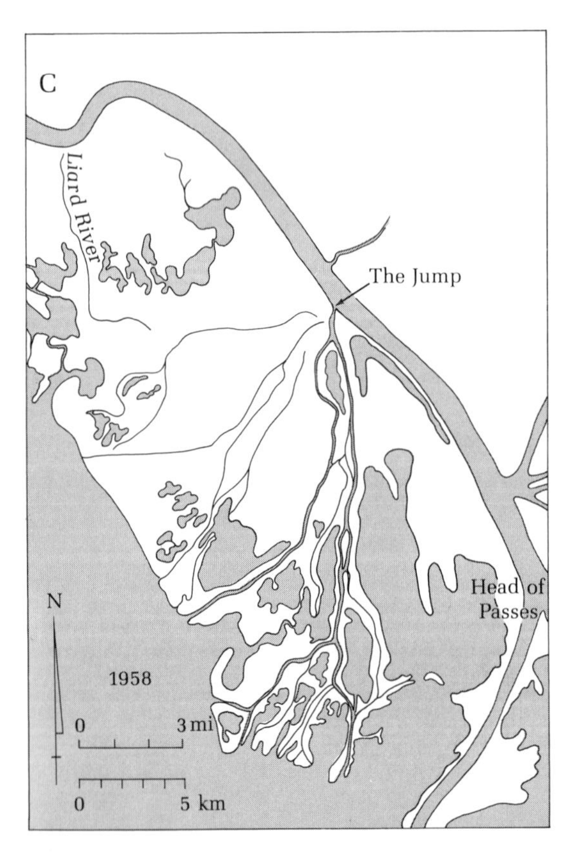

FIGURE 15.11

Three stages in the development of a microdelta by breaching of a natural levee of the Mississippi River: *A*, an embayment resulting from subsidence following the abandonment of the Mississippi River subdelta (1816); *B*, the microdelta near the height of depositional activity (1922); *C*, the microdelta showing subsidence effects during the present process of abandonment (1958). [After Morgan (1967), copyright 1967 by the American Association for the Advancement of Science.]

bay, called The Wash, at a rate that has advanced the broad shoreline 600 meters in the last century.

DELTAS

Deltas are very flat and marshy, and the only topographic relief may consist of natural levees that rivers build by deposition when they gently overspill their channels. If overspilling becomes deeper during a flood, the river will erode down through the levee and deposit a little microdelta. This happened four times between 1839 and 1891 at the mouth of the Mississippi River, and each of the resulting microdeltas was 10–20 kilometers in diameter. After it builds up for a while, a microdelta becomes higher than some nearby area and the next breakthrough in the levee occurs there. Because the whole delta is sinking, the edges of the microdelta are soon partially submerged; after a few decades, an overlapping microdelta may be deposited.

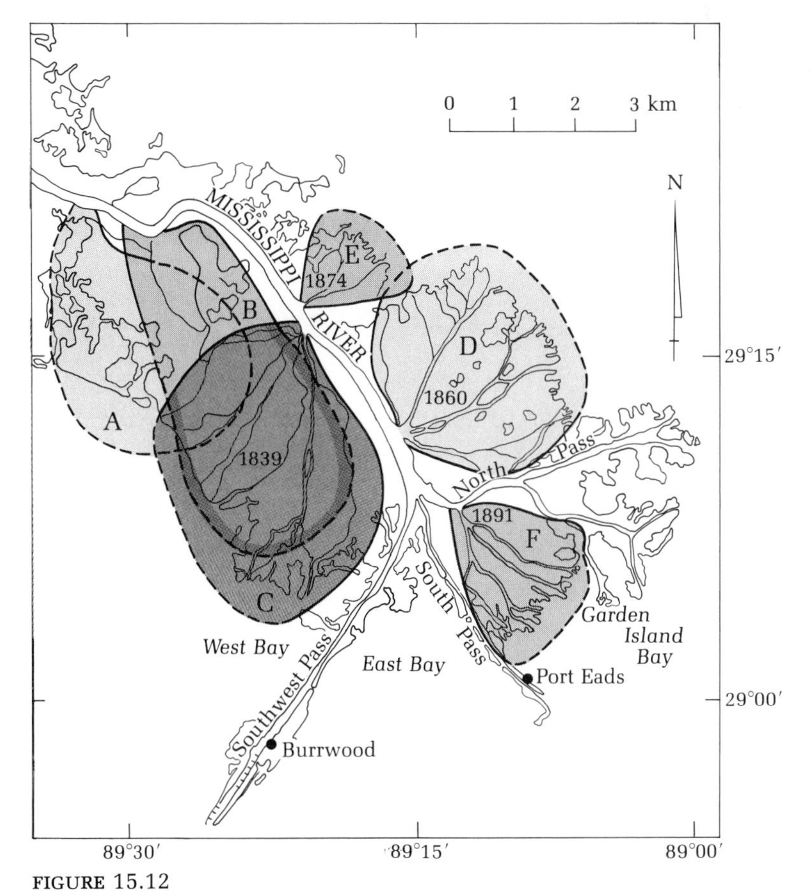

FIGURE 15.12

A century of microdelta deposition on the Mississippi delta front. Compare with Figure 15.11, which shows the evolution of microdelta *C*. [After Coleman and Gagliano (1964).]

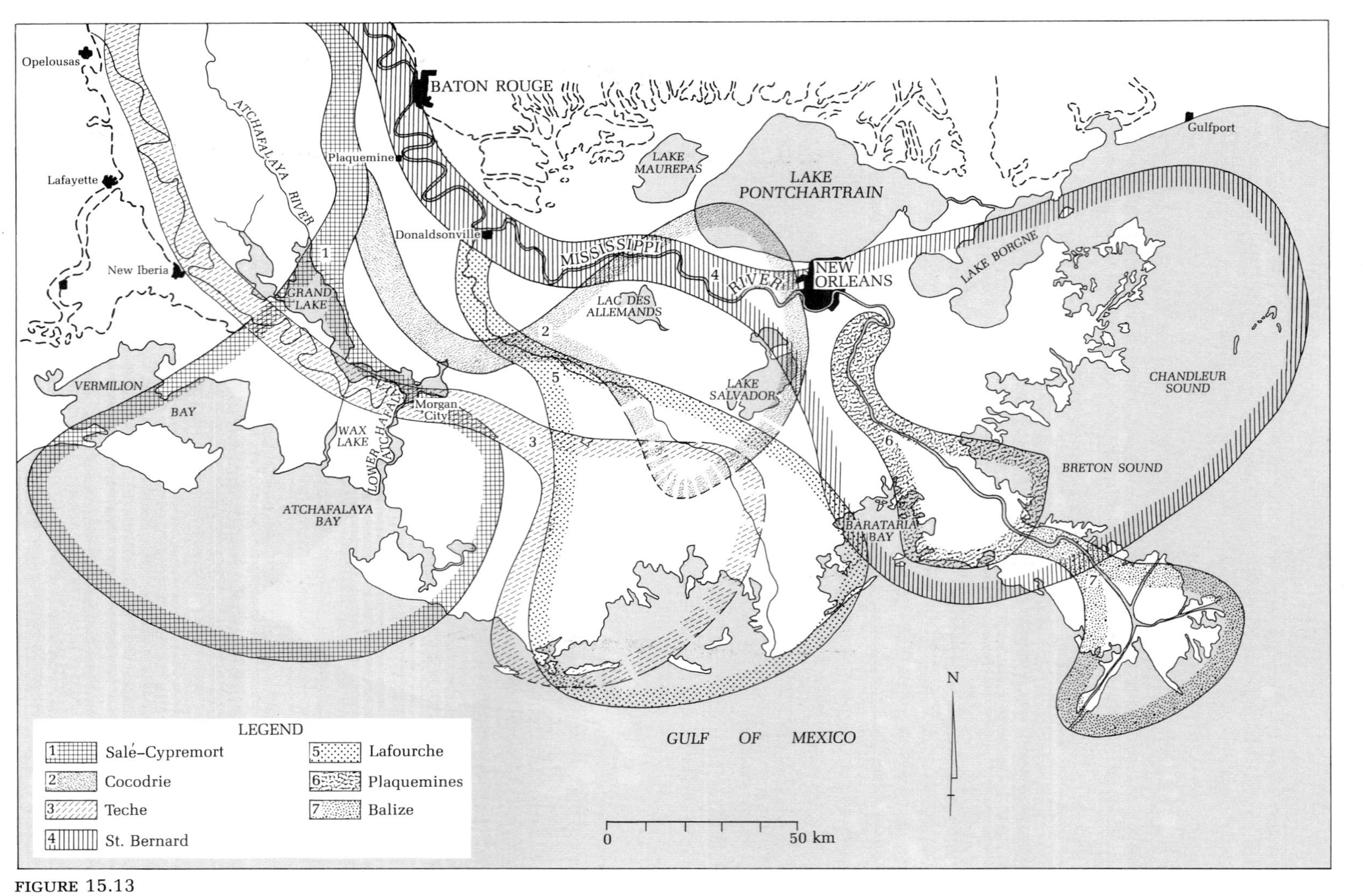

FIGURE 15.13
Location (*above*) and age (*below*) of discernible subdeltas of the Mississippi. [After Kolb and Van Lopik (1966).]

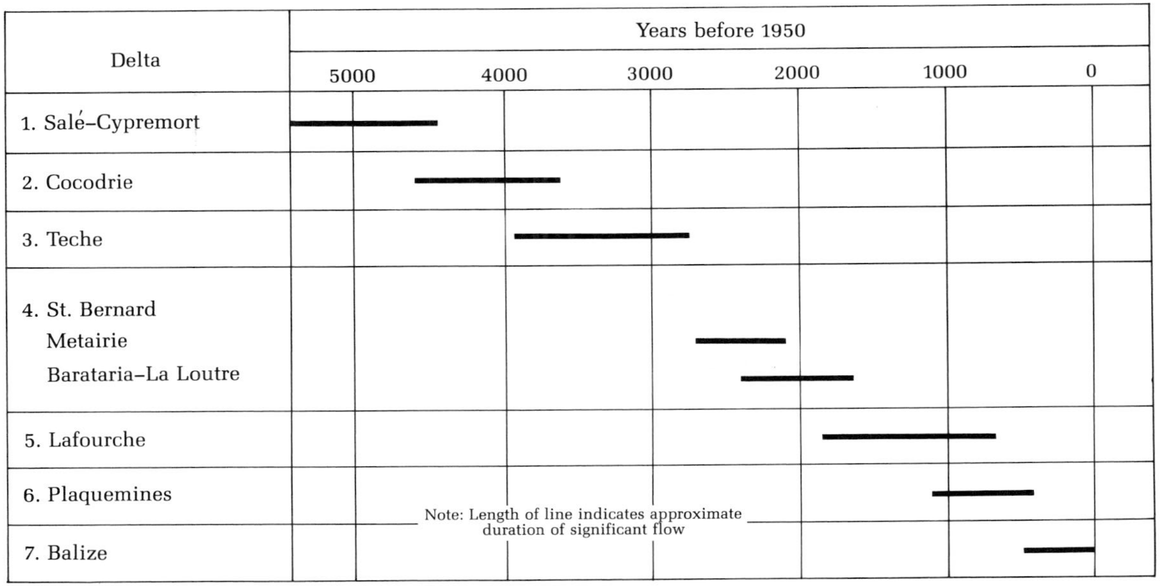

Delta	Years before 1950					
	5000	4000	3000	2000	1000	0
1. Salé–Cypremort						
2. Cocodrie						
3. Teche						
4. St. Bernard Metairie Barataria–La Loutre						
5. Lafourche						
6. Plaquemines						
7. Balize						

Note: Length of line indicates approximate duration of significant flow

After about 1000 years of deposition of overlapping microdeltas, the Mississippi has built a subdelta, about 65–100 kilometers in diameter, that is substantially above the surrounding region, which has been subsiding. Meandering inevitably causes the river to migrate into one of these low areas and begin to build another subdelta of overlapping microdeltas. One thousand years later, it may shift back to reoccupy an earlier site on the great growing delta. Seven main subdeltas have formed at the mouth of the Mississippi in the last 5000 years, and the river at the latitude of New Orleans has shifted sideways about 160 kilometers.

The active shifting of channels is irresistible on any long time scale, and this, plus subsidence, accounts for the scarcity of surviving ancient cities on large deltas.

It seems appropriate that the Menderes River in Turkey, which the ancient Greeks called the Meander, should provide examples of the fate of delta cities. Minoan and Mycenaean traders established trading posts along what is now the Turkish coast. Later, the Ionian Greeks founded permanent cities, several of which were larger than any in Greece itself, including Athens. Two, Miletus and Priene, were ports separated by a broad gulf 30 kilometers wide. About 500 B.C., Hippodamus of Miletus laid out the first "modern" rectangular street pattern, and Aristotle referred to him as the pre-eminent town planner of the times. Priene adopted the same orderly plan. Between them now winds the Menderes River "with a contrasting lack of system towards a receding sea," to quote Sir Mortimer Wheeler. The river has advanced its delta 16 kilometers and filled the estuary in 2500 years.

Ephesus, in the next embayment to the north, was once the greatest metropolis of Asia, and it has suffered the same fate. Founded by Amazons, according to Pindar, it was settled about 1100 B.C. by Ionian Greeks, and it grew to be both a great port and the principal shrine of the goddess Diana. The city's theater, which held 24,000 people, was the scene of the riot when St. Paul came to preach his Epistle to the Ephesians (Acts 14). The city's last great temple, built 300 years earlier in the time of Alexander the Great, was one of the seven wonders of the world. The great port is now completely silted. Subsidence of the ground has carried even the marble pavement of the harbor road below the level of the marshy waste that was the port. What of the wondrous temple, a great stone structure on the uncompacted soil of a subsiding delta? Mr. Wood of the British Museum searched from 1863 to 1869 and finally found it, sunk 9 meters under the level of the swamp.

Summary

1. Sediment flows in an essentially continuous "river of sand" from mountains to the deep sea. If the flow is changed at any point, the whole flow gradually responds.

2. Each river basin has various sizes of floods over a period of years. The greatest floods are the least frequent and, thus, those for which there is the least preparation.

3. Most sediment is transported during floods.

4. A normal river flowing on an unconfined sedimentary bed is in a delicate equilibrium, in which its capacity to do work equals the work that it actually does.

5. The equilibrium of a stream may be disturbed by natural or artificial causes. The stream then deposits or erodes until its slope and hydraulic depth are in a new equilibrium.

6. Lakes are transient, trapped bodies of water that are in an equilibrium even more delicate than that of rivers. Geologically, they have been rare.

7. Lakes change and are gradually filled whether man intervenes or not, but man can accelerate the changes enormously.

8. Rivers deposit sediment when they reach the sea and their potential energy is exhausted. The sediment stays put in a delta only if the energy of the ocean is unable to move it.

9. Deltas subside because of sediment loading, sediment compaction, and, in some places, because the region is sinking tectonically. Cities built on them also subside.

Discussion Questions

1. How is the potential energy of a river expended?

2. What happens to the sand in a river channel during a major flood?

3. Why do so many river valleys in the Northern Hemisphere have terraces?

4. How would you have controlled the Po River if you had vast resources and political power?

5. Why are there now so many lakes, if they have been rare during most of geologic time?

6. What happens to a lake that has no outlet, if the rivers that feed it are diverted to irrigate farms or provide water for cities?

7. Why are there so many estuaries on the east coast of the United States?

References

Coleman, J. M., and S. M. Gagliano, 1964. Cyclic sedimentation in the Mississippi River deltaic plain. *Gulf Coast Ass. Geol. Soc. Trans.* 14:67–80.

Emery, K., 1967. Estuaries and lagoons in relation to continental shelves. *In* G. H. Lauff, ed., *Estuaries* (AAAS Publ. 83), pp. 9–14. Washington, D.C.: American Association for the Advancement of Science.

Engel, A. E. J., 1969. Time and the earth. *Amer. Sci.* 57:458–483. [A wide-ranging iconoclastic view of the earth sciences today.]
Gillette, R., 1972. Stream channelization: conflict between ditchers, conservationists. *Science* 176:890–894. (Quoted from p. 890.)
Kolb, C., and J. Van Lopik, 1966. Depositional environments of the Mississippi River deltaic plain—southeastern Louisiana. In *Deltas,* pp. 17–62. Houston: Houston Geological Society.
Langbein, W., and L. Leopold, 1968. River channel bars and dunes—theory of kinematic waves. *U. S. Geol. Surv. Prof. Pap.* 422-L. 20 p.
Langlois, T., 1954. *The Western End of Lake Erie and its Ecology*. Ann Arbor: J. W. Edwards. (Quoted from p. 49.)
Leopold, L,, and W. B. Langbein, 1966. River meanders. *Sci. Amer.* 214(6): 60–70. (Available as *Sci. Amer.* Offprint 869.) [Rivers change their courses like a moving snake.]
——, M. G. Wolman, and J. P. Miller, 1964. *Fluvial Processes in Geomorphology*. San Francisco: W. H. Freeman and Company. [An exceptionally analytical and lucid account of a subject flooded with facts.]
Lyell, C., 1835. *Principles of Geology*. London: John Murray. (Quoted from vol. 1, p. 277.) [The best textbook of geology ever written.]
Mackin, H., 1948. Concept of the graded stream. *Geol. Soc. Amer. Bull.* 59: 463–512. (Quoted from p. 496.) [Rivers generally are in equilibrium and react if disturbed. This classic paper somewhat overemphasizes the importance of slope.]
Moore, J. G., Jr., responsible official, 1968. *Lake Erie Report*. Washington, D.C.: Federal Water Pollution Control Administration.
Morgan, J. P., 1967. Ephemeral estuaries of the deltaic environment. *In* G. H. Lauff, ed., *Estuaries* (AAAS Publ. 83), pp. 115–120. Washington, D.C.: American Association for the Advancement of Science.
Nelson, B., 1970. Hydrography, sediment dispersal, and recent historical development of the Po River Delta, Italy. *In* J. Morgan, ed., *Deltaic Sedimentation, Modern and Ancient* (Soc. Econ. Paleontol. Mineral. Spec. Publ. 15), pp. 152–184. Tulsa: Society of Economic Paleontologists and Mineralogists.
Powers, C. F., and A. Robertson, 1966. The aging Great Lakes. *Sci. Amer.* 215(5):94–104. (Available as *Sci. Amer.* Offprint 1056.)
Russell, R. J., 1967. Origins of estuaries. *In* G. H. Lauff, ed., *Estuaries* (AAAS Publ. 83), pp. 93–99. Washington, D.C.: American Association for the Advancement of Science.
Steers, J. A., 1967. Geomorphology and coastal processes. *In* G. H. Lauff, ed., *Estuaries* (AAAS Publ. 83), pp. 100–107. Washington, D.C.: American Association for the Advancement of Science.
Stratton, J., chairman, 1969. *Our Nation and the Sea* (Report of the Commission on Marine Science, Engineering, and Resources). Washington, D.C.: U.S. Government Printing Office.
Twenhofel, W. H., 1939. *Principles of Sedimentation*. New York: McGraw-Hill. [A systematic environmental analysis.]
Wheeler, M., ed., 1970. *Swans Hellenic Cruise Handbook*. Norwich: Jarrold and Sons. (Quoted from p. 183.)

16

THE SEDIMENTARY CYCLE AT SEA

The face of places, and their forms decay;
And that is solid earth, that once was sea:
Seas, in their turn, retreating from the shore,
Make solid land, what ocean was before.

Ovid, METAMORPHOSES

And this devouring sea, that naught doth spare,
The most part of my land hath washt away
And throwne it up unto my brothers share. . . .

Edmund Spenser, THE FAERIE QUEEN

21 MILES OF CALIFORNIA COAST PUT ON 5-YEAR 'DEATH NOTICE'

Headline in the LOS ANGELES TIMES, 21 November 1971

In the last few chapters, we have followed the material weathered from solid rock down to the sea, where the potential energy of rivers is exhausted. The movement of most of the material does not stop there, however, either because it still has potential energy itself or because it draws on the energy of the sea. It moves in three ways—in solution, in suspension, and along the bed of the sea. The fate of dissolved material depends on its geochemistry. Materials with long residence times, such as sodium, are mixed uniformly throughout the world ocean. Those with short residence times, such as calcium, tend to precipitate nearer river mouths or where organisms are concentrated, and are not uniformly mixed.

Clay remains finely suspended in river water. When it reaches the sea, however, it tends to form clumps, because salt water is a good electrical conductor and clay particles often have surface charges. The clumps are heavier and, thus, more difficult to keep in suspension; but they are also fluffy, and do not fall very rapidly. Consequently, ocean currents disperse fine sediment for hundreds of kilometers from the shore before the finest material has time to settle through 5 kilometers of water. It falls as a blanket on the sea floor, thickest near shore and gradually thinning farther out.

The coarser material—sand and silt—tends to be deposited in deltas and estuaries, but not necessarily for very long. In shallow water, it is exposed to the energy of waves and currents, which erode and disperse it along the shoreline and into deeper water. Fluctuations in sea level and submarine processes bring much of it to the top of the continental slope in anything from 10 to 1,000,000 years. There, its potential energy is readily available, and it slumps, cascades, or flows rapidly to the deep-sea floor. It spreads out as a thin sheet in the valleys and plains between submarine mountains, and forms distinctive deposits originating at the mouths of the submarine canyons of the continental slope.

The River of Sand

Standing in the surf, we commonly see sand suspended in the ocean after each wave plunges and breaks. This settles quickly—but, during the interval it is suspended, it moves with the water, and so it comes to rest a small distance from where it began. In this manner, wave by wave and in an endless series of small steps, rivers of sand drift along the beaches of the world.

The quantity of drifting sand varies markedly, ranging from 25,000 cubic meters per year at Atlantic Beach, North Carolina, to 150,000–380,000 cubic meters per year along the New Jersey shore, and up to 750,000 cubic meters

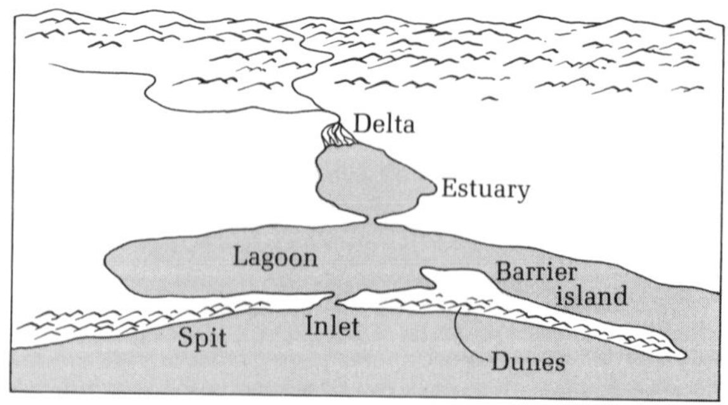

FIGURE 16.1
Features of a shoreline with a sand barrier and a lagoon.

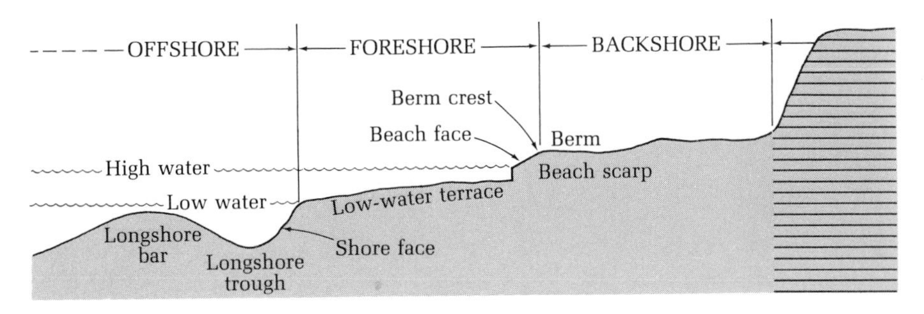

FIGURE 16.2
The principal subdivisions of a beach and of the adjacent shallow-water area. [From Shepard (1973).]

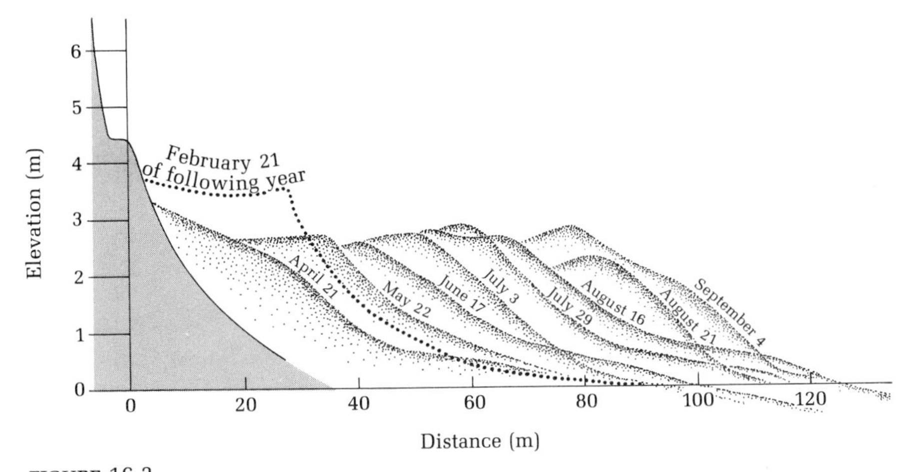

FIGURE 16.3
Growth of the berm at Carmel, California, during the spring and summer is indicated by this series of dated slopes, based on actual measurements. The vertical dimension is exaggerated 10 times. The dotted line shows how the berm was cut back during following winter. [After Bascom, "Beaches," copyright © 1960 by Scientific American, Inc. (all rights reserved).]

per year off Oxnard, California. This means that such amounts flow past each point on the beach in a year's time—or that, if a dam is constructed anywhere across the drift, such amounts will collect behind it in one year. The rate at which individual grains of sand move is even more variable, but the maximum may be 1500–3000 meters per day in the surf zone when sand waves are moving 15–30 meters per day. The daily rate averaged for a whole year is much less.

SOURCE AND DISPOSITION OF SAND

The ultimate source of the littoral drift, or the river of sand, is the erosion of the sea floor, the erosion of the shore, or the fresh deposits off river mouths,

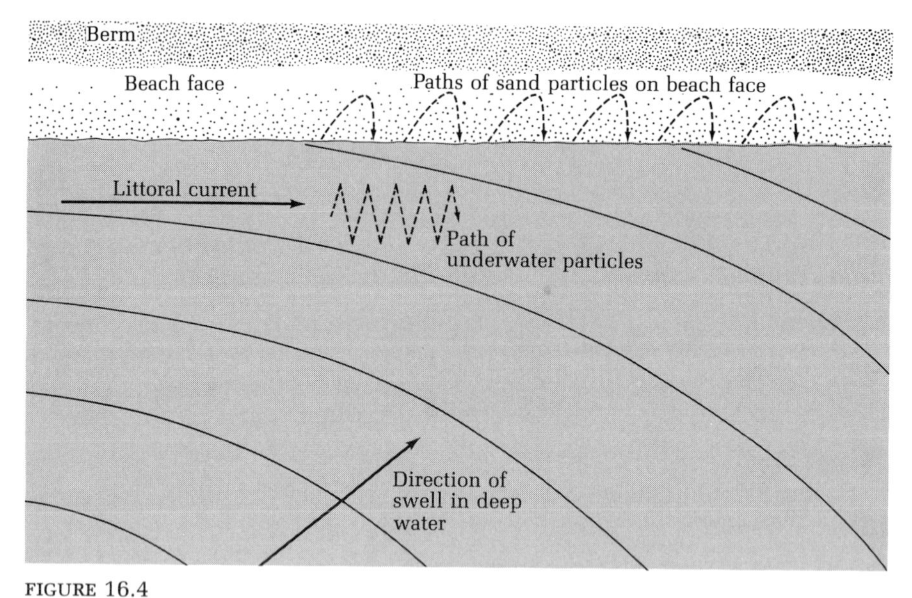

FIGURE 16.4

Littoral current, a current running parallel to the beach, is set up when waves move toward the beach at an oblique angle. Under such conditions, sand grains lifted by the surf, normally moved at right angles to the beach, are transported with the current. [From Bascom, "Beaches," copyright © 1960 by Scientific American, Inc. (all rights reserved).]

depending on the geography. An eroded delta is constantly replaced, and the drift may be in equilibrium in such an environment. Shore and sea-floor erosion products are not replaced, however; thus, there can be no long-term equilibrium of littoral drift. Many shorelines that are the source of littoral drift are the unconsolidated terminal moraines of the last glaciation. Cape Cod and Long Island are examples, and cliffs of the former have been worn back 300 meters in the last century. Solid rock is worn only very slowly by waves; thus, the littoral drift in the area may be minimal, once the morainal sand is eroded away. Likewise, erosion of the sea floor may only continue at the present rate until waves adjust the bottom profile to fit the recent rise in sea level. Once the waves have removed all the sand they can reach, the offshore source will be exhausted until sea level fluctuates again.

The river of sand comes from some source in a small area and flows along the shore. Where does it go? Once again, it depends on the geography. The eastern end of Long Island is being eroded, and sand drifts to the western end and extends the Rockaway beaches and Coney Island toward lower New York Bay. The coast of New Jersey to the south is also being eroded; a river of sand is moving north, and extending Sandy Hook toward New York Bay. Thus, the morainal sands are merely being shifted from more exposed points to a protected embayment. They remain on the edge of the land.

The geography of Southern California is entirely different. The sand is derived almost entirely from small rivers; the beaches are bounded by rocky

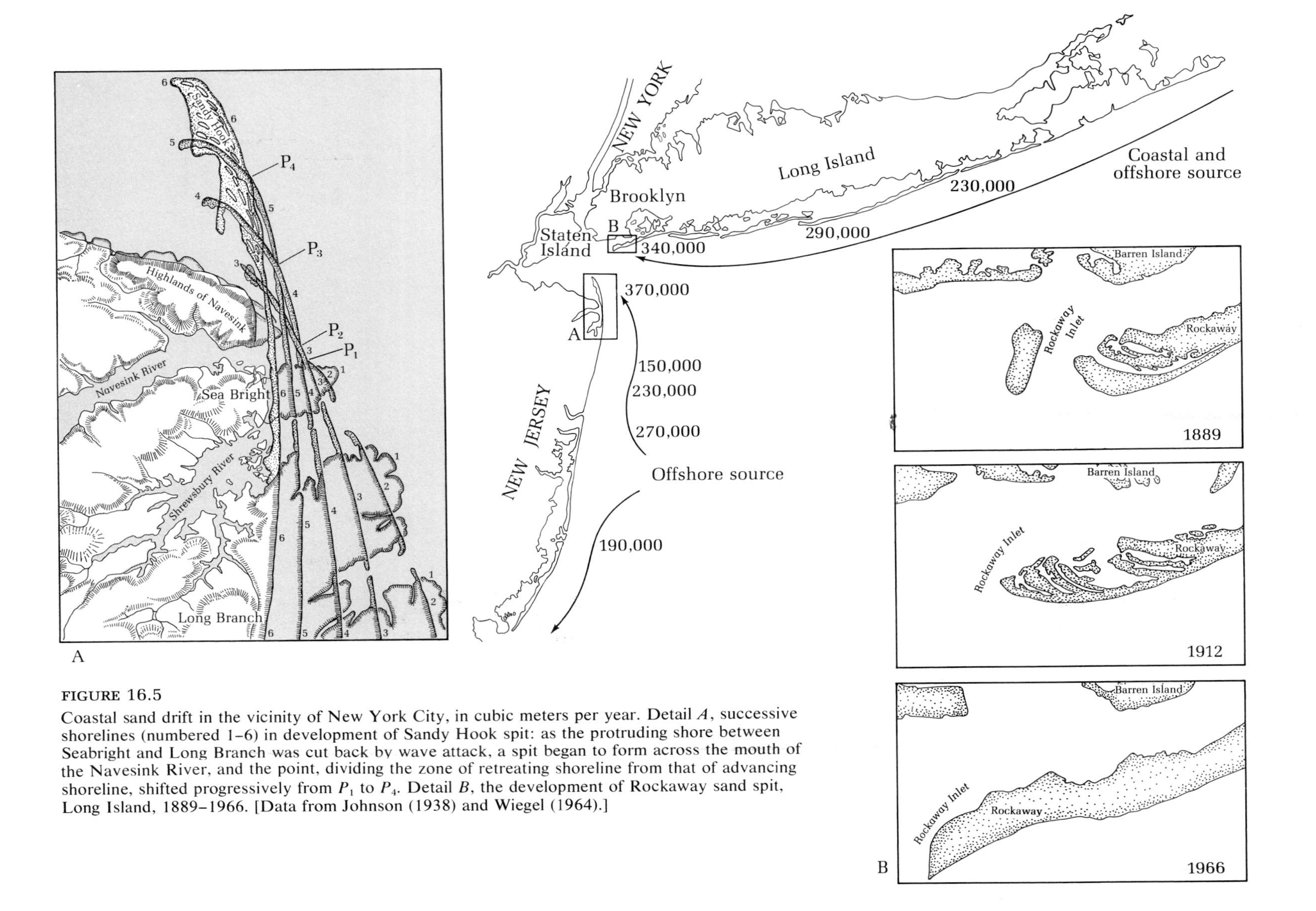

FIGURE 16.5

Coastal sand drift in the vicinity of New York City, in cubic meters per year. Detail *A*, successive shorelines (numbered 1–6) in development of Sandy Hook spit: as the protruding shore between Seabright and Long Branch was cut back by wave attack, a spit began to form across the mouth of the Navesink River, and the point, dividing the zone of retreating shoreline from that of advancing shoreline, shifted progressively from P_1 to P_4. Detail *B*, the development of Rockaway sand spit, Long Island, 1889–1966. [Data from Johnson (1938) and Wiegel (1964).]

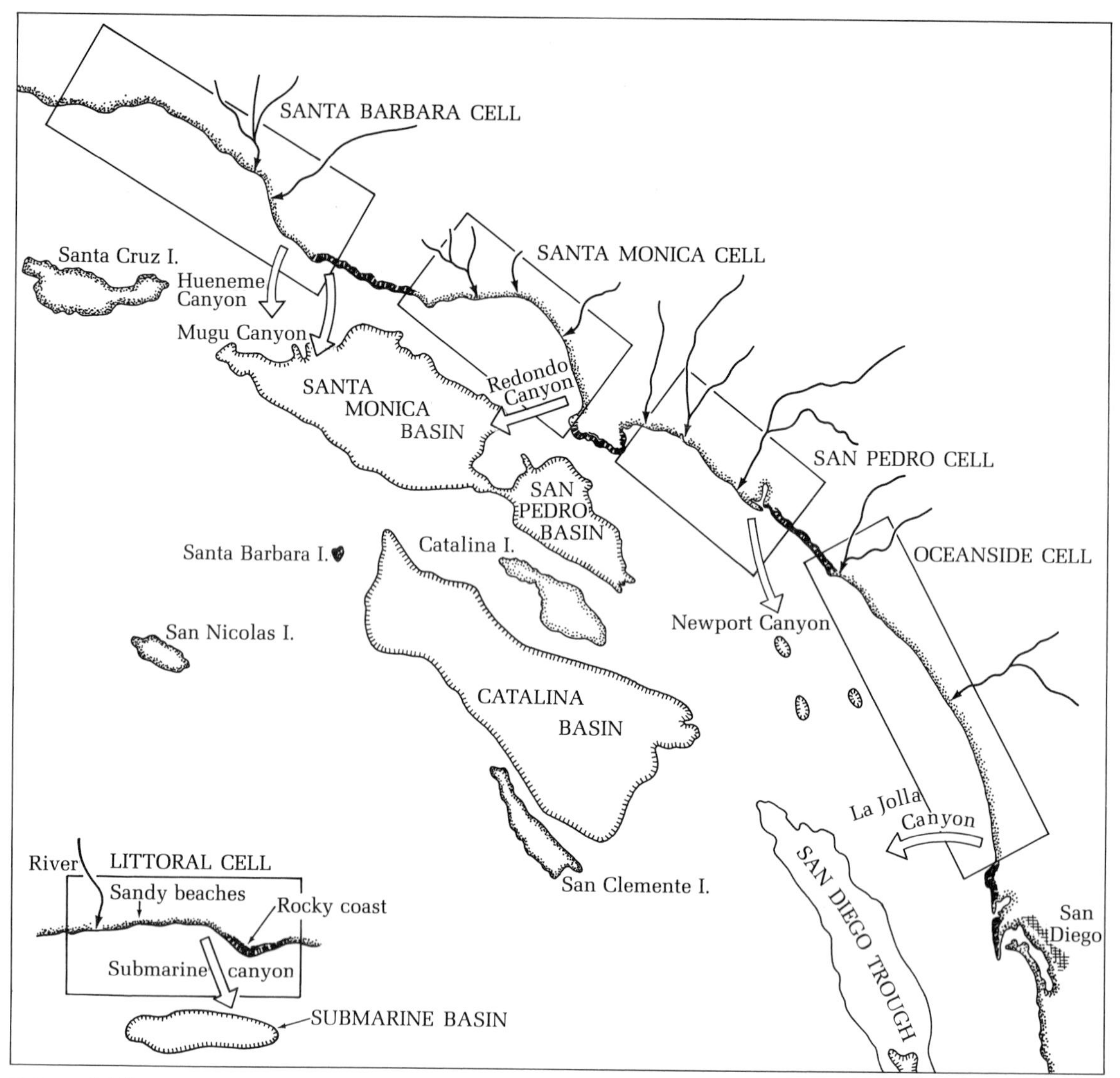

FIGURE 16.6

Sedimentary cells along the southern California coast; arrows indicate the loss of sand into the heads of submarine canyons at the southern end of each cell. [After Inman and Frautschy (1966).]

headlands; the continental shelf is narrow, and submarine canyons lead to deep water. Typically, sand is deposited at a river mouth, drifts along the beach, is deflected at a headland into a submarine canyon, and disappears into the sea.

GAPS IN THE RIVER OF SAND

Any long shoreline has gaps in the beaches that may also seem to be breaks in the river of sand—which is an absurdity. The gaps arise primarily because two fluid flows intersect at the shoreline—namely, the longshore or littoral drift and the cross-shore flow of rivers and tides. Storm surges also introduce an additional flow from the sea that breaks across the littoral drift into lagoons. The water mixes without difficulty, but the particulate sand cannot adjust in the same way. All it can do is go downstream in whatever the net direction of the different flows happens to be.

The simplest case to consider is that of a long barrier island maintained by littoral sand drift but cut by inlets through which much water flows but little sand. Fire Island and Jones Island, off Long Island in New York, provide an excellent example, They are separated by Fire Island Inlet, across which 120,000–460,000 cubic meters of sand drift each year. This drift occurs in two modes, of which the first is a relatively continuous drift which shifts offshore at the mouth of the inlet and forms a bar. This occurs because the jet of tidal water that moves in and out of the inlet loses its sediment-carrying capacity as it spreads out in the sea.

The other mode is for Fire Island, which is updrift, to be extended until it overlaps the tip of Jones Island by a few kilometers. Thus, the inlet, once perpendicular to the shore, becomes almost parallel. This is the present state of the inlet, in part because of artificial stabilization. Elsewhere, however, such inlets eventually jump back to near their earlier positions after the long inlets

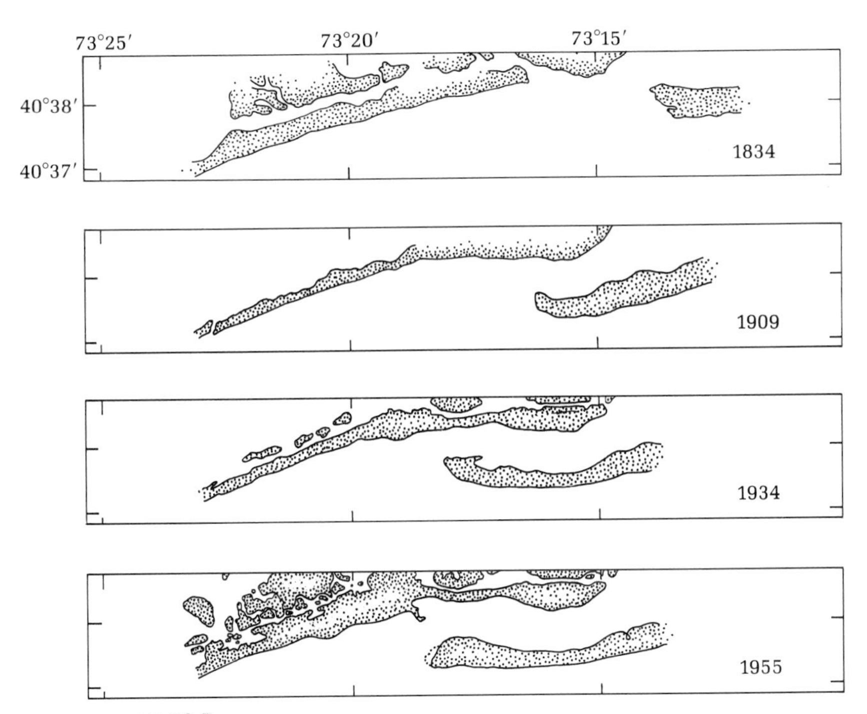

FIGURE 16.7
Migration of Fire Island inlet (Long Island, New York), as shown by high-water shore-line changes, 1834–1955. [After Taney (1961).]

become choked with sand. A storm surge might be the trigger for the jump. Once it occurred, the former tip of Fire Island would be joined to Jones Island, and, presumably, a new wave of sand would move down the beach. Such sand waves are known to drift along these beaches, although their source is obscure.

EQUILIBRIUM AND DISEQUILIBRIUM

The littoral drift is influenced by (1) the rate of sediment supply and (2) the availability of energy to transport it. If they are balanced, the river of sand is in equilibrium. If they are not, the system seeks to establish equilibrium by depositing or eroding. It is useful to visualize this equilibrating process in two components, one acting perpendicular, and the other parallel, to the beach. Consider first the perpendicular equilibrium. If waves of a certain size impinge on a sandy shore for very long, they establish a profile of equilibrium on the bottom. That is, the waves reshape the bottom until all their energy is used up in stirring sand and running up the beach slope. Any excess sand is deposited on the beach.

In general, winter waves are relatively large, and a wide, gentle slope is required on which they can expend their energy. Thus, the beach is eroded and sand waves move offshore to reduce the average slope of the shallow bottom and form bars when they come to rest. The summer waves are usually relatively small, and they erode the shallow sea floor and deposit surplus sand in a stockpile on the beach until a steeper, narrower profile of equilibrium is established.

In addition to this annual migration of sand, there are other cycles, both longer and shorter, connected with fluctuations of sea level and wave height. Most notably, the profile of the shore varies with the tidal cycle on a daily and monthly scale. In addition, the profile adjusts to the less frequent but more persistent changes of sea level associated with climatic variations and particularly the great fluctuations of glacial periods.

The sand movement parallel to the beach also fluctuates—during storms, annually, and over longer periods. The sand drift increases with stirring by large waves for the same reasons that the sediment load of a river increases during floods. The annual fluctuations are related to the onshore–offshore component. Much of the drift is stockpiled on the beach during the summer and, thus, moves only in winter.

Moving sand waves with a wavelength of 1500–3000 meters and a height of 1–2 meters occur in the North Sea. These waves apparently intersect the shoreline in Denmark, where they form horizontal undulations in the shoreline with about the same wavelength and a "height" of 60–90 meters. Similar undulations are observed along the shore of Long Island and doubtless are common elsewhere. They may arise when sand dunes or submarine

TABLE 16.1
Measured rates of sand drift along coasts

Location	Predominant direction of drift	Rate of drift (m^3/yr)	Method of measure of rate of drift	Years of record
ATLANTIC COAST				
Suffolk Co., N.Y.	W	230,000	Accretion	1946–1955
Sandy Hook, N.J.	N	376,000	Accretion	1885–1933
Sandy Hook, N.J.	N	334,000	Accretion	1933–1951
Asbury Park, N.J.	N	153,000	Accretion	1922–1925
Shark River, N.J.	N	230,000	Accretion	1947–1953
Manasquan, N.J.	N	275,000	Accretion	1930–1931
Barneget Inlet, N.J.	S	190,000	Accretion	1939–1941
Absecon Inlet, N.J.	S	306,000	Erosion	1935–1946
Ocean City, N.J.	S	306,000	Erosion	1935–1946
Cold Spring Inlet, N.J.	S	153,000	Accretion	–
Ocean City, Md.	S	115,000	Accretion	1934–1936
Atlantic Beach, N.C.	E	22,600	Accretion	1850–1908
Hillsboro Inlet, Fla.	S	57,500	Accretion	–
Palm Beach, Fla.	S	115,000–172,000	Accretion	1925–1930
Moriches Inlet, N.Y.	W	230,000	Accretion	–
Fire Island Inlet, N.Y.	W	120,000–460,000	Accretion	–
Rockaway Inlet, N.Y.	W	340,000	Accretion	–
GULF OF MEXICO				
Pinellas Co., Fla.	S	38,000	Accretion	1922–1950
Perdido Pass, Ala.	W	153,000	Accretion	1934–1953
Galveston, Texas	E	334,000	Accretion	1919–1934
PACIFIC COAST				
Santa Barbara, Calif.	E	214,000	Accretion	1932–1951
Oxnard Plain shore, Calif.	S	765,000	Accretion	1938–1948
Port Hueneme, Calif.	S	380,000	Accretion	1938–1948
Santa Monica, Calif.	S	206,000	Accretion	1936–1940
El Segundo, Calif.	S	124,000	Accretion	1936–1940
Redondo Beach, Calif.	S	23,000	Accretion	–
Anaheim Bay, Calif.	E	115,000	Erosion	1937–1948
Camp Pendleton, Calif.	S	76,500	Accretion	1950–1952
GREAT LAKES				
Milwaukee Co., Wis.	S	6,000	Accretion	1894–1912
Racine Co., Wis.	S	30,600	Accretion	1912–1949
Kenosha, Wis.	S	11,500	Accretion	1872–1909
Ill. state line to Waukegan	S	69,000	Accretion	–
Waukegan to Evanston, Ill.	S	43,600	Accretion	–
South of Evanston, Ill.	S	30,600	Accretion	–
HAWAII				
Waikiki Beach, Hawaii	–	7,650	Suspended load samples	–

Source: After Johnson (1956) and Wiegel (1964).

sand waves drift to the shoreline, or when rivers dump flood deposits on the beach, or by other means. The excess concentration of sand on the beach is eroded on the side toward the littoral drift and is deposited on the side away from the drift. Thus, a wave of sand moves along the beach.

Climatic fluctuations that last for centuries may cause the erosional habit of a river to change, with the result that quantities of sand may be deposited along the shore in new places. The littoral drift has a new source, and beaches may appear where none existed before. A remarkable example of a previously unknown, long-persistent littoral drift is available from Portugal (although the source of the drift has not been determined): Several prosperous ports, of which the largest was Aveiro, existed in the tenth century in a slight embayment of the central Portuguese coast, about 50 kilometers long and 8–12 kilometers wide. About that time, the forerunner of disaster appeared in the form of a sand spit that began to build southward from the northern end of the embayment. Two centuries later, it was 16 kilometers long, and the northern half of the embayment became a narrow bay. Aveiro was at its peak, and there were 150 ships based in the area. The spit extended itself inexorably south; after 800 years, the bay was completely sealed off and became a relatively stagnant lagoon. Commerce and fishing were destroyed. Agriculture, which was based on fertilizing with a compost of salt-marsh grass, declined sharply. The lagoon was a stagnant cesspool and became a source of pestilence. By 1797, deaths were twice as numerous as births. The people of the area did not accept these calamities placidly. Their attempts to control the sand spit are discussed in the next section after a general analysis of the effects of tampering with the river of sand.

FIGURE 16.8

Development of the lagoon of Aveiro, Portugal: *gray*, areas of sedimentation. [After Abecasis (1954).]

Tampering with the River of Sand

The river of sand is in equilibrium when the steadily replenished supply of sand to be moved is balanced by the energy steadily available to move it. Thus, disequilibrium can result from an increase or decrease in either the supply of sand or the available energy. On a long time scale, all possible combinations of changes occur naturally, and the end results show how the river of sand attempts to restore equilibrium and maintain continuity of flow. However, the efforts of man to control the sand drift are more illuminating, because the whole sequence of events is available for analysis.

CHANNELS AND JETTIES

Channels, stabilized by rocky jetties on each side, are commonly built through spits and beaches in the expectation that they will provide a safe entry to a lagoon or harbor. The experience of Aveiro, Portugal, after a spit closed off the lagoon entrance, is typical. Engineers were consulted by the city in 1685, 1756, 1777, 1780, and 1791; finally, construction of a channel began in 1801. In 1808, it was opened "forever." The drift of sand was naturally deflected outward to form a submerged sandbar around the gap in the spit. All remained well until 1859, when an excess of sand, presumably a sand wave moving along the beach, maintained a spit across the channel for some time. The channel was filled again in 1873–1874, and the engineers returned in 1886 to "improve" it for a decade. In 1908, a spit closed the gap temporarily. In 1927, the gap was "improved," and again, in 1932–1936, the jetties were extended. In 1937, the new channel became unfit for use. Further "improvement" began in 1949 and still continued when the history of the spit was published in 1954. In sum: (1) even primitive harbor engineering could open a channel, but it could not keep the channel open during the passage of sand waves; and (2) tampering with the river of sand, once initiated, required further tampering to maintain the desired change.

Modern harbor engineering in the United States often counters, stage by stage, the almost inevitable consequences of some initial tampering with littoral sand drift. In 1940, Port Hueneme was carved out of the edge of the Oxnard Plain in Southern California. A jetty impounded the river of sand west of, or updrift from, the harbor entrance. However, because wave energy for transport was still available to the east, 900,000 cubic meters of sand was eroded downdrift from the harbor in a single year. When the sand began to drift around the end of the jetty, it did not form the usual submarine bar leading to a beach on the downdrift side; instead, it was deflected into the head of a submarine canyon and disappeared into the deep sea. Consequently, a million cubic meters of sand was eroded each year east of the harbor. A seawall, built to limit the penetration of erosion, eventually extended 2100 meters downdrift. Its only effect was to shift the locus of erosion to just beyond the terminus of the wall. About 2,000,000 square meters of valuable land disappeared downdrift.

The erosion was countered by building a breakwater and new jetties to form a large sand trap short of the canyon. A program of dredging 1.5–2.3 million cubic meters of sand every two years began in 1960 and, presumably, will persist for as long as the harbor is in use. Sand is dredged from the trap and pumped through a pipeline to the downdrift side of the harbor entrance. The initial construction of the bypass cost roughly $4 million. Because the average annual cost for dredging is $0.5 million, it presumably also equalled $4 million by 1968.

Many new harbors are being cut into the California coast. Each crosses the river of sand and each must provide for a bypass. Thus, annual dredging burgeons. The port of Salina Cruz in Mexico has a more elegant system, which continuously sucks in sand from the updrift side and pumps it around the harbor to the downdrift side. Engineering technology has no difficulty in controlling sand, but the cost estimates in the United States in the past rarely provided for perpetual maintenance as they did at Salina Cruz.

OFFSHORE BREAKWATERS

An island near the coast shields the beach from the energy of the sea. Thus, the littoral drift has nothing to drive it in the lee of the island, and sand fills in the sheltered area with a characteristic spur of sand called a tombolo. The breakwater updrift from Port Hueneme is an artificial island specifically designed to trap an artificial tombolo. Thus, man takes advantage of an understanding of nature. Earlier, he paid the price of a lack of understanding when a breakwater was built parallel to the shore to provide a yacht harbor at Santa Monica, California: the desired harbor became an unwanted tombolo.

ELIMINATING THE SEDIMENT SOURCE

In many regions, rivers provide the source for littoral drift. We found, in the last chapter, that there are many reasons to dam river water and that the end result is sediment deposition upriver and erosion downriver. The effects of a dam far upstream may be accommodated in the river. However, flood-control dams on the short rivers of Southern California trap the sediment that used to feed the littoral drift. Thus, the beaches of the area are drifting toward submarine canyons without much prospect of natural replacement.

Similar problems may be anticipated along the great beaches of the east coast of the United States. The shoreline where erosion provides the source for littoral drift is generally as treasured as the shoreline to which the sand drifts and is deposited. However, the source region can be protected from erosion only by eliminating the drift and deposition. The energy of the sea still reaches the shore, and a new source region will arise where once there was equilibrium drift. The situation is just like extending the sea wall at Port Hueneme: if tampering affects any part of the river of sand, it will propagate along the whole length of the beach.

Even if the whole beach is stabilized with jetties and the shorter structures known as groins, the wave energy has to be used up somehow. The National Park Service has, perhaps only temporarily, stabilized the sand dunes on the beaches of the Outer Banks of North Carolina. The larger waves of winter no longer have as wide a mobile zone in which to adjust to an equilibrium profile. The waves tend to make a very steep beach and transport sand out into deep water. After 13 years, the beaches have lost 30% of their width. Eventually, the attempt at stabilization may have very undesirable consequences.

PROLONGED TAMPERING

On a piecemeal basis, man is tampering with almost every aspect of the flow of the river of sand and only slowly coming to understand the consequences. What are they? Reporter Dorothy Townsend has summarized some of them, based on a kilometer-by-kilometer inventory of 56,000 kilometers of coastline conducted by the Corps of Engineers. Some 43% of the coast is being eroded. The Corps estimates that protective measures with an estimated cost of $1.8 billion are justified for the 4300-kilometer portion of the coast that is "critically eroding." The people whose houses were undercut by the sea in Ventura County, California, in 1971 would probably agree. But "what forces diminished the 200-foot [60 meter] wide strand that once stood as a buffer between those homes and the sea? Was it the irrevocable etching of time and tide? Or was it 'the marina up at Ventura,' as the real estate man claimed, that blocked the littoral drift of replenishing sands, the 'nourishment' provided by nature?" Certainly, it was the entrance to the U.S. Naval Weapons Station harbor in Orange County to the south that cut off the sand to the house-sprinkled shore at Surfside Beach. The federal government accepted responsibility and pumps 1,500,000 cubic meters of sand on the beach every 6 years. Who will accept responsibility for the drifting centers of erosion and waves of clogging sand that will result from any but the most carefully planned efforts at controlling beach erosion? We have much to learn about the river of sand and its responses to forces that would unsettle its equilibrium.

Continental-Edge Effects

Ocean currents, large waves, and fluctuations of sea level mobilize the sediment of the continental shelf—the submerged margin of the continents. Like a slow littoral drift, sand and silt move parallel to shore and also to and fro perpendicular to it. In due course, no matter what the direction, the sediment reaches the outer edge of the continental shelf or a canyon leading to it. Below is the continental slope, which drops off 2–5 kilometers and resembles the front of an endless mountain range.

In a sense, we are back at stage one, where sediment begins to be eroded from the high, steep-sided peaks of continental mountains. There is little fresh hard rock, and much sediment, which is rather different from the mountain peaks. With regard to the sediment, however, the most important characteristic is the same in the two environments, whether under the air or under the sea. The height gives the sediment potential energy, and the steep slope allows this energy to be expended rapidly in great landslides, slumps, and other modes of gravity flow. The buoyancy of the water diminishes the potential energy of the sediment compared to what it would have in air, but this is countered by the fact that there tends to be much more sediment at the top of the continental shelf than at the top of a mountain.

Usually, the gravity flows are unobservable at the surface. For this reason, they were long unknown. In a few places, however, they have destroyed marine structures or equipment, and from these events we know some of their characteristics. The events that followed the Grand Banks earthquake of 1929 provided the best information available, although similar sequences of events have occurred in many other places and can be duplicated in the laboratory.

This earthquake, which shook eastern Canada, was the largest one ever recorded in the vicinity of the Grand Banks. Apparently, a large amount of sediment had accumulated since the last such earthquake, and an enormous slump, about 160 by 320 kilometers, dropped the whole face of the continental slope. This region was crossed by almost every submarine telegraph cable between Europe and the United States, and 16 breaks occurred essentially at the time of the quake and slump. This is the largest phenomenon of this kind known, and it may be that earthquakes are necessary to trigger slumps on this scale. However, smaller slumps occur commonly without quakes. If wet sediment accumulates rapidly, the excess water may not be able to escape from it, and a slump occurs underwater just as it does in improperly engineered fill on land.

There were many more submarine cables crossing the relatively smooth, channeled plain of the deep-sea floor where it sloped gently away from the steep area of slumping at the edge of the Grand Banks. One of these cables, the one nearest the slump, broke 58 minutes after the quake. Thereafter, for 13 hours, cable after cable broke downslope for a distance of 500 kilometers. The cables were broken by a turbidity current that formed as the slump collapsed and mixed with the surrounding water. A turbidity current is one of many phenomena that are driven by their own excess density within

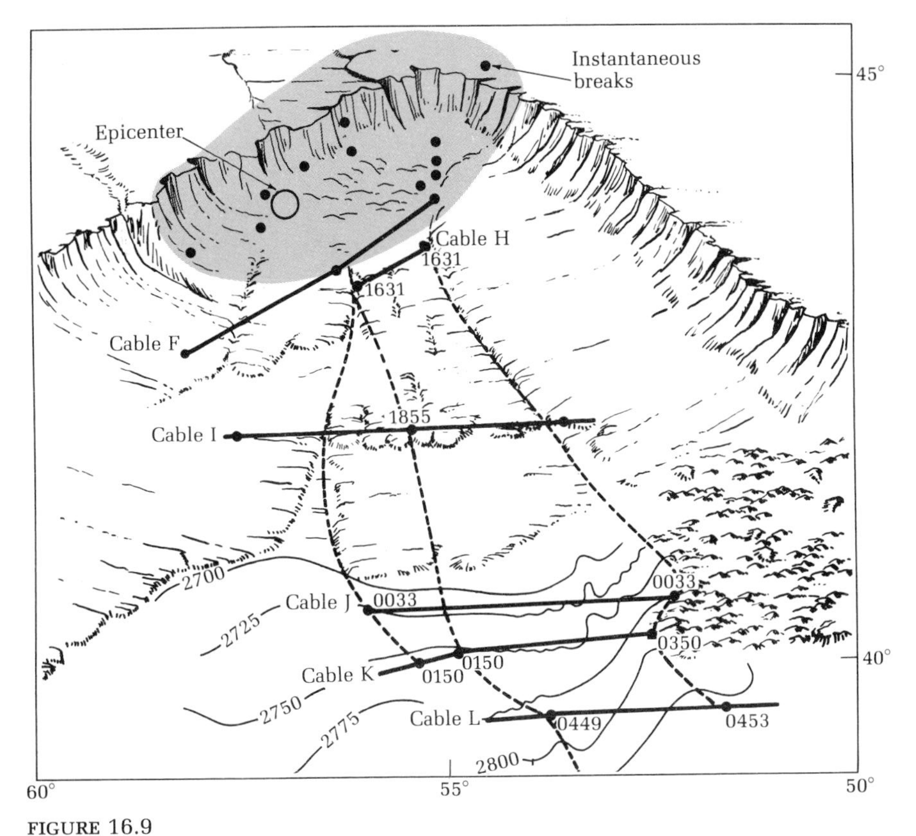

FIGURE 16.9
The epicenter of the Grand Banks earthquake of 1929 and the location and time of subsequent cable breaks. [From Menard (1964).]

rather similar fluids. Cold air masses and ocean currents flow downslope because they are denser than the air or water that surrounds them. The excess density of turbidity currents is the result of the sediment suspended in them. Instead of the sediment merely settling out of the water, the whole, unconfined mass moves downslope. Once again, this is not a unique phenomenon because dust storms, glowing avalanches from volcanoes, and some snow avalanches also move by this process of autosuspension. The important characteristic of autosuspensions is that they are extremely mobile compared to other mass movements. Thus, turbidity currents may flow at the rate of 30–50 kilometers per hour and extend as thin sheets for a thousand kilometers. They are the principal mechanism for spreading continental sand and silt over the sea floor.

SUBMARINE CANYONS

Relatively little sediment crosses the entire shelf and accumulates on the continental slope. Instead, most of it falls into submarine canyons, which intersect the continental shelf and slope in many places. The canyons may even intersect the shallow and vigorous littoral drift, as one does off Port Hueneme, California. The origin of these canyons is still conjectural. Some may be cut underwater by turbidity currents and slumps. Some may be cut subaerially, perhaps during the initial stages of continental drift, when splitting creates new margins that later sink.

Because of the paucity of observations, there is also some controversy about the details of what happens in submarine canyons. It appears that the sediment dumped in the head of a canyon moves as much as a meter in the process of self-compaction. Then, at intervals of between 1 and 1000 years, the sediment begins to slump and is transformed into a mobile turbidity current, which emerges as a jet from the mouth of the canyon onto the deep-sea floor. There, most of its potential energy is gone, and, on the average, it begins first to deposit its sand, then its silt, and later its clay. It flows partially in meandering, leveed channels and partially in sheets. The Coriolis acceleration, caused by the rotation of the earth, acts so as to divert the flow to the left in the Northern Hemisphere. This is the opposite of what might be expected, and it reflects a complex interaction between unimpeded flow

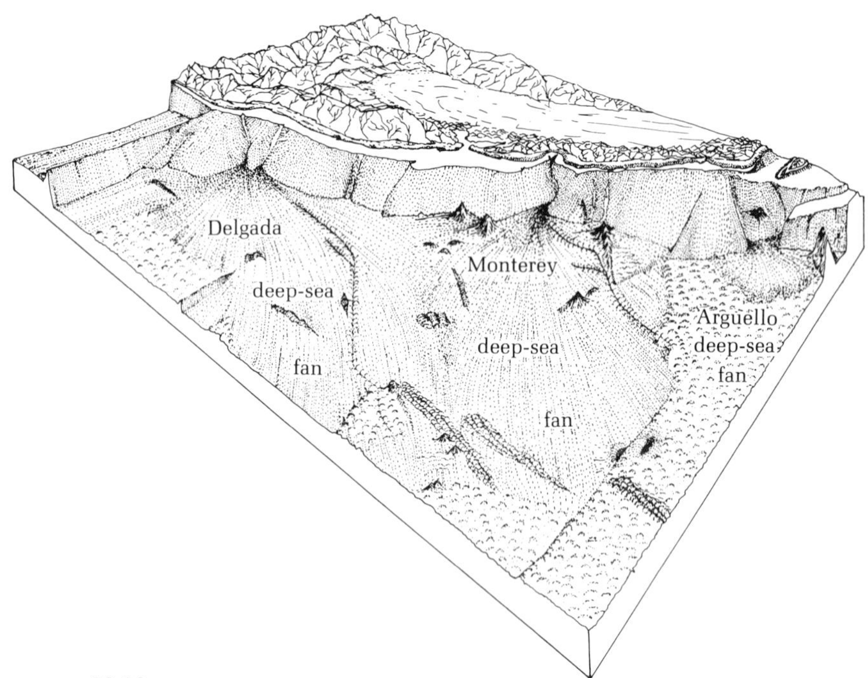

FIGURE 16.10

A block diagram of the sea floor and coast of central California. Deep-sea fans spread out from the mouths of Delgada, Monterey, and Arguello canyons. The first two fans are confined in a broad trough between the Mendocino and Murray fracture zones. [From Menard (1960), courtesy the Geological Society of America.]

and sedimentary processes that build higher levees on the right side and, thus, bias the location of "microdeltas" on the fans.

TECTONIC ENVIRONMENT

The tectonic environment of a continental margin dominates the accumulation of sediment nearby on the deep-sea floor. If the margin is a subduction zone, as it is off Chile at present, a deep-sea trench traps everything that flows on the bottom. Subduction then feeds the sediment back to the bottom of the continent, where magma forms and flows upward into the continental crust. Thus, the coarser material in the sedimentary cycle is soon restored to the source region. The finer, suspended, and dissolved materials in sea water escape for much longer.

If the continental margin is within a plate, as it is in the Atlantic, the coarse material remains where it is deposited by gravity flows. Thus, a long, narrow wedge of sediments accumulates to a thickness of 2–4 kilometers and depresses the sea floor with its load. This provides space for more sediment, and, gradually, the thickness increases to 10 kilometers or more. Bodies of sediment and sedimentary rock with these dimensions are modern, developing geosynclines (Chapter 4). They persist until something, perhaps the formation of a subduction zone, crumbles or squeezes them into mountain ranges. Almost all such ranges were once geosynclines.

The continental margin may be at or near a transform fault, as it is in various places on the west coast of North America. If so, the sediment deposited on the sea floor gradually drifts away from the submarine canyons or the continental hinterland from which it is derived. The sediment fans off central California probably contain sediment eroded from Mexico far to the south.

Dissolved Matter

The oceans receive 11 million cubic meters of river runoff per second, with about 100 parts per million of dissolved solids, the most common of which are bicarbonate (48.6%) and calcium (12.5%). These proportions contrast with those in the vast and salty ocean, in which the commonest constituents are chlorine (55.0%) and sodium (30.6%). "Thus," as the chemist Forchhammer said in 1865, "the quantity of the different elements in sea water is not proportional to the quantity of elements which river water pours into the sea, but inversely proportional to the facility with which the elements in sea water are made insoluble by general chemical or organo-chemical actions in the sea." The composition of the oceans apparently has varied little during geologic time. In chemical terms, it is approximately in a steady state in which the addition of an element is balanced by deposition in marine sediments. It is of interest to calculate the average time an atom of each element spends between its introduction into the sea and its deposition, and to compare it with the residence time of the water itself, which is about 40,000 years. Residence times vary from a few hundred years for beryllium, aluminum, titanium, chromium, iron, niobium, and thorium through more than 10 million years for lithium, magnesium, potassium, and rubidium, to about 100 million years for sodium.

These residence times help in the understanding of geochemical processes of importance in marine geology. The mixing time for the oceans is greater than the shorter residence times, and this reflects the fact that some of the elements enter the oceans as rapidly settling particles from the continents and oceanic volcanoes. Moreover, aluminum, titanium, and iron are examples of elements that are highly reactive in the formation of authigenic minerals, which are those minerals that form in place. If the elements fall and precipitate out before they are thoroughly mixed in the ocean, their concentration both in sea water and in sediments may be expected to vary from place to place. The thorium content of authigenic minerals provides an example. Thorium is accumulating in the Atlantic far more rapidly than in the Pacific, and in the North Atlantic more rapidly than in the South Atlantic. The variation in rates corresponds reasonably well with the volume of river discharge divided by the ocean volume in each basin. If a lot of material enters a small sea, more falls to the bottom than if a little enters a large sea.

This relationship also holds for elements with the longest residence times under certain circumstances that are rare but of great geological interest. Water leaves an ocean basin only by evaporation; therefore, sodium, chlorine, and other ions with very long residence times gradually become more concentrated if the basin is small and the influx and balancing evaporation are large—that is to say, if the residence time for the water is brief. Gradually a dense brine forms, rather like that of the Dead Sea (although the condition of the Dead Sea is mainly the result of simple evaporation decreasing its volume).

Simple evaporation of even the deepest ocean is incapable of producing the thickness and proportions of various salts preserved in the great salt deposits of Germany and Michigan. Thus, the ordinary condition for deposition of salt is inflow balancing evaporation, but with a brief residence time. This is just the condition that may be anticipated when an ocean basin begins to form by the initial splitting of a large continent. A crack as long as the Atlantic coast of South America once began to open and a very narrow ocean basin formed. The formation of the Andes by plate convergence may have reversed the continental drainage from westward to eastward, and the flow of water may have been great. It is not surprising, according to this model, to find that salt deposits are common in small basins along the continental shelf of the South Atlantic.

Another geologically important factor affecting the concentration of chemicals in sea water is biological activity. Sunlight penetrates only a few hundred meters, and photosynthesis and plant life are restricted to the top 200 meters. Most animal life also stays near the surface to feed on the plants, although scavengers exist at all depths and even in bottom sediment. Phosphorus and nitrogen are removed from all productive surface waters and calcium and silicon are preferentially removed in regions where they are incorporated into the living skeletons of plankton. In this manner, calcium carbonate is depleted from the surface equatorial waters by sedimentation of dead foraminifers and coccoliths, and silicon dioxide is removed from the waters of high latitudes by diatoms (see Figure 16.12).

Because of oxygen's role in life processes, its concentration in sea water is highly variable. The concentration is highest near the surface because of photosynthesis, and it passes through a minimum at an intermediate depth because of the decay of sedimenting organic matter. Mixing keeps the oxygen level relatively constant at greater depths.

Another gas whose concentration in sea water is highly variable is hydrogen sulfide. This gas is locally important in the bottom waters of stagnant basins in which all the oxygen is consumed. These are the sorts of basins now forming as a result of pollution with excess nutrients. They are very important in geology because they are the environment of deposition of extensive deposits of black shales containing the iron sulfide mineral pyrite. These shales also sometimes contain perfect fossils of organisms that are preserved when they fall into a sterile environment. The Black Sea and some deep narrow fjords in Norway are modern basins of this type.

It appears that sea water is undersaturated with respect to most of the common metal ions. The exception is calcium, which may precipitate inorganically as calcium carbonate, especially in warm, shallow waters. At greater depths, it is the solution of calcium carbonate that has the most interesting geologic effects, because it is one of the principal controls on the distribution of calcareous oozes. In an elegant experiment, Melvin Peterson of Scripps Institution of Oceanography placed carefully weighed spheres of calcite in sea water at various depths along a moored buoy in the central Pacific. Solution occurred at all depths, but it was much more rapid below 3700 meters, which corresponds to the maximum depth at which carbonate sediments are found in this region.

Pelagic Sediments

Pelagic sediments are those derived from the dissolved or suspended matter in the ocean. They are deposited at rates of about a centimeter per 1000 years; thus, they are identifiable only where other types of more rapid sedimentation

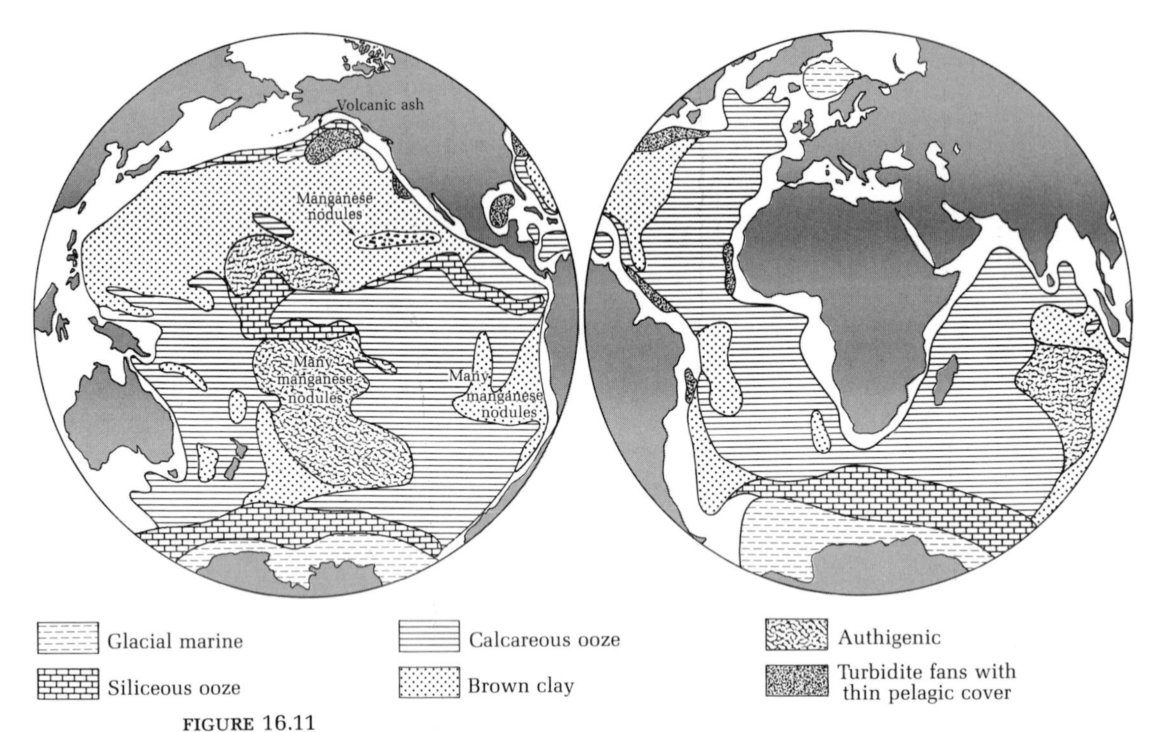

FIGURE 16.11
Distribution of deep-sea sediments. (All boundaries are subject to extensive changes as information becomes more abundant.) [From Shepard (1973).]

do not occur. Because they accumulate so slowly, they contain highly exotic components, including many from outer space, which, although they may fall everywhere, cannot be located in other sediments because of dilution. Likewise, seemingly minor variations in the generally constant environment of the sea are enough to completely change the character of the pelagic sediments.

BIOGENIC SEDIMENTS

Pelagic sediments have both biogenic (or organic) and nonbiogenic sources. The biogenic components include the teeth of sharks and ear bones of whales, which ultimately are all that remain of the terrors of the deep. Most of the biogenic material consists of the skeletons of harmless microscopic plants and animals that are made of resistant materials. Those organisms with skeletons of silica are the plants called diatoms and the animals called radiolarians; those with skeletons of calcium carbonate are the plants known as coccoliths and the animals called foraminifers.

The microscopic plants derive their skeletons from solution in sea water; the animals build theirs partly from the same source, but also from the microorganisms on which they prey. Thus, the abundance of organisms in the surface waters is largely determined by the availability of nutrients, of which nitrogen is usually the limiting one. Because nutrients are brought to the surface by upwelling, they are abundant along the equator, along the west coasts of continents, and in zones of current interactions.

The abundance of biogenic components in sediments, however, depends on the rate of solution of the skeletons after the organisms die. Skeletons fall relatively rapidly through the water; once on the bottom, however, they are exposed to solution for decades or longer before they are buried. In favorable circumstances, calcium carbonate skeletons accumulate most rapidly of all the components of pelagic sediment. However, they dissolve relatively rapidly below about 4000 meters, in most places, so not much of the calcium carbonate is preserved at greater depths. The resistant carbonate residue is usually the skeletons of the large foraminifers. At much greater depths, the calcium carbonate dissolves completely and the less voluminous silica skeletons tend to dominate the sediment. Bands of diatomaceous ooze occur at high latitudes in each hemisphere, and radiolarian ooze is found where the equatorial waters are too deep for carbonates.

NONBIOGENIC SEDIMENTS

There is a hierarchy of rates even among these sedimentary components that are deposited so very slowly. The nonbiogenic components are derived mainly from dissolved matter, but also from particulate matter, which is not as uniformly distributed. Thus, these components are much more variable than the biogenic ones. They depend on the character and location of the source, the distance from it, the mode of transport, and rates of sedimentation. They are relatively independent of depth, either because they are the insoluble minerals that have survived the whole sedimentary cycle, or because they form on the sea bottom.

Wind-blown dust, which is a significant component of pelagic sediments, is concentrated in the midlatitudinal bands in which deserts are most widespread. Dust also blows west from the Sahara Desert in such abundance that ships observe hazy skies 25–35% of the time, and fine sand is often found on the decks.

Volcanic ash is also an important component of pelagic sediment, and it is a particularly useful and important one because the ash layer from a single dated eruption is often identifiable. It provides a time marker that is both widespread and distinctive. Near subduction zones or around individual volcanic islands there are often so many ash emitters that a volcanic mud accumulates instead of a pelagic sediment. One grades into another.

The most common type of nonbiogenic sediment is red clay, which includes components from many sources. If wind-blown dust and ash and clay from

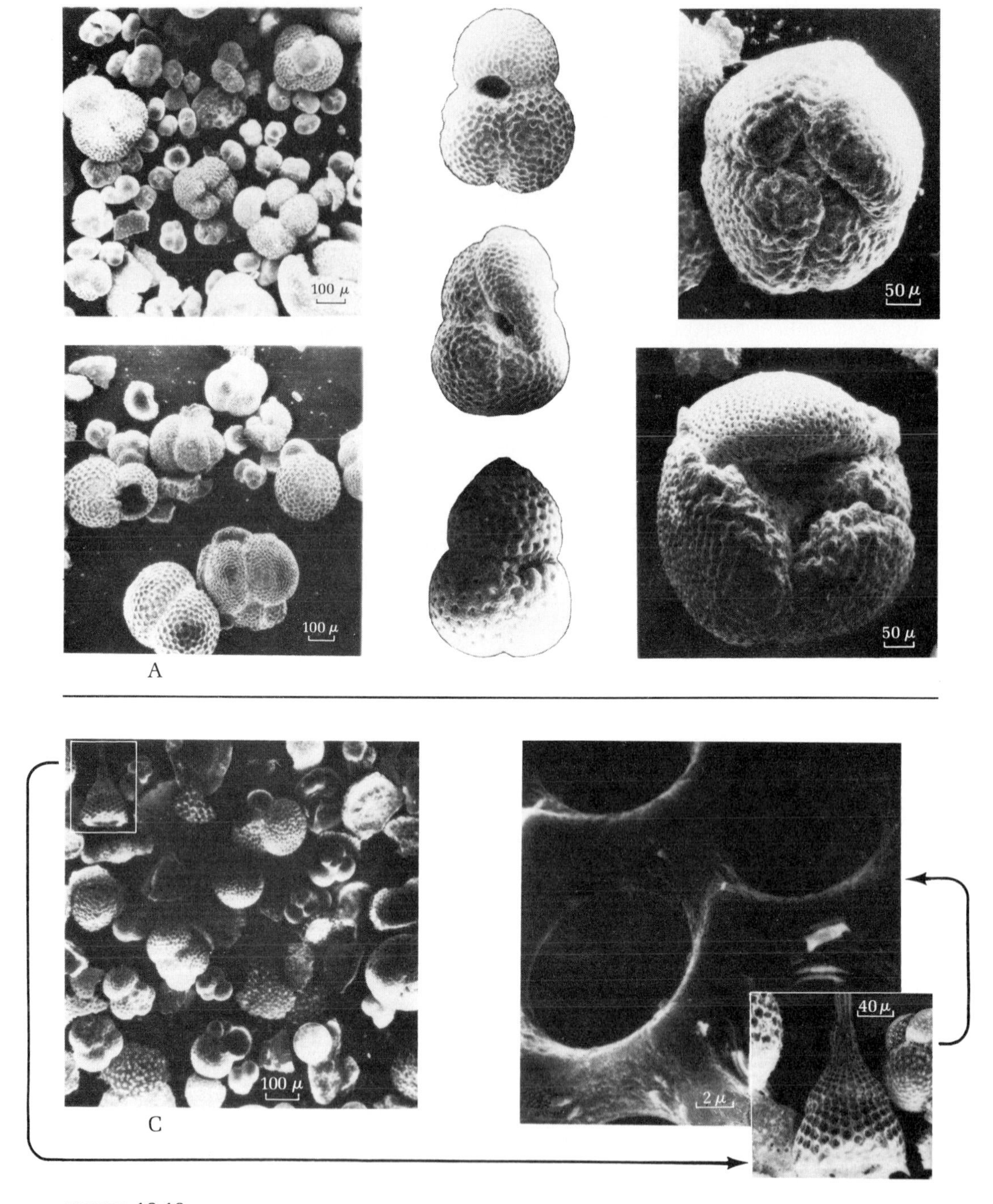

FIGURE 16.12

Enlarged views of the skeletons of marine microorganisms that occur commonly in pelagic sediment, including calcareous animals (*A*, foraminiferans), calcareous plants (*B*, coccoliths), siliceous animals (*C*, radiolarians), and siliceous plants (*D*, diatoms). The same skeletons are shown at different enlargements to display the complex but orderly structures. Note the scale indicated for each.

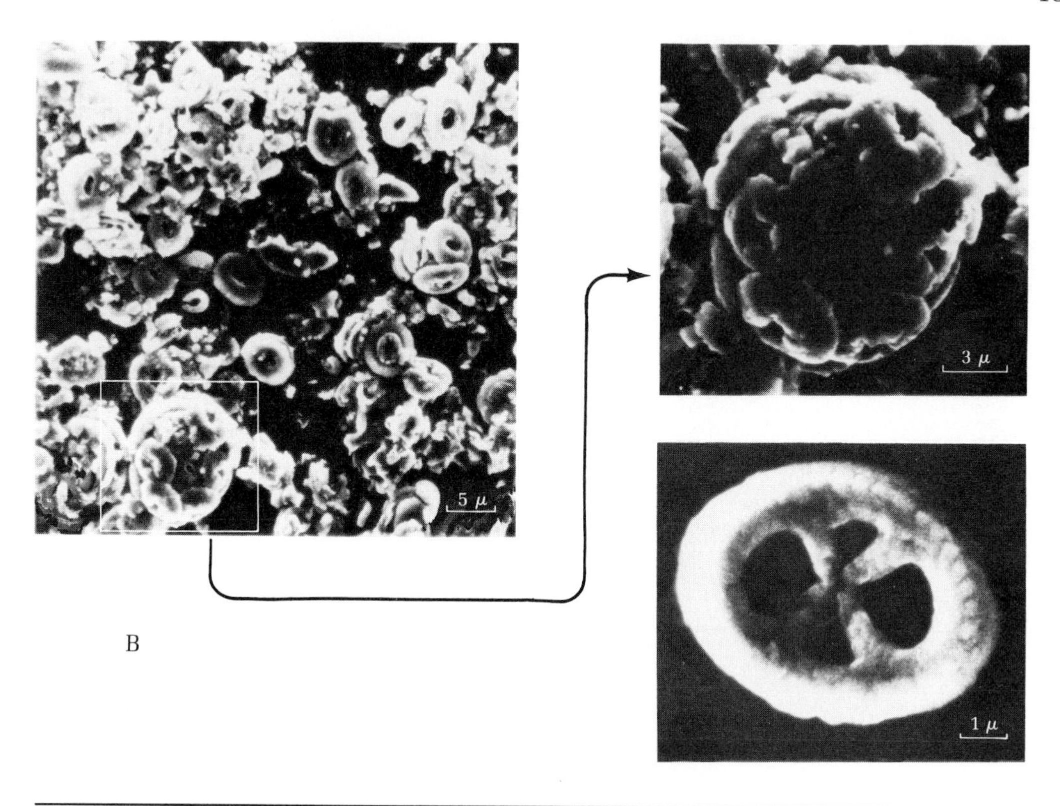

B

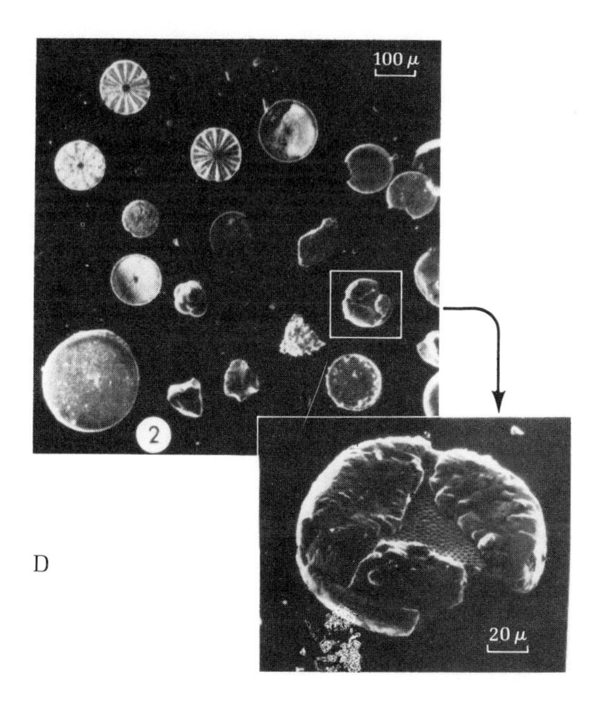

D

A

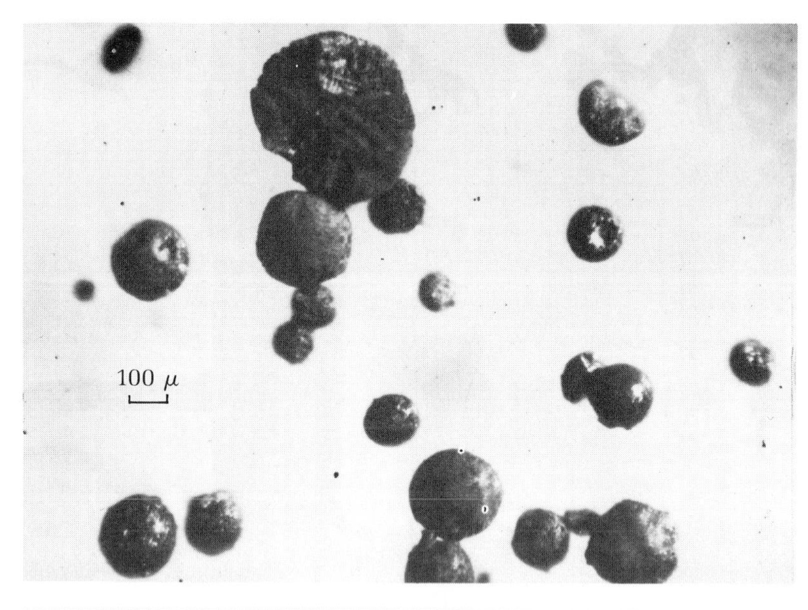

B

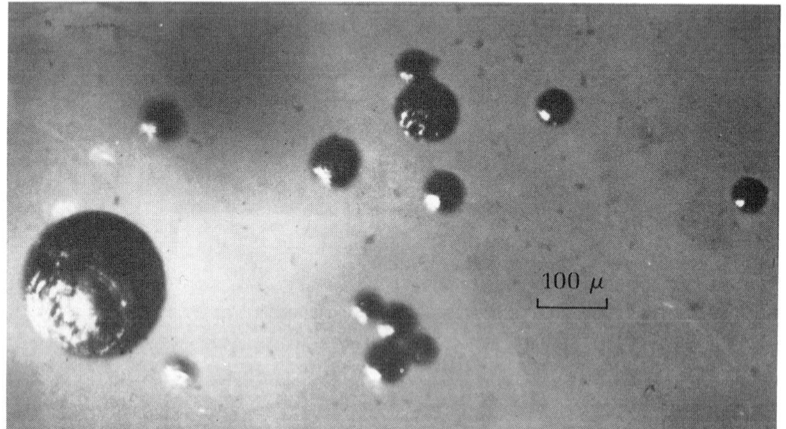

FIGURE 16.13

Cosmic spherules from deep-sea sediments: *A*, silicate spherules; *B*, magnetite-coated nickel-iron spherules. [From Hunter and Parkin (1960).]

rivers are scarce, the even rarer but more evenly spread cosmic components are more detectable. Most notable are the microscopic spheres of nickel-iron that ablate as droplets from the leading surfaces of meteorites as they burn up in the air. Presumably, there are also stony spheres from ablating stony meteorites—but they may weather into clay, and, in any event, cannot be collected and concentrated easily with a magnet like the iron ones.

All else failing, the deep-sea floor becomes covered with manganese nodules, which are precipitated from the solutions in sea water and the surface layer of sediment. These contain oxides of manganese and iron in a layered

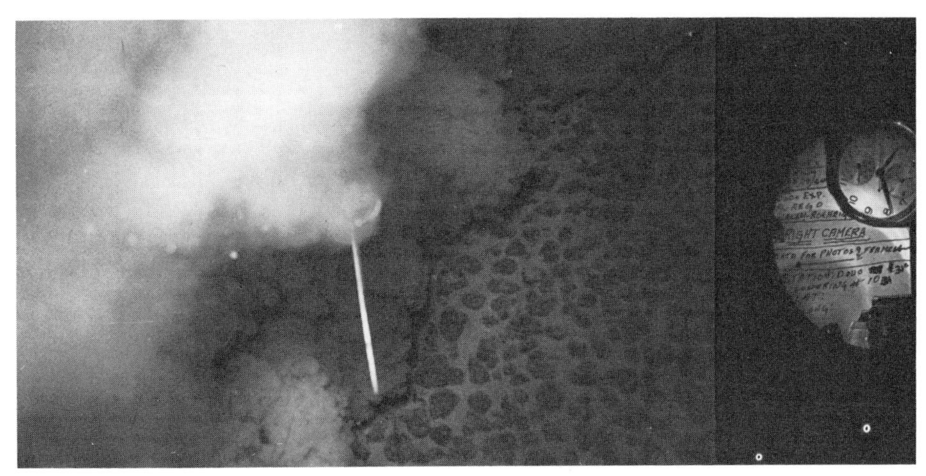

FIGURE 16.14

Manganese nodules covering the sea floor at a depth of 5235 meters (lat. 9°10′N, long. 168°50′ W). At the left is suspended sediment stirred up by an instrument that is in the field of view. At the right is information photographed to identify the exposure number. [Photo by Tom Walsh. Dodo Expedition, Station 31C (camera).]

FIGURE 16.15

A very large piece of white calcareous ooze, thickly coated with black ferromanganese oxides, cut into three sections to expose the interior. The scale is in inches. [U.S. Navy photo.]

crystalline structure that readily captures other heavy metals. The iron and manganese commonly reach concentrations of 20–30%, and nickel, cobalt, and copper reach 0.5–2%. In favorable circumstances, all of these metals, except iron, could become mineable ores within the next few decades. This possibility will be considered further in Chapter 19.

STRATIGRAPHIC EFFECTS OF PLATE TECTONICS

The JOIDES drilling program in many places penetrates a sequence of red clay over carbonate ooze over basalt. This is to be expected because of the mode and depth at which the sea floor is created and the way in which it subsides. Sea-floor spreading forms a strip of bare volcanic rock that lies generally at a depth of roughly 3000 meters. Calcium carbonate covers the rock at a rate of a few centimeters per 1000 years until sinking reaches the depth at which carbonate begins to dissolve. All but the very surface of the carbonate ooze is preserved because it is coated with a protective layer of red clay. This accumulates very slowly but steadily until the plate plunges into a subduction zone and the cycle is ended.

Pelagic oozes tend to have very different chemistry from place to place, so the material restored to a continent by subduction rarely has the average composition of the rocks of the source area. In this process, a mechanism exists to maintain local inhomogeneity of the continental crust and upper mantle, even though it is cycled again and again by the energy of the sun and earth.

Abyssal Processes

Not even the sediment on the deep-sea floor is eternally at rest. The top few centimeters are stirred by slow, gentle, but steady processes and also by fast, relatively intense, but very rare ones. One of the most widespread mechanisms for stirring is by burrowing organisms, which are various types of worms and other invertebrates. These ingest sediment, just as earthworms ingest soil, and pile it on the surface at the burrow entrances. Some animals move across the bottom, scavenging randomly for food that falls from the productive surface waters.

Gentle contour currents, which are concentrated on the western sides of ocean basins, slowly sculpture the bottom sediment there into ripples and drifting dunes. Elsewhere, these bottom-seeking currents appear to have much less effect, except where they are accelerated as they move through narrow passes in the bottom relief. The fracture zones that transect the Mid-Atlantic Ridge are examples of such passes that are typically scoured by intensified bottom currents. Similar scouring occurs in the passes between the great volcanic ridges of the western Pacific. Very careful study shows that less intense

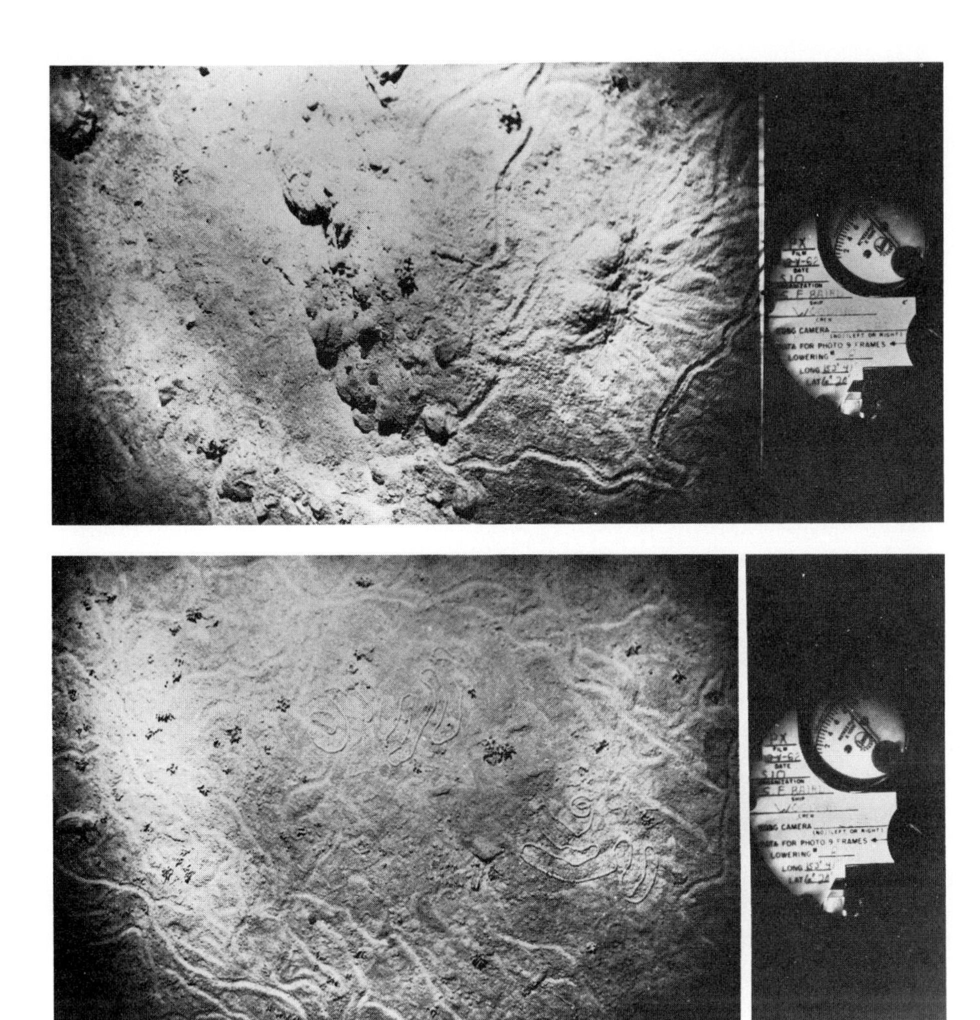

FIGURE 16.16
Two photos showing animal tracks, burrowings, and trails of excrement on the floor of the Solomon Trench (lat. 6°16.7′S, long. 153°43.2′E) at a depth of 8500 meters. [S.I.O. photos courtesy R. L. Fisher.]

erosion takes place even in the minor valleys among the endless abyssal hills in the western equatorial Pacific. The scoured sediment in each place is transported only a few kilometers through the pass. Once the current can spread out beyond the pass, it slows and deposits sediment.

Stirring of a more intense but rarer sort occurs when the giant waves of tsunamis and meteorite splashes spread through the oceans. Near the wave sources, stirring may be briefly intense; farther away, however, it may only be strong enough to erode sediments where it is concentrated in valleys.

The probability of erosion by waves or currents also varies with the type of sediment. Red clay has a smooth surface that is difficult to erode—like all clays—but it consists of very fine particles that are easy to transport. These factors make the role of stirring organisms unusually important. The organisms make irregular mounds that are easier to erode, and many of them also squirt waste clay into the water when they excrete. Thus, *they* do the eroding, and very gentle currents can then transport the clay for long distances. By contrast, the carbonate oozes contain coarse grains of foraminiferal skeletons that are easy to erode but difficult to transport. Therefore, compared to the red clays, these oozes may be stirred more by intense abyssal processes and less by burrowers.

Return to Continents

Continental material has a low density and floats on the denser rock of the mantle. There is no way to destroy large pieces of it by pulling them into the mantle. They merely bump into other pieces of continent at subduction zones, instead of plunging like oceanic crust. What of the individual grains of sand and the crystals of clay eroded from continents? Do they plunge downward with the oceanic crust on which they rest? Is this a way to recycle the continents back into the interior of the earth? The evidence is equivocal.

The abyssal clays and oozes are thin and have no strength, and it is hard to conceive that they can be pulled down between two hard, rigid, converging plates. It would be no more likely than, say, that jelly spilled on a smooth table would cling to the table when you tried to scrape it off with a sharp knife. Although it is seemingly clear what ought to happen, it is difficult to observe what actually does. In only a few places can it be demonstrated that abyssal sediment is actually scraped off a plunging plate and plastered against the edge of the next plate. The best evidence is that blocks or wedges of sediment slide upward on the continental slope without being destroyed. In some places, it appears that the sediment is contorted and literally smeared on the slope. The geologic record also suggests that sometimes the volcanic and oceanic crustal layers become detached from the plunging plate and slide up the slope and onto the next plate. The overlying abyssal sediment, in such a situation, may remain relatively intact.

In California and other sites of former subduction zones, abyssal sediments and oceanic crust have had a more complex history at the plate edge. They plunged down 15–30 kilometers with the lithosphere and were metamorphosed before their low density or heating caused them to penetrate back up to the surface through the overlying plate. This is enough to show that, somehow, some of the weak abyssal sediment is not scraped off at the edge. Perhaps the plates have undulations that prevent some areas from being scraped; perhaps they can sometimes push past each other without scraping together.

The quantity of ancient abyssal sediment that can be identified in association with the subduction zone under western North America seems much too small, considering the amount of oceanic crust that appears to have plunged there. It is possible that, somehow, much of the sediment escaped the scraping edge and has been lost to the mantle. If so, the sedimentary cycle may be more complex than is now thought, but presumably the lighter elements will return to the surface in due course to complete the cycle.

Summary

1. The weathering products that reach the sea are distributed in ways and for distances that depend on their size, if they are solids, or on their geochemistry, if they are in solution.
2. Sand migrates along the shore. Beaches, spits, and barrier islands are temporary features that exist where sand is stockpiled in the course of the migration.
3. Any natural or manmade alteration of the river of sand along the shore causes gradual changes elsewhere in the flow.
4. Even though it is underwater, sediment at the edge of the continental shelf possesses potential energy because it is high above the deep-sea floor. Consequently, the steep continental slope is a dynamic environment of slumping and fast-moving turbidity currents.
5. Turbidity currents spread relatively coarse sediment for hundreds to thousands of kilometers from continental margins.
6. Dissolved matter mixes in the ocean. Soluable elements, with long residence times, are thoroughly mixed and uniformly distributed. Less soluble elements, with short residence times, are only mixed locally and are unevenly distributed, both in sea water and in marine sediments.
7. Pelagic sediments are deposited very slowly from material in suspension or solution. They consist of cosmic, terrestrial, and volcanic dust, plus the skeletons of marine microorganisms.
8. Although abyssal processes move sediment, most are very weak, except where currents are locally focused by topography.
9. The sediment on the deep-sea floor returns to the continents in subduction zones.

Discussion Questions

1. What happens to a beach if the supply of sand is interrupted by a breakwater or groin?
2. What happens if a channel is cut through an offshore bar or spit?
3. What do you suppose was the origin of the sand that began to move south along the Portuguese coast in the tenth century?

4. Why is it unwise to lay submarine cables along the continental slope?
5. How is a highly soluble compound like salt (NaCl) ever removed from the sea in appreciable quantities by natural means?
6. What kind of sediment occurs under equatorial waters 4000 meters deep? Why does it occur there?

References

Abecasis, C., 1954. The history of a tidal lagoon inlet and its improvement. *In* J. W. Johnson, ed., *Proceedings of the Fifth Conference on Coastal Engineering, Grenoble, France,* pp. 329–363. Richmond, Calif.: Council on Wave Research, University of California. [A history of shoreline changes spanning 900 years.]

Arrhenius, G., 1963. Pelagic sediments, *In* M. Hill, ed., *The Sea,* v. 3, pp. 655–727. New York: John Wiley & Sons.

Bascom, W., 1960. Beaches. *Sci. Amer.* 203(2):80–94. (Available as *Sci. Amer.* Offprint 845.) [Excellent photographs of effects of tampering with littoral drift.]

Berger, W. H., and U. von Rad, 1972. Cretaceous and Cenozoic sediments from the Atlantic Ocean. In *Initial Reports of the Deep Sea Drilling Project,* vol. 14, pp. 787–954. Washington, D.C.: U.S. Government Printing Office.

Bruun, P., 1954. Migrating sand waves or sand humps. *In* J. W. Johnson, ed., *Proceedings of the Fifth Conference on Coastal Engineering, Grenoble, France,* pp. 269–295. Richmond, Calif.: Council on Wave Research, University of California.

Dolan, R., 1972. Barrier dune system along the Outer Banks of North Carolina: a reappraisal. *Science* 176:286–288. [Attempts to stabilize the beaches may destroy them.]

Forchhammer, G., 1865. On the composition of sea water in the different parts of the ocean. *Phil. Trans. Roy. Soc. London* 155:203–262. (Quoted in Goldberg, 1963.)

Gilluly, J., J. Reed, Jr., and W. Cady, 1970. Sedimentary volumes and their significance. *Geol. Soc. Amer. Bull.* 81:353–376. [There is much more sediment off the Atlantic coast than the Pacific coast of the United States because of subduction under the Pacific.]

Goldberg, E. D., 1963. The oceans as a chemical system. *In* M. N. Hill, ed., *The Sea,* vol. 2, pp. 3–25. New York: John Wiley & Sons.

Heezen, B., and C. Hollister, 1964. Deep-sea current evidence from abyssal sediments. *Mar. Geol.* 1:141–174. [The floor of the deep sea is stirred in many ways.]

———, ———, and W. Ruddiman, 1966. Shaping of the continental rise by deep geostrophic contour currents. *Science* 152:502–508. [In some places they reshape the sedimentary fans at the continental margin.]

Herron, W., and R. Harris, 1966. Littoral bypassing and beach restoration in the vicinity of Port Hueneme, California. In *Proceedings of the Tenth Conference on Coastal Engineering, Tokyo, Japan,* pp. 651–675. New York: American Society of Civil Engineers. [The bypassing goes on forever.]

Hunter, W., and D. W. Parkin, 1960. Cosmic dust in Recent deep-sea sediments. *Proc. Roy. Soc. London* A255:382–397.

Inman, D. L., and J. D. Frautschy, 1966. Littoral processes and the development of shorelines. In *Coastal Engineering* (Proc. Santa Barbara Specialty Conf., Amer. Soc. Civ. Eng.), pp. 511–536. New York: American Society of Civil Engineers.

Johnson, D., 1938. *Shore Processes and Shoreline Development* (rev. ed.). New York: John Wiley & Sons. [A classical treatise based on a theory of evolutionary changes leading from one type of shoreline to another.]

Johnson, J. W., 1956. Dynamics of nearshore sediment movement. *Amer. Ass. Petrol. Geol. Bull.* 40(9):2211–2232.

Menard, H., 1960. Possible pre-Pleistocene deep-sea fans off central California. *Geol. Soc. Amer. Bull.* 71:1271–1278.

——, 1964. *Marine Geology of the Pacific*. New York: McGraw-Hill.

Peterson, M. N. A., 1966. Calcite: rates of dissolution in a vertical profile in the central Pacific. *Science* 157:1542–1544. [Experiments show a marked vertical change in rate of dissolution.]

Shepard, F., 1973. *Submarine Geology* (3rd ed.). New York: Harper & Row. [The standard modern text.]

——, and H. Wanless, 1971. *Our Changing Coastlines*. New York: McGraw-Hill. [Up-to-date photographic coverage of U.S. coastlines, with many historical photographic comparisons and personal anecdotes.]

Taney, N., 1961. Geomorphology of the south shore of Long Island, New York. *U.S. Army Corps Eng. Beach Erosion Bd. Tech. Mem.* (128).

Townsend, D., 1971. 21 miles of California coast put on 5-year 'death notice.' *Los Angeles Times*, 21 November, sect. C. pp. 1, 4. (Quoted from sect. C, p. 4.)

Turekian, K., 1968. *Oceans*. Englewood Cliffs, N.J.: Prentice-Hall. [An up-to-date paperback that emphasizes marine sedimentation and geochemistry.]

Wiegel, R. L., 1964. *Oceanographical Engineering*. Englewood Cliffs, N.J.: Prentice-Hall.

Zenkovich, V. P., 1967. *Processes of Coastal Development* (translated from the 1962 Russian edition and edited by J. A. Steers). New York: John Wiley & Sons. [Russian examples of beach phenomena.]

RESOURCES

JOHN S. SHELTON

17

WATER AS A RESOURCE

. . . as it has been well said, not one of us, if we are scientists at heart, can afford to ignore any branch of our science, "even though it be conspicuously—and even glaringly—useful."

Charles Lapworth, in GEOLOGICAL MAGAZINE (1899)

Water is the best of all things.

Pindar (522–448 B.C.), OLYMPIAN ODES

We now begin the first of three chapters concerned with natural resources, and we are immediately confronted with a difficulty in defining the term "resource." In a general way, it means an obtainable reserve supply of some desirable thing—but "obtainable" at what cost, and "desirable" compared to what?

Some natural resources, such as air and water, are renewable, meaning that, in general, they restore themselves. Fish and forests are also considered to be renewable resources, but it is clear that they *can* be exhausted locally—or even entirely. This is true of all renewable resources, if their quality is taken into account. Pure air and water do not automatically renew themselves in a polluted city.

Metal ores, such as the ores of copper, are not naturally renewable, except on a geologic time scale, but they can be obtained again and again by recycling. This may not be done because the cost does not equal the value, but it is technologically possible, and both cost and value vary with time. At one time, for example, copper was mined from very concentrated ores, and rock with 1%

copper was worthless. Now the high-grade ores are exhausted, and 1% ores are valuable if they occur in large bodies. When the low-grade ores are gone, copper could be mined by processing ordinary rock, soil, or sea water, but the cost would increase. Already, some copper is being recycled from waste. Alternatively, some other material could be substituted for copper. Very valuable materials, such as diamond and emerald, have already been synthesized, and others could be, were there enough demand for them.

Viewed in the simplest way, almost all natural resources could be obtained in any desired quantity, provided they were considered worth the trouble and expense of obtaining them. Few, if any, important materials have limits as resources, provided the desire for them is great enough and that energy is available to mine, modify, concentrate, or manufacture them.

Albert Einstein discovered that energy and matter are interchangeable, and, in a general way, that applies to resources. Ultimately, therefore, the cost and availability of material resources depend on the energy resources. At present, the latter are quite limited. In the past, society has lived off the stored energy of the sun in the form of wood, coal, oil, and gas. These will not last for more than a few centuries, according to realistic estimates. Now we are beginning to use the energy of the atom in nuclear power plants, but, like coal or oil, the supply of suitable radioactive minerals may not last very long.

In the immediate future, the limit on material natural resources depends on arbitrary priorities set by society. The energy to extract, refine, or manufacture materials is now available. Once it is gone, however, man may no longer be capable of mining any but concentrated ores, and such ores will all have been exhausted. Thus, material natural resources may be very limited in the foreseeable future.

With limitless power, other resources might also be virtually limitless, except for the limits posed by pollution. It appears that this dream may conceivably be realized. Some planned types of nuclear power plants—namely, breeder reactors—would produce even more radioactive fuel than they consume. They are now in the testing stage. Experiments are also underway to try to develop thermonuclear power by controlling the reactions that occur in hydrogen bombs. If this power source can ever be tapped, its fuel is inexhaustible and the future of world mineral resources will be much brighter. However, the price to be paid in pollution, particularly thermal pollution, is unknown. The world and its resources will still be limited, even if power is not.

We have already considered water in its eternal circulation from ocean to air to land and back (Chapter 10), in its controlling influence on climate and weather (Chapters 11 and 12), in its influence on soil formation and stability (Chapters 13 and 14), and in its role in the erosion and transportation of sediment (Chapters 15 and 16). It also has a unique importance as a resource, and it is to that aspect of water that we now turn. Water is abundant and cheap, and it has an extraordinary range of valuable properties that may be put to use. We drink it, we nourish our food with it, and we use it to dissolve away our waste products. We use it for heating, for cooling, and for transporting goods. We even use its energy, derived from the sun, to produce hydroelectrical power and to transport waste materials to the sea. The emerging local problems of water shortages in the United States tend to arise from these conflicting uses of this remarkable resource that is both so cheap and so valuable.

We shall consider increasingly complex environmental problems and related social problems in the course of this chapter, and it may be helpful to outline the underlying plan: First, we shall consider what water is used for in different parts of the United States, and then the existing supply and demand in the east and west. The opening sections are concerned largely with surface water, which has already been discussed in earlier chapters. The next sections deal with ground water and some factors that affect underground flowage and the chemistry of water. The final sections discuss some problems related to dividing a fluctuating water supply and to geologic hazards associated with distant transfers of water.

Water Uses

The possible sources and supply of water for a particular use depend intimately upon the nature of that use and the number of its properties that are consumed by that use. The uses of water are very diverse and include some that may hardly affect other, later uses. Foremost of these is the generation of hydroelectrical power by storing water behind a dam and releasing its potential energy by letting it fall to turn electrical generators. The water emerges from the power plant as pure and almost as abundant as when it entered. Even so, a surprising number of values and potentially useful properties are lost when a river is dammed, namely: (1) part of the potential energy; (2) part of the water, because of increased evaporation from the lake, which has a larger surface than that of the original river; (3) the sediment that is trapped by the dam; and (4) the use of the river for shipping, fish migration, and certain types of recreation. Some of these lost values can be restored by canal locks to raise ships or by bypassing channels for fish.

Power generation hardly detracts from other uses of water, but, carried to extreme, the additional evaporation could seriously deplete the volume of the river, which could destroy its utility for one of water's most critical functions—namely, the removal of wastes. It might also reduce the supply below the requirements for industrial cooling and other processes, and below the amount necessary to maintain low salinity in estuaries. In the latter case, most of the typical organisms of estuaries would die because of encroachment of sea water.

Generally, the least desirable use of water is one that causes the largest fraction of it to evaporate and, thus, be lost for all other uses. This weakens the flow, decreases the potential energy, and concentrates all of the pollutants that are in the water. The addition of pollutants to water also destroys many of its potential uses, but by no means all of them. Pollution by water users has many sources, each of which affects especially the same type of potential user downstream. Thus, upstream farmers are responsible for the increased leaching of salts into rivers, and other farmers downstream cannot use the salty water; upstream power plants pollute the water with heat, and others downstream cannot use the warmed water for cooling; and cities dump their sewage upstream of other cities' intakes for domestic water. Fortunately, these abuses are being corrected, albeit slowly.

Major Uses of Water in the United States

The major uses of water in the United States can be divided into five categories that tend to have different sources, that lose markedly different proportions of their water to evaporation, and that produce quite different levels of pollution. The total use in this country is about 9860 million cubic meters per day (mcmd) and the total consumption is about 429 mcmd. The greatest individual use by far is 8720 mcmd for generating electrical power. The accompanying loss to evaporation of lake water is uncertain. It is known that 42 mcmd is evaporated from all the lakes of the 17 western states, and most of the large ones are artificial reservoirs. Thus, in the west, a major factor in water consumption is evaporation from water behind dams.

The next most important user is industry, with 645 mcmd that is self-supplied and another 26.6 mcmd from public supplies. The self-supplied water is used almost entirely for cooling power plants that burn coal, gas, or oil, and 25% of the water is saline, as many plants have been built along the seashore. Only 2%, or 12.9 mcmd, of the self-supplied water is consumed; except for the heat that it acquires, it is largely unpolluted and, thus, is fit for most other uses. The remainder, however, includes water heavily polluted with industrial wastes. Most of the self-supplied industrial water is used in the eastern states, where thermal power plants are concentrated.

Agriculture, which requires 455 mcmd for irrigation, is only the third largest user of water, but it accounts for 80% of all water lost and is incomparably the greatest consumer. About 92 mcmd is lost to seepage and evaporation in the process of moving water to the fields, and 282 mcmd is consumed as a consequence of spreading the water in shallow furrows. Thus 75% of all irrigation water is consumed, and, as a consequence of evaporation, the drainage from farm land tends to be excessively salty, so it may not be reusable for most purposes. Irrigation is concentrated in the western states, but it is increasingly popular in the east as well.

The "public supply" of water is largely for urban uses and totals 89.5 mcmd, of which a third is for industrial use. Slightly less than 20 mcmd, or 22%,

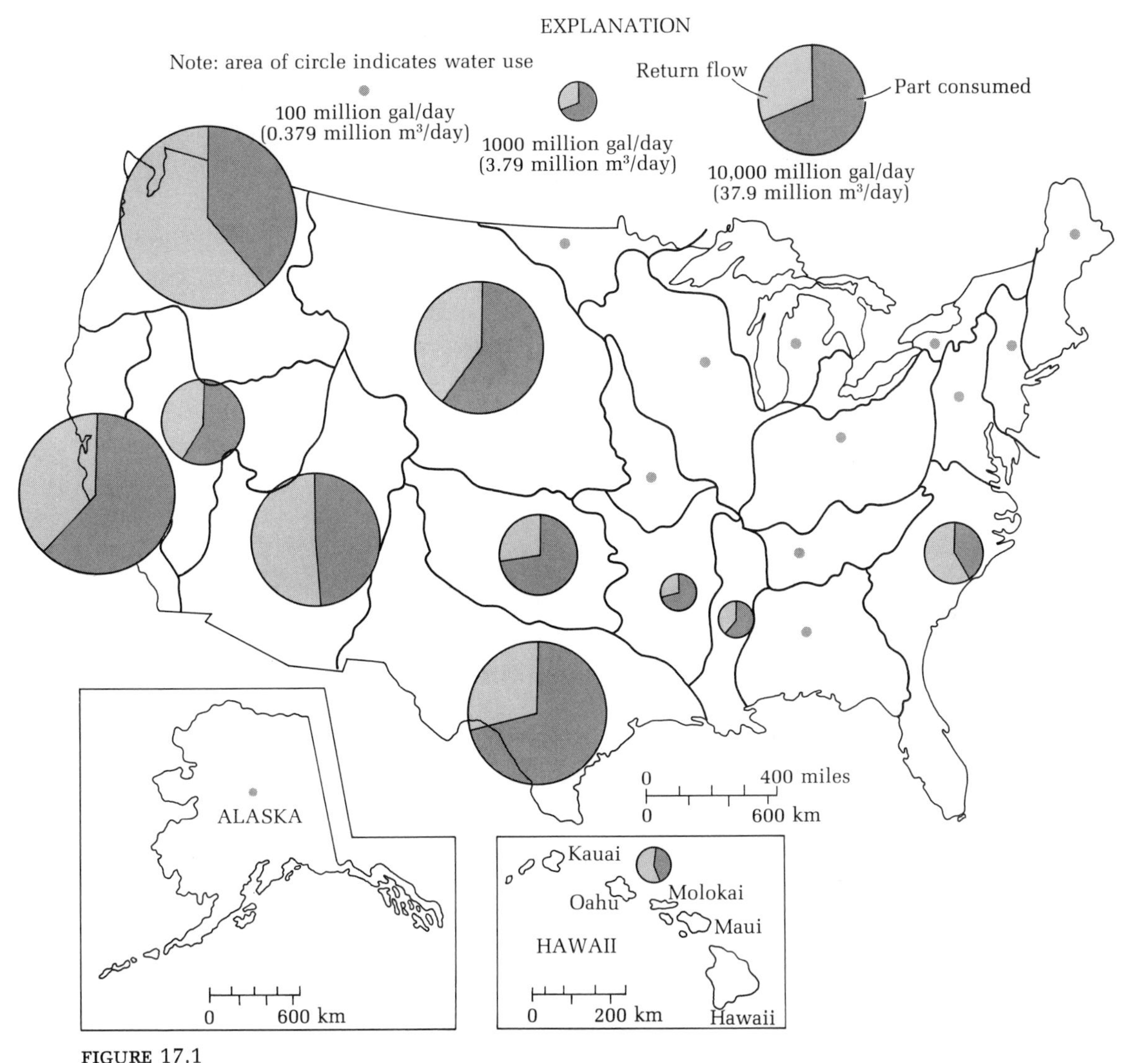

FIGURE 17.1

Use and consumption of irrigation water in "water-use regions" in 1965. The regions are roughly equivalent to river-drainage basins, except that the Mississippi basin is subdivided. Irrigation, which is practiced mainly in the western states, is by far the largest consumer of water. [From Murray (1968).]

is consumed, and the remainder is more or less polluted with human, animal, and industrial wastes.

"Rural domestic and livestock" require only 15.2 mcmd, which is drawn largely from local wells and is 80% consumed. Much of the highly polluted remainder returns to the ground.

In sum, water is used mostly for purposes that do not consume it. Water is used predominantly for different purposes in the western and eastern states. The eastern uses, in general, are nonconsumptive, whereas those in the west—mainly irrigation and hydroelectric power—are consumptive. Consequently, in the 48 contiguous states, 85% of the water consumption takes place in the 17 western states and only 15% in the 31 eastern states. The difference can also be expressed in terms of average consumption per person: This is only 11.5–15 cubic meters (30–40 gallons) per day in the urban, industrialized northeastern triangle from Chicago to Washington to Boston, but it is 379–1325 cubic meters (1000–3500 gallons) per day in the western states.

Supply and Demand

The annual supply of water is the sum of the amount that falls from the sky plus the amount withdrawn from ground water. In the long run, however, the ground water must be replenished from the surface, and, for some purposes, it can be ignored in assessing the supply. For most purposes, the runoff of river water is a reasonable approximation of the supply of water. It amounts to 4550 mcmd for the 48 states; about a third of this is in the 17 western states, which have half the area. Thus, this supply is markedly different in the two "halves" of the country. Moreover, much of the runoff in the western half drains into the Missouri, Arkansas, and Red rivers; and these, in turn, empty into the Mississippi, which is in the eastern half. This water becomes available for many purposes, such as recreation, removing wastes, and generating power, and it adds to the general abundance in the east.

The runoff varies with seasonal rainfall, spring melting, and climatic fluctuations. For this reason, a distinction is made between the average annual supply and the dependable supply, which is usually defined as a flow that is exceeded in 9 years out of 10. The dependable supply can be increased by various means, such as constructing reservoirs and recharging ground water. The addition of reservoirs will increase evaporation, however, so a dependable supply is augmented at the cost of the total water supply. The total dependable supply in 1955 was roughly 1138 mcmd, but, by 1980, it is expected to be more than 1890 mcmd. For the 48 contiguous states considered as a whole, this was adequate for all uses as of 1965. The picture is grimmer for the 17 western states, however: use in 1965 exceeded the dependable supply, and even the consumptive use approached half the dependable supply. The situation may be better by 1980 because of planned measures to increase reservoir capacity. However, to the extent that this increases evaporation, other problems may arise.

What of projected supply and demand? The population is steadily increasing, and the per capita use increases even faster. What will happen is still uncertain, but the trends from 1950 to 1965 provide some basis for speculation. In rural supplies, for example, there was hardly any change at all in either

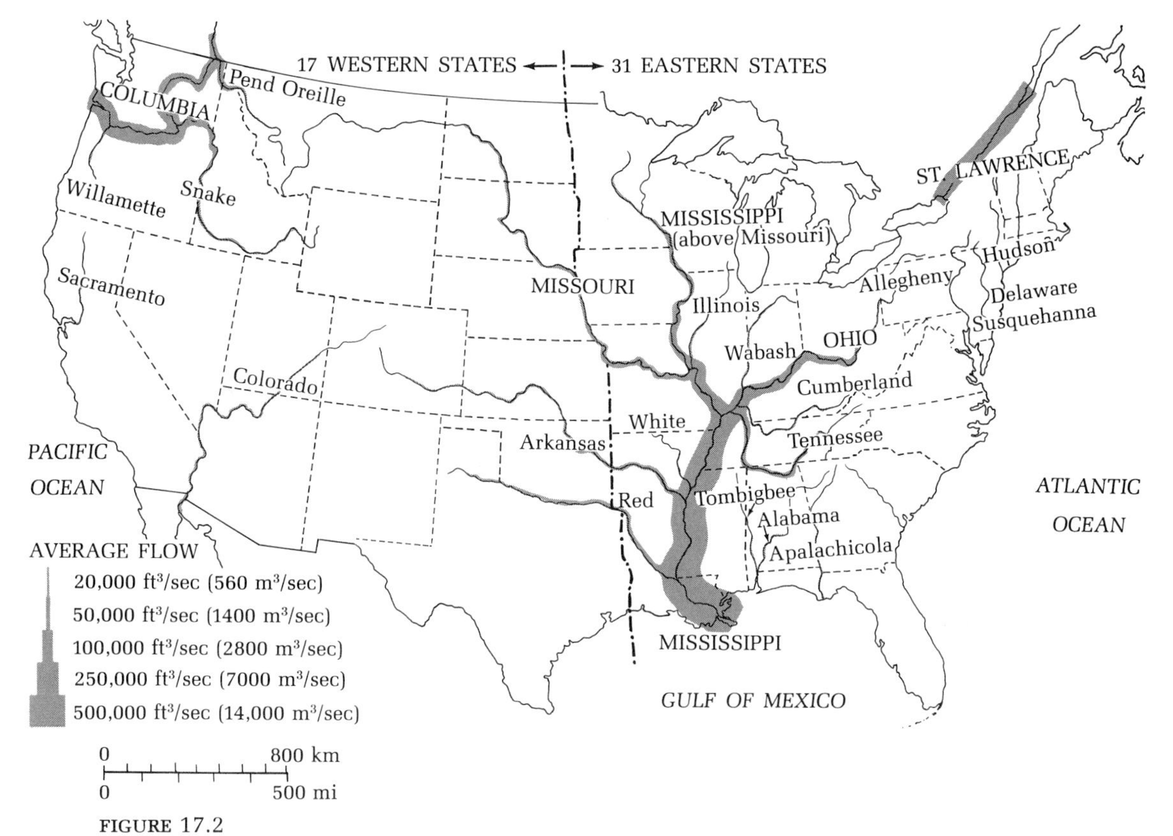

FIGURE 17.2

Large rivers of the United States. Those shown have an average flow at the mouth of at least 19,000 cubic feet per second (530 cubic meters per second), based on data for the years 1921–1945. Note the importance of the Mississippi and the Columbia, and that much of the area of the 17 western states drains into the eastern region. [After Miller et al. (1963).]

surface or ground-water withdrawals during the period. This may reflect the fact that the population is shifting to the cities and that farms are being abandoned in the process. The population of California, for example, was already 86.4% urban in 1960. The shift toward the cities may also account for a relatively small increase in water for irrigation – from 397 mcmd to 440 mcmd – in this period. However, in 1950, the use of water for irrigation may already have approached its possible maximum. Because irrigation is the chief consumer of water, it appears that total consumption may grow relatively slowly compared to use. Industrial use, for example, roughly doubled from 1950 to 1965, but little of that water is actually consumed. The use of water for hydroelectric power also doubled, and this trend is apt to continue as pollution-free power becomes more desirable. Unfortunately, such growth will be accompanied by increasing losses to evaporation, which are already very large.

A noteworthy trend is the increasing use of ground water for most purposes. It has doubled in 15 years from about 114 mcmd to 233 mcmd. Almost all

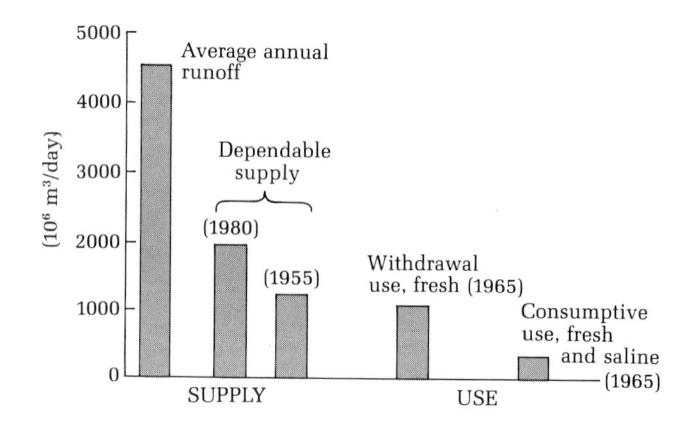

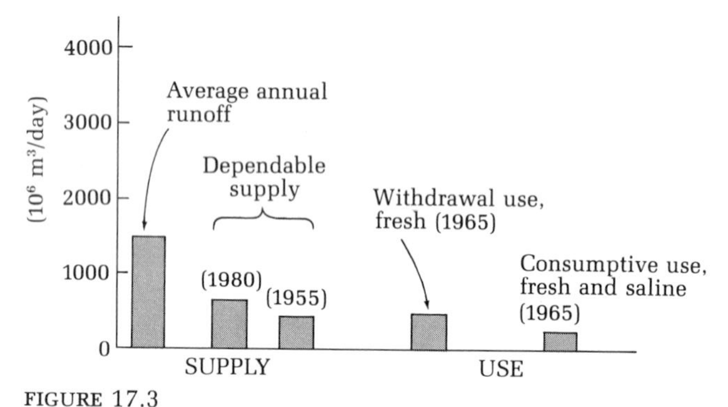

FIGURE 17.3
Water supply and demand in the 48 conterminous states (*above*) and in the 17 western states (*below*). [From Murray (1968).]

of the growth has occurred in the categories of irrigation and public supplies, which totalled 182 mcmd in 1965. Much of this was in California and Texas, which used 102 mcmd for these purposes. This has had some side effects, such as subsidence, which we have already considered. There are other effects, including exhaustion of the resource, which will be discussed after we study the geology of ground water.

IS THERE ENOUGH WATER?

The official position of the U.S. Geological Survey, expressed by geologist Raymond Nace, is that we are not now running out of water. Acute local problems exist, but they are generally solvable at a cost that would appear trivial were water not originally so cheap and were it not so widely used. As the costs go up, the water is used for more important purposes, particularly for public supplies. If present trends continue, we would generally "run out of water" (that is, our activities would be limited by the water supply). However, to quote Nace:

> Predictions that point to a crisis in 1980, 2000, 2050 or some other year have no value because they are misleading and therefore may be dangerous. They are merely straight-line projections of historical and current trends, and they seem to assume that withdrawal use of water will be as irrational in the future as it has been in the past. They are equally misleading and dangerous in that they ignore completely the nonwithdrawal use of water that maintains the environment and grows most of the food.

Nace's last sentence deserves emphasis. River runoff is generally taken as equivalent to annual water supply, but, in fact, the supply also includes all the rain that does not run off. Farms and forests receive rainfall, and by far the largest fraction of them function without irrigation or runoff. Cities are watered by rainfall; rain replaces evaporation in lakes; and the surface of the land is cleaned by raindrops.

The present supply of water is generally adequate, and society has many political, economic, and social means of accommodating to any limit that may be approached. Ground-water storage, rainmaking, and desalination are a few technical possibilities for solving the local problems that exist within the national abundance.

One major possibility for a broad concern remains—namely, a climatic change toward aridity. Such a change should be slow, but, to the extent that we allow use to approach supply, we lose any ability to accommodate to long-term changes. Our earlier analysis of river flow showed that the amount of water available in a whole drainage basin may vary by as much as 100% from one decade to another. Then we were considering the hazards of flooding when society adjusts construction on flood plains to low levels of river flow. The social effects of climatic change might be much greater if society instead expanded to dependency on a high level of river flow.

Ground Water

Some rainfall runs off at the ground surface and flows downhill in familiar ways—that is, as streams and rivers. Some penetrates the surface and flows in more complex ways as ground water. The two flows tend to intermix as they move toward the sea. Rivers and lakes may add or subtract from the ground water, depending on many factors. A mountain spring, for example, marks a transition from ground water to surface water. A brook that flows from the spring will return the water to the ground if it flows into dry desert sand of the sort that surrounds many mountains in the arid western states.

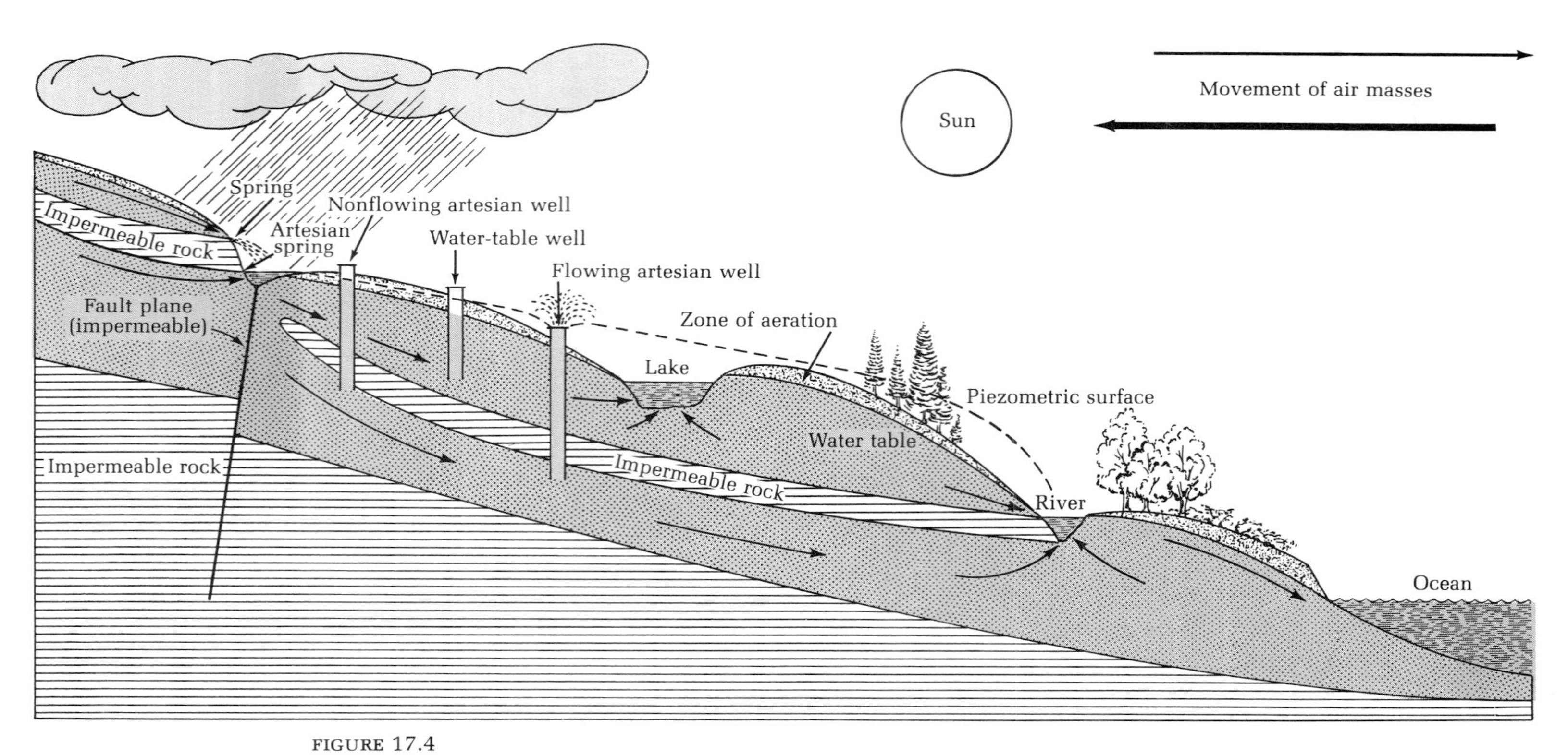

FIGURE 17.4

Ground water is part of the hydrologic cycle, in which water is ceaselessly evaporated from, and precipitated on, the surface of the earth. When a well is sunk into permeable rocks, the water stands at the level of the water table. Impermeable barriers, such as the fault plane shown, may bring the water table to the surface. When a well is sunk into permeable rocks beneath an impermeable formation, the water rises. Such a well is termed artesian. The level to which artesian water will rise is called the piezometric surface. [After Sayre, "Ground Water," copyright © 1950 by Scientific American, Inc. (all rights reserved).]

AQUIFERS AND AQUITARDS

The porosity of a sediment or rock determines the quantity of water it can hold, but its permeability is far more important because it determines how much can move in or out. If water moves relatively freely through it, the material is called an aquifer; if water is relatively retarded by it, the material is called an aquitard. Common permeable materials are unconsolidated sand and gravel and their equivalent rocks, sandstone and conglomerate. Ground water moves through such rocks by flowing in the tiny pore spaces among the grains. Lava flows and limestones also tend to be highly permeable, even though there may be no connected open spaces comparable to pores. This is because such rocks tend to contain large, intricate passages comparable to a capriciously designed sewer system. The volcanic conduits are a consequence of the original flows, but the limestone ones result from solution by ground water and, thus, develop gradually.

Almost any aquitard is at least slightly permeable because of the same kinds of cracks that localize weathering. However, many materials, such as clay and shale, effectively prevent the flow of water. Many sandstones are also aquitards if the pore spaces become filled with minerals deposited by ground water. Other common barriers to water flow are intrusive igneous rocks, and faults that are sealed by the clay that is produced on them by friction. Thus, the paths of ground-water flow tend to be complex. Water may move horizontally through an aquifer that is bounded above and below by aquitards; but the horizontal flow may be dammed by the almost impermeable wall of a fault; and so on. However, the phenomena are well known and, after detailed mapping, it is even possible to utilize aquifers as natural underground reservoirs in which to store ground water.

The residence time of ground water is about 5000 years (Chapter 10), so, on the average, it moves much more slowly to the sea than surface water does. Rates of flow probably average about 15–20 meters per *year*, which means that ground water can be replenished only very slowly. However, water in some aquifers moves relatively rapidly, at least for small distances. Ground water may flow through highly permeable sands at rates of 10–100 meters per *day*. In the extreme case of open channels in limestone or lava caves, underground streams flow just as fast as those on the surface.

WATER TABLE

In any particular area, the top surface of water-saturated soil or rock, the water table, corresponds to the level of water in open wells. The water table is ordinarily underground, but it always intersects the surface at springs, rivers, and lakes. In addition, it may temporarily intersect the surface during periods of flooding or during a spring thaw—or, in flat regions with slow runoff, where swamps are common, after every rain.

Ground water either moves downhill or is ponded—just like surface water. It is replenished by rain, except in periods of drought, when the water table slowly sinks as the water flows away. Thus, the level of water in a well naturally fluctuates in response to variations in supply.

Drilling a well introduces a new element into a ground-water system. The increased withdrawal that follows inevitably lowers the water table, although the long-term effect may be trivial for small withdrawals from a copiously flowing aquifer. At the opposite extreme, the water table drops permanently in response to large withdrawals from a ponded, almost static store of ground water. There is also a short-term effect caused by pumping. A conical depression of the water table, hundreds of meters wide and a few meters deep at the center, forms around the well. It usually vanishes within a few days after pumping stops as ground water flows in to restore equilibrium.

ARTESIAN WATER

A soil zone, with its pores partially filled with air, exists above the water table; thus, the pressure at the water table is atmospheric. If a well is drilled through the water table, the water level is unchanged: the pressure is the

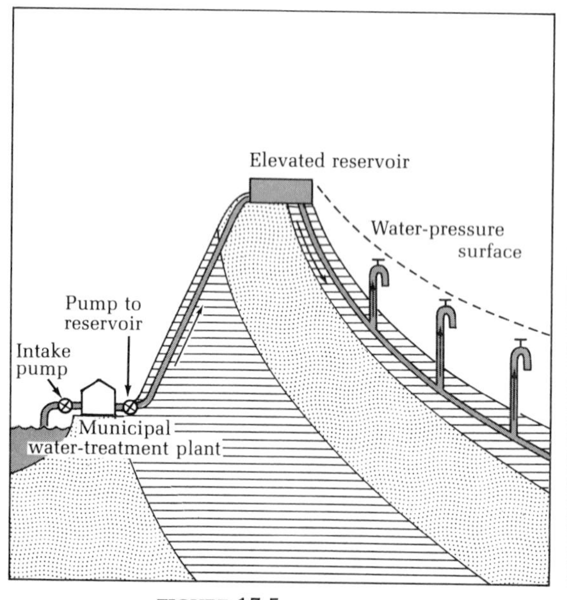

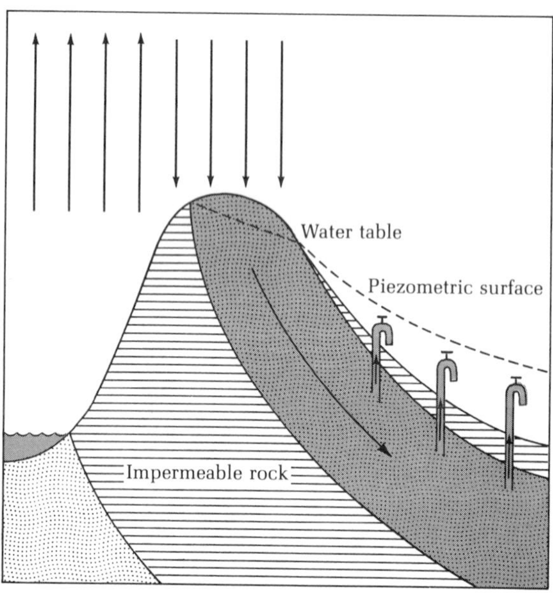

FIGURE 17.5

A nonartesian water system (*left*) draws water from a low-lying well or reservoir, treats it, and pumps it into an elevated tank. The water than flows freely from the outlets that are lower than the water-pressure surface. An artesian water system (*right*) requires no pumps or tanks. The water is precipitated on a "recharge area," and is drawn from the same stratum below it. (The piezometric surface is analogous to the water-pressure surface.) [From Sayre, "Ground Water," copyright © 1950 by Scientific American, Inc. (all rights reserved).]

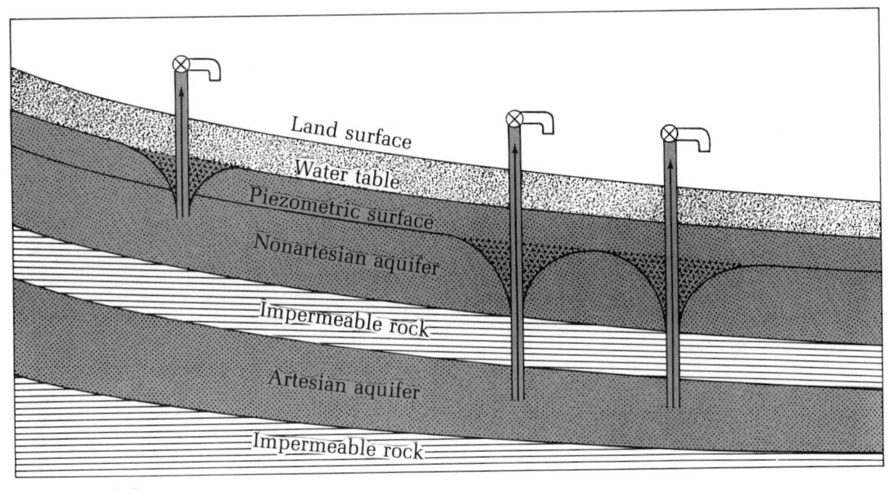

FIGURE 17.6
Artesian and nonartesian systems may be used in the same area without interfering with each other. In this diagram, water is pumped both from artesian and from nonartesian aquifers. The nonartesian well at the left causes a drop in the water table, but does not affect the piezometric surface. The artesian wells at the right cause drops in the piezometric surface, but do not affect the water table. [From Sayre, "Ground Water," copyright © 1950 by Scientific American, Inc. (all rights reserved).]

same before and after drilling. Consider a very different configuration in which a sloping aquifer is bounded, top and bottom, by aquitards: Water feeds into the aquifer at the upper, exposed end, and the water table is not far below the ground surface. The water in the aquifer at greater depths is under pressure like the water in an urban water main. A well that taps such an aquifer produces the same effect as opening a fire hydrant. The released hydrant water rises—as high as a house top—toward the level of the reservoir that feeds the water system, but valves and friction losses in small pipes restrict the lift. In the same way, artesian water rises in a well because it is under pressure. Likewise, the difficulties of flowing through the ground prevent the rising water from reaching the level of the water table in the source area. However, artesian water may be under very high pressure and may emerge above the well as a spouting fountain. The level to which it rises, the piezometric surface, fluctuates like the water table but in response to different factors (see Figures 17.4–17.6).

Artesian wells were developed in the French district of Artois in 1126—hence the name. They are very important in many regions, including Australia and the United States. The Dakota sandstone, an artesian aquifer under the Dakotas, Nebraska, and nearby areas, has its source area hundreds of kilometers away in the foothills of the mountains to the northwest. It is only about 30 meters thick, but it supplies water to more than 15,000 wells. Heavy withdrawals have steadily reduced pressure since the aquifer was tapped in 1882. The supply of water has not limited its use, but pumping may become necessary when the pressure drops—and that costs money. Thus, the supply of water costing less than a certain amount may be limited.

BOX 17.1 WATER WITCHING

The transitions from common sense to technology and from magic to science occurred gradually over several centuries, and some are not yet completed. We still have witch doctors and faith healers as well as medical doctors. We still have builders following traditional methods of construction as well as civil engineers designing bridges with new materials. In many respects, these various practitioners are economically, as well as ideologically, competitive. An experienced witch doctor, for example, certainly can be helpful in alleviating fear and discomfort and even in curing some ailments. He may not be as effective as an M.D., but, in some areas, he may be more available and cheaper.

Water witcher or dowser of the sixteenth century, styled after old woodcuts. [From Gilluly, Waters, and Woodford, *Principles of Geology*, W. H. Freeman and Company, copyright © 1968.]

For these same reasons, magicians are still competitive with scientists in many circumstances. One such group of magicians are water witchers or dowsers. Commonly, although they wear modern costume, dowsers still follow the same practices that were followed by water witches in medieval times. Past and present farmers have wanted to know where to dig a well that would strike water. Grasping a forked stick with both hands, a dowser walks about until an "irresistible force" pulls on the end of the stick and causes it to point downward when it is over a supply of ground water. That is the story, although it would be difficult to find a scientist who believes in the mysterious "force."

How, then, can one explain the fact that a well drilled where the stick points will usually produce water? The scientific explanation is threefold: (1) aquifers are common, and most wells locate at least some water; (2) people tend to emphasize successes and to ignore or explain away failures; and (3) a practicing magician is a transitional scientist and draws on his experience of successes in seeking water. (Professional training is essential to solve difficult problems but not to solve easy ones. It just improves the odds for success.)

The last point may be the most important. Dowsers may be relatively available and cheap in rural areas, compared to ground-water scientists, even though they are not as skillful in finding water in difficult circumstances. In addition, because they are generally more skillful than most people, their services may actually be of value.

An interesting account of water dowsing, based on personal experience and written from a viewpoint sympathetic to the practitioners, is given by the well-known novelist Kenneth Roberts.

Solution and Deposition

The intense processes of solution, migration, and deposition of dissolved material that are characteristic of soil formation at the rock surface also occur deeper in the earth in aquifers. The ground water slowly dissolves materials where it is undersaturated with certain elements and deposits materials where it is supersaturated. In between, it may transport calcium, silicon, and various more soluble substances for long distances.

The principal effect of ground-water deposition is probably the conversion of sediment into sedimentary rock by filling pore spaces with calcium carbonate or silica. The most obvious effect of ground-water solution is the formation of karst topography in limestone. However, the deposition of silica requires its solution somewhere else, and a slow interchange of many elements may be commonplace.

The history of ground-water solution and deposition may be very complex in a given locality, particularly in the shallow zone that is sometimes above and sometimes below the water table. Limestone caves produced by solution are often partially refilled by curious formations deposited from ground water. Moreover, solution followed by deposition is responsible for the existence of many fossils. Calcium carbonate shells are slowly dissolved away and may be replaced by silica, which, being resistant, is preserved.

Ground water exists in thermal and volcanic areas, and it is heated before returning to the surface as hot springs and the intermittent fountains called

FIGURE 17.7
Hot-spring terraces of calcium carbonate, Mammoth Hot Springs, Yellowstone National Park, Wyoming. [Photo by Tad Nichols, Tucson, Arizona, from Gilluly, Waters, and Woodford, *Principles of Geology*, W. H. Freeman and Company, copyright © 1968.]

FIGURE 17.8
Old Faithful Geyser in eruption, Yellowstone National Park, Wyoming. [Photographed in 1872 by William H. Jackson, official photographer of the Hayden Survey, 1870–1879.]

geysers. Hot water is a much better solvent than cold; hence, when the hot solutions cool, they deposit material in abundance. This produces the very interesting mineral pools and cascades that are characteristic of such areas as Yellowstone National Park.

Hot mineral springs have been used as health spas since the dawn of history. The town of Bath in England has natural hot water and was a Roman spa. Likewise, Marienbad (*bad* is the German word for "bath") and other such resorts are located at hot springs. Mineral waters, containing small amounts of dissolved material, are commonly bottled for sale. Vichy water and Perrier water are among the popular French brands that are sold all over the world.

People value mineral water for bathing, if they find it at a spa, and they will buy it for drinking if it is bottled. However, mineral water is usually very unpopular if it is free and comes out of a household water tap. This is because most mineralized water does not taste very good, and it damages plumbing as well.

One important component of mineralized tap water is calcium carbonate, and its concentration, in parts per million of water, is expressed as hardness. Hard ground water is of little consequence in most of the eastern United States, because most ground water there is soft and, in any event, surface water is far more important. Ground water finds more use in the Midwest, but, because it is used mainly for agriculture, the fact that it is hard is easily endurable. As with most water problems in the United States, it is only in the southwest that water hardness is critical. Urban water in southern California is drawn from hard ground water and hard surface water. Calcium carbonate precipitates from water as it evaporates—or, in some places, even in the pipes. In time, the pipes are damaged. Noteworthy deposition occurs around dripping faucets and shower heads, so these fixtures are ordinarily repaired. This contrasts with the situation in New York City, where water is soft and the amount used is not metered as it is in California. In New York City, a significant amount of water is lost because of leaky plumbing that probably would be quickly repaired in Los Angeles. Thus, when an "acute water crisis" occurs in one locality, there may be solutions that do not apply to similar "crises" in other localities.

Contact with other Fluids

Fresh water is not the only fluid under ground. There are also important amounts of oil and salt water that interact with it in various ways. If the fluids are not in motion, the interaction is simple. The oil, which is least dense, floats in a horizontal layer on the fresh water, which, in turn, floats on the yet denser salt water.

The distribution of underground fluids is rarely so simple, however, because the ground water is usually moving and the configuration of permeable rocks is complex. As to the latter, oil fields are common where oil-bearing rocks are folded upward into anticlines but rare where they are folded down into synclines (Chapter 18). Being lighter than ground water, the oil floats up into the anticlines and is trapped. The rare oil in synclines has been prevented from floating away by some other phenomenon, such as an impermeable fault zone.

The effect of ground-water movement is to generate a slope between two fluids of different density. The distribution of water in an oceanic island provides an illuminating example. Consider a small, low island of coral sand on an atoll. Without rain, sea water would form a water table extending at sea level through the island. However, rain falls on the island, soaks into the sand, and floats on the sea water. Like an iceberg, the part above sea level is supported by a root of fresh water that extends down into the sea water below the surface of the island (Figure 17.9). The pool of fresh water constantly spreads

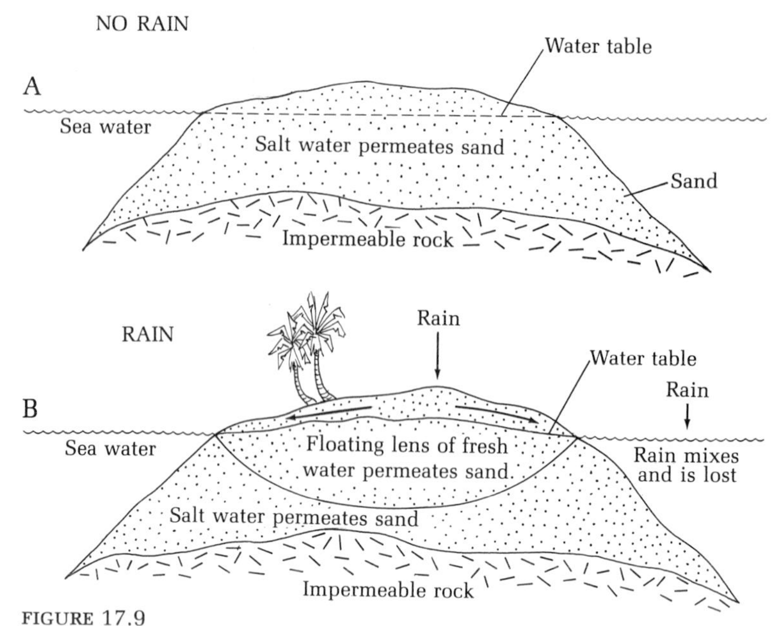

FIGURE 17.9

In a hypothetical region without rain (A), a sandy island is permeated by sea water, which even moves up and down in response to the tides. In a region with rain (B), a lens-shaped layer of fresh water is restrained from mixing by the sand of the island, and floats, like ice, on the salt water. The fresh-water table stands above sea level just as the top of an iceberg stands above the sea.

out from the island and is lost by mixing with the sea. At the same time, it is constantly replenished by rain; thus, it exists in dynamic equilibrium. The upper surface of the pool slopes because the water is flowing like a river. The lower surface slopes because each column of fresh water is floating, and the higher upper surface is balanced by a depression of the lower one. With regard to ground water, the edges of continents are like one-sided islands. Rain is a source for a steady flowage of fresh ground water toward the sea. Because the continental edges are relatively high, the water table is high and the sloping interface with sea water is correspondingly deep. Usually, fresh-water aquifers form a tongue between the salt-water aquifers below and the sea itself above. Thus, springs of fresh water commonly flow up through coastal sea water, and, if copious enough to minimize mixing, are used as a source of drinking water by explorers, who then need not come ashore.

Ground water is also the source of urban water for many coastal cities, and the contact with saline ground water gives rise to serious problems. As urban use grows, the wells pump harder, and fresh water withdrawal exceeds replacement. Salt water then moves in and up to fill the gap, and the water supply may be ruined unless expensive recharging of fresh water begins.

Mining and Recharging Water

Ground water commonly is in short supply in areas where surface water flows, wasted, into the sea. Thus, many cities have introduced measures to recharge ground water by spreading surface waters where they will soak into the urban aquifers. North of Miami are vast water-conservation districts where the water of the Everglades is ponded and allowed to soak into the ground instead of flowing south into Everglades National Park. The Miami water supply is drawn from shallow wells. Experience has shown that saltwater incursion may endanger the water supply unless many water uses are constantly balanced. In southern California, the coarse sands and gravels that surround the mountains are ideal feeders for aquifers, which are recharged by storing water behind low dams that are primarily intended for flood control.

Ground water may also be in short supply simply because it goes to high-volume uses like industrial cooling and air conditioning. Such uses merely warm the water and do not pollute it, so it is ordinarily feasible to return water to the ground. This increases the cost of water, but New York now requires that ground water used for cooling on Long Island must be restored to the aquifer from which it is drawn. The danger of salt-water incursion is so great that recharge basins have also been constructed inland on the island. The average temperature of the water has increased a few degrees as a result of this recycling, and its value for cooling, consequently, has decreased while the need is still growing.

The supply of urban ground water is also adversely affected by the mere growth of a city. Where once the water soaked into the ground, it now races to the ocean on the impermeable surfaces of roofs, sidewalks, and streets and through drainage pipes. Although relatively unimportant at present, this factor will become increasingly important as isolated cities merge into the great conurbations that are expected to take shape in the near future.

In some regions, the ground water is being withdrawn much faster than it can be resupplied by surface water. Moreover, because the withdrawal is mainly for irrigation, most of the water either evaporates or is incorporated into crop tissues and cannot be used to recharge the ground, regardless of the cost or the law. In such regions, the water is being mined and exported as crops, just as the soil fertility is mined and exported in other places.

The High Plains of eastern New Mexico and western Texas provide a typical example. Rainfall is low, evaporation is high, and the aquifers have been isolated from an abundant water source by downcutting of river valleys. There was hardly any interest in, or occupation of, the area until soon after the turn of the century. Then it was realized that the water could be pumped to the surface to utilize a climate that was ideal for farming. Sixty years later, a prosperous farming community is perched on a perched water supply that has decreased by more than 40% in some places. The water table has dropped by as much as 30 meters, however, and the cost of pumping steadily increases. Thus, as with any mining operation, the withdrawal is self-limiting. Marginal farms will become uneconomic, and the intensity of agriculture will wane. This will not lead to a rapid rise of the water table and a resurgence of farming.

Present estimates are that it will take 4000 years to restore the level of the water table. However, this assumes that there will be no artificial recharge. It is possible that the very rivers that isolated the aquifers can be used to recharge them—provided that the yield from farming is worth the cost of construction and pumping.

BOX 17.2 CAVES AND SPELUNKING

Caves have the same fascination for some people today as they had for their distant ancestors at the dawn of time. Most people who venture into caves do so in such convenient places as Carlsbad Caverns National Park in New Mexico, where an elevator takes them down 250 meters to the main caverns. There they see such spectacles as the "Big Room," 550 meters long by 70 meters high and full of peculiar evidences of ground water solution and deposition.

There are numerous other famous caves that are public or commercial attractions, but their very popularity drives out everything but people. In their primitive state, many contain fascinating animals, such as species of blind albino salamanders, fish, or crayfish, that are found nowhere else. Perhaps most common of all are various species of bats, which like to roost in safety on the dark ceilings and emerge at dusk to forage. They deposit guano, which is mined for use as a fertilizer because of its high nitrogen content. More than 100,000 tons of bat guano were mined from Carlsbad Caverns alone.

For some people, the popular, easily accessible caves are not enough, and they go exploring in the depths. They are called *spelunkers,* a word derived from one of the Latin words for cave (which, in turn, was derived from the Greek). In 1941, the National Speleological Society was organized for the purposes of studying and preserving caves. Local chapters, called grottoes, exist throughout the United States.

Spelunkers, like scuba divers, have one cardinal rule—*Never go into a cave alone.* Another important rule is *Always mark your*

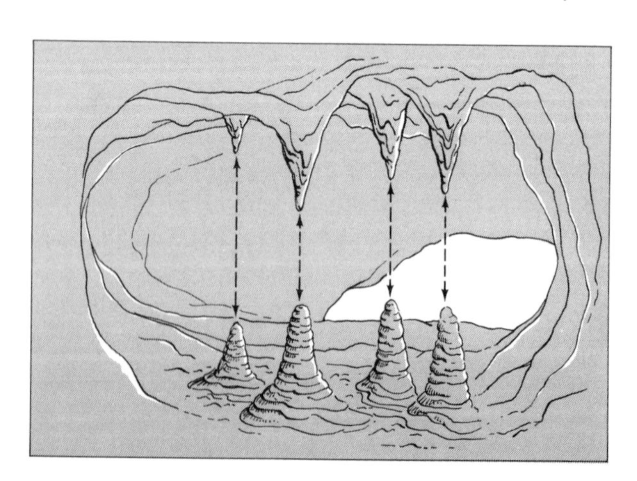

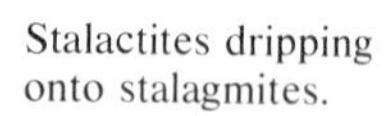

Stalactites dripping onto stalagmites.

Dividing Basin Water

Water is still available almost anywhere and at almost any time. It is just becoming more expensive locally, more dependent on artificial transfers, and more vulnerable to natural fluctuations and hazards as the limit of the supply is approached.

path so you can return (as Theseus did when he entered the labyrinth to kill the Minotaur). Diving in caves, which has been a popular sport on the French Riviera for some time, is becoming increasingly widespread. Among other things, diving or swimming can be a relatively easy way to move around in an otherwise dirty, slippery, and precipitous cave. Booted or finned, exploring a cave can be an exciting way to study the earth.

Stalactites form on cave ceilings where ground water, saturated with calcium carbonate, drips down and part of it evaporates. Stalagmites form on the ground below the stalactites, where additional evaporation and precipitation occur. The stalactites tend to have pointed tips, because the dripping is focused by gravity. The stalagmites are blunted, by contrast, because the falling drops splash. Ceiling jointing or faulting may channel ground water along cracks, thereby producing thin vertical sheets of deposits rather than isolated cones.

In time, stalactites and stalagmites tend to grow together to form columns, several of which may fuse together if the process continues. The columns, or the walls of caves, may have simple smooth surfaces or they may by fantastically fluted or draped, depending on variations in ground water flowage.

Deposition from ground water also occurs in pools on the floors of caverns. This deposition may produce such oddities as "cave pearls" of calcium carbonate, which form layer by layer on sand grains that are agitated by water flow.

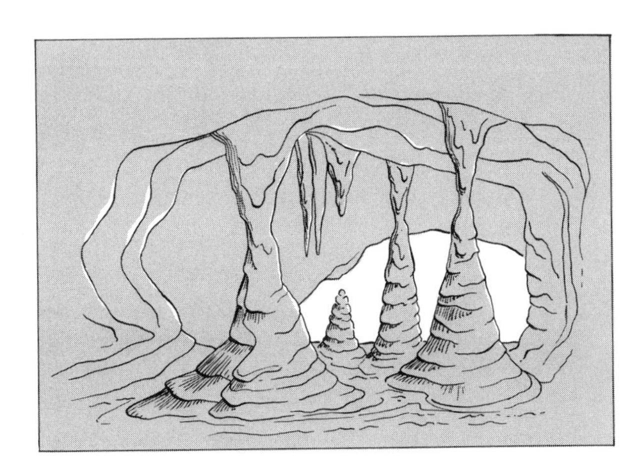

Column formed of $CaCO_3$.

Ultimately, if present trends continue, most of the western states will share a common problem of water shortages for some purposes. At present, however, the problems—more social than physical—are how to divide the water within a drainage basin, and whether to transfer it from one basin to another.

The Colorado River basin provides the outstanding example of the complexities of attempting a legal division of a water supply of unknown but certainly fluctuating volume. The river and its tributaries are fed by the rain and melting snows of the Rocky Mountains, mainly in Wyoming, Colorado, and New Mexico. It flows on through Utah and Arizona and along the borders of California and Nevada. It is blocked by numerous dams, and simultaneously supplies power and agricultural and urban water to a vast region that extends beyond the basin to southern California.

Senator Frank Moss of Utah has summarized the efforts of the seven concerned states to divide the properties of the river. In 1922, the Colorado River Compact divided the river into an Upper Basin and a Lower Basin with the boundary at Lees Ferry in northernmost Arizona. The Lower Basin was allotted—for all time!—9.250 billion cubic meters per year (bcmy) plus 1.232 bcmy when available. By later treaty, Mexico was allotted 1.850 bcmy. The Upper Basin received 9.250 bcmy, minus whatever might be necessary to supply the Lower Basin and Mexico in a dry year, plus whatever might be surplus in a wet year.

The seven states that signed the Compact in 1922 were under the impression that the average flow at Lees Ferry would be 22.800 bcmy because that is what it had been in the past. That would have supplied all the agreed allotments, plus 0.862–1.230 bcmy for evaporation. Unfortunately, the climate entered a dry cycle—or began a long-term change—and the average flow was only 16.0 bcmy from 1930 to 1964. Meanwhile, Los Angeles grew, and agriculture flourished in the whole basin, The Upper Basin states divided their reduced share of water among themselves in 1948, but needs, acrimony, and legal proceedings continue to grow in the Lower Basin. Mexico, for example, has found that its allotment is being met, but, by 1972, the water was so salty that it was unusable. A formal protest caused the salinity to be rectified, but it is clear that future water compacts will be concerned with matters of water quality as well as quantity.

Interbasin Transfers

If the governments within a basin have such difficulties with dividing the water, it might seem impossible to reach agreement to transfer water from one basin to another. However, transfers have been commonplace for half a century, and plans for more grow increasingly grandiose. It appears that

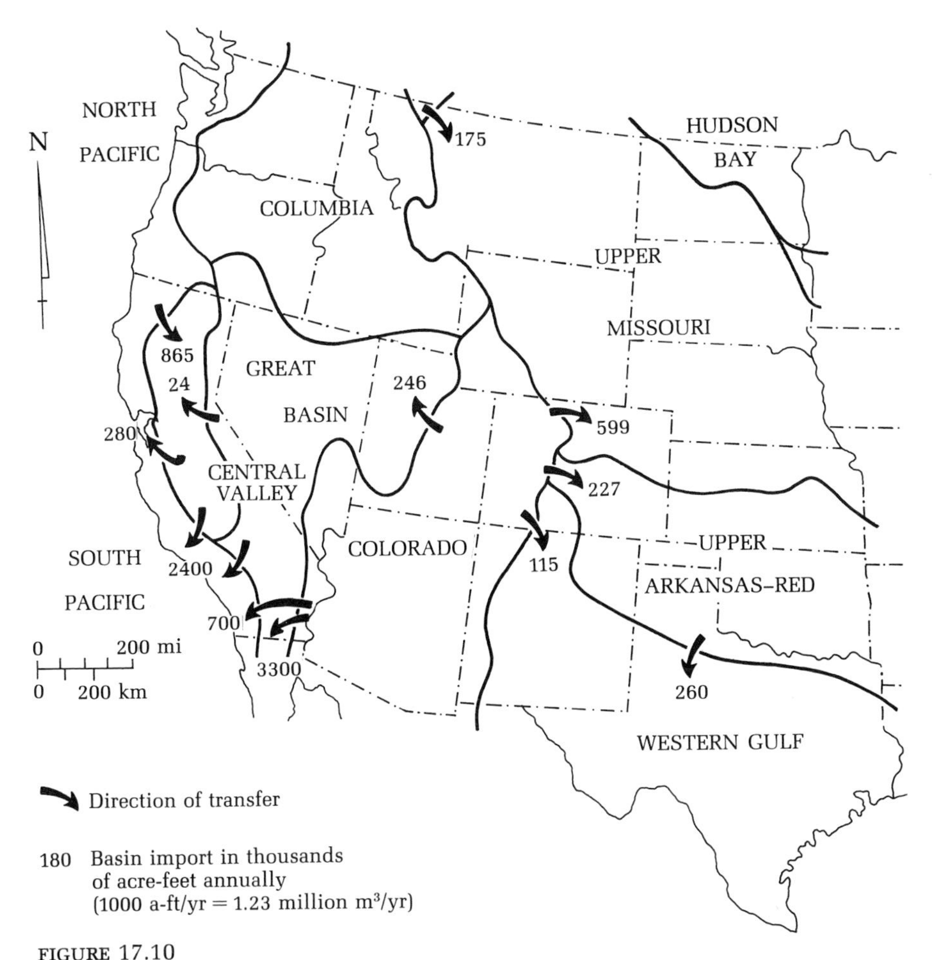

FIGURE 17.10
Aggregate water transfers between major river basins according to Quinn (personal communication). By far the largest transfer is to support agriculture in the lower parts of the Great Basin. The California State Water Project also transfers Central Valley water to the South Pacific basin.

basins with growing water shortages tend to plan far ahead to alleviate them, and basins with large surpluses care little about conserving water. Moreover, basins with shortages of water tend to have abundant people, and our habit is to take water to people rather than the reverse. In 1968, 20% of the population in the 17 western states was supplied by water imported from more than 160 kilometers away.

The general situation in the west is that water is commonly transferred from one drainage basin to another but not across state boundaries. Los Angeles, Denver, Colorado Springs, Laramie, and Salt Lake City draw water out of the Colorado River basin, but only from the part that lies in their respective states.

The total amount transferred between basins in 1968 was more than 22.2 bcmy, or 13% of all withdrawals from streamflow in the West. Since that time, the California State Water Project has started to drain an enormous amount of water from northern to southern California. Even then, southern California was by far the greatest beneficiary of interbasin transfers. The urban coastal strip already drew more than 1.230 bcmy from the north and east. The sparsely settled eastern agricultural part of the state, however, imported three times as much as the cities.

What will happen in the future? There are seemingly fantastic proposals—or dreams—of bringing water to the dry southwest from the Columbia River—or even from the Mackenzie River in the Arctic Basin. There are plans for a

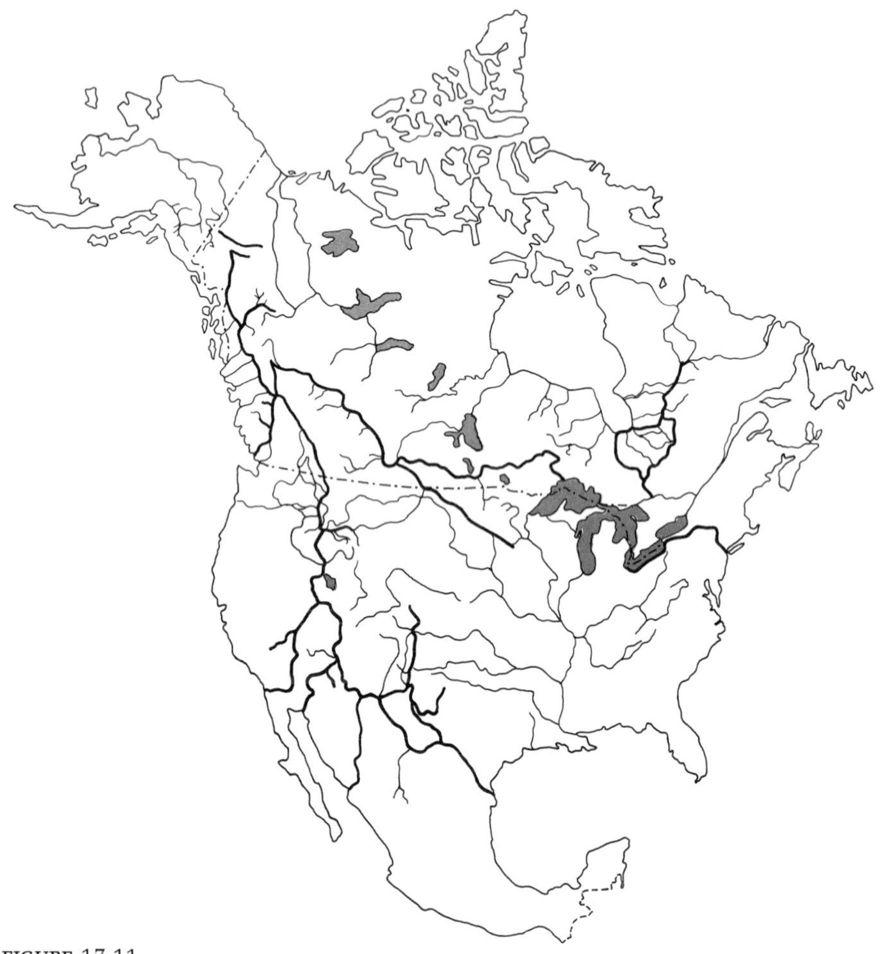

FIGURE 17.11

The North American Water and Power Alliance proposals: *heavy lines,* canals or linking rivers. [From Simons (1969), courtesy the Ralph M. Parsons Company, Los Angeles and New York.]

pipeline on the continental shelf from northern to southern California. Moreover, the proposals are not confined to the arid west. One plan would shift water from rivers in eastern Texas toward the west by construction of a canal parallel to the coast. Another would convert Long Island Sound into a freshwater reservoir for New York, particularly for Long Island.

These plans are of interest to us here because of the increasing likelihood that the complexity of the projects will make them vulnerable to natural hazards. This is not meant to suggest that people are likely to die of thirst, or even that they are likely to be thirsty. Water to support human life is really superabundant. However, other water uses may become limited by supplies or cost. For example, southern California cities pay a wholesale cost for Colorado River water that is ten times greater than the price that some farmers in the same area pay for the same water. Much of the irrigation water is used to grow "low-value crops such as alfalfa," according to Quinn. If the farmers paid the same price as city dwellers, there would be no need to plan more interbasin transfers to southern California.

A great system of aqueducts already exists, however, and there is every suggestion that it may be expanded until water consumption in western North America matches the supply during one of the rainier cycles of climate. We have briefly discussed the distress that can result locally when the climate fluctuates. Enlarging population and the area affected by a fluctuation merely multiplies the potential distress.

Such natural hazards as earthquakes and volcanic eruptions will also take their toll of the vast and complex water system. The canals and conduits to southern California cross a tectonic-plate boundary and inevitably will suffer for it. Moreover, earthquakes are common in much of the West, even away from plate boundaries. It is not alarmist, merely realistic, to expect that the aqueducts will be broken repeatedly. Likewise, it is realistic to anticipate that they will be speedily repaired and that prudent measures will avert even temporary thirst.

What will probably occur is that short-term natural hazards will cause episodic damage to the water system and that the necessity for repairs will add to the cost of water. Relatively slow and unnoticed geologic changes may cause much more difficulty and add much more to costs than earthquakes. A water system is powered almost entirely by gravity; thus, it is extremely susceptible to changes in level. All the changes caused by nature and man that are discussed in earlier chapters are slowly warping the surface up and down in the western states. Part of one of the canals in California has already been rebuilt for this reason. The local cause has been taken into account in further canal building in the area. Other problems may be anticipated elsewhere. The great western canyons and the thick sedimentary fill in basins show that rivers are actively adjusting their grade to changes in the level of the land. The aqueducts are not.

Summary

1. Water is abundant and cheap, but shortages arise because of conflicting uses.
2. Water is used in the United States mainly to generate electrical power, but it is then reusable.
3. Water is consumed mainly by irrigation.
4. Average consumption of water per person is 50–100 times higher in the western states than the eastern ones because of losses to irrigation and evaporation behind dams.
5. The total supply of water is twice as great in the eastern states as in the western ones. Supplies are generally adequate, at present, for almost all purposes, but the population, water uses, and climate may change.
6. Ground water moves downhill like river water, but more slowly. It is also more apt to dissolve or to deposit minerals and to be trapped behind naturally impermeable dams.
7. Various liquids, in addition to fresh water, exist underground. If they are not flowing, they are sorted by density, with the least dense floating on top. If the fluids are moving laterally, the contact between them slopes instead of being horizontal.
8. Problems arise in dividing basin water because of the difficulty of determining the reliability of the supply.
9. Interbasin transfers of water are increasingly common in the western states. The complexity of the projects will make them increasingly vulnerable to natural hazards.

Discussion Questions

1. What are eight different uses for the water of a single river? How does each affect the river downstream?
2. Why are volcanic rocks generally highly permeable?
3. What causes a spring?
4. What is the principal effect of deposition of minerals from ground water?
5. What happens to the water table if ground water is removed faster than it is recharged?
6. What must be done if the ground rises or sinks under an aqueduct?

References

Gilluly, J., A. C. Waters, and A. O. Woodford, 1968. *Principles of Geology* (3rd ed.). San Francisco: W. H. Freeman and Company.

Miller, D., J. Geraghty, and R. Collins, 1963. *Water Atlas of the United States.* Port Washington, N.Y.: Water Information Center. [Forty plates plus matching text. A very useful compilation.]

Moore, C., and T. Poulson, 1966. *The Life of the Cave.* New York: McGraw-Hill. [A well-illustrated and informative account for spelunkers and other amateur geologists.]

Moss, F., 1967. *The Water Crisis.* New York: Frederick A. Praeger. [An informative and reasoned analysis of United States water problems and possible solutions by the Senator from Utah.]

Murray, C. R., 1968. Estimated use of water in the United States, 1965. *U.S. Geol. Surv. Circ.* (556). [The official guide.]

Nace, R., 1967. Are we running out of water? *U.S. Geol. Surv. Circ.* (536). (Quoted from p. 5.)

Quinn, F., 1968. Water transfers. *Geogr. Rev.* 58:108–132. (Quoted from p. 123.) [They are increasing rapidly in the west and more are planned.]

Sayre, A. N., 1950. Ground water. *Sci. Amer.* 183(5):14–19. (Available as *Sci. Amer.* Offprint 818.)

Simons, M., 1969. Long-term trends in water use. *In* Richard Chorley, ed., *Water, Earth, and Man*, pp. 535–544. London: Methuen.

Waring, G. A., 1965. Thermal springs of the United States and other countries of the world – a summary (revised by R. R. Blankenship and R. Benthal). *U.S. Geol. Surv. Prof. Pap.* (492).

18

ENERGY AND FUELS

It is difficult for people living now, who have become accustomed to the steady exponential growth in the consumption of energy from the fossil fuels, to realize how transitory the fossil-fuel epoch will eventually prove to be when it is viewed over a longer span of human history.

M. King Hubbert, in SCIENTIFIC AMERICAN (1971)

The availability of sources of energy and human successes in developing them make it possible for so many of our species to live on the earth. At times in the past, energy has been supplied by human slaves, but these are now replaced by machines or energy slaves. The rich have many, but even the poor have more than they might think. It takes energy to mine the iron ore and fabricate the steel for a farmer's plow, energy to distribute food and water, and energy to build shelter. These are simple but essential things. It takes far more energy per capita to supply people with the automobiles, televisions, jet planes, telephones, and other devices of technological civilization that are desired by most of humanity.

Thus it is that fuels and other energy sources are among the most eagerly sought of all things. There are many sources of energy, such as oil, gas, radioactivity, gravity, geothermal heat, and sunlight, and we utilize them with different degrees of success. Some of these sources are limited, and others are relatively limitless, but what are the limits? Some lead to moderate pollution, some to much, and none can be used in quantity without local environmental changes. Is the production of extra energy more important than the preservation of the environment? Are the Indians right in objecting to the construction of power pylons across their reservations in order to allow urban housewives to open tin cans electrically? These are but a few of the questions that must be answered during the future development of energy resources. In this chapter, we confine ourselves to narrower but still difficult questions—namely, the extent of existing energy resources and the prospects for finding more. This is essential information for meaningful understanding of energy problems at any level. For example, intensive farming with gasoline-powered tractors might seem to offer a solution to food problems even a century from now—but not if there will be no gasoline.

Estimating Consumption of Energy

The world consumption of energy in 1967 was about 45 million million (45×10^{12}) kilowatt-hours per year, according to United Nations sources. This rather meaningless number may be compared with an estimated consumption of 90 million million kilowatt-hours per year in 1980; the remarkable fact emerges, then, that consumption is expected to double in only thirteen years. This is a conservative estimate. Consumption in the United States has doubled every 10 years during this century, and world consumption, at present, is growing even faster. In any event, this is a major factor in the immediate future of the world. New oil fields will be needed by the hundreds; likewise, gas, coal, and uranium must be discovered; tank ships, refineries, and power plants must be built; and pipelines and power pylons must form a net across the surface of the earth. What is the basis for this crucial estimate?

In essence, we know the present consumption of energy, we know the history of the growth of consumption, and we extrapolate that history forward to make the estimate. This seems reasonable, but it incorporates a fundamental flaw. The growth is exponential—that is, it increases by a constant fraction each year, like compound interest. In a finite world, this cannot possibly continue for long. There is always some limit. Consider, for example, the interest paid by a bank on savings deposits. It is only a few percent, but, if the interest is compounded (and taxes are not taken into account), the principal multiplies twofold in about 20 years, eightfold in 60, and 32-fold in a century. In 400 years, it multiplies a million times, and in 800 years, the savings account *should* contain all the money in the world. Of course, it wouldn't: something would happen to prevent it. Something will also happen to slow the growth of energy consumption. Meanwhile, however, it continues to grow.

At present, the relatively few people in the developed countries use 86% of the world's energy, and the relatively many elsewhere use only 14% (see

TABLE 18.1
World energy consumption, 1967

	Solid (coal)		Liquid (oil)		Gas	
	10^{12} kw-hr	Percentage of world consumption	10^{12} kw-hr	Percentage of world consumption	10^{12} kw-hr	Percentage of world consumption
DEVELOPED COUNTRIES						
United States	3.52	20.2	6.33	35.7	5.58	64.1
Canada	0.18	1.0	0.63	3.6	0.38	4.4
Western Europe	3.67	21.1	4.42	24.9	0.33	3.8
Eastern Europe	2.42	13.9	0.37	2.1	0.26	3.0
USSR	3.47	19.9	2.23	12.6	1.66	19.1
Japan	0.61	3.5	1.11	6.3	0.02	0.2
Oceania	0.26	1.5	0.24	1.4	–	–
Total	14.13	81.1	15.33	86.6	8.23	94.6
DEVELOPING COUNTRIES						
Communist Asia	1.97	11.3	0.13	0.7	–	–
Other Asia (exc. Japan)	0.76	4.4	0.85	4.8	0.12	1.4
Africa	0.43	2.5	0.29	1.6	0.02	0.2
Other America	0.10	0.6	1.14	6.4	0.33	3.8
Total	3.26	18.7	2.41	13.5	0.47	5.4
World Total	17.39	100.	17.74	100.	8.70	100.

Source: Data from Matthews (1970).

Table 18.1). This relation is equivalent to a definition of "developed country." Most conspicuously, the very rich United States with roughly 6% of the world's population, consumes 35% of the world's energy. Presumably, "developing" countries are attempting to become "developed" and, thus, consume increasingly more energy per person. This provides a basis for predicting future consumption. Conceivably, everyone in the world wants to have as many mechanical and energy slaves as the average resident of the United States has at present (see Figure 18.1).

What about the types of energy consumed? In the world as a whole, and in round terms, 40% of the energy comes from coal, 40% from oil, and 20% from gas—which means that very little comes from other sources. Coal, oil, and gas are fossil fuels, the remains of ancient plants and animals that once converted the energy of the sun into organic compounds that have since been fossilized and preserved. The exact amount of fossil fuel in the world is uncertain, but there is no slightest doubt that it has a limit. Almost all the energy now being used is produced by burning this fossil fuel. Consequently, the whole pattern of civilization must adjust to a new source of energy *before* fossil fuels are exhausted.

Hydroelectric		Nuclear		Overall	
10^{12} kw-hr	Percentage of world consumption	10^{12} kw-hr	Percentage of world consumption	10^{12} kw-hr	Percentage of world consumption
0.22	21.8	0.03	25.	15.68	34.8
0.13	12.9	–	–	1.32	2.9
0.32	31.7	0.09	75.	8.83	19.6
0.01	1.0	–	–	3.06	6.8
0.09	8.9	n.a.	n.a.	7.45	16.6
0.07	6.9	–	–	1.81	4.0
0.02	2.0	–	–	0.52	1.2
0.86	85.2	0.12	100.	38.67	85.9
0.04	4.0	–	–	2.14	4.8
0.03	3.0	–	–	1.76	3.9
0.02	2.0	–	–	0.76	1.7
0.06	5.9	–	–	1.63	3.6
0.15	14.9	–	–	6.29	14.0
1.01	100.	0.12	100.	44.96	100.

Some fossil fuels will be consumed before others, and the history of consumption in the United States suggests what may happen. Until 1900, the total annual consumption of energy merely paralleled population growth (Figure 18.2). Thereafter, except during the great depression, the consumption of energy grew about three times as fast as the population.

What do we do with it all? About 30% goes for household and commercial uses, 40% for industry, and 30% merely to transport people and things from place to place. This contrasts with developing countries, in which transportation is relatively trivial: people stay home, for the most part, and things are mainly produced locally. The relative importance of transportation accounts for the critical role of oil in the economy of the developed world. Coal and oil are the sources of equal quantities of American energy, but almost all the transportation is fueled by oil (Table 18.2). Likewise, electrical utilities are predominantly fueled by coal, and houses and offices are heated by gas. A shortage of any one of these will necessitate extensive technological changes.

The sources and uses of energy in the United States will change in the future as they have in the past (Figure 18.4). As late as 1850 almost all our energy came from wood and this continued until the cheap forests were gone. By

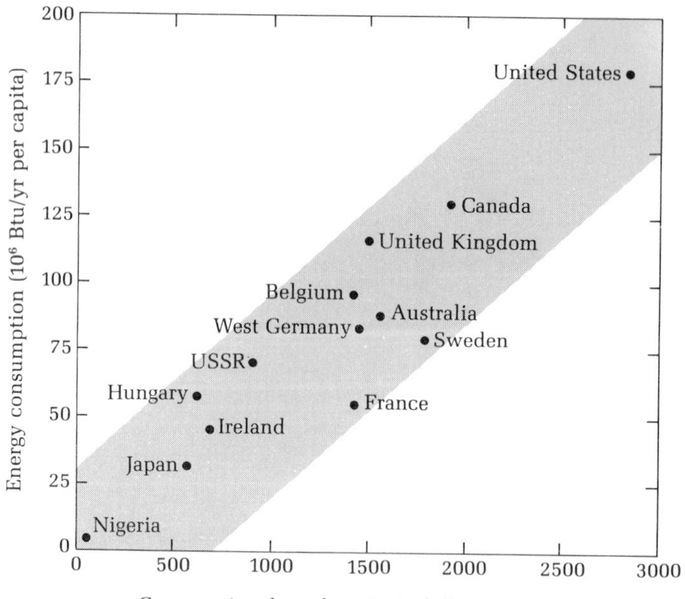

FIGURE 18.1

Commercial energy use and the gross national product show a reasonably close correlation. The average citizen of the United States uses roughly 40 times as much energy as the average Nigerian, but only twice as much as an Australian. [From Starr, "Energy and Power," copyright © 1971 by Scientific American Inc. (all rights reserved).]

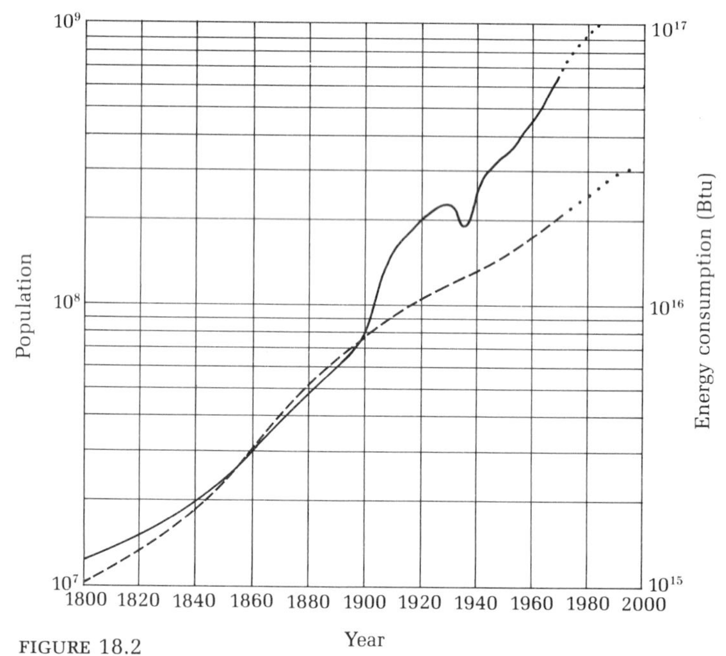

FIGURE 18.2

Growth of U.S. energy consumption (*solid line*) has outpaced the growth of U.S. population (*dashed line*) since 1900, except during the energy cutback of the depression years. [After Cook, "The Flow of Energy in an Industrial Society," copyright © 1971 by Scientific American, Inc. (all rights reserved).]

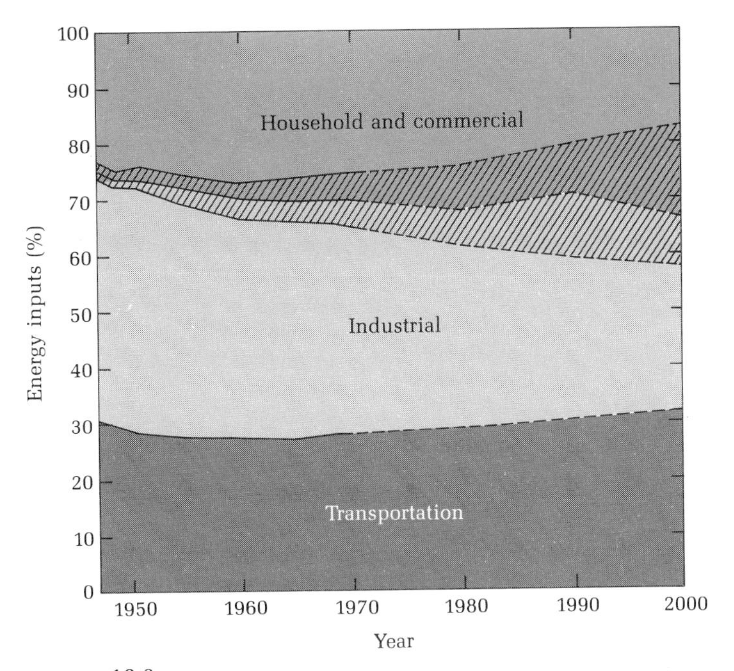

FIGURE 18.3

Useful work is distributed among the various end-use sectors of the U.S. economy as shown in this graph. The trend has been for industry's share to decrease, with household and commercial uses (including air conditioning) and transportation growing. Electricity (*hatched area*) accounts for an ever larger share of the work. U.S. Bureau of Mines figures in this chart include nonenergy uses of fossil fuels, such as for chemicals, which constitute about 7% of total energy inputs. [From Cook, "The Flow of Energy in an Industrial Society," copyright © 1971 by Scientific American, Inc. (all rights reserved).]

1900, coal had replaced wood, except for a few percent of energy supplied by hydroelectric dams and oil-gas. Since then, all power sources except wood have been used increasingly (Figure 18.5) and nuclear energy has been introduced. The *proportion* of coal use has decreased, however, and hydroelectric power has retained only a constant proportion. Thus the great expansion since 1900 has been at the expense of oil and gas which have assumed a correspondingly crucial role in our economy and our lives.

Estimating Mineral Reserves

The future of the human race and of the world depends on the reserves of minerals that still lie in the earth, but we are not yet skilled at estimating how great they may be. There are many ways of making estimates, but none is universally, or even generally, accepted. In part, this reflects the brevity of our appreciation of the problem of global mineral reserves. Until the population and industry began to expand so rapidly, there was never any question

TABLE 18.2
U.S. energy consumption, 1968

Fuel sector	Energy consumed (10^9 kw-hr) Utilities	Industry	Transport	Household, commercial, miscellaneous	Total
SOLID					
Bituminous coal and lignite	2.07	1.58	0.00	0.13	3.78
Anthracite coal	0.02	0.02	–	0.04	0.08
Total solid fuels	2.09	1.60	0.00	0.17	3.86
LIQUID					
Liquified gases	–	0.02	0.04	0.21	0.27
Jet fuel: naphtha	–	–	0.20	–	0.20
Jet fuel: kerosene	–	–	0.37	–	0.37
Gasoline	–	–	3.01	–	3.01
Kerosene	–	0.04	–	0.13	0.17
Distillate fuel oil	0.01	0.11	0.36	1.00	1.48
Residual fuel oil	0.34	0.31	0.23	0.36	1.24
Still gas	–	0.26	–	–	0.26
Petroleum coke	–	0.10	–	–	0.10
Total liquid fuels	0.35	0.84	4.21	1.70	7.10
OTHERS					
Natural gas, dry	0.95	2.58	0.18	1.89	5.60
Hydropower	0.22	–	–	–	0.22
Nuclear power	0.04	–	–	–	0.04
Total all fuels	3.65	5.02	4.39	3.75	16.82*

Source: Data from Matthews (1970).
**Note*: It is estimated that, by the year 2000, total energy consumption in the U.S. will be 45–51 billion thermal kilowatt-hours.

about the limits of the world. The question was merely one of who would be lucky enough to find the ore and strike it rich.

Now the existence of limits to resources is recognized, but measuring, estimating, or even defining mineral reserves remains extraordinarily difficult. With unlimited cheap energy it might be possible to think of mining ordinary rock. Thus, the mineral reserves for any given material are the same as the amount in the crust of the earth, as measured by geochemists. However, thermal or other pollution puts a restriction on unlimited energy use, so the degree of concentration of ores or fuels enters into the definition of mineral reserves. These matters are the subject of economic geology. Further considerations are the possibility of developing a reserve (technology), and the cost of development versus the value of the material mined (economics). But these factors are by no means all important—or even the most important.

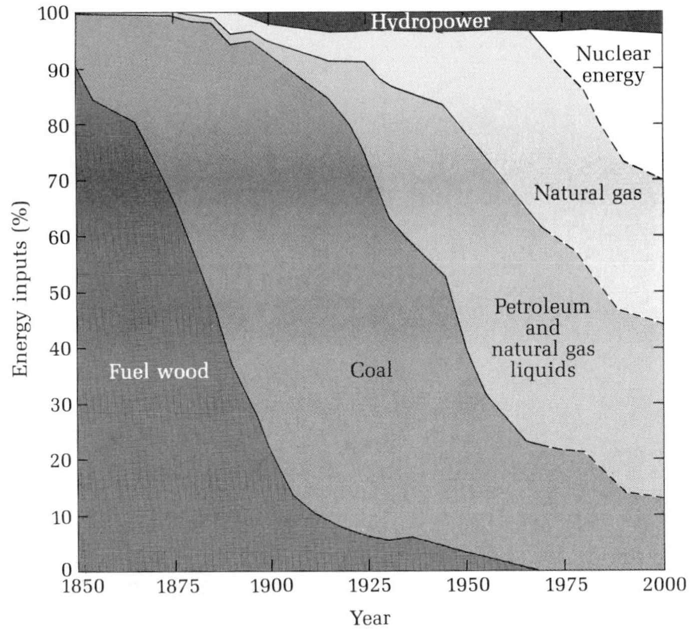

FIGURE 18.4

Fossil fuels now account for nearly all the energy input into the U.S. economy. Coal's contribution has decreased since World War II; that of natural gas has increased most in that period. Nuclear energy should contribute a substantial part of the input within the next 20 years, unless difficulties with technology and pollution cannot be surmounted in that time. [After Cook, "The Flow of Energy in an Industrial Society," copyright © 1971 by Scientific American, Inc. (all rights reserved).]

A country at war so highly values steel and aluminum for their utility in weapons that they may be priceless by ordinary standards. Thus, the definition of a mineral reserve varies with necessity and, ultimately, is controlled by politics. With due consideration for these and other factors, Vincent McKelvey of the U.S. Geological Survey has defined reserves graphically in terms of two factors—namely, increasing degree of certainty that they exist in the ground, and increasing feasibility of economic recovery. Reserves are recoverable with present technology at present costs and prices. Proved reserves are those known to exist; they are the only certain reserves. Minerals that are not recoverable under present economic conditions are called resources. Large quantities of resources are known to exist and are classified as marginal. Moreover, parts of the world are unexplored. Thus, new proved reserves will be discovered; meanwhile, on the basis of geology, they are considered as probable, possible, or undiscovered. Finally, there are expectable, but as yet undiscovered, resources that might become reserves if there is a change in economics. All these definitions are qualitative.

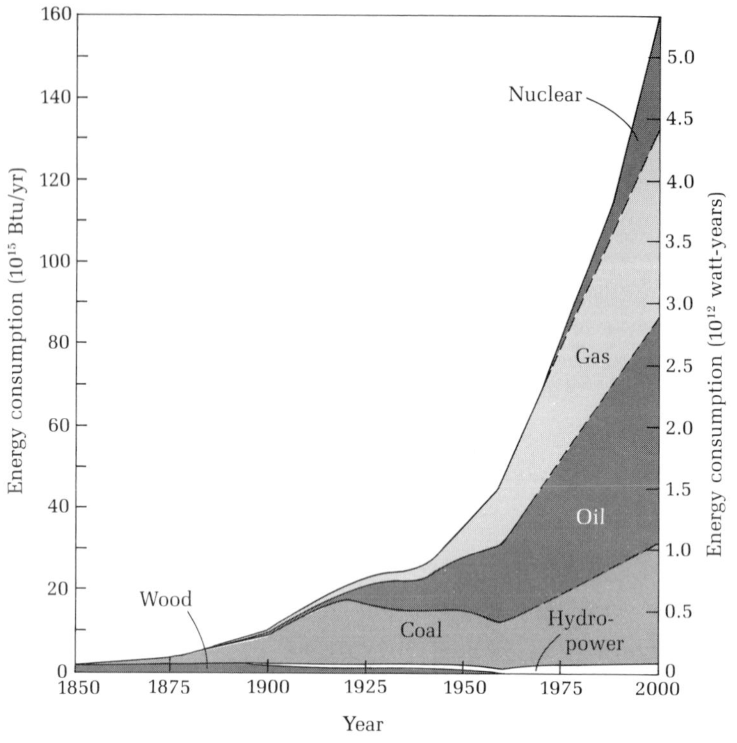

FIGURE 18.5

U.S. energy consumption has been multiplied some 30 times since 1850, when wood supplied more than 90% of all the energy units. By 1900, coal had become the dominant fuel; since then, however, oil and gas have become more important. [After Starr, "Energy and Power," copyright © 1971 by Scientific American, Inc. (all rights reserved).]

Quantitative estimates of reserves and resources are necessary for rational planning, but they are rather unsatisfactory, as might be anticipated by the many qualitative factors involved. One method of estimating is to compare reserves of minerals in some areas with abundances in the crust. Minable reserves, expressed in tons, of long-sought and well-explored elements in the western United States are about 1–10 billion times their crustal abundances expressed as percentages (Figure 18.6). This relation provides a basis for predicting the minable reserves in relatively unexplored areas.

The geologic occurrence of some valuable fuels—particularly coal, oil, and natural gas—is such that more refined estimates can be made for them than for most minerals. The fossil fuels occur only in sedimentary rock; moreover, only certain types of such rocks are likely to contain them. In addition, oil and gas migrate in the same manner as ground water (Chapter 10), and their location is controlled by the permeability of rocks and the existence of structural traps.

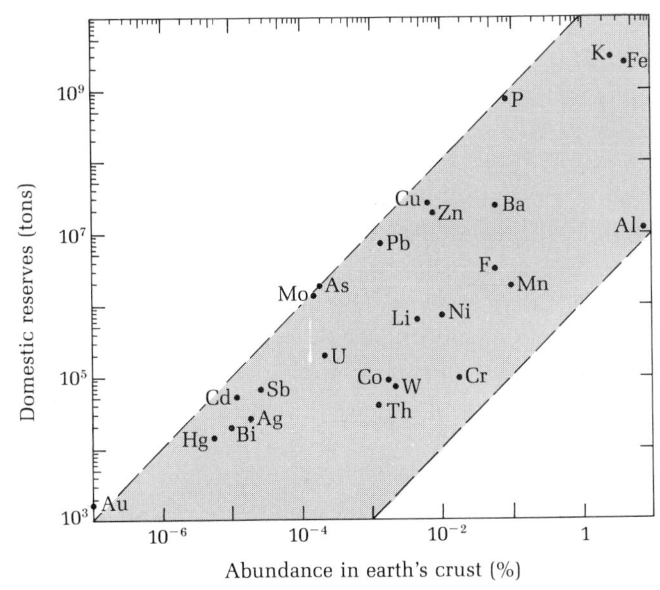

FIGURE 18.6
Domestic reserves of elements increase in proportion to their abundance in the earth's crust. Thus, gold (Au) is very rare in the crust, and the reserves of gold are low; iron (Fe) is abundant and, the reserves of iron are large.

Because oil and gas are so crucial in industrial societies, many estimates of the ultimate amount of reserves of these materials have been made. One is closely related to the reserves–abundance relation for elements, which was discussed above. Petroleum consultant Lewis Weeks and others have observed that, on the average, a cubic mile of sediment contains 50,000 barrels of oil. Thus, by measuring the volume of sediment in the world, the reserve of oil can be calculated. This method of estimating reserves points to the growing importance of the continental shelves in the oil industry. The shelves are only a small fraction of the total area of the continents, but they are underlain almost entirely by sedimentary rock. Moreover, the very promising shelves were not available for drilling – for technological reasons – until many of the best regions on land had been developed.

Another method of estimating reserves is to graph the history of effectiveness of exploration for them. This can be done by analyzing the historical relation between the length of wells drilled and the amount of oil discovered. The proportion of successful drilling has decreased over the years; consequently, this method of estimating indicates that relatively limited resources remain to be discovered.

A third method is a modification of the first two. It considers both the average amount of oil that is now estimated to exist in a volume of sedimentary rock, and the amount of drilling that will probably be necessary to discover the total average amount.

A fourth method takes into account the frequency of oil fields of different sizes. In the United States, 5% of the fields contain 50% of the oil reserves, and 50% of the fields contain 95% of the oil. In the less developed USSR, 5% of the fields contain fully 75% of all the reserves. Thus, many small oil fields hold only a minor fraction of the reserves, fewer intermediate fields contain more, and a very few large fields contain most of the oil. A graph of this relation can be extrapolated to predict future discoveries and reserves. Other things being equal, the method predicts that many more small fields will be discovered in the USSR but that this will not greatly increase oil reserves.

These different methods of estimating reserves give values that differ by large factors. For the relatively well-known area of the United States, estimates of recoverable oil reserves range from 190 to 550 billion barrels. Estimates for other regions or other minerals are correspondingly uncertain—and will remain so, until an adequate effort is expended on mineral exploration and resource prediction.

Lifetimes of Mineral Resources

The lifetime of a nonrenewable resource is the interval of time between when it is first used and when it is exhausted. For many social purposes, the lifetime is one of the most important bits of information about a resource. For example, knowledge of the time remaining before the resource is gone is essential for all types of rational planning by government or industry.

Inevitably, many attempts are made to estimate the lifetimes of mineral resources, especially those of the crucial energy resources. One basis for estimation is simply that the lifetime is equal to the total original amount of the resource divided by the average rate of consumption. As we have already seen, however, the amount of any resource is only roughly known. It might appear that, as a consequence, estimates of the lifetimes of mineral resources are of little value.

This would be true if the rate of consumption were constant. In the language of childhood, if you have 15 oranges and eat one per day, they will last for 15 days. Thus, if you didn't know whether you had 15 or 31 oranges, it would be difficult to plan. However, everything we know about fuel and energy consumption indicates that they are each increasing by a constant proportion each year. This changes the arithmetic dramatically. Suppose you eat one orange the first day, two the second, four the third, and so on. On the fourth day, you consume 8, and all 15 oranges are gone. If you actually had 31 oranges, it would make little difference in the lifetime of the resource. On the fifth day, you would eat 16 oranges, and all 31 would be gone. The life of the resource would be only one more day—in other words, one more doubling period.

TABLE 18.3
The calculated effects of increases in global fuel reserves

Resource	*Known reserves*	*Lifetime at present consumption (yr)*	*Estimated increase in consumption (%/yr)*			A: *lifetime at average estimated rate of increase*	B: *lifetime as in* A *but assuming 5 times the known reserves*
			Minimum	*Average*	*Maximum*		
Coal	5×10^{12} tons	2300	3.0	4.1	5.3	111	150
Natural gas	1.1×10^{15} ft^3	38	3.9	4.7	5.5	22	49
Oil	455×10^9 bbl	31	2.9	3.9	4.9	20	50

Source: Meadows et al. (1972).

The doubling period for the consumption of mineral fuels is about ten years. The estimates of oil reserves in the United States differ by somewhat more than a factor of two. Thus, the difference between the minimum and maximum estimates of the lifetime differ by only about ten years, if consumption does not first change.

This is the most significant conclusion, with regard to minerals, made by the Club of Rome, an environmentally concerned group, after a computerized analysis of the "predicament of mankind." The uncertainties of estimates of mineral resources matter little *if the rates of consumption continue to grow as they are doing now.* Donella and Dennis Meadows and others calculate the effects of enormous increases in global reserves of fuels as shown in Table 18.3. The conclusion that the present rates of consumption cannot long continue to grow seems inescapable. The amount, distribution, and availability of fuel resources will be of paramount importance to the future of mankind. Even if the

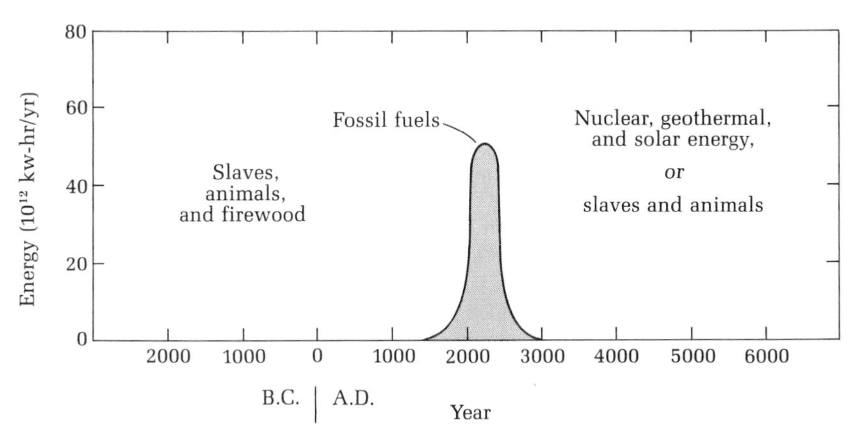

FIGURE 18.7
The energy sources of mankind, illustrating the brief duration of the fossil-fuel epoch. [After Hubbert (1967).]

increase in consumption is brought under control, major changes in fuel consumption will occur. People talk of doubling the present world population and of providing everyone with the same number of energy slaves as the average American expects to have in a decade or two. They are not talking about the pattern of energy resources that we draw upon today.

Doubling consumption cannot continue because fossil fuels are finite. What will happen? It may be helpful to visualize the history of discoveries of a finite resource. The original discovery rate is zero, the final rate is zero, and therefore the rate reaches a maximum at some intermediate point. Extensive government records show that discovery increases by a constant fraction; that is, the *rate* of discovery increases exponentially. This is reasonable because more and more uses are developed, and more of the resource can be sold.

Many—indeed, most—things that grow follow a pattern like the growth of discovery. A bean sprout has a zero rate of growth at the beginning and end, and the rate of growth is exponential for part of the time in between. The same growth pattern is followed by other plants and animals, and by social organizations such as armies or economic activities such as the production of steel.

The general mathematical character of such growth is consequently well known. A graph of the history of cumulative growth displays a characteristic S-shaped curve, called a logistic curve, that results from exponential growth that approaches a limit. A notable characteristic of the logistic curve is that the second half of the curve is the reverse of the first. Thus, if the early growth is known, the whole history can be calculated.

American geophysicist M. King Hubbert has applied the logistic curve to the discovery and production of oil. Each barrel produced leaves one less in the

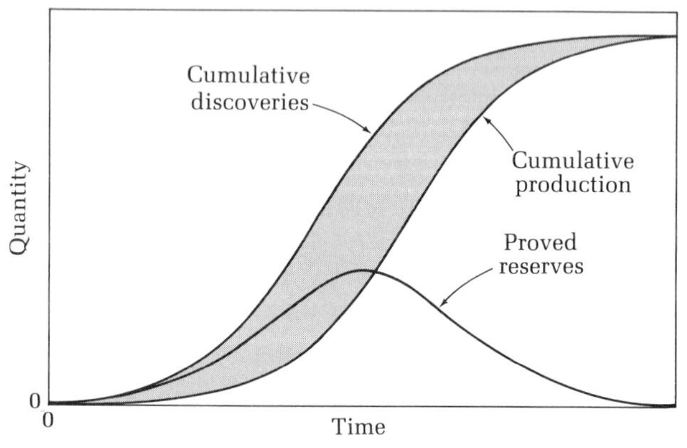

FIGURE 18.8

Generalized form of curves of cumulative discoveries, cumulative production, and proved reserves for a nonrenewable resource during a full cycle of production. *Shaded area*, the time lapse between discovery and production. The proved reserves are obtained at any time by subtracting cumulative production from cumulative discoveries. [After Hubbert (1962).]

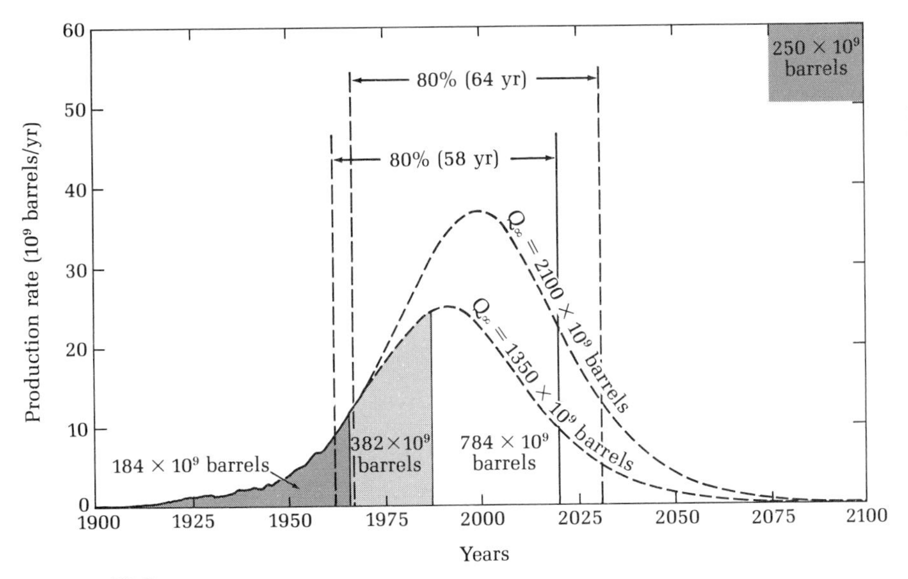

FIGURE 18.9
Complete cycles of world crude-oil production for two values of ultimate quantity of oil produced (Q_∞). [From Hubbert, "Energy Resources," in Cloud, *Resources and Man*, W. H. Freeman and Company, copyright © 1969.]

ground. Consequently, if the discoveries are predictable, so are the remaining reserves of oil (Figure 18.8).

Once again, the actual amount of the original resource is of surprisingly little importance with regard to its lifetime. The exponential increase in discoveries (and production and consumption) speedily disposes of whatever may exist. Hubbert has taken two estimates of the original world oil resource and used discovery data plus the logistic-curve analysis to calculate lifetimes. If the resource was originally 1350 billion barrels, 90% of it will be gone by about 2020 A.D. If it was about 50% greater—namely, 2100 billion barrels—it will be gone by about 2030 A.D. In short, the (highly probable) consumption according to the logistic curve gives only a few more decades to the life of a resource than the (impossible) uncontrolled exponential consumption.

Fossil Fuels

Ancient sunshine has been preserved in several forms of fossil fuel, of which the most important are coal, natural gas, oil, tar sand and oil shale. Of these, by far the most abundant is coal, which amounts to 88% of all fossil fuel (Figure 18.10). However, it is neither as cheap nor as versatile as oil and gas. Coal is solid and is generally acquired by the expensive process of mining, although it can be burned in place, underground, and the gas can be collected at a lower cost. Oil and gas, by contrast, flow naturally or can be pumped out of pipes with ease and at low cost. Moreover, the volatile liquids and gases are relatively

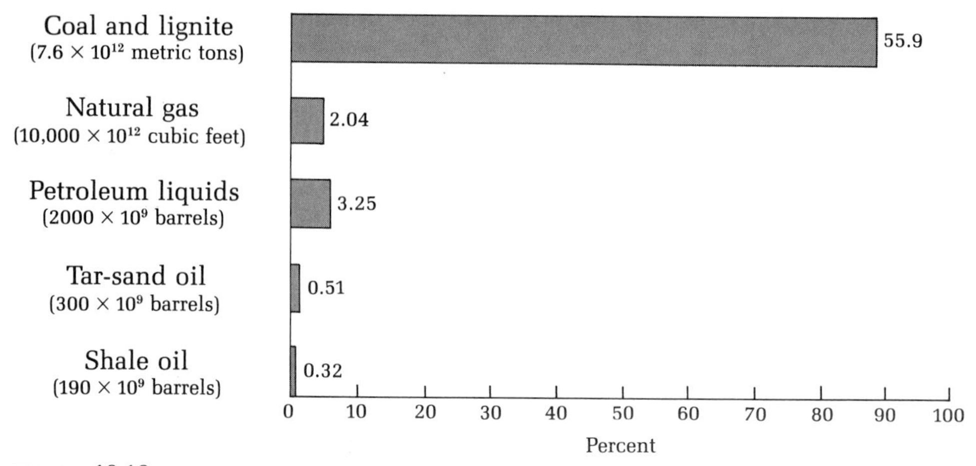

FIGURE 18.10

Energy content of the world's initial supply of recoverable fossil fuels is given in percentages of the total and in units of 10^{15} thermal kilowatt-hours (numbers after bars). Coal and lignite, for example, contain 55.9×10^{15} thermal kilowatt-hours of energy and represent 88.8% of the recoverable energy. [After Hubbert, "The Energy Resources of The Earth," copyright © 1971 by Scientific American, Inc. (all rights reserved).]

easy to transport and use. Thus, they are heavily exploited. Oil in solid form occurs in tar sands, and oil in nonflowing form exists in shale rock. In these forms, however, it has yet to find much use, because its extraction is more expensive; but as oil and gas are consumed, the solid fossil fuels will become more important, and so will the countries in which they are found.

COAL

Coal consists essentially of solid remains of plants, concentrated in layers and more or less modified by temperature, pressure, and chemical changes that have driven out most volatiles. We may visualize soggy Irish peat as similar to the original material. As it is buried, the pressure of the overlying rocks may convert it to lignite or brown coal. Continuing changes produce first bituminous and then anthracitic coal, which are hard and black and contain much more available energy per gram. The various grades of coal have different properties (such as the ability to form coke—which is used in steel-making—or to burn without smoke) that make them desirable for different purposes.

The availability of coal has been essential to the development of such great industrial nations as Britain, Germany, and the United States. Coal diminished in importance as the consumption of oil and gas grew, but it remains the staple energy source for power plants that generate electricity, and it is still highly important in industry, particularly in the chemical and steel industries. Because it will last much longer than oil and gas, its distribution will become increasingly important in international affairs, unless new energy sources are developed. Present estimates indicate that the two existing superpowers, the United

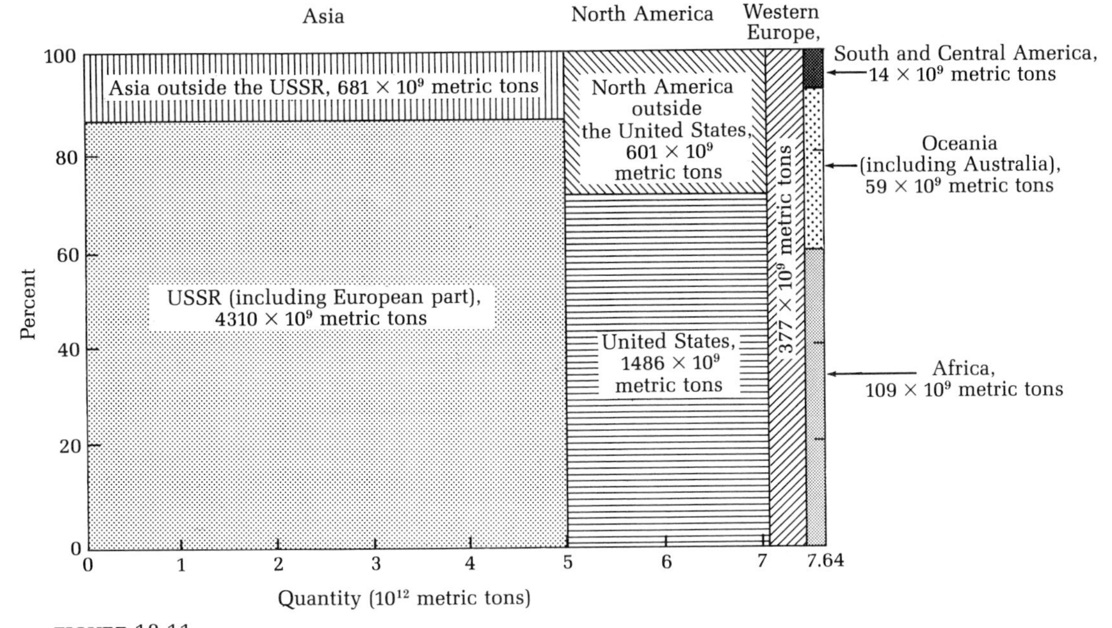

FIGURE 18.11

Coal resources of the world are indicated on the basis of data compiled by Paul Averitt of the U.S. Geological Survey. The figures represent the total initial resources of minable coal, which is defined as 50% of the coal actually present. The horizontal scale gives the total supply. Each vertical block shows the apportionment of the supply in a continent. From the first block, for example, one can ascertain that Asia has some 5×10^{12} metric tons of minable coal, of which about 86% is in the USSR. [From Hubbert, "Energy Resources," in Cloud, *Resources and Man,* W. H. Freeman and Company, copyright © 1969; based on data from Averitt (1969).]

States and the USSR, have had the good fortune to include most of the known reserves of coal within their boundaries (Figure 18.11). Thus, most developing countries are not likely to develop by selling—or even using—any vast reserves of coal.

Technology may play an important role in the future use of coal. Minable reserves are generally defined as half of the coal present in beds no thinner than 36 centimeters and generally no deeper than 1.2 kilometers. The reserves can be greatly increased if underground burning can be developed economically and without unacceptable environmental effects. Likewise, transportation of coal slurries in pipelines is opening many low-grade coal deposits to development. Some of the controversial new power plants of the southwestern states are the result of inexpensive transportation of lignite for hundreds of miles to power plants that are located on rivers to provide water for cooling.

OIL

Liquid oil, which is the result of anaerobic decay of marine organic matter, is much less volatile than natural gas. Because it is relatively viscous, it flows with difficulty compared to ground water or gas. In California and some other places, the oil grades into asphalt or tar, which is a solid that may require heating to make it flow. However, it does flow, and this characteristic is one of the main reasons why it is cheap.

Fresh organic matter typically accumulates on the sea floor mixed in with slowly accumulating, fine-grained mud. As time passes, the mud is buried. Perhaps alternate layers of mud and sand are deposited one above another. The mud is compressed into shale and the fluids in the mud—water, oil, and gas—are driven out. Some escape into the sea above, but large quantities merely migrate to the relatively incompressible sands around them. The shales tend to become impermeable. If the sands do not become too cemented in the process of becoming sandstone, the gaps between grains may be filled with fluids, and the sandstone is a source bed for oil.

The existence of such a source bed, full of a mixture of water, oil, and gas, is not enough in itself to permit production of oil. The oil must first be naturally separated from the other fluids and concentrated somewhere. This occurs underground, in favorable circumstances, by virtue of the differences in density of the fluids. In an anticlinal fold, to pick a simple example, the gas migrates to the top of the fold in the source rock, where it is trapped by an impermeable shale. The oil floats upward through the denser water and accumulates in a layer between the gas and water. Such a layer of oil concentrated in a reservoir is at last ready to be removed. In some circumstances, the pressure of the gas forces the oil to the well head in a gusher like an artesian well. Otherwise, the oil is pumped; only a few wells are needed, however, because the oil flows toward them from the surrounding layer. Eventually, the oil ceases to flow,

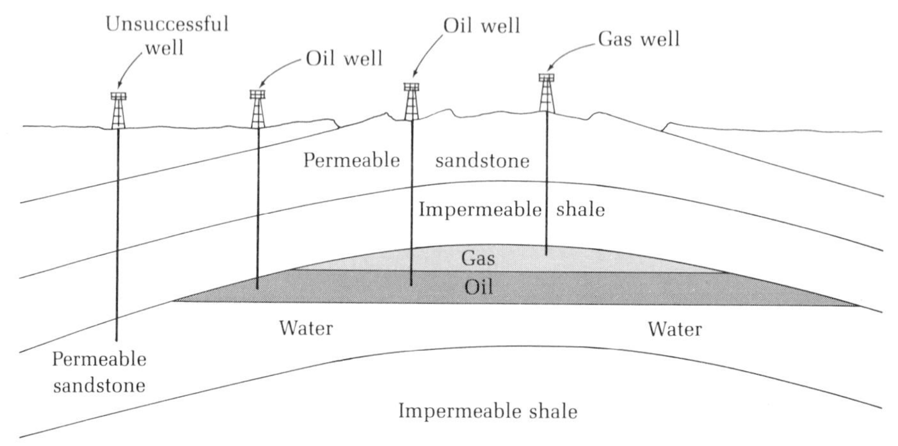

FIGURE 18.12
Oil and gas may migrate from shale into permeable sandstone and become trapped in an anticlinal fold, as shown, or in other permeability traps. In the trap, the gas, oil, and water become separated and concentrated in layers because of the differences in their densities.

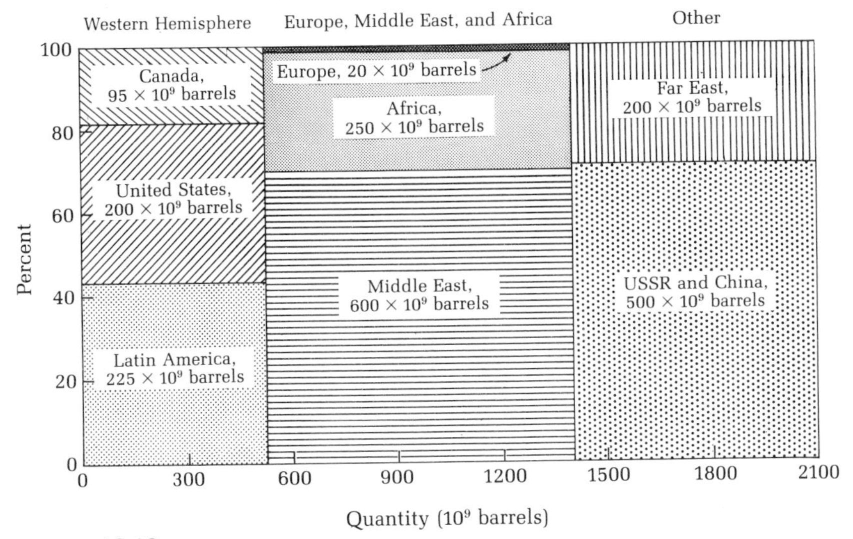

FIGURE 18.13

Petroleum resources of the world are derived from estimates made in 1967 by W. P. Ryman of the Standard Oil Company of New Jersey. They represent ultimate crude-oil production, including oil from offshore areas, and consist of oil already produced, proved and probable reserves, and future discoveries. Estimates as low as 1350×10^9 barrels have also been made. [After Hubbert, "Energy Resources of the Earth," copyright © 1971 by Scientific American, Inc. (all rights reserved).]

and water commonly replaces it in the well (much as a salt-water intrusion into a coastal water well). Half or two-thirds of the oil normally remains in the sandstone when the first phase of production ceases. The extraction of this residual oil grows more desirable as the supply dwindles, even though it costs more than simple pumping. The second phase of extraction commonly involves pumping water down into the periphery of the reservoir rock. This flows toward the pumping wells in the center of the oil field and flushes some of the residual oil with it.

By far the greatest concentration of oil reserves is in the vast fields of the Middle East—which is why a few small, arid, underpopulated, nonindustrial, militarily impotent nations are so important in world affairs (Figure 18.13). The Middle East oil is exported to the whole world, but largely to Japan and Europe. It is not used locally to any great extent. Judging by present trends, these circumstances will not last very long. A few scores of years, and the endless lines of tank ships will no longer ply from the deserts to the centers of industrial power. Meanwhile, the small, arid countries will have become very rich.

GAS

Natural gas consists of the more volatile components that result from the same processes that yield oil. It is a clean fuel and a cheap one, provided it can be transported from the well to the user at a reasonable cost. In North America, this is easily accomplished by pipeline, and gas is important in our economy. However, there are no pipelines from the enormous gas reserves of South America, the Middle East, and North Africa to the potential consumers of North America, Europe, and Japan. Thus, most of the gas that is produced incidentally to pumping oil in those areas is merely burned at the well head—as it used to be in the United States.

It is generally assumed, for want of data, that approximately as much fossil-fuel energy is in the form of gas as is in the form of oil. Consequently, a very powerful incentive exists to produce local uses for gas in developing countries or to devise means of transporting gas for large distances. A promising method, now in limited but expanding use, is to liquify the gas by cooling it, and then to move it in refrigerated tank ships.

Gas is volatile, relatively mobile underground, and tends to flow freely into wells. In some places in which it could easily be utilized, however, it is trapped in impermeable rocks. Experiments have been carried out to test the feasibility of shattering the rocks with underground nuclear explosions and, thereby, creating artificial permeability. It appears that the experiments have been successful and that new reserves may thus be created, but the tests cannot be completely evaluated until the rock cools, the short-lived radioactivity decays, and the gas is actually put into pipelines.

TAR SANDS

Tar, a crude oil so thick and viscous that it will not flow, can be mined economically if it is concentrated enough. Little effort has been made to develop most tar sands because they cannot compete with flowing oil. However, the Athabasca sands in Alberta, Canada, contain an estimated 300 billion barrels of tar—or as much as all the oil resources in Canada and the United States together. This is a resource worthy of development, and several oil companies have begun large-scale mining and heating to extract the tar. Chemically, the tar is like oil; thus, it can be refined in more or less conventional ways.

OIL SHALES

The "oil" in oil shale has two unfavorable characteristics; it must be mined and heated, and the organic materials cannot be refined without special equipment. Oil shales exist on most continents, but the total proven reserves are at best about 200 billion barrels. Oil shale may someday become important for its organic compounds, which are of value in the chemical industry, even if they are not feasible fuels. Development of Colorado oil shales may be limited for lack of uncommitted water in the region.

Future Fuels

It appears that most oil and gas will be gone within a few score years, and most coal within a few hundred. What will replace the fossil fuels in mankind's search for energy? Predictions are illuminated by flashes of optimism and darkened by clouds of gloom. Everyone concerned agrees that something *must* happen, but few agree about what it might be. The uncertainties are many and on a grand scale. For any new global power system there are uncertainties: pollution, destruction in the course of construction, exhaustion of related resources, and technological, social, climatic, and geologic change.

It might seem that we are already committed to nuclear power and that the matter is settled. Yet, hardly a month passes without reports of some new difficulty in nuclear power plants or in the disposal of radioactive wastes, or some new objections to new power plants on environmental grounds.

There is also the curious fact that an irretrievable mistake may be made if we proceed too rapidly. Nuclear reactors are of three types—burners, converters, and breeders. Existing reactors are burners which use up U-235. This is the only naturally occurring isotope that is capable of sustaining a chain reaction; that is, it is spontaneously fissionable by capture of slow or thermal neutrons. Other naturally occurring isotopes, such as U-238, fission naturally by decay, but so slowly that they are not useful for generating power. Unfortunately, U-235 occurs as only 1 part in 141 parts of natural uranium; the remainder, except for a trivial amount of U-234, is the less useful U-238.

According to McKelvey, the proven reserves and likely resources of U-235 in the United States are

> . . . just about enough to supply the lifetime needs of reactors in use . . . in 1968 and only half that required for reactors expected to be in use by 1980. . . . Plainly, the significance of uranium as a commercial fuel lies in its use in the breeder reactor, and one may question, as a number of critics have, . . . the advisability of enlarging nuclear generating capacity until the breeder is ready for commercial use.

A breeder reactor theoretically can convert abundant U-238 and Th-232 to isotopes that are usable in reactors. Thus, more nuclear fuel is "bred" than is used. If all reactors were breeders from the start, the supply of usable radioactive isotopes would be almost limitless—except for the limits set by pollution. However, no commercial breeder reactor exists and no one knows what the effects of a vast new reactor technology will be.

Stars are powered by the fusion of light-weight elements, and it is possible that the same reactions can generate unlimited power on earth using sea water as a source. If so, the technology for controlling it has yet to be invented, although we can release uncontrolled fusion energy in the form of thermonuclear (hydrogen) bombs.

Many natural sources of energy remain to be developed—namely, solar, tidal, geothermal, and hydroelectric. They have various capabilities and disadvantages, but all have the enormous advantages that they are nonpolluting, except for the heat they generate, and renewable.

Solar radiation (Chapter 3) is capable of producing energy beyond our present dreams. If a small percentage of the area of the deserts of the United States were converted to solar-energy absorbers, we could generate all the energy we would need until the year 2000. Other sunny lands would have equal success. The cost of the installation would be enormous, of course, and the problems of energy storage at night and in cloudy periods would not be trivial. Nonetheless, solar energy has possibilities.

Some of the energy of the sun that reaches the earth is expended in imparting potential energy to water by elevating it during the global atmospheric circulation. Thus, we already capture a fraction of inexhaustible solar energy by building hydroelectric dams, lakes, power plants, and electrical transmission lines in the style of southern California.

At present, about 45,000 megawatts of water power are available in the United States, and the ultimate capacity is estimated by the Federal Power Commission to be about 160,000 megawatts, or more than three times as much as now exists. Western Europe has developed even more of its potential water power, but elsewhere development has hardly begun. In 1962, the total developed was 152,000 megawatts out of a potential 2,857,000 megawatts (Table 18.4). This was about four times as great as the installed electrical-power capacity of the world in 1964. Thus, hydropower cannot readily replace fossil fuels, especially in transportation, but it can fill some of the gap until technology has a chance to produce safe and abundant nuclear power.

TABLE 18.4
World water-power capacity as of 1962

Region	*Potential (10^3 Mw)*	*Percent of total*	*Development (10^3 Mw)*	*Percent developed*
North America	313	11	59	19
South America	577	20	5	
Western Europe	158	6	47	30
Africa	780	27	2	
Middle East	21	1	–	
Southeast Asia	455	16	2	
Far East	42	1	19	
Australasia	45	2	2	
USSR, China, and satellites	466	16	16	3
Total	2,857	100	152	

Source: Hubbert (1969).

The distribution of hydropower is particularly striking compared to that of other energy sources. Most of the capacity is in Africa, which is relatively deficient in fossil fuels, and in South America, which has little coal. If the capacity is put to use, the developing nations will have the energy that is now available to the developed ones—unless electricity can be exported for intercontinental distances.

Hydropower is clean and abundant, but it is not developed without paying a price and it is not as reliable as it might seem. Reservoirs fill with sediment, and the flow of rivers often can be disrupted only at great cost to navigation and agriculture (Chapter 17). The Aswan High Dam in Egypt provides examples of the multitude of ecological changes that can accompany construction of dams. Moreover, climates fluctuate, and so do stream flows. A prolonged drought could play havoc with a hydropower system.

The gravitational attraction of the tides is another source of energy. It is already being utilized in France and Russia, and there are plans for utilizing it in many other places. Locally, such energy may be important; on a global scale, however, it is relatively trivial: at best, it will amount to about 2% of the potential of hydropower.

The heat escaping from the earth is equivalent to an enormous amount of renewable energy and might seem to be a promising source for geothermal power. Most of it is too widely diffused to be captured, however; thus, only the small amount released in thermal areas has much promise. In fact, many thermal areas are already being utilized. Italy began to develop geothermal power in 1904 at Larderello, and now geothermal power plants exist in half a dozen countries. Many areas remain to be developed, and the western United States appears promising in this respect. Nonetheless, on the global scale, geothermal energy probably will not amount to much. Like tidal power, it has a potential of only about 2% of hydropower.

Sources of Energy

For ages, man drew upon the heat of wood fires and the muscles of animals and slaves for energy. Then industry began to expand, and more and cheaper fuel was needed. Hardly a century ago, man began to burn the fossil fuels that had taken 600 million years to accumulate. It appears that, in a few centuries, they will be gone forever. Some new change must occur by then. Either the number of people or the energy consumed by each person must decrease, unless new sources of energy are found. What these sources may be remains to be seen. No matter what energy sources may become available, however, it appears that development will ultimately be limited by pollution.

Summary

1. World consumption of energy is expected to double in the next thirteen years and to continue to expand rapidly for some time thereafter. Thus, new energy resources will be needed to meet the demand.
2. The United States, with 6% of the population, uses 35% of the total world consumption of energy. It produces far less.
3. Estimates of natural resources contain uncertainties. The amount of a mineral reserve varies with necessity and is controlled by political considerations. A reserve can be defined in terms of the degree of certainty that it exists and the feasibility of its economic recovery.
4. Oil, gas, and coal have been the subject of prolonged and intense prospecting. Various correlations between types and volumes of rock and types and volumes of fuel resources have emerged. These can be used to estimate the amounts that remain to be discovered.
5. The estimated lifetime of fuel resources is affected more by extrapolations of rapidly increasing consumption than by the amount of fuel in the ground.
6. Practically all energy is now derived from fossil fuels, but they will probably be exhausted before 3000 A.D.
7. Oil—which fuels transportation, among other things—will probably be almost exhausted within 75 years and will reach peak production by the end of the century.
8. Nuclear power offers much for the future, but its hazards are not yet fully assessed.
9. The possibilities of utilizing solar, geothermal, and thermonuclear power are being explored. Something will be needed to replace fossil fuels.

Discussion Questions

1. Why will more sources of energy be needed in the future?
2. In what way does the pattern of energy consumption in developed countries differ from that in underdeveloped ones?
3. How is it possible to estimate the lifetime of the world's supply of oil or coal when neither the total amount nor the future rate of consumption is known?
4. Why have oil and gas been extracted more quickly than coal?
5. What is the role of gas in the extraction of oil?
6. How will the energy of geothermal or solar heat become usable in New York City or Chicago?
7. What problems may accompany the development of new energy resources?

References

Averitt, P., 1969. Coal resources of the United States, Jan. 1, 1967. *U.S. Geol. Surv. Bull.* (1275).

Cloud, P., chairman, 1969. *Resources and Man.* San Francisco: W. H. Freeman and Company. [A study by the Committee on Resources and Man of the National Academy of Sciences–National Research Council. Some of the (gloomier? more realistic?) conclusions by these distinguished scientists have become controversial.]

——, 1971. Resources, population, and the quality of life. *In* S. F. Singer, ed., *Is There an Optimum Level of Population?* pp. 8–31. New York: McGraw-Hill.

Cook, E., 1971. The flow of energy in an industrial society. *Sci. Amer.* 224(3): 134–147. (Available as *Sci. Amer.* Offprint 667.)

Hubbert, M. K., 1962. *Energy Resources: A Report to the Committee on Natural Resources.* Washington, D.C.: National Academy of Sciences–National Research Council.

——, 1969. Energy resources. *In* P. Cloud, chairman, *Resources and Man,* pp. 157–241. San Francisco: W. H. Freeman and Company.

——, 1971. The energy resources of the earth. *Sci. Amer.* 224(3):60–87. (Available as *Sci. Amer.* Offprint 663.)

Lovering, T., 1969. Mineral resources from the land, *In* P. Cloud, chairman, *Resources and Man,* pp. 109–134. San Francisco: W. H. Freeman and Company.

Luten, D., 1971. The economic geography of energy. *Sci. Amer.* 244(3):165–175. (Available as *Sci. Amer.* Offprint 669.)

McKelvey, V., 1972. Mineral resource estimates and public policy. *Amer. Sci.* 60(1):32–40. (Quoted from p. 35.)

Matthews, W. H. editor, 1970. *Man's Impact on the Global Environment: Assessments and Recommendations for Action* (report of the Study of Critical Environmental Problems [SCEP]). Cambridge, Mass.: The MIT Press.

Meadows, D. H., D. L. Meadows, J. Randers, and W. Behrens, III, 1972. *The Limits to Growth.* New York: Universe. [The controversial predictions of the Club of Rome's Project on the Predicament of Mankind.]

Starr, C., 1971. Energy and power. *Sci. Amer.* 224(3):36–49. (Available as *Sci. Amer.* Offprint 661.)

Weeks, L. G., 1958. Fuel reserves of the future. *Amer. Ass. Petrol. Geol. Bull.* 42:431–438.

19

MINERAL RESOURCES

Having now refuted the opinions of others, I must explain what it really is from which metals are produced.

Agricola

DE ORTU ET CAUSIS SUBTERRANEORUM (1546)

A commercial mineral exploration house is composed of realists who do not pay their geologists for theories.

B. B. Brock

A GLOBAL APPROACH TO GEOLOGY (1972)

In the last two chapters, we have considered some essential natural resources—water and energy. Now we concern ourselves with many of the important mineral resources that shape the character of our civilization. For most purposes, these resources are not unique—that is, one can be substituted for another. For example, various metals are good conductors and therefore, are suitable for telephone or power cables. We use copper because it is relatively abundant and cheap, but aluminum and silver have served in special circumstances. Likewise, lead and copper have been used for cooking pots, water pipes, and roofing. The danger of poisoning has eliminated lead from domestic use, and high cost has confined lead roofing mainly to large public buildings. At the same time, galvanized iron has become so inexpensive that it is used as roofing in the cheapest dwellings. Thus, we see that the development of most mineral resources is extraordinarily dependent on economics, technology, and other social factors.

The natural materials of interest here can be classified for different purposes in many ways. According to chemistry, they are metals or nonmetals. According to use, they are building materials, ceramics, chemical sources, precious metals, iron and its alloys, and so on. With regard to origin, they are variously: (1) primary and secondary; (2) igneous, metamorphic, and sedimentary; (3) shallow or deep; and (4) weathered or not. Considering economics, some mineral resources, such as gravel, are so common, widespread, and cheap that the cost of transportation may dominate production. Others, such as diamond, are so rare and valuable that it is worthwhile to synthesize them.

We may profitably view mineral resources as participating in the tectonic, igneous, metamorphic, sedimentary, and water cycles. They rise from the depths of the earth and return to them. They differ from other rocks and minerals solely in that they temporarily contain a concentration of some desired material that it is economical to extract while certain social conditions prevail. We can consider men and their machines as exceptionally effective and selective agents of erosion in the sedimentary cycle. We can consider manufacture, transportation, and merchandizing as dispersion in the sedimentary cycle.

Thus, the origin and history of mineral deposits present a small sample of the earth sciences and their social implications. As a consequence, we shall use this chapter not only to discuss mineral deposits per se, but also to illuminate the fundamental unity of the earth sciences and, thereby, to summarize this book.

Resources and Their Expected Lives

Estimates of the expectable lives of the mineral resources of interest here suffer from the same uncertainties as those of coal and oil, which were discussed in the last chapter. However, they do merit analysis because they indicate the character of future shortages.

IRON AND ITS ALLOYS

The most important of these resources to our civilization is iron, and it is relatively abundant. At the present rate of consumption, there is enough for several hundred years (Table 19.1). Even with likely increases, there is at least enough iron in the ground to last a hundred years, and scrap iron is relatively easy to recycle. Thus, iron itself is not one of the most critical mineral resources. However, its use is highly dependent on the continuation of existing patterns of world commerce and upon the availability of the other metals that are used to make the alloys of iron, or steel.

We shall consider the importance of commerce in the next section. Table 19.1 indicates the reserves and expectable lives of the principal alloying metals: chromium, cobalt, manganese, molybdenum, nickel, and tungsten. Of these, apparently, chromium reserves will last longer than iron, but all the others will be exhausted long before it, even at present rates of consumption. Moreover,

TABLE 19.1
Quantity, location, and expectable life of some mineral resources

Resource	Known global reserves	Life at present rate of use (yr)	Average projected rate of growth (%/yr)	Exponential index (yr)*	Exponential index calculated using 5 times known reserves (yr)
Aluminum	1.17×10^9 tons	100	6.4	31	55
Chromium	7.75×10^8 tons	420	2.6	95	154
Cobalt	4.8×10^9 lb	100	1.5	60	148
Copper	308×10^6 tons	36	4.6	21	48
Gold	353×10^6 troy oz	11	4.1	9	29
Iron	1×10^{11} tons	240	1.8	93	173
Lead	91×10^6 tons	26	2.0	21	64
Manganese	8×10^8 tons	97	2.9	46	94
Mercury	3.34×10^6 flasks	13	2.6	13	41
Molybdenum	10.8×10^9 lb	79	4.5	34	65
Nickel	147×10^9 lb	150	3.4	53	96
Platinum group	429×10^6 troy oz	130	3.8	47	85
Silver	5.5×10^9 troy oz	16	2.7	13	42
Tin	4.3×10^6 lg tons	17	1.1	15	61
Tungsten	2.9×10^9 lb	40	2.5	28	72
Zinc	123×10^6 tons	23	2.9	18	50

Source: Meadows et al. (1972).
*Expected life if use continues to grow at projected average rate.

Countries or areas with highest reserves (% of world total)	*Prime producers (% of world total)*	*Prime consumers (% of world total)*	*U.S. consumption (as % of world total)*
Australia (33), Guinea (20), and Jamaica (10)	Jamaica (19) and Surinam (12)	U.S. (42) and USSR (12)	42
S. Africa (75)	USSR (30) and Turkey (10)		19
Rep. Congo (31) and Zambia (16)	Rep. Congo (51)		32
U.S. (28) and Chile (19)	U.S. (20), USSR (15), and Zambia (13)	U.S. (33), USSR (13), and Japan (11)	33
S. Africa (40)	S. Africa (77) and Canada (6)		26
USSR (33), S. America (18), and Canada (14)	USSR (25) and U.S. (14)	U.S. (28), USSR (24), and W. Germany (7)	28
U.S. (39)	USSR (13), Australia (13), and Canada (11)	U.S. (25), USSR (13), and W. Germany (11)	25
S. Africa (38) and USSR (25)	USSR (34), Brazil (13), and S. Africa (13)		14
Spain (30) and Italy (21)	Spain (22), Italy (21), and USSR (18)		24
U.S. (58) and USSR (20)	U.S. (64) and Canada (14)		40
Cuba (25), New Caledonia (22), USSR (14), and Canada (14)	Canada (42), New Caledonia (28), and USSR (16)		38
S. Africa (47) and USSR (47)	USSR (59)		31
Communist countries (36) and U.S. (24)	Canada (20), Mexico (17), and Peru (16)	U.S. (26) and W. Germany (11)	26
Thailand (33) and Malaysia (14)	Malaysia (41), Bolivia (16), and Thailand (13)	U.S. (24) and Japan (14)	24
China (73)	China (25), USSR (19), and U.S. (14)		22
U.S. (27) and Canada (20)	Canada (23), USSR (11), and U.S. (8)	U.S. (26), Japan (13), and USSR (11)	26

the use of all the alloying metals except cobalt is expanding more rapidly than the use of iron. If these trends continue, most reserves of alloying metals will be consumed in less than half a century. It appears that the technology of steelmaking may change radically long before we must face a shortage of iron ore. The nickel, cobalt, and manganese in sea-floor manganese nodules have a potential for alleviating acute shortages, but the deposits are very low-grade compared to existing reserves.

OTHER INDUSTRIAL METALS

The other principal industrial metals are aluminum, copper, lead, mercury, tin, and zinc, and the known reserves of all but the first one are extremely small. Even at present rates of consumption, the known reserves of the last four will be mined out before the end of the century. It may be anticipated that lower grades of ore will be developed as the cost of these metals rises in response to shortages. Even so, those four metals seem on the verge of limiting the production of some types of technology.

Aluminum is abundant in the earth's crust and is concentrated in lateritic soils. Consequently, the supply is relatively unlimited, provided the value of the metal warrants the cost of extracting it. This is shown by the history of U.S. production of bauxite—the main ore of aluminum. U.S. domestic production was relatively constant at a low level until 1940, while imports increased to about the same level (Figure 19.1). Then, during World War II, aluminum

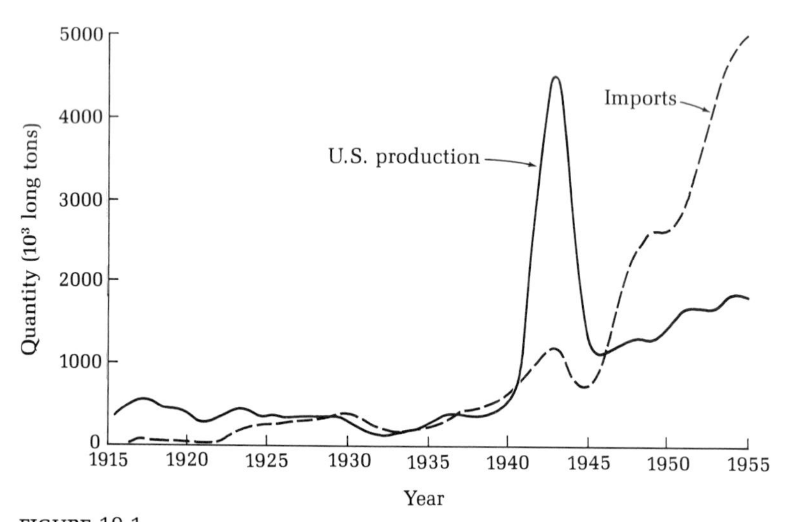

FIGURE 19.1

Bauxite produced in and imported to the United States, 1915–1955. High-cost domestic production increased rapidly to meet the needs for airplanes during World War II. [From Lovering, "Mineral Resources from the Land," in Cloud, *Resources and Man*, W. H. Freeman and Company, copyright © 1969.]

was urgently needed for aircraft and was produced domestically from low-grade, high-cost ores. After the war, these ores were largely replaced by lower-cost imports. Domestic production could be resumed at any time, and many other low-grade deposits exist elsewhere.

Aluminum has another flaw that may limit it as a replacement for iron. It takes an enormous amount of electrical energy to transform the ore into metal. For this reason, the mainly tropical ores are shipped to smelters in such places as British Columbia and Norway, where cheap hydroelectric power has been developed. Perhaps the undeveloped hydropower of Africa and South America might be used for this purpose in the future.

Copper is another metal with very large potential reserves, provided technology is developed for mining the sea floor. Manganese nodules commonly contain 0.3–0.5% copper. This concentration is not much below that of ores that are now mined, and so it seems possible that sea-floor mining will replace exploitation of the diminishing resources of the land.

PRECIOUS METALS

Of the precious metals, gold, silver, and platinum, the first two are relatively important because they are the foundation of world currency. As such, their prices are largely controlled by agreements among nations rather than their value in an open market. Much larger reserves would exist if the price were higher. However, the present known reserves will be exhausted in a few decades.

The need for money will rise with the population. One of two things can happen. Either the currencies will be devalued compared to gold, or precious metals will be abandoned as a standard.

FERTILIZERS

The high agricultural yields necessary to feed even the present population depend on the availability of the chemicals nitrogen, phosphorus, and potassium, which make the soil fertile. Because nitrogen can be fixed as soluble nitrate from nitrogen gas in the air, it is not a limited resource. The process was developed by German scientists when normal supplies were cut off by the blockade during World War I. It requires a large amount of energy, however, so the nitrate minerals of the Chilean deserts may again be more important, briefly, at some time.

Potassium (or potash) salts precipitate from concentrated brines when sea water is almost completely evaporated. Thus, layers of potassium salts occur with common salt (sodium chloride) in beds that are now incorporated in ancient sedimentary rocks. Until World War I, Germany had a virtual monopoly, and it denied potassium to the Allies just as they cut off nitrogen from Germany. Necessity broke the monopoly when new deposits were found in

New Mexico. Now deposits of potassium are widely known, although the present reserves are ill defined and presumably do not approach the ultimate limits. It is anticipated that about half the world's production will come from Canada by 1980, although the Canadian deposits were first developed only in the 1960s. Presumably, other major discoveries may be expected. In any event, because potassium is available in unlimited supply in sea water, it can always be obtained if it is considered worth the cost of the energy necessary to concentrate it.

Thus, phosphorus emerges as the potentially limiting fertilizer. Unlike potassium, it forms insoluble compounds, and it is concentrated generally in the form of phosphates on the sea floor. Ancient marine phosphates are widespread in high-grade deposits in the United States and North Africa. High-grade and low-grade reserves occur elsewhere on land and also on shallow marine banks.

At the present rate of production, reserves of phosphate on land are adequate for about 2500 years, and submarine supplies are also extensive. Therefore, there is no immediate likelihood of a crisis resulting from a shortage of any chemical fertilizer. The only obvious problem is that use of the fertilizers is growing at 6–8% per year, which means that it is doubling in about 10 years. Fortunately, even a quadrupling of consumption will not approach the available supply. The limits of production are more likely to be posed by cost and pollution.

Mineral Production and World Trade

Local mineral wealth has made first one nation rich and then another. Athens financed its ancient wars with the silver of Laurium, and Alexander funded his with the gold of Macedon. During the Middle Ages, Germany was the center of lead, zinc, and silver production and a focus for the first training of geologists. Great Britian moved into the forefront during the industrial expansion of the nineteenth century; the tiny realm was successively the world's leading producer of lead, copper, iron, and then coal. Bolstered as well by the resources of the empire, Great Britian was the wealthiest nation in the world.

Now most of the Greek, German, and British mines are exhausted. The far greater mineral resources of the United States have supported its advance to become the richest nation in the world today. However, the future is already foreshadowed. The gold of California and Alaska is gone, so is the silver of Colorado. The oil of Texas and the iron of Minnesota are going fast, and the United States is becoming ever more dependent on imports and the preservation of peaceful world trade.

The industrial nations, with the possible exception of the USSR with its enormously extensive territory, are all becoming increasingly dependent on peaceful commerce. The power plants, smelters, and factories are in a few places and the resources are spread throughout the world. The importance of commerce is obscured in those countries, such as the United States, that have domestic resources. It is remarkably clear, however, in Japan, which is essentially an industrial nation that lacks mineral resources. In 1951, Japan imported a few million tons of iron ore, about 60% from Asia and 35% from North America (Figure 19.2). By 1970, the imports exceeded 50 million tons and the

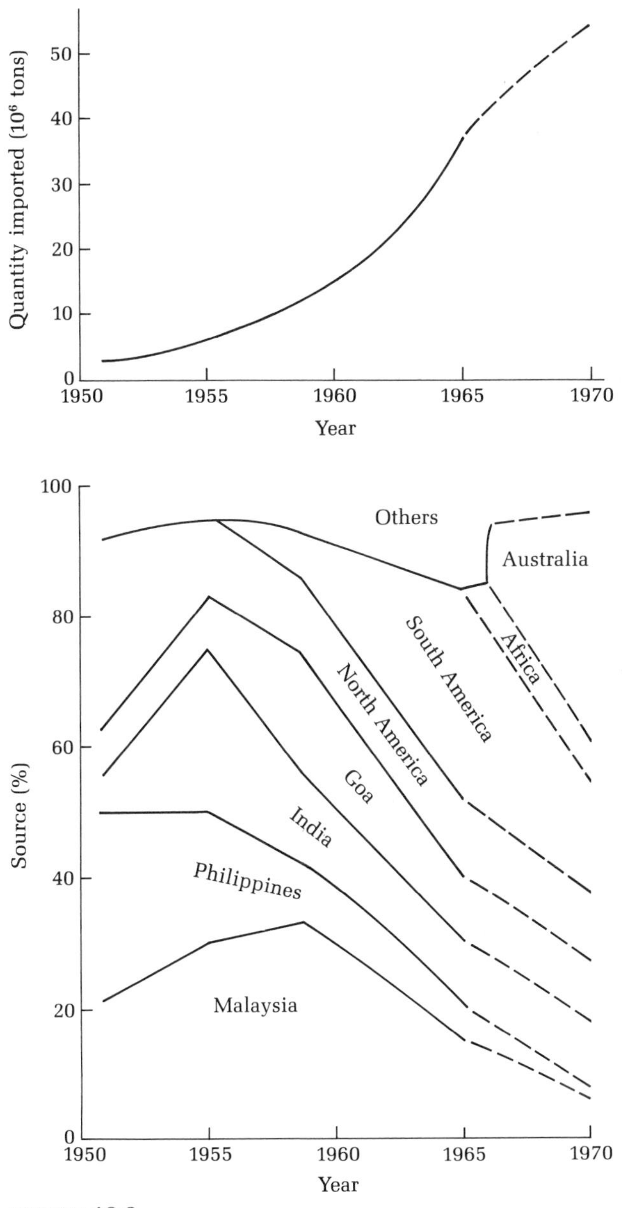

FIGURE 19.2
Past and projected imports and sources of Japanese iron ore.

pattern of suppliers had changed markedly. Asia supplied ten times as much ore as in 1951, but its proportion of the total had declined to less than 30%. Meanwhile, Peru and Chile were supplying more than 10 million tons per year, even though they became suppliers only in 1955. Even this new source could not meet the demand, and mountains of newly discovered ore in Australia began to flow to Japan. Within a few years, they approached 20 million tons per year, or 40% of the total. Industrial Japan was in the process of buying up the natural resources it sought to acquire by conquest during World War II. Both the importers and the exporters became more and more committed to continuing free commerce as a result.

During the coming centuries of growing shortages of many minerals, world commerce will become increasingly critical if world resources are to be utilized in any rational way. The resources are unevenly distributed. Only commerce can move them so that the iron of one region can be alloyed with the molybdenum of another, using the coal of a third region to power the factories of a fourth.

Some of the patterns of the change are already evident in the distribution and the history of production. South Africa, for example, has the world's largest reserves of chromium, gold, manganese, and platinum, and this fact has had a significant effect upon international affairs. However, it is also the leading producer (77%) of gold outside of Russia—whose production is secret but growing. Charles Park of Stanford University has estimated that South African production peaked in the late 1960s and will drop, by 1980, to a third of its maximum. The influence of this source area may decline with the gold production, although the other mineral resources will remain.

Gold deposits typically have been exhausted within a few years of their discovery; thus, the enormous South African reserves have lasted far longer than

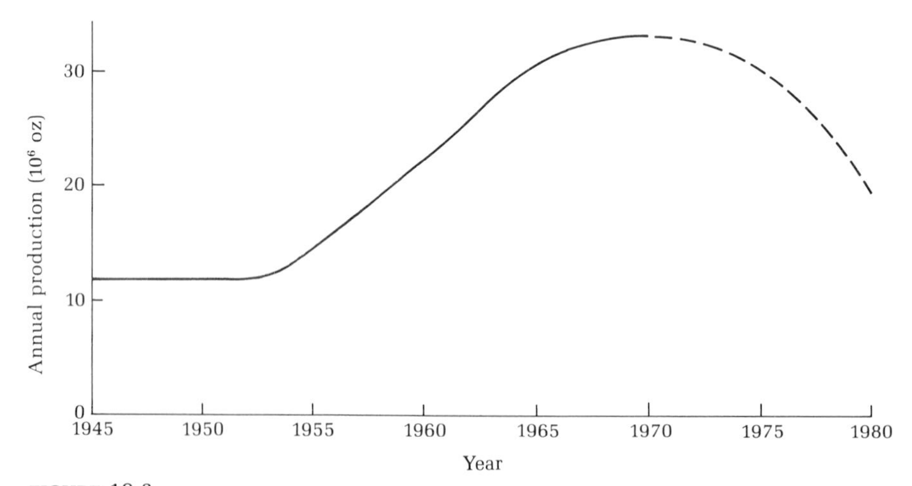

FIGURE 19.3
Past and projected annual production of gold in South Africa. The value of gold increased in the spring of 1971, and this may influence production.

most. Deposits of other, less precious, minerals have generally lasted for a few decades—or long enough to establish some stable pattern of commerce. Now industry has expanded to the point where it can gobble up major new mineral discoveries almost as fast as they can be developed. Eventually, we shall have to change our ways. Meanwhile, commerce and society itself depend on the successful application of the earth sciences to the discovery of new reserves of most mineral resources.

Igneous and Metamorphic Ores

The rising silicate magmas of the igneous cycle contain traces of various metals that are concentrated in different ways to become ore bodies. The histories of most ores are very complex and controversial. Consequently, there is much more agreement about the principles of ore formation than about the origin of any particular ore body. Thus, it seems useful, in this section, to minimize reference to examples and to concentrate on principles.

Concentration of metals occurs on all scales. The magmatic intrusions are themselves concentrated regionally—often at plate boundaries or above plunging plates. They are further concentrated locally along individual faults or fault zones. General geologic knowledge thus enters into the scientific exploration for ores.

Within the magma, metallic ores are concentrated—in the same ways as silicate minerals—as a consequence of crystal structure of the major minerals, differences in order of crystallization, density, and so on.

Some elements that form valuable mineral deposits at some stage in the geologic cycles, such as aluminum and potassium, are incorporated, in the igneous cycle, into the major rock-forming minerals and, thus, are initially dispersed. Many elements, however, do not readily form silicates or fit into silicate crystal lattices as impurities. Generally, they tend to become concentrated during the cooling and crystallization of the magma.

Concentration of rarer elements may occur by early formation of crystals while most of the magma is liquid or a crystalline mush. Alternatively, it may result from late crystallization, in which case the concentrated solutions may remain in the magma chamber or escape to the surrounding rocks.

Generally, minerals that form early must be further concentrated by gravitational settling through the remaining liquid. They would not be worth mining if they were disseminated uniformly. Diamonds are an exception. They are so valuable that the whole mass of rock in a diamond pipe can be mined to obtain the disseminated crystals. Chromite ($FeCr_2O_4$) is an example of an important ore mineral that is concentrated during the early stage of magma solidification. It is one of the first minerals to crystallize, and, because it has a density of 4.1–4.9 grams per cubic centimeter, it settles out to form layers of ore. Some ores of chromite and magnetite that might be expected to have accumulated placidly by settling have apparently been injected as liquids or mush after the rock-forming minerals have solidified. It is uncertain how the ores are stockpiled or remobilized during and after the solidification of the rocks.

Most metallic ores consist of minerals that crystallize late (at low temperatures) and, thus, are concentrated as residual liquids. Commonly, the ores are sulfides, including those of arsenic, cadmium, cobalt, copper, lead, mercury, molybdenum, nickel, silver, and zinc. Some are emplaced by liquid injections that cut across the already solidified parts of the intrusion or the surrounding rock.

The injections are little different from ordinary igneous dikes and sills except in their chemistry and mineralogy. The other modes of ore emplacement late in the cooling of the magma are quite different. The nature of the phenomena are complex, but it is clear that hot gases and solutions spread outward and upward from the magma and alter the solid rock or fill cracks and faults with minerals. The phenomena are generally gradational.

The hot emanations that emerge from volcanic vents probably resemble those in the interior. They tend to be highly corrosive acid solutions rich in metals. The 300°C brine discovered in a hole drilled near the plate boundary in California contains 30% dissolved solids, including copper, iron, lead, lithium, manganese, silver, and zinc, which indicates the type of hot solutions that circulate above igneous intrusions and that deposit ores.

TABLE 19.2
List of the common ore minerals

Metal	*Ore Mineral*	*Composition*	*Percentage of metal*
Gold	Native gold	Au	100
	Calaverite	$AuTe_2$	39
	Sylvanite	$(Au,Ag)Te_2$	–
Silver	Native silver	Ag	100
	Argentite	Ag_2S	87
	Cerargyrite	$AgCl$	75
Iron	Magnetite	$FeO \cdot Fe_2O_3$	72
	Hematite	Fe_2O_3	70
	"Limonite"	$Fe_2O_3 \cdot H_2O$	60
	Siderite	$FeCO_3$	48
Copper	Native copper	Cu	100
	Bornite	Cu_5FeS_4	63
	Brochantite	$CuSO_4 \cdot 3Cu(OH)_2$	62
	Chalcocite	Cu_2S	80
	Chalcopyrite	$CuFeS_2$	34

TABLE 19.2 (continued)

Metal	*Ore Mineral*	*Composition*	*Percentage of metal*
Copper *(contd.)*	Covellite	CuS	66
	Cuprite	Cu_2O	89
	Enargite	$3Cu_2S \cdot As_2S_5$	48
	Malachite	$CuCO_3 \cdot Cu(OH)_2$	57
	Azurite	$2CuCO_3 \cdot Cu(OH)_2$	55
	Chrysocolla	$CuSiO_3 \cdot 2H_2O$	36
Lead	Galena	Pbs	86
	Cerussite	$PbCO_3$	77
	Anglesite	$PbSO_4$	68
Zinc	Sphalerite	ZnS	67
	Smithsonite	$ZnCO_3$	52
	Hemimorphite	H_2ZnSiO_5	54
	Zincite	ZnO	80
Tin	Cassiterite	SnO_2	78
	Stannite	$Cu_2S \cdot FeS \cdot SnS_2$	27
Nickel	Pentlandite	$(Fe,Ni)S$	22
	Garnierite	$H_2(Ni,Mg)SiO_3 \cdot H_2O$	–
Chromium	Chromite	$FeO \cdot Cr_2O_3$	68
Manganese	Pyrolusite	MnO_2	63
	Psilomelane	$Mn_2O_3 \cdot xH_2O$	45
	Braunite	$3Mn_2O_3 \cdot MnSiO_3$	69
	Manganite	$Mn_2O_3 \cdot H_2O$	62
Aluminum	Bauxite	$Al_2O_3 \cdot 2H_2O$	39
Antimony	Stibnite	Sb_2S_3	71
Bismuth	Bismuthinite	Bi_2S_3	81
Cobalt	Smaltite	$CoAs_2$	28
	Cobaltite	$CoAsS$	35
Mercury	Cinnabar	HgS	86
Molybdenum	Molybdenite	MoS_2	60
	Wulfenite	$PbMoO_4$	39
Tungsten	Wolframite	$(Fe,Mn)WO_4$	76
	Huebnerite	$MnWO_4$	76
	Scheelite	$CaWO_4$	80

Source: After Bateman (1950).

Alteration (metasomatism) of surrounding rocks and of the intrusion itself occurs on an enormous scale. Waldemar Lindgren of the U.S. Geological Survey showed that every cubic meter (about 3 metric tons) of the altered limestones at Morenci, Arizona, had lost 1.7 tons of CaO and CO_2 and had gained 2.5 tons of SiO_2 and Fe_2O_3. Limestones are particularly susceptible to alteration by hot acid solutions, and this scale of change may not be typical of other rocks. The metallic emanations, derived from the magma and perhaps from leaching of the surrounding rock, also alter and replace rock to form ore bodies.

The decreasing intensity of alteration and temperature of crystallization around the igneous intrusion tends to form mineral zones. However, the zones are irregular where permeable beds and joints influence the movement of emanations. The alteration of the whole rock provides a basis for distinguishing these deposits from hydrothermal ones, which also result from migration of hot solutions.

Volcanoes and thermal areas discharge enormous quantities of steam, and mining geologists took this as evidence that magma emits similar quantities. We now know that almost all the steam is simply reheated rain that has flowed down as ground water and become heated. That much is clear. How and where the water is charged with solutions of ores remains obscure. The evidence of the Valley of Ten Thousand Smokes, in Alaska, is that metals and many other elements can reach the surface as vapors in volcanic zones. Presumably, they can produce ore solutions at depth, even though the magma contains little water.

Whatever their origin, hydrothermal solutions produce many of the most important metalliferous ore bodies. Many of these are extremely high-grade veins that fill cracks in essentially barren and unaltered rock. Others fill the pores in sand and thus provide an unusual cement—deposited like more common cements by precipitation from moving ground water. Yet others, the "porphyry" or "disseminated" copper ores, consist of a mesh of intersecting veinlets that are so closely spaced that the intervening rock is intensely altered.

The distribution of hydrothermal ores in veins is related to the structural history of an area. The underground search for high-grade but discontinuous veins has provided much of the impetus for the study of small-scale geologic structures. These tend to be increasingly complex as the scale of deformation grows smaller. The great dislocations of plate tectonics are no longer of more than background interest. On the scale of the mining geologist, tiny faults may change direction, as though refracted, as they pass from shale to sandstone. Faulting may occur repeatedly along different, although related, directions. All these events may influence the paths followed by ore-bearing solutions and, therefore, the ultimate location of veins. Thus, to be effective in finding and developing igneous and metamorphic mineral deposits, the geologist must have a wide range of knowledge. Additional understanding is necessary in order to reconstruct the history of the ores as they become exposed to weathering and to the sedimentary cycle.

Weathering

The physical and chemical weathering processes that affect ordinary rocks also act on ore deposits that are in the surficial zone of oxidation. Thus, some elements are oxidized in place, some are dissolved and transported away, and some remain as inert residues. The difference between metalliferous ores and common rocks is that the metallic elements—such as lead and zinc—that are mobilized in concentrated solutions are just as poisonous to plants as to people. Without plants, ordinary soils do not form, despite weathering. In such places as New Caledonia, strange soils develop on metalliferous rocks and support growth of equally strange plants.

This section is concerned largely with the chemical effects of exposure to the zone of oxidation, whereas the next section deals with materials in the sedimentary cycle. Rain and ground water carry oxygen to ores and generally leach the surficial minerals and concentrate them as precipitates at lower levels. Most soil formation extends down only a meter or so, and weathering is generally confined to the upper 100 meters. However, the sulfides, especially the common mineral pyrite (FeS_2), readily decompose to oxides and sulphuric acid. In this reactive environment, sulfides have been found to oxidize to depths of almost 1000 meters.

Without erosion, the surface of the initial ore deposit becomes leached, and a layer of enriched oxidized ores accumulates at some generally shallow depth. In addition, new sulfides precipitate from solutions that migrate below the oxidation zone. These enriched zones may be economically minable even when the original sulfide ore is not. With the passage of time, erosion removes some of the leached surface rock, and the oxidized zone migrates downward like a soil zone. Thus, even greater enrichment occurs.

Most sulfide ores are at least partially enriched in this way. Another class of ores is formed only in the zones of deep leaching. In temperate climates, this produces clays, which can be valuable for construction and pottery. In tropical climates, some lateritic soils, leached of silica, become ores of bauxite (hydrous aluminum oxide).

Bauxite is named after the Baux district in France, which is not in the tropics. Likewise, bauxite ores occur in Arkansas and Georgia. How is this distribution reconciled with an origin by tropical weathering? The deposits were formed during Tertiary times, when the weather was warmer than now. Moreover, continental drift is moving these deposits northward, and they were closer to the equator when they formed.

The age distribution of bauxite merits interest. Most deposits of bauxite are at least 10 million years old, which probably reflects very slow formation. At the opposite extreme, few are more than 60 million years old. This may reflect the difficulty of preserving soft, thin, surficial bauxite layers during normal cycles of erosion and deposition. For this reason, some bauxite ores are confined to solution depressions in the karst topography of Hungary,

Yugoslavia, and other areas of limestone. The intervening lateritic soil has been eroded away.

Sedimentary Ores

Sedimentary ores, like other sedimentary rocks, can be divided into two classes, depending on whether they result primarily from chemical processes or from physical ones. The former generally consist of elements that are readily soluble and highly mobile until they are deposited on the sea floor. The latter consist of relatively insoluble, dense, durable materials that are sorted and concentrated locally by rivers and waves as they move toward the deep sea.

Consider first the placer ores that are physically concentrated. Diamond is an excellent example. It is chemically inert under ordinary conditions, it is dense, and it occurs as large crystals that are difficult to move. Consequently, as the diamond pipes are eroded away, big diamonds tend to remain behind as a surficial lag gravel. If erosion is more vigorous, all sizes of diamonds are transported away by rivers. Diamond is the hardest natural substance, so it is hardly worn during transport, regardless of the distance. It does have a cleavage, however, so it may be fractured in unusual circumstances.

Diamonds move with the sand bed, and it moves mainly during floods. Because of their great density, the diamonds are among the first grains to settle out of suspension as a flood wanes. Thus, they tend to become concentrated on bedrock surfaces under the sand that comes to rest more slowly. Diamonds have been traced from the source pipes in central South Africa along the Orange River to the sea. Along the coast, they have been concentrated by wind and waves, which are highly selective winnowers of low-density materials. The diamonds tend to remain as residual or lag materials on beaches, but they also are dredged from the continental shelf. Presumably, this is because beaches existed on the shelf when Pleistocene glaciation lowered sea level.

Gold is another material that occurs in placer ores. It is durable and inert. Its density is 19.3 grams per cubic centimeter, and this makes it very difficult for running water to transport. It not only occurs on the bedrock of river channels, like diamonds, but tends to be concentrated in cracks and crevices in the rock. Gold occurs abundantly as very fine flakes that are transported along with the less dense but coarser sand. This is the type of gold that can be panned in small streams in gold districts.

There are many other less valuable minerals that form placer ores by physical concentration. One of these is cassiterite (SnO_2), a tin ore that is mined extensively from sand and gravel, particularly in southeast Asia. Notable deposits occur in valleys on small islands in Indonesia, and they have been followed out into the submerged Pleistocene valleys of the continental shelf. This "sea tin" is scooped up by floating dredges.

Another type of deposit, called "black sand," occurs in layers on many modern and fossil beaches. It generally contains a group of minerals that have been mined in some places for iron, titanium, and zirconium. It is most interesting, however, for its content of monazite, which is a phosphate mineral containing the radioactive element thorium. If breeder reactors are developed in time, this thorium can be converted to uranium fuel. For this reason, the governments of India and Brazil have ceased exporting the black sands, which were being mined for the other materials they contain.

Elements that are easily dissolved generally have a long residence time in the ocean and are difficult to remove from it. Nonetheless, slow chemical or biochemical precipitation does occur and mineral reserves or resources do form in special circumstances.

Precipitation is very slow, but a layer of ore can form if other forms of sedimentation are even slower. This seems to account for the formation of layers of phosphate nodules and rock on shallow banks that are swept free of sediment even though they are near land. Very slow deposition also may contribute to the existence of an extensive carpet of black nodules of ferromanganese oxides on some areas of abyssal red clay. The nodules contain only small proportions of nickel, cobalt, and copper, but the amounts are large enough to make them significant resources.

The elements potassium and sodium, with residence times in sea water of 10–100 million years, are deposited only in extraordinary circumstances, such as those that accompanied the initial opening of the Atlantic Ocean. However, these circumstances arise often enough to provide extensive deposits of sodium salts and much smaller ones of potassium.

Low-grade Minerals from the Sea

From the land we mine the salts and phosphates that were once deposited or precipitated from ancient seas.

From under the sea we extract oil, gas, sulfur, tin, diamonds, and other continental mineral resources that happen to be slightly submerged near the edges of continents. In the foreseeable future, we may begin to dredge the manganese nodules of the deep sea. However, the sea and the sediment below it have an additional potential as resources for the distant future.

Many industrial metals, including manganese, nickel, copper, cobalt, lead, and molybdenum, are as much as fifty times more abundant in pelagic sediments than in average igneous rocks (Figure 19.4). Similar low-grade concentrations of these elements also occur on land and may be more logical candidates for development than the sea-floor sediments. Consider, however,

BOX 19.1 DIVING FOR GOLD

It is sometimes possible to obtain very fine gold "flour" by panning sand that is merely scooped out of the bed of a stream in a gold district. However, gold mixed with sediment is apt to be too limited in quantity to have a significant value. This is because the high density of gold causes all but the most minute particles to sink to the bedrock when floods mobilize the sediment in a stream bed.

For this reason, professional miners, and those who would be, commonly dredge out all the sand from the bed, thereby exposing the gold-bearing zone. Alternatively, they may deflect the stream and expose the bed to mining. The forty-niners in California were fortunate because much of the gold was at the base of stream gravels deposited in Tertiary time and the streams had since been deflected naturally.

Deflecting streams and mining are relatively expensive and time consuming, and floating dredges can be used only on the largest ore deposits because of the high cost of building and operating them. Recently, California divers have developed a technique that is inexpensive but has a reasonable chance of obtaining worthwhile amounts of gold from streams.

Divers can use a scuba, but it is generally more convenient to use a "hookah" apparatus with a conventional demand valve like that of a scuba but connected to a hose leading to a pump floating at the surface. The diver inspects the bottom for crevices and directs a small suction dredge into them to suck out coarse gravel and gold. The same engine that drives the air compressor can pump out the gravel and gold in the crevice. At the surface, a buddy diver monitors a floating sluice that separates the gold from the gravel. He also makes sure that the diver is safe.

An article by William Clark in *California Geology* describes gold diving in some detail.

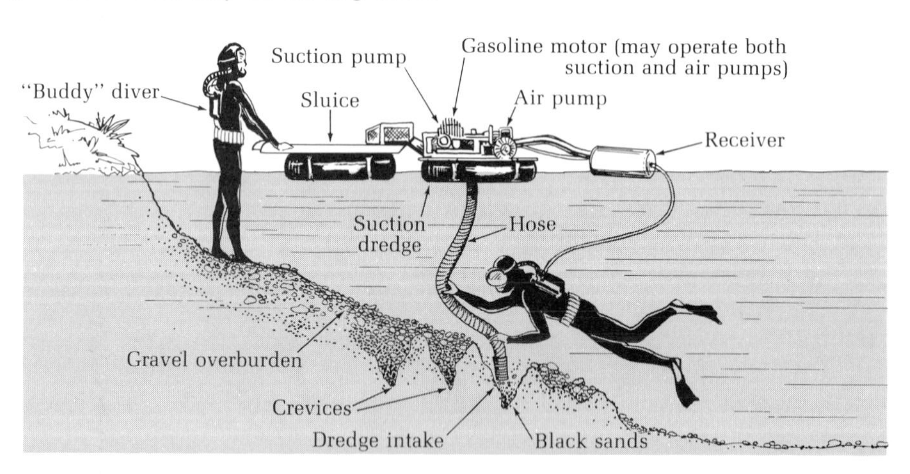

A typical gold-diving operation. [From Clark (1972).]

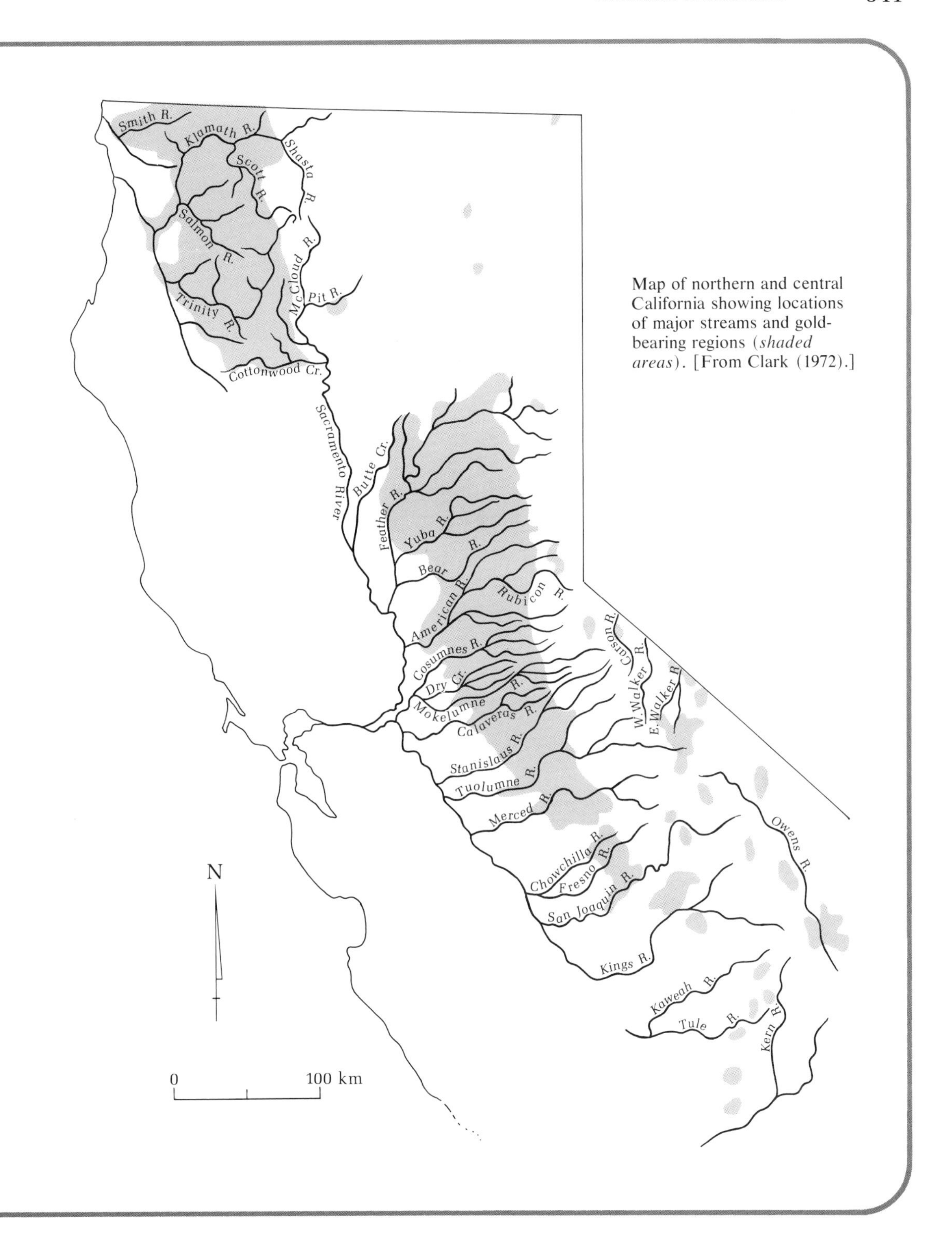

Map of northern and central California showing locations of major streams and gold-bearing regions (*shaded areas*). [From Clark (1972).]

the economic and social conditions that might be expected to lead to the mining of low-grade resources: First, the high-grade ores will be exhausted, and those of the industrial nations will be consumed first. Second, power will be cheap. Third, industry and the general population will have expanded and pollution and preservation of land will be more critical. Fourth, the new low-grade ore bodies will need to be enormous in order to merit development.

If these conditions apply, the virtually unlimited red clay may be more attractive than it is now. Industrial nations will find its exploitation free of political and environmental problems—which can be more limiting than economics.

Sea water contains at least 64 elements and is abundant and free; moreover, it is easy to process in ways that hardly affect the environment. Magnesium, bromine, sodium, and chlorine (the last two as common salt, NaCl) are already being extracted from sea water. However, they are among the elements most concentrated in sea water, and most of the remainder are extremely dilute. Only a quarter of the elements reach even the low concentration of 1.8 kilograms per million liters. Preston Cloud has calculated the effort necessary to obtain the valuable metal zinc from the sea. Only 400 tons, worth $120,000 at 1968 prices, could be stripped from a volume of sea water equal to the annual flow of the Delaware and Hudson rivers. The pumping charges alone would make the operation questionable.

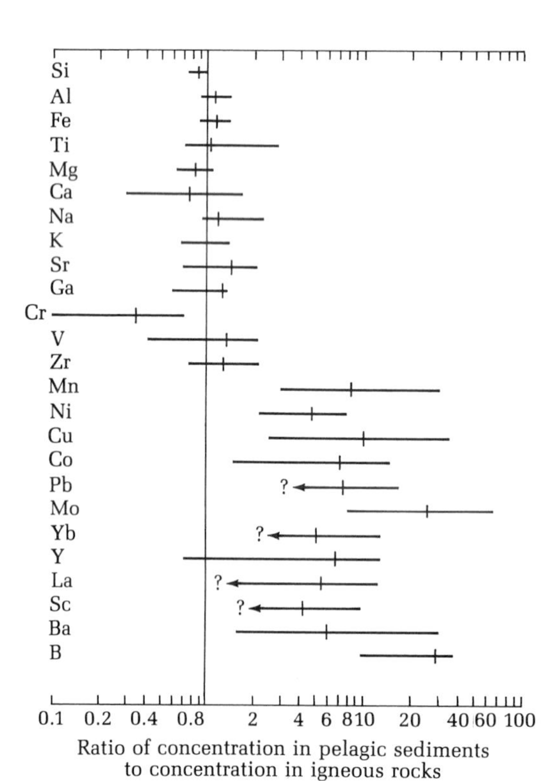

FIGURE 19.4
Ratios of average elemental abundances in Pacific pelagic sediments and igneous rocks. Variations in ratios are indicated by the lengths of the horizontal lines, and averages by a short vertical line. Copper (Cu), for example, is about 2–40 times more concentrated in pelagic sediment than in igneous rock; its average concentration in such sediment is about 10 times that in igneous rock. [From Goldberg and Arrhenius (1958).]

Appropriate processing could remove many elements at the same time, and this might be more economical at some time. At present, however, only a few metals are needed, and the apparently valuable by-products would oversupply the need and reduce the price accordingly. This objection may be less important as high-grade reserves of more metals are exhausted.

Origin of Igneous Mineral Provinces

Most of the major ore deposits of the world are notably concentrated in small regions and were emplaced during geologically brief periods. Thus, they approach uniqueness; accordingly, they may be attributed to exceptional circumstances or even to unique phenomena. The late history of many ore deposits is known, but the cause of the concentrations in provinces remains obscure. These are not minor concentrations: 40% of the world's reserve of molybdenum is in one intrusion in Colorado, 77% of the tungsten reserve is in China, half the reserve of tin is in southeast Asia, and 75% of the chromium is in South Africa. Likewise, most of the banded iron ores were deposited at about the same time, although they are widely distributed.

We conclude this chapter with some speculations about the origin of mineral provinces, mainly igneous ones, and their relation to the major geologic cycles and the history of the earth. There are three general explanations for the existence, in the relatively homogeneous crust of the earth, of a concentration of some element or mineral (inhomogeneity). The concentration may (1) have existed from the beginning, or (2) have been produced by some unique and irreversible global event during the history of the earth, or (3) have been produced by repeated local events. All of these explanations may apply to the origin of mineral provinces.

The accretion model for the formation of the earth offers an easy explanation for the initial existence of concentrations. The meteorites that strike the earth do not have uniform compositions, and, presumably, the planetesimals that formed the earth did not either. Thus, the problem is one of preserving some inhomogeneities while the earth accretes and while the mantle is stirred by convection, and then bringing them to the surface in ore bodies. The likelihood of preservation depends on the intensity of stirring. The recent discovery that the mantle is not entirely homogeneous suggests that mixing has never been completed and that inhomogeneities comparable in size to igneous intrusions may have been preserved. Bringing them to the surface requires the localization of heat, and the existence of mantle hot spots indicates that heat is indeed localized from time to time.

"Initial," as it is used above, refers to the time of origin of the earth. However, meteorites, large and small, are still hitting the earth, and their "initial" composition refers to the time when they arrive. Some large concentrations of metallic ores may not have been brought up from the mantle, but may merely be manifestations of astroblemes. They are preserved remnants of meteorite impacts that occurred on the surface where they are now found.

These are not idle speculations, because they illustrate the point that theory and practice are intimately related. The observed distribution of elements must be explained by, or be compatible with, the theory of the origin of the earth. The thinking of the ore-seeker is fruitfully guided by concepts of the origin and history of the earth.

Let us turn to the role of unique and irreversible events in the formation of mineral provinces. These are events in the differentiation of the interior and the evolution of the air and sea. The most significant "event" with regard to ores occurred during the interval from 3200 to 1800 million years ago when almost all banded iron formations were deposited. It was during this interval that oxygen, released by the first plants, became available and began to change the air from its initial reducing state to its present oxidizing one.

The preservation of initial inhomogeneities and the evolution of the earth can account for some mineral types and provinces, but it can offer little hope of explaining many others. It appears that some mineral provinces are related to plate boundary phenomena and, thus, to repeating processes that concentrate elements locally. Many observations show that vulcanism of all sorts is accompanied by emanations of metals. Evidence of these has been detected in the sediments of marine spreading centers in the form of unusual minerals, such as barite ($BaSO_4$), and unusual concentrations of metallic elements. It appears, therefore, that spreading centers are also mineralization centers and that a thin, metalliferous layer lies at the base of oceanic sediment. Presumably, this layer is of little value in itself, but it is a potential source for ores if the metals can be concentrated.

Extraordinary deposits of metalliferous sediment have been discovered in association with hot brines in small basins in the center of the Red Sea (Figures 19.5 and 19.6). The hot brines do not rise to the sea surface because they are so saline that their density is greater than the overlying water. Among other elements, they contain more than 1500 times as much zinc as normal sea water and 30,000 times as much lead. If the time comes to extract metals from sea water, this may be the place to begin.

Below the brines is sediment that contains layers rich in sulfides, oxides, and carbonates of various metals, mainly iron, manganese, zinc, and copper accompanied by significant amounts of gold and silver.

Only the top 10 meters of sediment in these small basins has been sampled, but it contains an estimated $2 billion worth of gold, silver, copper, and zinc, at 1970 prices. What it would be worth after the expense of mining and extracting it is unknown, but it is soft mud and easy to remove. Moreover, the sediment is almost 100 meters thick in places, so the resource is very large. To whom it belongs is highly debatable within the framework of existing international law.

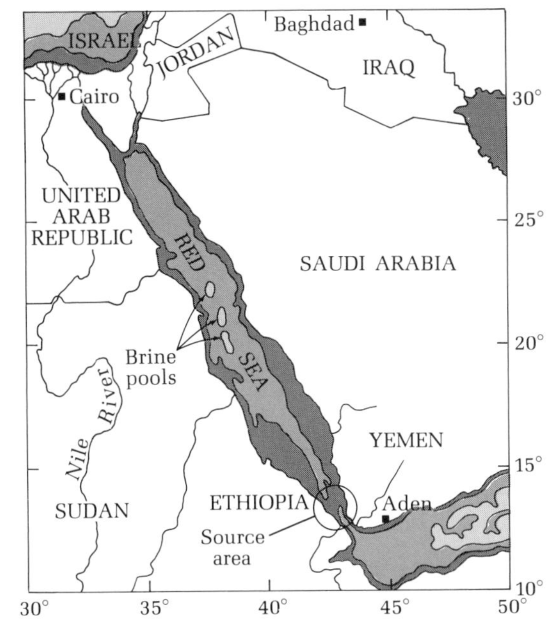

FIGURE 19.5
Location and source of the Red Sea brine pools. [After Degens and Ross, "The Red Sea Hot Brines," copyright © 1970 by Scientific American, Inc. (all rights reserved).]

The origin of the brines suggests that ore deposits of unknown extent exist below the muds. From an analysis of the isotopic composition of the brines, it appears that they are derived from sea water in the southernmost part of the Red Sea. This water migrates downward and flows horizontally for hundreds of kilometers before it emerges. The environment is a spreading center; hence, the crust is relatively hot and, presumably, is emitting metallic and acidic solutions. Thus, the percolating sea water becomes a hot, chemically active solution that apparently leaches salts and metals from sedimentary rock. When it begins to rise and cool, it probably deposits most of the metals as carbonate and sulfide veins in bedrock. The fraction that reaches the sediment and brine depends on the temperature and the rate of flow of the brines.

A consideration of the rate of sediment accumulation makes it seem likely that these deposits have formed in only the last 200,000 years. The circumstances are unusual because of the peculiar geography of the Red Sea. Spreading centers are common, however, and they are characterized by emissions of potent, hot solutions. Other important ore deposits may accumulate upon them. This is an intriguing speculation because the deposits will split naturally, in due course, and become attached to different plates. How many ore deposits have far-travelling twins?

Ultimately, oceanic plates are consumed at subduction zones, and they plunge into the mantle bearing whatever metals they may have accumulated. These are not limited to those that emanate at spreading centers. Pelagic

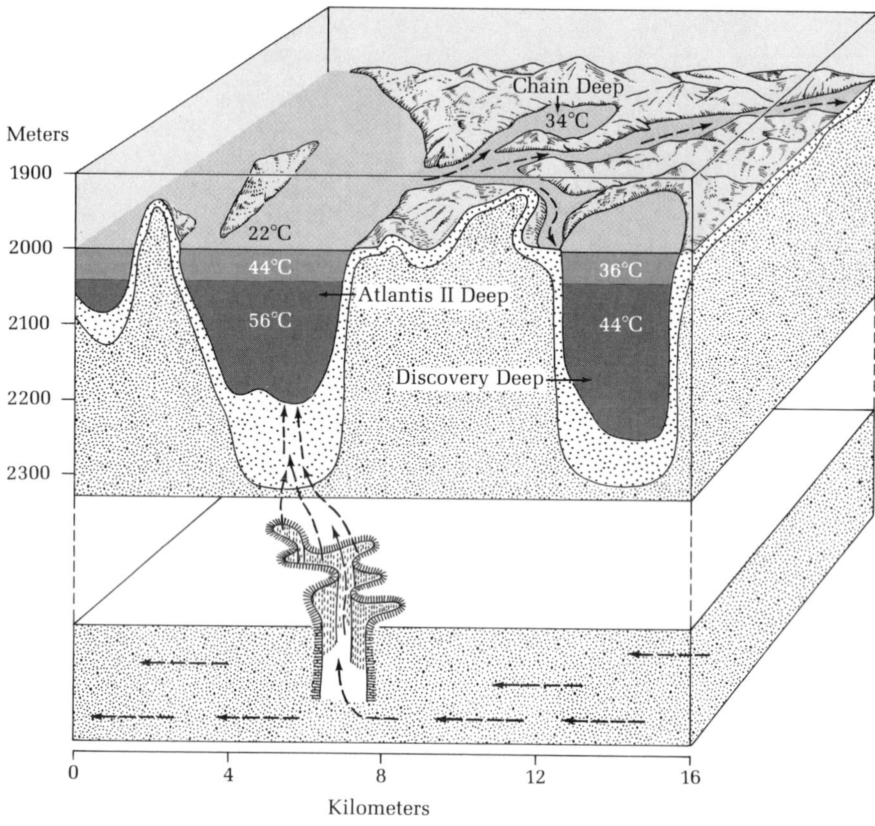

FIGURE 19.6

Bottom topography in the area of the three deeps in the Red Sea is seen near the top of this exploded cross section of the basement underlying the spreading center. The view is from the west; north is to the left. The long axis of the block measures 16 kilometers, but depth indications below 2300 meters are arbitrary. Top arrows show how overflow of the Atlantis II Deep can periodically replenish the two other pools. The winding "canyon" visible in the lower block represents zones in the basement rock where the percolating brine, cooling as it rises toward the floor of the Atlantis II Deep, has probably deposited rich veins of various metallic ores in the rock below. [From Degens and Ross, "The Red Sea Hot Brines," copyright © 1970 by Scientific American, Inc. (all rights reserved).]

oozes, and particularly ferromanganese nodules, contain low-grade concentrations of metals as well.

As a plate plunges, the temperature and pressure gradually increase. Various components melt and move upward. The most voluminous are the siliceous igneous rocks associated with subduction zones, but the metals are also mobilized at various depths. Because a plate plunges at an angle, different depths are at different distances from the line of subduction. Thus, plate tectonics leads to a model capable of explaining the existence of large concentrations of a few elements in a mineral province of limited age and extent.

Summary

1. Few of the mineral resources of importance to society are unique: often, one can be substituted for another. Thus, the development of mineral resources is extraordinarily dependent on economics, technology, and other social factors.
2. Supplies of iron and most other major industrial metals appear adequate for decades or centuries, provided there is political stability so that the resources of the whole world are available.
3. Sources of chemical fertilizers do not appear to pose a limit to agriculture.
4. Many metallic ores, especially sulfides, are the direct result of the igneous cycle.
5. Other ores, especially oxides, are concentrated by weathering of igneous ores or rocks.
6. Sedimentary ores result from concentration of resistant solid materials during transportation or from precipitation of dissolved materials.
7. Many industrial metals are 50 times more abundant in pelagic sediments than in the average igneous rock. Thus, the sediments may be a vast resource for the future.
8. Mineral provinces are notably concentrated in small regions and emplaced during geologically brief periods. Their origin may be related to exceptional geologic circumstances.

Discussion Questions

1. Do mineral resources participate in the major geologic cycles like other materials of the crust?
2. What is the major problem in the utilization of increasing amounts of plentiful aluminum?
3. Other things being equal, why are mine tailings from igneous ores likely to yield more stream pollution than those from ores produced by weathering?
4. Why do gold and diamonds occur as placer ores?
5. What are some desirable aspects of extracting elements from sea water?
6. What is the role of sea-floor spreading in the formation of the Red Sea metalliferous brines?

References

Bateman, A., 1950. *Economic Mineral Deposits* (2nd ed.). New York: John Wiley & Sons. [A widely used, comprehensive treatise.]

Brock, B. B., 1972. *A Global Approach to Geology*. Cape Town: A. A. Balkema. (Quoted from p. 2.) [An economic geologist analyzes global fault patterns.]

Clark, W., 1972. Diving for gold in California. *Calif. Geol.* 25(6):123–125. [A hobby can pay for itself in delightful surroundings.]

Cloud, P., 1969. Mineral resources from the sea. *In* P. Cloud, chairman, *Resources and Man,* pp. 135–156. San Francisco: W. H. Freeman and Company. [They tend to enlarge resources but not proven reserves.]

Degens, E., and D. Ross, 1970. The Red Sea hot brines. *Sci. Amer.* 222(4):32-42.

Flawn, P., 1966. *Mineral Resources.* Chicago: Rand McNally [Focuses on social and economic aspects of resources.]

Goldberg, E., and G. Arrhenius, 1958. Chemistry of Pacific pelagic sediments. *Geochim. Cosmochim. Acta* 13:153–212. [Industrial metals tend to be concentrated in pelagic sediments compared to igneous rock.]

Lindgren, W., 1905. Clifton-Morenci District, Arizona. *U. S. Geol. Surv. Prof.* (43). [Quantitative analysis of metasomatism.]

Lovering, T., 1969. Mineral resources from the land. *In* P. Cloud, chairman, *Resources and Man,* pp. 109–134. San Francisco: W. H. Freeman and Company. [Mineral production is only approximately related to mineral price, and other myths exploded.]

Meadows, D. H., D. L. Meadows, J. Randers, and W. Behrens, III, 1972. *The Limits to Growth.* New York: Universe. [The horrors of exponential expansion.]

Park, C., Jr., 1968. *Affluence in Jeopardy.* San Francisco: Freeman, Cooper. [Identifies American as well as global mineral shortages.]

Skinner, B., 1969. *Earth Resources.* Englewood Cliffs, N.J.: Prentice-Hall. [Resources from the perspective of the geochemist.]

APPENDIX 1

WORLD POPULATION

The following table gives United Nations estimates of future world population, as of 1968, rounded to the nearest million. High and low estimates are given for less-developed regions.

	Population (millions)			
	1970	*1980*	*1990*	*2000*
More-developed regions	1090	1210	1336	1454
Less-developed regions	2523–2563	3137–3379	3820–4425	4523–5650
East Asia	814–846	932–1062	1038–1285	1142–1493
South Asia	1121–1126	1439–1518	1786–2032	2119–2617
Europe	462	497	533	568
USSR	243	271	302	330
Africa	344–346	448–466	583–649	734–906
Northern America	228	261	299	333
Latin America	243–245	323–337	424–469	540–645
Oceania	19	24	31	36
World total	3632	4457	5438	6494

Source: Data from U.N. Document ESP/P/WP 37.

APPENDIX 2

MEASUREMENT CONVERSION TABLE

	U.S. customary units	Metric equivalent	Metric units	U.S. customary equivalent	Equivalents in same system
LENGTH					
	1 in	= 2.54 cm, 0.0245 m	1 μm	= 0.00003937 in	1 mi = 5280 ft, 1760 yd
	1 ft	= 0.3048 m	1 mm	= 0.03937 in	1 rod = 16.5 ft
	1 yd	= 0.9144 m	1 cm	= 0.3937 in	1 yd = 3 ft
	1 mi	= 1609 m, 1.609 km	1 m	= 39.37 in, 3.281 ft, 1.094 yd	1 ft = 12 in
			1 km	= 0.6214 mi, 1094 yd, 3281 ft	1 km = 1000 m
					1 mm = 0.001 m
					1 cm = 0.01 m
					1 μm = 0.000001 m
AREA					
	1 in^2	= 6.452 cm^2	1 cm^2	= 0.1550 in^2	1 acre = 43,560 ft^2, 4840 yd^2, 160 rod^2
	1 ft^2	= 929 cm^2	1 m^2	= 10.76 ft^2, 1.20 yd^2	1 mi^2 = 640 acres
	1 acre	= 4047 m^2	1 km^2	= 247.1 acre, 0.3861 mi^2	1 km^2 = 1,000,000 m^2
	1 mi^2	= 2.590 km^2, 2,590,000 m^2			1 ft^2 = 144 in^2
					1 yd^2 = 9 ft^2
VOLUME					
	1 in^3	= 16.39 cm^3	1 cm^3	= 0.06102 in^3	1 cm^3 = 0.000001 m^3
	1 ft^3	= 0.02832 m^3	1 m^3	= 35.31 ft^3, 1.308 yd^3	1 km^3 = 1,000,000,000 m^3
	1 yd^3	= 0.7646 m^3	1 km^3	= 0.239 mi^3	1 yd^3 = 27 ft^3
	1 mi^3	= 4.17 km^3			1 ft^3 = 1728 in^3

U.S. customary units	*Metric equivalent*	*Metric units*	*U.S. customary equivalent*	*Equivalents in same system*	
CAPACITY					
Liquid measure					
1 fl oz	= 29.573 ml	1 ml	= 0.0338 fl oz	1 bbl	= 31.5 gal, 42 gal (oil)
1 pt	= 0.473 liter	1 liter	= 2.114 pt, 1.057 qt, 0.264 gal	1 gal	= 4 qt, 231 in^3
1 qt	= 0.946 liter			1 qt	= 2 pt, 57.749 in^3
1 gal	= 3.785 liter			1 pt	= 16 oz, 28.875 in^3
1 bbl	= varies from 117.4 to 159.0 liters, established by law or usage			1 liter	= 1000.027 cm^3
Dry measure					
1 pt	= 0.551 liter	1 ml	= 0.0610 in^3	1 bu	= 4 pk, 2150.42 in^3
1 qt	= 1.101 liter	1 liter	= 1.8162 pt, 0.9081 qt	1 pk	= 8 qt, 537.605 in^3
1 pk	= 8.810 liter			1 qt	= 2 pt, 67.2006 in^3
1 bu	= 35.24 liter			1 pt	= 33.6003 in^3
MASS					
1 oz av	= 28.35 g	1 g	= 0.03527 oz av	1 ton	= 2000 lb av
1 lb av	= 453.59 g, 0.45359 kg	1 kg	= 2.205 lb av	1 lb av	= 16 oz av
1 ton (short)	= 907.18 kg	1 metric ton	= 2204.6 lb av, 1.102 ton (short)	1 metric ton	= 1000 kg
MISCELLANEOUS UNITS					
1 acre-ft	= 1235 m^3				
1 acre-ft/yr	= 3.38 m^3/day				
1 ft^3/sec	= 2450 m^3/day				
1 bgd	= 3,790,000 m^3/day				

ABBREVIATIONS USED IN TABLE

bbl: barrel	in: inch	μm: micron
bgd: billion gallons per day	kg: kilogram	oz av: avoirdupois ounce
bu: bushel	km: kilometer	pk: peck
cm: centimeter	lb av: avoirdupois pound	pt: pint
fl oz: fluid ounce	m: meter	qt: quart
ft: foot	mi: mile	sec: second
g: gram	ml: milliliter	yd: yard
gal: gallon	mm: millimeter	yr: year

Note: Area and volume units are abbreviated by using the exponents 2 or 3 with the proper units; for example, cm^2 denotes square centimeters and km^3 denotes cubic kilometers.

APPENDIX 3

THE CHEMICAL ELEMENTS

The following table names the chemical elements and gives their symbols, atomic numbers, and atomic weights. The atomic weights are from the report of the International Commission on Atomic Weights, published in the *Journal of the American Chemical Society*, August 20, 1958.

Name	*Symbol*	*Atomic number*	*Atomic weight*
Actinium	Ac	89	227.*
Aluminum	Al	13	26.98
Americium	Am	95	243.*
Antimony	Sb	51	121.76
Argon	Ar	18	39.944
Arsenic	As	33	74.92
Astatine	At	85	210.*
Barium	Ba	56	137.36
Berkelium	Bk	97	249.*
Beryllium	Be	4	9.013
Bismuth	Bi	83	208.99
Boron	B	5	10.82
Bromine	Br	35	79.916
Cadmium	Cd	48	112.41
Calcium	Ca	20	40.08
Californium	Cf	98	251.*
Carbon	C	6	12.011
Cerium	Ce	58	140.13
Cesium	Cs	55	132.91
Chlorine	Cl	17	35.457
Chromium	Cr	24	52.01
Cobalt	Co	27	58.94
Copper	Cu	29	63.54
Curium	Cm	96	247.*
Dysprosium	Dy	66	162.51
Einsteinium	Es	99	254.*
Erbium	Er	68	167.27
Europium	Eu	63	152.0
Fermium	Fm	100	253.*
Fluorine	F	9	19.00
Francium	Fr	87	223.*
Gadolinium	Gd	64	157.26
Gallium	Ga	31	69.72
Germanium	Ge	32	72.60
Gold	Au	79	197.0
Hafnium	Hf	72	178.50

Name	Symbol	Atomic number	Atomic weight
Helium	He	2	4.003
Holmium	Ho	67	164.94
Hydrogen	H	1	1.0080
Indium	In	49	114.82
Iodine	I	53	126.91
Iridium	Ir	77	192.2
Iron	Fe	26	55.85
Krypton	Kr	36	83.80
Lanthanum	La	57	138.92
Lead	Pb	82	207.21
Lithium	Li	3	6.940
Lutetium	Lu	71	174.99
Magnesium	Mg	12	24.32
Manganese	Mn	25	54.94
Mendelevium	Md	101	256.*
Mercury	Hg	80	200.61
Molybdenum	Mo	42	95.95
Neodymium	Nd	60	144.27
Neon	Ne	10	20.183
Neptunium	Np	93	237.*
Nickel	Ni	28	58.71
Niobium	Nb	41	92.91
Nitrogen	N	7	14.008
Nobelium	No	102	253.*
Osmium	Os	76	190.2
Oxygen	O	8	16.000
Palladium	Pd	46	106.4
Phosphorus	P	15	30.975
Platinum	Pt	78	195.09
Plutonium	Pu	94	242.*
Polonium	Po	84	210.
Potassium	K	19	39.100
Praseodymium	Pr	59	140.91
Promethium	Pm	61	147.*
Protactinium	Pa	91	231.
Radium	Ra	88	226.05
Radon	Rn	86	222.
Rhenium	Re	75	186.22
Rhodium	Rh	45	102.91
Rubidium	Rb	37	85.48
Ruthenium	Ru	44	101.1
Samarium	Sm	62	150.35
Scandium	Sc	21	44.96
Selenium	Se	34	78.96
Silicon	Si	14	28.09
Silver	Ag	47	107.880
Sodium	Na	11	22.991
Strontium	Sr	38	87.63
Sulfur	S	16	32.066
Tantalum	Ta	73	180.95
Technetium	Tc	43	99.*
Tellurium	Te	52	127.61
Terbium	Tb	65	158.93
Thallium	Tl	81	204.39
Thorium	Th	90	232.05
Thulium	Tm	69	168.94
Tin	Sn	50	118.70
Titanium	Ti	22	47.90
Tungsten	W	74	183.86
Uranium	U	92	238.07
Vanadium	V	23	50.95
Xenon	Xe	54	131.30
Ytterbium	Yb	70	173.04
Yttrium	Y	39	88.91
Zinc	Zn	30	65.38
Zirconium	Zr	40	91.22

*The atomic weights marked with asterisks are those of radioactive elements whose atomic weights depend upon the method of manufacture. The listed isotope may be either the one of longest known half-life or a better-known one.

APPENDIX 4

IDENTIFICATION OF MINERALS

An attempt has been made to keep mineral names in the text to a minimum. A more extensive list of minerals and their properties is appended here as a guide to their field identification, and as a brief, unified glossary of mineral names that may appear in publications about geology—even those intended for the general reader.

Some minerals can be identified only with advanced laboratory equipment; many others may be identifiable by physical properties that can be determined by simple and inexpensive techniques. The most useful of these properties, generally, are those related to appearance, hardness, and density. "Appearance" may be further subdivided according to crystal form, cleavage, fracture, color, streak, and luster.

Crystal Form

As discussed in Chapter 4, minerals tend to solidify as regular, characteristically shaped crystals bounded by smooth planes called crystal faces. When these faces are present, their shapes and the angles between them are useful for identification. However, many minerals occur in shapeless granular forms or as crystals so tiny that the crystal faces are not visible. Crystal faces should be carefully distinguished from cleavage surfaces, which may or may not be parallel to the faces. The distinction is that crystal faces appear only on the unbroken outside of a crystal, whereas cleavage surfaces appear only on broken or cracked fragments of a crystal.

Cleavage

As discussed in Chapter 4, many minerals cleave (break) along smooth cleavage planes determined by the structure of the crystals. Broken fragments of these minerals have characteristic shapes, which aid in identifying the minerals. This is because the spacing and angles between cleavage surfaces are characteristic of individual minerals or groups of them. The mica minerals provide us with a useful example of cleavage: because they have only one cleavage plane and the cleavage surfaces are very close, mica minerals can be split into countless thin flakes or sheets.

Fracture

Many minerals fracture (break) along irregular surfaces, and this characteristic may be as diagnostic as cleavage. Quartz, for example, breaks along characteristic curved surfaces called conchoidal fractures.

Color

Some minerals have a constant color that is diagnostic. Others, including the common minerals calcite and quartz, have a wide range of colors, depending on the minor impurities present. Additionally, the color of many minerals is altered by weathering. Thus, it may be necessary to obtain a freshly broken surface in order to determine color. Both the fact that alteration occurs and that the color changes may be diagnostic characteristics.

Streak

Many materials, such as chalk, leave a streak of their powder when they are rubbed on a hard, slightly rough surface. In mineralogy, the streak of a mineral is obtained by rubbing it against a piece of unglazed porcelain. The streak of a mineral is more constant in color than is the mineral itself; the mineral and its streak may be the same color or the two may be entirely different. The streak may be useful for identification by itself. In addition, the difference between the color of the specimen and the color of the streak may help in identifying such minerals as hematite, which has a black or red color and a characteristic brownish-red streak.

Luster

Luster refers to the appearance of a mineral when ordinary light is reflected from it. Several types of luster may be distinguished by their resemblance to the distinctive lusters of certain materials. Most terms to describe luster are self-explanatory. They include: dull, earthy, metallic, pearly, resinous, and silky. Vitreous luster resembles that of glass; adamantine luster is like that of diamond.

Hardness

The property of hardness is determined by the ability of one material to scratch another. The relative hardness of two minerals may be established by firmly scraping a pointed corner of one across the flat surface of another. If the mineral with the pointed corner is harder, it will leave a scratch or cut on the flat surface. In mineralogy, a standard set of minerals is used to establish a numerical scale of hardness (known as the Mohs scale, after the German mineralogist Friedrich Mohs, who devised it):

1. talc
2. gypsum
3. calcite
4. fluorite
5. apatite
6. orthoclase
7. quartz
8. topaz
9. corundum
10. diamond

Any one of these minerals will scratch all of those with lower numbers. An unknown mineral that is scratched by topaz but itself scratches quartz has, thus, a hardness between 7 and 8 on this scale. The blade of a pocket knife has a hardness of about 5.5, a copper coin, about 3.5, and a thumbnail, about 2.5.

Specific Gravity

The density of a mineral is measured as specific gravity, which is expressed as the ratio of the weight of a substance to that of an equal volume of water.

Specific gravity is defined as:

$$\text{specific gravity} = \frac{(\text{weight in air})}{(\text{weight in air}) - (\text{weight in water})}$$

Each of the terms in parentheses can be easily measured with a laboratory balance. For field purposes, the relative density of a mineral specimen can often be estimated merely by hefting it up and down in the hand.

Other Properties

Minerals have many other properties that may aid in their identification. Magnetite and pyrrhotite, for example, are the only common minerals that are attracted by a magnet. Other minerals, such as calcite, can be dissolved in a weak acid; thus, a few drops of lemon juice can help to distinguish it from some minerals with similar appearance. Other minerals, notably radioactive ones, are commonly identified with special equipment, even in the field.

Table of Minerals and Diagnostic Properties

Mineral	*Form*	*Cleavage*	*Hardness*	*Specific gravity*	*Other properties*
Actinolite. Calcium iron silicate, $Ca_2(Mg,Fe)_5Si_8O_{22}(OH)_2$.	Slender crystals, usually fibrous.	Two good cleavages meeting at angles of 56° and 124°.	5–6	3.0–3.3	Color white to light green. Transparent to translucent. Colorless streak. Vitreous luster.
Albite (sodic plagioclase feldspar). Sodium aluminum silicate, $NaAlSi_3O_8$.	Tabular crystals.	Good in two directions at 93° 34′.	6	2.62	Colorless, white, or gray. transparent to translucent. Streak colorless. Opalescent variety is moonstone.
Amphibole. A group of complex, solid-solution silicates, chiefly of calcium, magnesium, iron, and aluminum. Similar to pyroxene in composition, but containing a little hydroxyl (OH^-) ion. The commonest of the many varieties of amphibole is hornblende.	Long, prismatic, 6-sided crystals; also in fibrous or irregular masses of interlocking crystals and in disseminated grains.	Two good cleavages meeting at angles of 56° and 124°.	5–6	2.9–3.2	Color black to light green; or colorless. Opaque. Highly vitreous luster on cleavage surfaces. Distinguished from pyroxene by the difference in cleavage angle and in crystal form. Amphibole also has much better cleavage and higher luster than pyroxene.
Anhydrite. Anhydrous calcium sulfate, $CaSO_4$.	Crystals rare. Commonly in massive fine aggregates.	Three directions at right angles, forming rectangular blocks.	3–3.5	2.89–2.98	White with various tinges. Transparent to translucent. Streak colorless.
Anorthite (calcic plagioclase feldspar). Calcium aluminum silicate, $Ca(Al_2Si_2O_8)$.	Prismatic crystals, also massive.	Good in two directions at 94° 12′.	6	2.76	Colorless, white, gray, green, red, or yellow. Vitreous to pearly luster. Colorless streak. Striations.
Aragonite. Calcium carbonate, $CaCO_3$.	Needle-shaped crystals with rectangular cross section.	One distinct cleavage.	3.5–4	2.93–2.95	Color usually white. Vitreous luster. Colorless streak. Forms compound (twinned) crystals.

(*continued*)

Table of Minerals and Diagnostic Properties (*continued*)

Mineral	*Form*	*Cleavage*	*Hardness*	*Specific gravity*	*Other properties*
Augite. A ferromagnesian silicate, $Ca(Mg,Fe,Al)(Al,Si_2O_6)$.	Short stubby crystals.	Prismatic along two planes nearly at right angles.	5–6	3.2–3.4	Dark green to black. Translucent only on thin edges. Streak greenish-gray. Vitreous luster.
Azurite. Blue copper carbonate, $Cu_3(CO_3)_2(OH)_2$.	Complex crystals, sometimes in radiating groups.	Fibrous.	4	3.77	Intense azure blue. Opaque. Vitreous to dull earthy luster. Streak pale blue. Effervesces with hydrochloric acid.
Bauxite. Hydrous aluminum oxides, indefinite composition.	Rounded grains, earthy masses.	Uneven.	1–3	2–3	Yellow, brown, gray, white. Opaque. Dull to earthy luster. Colorless streak.
Beryl (emerald). Beryllium aluminum silicate, $Be_3Al_2(SiO_3)_6$.	Hexagonal prisms.	Conchoidal to uneven fracture.	7.5–8	2.63–2.80	Many colors. Chiefly green. Transparent to subtranslucent. Vitreous or resinous luster. Streak white.
Biotite (black mica). A complex silicate of potassium, iron, aluminum, and magnesium, variable in composition but approximately $K(Mg,Fe)_3AlSi_3O_{10}(OH)_2$.	Thin, scalelike crystals, commonly 6-sided, and in scaly, foliated masses.	Perfect in one direction, yielding thin, flexible scales.	2.5–3	2.7–3.2	Black to dark brown. Translucent to opaque. Pearly to vitreous luster. White to greenish streak.
Bornite (peacock ore). Copper iron sulfide, Cu_5FeS_4.	Some cubic crystals. Usually massive.	Uneven.	3	5.06–5.08	Brownish-bronze on fresh fracture. Tarnishes to variegated purple, blue, and black. Metallic luster. Opaque. Streak grayish black.

Calcite. Calcium carbonate, $CaCO_3$.	"Dog-tooth" or flat crystals showing excellent cleavages; granular, showing cleavages; also masses too fine-grained to show cleavages distinctly.	Three highly perfect cleavages at oblique angles, yielding rhomb-shaped fragments.	3	2.72	Commonly colorless, white, or yellow, but may be any color owing to impurities. Transparent to opaque, transparent varieties showing strong, double refraction (e.g., 1 dot seen through calcite appears as 2). Vitreous to dull luster. Effervesces readily in cold dilute hydrochloric acid.
Carnotite. Potassium uranyl vanadate, $K_2(UO_2)_2(VO_4)_2 \cdot 8H_2O$.	Earthy powder.	Not apparent.	Very soft	4.1, approx.	Brilliant canary-yellow color. An ore of vanadium and uranium.
Cassiterite. Tin dioxide, SnO_2.	Well-formed, 4-sided prismatic crystals terminated by pyramids; 2 crystals may be intergrown to form knee-shaped twins; also as rounded pebbles in stream gravels.	None; curved to irregular fracture.	6–7	7	Brown to black. Adamantine luster. White to pale-yellow streak. Chief ore of tin.
Chalcedony (cryptocrystalline quartz). Silicon dioxide, SiO_2.	Crystals too fine to be visible; some are conspicuously banded, or in masses.	None; conchoidal fracture.	6–6.5	2.6	Color commonly white or light gray, but may be any color owing to impurities. Distinguished from opal by dull or clouded luster.
Chalcocite (copper glance). Cuprous sulfide, Cu_2S.	Massive, rarely in crystals of roughly hexagonal shape. May be tarnished and stained to blue and green.	Indistinct, rarely observed.	2.5–3	5.5–5.8	Blackish-gray to steel gray, commonly tarnished to green or blue. Dark gray streak. Very heavy. Metallic luster. An important ore of copper.
Chalcopyrite Copper iron sulfide, $CuFeS_2$.	Compact or disseminated masses, rarely in wedge-shaped crystals.	None; uneven fracture.	3.5–4	4.1–4.3	Brassy to golden-yellow. Tarnishes to blue, purple, and reddish iridescent films. Greenish-black streak. Distinguished from pyrite by deeper yellow color and softness. A common copper ore.

(*continued*)

Table of Minerals and Diagnostic Properties (*continued*)

Mineral	*Form*	*Cleavage*	*Hardness*	*Specific gravity*	*Other properties*
Chlorite. A complex group of hydrous magnesium aluminum silicates containing iron and other elements in small amounts.	Commonly in foliated or scaly masses; may occur in tabular, 6-sided crystals resembling mica.	One perfect cleavage, yielding thin, flexible, but inelastic scales.	1–2.5	2.6–3.0	Grass-green to blackish-green color. Translucent to opaque. Greenish streak. Vitreous luster. Very easily disintegrated.
Chromite. Iron chromium oxide, $FeCr_2O_4$.	Massive.	Uneven fracture.	5.5	4.6	Black to brownish-black. Subtranslucent. Metallic luster. Dark brown streak.
Copper (native copper). An element, Cu.	Twisted and distorted leaves and wirelike forms; flattened or rounded grains.	None.	2.5–3	8.8–8.9	Characteristic copper color, but commonly stained green. Highly ductile and malleable. Excellent conductor of heat and electricity. Very heavy.
Corundum (ruby, sapphire). Aluminum oxide, Al_2O_3.	Barrel-shaped crystals.	Basal or rhombohedral parting.	9	4.02	Many colors, depending on impurities. Transparent to translucent. Adamantine to vitreous luster. Colorless streak.
Diamond. High-density form of the element carbon, C.	Octahedral crystals.	Octahedral.	10	3.5	Many colors, depending on impurities. Transparent. Adamantine luster. Colorless streak.
Dolomite. Calcium magnesium carbonate, $CaMg(CO_3)_2$.	Rhomb-faced crystals showing good cleavage; also in fine-grained masses.	Three perfect cleavages at oblique angles, as in calcite.	3.5–4	2.9	Variable in color, but commonly white. Transparent to translucent. Vitreous to pearly luster. Powder will effervesce slowly in cold dilute hydrochloric acid, but coarse crystals will not.

Epidote. A complex calcium iron aluminum silicate, $Ca_2(Al,Fe)_3(SiO_4)_3(OH)$.	Short, 6-sided crystals or radiate crystal groups, and in granular or compact masses.	One good cleavage; in some specimens, a second poorer cleavage at an angle of 115° with the first.	6–7	3.4	Characteristic yellowish-green (pistachio-green) color. Vitreous luster.
Fluorite. Calcium fluoride, CaF_2.	Cubic crystals, also massive.	Four good cleavages parallel to faces of an octahedron.	4	3.18	Many colors. Transparent to translucent. Vitreous luster. Colorless streak.
Galena. Lead sulfide, PbS.	Cubic crystals common, but mostly in coarse to fine granular masses.	Perfect cubic cleavage (three cleavages mutually at right angles).	2.5	7.3–7.6	Silvery-gray color. Metallic luster. Silvery-gray to grayish-black streak. Chief ore of lead.
Garnet. A group of solid solution silicates having variable proportions of different metallic elements. The most common variety contains calcium, iron, and aluminum, but garnets may contain many other elements.	Commonly in well-formed equidimensional crystals, but also massive and granular.	None; conchoidal or uneven fracture.	6.5–7.5	3.4–4.3	Commonly red, brown, or yellow, but may be other colors. Translucent to opaque. Resinous to vitreous luster.
Gold. An element, Au.	Massive or in thin plates; also in flattened grains or scales; distinct crystals very rare.	None.	2.5–3	15.6–19.3	Characteristic gold-yellow color and streak. Rarely in crystals. Extremely heavy. Very malleable and ductile. Variable density reflects impurities.
Graphite. An element, C.	Foliated or scaly masses.	Good in one direction.	1–2	2.3	Black to steel-gray. Opaque. Metallic or earthy luster. Black streak. The "lead" of pencils.

(continued)

Table of Minerals and Diagnostic Properties (*continued*)

Mineral	*Form*	*Cleavage*	*Hardness*	*Specific gravity*	*Other properties*
Gypsum. Hydrous calcium sulfate, $CaSO_4 \cdot 2H_2O$.	Tabular crystals, and cleavable, granular, fibrous, or earthy masses.	One perfect cleavage, yielding thin, flexible folia; 2 other much less perfect cleavages.	2	2.2–2.4	Colorless or white, but may be other colors when impure. Transparent to opaque. Luster vitreous to pearly or silky. Cleavage flakes flexible but not elastic like those of mica.
Halite (rock salt). Sodium chloride, NaCl.	Cubic crystals, granular masses.	Excellent cubic cleavage (3 cleavages mutually at right angles).	2–2.5	2.1	Colorless to white, but of other colors when impure. The color may be unevenly distributed through the crystal. Transparent to translucent. Vitreous luster. Salty taste.
Hematite. Ferric iron oxide, Fe_2O_3.	Highly varied, compact, granular, fibrous, or earthy, micaceous; rarely in well-formed crystals.	None, but fibrous or micaceous specimens may show parting resembling cleavage; splintery to uneven fracture.	5–6.5	4.9–5.3	Steel-gray, reddish-brown, red, or iron-black in color. Metallic to earthy luster. Characteristic brownish-red streak. Hematite is the most important iron ore.
Hornblende. A complex ferromagnesian silicate.	Long, prismatic crystals. Fibrous masses.	Perfect prismatic at 56° and 124°.	5–6	3.2	Dark green to black. Translucent on thin edges. Vitreous luster; fibrous variety, silky. Colorless streak.
Kaolinite. Hydrous aluminum silicate, $H_4Al_2Si_2O_9$. Representative of the 3 or 4 similar minerals common in clays.	Commonly in soft, compact, earthy masses.	Crystals always so small that cleavage is invisible without microscope.	1–2	2.2–2.6	White color, but may be stained by impurities. Greasy feel. Adheres to the tongue, and becomes plastic when moistened. "Claylike" odor when breathed upon.

Kyanite (disthene). Aluminum silicate, Al_2SiO_5.	Long, bladelike crystals.	One perfect, and one poor cleavage, both parallel to length of crystals; and a crude parting across the crystals.	4–7	3.5–3.7	Colorless, white, or a distinctive pale blue color. Can be scratched by knife parallel to cleavage, but is harder than steel across cleavage.
"Limonite." Microscopic study shows that the material called limonite is not a single mineral. Most "limonite" is a very finely crystalline variety of the mineral **Goethite** containing absorbed water. Hydrous ferric oxide with minor amounts of other elements, roughly $Fe_2O_3 \cdot H_2O$.	Compact or earthy masses; may show radially fibrous structure.	None; conchoidal or earthy fracture.	1–5.5	3.4–4.0	Yellow, brown, or black in color. Dull earthy luster, which distinguishes it from hematite. Characteristic yellow-brown streak. A common iron ore.
Magnetite. A combination of ferric and ferrous oxides, Fe_3O_4.	Well-formed, 8-faced crystals, more commonly in compact aggregates, disseminated grains, or loose grains in sand.	None; conchoidal or uneven fractures; may show a rough parting resembling cleavage.	5.5–6.5	5.0–5.2	Black. Opaque. Metallic to submetallic luster. Black streak. Strongly attracted by a magnet. Magnetite is an important iron ore.
Muscovite. (white mica; isinglass). A complex potassium aluminum silicate, $KAl_3Si_3O_{10}(OH)_2$ approximately, but varying.	Thin, scalelike crystals and scaly, foliated aggregates.	Perfect in one direction, yielding very thin, transparent, flexible scales.	2–3	2.8–3.1	Colorless, but may be gray, green, or light brown in thick pieces. Transparent to translucent. Pearly to vitreous luster.
Olivine. Magnesium iron silicate, $(Fe,Mg)_2SiO_4$.	Commonly in small, glassy grains and granular aggregates.	So poor that it is rarely seen; conchoidal fracture.	6.5–7	3.2–3.6	Various shades of green, also yellowish; opalescent and brownish when slightly altered. Transparent to translucent. Vitreous luster. Resembles quartz in small fragments but has characteristic greenish color, unless altered.

(continued)

Table of Minerals and Diagnostic Properties (*continued*)

Mineral	*Form*	*Cleavage*	*Hardness*	*Specific gravity*	*Other properties*
Opal. Hydrous silica, with 3% to 12% water, SiO_2nH_2O. Because it does not have a definite geometric internal structure, it is a mineraloid, not a true mineral.	Amorphous. Commonly in veins or irregular masses showing a banded structure. May be earthy.	None; conchoidal fracture.	5.0–6.5	2.1–2.3	Color highly variable, often in wavy or banded patterns. Translucent or opaque. Somewhat waxy luster.
Orthoclase (potassium feldspar). Potassium aluminum silicate, $KAlSi_3O_8$.	Prismatic crystals and formless grains.	Good in two directions at or near 90°.	6	2.57	White, gray, pink. Translucent to opaque. Vitreous luster. White streak.
Potassium feldspar (orthoclase, microcline, and sanidine). Potassium aluminum silicate, $KAlSi_3O_8$.	Boxlike crystals, massive.	One perfect and 1 good cleavage, making an angle of 90°.	6	2.5–2.6	Commonly white, gray, pink, or pale yellow; rarely colorless. Commonly opaque but may be transparent in volcanic rocks. Vitreous. Pearly luster on better cleavage. Distinguished from plagioclase by absence of striations.
Plagioclase feldspar (soda-lime feldspars). A solid solution group of sodium calcium aluminum silicates, $NaAlSi_3O_8$ to $CaAl_2Si_2O_8$.	In well-formed crystals and in cleavable or granular masses.	Two good cleavages nearly at right angles (86°). May be poor in some volcanic rocks.	6–6.5	2.6–2.7	Commonly white or gray, but may be other colors. Some gray varieties show a play of colors called opalescence. Transparent in some volcanic rocks. Vitreous to pearly luster. Distinguished from orthoclase by the presence on the better cleavage surface of fine parallel lines or striations.

Pyroxene. A solid-solution group of silicates, chiefly silicates of calcium, magnesium and iron, with varying amounts of other elements. The commonest varieties are augite and hypersthene.	Commonly in short, 8-sided, prismatic crystals; the angle between alternate faces nearly 90°. Also as compact masses and disseminated grains.	Two cleavages at nearly 90°. Cleavage not always well developed; in some specimens, conchoidal or uneven fracture.	5–6	3.2–3.6	Commonly greenish to black in color. Vitreous to dull luster. Gray-green streak. Distinguished from amphibole by the right-angle cleavage, 8-sided crystals, and by the fact that most crystals are short and stout, rather than long, thin prisms, as in amphibole.
Pyrite ("fool's gold"). Iron sulfide, FeS_2.	Well-formed crystals, commonly cubic, with striated faces; also granular masses.	None; uneven fracture.	6–6.5	4.9–5.2	Pale brassy-yellow color; may tarnish brown. Opaque. Metallic luster. Greenish-black or brownish-black streak. Brittle. Not a source of iron, but used in the manufacture of sulfuric acid. Commonly associated with ores of several different metals.
Quartz (rock crystal). Silicon dioxide, SiO_2.	Six-sided prismatic crystals, terminated by 6-sided triangular faces; also massive.	None or very poor; conchoidal fracture.	7	2.65	Commonly colorless or white, but may be yellow, pink, amethyst, smoky-translucent brown, or even black. Transparent to opaque. Vitreous to greasy luster.
Serpentine. A complex group of hydrous magnesium silicates, roughly $H_4Mg_3Si_2O_9$.	Foliated or fibrous, usually massive.	Commonly only one cleavage, but may be in prisms. Fracture usually conchoidal or splintery.	2.5–4	2.5–2.65	Feels smooth, or even greasy. Color leek-green to blackish-green but varying to brownish-red, yellow, etc. Luster resinous to greasy. Translucent to opaque. Streak white.

(*continued*)

Table of Minerals and Diagnostic Properties (*continued*)

Mineral	*Form*	*Cleavage*	*Hardness*	*Specific gravity*	*Other properties*
Sillimanite (fibrolite). Aluminum silicate, Al_2SiO_5.	In long slender crystals, or fibrous.	Parallel to length, but rarely noticeable.	6–7	3.2	Gray, white, greenish-gray, or colorless. Slender prismatic crystals or in a felted mass of fibers. Streak white or colorless.
Silver. An element, Ag.	In flattened grains and scales; rarely in wirelike forms, or in irregular needle-like crystals.	None.	2.5–3	10–11	Color and streak are silvery-white, but may be tarnished gray or black. Highly ductile and malleable. Very heavy. Mirrorlike metallic luster on untarnished surfaces.
Sphalerite. Zinc sulfide (nearly always containing a little iron), ZnS.	Crystals common, but chiefly in fine to coarse-granular masses.	Six highly perfect cleavages at 60° to one another.	3.5–4	3.9–4.2	Color ranges from white to black but is commonly yellowish-brown. Trans-lucent to opaque. Resinous to adamantine luster. Streak white, pale yellow or brown. Most important ore of zinc.
Staurolite. Iron aluminum silicate, $Fe(OH)_2(Al_2SiO_5)_2$.	Stubby prismatic crystals, and in cross-shaped twins.	Poor and inconspicuous.	7–7.5	3.7	Red-brown or yellowish-brown to brownish-black. Generally in well-shaped crystals larger than the minerals of the matrix enclosing them.
Talc. Hydrous magnesium silicate, $Mg_3(OH)_2Si_4O_{10}$.	In tiny foliated scales and soft compact masses.	One perfect cleavage, forming thin scales and shreds.	1	2.8	White or silvery-white to apple-green. Very soft, with a greasy feel. Pearly luster on cleavage surfaces.

Topaz. Aluminum fluorosilicate, $Al_2SiO_4(F,OH)_2$.	Prismatic crystals.	Good in one direction.	8	3.4–3.6	Many colors, yellow common. Transparent to translucent. Vitreous luster. Colorless streak.
Tourmaline. A complex silicate of boron, aluminum, and other elements.	Crystals commonly with curved triangular cross section.	Poor.	7–7.5	3–3.25	Many colors, green common. Translucent. Vitreous to resinous luster. Colorless streak.
Uranite (pitchblende). Uranium oxide, UO_2 to U_3O_8.	Regular 8-sided or cubic crystals; massive.	None, fracture uneven to conchoidal.	5–6	6.5–10	Color black to brownish-black. Luster submetallic, pitchlike, or dull. Chief mineral source of uranium.
Zircon. Zirconium neosilicate, $Zr(SiO_4)$.	Tetragonal pyramid and dipyramid.	None.	7.5	4.68	Many colors. Translucent, some transparent. Adamantine luster. Colorless streak.

Source: After J. Gilluly, A. C. Waters, and A. O. Woodford, 1968, *Principles of Geology* (3rd ed.), W. H. Freeman and Company, San Francisco; and C. Hurlbut, Jr., 1961, *Dana's Manual of Mineralogy* (17th ed.), John Wiley & Sons, New York.

APPENDIX 5

IDENTIFICATION OF ROCKS

Rock names in the text have been kept to a minimum. A more comprehensive classification is presented here as a guide to the field identification of rocks, and as a brief, unified glossary of rock names that may appear in publications intended for the general reader.

Far more elaborate classifications of rocks exist. Indeed, the number of possible classifications is infinite, because rocks grade into one another without regard for the desire of geologists to distinguish them. Even so, a large fraction of rocks lend themselves to identification and classification according to the characteristics that are outlined below.

The basic procedure in field identification of a rock consists in answering three questions:

1. Is the rock igneous, sedimentary or metamorphic?
2. Of what minerals is it composed?
3. What is its texture?

For a guide to the first and second questions, the student should refer to Chapter 4 as well as the material in this appendix.

Igneous Rocks

TEXTURE

The texture of an igneous rock can be described in terms of the degree of crystallinity, the size of crystals, the variation in mineral size, and the presence and size of fragments of glass and rock. These textures reflect the origin of the rocks.

If a mixture of molten silicates is chilled quickly, it forms glass. If it cools slowly, it forms larger and larger crystals until it is solid. Fast cooling tends to occur at the surface, slow cooling at depth below the surface. A mixture of mineral sizes generally indicates that some minerals had time to grow to large size in a magma, which then moved up to a cooler, near-surface environment. There, the remainder of the melt chilled rapidly and formed glass or small mineral grains. Such textures result from cooling of a relatively continuous liquid; but if the liquid froths or is blasted apart by contained gas, the molten fragments more or less cool in the air and more or less stick together when they fall to the ground.

Granular texture Composed of interlocking crystals that are of relatively uniform size and large enough to be identified without magnification. The average mineral size in different rocks may vary from 0.5 millimeter to more than 1 centimeter in diameter. Minerals in common rocks, such as granite, average 3 to 5 millimeters in diameter. Large crystals require time to grow slowly.

Aphanitic texture Composed largely of tiny crystals, less than 0.5 millimeters in diameter, that cannot individually be identified. The crystals give a rock a stony luster, in contrast to the vitreous luster of glassy rocks. Most lava flows have aphanitic texture. In the last stage of flow, the minerals may be aligned in streaks along flow lines. Thus, they may bear some resemblance to layered sedimentary rocks.

Glassy texture Composed largely of massive or streaky volcanic glass. Small minerals may be scattered through the glass, and bubbles of gas may be entrapped in it. If the bubbles predominate, the texture is pumiceous and the rock is called pumice.

Porphyritic texture Composed of two widely different sizes of minerals or glass. The large crystals are called phenocrysts and the remainder is the groundmass, which may be granular, aphanitic, or glassy. A complete description of the texture of a porphyritic rock includes the texture of the groundmass. An example is "porphyritic-aphanitic texture."

Pyroclastic texture Composed of slivers of volcanic glass, minerals, and broken fragments of volcanic rock. The glassy slivers and fragments of pumice may weather rapidly to clay. Pyroclastic rocks may be further subdivided according to the size of the fragments: volcanic tuff includes fragments no larger than 4 millimeters in diameter; volcanic breccia includes larger fragments.

TABLE OF IGNEOUS ROCKS

	Predominant minerals			
Textures	*Feldspar and quartz*	*Feldspar predominates (no quartz)*	*Ferromagnesian minerals and feldspar (no quartz)*	*Ferromagnesian minerals (no quartz or feldspar)*
Granular	**Granite** (potassium feldspar predominates) and **Granodiorite** (plagioclase feldspar predominates)	**Diorite**	**Gabbro** **Dolerite** or **Diabase** (if fine grained)	**Peridotite** (with both olivine and pyroxene) **Pyroxenite** (with pyroxene only) **Serpentine** (with altered olivine and pyroxene)
Aphanitic (generally porphyritic-aphanitic)	**Rhyolite** and **Dacite**	**Andesite**	**Basalt**	[Rocks of the texture and composition represented by this part of the table are rare.]
Glassy	**Obsidian** (if massive glass) **Pumice** (if a glass froth)		**Basalt glass**	
Pyroclastic	**Volcanic tuff** (fragments up to 4 mm in diameter) **Volcanic breccia** (fragments more than 4 mm in diameter)			

Decreasing silica content →

Decreasing grain size ↓

Granite. Granular texture. Generally light colored. Feldspar and quartz are most abundant minerals. Biotite or hornblende, or both, present in most granite. The name granite is generally reserved for rocks of these characteristics that contain potassium feldspar as the chief mineral. *Granodiorite* is the term for similar rocks in which plagioclase predominates. Granodiorite can often be distinguished from granite by the fine striations on cleavage planes of plagioclase. Some granites are metamorphic rather than igneous rocks.

Diorite. Granular texture. Composed of plagioclase and lesser amounts of ferromagnesian minerals, including hornblende, biotite, and pyroxene.

Gabbro. Granular texture. Generally dark color. Plagioclase and pyroxene dominant. Small amounts of other ferromagnesian minerals. Very fine-grained gabbro, called *dolerite* or *diabase*, is common in dikes and thin sills.

Peridotite. Granular texture. Dark color. Olivine a dominant mineral. Alters to the rock *serpentine*.

Pyroxenite. Granular texture. Dark color. Pyroxene a dominant mineral. Alters to *serpentine*.

Rhyolite. Aphanitic equivalent of granite. Commonly with phenocrysts of quartz and potassium feldspar. Color variable, generally light. Flow lines or streaks common.

Dacite. Aphanitic equivalent of granodiorite. Commonly with crystals of quartz and plagioclase.

Andesite. Aphanitic equivalent of diorite. Generally porphyritic with plagioclase phenocrysts. Flow lines. Color generally dark gray or greenish-gray.

Basalt. Aphanitic equivalent of gabbro. Few phenocrysts. Black to medium gray. The most abundant rock on the surface of the earth.

Basalt glass. Jet-black natural glass that is hardly transparent, even on thin edges.

Obsidian. Black, red, or brown natural glass that is almost transparent in thin pieces. Vitreous luster. Breaks with a curved fracture. *Pitchstone* is obsidian with a dull luster like pitch.

Pumice. Obsidian froth that will float in water.

Volcanic tuff. Pyroclastic rock composed of fragments less than 4 millimeters in diameter. Grades into sedimentary rock. If the fragments were hot enough to fuse together, the rock is an *ignimbrite* or *welded tuff*.

Volcanic breccia. Composed predominantly of fragments more than 4 millimeters in diameter. May also contain spatterings of lava that have cooled in the air as large bombs or smaller lapilli. May be deposited by lahars as mud flows.

Sedimentary Rocks

TEXTURE

The texture of a sedimentary rock is largely determined by the size and nature of the constituent particles and how they are bound together.

Clastic texture Composed of broken and worn fragments of older minerals, rocks, or organic solids that have been cemented together.

Organic texture Composed of cemented organic debris (shells, plant remains, and so forth) in which the individual fragments are preserved so well that organic features are conspicuous or dominant.

Crystalline texture Composed of interlocking crystals precipitated from relatively cool solutions compared to magma.

TABLES OF SEDIMENTARY ROCKS

Clastic rocks

Consolidated rock	*Original unconsolidated debris*	*Chief mineral or rock components*	*Diameter of fragments*
Conglomerate	Gravel (rounded pebbles)	Quartz, and rock fragments	More than 2 mm
Breccia	Rubble (angular fragments)	Rock fragments,	More than 2 mm
Sandstone	Sand		2 to $\frac{1}{16}$ mm
Quartz sandstone	Quartz-rich sand	Quartz	
Arkose	Feldspar-rich sand	Quartz and feldspar	
Graywacke	"Dirty sand," with clay and rock fragments	Quartz, feldspar, clay, rock fragments, volcanic debris	
Shale	Mud, clay, and silt	Clay minerals, quartz	Less than $\frac{1}{16}$ mm
Clastic limestone	Broken and rounded shells and calcite grains	Calcite	Variable

Organic rocks

Consolidated rock	*Original nature of material*	*Chief mineral or rock components*	*Chemical composition of dominant material*
Limestone	Shells; chemical and organic precipitates	Calcite	$CaCO_3$
Dolomite	Limestone, or unconsolidated calcareous ooze, altered by solutions	Dolomite	$CaMg(CO_3)_2$
Peat and **Coal**	Plant fragments	Organic materials	C, plus compounds of C, H, O
Chert	Siliceous shells and chemical precipitates	Opal, chalcedony	SiO_2 and SiO_2nH_2O

Chemical rocks

Consolidated rock	*Source*	*Chief minerals*	*Chemical composition of dominant material*
Evaporates, or **Salt deposits**	Evaporation residues from the ocean or saline lakes	Halite, gypsum, anhydrite	Varied, chiefly NaCl and $CaSO_4 \cdot 2H_2O$

Conglomerate. Cemented gravel with rounded pebbles. Generally with a wide range of particle sizes.

Breccia. Like conglomerate, only with angular pebbles. Grades into volcanic or fault breccia.

Sandstone. Cemented sand. Sand is defined as consisting of particles from $\frac{1}{16}$ millimeter to 2 millimeters in diameter. Grades into shale or conglomerate.

Quartz sandstone. A sandstone composed mainly of grains of quartz such as would result if a typical beach sand were cemented. Grains generally rounded. Well sorted.

Arkose. Sandstone with abundant feldspar as well as quartz. Grains generally angular. Moderately sorted.

Graywacke. "Dirty" sandstone. Poorly sorted. Contains clay and rock fragments. Grains generally angular.

Shale. Hardened or cemented mud. Mud, in the geological sense, is composed of very small mineral and organic particles. Clays predominate; micas, quartz, and other minerals often present. Shale splits easily along closely spaced planes parallel to the bedding planes, or nearly so. Some cemented muds show little layering, and split into angular blocks. They are *mudstones.* Mudstones with grains in the silt size ($\frac{1}{256}$ millimeter to $\frac{1}{16}$ millimeter in diameter) are *siltstones.* Black shales may contain brass-colored crystals of the iron sulfide called pyrite.

Limestone. A common sedimentary rock composed almost wholly of calcium carbonate. The chief mineral is calcite, although aragonite (see Appendix 4) may occur.

Organic limestones. Rocks composed of fragments of calcareous organisms. Examples are *coral limestone* and *foraminiferal limestone. Coquina* is limestone composed of the coarse shells of mollusks. *Chalk* is composed mainly of the fragmented skeletons of coccoliths.

Clastic limestones. Cemented sands composed of grains of calcite worn largely from shells of organisms.

Chemical limestone. Limestone deposited directly from solution in shallow, warm seas, or in hot springs (where it is called *travertine*).

Dolomite. Resembles and grades into limestone but consists predominately of the mineral dolomite, which is calcium magnesium carbonate. Most dolomite is formed by alteration of limestone.

Chert. A very fine-grained, hard rock, rich in silica and commonly containing remains of the siliceous skeletons of organisms. Occurs abundantly as nodules shaped like potatoes or pillows within shale or limestone. Dark-colored chert is *flint. Petrified wood* is buried wood replaced, molecule by molecule, by silica-bearing ground-water solutions.

Coal. An organic rock consisting of fragments or remains of plants that have been fossilized by the heat and pressure accompanying burial and deformation. Black or dark brown. Low density.

Salt deposits or evaporites. Layers of rock formed by chemical precipitation from concentrated brines. *Rock gypsum,* composed of the mineral gypsum ($CaSO_4 \cdot 2H_2O$) or anhydrite ($CaSO_4$), is abundant. *Rock salt,* or halite ($NaCl$), is less abundant but common.

Metamorphic Rocks

TEXTURE

Because metamorphic rocks grade into igneous and sedimentary ones, they may be impossible to distinguish without laboratory tests. Texture is often a help, however, not only in identifying metamorphic rocks as such, but also in distinguishing among them.

Metamorphic textures may be divided, on the basis of crystal size and orientation, into two groups. In foliated textures, such flat, platy minerals as mica are aligned nearly parallel to each other. A foliated rock splits easily in the plane of the flat sides of the minerals. Rocks with unfoliated textures are composed of equidimensional or randomly oriented platy minerals. Such rocks break into angular fragments.

Gneissose texture Coarse mineral grains. Coarsely foliated with folia one millimeter to several centimeters apart. Folia straight to crenulated.

Schistose texture Minerals visible. Finely foliated. Rock splits easily along thin, parallel foliations. Equidimensional minerals, such as garnet, may be present, in addition to abundant platy or rodlike minerals.

Slaty texture Very fine foliation. Easy splitting along nearly perfect, parallel planes that correspond to the orientation of platy minerals.

Granoblastic texture Roughly similar to granular texture in igneous rocks. Generally unfoliated. Interpenetrating, equidimensional mineral grains. Largely feldspar, quartz, garnet, and pyroxene.

Hornfelsic texture Unfoliated. Mineral grains mostly too small to be visible without magnification. Rock breaks into sharply angular pieces with curved sides.

TABLES OF METAMORPHIC ROCKS

Foliated rocks

Name	*Commonly derived from*	*Texture*	*Chief minerals*
Migmatite	Mixtures of igneous and metamorphic rocks	Coarsely banded, highly variable	Feldspar, amphibole, quartz, biotite
Gneiss	Granite, shale, diorite, mica schist, rhyolite, etc.	Gneissose	Feldspar, quartz, mica, amphibole, garnet, etc.
Mica schist	Shale, tuff, rhyolite	Schistose	Muscovite, quartz, biotite
Chlorite schist	Basalt, andesite, tuff	Schistose to slaty	Chlorite, plagioclase, epidote
Amphibole schist	Basalt, andesite, gabbro, tuff	Schistose	Amphibole, plagioclase
Slate and **Phyllite**	Shale, tuff	Slaty	Mica, quartz

Unfoliated or faintly foliated rocks

Name	*Commonly derived from*	*Texture*	*Chief minerals*
Hornfels	Any fine-grained rock	Hornfelsic	Highly variable
Quartzite	Sandstone	Granoblastic, fine-grained	Quartz
Marble	Limestone, dolomite	Granoblastic	Calcite, magnesium and calcium silicates
Tactite	Limestone or dolomite plus magmatic emanations	Granoblastic, but coarse and variable	Varied; chiefly silicates of iron, calcium, and magnesium, such as garnet, epidote, pyroxene, amphibole
Amphibolite	Basalt, gabbro, tuff	Granoblastic	Hornblende and plagioclase, minor garnet and quartz
Granulite	Shale, graywacke, or igneous rocks	Granoblastic	Feldspar, pyroxene, garnet, kyanite, and other silicates

Migmatite. Small-scale mixtures of igneous and metamorphic rocks characterized by rough bands and veins. Most contain abundant feldspar and quartz, but mineral composition is highly variable.

Gneiss. Coarse-grained gneissose rock. Feldspar abundant. Quartz, amphibole, garnet, and mica common. Derived from many types of rocks.

Mica schist. Schistose rock composed mainly of biotite, muscovite, and quartz. Derived mostly from shale.

Chlorite schist or *greenschist.* Green, very fine-grained rock. Schistose to slaty texture. Composed of chlorite, plagioclase, and epidote. Derived mainly from basalt and andesite. Remnants of original volcanic structures may remain.

Amphibole schist. Schistose texture. Composed chiefly of amphibole and plagioclase. Derived mainly from basalt, gabbro, and chlorite schist.

Slate. Very fine-grained, slaty texture. Dull on cleavage surfaces. Mostly derived from shale. Remnants of original sedimentary structures may be preserved.

Phyllite. Like slate, except slightly less fine-grained, and shiny on cleavage surfaces.

Tactite. Granoblastic, but variable texture, variable grain size, and variable mineral composition. Generally contains amphibole, pyroxene, garnet, and epidote. Tactite is a broad term for rocks that form where calcareous sedimentary rock is intruded by granite or granodiorite. Metallic ores may be associated with tactite.

Amphibolite. Granoblastic texture. Consists mainly of plagioclase and amphibole. Formed from various rocks.

Granulite. Granoblastic texture. Medium-grained to coarse-grained rock consisting mainly of feldspar, pyroxene, and garnet. Formed by intense metamorphism of various rocks.

Hornfels. A broad name for hard, unfoliated, very fine-grained rock that breaks into angular pieces. Variable mineral composition. Formed by recrystallization of rocks around an igneous intrusion.

Quartzite. Granoblastic textures. Very hard rock consisting mainly of interlocking grains of quartz. Breaks across the grains, unlike most sandstones, which break around them. Formed by metamorphism of quartz sandstone.

Marble. Granoblastic. Fine-grained to coarse-grained. Composed mainly of calcite or dolomite, or both. Formed by the metamorphism of limestone and dolomite.

APPENDIX 6

MAPS AND CHARTS

Map Projections

A map is a representation on a plane surface of all or part of the curved surface of the earth. The surface of a sphere cannot be peeled and spread out flat without some distortion occurring. Consequently, as the following figure shows, it is impossible to construct a map that accurately portrays both angles and distances on the earth.

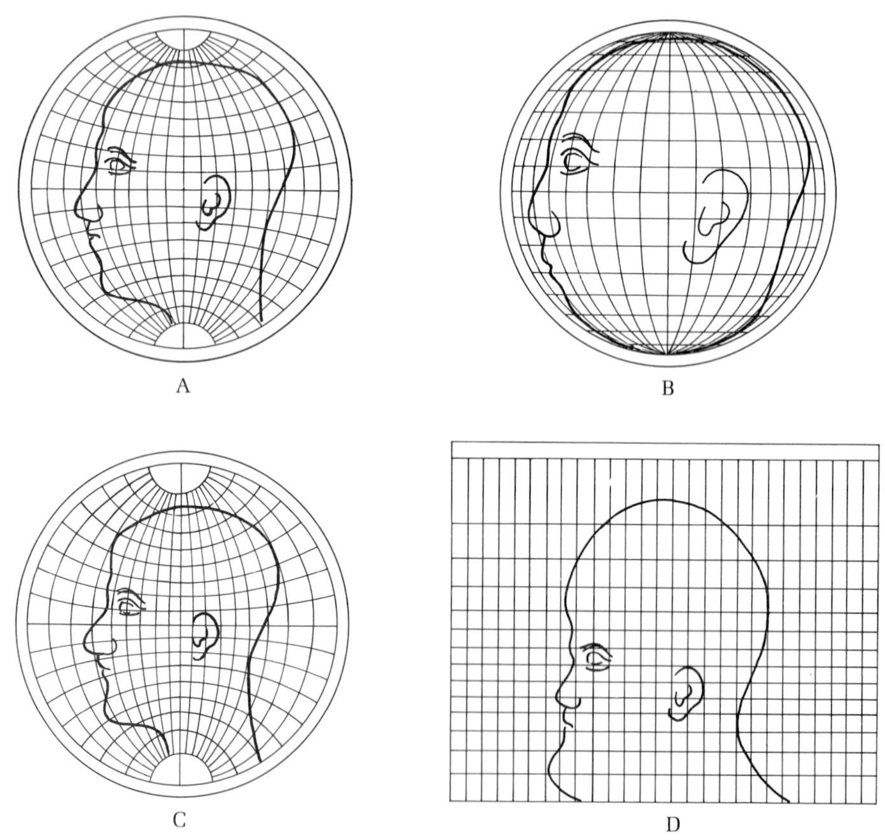

The distortion and exaggerations inherent in various systems of projection: *A*, a man's head is drawn on a globular projection of a hemisphere with lines of latitude and longitude; *B*, the same head on an orthographic projection; *C*, the same head on a stereographic projection; *D*, the same head on a Mercator projection. [From C. H. Deetz and O. S. Adams, 1945, *Elements of Map Projection* (5th ed.), U.S. Coast and Geodetic Survey Spec. Publ. 68.]

Cartographers prepare maps by transforming the spherical surface to a plane through the device of map projection. This device can be illustrated by optical analogies. Consider a small, bright point of light shining through (or from within) a transparent sphere marked with the outlines of the continents, so that those outlines cast shadows on a screen. The projected shadows are a map. Many types of maps are prepared for special purposes by varying the location of the (imaginary) light source or the location and shape of the screen. In practice, cartographers compute map projections mathematically, rather than actually projecting them optically.

Different types of maps have different desirable properties. The most common is called a Mercator projection, after the inventor. Preparation of a Mercator map can be visualized as follows:

1. Imagine a light source at the center of the earth.
2. Imagine a cylindrical screen, covered with photographic emulsion, wrapped around the earth and tangent at the equator.
3. Project the surface of the earth on the cylinder, and then cut the cylinder and spread it flat. On the map, lines of longitude appear as straight, parallel, equally spaced vertical lines. Lines of latitude appear as straight, parallel, unequally spaced horizontal lines.

The following figure illustrates this.

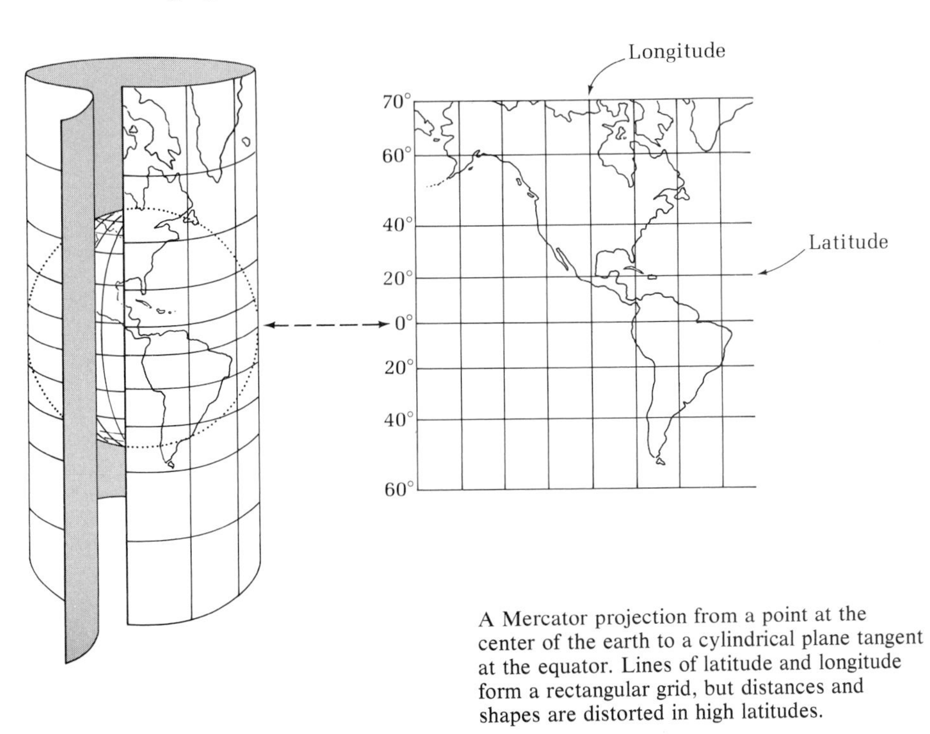

A Mercator projection from a point at the center of the earth to a cylindrical plane tangent at the equator. Lines of latitude and longitude form a rectangular grid, but distances and shapes are distorted in high latitudes.

A Mercator map, projected in this way, is grossly distorted in high latitudes and is incapable of showing the North Pole or South Pole. It distorts angles and areas, and is of little use for many purposes. However, it has the useful characteristic for navigators that a line connecting two points on the map crosses all lines of longitude at a constant angle. Such a line is called a rhumb line. The angle of the line determines the course that a ship or airplane would follow in order to hold a constant compass heading from one point to the other.

The shortest distance on a sphere is a great circle, not a rhumb line. Thus, to travel between two distant points, it is normal to change compass course repeatedly. These changes can be established on a different type of gnomonic projection called a great-circle sailing chart (see the following figure). A great circle is a straight line on such a chart, but it intersects lines of latitude and longitude at highly variable angles depending on the location.

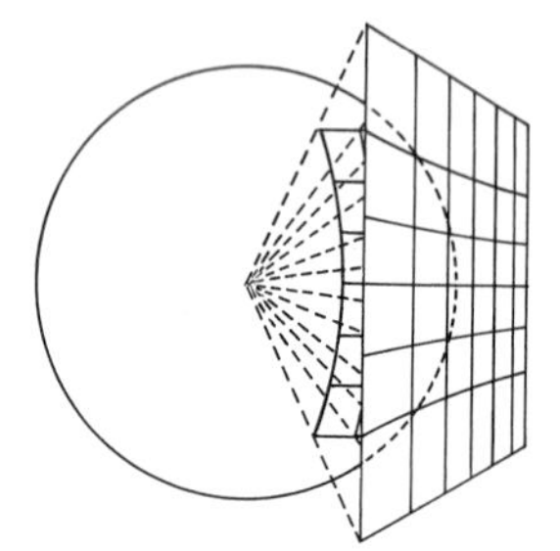

A gnomonic projection from a point at the center of the earth to a flat plane tangent to the surface. All straight lines on the plane are great circles on the globe.

Other map projections also have their special uses, and none is perfect for all purposes. The projection selected may depend on several factors, including the scale of the map, the size of the area mapped, and the nature of the features mapped. In the earth sciences, many types of maps begin with the shape of the physical features of the territory—topography on land or bathymetry at sea. Various other data—such as the distribution of sediment, or rocks of a given age, or volcanoes, or plate boundaries—are then superimposed on maps of physical features.

Topographic Maps and Bathymetric Charts

Certain special terms are useful in describing maps (land) and charts (sea):

Datum plane is the reference surface to which all observations are referred on the map. It is usually mean sea level.

Elevation or altitude on land is the vertical distance between a point and the datum plane. Land points below sea level, such as Death Valley, have a negative altitude.

Depth, at sea, is the vertical distance to a point below the datum plane. There are no negative depths.

Relief of an area of land or sea floor is the difference in elevation between the tops of hills and the bottoms of valleys.

Height is the vertical distance between a point and its immediate surroundings.

Contour line is a line, on a map or chart, connecting points of equal elevation or depth. A contour marks the intersection of the surface of the earth with a plane parallel to the datum plane but at a different elevation or depth. Contours represent the third dimension on a two-dimensional map surface. They show hills and valleys in symbolic and simplified form.

Contour interval is the vertical distance between successive contour lines.

Bench mark is a surveyed point of known elevation and position on land.

Scale of a map or chart is the ratio of the distance between two points on the surface of the earth to the distance between the same two points on the map or chart. It may be expressed in several ways, all of which may appear on the same map: (1) Verbal scale, such as "one mile to the inch" or "one centimeter to two kilometers." (2) Graphic scale, which is a line subdivided into units equivalent to some standard distance, such as a kilometer. On nautical charts, the graphic scale is given by the distance between lines of latitude. One minute of latitude is one nautical mile. (3) Fractional scale gives the proportion of a distance on the map to one on the surface of the earth. Thus, if 1 centimeter represents 1 kilometer (100,000 centimeters), the scale is 1:100,000.

A topographic map or bathymetric chart is a representation of the physical features of an area by means of contour lines. The scale and contour interval are selected to display features of interest.

Geologic Maps

Geologic maps show the distribution of sediments and rocks on the earth's surface. The rocks are commonly divided on the basis of relative or absolute age, or physical characteristics. The rock unit that is mapped is the formation. A formation has upper and lower boundaries that can be recognized easily in the field, and is large enough to be shown on the map.

A contact is the surface separating two formations or minor rock units. It appears on a geologic map as the line that marks the intersection of the contact surface with the ground surface.

Many sedimentary formations approach an ideal tabular form, and the contacts are approximately parallel. If an ideal formation is horizontal, the contact lines follow topographic contour lines. If the formation is not horizontal, its contacts intersect contour lines.

The orientation of a formation is described in terms of its dip and strike. The dip is the acute angle between a formational contact and an intersecting imaginary horizontal plane. The strike is the bearing or compass direction of the line of intersection between the contact and the plane.

Geophysical Maps

Geophysicists commonly prepare two types of maps. One type shows the observations without corrections. Examples are maps of the observed gravity or magnetic field. For the latter, the magnetic field is observed at many points, and lines are drawn through observations with the same value. These lines are contours that show the position of the magnetic field at the surface of the earth relative to a datum of a zero field.

For many purposes, it is interesting to make a map of anomalies rather than the observed field. In such a case, the smoothed magnetic (or gravity) map itself is taken as a datum surface, and local variations from it are plotted on a map. The values of the variations are then contoured once more. These anomaly contours often can be related quite easily to relief or geologic structures or to other types of anomalies.

Geochemical Maps

Geochemical observations may lend themselves to mapping by procedures similar to those used for geophysical mapping. One technique is increasingly useful in preliminary prospecting for low-grade ores. Rock samples from an area are analyzed for desired elements. The concentrations are plotted on a map, and lines of equal concentration are drawn. Promising areas tend to occur within the contour of highest values, and further prospecting can be concentrated there.

Air and Water Maps

Meteorologists and oceanographers make maps of surfaces of equal pressure, temperature, or salinity in order to understand and predict fluid motions. If the data are to be comparable, they must be collected at the same time. In most respects, the maps are prepared and presented in much the same way as other geophysical maps. Observations or anomalies are plotted, contoured, and interpreted. However, each map applies only to the time when the observations were made. Predictions can be made by comparing successive maps and observing the direction and rate of changes.

GLOSSARY

This glossary is based on *Glossary of Geology and Related Sciences with Supplement* (2nd ed.), published by the American Geological Institute, Washington, D.C., under the auspices of the National Academy of Sciences–National Research Council.

A horizon: The soil zone immediately below the surface, from which soluble material and fine-grained particles have been moved downward by water seeping through the soil. Varying amounts of organic matter give the A horizon a color ranging from gray to black. Also called *A zone*.

aa: Lava whose surface is randomly broken into angular jagged blocks.

ablation: Wearing away by evaporation and melting.

abrasion: Erosion of rock material by friction of solid particles.

absolute time: Geologic time measured in years. Compare *relative time*.

abyssal hill: A deep-sea hill with less than 500 meters of relief and usually very elongate. Generally formed at spreading centers.

abyssal plain: A deep-sea plain with a slope of less than 1 meter per kilometer. Commonly produced by turbidity currents.

acceleration: A change in velocity.

adiabatic rate: In a body of air moving upward or downward, the change in temperature that occurs without exchange of heat with the air through which it moves.

advection: A localized flow caused by density differences in a fluid. Examples: a thunderhead in the air or a plume in the mantle.

aerosol: A gaseous suspension of fine solid or liquid particles.

aftershock: An earthquake that follows a larger earthquake and originates near it.

aggradation: Upward building of a land surface or the bottom of a body of water by deposition of sediment.

albite: One of the plagioclase feldspars, $NaAlSi_3O_8$.

air mass: A large body of transient but distinctive air that is colder or warmer than the surrounding air and moves accordingly.

alkali feldspar: A feldspar rich in sodium and potassium.

alluvial fan: An assemblage of sediments marking the place where a stream moves from a steep gradient to a flatter gradient and suddenly loses much of its transporting power. Typical of arid and semiarid climates on land and of the base of continental slopes. Compare *delta*.

alluvial soil: A soil derived from distant sources rather than from the underlying bedrock.

alluvium: Sediment deposited by rivers in relatively recent time.

amorphous: Formless. Descriptive of matter in which there is no orderly arrangement of atoms.

amphibole group: Ferromagnesian silicates with a double chain of silicon–oxygen tetrahedra. Common example: hornblende. Compare *pyroxene group*.

anaerobic: Living in an environment lacking oxygen. The products of such an environment.

anaerobic decay: Decay in the absence of oxygen.

andesite: A fine-grained igneous rock with no quartz or orthoclase, consisting of about 75% plagioclase feldspars and the balance ferromagnesian silicates. Characteristic of subduction zones of the Pacific Ocean, and confined to continental sectors.

angular momentum: A vector quantity, the product of mass times radius of orbit times velocity. The energy of motion of the solar system.

angular unconformity: An unconformity in which the older strata dip at an angle different from that of the younger strata.

anhydrite: The mineral calcium sulfate, $CaSo_4$, which is gypsum without the water of crystallization.

anion: A negatively charged ion.

anomaly: Something different from what is expected or calculated, such as a gravity anomaly or magnetic anomaly.

anorthite: Calcic feldspar, $CaAl_2Si_2O_8$. One of the plagioclase feldspars.

anthracite: Metamorphosed bituminous coal containing 95–98% carbon.

anticline: A configuration of upward folded, stratified rocks in which the strata dip in two directions away from a crest just as the sides of a common gable roof dip away from the ridgepole. The reverse of a *syncline*.

anticyclone: Mass of high-pressure air rotating clockwise (in the Northern Hemisphere) as seen from space.

Apollo Mission: A series of voyages to the moon by U.S. astronauts.

Aquarian Expedition: For the purposes of this book, an imaginary voyage to the earth by beings from another solar system.

aquifer: A permeable material through which ground water moves.

aquitard: A rock layer that inhibits the flow of ground water.

archipelagic apron: Sediment and volcanic rock around a group of islands in the deep sea.

arc–trench system: A system of structures that occurs over a subduction zone. The arc is commonly a line of andesitic volcanoes and islands.

arête: A narrow, saw-toothed ridge between two cirques resulting from their having developed in close proximity.

argentite: A mineral, silver sulfide, Ag_2S. An ore of silver.

arroyo: A flat-floored, vertically walled channel of an intermittent stream typical of semiarid climates.

artesian water: Water that is under pressure when tapped by a well and thus is able to rise above the depth at which it was first encountered. It may or may not flow out at ground level.

ash: Volcanic fragments consisting of sharply angular glass particles, smaller than cinders.

asphalt: A brown to black solid or semisolid bituminous substance. Occurs in nature but is also obtained as a residue from the refining of certain hydrocarbons (then known as "artificial asphalt").

asteroids: Small bodies in orbit around the sun.

asthenosphere: A layer of the mantle below the lithosphere that is distinguished by being weak although not liquid.

astrobleme: A crater or rock structure resulting from the impact of a meteorite or larger body from space.

astrogeology: The geology of celestial bodies other than the earth. Generally restricted in meaning to the geology of the extraterrestrial bodies of our planetary system.

asymmetric fold: A fold in which one limb dips more steeply than the other.

atoll: A ring of low coral islands arranged around a central lagoon.

atom: The smallest unit of an element, consisting of protons, neutrons, and electrons.

atomic energy: Energy associated with the nucleus of an atom. It is released when the nucleus is split.

atomic mass: The nucleus of an atom contains 99.95% of its mass. The total number of protons and neutrons in the nucleus is called the *mass number*.

atomic number: The number of positive charges on the nucleus of an atom; the number of protons in the nucleus.

atomic reactor: A huge apparatus in which a radioactive core heats water or some other material under pressure and passes it to a heat exchanger.

atomic size: The radius of an atom (average distance from the center to the outermost electron of the atom).

augite: A dark-green pyroxene mineral, principally $(Ca,Na)(Mg,Fe,Al)(Si,Al)_2O_6$.

australopithecine hominids: Precursors of modern men.

authigenic: Constituents or minerals generated on the spot. Not transported.

autosuspension: A process in which a fluid mass moves because of the excess density produced by material suspended within it.

autotroph: A plant or plantlike organism capable of manufacturing its own food by synthesis from inorganic materials.

B horizon: The soil zone of accumulation that lies below the A horizon. Here is deposited some of the material that has moved downward from the A horizon. Also called *B zone*.

bar: A submerged, linear body of sand.

barchan: A crescent-shaped dune with wings or horns pointing downwind. Has a gentle windward slope and steep lee slope inside the horns.

barrier island: A low, sandy island near the shore and parallel to it.

barrier reef: A reef separated from a land mass by a lagoon of varying width and depth that opens to the sea through passes in the reef.

basalt: A fine-grained igneous rock dominated by dark-colored minerals, consisting of plagioclase feldspars (more than 50%) and ferromagnesian silicates. The typical rock of spreading centers.

batholith: A discordant pluton that increases in size downward, has no determinable floor, and shows an area of surface exposure exceeding 100 square kilometers.

bauxite: A mixture of hydrous aluminum oxides that is the chief ore of commercial aluminum.

bcmy: For the purposes of this book, billion cubic meters per year.

beach: A linear body of sand backed by land and fronted by water.

bedding: (1) A collective term used to signify the existence of beds or layers in sedimentary rocks. (2) Sometimes synonymous with *bedding plane*.

bedding plane: A surface separating layers of sedimentary rocks. Each bedding plane marks the termination of one deposit and the beginning of another of different character, such as the surface separating a sand bed from a shale layer. Rocks tend to separate or break readily along bedding planes.

bed load: Material in movement along a stream bottom, or, if wind is the moving agency, along the surface. Compare *solution, suspension, traction*.

berm: In the terminology of coastlines, berms are storm-built beach features that resemble small terraces; on their seaward edges are low ridges built up by storm waves.

biotite: "Black mica," actually ranging from dark brown to green. A rock-forming ferromagnesian silicate mineral, $K(Mg,Fe)_3AlSi_3O_{10}(OH)_2$, with tetrahedra arranged in sheets.

bituminous coal: Soft coal, containing about 80% carbon and 10% oxygen.

black body: In physics, a theoretical body that absorbs all the energy that falls upon it.

black-body radiation: Thermal radiation emitted by a black body at a specific temperature.

body wave: A wave that travels through the body of a medium, as distinguished from a wave that travels along a free surface.

boulder: A rock fragment with a diameter of 256 millimeters or more.

boulder train: A series of glacial erratics from the same bedrock source (usually with some property that permits easy identification), arranged across the country in the shape of a fan with the apex at the source and widening in the direction of glacier movement.

boundary layer: In fluid dynamics, a layer at the boundary of a fluid in which motion is significantly influenced by the roughness or other properties of the bounding material.

braided stream: A complex tangle of converging and diverging stream channels separated by sand bars or islands. Characteristic of flood plains on which the amount of debris is large in relation to the discharge of the stream.

breaking strength: The stress difference at fracture.

breccia: A clastic rock made up of angular fragments of such size that an appreciable percentage of the volume of the rock consists of particles of granule size or larger.

brittle: Breaking easily.

C horizon: The soil zone that contains partially disintegrated and decomposed parent material. It lies directly under the B horizon and grades downward into unweathered material. Also called *C zone*.

calcite: A mineral composed of calcium carbonate, $CaCo_3$.

caldera: A roughly circular, steep-sided volcanic basin with a diameter at least three or four times its depth. Commonly at the summit of a volcano. Compare *crater*.

caliche: A whitish accumulation of calcium carbonate in the soil profile.

calving: As applied to glacier ice, the process by which a glacier that terminates in a body of water breaks away in large blocks. Such blocks form the icebergs of polar seas.

carat: A unit used in measuring gems, equal to 205 milligrams (the metric carat is 200 miligrams). Compare *karat*.

carbonate mineral: A mineral formed by the combination of the complex ion $(CO_3)^{2-}$ with a positive ion. Common example: calcite, $CaCO_3$.

carbon-14: Radioactive isotope of carbon, $^{14}_{6}C$, which has a half-life of 5730 years. Used to date events back to about 50,000 years ago.

cassiterite: A mineral, tin dioxide, SnO_2, that is the principal ore of tin.

cataclasis: The crushing of a rock by fracture and rotation of mineral grains.

cation: A positively charged ion.

cave pearl: A smooth, rounded body of calcium carbonate formed by concentric precipitation around a nucleus.

cementation: The process by which a binding agent is precipitated in the spaces between the individual particles of an unconsolidated deposit. The most common cementing agents are calcite, dolomite, and quartz.

central vent: An opening in the earth's crust from which magmatic products are extruded. A volcano is an accumulation of material around a central vent.

chalcocite: A mineral, copper sulfide, CuS_2.

chalk: A variety of limestone made up, in part, of biochemically derived calcite, in the form of the skeletons or skeletal fragments of microscopic oceanic plants and animals, mixed with very fine-grained calcite deposits that may be either biochemical or inorganic in origin.

chemical weathering: The weathering of rock material by chemical processes that transform the original material into new chemical combinations.

chernozem: A very dark soil, rich in organic matter, found in subhumid climates. Typically planted to wheat and other grain crops.

chert: Granular cryptocrystalline silica similar to flint but usually light in color. Occurs as a compact massive rock, or as nodules.

chestnut soil: A dark brown soil characterized by a growth of short grasses and found in subhumid to semiarid climates.

cinders: Volcanic fragments that are small, slaglike, solidified pieces of magma 0.5–2.5 centimeters across.

cinder cone: Structure built exclusively or predominantly of pyroclastic ejecta dominated by cinders. Parasitic to major volcano, it slopes up 30–40° and seldom exceeds 500 meters in height.

cinnabar: A mineral, mercuric sulfide, HgS. The principal ore of mercury.

cirque: A steep-walled hollow in a mountainside at high elevation, formed by ice-plucking and frost action, and shaped like a half-bowl or amphitheater. Serves as principal gathering ground for the ice of a valley glacier.

clastic texture: Texture shown by sedimentary rocks formed from deposits of mineral and rock fragments.

clay: Particles of *clay size*.

clay minerals: Finely crystalline, hydrous silicates that form as a result of the weathering of such silicate minerals as feldspar, pyroxene, and amphibole. The most common clay minerals belong to the *kaolinite*, *montmorillonite*, and *illite* groups.

clay size: Refers to particles with a diameter of 0.004 millimeter or less.

cleavage: (1) Mineral cleavage, a property possessed by many minerals of breaking in certain preferred directions along smooth plane surfaces; the planes of cleavage are governed by the atomic pattern, and represent surfaces across which atomic bonds are relatively weak. (2) Rock cleavage, a property possessed by certain rocks of breaking with relative ease along parallel planes or nearly parallel surfaces.

climate: The average weather or summation of weather for years or decades in a particular place.

coal: A sedimentary rock composed of combustile matter derived from the partial decomposition and alteration of cellulose and lignin of plant materials.

coesite: High-pressure form of quartz, with a density of 2.92. Associated with many impact craters.

colloidal size: Refers to particles 0.2–1.0 micron (0.0002–0.001 millimeter), in diameter.

columnar jointing: A pattern of jointing that blocks out columns of rock. Characteristic of tabular basalt flows or sills.

compaction: Reduction in pore space between individual grains as a result of pressure of overlying sediments or pressures resulting from earth movement.

competent rock: Strata or beds, commonly sandstone, that, during folding, are able to bend without much internal flowage. They support incompetent strata, such as shale, that flow.

composite volcano: A volcano composed of interbedded lava flows and pyroclastic material, and characterized by slopes of close to 30° at the summit, reducing progressively to 5° near the base.

compound: A combination of atoms or ions of different elements. The mechanism by which they are combined is called a bond.

compression: A squeezing stress that tends to decrease the volume of a material.

computer modeling: A technique whereby a mathematically simulated material or organization is subjected to mathematically simulated stresses, and the consequences are calculated.

conduction: Transmission of energy, notably heat, by moving it from atom to atom. Compare *convection, radiation*.

cone of depression: A dimple in the water table that forms as water is pumped from a well.

cone sheet: A dike that is part of a concentric set that dips inward, like an inverted cone.

conglomerate: A detrital sedimentary rock containing fragments that are more or less rounded; an appreciable percentage of the volume of the rock consists of particles of granule size or larger.

contact metamorphism: Metamorphism at or very near the contact between magma and rock during intrusion.

continental crust: See *crust*.

continental drift: The hypothesis that earlier continents split into pieces that drifted laterally to form the present-day continents.

continental glacier: An ice sheet that obscures mountains and plains of a large section of a continent. Continental glaciers exist presently on Greenland and Antarctica.

continental rise: In some places, the base of the continental slope is marked by the somewhat gentler continental rise, which leads downward to the deep-ocean floor.

continental shelf: A shallow, gradually sloping zone extending from the sea margin to a depth at which there is a marked or rather steep descent into the depths of the ocean down the continental slope. The seaward boundary of the shelf averages about 130 meters in depth but may be more or less.

continental slope: Portion of the ocean floor extending downward from the seaward edge of the continental shelf.

contour current: A subsurface current that flows parallel to bottom contours. Geologically important, chiefly in deep water on the western sides of ocean basins.

conurbation: An essentially continuous urban area containing several large cities.

convection: Transmission of energy (and matter) by motion of material. A result of differences in density that are gravitationally unstable. Compare *conduction, radiation.*

convection current: A current caused by convection.

convection plume: See *advection.*

cordillera: A broad belt of mountain ranges.

core: The innermost zone of the earth, which is surrounded by the mantle.

Coriolis effect (or Coriolis acceleration): The change in direction of motion, or tendency to change, caused by the rotation of the earth. In the Northern Hemisphere, the Coriolis effect causes a moving body to veer, or try to veer, to the right of its direction of forward motion; in the Southern Hemisphere, to the left. The magnitude of the effect is proportional to the velocity of a body's motion.

cosmic dust: Very fine material that falls on the earth from space.

cosmic ray: The nuclei of atoms that reach the earth from space with enormous energy because of their great velocity.

crater: (1) A roughly circular, steep-sided volcanic basin, with a diameter less than three times its depth, commonly at the summit of a volcano. Compare *caldera.* (2) A depression caused by a meteorite or a bomb, either by direct impact or by explosive impact.

creep: As applied to soils and surficial material, slow downward movement of a plastic type. As applied to elastic solids, slow permanent yielding to stresses that are less than the yield point if applied for a short time only.

crevasse: A deep crevice or fissure in glacial ice.

crust: The outermost layer of the earth, defined by chemistry and some properties of seismic waves. Composed of solid rock 35–50 kilometers thick under continents and 4–6 kilometers thick under ocean basins. It rests on the mantle. Compare *lithosphere.*

cryptocrystalline: A state of matter in which there is actually an orderly arrangement of atoms characteristic of crystals, but in which the units are so small (that is, the material is so fine-grained) that the crystalline nature cannot be determined with the aid of an ordinary microscope.

crystal: A solid with orderly atomic arrangement that may or may not develop external faces that give it crystal form.

crystal form: The geometrical form taken by a mineral as an external expression of the orderly internal arrangement of atoms.

crystalline structure: The orderly arrangement of atoms in a crystal. Also called crystal structure.

crystallization: The process through which crystals develop from a fluid, viscous, or dispersed state.

cumulus cloud: A detached dense cloud that rises in the shape of a dome or tower from a level low base.

Curie temperature (or Curie point): The temperature above which magnetic material ordinarily loses its magnetism. On cooling below this temperature, it regains its magnetism. Example: iron loses its magnetism above 760°C and regains it as it cools below this temperature.

current ripple marks: Ripple marks, asymmetric in form, formed by air or water moving more or less continuously in one direction.

cycle of erosion: A qualitative description of river valleys and regions passing through the stages of youth, maturity, and old age with respect to the amount of erosion that has occurred.

cyclone: A mass of low pressure air rotating counterclockwise (in the Northern Hemisphere) as seen from space.

debris slide: A small, rapid movement of largely unconsolidated material that slides or rolls downward to produce an irregular topography.

decomposition: Synonymous with *chemical weathering.*

deep-sea fan: An alluvial fan at the base of a continental slope.

deep-sea trench: A linear depression, 6–10 kilometers deep, produced where a lithospheric plate plunges in a subduction zone.

deformation of rocks: Any change in the original shape or volume of a rock mass. Folding, faulting, and plastic flow are common modes of rock deformation.

degradation: Lowering of the surface of the land by erosion.

delta: A plain underlain by an assemblage of sediments that have accumulated where a stream,

delta *(continued)* flowing into a body of standing water, has had its velocity and transporting power suddenly reduced. Originally so named because many deltas are roughly triangular in plan, like the Greek letter "delta" (Δ) with the apex pointing upstream.

density: A measure of the concentration of matter, expressed as the mass per unit volume.

deposition: Emplacement of material.

detrital sedimentary rocks: Rocks formed from accumulations of materials derived either from erosion of previously existing rocks or from the weathered products of such rocks.

diamond: A mineral composed of a high-pressure form of the element carbon.

diapir: See *piercement structure.*

dike: A tabular discordant intrusion, usually vertical, typical of spreading centers.

diorite: A coarse-grained igneous rock with the composition of andesite (no quartz or orthoclase, consisting of about 75% plagioclase feldspars and the balance of ferromagnesian silicates).

dip: The acute angle that a surface or line makes with a horizontal plane. Compare *strike.*

dipole: Any object that is oppositely charged at two points.

dipole magnetic field: The portion of the earth's magnetic field that can best be described by a dipole passing through the earth's center and inclined to the earth's axis of rotation.

dip-slip fault: A fault on which the displacement is in the direction of the fault's dip.

discharge: With reference to stream flow, the quantity of water that passes a given point in a given time.

discordant pluton: An intrusive igneous body with boundaries that cut across the surfaces of layering or foliation of the rocks into which it has been intruded.

distributary channel (or distributary stream): A river branch that flows away from the main stream and does not rejoin it. Characteristic of deltas and alluvial fans.

divide: A line separating two drainage basins.

DNA: Deoxyribonucleic acid, a long, helical molecule that contains coded genetic information, which determines the individual hereditary characteristics of organisms.

dolomite: A mineral composed of the carbonate of calcium and magnesium, $CaMg(CO_3)_2$. Also used as a rock name for formations composed largely of the mineral dolomite.

dome: An anticlinal fold without a clearly developed linearity of crest, so that the beds involved dip in all directions from a central area, like an inverted cup. The reverse of a *basin.*

Doppler effect: A change in the observed frequency of a wave resulting from relative motion between source and receiver. In astronomy, causes the *red shift.*

drainage basin: The area from which a given stream and its tributaries receive their water.

drift: Any material laid down directly by ice, or deposited in lakes, oceans, or streams, as a result of glacial activity. Unstratified glacial drift is called *till* and forms *moraines.* Stratified glacial drift forms *outwash plains, eskers, kames,* and *varves.*

dripstone: Mineral matter, usually calcium carbonate, deposited from solution by dripping, as in a cave. See *stalactite, stalagmite.*

drumlin: A smooth, streamlined hill composed of glacial till.

ductile: Capable of undergoing large permanent deformation without fracture.

dune: A mound or ridge of sand piled up by wind.

dust: Solid particles with a diameter less than 0.06 millimeter that are or have been carried in suspension in the air.

dust-cloud hypothesis: The hypothesis that the solar system was formed from the condensation of intersteller dust clouds.

dust dome: A dome-shaped concentration of dust that forms in the air over a city.

dynamo: A generator for producing electrical current.

earthflow: A combination of slump and mudflow.

earth-power unit: Power equivalent to that of the internal heat of the earth (30×10^{12} watts).

ecliptic: The apparent path of the sun in the heavens; the plane of the planets' orbits.

elastic energy: The energy stored within a solid during elastic strain and released during elastic rebound.

elasticity: A property of materials that defines the extent to which they resist elastic strain.

elastic rebound: The recovery of elastic strain when a material breaks or when the deforming force is removed.

elastic strain: Deformation that is restored when the force is removed. The amount of yield is proportional to the force.

electron: A fundamental particle of matter, the elementary unit of negative electrical charge.

element: A unique combination of protons and electrons that cannot be broken down by ordinary chemical methods. The fundamental chemical properties of an element are determined by the number of its protons, or its atomic number. The number of neutrons varies, thus allowing formation of isotopes or different weights for the same element.

end moraine: A ridge or belt of till marking the farthest advance of a glacier. Sometimes called *terminal moraine.*

energy: The capacity for doing work. It takes such forms as kinetic, potential, heat, chemical, electrical, and atomic energy, which are interchangeable. It is also interchangeable with mass.

entrenched meander: A meander cut into underlying bedrock when regional uplift allows the originally meandering stream to resume downward cutting.

epicenter: An area on the surface directly above the focus of an earthquake.

equinoxes, precession of: A circular motion of the earth's axis of rotation with a period of 26,000 years.

erosion: The removal of material by *abrasion, mechanical weathering, chemical weathering,* and similar processes. The term also includes *transportation* of the products of such processes.

erratic: In the terminology of glaciation, a stone or boulder carried by ice to a place where it rests on or near bedrock of different composition.

escape velocity: The minimum velocity an object must have in order to escape from a gravitational field. For the moon, this is about 2.38 kilometers per second, and for the earth, about 11.2 kilometers per second.

esker: A steep-sided ridge of stratified glacial drift.

estuary: The drowned mouth or lower valley of a river.

eucaryotic cells: In biology, cells with nuclear walls and the capacity for reproduction by replication of DNA.

eugeosynclinal rocks: Rocks occurring in long thick deposits in tectonic environments roughly equivalent to subduction zones.

Euler latitude: The latitude relative to an Euler pole instead of the earth's pole of rotation.

Euler pole: A geometrical point on the surface of the earth around which a lithospheric plate rotates as it moves over the spherical surface. The relative motion between two plates is also a rotation around an Euler pole.

eustatic change of sea level: A worldwide change in sea level.

eutrophic: Refers to a lake or sea in which excess nutrients have fueled excess biological productivity and reduced or exhausted the available oxygen.

evaporation: The process by which a liquid becomes a vapor even at a temperature below its boiling point.

evaporite: A rock composed of minerals that have been precipitated from solutions concentrated by the evaporation of solvents. Examples: rock salt, gypsum, anhydrite.

exfoliation: The process by which plates of rock are stripped from a larger rock mass by physical and chemical forces.

exfoliation dome: A large, rounded domal feature produced in homogeneous coarse-grained igneous rocks and sometimes in conglomerates by the process of exfoliation.

extrusive rock: A rock that has solidified from a mass of magma that poured or was blown out upon the earth's surface.

facies: One of two or more assemblages of minerals, rocks, textures, or fossils reflecting varied environments.

fan: See *alluvial fan* and *deep-sea fan.*

fault: A surface of rock rupture along which there has been differential movement.

feldspar: Silicate minerals composed of silicon–oxygen and aluminum–oxygen tetrahedra linked together in three-dimensional networks with positive ions fitted into the interstices of the negatively charged framework of tetrahedra.

ferromagnesian mineral: A silicate mineral in which the positive ions are chiefly iron, magnesium, or both.

ferromagnetism: A property of certain metal alloys and compounds in which, below the Curie temperature, the magnetic domains tend to line up in a magnetic field.

fire storm: A fire (as in a bombed city) so great that it produces major changes in atmospheric circulation around and above it.

fissure eruption: Extrusion of lava from a linear fissure. Typical of spreading centers.

fjord: A glacially deepened valley that is now flooded by the sea to form a long, narrow, steep-walled inlet.

flood plain: Area bordering a stream over which water spreads in time of flood.

flood plain of aggradation: A flood plain formed by the building up of the valley floor by sedimentation.

flow: In rocks, any deformation, not instantly recoverable, without permanent loss of cohesion.

fluid: Material that offers little or no resistance to forces tending to change its shape.

flux: In physics, a flow of matter or energy as a fluid, or considered as a fluid.

focus: *n.* The source of a given set of earthquake waves. *v.* To concentrate.

fog: A suspension of very fine droplets in air.

fold: A bend, flexure, or wrinkle in rock.

fold mountains: Mountains consisting primarily of elevated, folded sedimentary rocks.

foliation: A layering in some rocks caused by parallel alignment of minerals. A textural feature of some metamorphic rocks. Produces rock cleavage.

footwall: One of the blocks of rock involved in fault movement. The one that would be under the feet of a person standing in a tunnel along or across the fault. Opposite the *hanging wall.*

foraminiferans: Microscopic animals of the order Foraminifera of the phylum Protozoa, which are abundant in some regions of the ocean and whose skeletons, generally calcium carbonate, are a significant part of some pelagic sediments.

force: Mass times acceleration.

foreset beds: Inclined layers of sediment deposited on the advancing edge of a growing delta or along the lee slope of an advancing sand dune.

foreshock: A relatively small earthquake that precedes a larger earthquake by a few days or weeks and originates at or near the focus of the larger earthquake.

fossil: Geologic evidence of past life, such as the shell of an ancient clam, the footprint of a long-extinct animal, or the impression of a leaf in a rock.

fossil fuels: Fuels consisting of organic remains, such as oil, natural gas, and coal.

fractionation: A process whereby crystals that formed early from a magma are effectively removed from the environment in which they formed.

fracture: In rocks, a complete loss of cohesion and resistance to stress; separation into two or more parts. As a mineral characteristic, the way in which a mineral breaks when it does not have cleavage.

fracture cleavage: A system of joints spaced a fraction of a centimeter apart.

fracture zone: A linear zone of ridges and troughs generally separating regions of different depth. The topographic expression of a transform fault.

friction: A force at the boundary between two solids that resists tangential motion.

front: In meteorology, the line at which the surface between two air masses intersects the ground.

frost heaving: The heaving of unconsolidated deposits as lenses of ice grow below the surface by acquiring capillary water from below.

fumarole: A vent for volcanic steam and gases.

***G*:** "Big *G*," the symbol for the gravitational constant.

gabbro: A coarse-grained igneous rock with the composition of basalt.

galena: A mineral, lead sulfide, PbS. The principal ore of lead.

garnet: A family of silicates of iron, magnesium, aluminum, calcium, manganese, and chromium, which are built around independent tetrahedra and appear commonly as distinctive twelve-sided, fully developed crystals. Characteristic of metamorphic rocks. Generally cannot be distinguished from one another without chemical analysis.

gas: (1) A state of matter that has neither independent shape nor volume, can be compressed readily, and tends to expand indefinitely. (2) In geology, the word is sometimes used to refer to *natural gas,* the gaseous hydrocarbons, predominately methane, that occur in rocks. (Similarly, the word *oil* is sometimes used to refer to *petroleum*).

geode: A roughly spherical, hollow or partially hollow accumulation of mineral matter from a few centimeters to nearly half a meter in diameter. An inner layer is lined with crystals that project inward toward the hollow center. Geodes are most commonly found in limestone and more rarely in shale.

geodesy: The study of the dimensions of the earth.

geodetic network: An array of very accurately located points. Changes in distance and angle between them indicate tectonic strain.

geographic poles: The points on the earth's surface marked by the ends of the earth's axis of rotation.

geoid: The figures of the earth considered as a mean sea level surface extended continuously through the continents.

geologic column: A chronological arrangement of rock units in columnar form with the oldest units at the bottom and the youngest at the top.

geologic cycles: The major (predominantly inorganic) cycles in which air, water, and rock circulate in and on the earth.

geologic time scale: A chronologic sequence of units of earth time.

geology: An organized body of knowledge about the earth. Equivalent to earth sciences.

geophysical fluid dynamics: The application of the physics of fluid motion to the motion of the atmosphere, ocean, and interior of the earth.

geophysical prospecting: Mapping rock structures by methods of experimental physics. Includes measuring magnetic fields, the force of gravity, electrical properties, seismic wave paths and velocities, radioactivity, and heat flow.

geophysics: The physics of the earth.

geostrophic wind: A wind that flows parallel to lines of equal pressure in the atmosphere because of a balance between a pressure gradient force and the Coriolis effect.

geosyncline: A basin in which thousands of meters of sediments has accumulated, with accompanying progressive sinking of the basin floor. Common usage of the term includes both the accumulated sediments themselves and the geometrical form of the basin in which they are deposited. Most folded mountain ranges were built from geosynclines, but not all geosynclines have become mountain ranges.

geothermal field: An area in which wells are drilled in order to obtain elements contained in solution in hot brines, and to tap heat energy.

geyser: A special type of thermal spring that intermittently ejects its water with considerable force.

glacier: A mass of ice, formed by the recrystallization of snow, that flows under the influence of gravity.

glacier ice: A unique form of ice developed by the compression and recrystallization of snow and consisting of interlocking crystals.

glass: A form of matter that exhibits most properties of a solid but has the atomic arrangements, or lack of order, of a liquid.

***Glossopteris* flora:** A late Paleozoic assemblage of fossil plants named for the seed-fern *Glossopteris,* one of the plants in the flora. Widespread in South America, South Africa, Australia, India, and Antarctica.

glowing avalanche: A *nuée ardente,* an avalanche of incandescent volcanic ash and debris that flows down the side of a volcano.

gneiss: A metamorphic rock with coarse cleavage. Commonly resembles granite.

Gondwanaland: A Mesozoic continent that split into present-day South America, Africa, Australia, India, and Antarctica.

graben: An elongated, trenchlike structural form bounded by parallel normal faults created when the block that forms the trench floor moved downward relative to the blocks on either side.

gradation: Leveling of the land. This is constantly being brought about by the forces of gravity and such agents of erosion as surface and underground water, and wind, glacier ice, and waves.

graded bedding: The type of bedding shown by a sedimentary deposit in which particles become progressively finer from bottom to top.

graded stream: A stream in which the capacity to do work is balanced by the work being done.

gradient: A slope, as of a stream bed.

granite: A coarse-grained igneous rock dominated by light-colored minerals, consisting of about 50% orthoclase, 25% quartz, and the balance plagioclase feldspars and ferromagnesian silicates.

granodiorite: A coarse-grained igneous rock intermediate in composition between granite and diorite.

granular texture: Composed of mineral grains large enough to be seen by the unaided eye.

graphite: A mineral composed entirely of carbon. "Black lead." Very soft because of its crystalline structure.

gravimeter: An instrument for measuring the force of gravity. Also called gravity meter.

gravitation: The attraction of all matter in the universe for all other matter.

gravity anomaly: A difference between the observed and computed values of gravity.

gravity prospecting: Mapping the force of gravity at different places to determine differences in specific gravity of rock masses, and, through this, the distribution of masses of different specific gravity. Done with a *gravimeter* (gravity meter).

gray-brown podzol: Acid soil under broadleaf deciduous forest.

greenhouse effect: The heating that results because the air is relatively transparent to incoming shortwave solar radiation but not to outgoing, reflected, long-wave radiation.

groundmass: The finely crystalline or glassy portion of a porphyry.

ground moraine: Till deposited from a glacier as a veneer over the landscape and forming a gently rolling surface.

ground water: Underground water within the zone of saturation.

ground-water table: The upper surface of the zone of saturation for underground water. It is an irregular surface with a slope or shape determined by the quantity of ground water and the permeability of the earth materials. In general, it is highest beneath hills and lowest beneath valleys. Also referred to as *water table.*

guyot: A flat-topped *seamount* rising from the floor of the ocean like a volcano but planed off on top and covered by 200 meters or more of water. Also called tablemount.

gypsum: Hydrous calcium sulfate, $CaSO_4 \cdot 2H_2O$. A common, soft mineral in sedimentary rocks. Sometimes occurs as a layer under a bed of rock salt, because it is one of the first minerals to crystallize when sea water evaporates. Alabaster is a fine-grained massive variety of gypsum. Compare *anhydrite.*

Hadley cells: Latitudinal cells in the atmosphere arising from heating at the equator and cooling at the poles.

half-life: The time needed for one half of a sample of a radioactive element to decay.

halite: A mineral, rock salt or common salt, NaCl. Occurs widely disseminated, or in extensive beds and irregular masses, precipitated from sea water and interstratified with rocks of other types as a true sedimentary rock.

hanging valley: A tributary valley whose mouth has a greater elevation than the floor of the valley it joins. Often (but not always) created by a deepening of the main valley by a glacier.

hanging wall: One of the blocks involved in fault movement. The one that would be above the head of a person standing in a tunnel along or across the fault. Opposite the *footwall.*

hardness: A property of minerals: the resistance of a smooth surface of a mineral to scratching. The Mohs scale of relative hardness consists of ten minerals. Each of these will scratch all those below it in the scale and will be scratched by all those above it: (1) talc, (2) gypsum, (3) calcite, (4) fluorite, (5) apatite, (6) orthoclase, (7) quartz, (8) topaz, (9) corundum, (10) diamond.

head: The difference in elevation between intake and discharge points for a liquid.

heat: The capacity to elevate the temperature of a mass.

heat flow: The product of the thermal gradient and the thermal conductivity of earth materials. Its average over the whole earth is 1.2 ± 0.15 microcalories per square centimeter per second.

hematite: Iron oxide, Fe_2O_3. The principal ore mineral for about 90% of the commercial iron produced in the United States. Has a characteristic red color when powdered.

historical geology: The branch of geology that deals with the history of the earth, including a record of life on the earth as well as physical changes in the earth itself.

horn: A spire left where cirques have eaten into a mountain from more than two sides around a central area. Example: the Matterhorn in the Swiss Alps.

hornblende: A rock-forming ferromagnesian silicate mineral with double chains of silicon–oxygen tetrahedra. An amphibole.

horst: An elongated block bounded by parallel faults that has been elevated relative to the blocks on both sides.

hot spot: A small area in the mantle where heating produces volcanoes in the lithosphere as it drifts over the spot.

hot spring: A *thermal spring* that brings hot water to the surface.

hurricane: A severe, tropical cyclonic storm, with winds exceeding 135 kilometers per hour, originating in the Atlantic or the Caribbean.

hydration: The process by which water combines chemically with other molecules.

hydraulic depth: In a river, the ratio of the cross-sectional area to the wetted perimeter. A factor in the resistance to flow.

hydraulic mining: Use of a strong jet of water to move deposits of sand and gravel from their original site to separating equipment, where the sought-for mineral is extracted.

hydraulic slope: In a river, the slope of the surface of the water, as distinguished from the slope of the bed.

hydrocarbon: A compound of hydrogen and carbon that burns in air to form water and oxides of carbon. There are many hydrocarbons. The simplest, methane, is the chief component of natural gas. Petroleum is a complex mixture of hydrocarbons.

hydroelectric power: Conversion of energy to electricity by the fall of water.

hydrostatic uplift: The uplift on a submerged object equivalent to the weight of the fluid displaced.

hydrothermal solution: A hot, watery solution that usually emanates from a magma in the late stages of cooling. It may deposit minor elements in economically workable concentrations.

hypsithermal time: A period about 5000 years ago when the earth was generally warmer than now. Also called *megathermal time* and *postglacial optimum.*

ice sheet: A broad moundlike mass of glacial ice of considerable extent that has a tendency to spread radially under its own weight. Small ice sheets are sometimes called *icecaps.*

igneous rock: An aggregate of interlocking silicate minerals formed by the cooling and solidification of magma.

illite: A clay-mineral family of hydrous aluminum silicates. Structure is similar to that of montmorillonite, but aluminum is substituted for 10–15% of the silicon. Illite is the commonest clay mineral in clayey rocks and recent marine sediments, and is present in many soils.

incompetent rock: A weak rock that is unable to lift its own weight or maintain its internal structure during folding.

insolation: Exposure to solar radiation.

intensity: Of an earthquake, a measure of the effects of its waves on man, structures, and the earth's surface at a particular place. Compare *magnitude.*

intermittent stream: A stream that carries water only part of the time.

intraplate phenomena: Faulting, folding, and igneous activity that occur within a tectonic plate but are not intense enough to break it into two plates.

intrusive rock: A rock that solidified from a mass of magma that invaded the earth's crust but did not reach the surface.

inversion: In meteorology, a condition in which a layer of warm air lies over a cold layer.

ion: An electrically unbalanced form of an atom, or group of atoms, produced by the gain or loss of electrons.

ionic bond: A bond in which ions are held together by the attraction of opposite electrical charges.

ionic radius: The average distance from the center to the outermost electron of an ion.

isobar: In meteorology, a line connecting points at which atmospheric pressure is equal.

isobaric surface: In meteorology, a surface along which atmospheric pressure is constant.

isoseismic line: A line connecting all points on the surface of the earth at which the intensity of shaking produced by an earthquake is the same.

isostasy: The ideal condition of balance that would be attained by earth materials of differing densities if gravity were the only force governing their heights relative to each other.

isothermal liquid: A liquid that has the same temperature throughout its volume.

isotopes: Different forms of an element produced by variations in the number of neutrons in the nucleus and, thus, variations in the atomic weight.

isotopic dating: Determination of the age of a body by the ratio or ratios of isotopes that it contains.

jasper: A red, granular, cryptocrystalline silica usually colored by hematite inclusions.

JOIDES: Acronym for Joint Oceanographic Institutions for Deep Earth Sampling. A program of the National Science Foundation.

joint: A break in a rock mass along which there has been no relative lateral movement of the rock on opposite sides.

juvenile water: Water brought to the surface or added to underground supplies by magma.

kame: A steep-sided hill of stratified glacial drift. Distinguished from a *drumlin* by lack of unique shape and stratification.

kaolinite: A clay-mineral, a hydrous aluminum silicate, $Al_4Si_4O_{10}(OH)_8$.

karat: A measure comprising 24 units used to specify the proportion of gold in an alloy. Thus, 24-karat gold is pure, and 12-karat gold is 50% other metals.

karst topography: Irregular topography characterized by sinkholes, streamless valleys, and streams that disappear underground, all developed by the action of surface and underground water in soluble rock such as limestone.

kettle: A depression in the ground surface formed by the melting of a block of ice buried or partially buried by glacial drift, either outwash or till.

kinematic wave: A wave, generally in particulate material, in which the velocity of flow decreases when the quantity of material exceeds a certain value. Thus, automobile traffic on a highway comes to a halt when there are too many cars.

kinetic energy: Energy of movement. The amount possessed by an object or particle depends on its mass and speed.

lahar: A volcanic mudflow, often catastrophic.

laminar flow: The mechanism by which a fluid such as water moves slowly along a smooth channel, or through a tube with smooth walls, with fluid particles following paths parallel to the channel or walls. Compare *turbulent flow*.

landslide: A general term for relatively rapid mass movements, including *slump, rock slide, debris slide, mudflow*, and *earthflow*.

landslide splash: A large splash generated when a landslide falls on a sufficient body of water.

lapilli: Small, solidified fragments of lava (*sing*. lapillus).

laser: A device for converting incident electromagnetic energies of mixed frequencies to highly amplified and coherent energy at a discrete frequency in the range of visible light.

latent heat: The number of calories per unit mass that must be added to a material at the melting point to cause melting. These calories do not raise the temperature.

lateral fault: See *strike-slip fault*.

lateral moraine: A ridge of till along the edge of a valley glacier, composed largely of material that fell to the glacier from valley walls.

laterite: Tropical soil rich in hydroxides of aluminum and iron formed under conditions of good drainage.

latitude: A determination of distance from the equator on the earth's surface. Measured as the angle between the zenith (celestial pole) and the equatorial plane.

latitudinal motion: Motion toward or away from the equator.

lattice: A three-dimensional array of points or lines.

Laurasia: The ill-defined ancient continent from which the present Eurasian and North American continents have been derived.

lava: Magma that has poured out onto the surface of the earth, or rock that has solidified from such magma.

leaching: The removal by water of soluble matter from soil and rocks.

left-lateral fault: A strike-slip fault along which the ground on the opposite side appears, to an observer facing it across the fault, to have moved leftward.

levee: A broad, low ridge of sediment deposited beside a channel by overspilling during floods. Artificial levees are built to prevent overspilling. Natural levees are common on river deltas and deep-sea fans, where they are deposited by turbidity currents.

lignite: A low-grade coal with about 70% carbon and 20% oxygen. Intermediate between *peat* and *bituminous coal*.

limestone: A sedimentary rock composed largely of the mineral calcite, $CaCO_3$, that has been formed either by organic or by inorganic processes.

limonite: Iron oxide with no fixed composition or atomic structure. Encountered as ordinary rust, or the coloring material of yellow clays and soils.

liquid: A state of matter that flows readily so that the mass assumes the form of its container but retains its independent volume.

lithification: The process by which unconsolidated rock-forming materials are converted into a consolidated or coherent state.

lithosphere: The outermost, rigid layer of the earth, commonly 100 kilometers thick, including the crust and the uppermost part of the mantle. It is broken into moving tectonic plates and rests on the asthenosphere. Compare *crust.*

lithospheric plate: One of the pieces into which the lithosphere is broken by tectonic processes.

littoral ash cone: A cone of volcanic ash that forms where a lava flow meets a sizable body of water.

littoral drift: The movement of sediment, largely sand, along a shore.

load: The amount of material that a transporting agency, such as a stream, a glacier, or the wind, is actually carrying at a given time.

loam: A mixture of sand, silt, and clay. Usually a desirable soil.

loess: An unconsolidated, unstratified aggregation of small, angular mineral fragments, usually buff in color and generally wind-deposited. Characteristically able to stand on very steep to vertical slopes.

logistic curve: The shape of the line in a graph of exponential growth that is saturated or limited in some way—as all natural exponential growth is. The curve is S-shaped or sigmoid.

longitude: On the earth's surface, an angular distance east or west measured relative to some standard meridian, such as Greenwich.

longitudinal motion: Motion east or west and, thus, along a constant latitude.

mafic rock: A dark igneous rock rich in ferromagnesian minerals. Most common on the sea floor.

magma: A naturally occurring silicate melt, which may contain suspended silicate crystals or dissolved gases, or both.

magnetic anomaly: The difference between observed value of the earth's magnetic field and computed value at a given point.

magnetic declination: The angle of divergence between a geographic meridian and a magnetic meridian. It is measured in degrees east and west of geographic north.

magnetic dip: The vertical angle between a freely suspended compass needle and the horizontal. The dip of the earth's magnetic field at a given point.

magnetic domain: A very small, spontaneously magnetized region within a magnetized crystal or body. Within each domain, the elementary atomic magnetic moments are essentially aligned.

magnetic field of the earth: The magnetic lines of force in and around the earth and generated within it.

magnetic inclination: The angle that the magnetic needle makes with the surface of the earth. Also called the dip of the magnetic needle.

magnetic polarity epoch: One of the time intervals between reversals of the earth's magnetic field. See *polarity epoch.*

magnetic pole: The north magnetic pole is the point on the earth's surface at which the north-seeking end of a magnetic needle free to move in space will point directly down. At the south magnetic pole, the same end points directly up. These poles are also known as dip poles.

magnetic reversal: A shift of 180° in the earth's magnetic field such that a north-seeking needle of a magnetic compass points to the south rather than to the north magnetic pole.

magnetite: A mineral, iron oxide, Fe_3O_4. Black, strongly magnetic, and an important ore of iron.

magnitude: Of an earthquake, a measure of the total energy released. Compare *intensity.*

mantle: The intermediate zone of the earth. Surrounded by the crust, it rests on the core at a depth of about 2900 kilometers.

mantle plume: A localized flow of material in the mantle. See *advection.*

marble: A metamorphic rock of granular texture, no rock cleavage, and composed of calcite, dolomite, or both.

maria: The dark-toned "seas" of the moon. They mark the topographically low areas of the moon.

marsh gas: Methane, CH_4, the simplest paraffin hydrocarbon and the predominant component of natural gas.

mass: The quantity of matter. It is obtained on the earth's surface by dividing the weight of a body by its acceleration due to gravity.

mass movement: Surface movement of earth materials induced by gravity.

matter: Anything that occupies space. Usually defined by describing its states and properties: solid, liquid, or gaseous; possesses mass, inertia, color, density, melting and boiling points, hardness, crystal form, mechanical strength, or chemical properties. Composed of atoms.

mcmy: For the purposes of this book, million cubic meters per year.

meander: (1) *n.* A turn or sharp bend in a stream's course. (2) *v.i.* To turn, or bend sharply.

meander belt: The zone along a valley floor that encloses a meandering river.

mechanical weathering: The process by which rock is broken down into smaller and smaller fragments as the result of energy developed by physical forces. Also called disintegration.

medial moraine: A ridge of till formed by the junction of two lateral moraines when two valley glaciers join to form a single ice stream.

megathermal time: See *hypsithermal time.*

meridian: A line of longitude.

metal: An element or alloy that is a good conductor of electricity and heat. Examples: gold, silver, aluminum, bronze, brass. Seventy-seven of the elements are metals.

metamorphic rock: Any rock that has been changed in texture or composition by heat, pressure, or chemically active fluids after its original formation.

metamorphism: A process whereby rocks undergo physical or chemical changes, or both, to achieve equilibrium with conditions other than those under which they were originally formed. Weathering is arbitrarily excluded from the meaning of the term. The agents of metamorphism are heat, pressure, and chemically active fluids.

metasomatism: A process whereby rocks or minerals are altered when liquids and volatiles exchange ions with them.

meteor: The streak of light emanating from a transient celestial body that enters the earth's atmosphere with great speed and becomes incandescent from heat generated by resistance of the air. Compare *meteorite, meteoroid.*

meteoric water: Ground water derived primarily from precipitation.

meteorite: A stony or metallic body that has fallen to the earth from outer space.

meteoroid: Collective name for relatively small stony or metallic particles in, or arriving from, space.

methane: The simplest paraffin hydrocarbon, CH_4. The principal constituent of natural gas. Sometimes called *marsh gas.*

micas: A group of silicate minerals characterized by perfect sheet or scale cleavage resulting from their atomic pattern, in which silicon–oxygen tetrahedra are linked in sheets. Biotite is the ferromagnesian black mica. Muscovite is the potassic white mica.

microseism: A small shaking. Specifically limited in technical usage to earth waves generated by sources other than earthquakes, and most frequently to waves with periods of 1–9 seconds from sources associated with atmospheric or marine storms.

midocean ridge: A broad, bilaterally symmetrical, elongate submarine ridge with sloping sides, not necessarily in the middle of an ocean. Midocean ridges form at spreading centers and split. The materials formed there sink as each side ages and moves laterally. This sinking gives the ridge its form. The central part of the ridge does not sink unless spreading stops.

mineral: A naturally occurring element or compound. For some purposes, the definition is restricted to inorganic crystalline solids. Other definitions, especially ones concerned with resources, include organic solids, liquids, and gases.

mineral reserve: The amount of a mineral that is recoverable with present technology at present costs and prices. *Proved reserves* are those known to exist. Compare *mineral resource.*

mineral resource: The amount of a mineral that exists or probably exists but may not be recoverable under present economic conditions. Compare *mineral reserve.*

mining: Removing material from the earth.

moat: In marine geology, a deep wide trench around the submerged base of a large oceanic volcano or group of volcanoes.

molecule: The smallest unit of a compound that displays the properties of that compound.

momentum: The product of mass and velocity.

monadnock: A hill left as a residual of erosion, standing above the level of *peneplain.*

monocline: A double flexure connecting strata at one level with the same strata at another level.

monsoon: A seasonal wind that blows on or off a continent bringing rain or drier weather, respectively.

montmorillonite: A clay-mineral family of hydrous aluminum silicates with one ionic sheet of aluminum and hydroxyl between two Si_4O_{10} sheets.

moon: A natural satellite.

moraine: A general term applied to certain landforms composed of sediment deposited by glaciers.

mountain: Any part of a landmass that projects conspicuously above its surroundings.

mountain chain: A series or group of connected mountains having a well-defined trend or direction.

mountain range: See *range.*

mud: A wet detrital material made up of silt- and clay-sized particles.

mudflow: Flow of a well-mixed mass of rock, earth, and water that behaves like a fluid and flows down slopes with a consistency similar to that of newly mixed concrete.

mudstone: Fine-grained, detrital sedimentary rock made up of silt- and clay-sized particles. Distinguished from shale by lack of fissility.

muscovite: "White mica," a nonferromagnesian rock-forming silicate mineral, $KAl_2(AlSi_3O_{10})(OH)_2$, with its tetrahedra arranged in sheets. Sometimes called potassic mica.

nappe: A sheet of rock that has been transported a large distance along the thrust fault that forms its lower boundary.

natural gas: A mixture of gaseous hydrocarbons, predominately methane, that occurs in rocks.

natural resources: Energy and materials available in nature.

neap tide: A tide of lowest range, produced when sun, earth, and moon define a right angle. Compare *spring tide*.

neutron: An electrically neutral, fundamental particle of matter found in the nucleus of an atom.

nodule: An irregular, knobby-surfaced body of mineral that differs in composition from the rock or sediment in which it is formed.

nondipole magnetic field: That portion of the earth's magnetic field that would remain were the dipole field and the external field removed.

normal fault: A fault in which the hanging wall appears to have moved downward relative to the footwall. Compare *reverse fault*.

nucleation: The initial formation of a droplet or crystal around a particle in the medium.

nucleus: The protons and neutrons constituting the central part of an atom.

nuée ardente: Literally "hot cloud." A French term (pl. nuées ardentes) for a *glowing avalanche*.

nutation: An irregular motion of the earth's pole of rotation with a period of 18.6 years. Caused by the attraction of the sun and moon on the equatorial bulge.

oblique slip fault: A fault with components of relative displacement along both strike and slip.

obsidian: Volcanic glass.

oil: In geology, refers to *petroleum*.

oil shale: Shale containing such a high proportion of hydrocarbons as to be capable of yielding petroleum on distillation.

olivine: A rock-forming ferromagnesian silicate mineral that crystallizes early from a magma and weathers readily at the earth's surface. A solid solution with the general formula $(Mg,Fe)_2SiO_4$.

ooze: A deep-sea deposit consisting 30% or more by volume of the hard parts of very small organisms. If a particular organism is dominant, its name is used as a modifier, as in *radiolarian ooze*.

onshore–offshore breezes: The daily cycle of coastal winds caused by the different rates of heating and cooling of land and water.

opal: Amorphous silica with varying amounts of water. A mineral gel.

open-pit mining: Surface mining represented by sand and gravel pits, stone quarries, copper mines in some of the western states, and diamond mines in Africa.

orbital velocity: The velocity of a particle (molecule) moving in a circular or elliptical path as a wave form moves through a substance. Compare *phase velocity*.

order of crystallization. The chronological sequence in which crystallization of the various minerals of an assemblage takes place.

ore deposit: Metallic minerals in concentrations that can be worked at a profit.

orogeny: The process by which mountain structures develop.

orographic effect: In meteorology, the effects that result from an air mass being displaced upward and concentrated as it flows over a mountain.

orthoclase: A mineral, potassic feldspar, $K(AlSi_3O_8)$.

outwash: Material carried from a glacier by meltwater and laid down in stratified deposits.

outwash plain: A flat or gently sloping surface underlain by outwash.

overbank deposits: The sediments (usually clay, silt, and fine sand) deposited on a flood plain by a river overflowing its banks.

overturned fold: A fold in which at least one limb is overturned—that is, has rotated through more than 90°.

oxbow: An abandoned meander—that is, a meander cutoff from the stream that formed it.

oxbow lake: An abandoned meander isolated from the main stream channel by deposition, and filled with water.

oxide mineral: A mineral formed by the direct union of an element with oxygen. Examples: ice, corundum, hematite, magnetite, cassiterite.

ozone: A molecule of oxygen with three atoms, O_3. Formed from ordinary oxygen gas, O_2, by exposure to ultraviolet radiation or lightning.

P wave: See *primary wave*.

pahoehoe: Lava whose surface is smooth and billowy, frequently molded into forms that resemble huge coils of rope.

paleomagnetism: The study of the earth's magnetic field as it has existed during geological time.

Pangaea: A hypothetical Mesozoic continent from which all others are postulated to have originated through a process of fragmentation and drifting. Compare *Gondwanaland, Laurasia*.

Panthalassa: The world ocean that surrounded the one or two continents that existed in Mesozoic time.

peat: Partially reduced plant or wood material containing approximately 60% carbon and 30% oxygen. An intermediate material in the process of coal formation.

pebble: A particle of *pebble size.*

pebble size: A volume greater than that of a sphere with a diameter of 4 millimeters and less than that of a sphere with a diameter of 64 millimeters.

pedology: The science that treats of soils—their origin, character, and utilization.

pegmatite: A small intrusion of exceptionally coarse texture, with crystals up to 12 meters in length, commonly formed within, or at the margin of, a batholith.

pendulum: An object so suspended that, after it is displaced, a restoring force will return it to the starting position. If displaced, then released, it oscillates, completing one to-and-fro swing in a time called its period.

pelagic sediment: Slowly deposited sediment of the deep-sea floor.

peneplain: A nearly flat plain representing an advanced stage of erosion.

perched water table: The top of a zone of saturation that bottoms on an impermeable horizon above the level of the general water table in the area. Is generally near the surface, and frequently supplies a hillside spring.

peridotite: A coarse-grained igneous rock dominated by dark-colored minerals, consisting of about 75% ferromagnesian silicates and the balance plagioclase feldspars.

period: For an oscillating system, the length of time required to complete one oscillation.

permafrost: Permanently frozen ground.

permeability: For a rock or an earth material, the ability to transmit fluids.

petroleum: A complex mixture of hydrocarbons, accumulated in rocks, and dominated by paraffins and cycloparaffins. Crude petroleums are classified as paraffin-base if the residue left after the volatile components have been removed consists principally of a mixture of paraffin hydrocarbons, and as asphalt-base if the residue is primarily cycloparaffins.

petrology: The study of rocks.

phase: A homogeneous, physically distinct portion of matter in a system that is not homogeneous, as in the three phases ice, water, and aqueous vapor.

phase velocity: The velocity at which the form of a wave moves forward. Compare *orbital velocity.*

phenocryst: A crystal significantly larger than the crystals of surrounding minerals.

phosphate rock: A sedimentary rock containing calcium phosphate.

photosynthesis: The process by which carbohydrates are compounded from carbon dioxide and water in the presence of sunlight and chlorophyll.

physical geology: The branch of geology that deals with the nature and properties of the materials constituting the earth; the distribution of materials throughout the globe; the processes by which they are formed, altered, transported, and distorted; and the nature and development of the landscape.

piercement structure: A structure, such as a salt dome, that penetrates the strata above it. Also called *diapir.*

piezometric surface: An imaginary surface that coincides with the static level of the water in an aquifer.

pillow lava: Volcanic rock in sacklike bulbous masses about a meter long that develop only when lava flows underwater.

pingo: A hill of sediment pushed up by the expansion of ice beneath it.

pirate stream: One of the two streams in adjacent valleys that extended its valley headward until it breached the divide between them, and that captured the upper portion of the neighboring stream.

placer ore: A sedimentary ore generally consisting of dense, resistant minerals.

plagioclase feldspars: *Albite, anorthite,* and minerals of intermediate composition.

plain: A region of low relief composed of nearly horizontal sedimentary rocks.

planet: A heavenly body that changes its position from night to night with respect to the background of stars.

planetesimals: Small solid astronomical bodies that are collected by gravity to form a planet.

plasma: Hot ionized gas.

plastic deformation: A permanent change in shape or volume that does not involve failure by rupture, and that, once started, continues without increase in the deforming force.

plate, tectonic: See *tectonic plate.*

plate tectonics: The study of the motions and interactions of tectonic plates.

plateau basalt: Basalt poured out from fissures in floods that tend to form great plateaus.

playa: The flat-floored center of an undrained desert basin.

playa lake: A temporary lake formed in a playa.

plume: See *mantle plume* and *advection.*

pluvial lake: A lake formed during a pluvial period.

pluvial period: A period of increased rainfall and decreased evaporation, which prevailed in nonglaciated areas during the time of ice advance elsewhere.

podzol: An ashy gray or gray-brown soil, low in iron and lime, that is formed under coniferous forest in moist and cool conditions.

polar wandering (or polar migration): A movement of the position of the magnetic pole during past time in relation to its present position.

polarity: The orientation of the earth's magnetic field, whether normal, as it is now, or reversed. See *magnetic reversal.*

polarity epoch: A period of time during which the earth's magnetic field has been oriented dominantly in either a normal or a reverse direction. A polarity epoch may be marked by shorter intervals of the opposite sign, called polarity events. See *magnetic reversal.*

porosity: The percentage of open space or interstices in a rock or other earth material. Compare *permeability.*

porphyritic: A textural term for igneous rocks in which larger crystals, called phenocrysts, are set in a finer groundmass, which may be crystalline, glassy, or both.

porphyry: An igneous rock containing conspicuous phenocrysts in a fine-grained or glassy groundmass.

portland cement: A hydraulic cement consisting of compounds of silica, lime, and alumina. It is widely used as the cementing ingredient in mortar and concrete.

positive charge: An electrical property of matter in which protons are not balanced by electrons.

postglacial optimum: See *hypsithermal time.*

potassic feldspar: The mineral orthoclase, $K(AlSi_3O_8)$.

potential energy: Stored energy waiting to be used. The energy that a piece of matter possesses because of its position or the arrangement of its parts.

pothole: A hole ground into the solid rock of a stream channel by sand, gravel, and boulders caught in an eddy of turbulent flow and swirled for a long time over one spot.

power: The rate of doing work.

prairie: A nearly flat grassy upland.

precession of equinoxes: See *equinoxes.*

precipitation: (1) The discharge of water from the atmosphere in the form of rain, snow, hail, sleet, fog, or dew. (2) The process of separating dissolved solids from a solution by evaporation, cooling, or chemical change.

pressure: Force per unit area.

primary wave: The earthquake body waves that travel fastest and advance by a push-pull mechanism. Also known as longitudinal, compressional, or P waves. Compare *secondary wave.*

prograde orbit: Of a moon, motion in the same direction as the planet being orbited.

proton: A fundamental particle of matter with a positive electrical charge, found in the nucleus of an atom.

proved reserve: See *mineral reserve.*

pumice: Pieces of lava up to several centimeters across that have trapped bubbles of steam or other gases as they were thrown out. Some pieces have enough buoyancy to float on water.

pyrite: A mineral, iron sulfide, FeS_2. Colloquially called fool's gold.

pyroclastic debris: Fragments blown out by explosive volcanic eruptions and subsequently deposited on the ground. Include *ash, cinders, lapilli, volcanic bombs,* and *pumice.*

pyroxene group: Ferromagnesian silicates composed of single chains of silicon–oxygen tetrahedra.

quartz: A mineral, SiO_2, composed exclusively of silicon–oxygen tetrahedra with all oxygens joined together in a three-dimensional network. Its crystal form is a six-sided prism tapering at the end, with the prism faces striated transversely. An important rock-forming mineral.

quartzite: A metamorphic rock composed of quartz and commonly formed by the metamorphism of sandstone. Has no rock cleavage. Breaks through sand grains, whereas sandstone breaks around the grains.

quick clay: A water-bearing clay that readily liquifies when disturbed.

quicksand: Saturated sand with grains in very loose array or with grains partially supported by the upward motion of water, as from a spring.

radar: Ultrahigh-frequency electromagnetic radiation.

radiation: Transmission of energy as waves. Example: light passing through air or window glass. Compare *conduction, convection.*

radiation damage: Microscopic damage to the structure of a nonmetal caused by penetration of high-energy neutrons or cosmic rays.

radioactivity: The spontaneous breakdown of an atomic nucleus, with radiation of energy.

radiocarbon: See *carbon-14.*

radiolarians: Minute, single-celled animals. Their siliceous skeletons dominate some types of pelagic sediment.

rain shadow: An area with less than normal rain because it is downwind of a mountain or other object that strips much of the water out of passing air masses.

rain wash: The water from rain after it has fallen to the ground and before it has been concentrated in definite stream channels.

range, mountain: An elongated series of mountain peaks considered to be a part of one connected unit, such as the Applachian Range or the Sierra Nevada.

recessional moraine: A ridge or belt of till formed during a period of temporary stability or slight readvance while a glacier is generally shrinking.

recrystallization: Deformation or change on a molecular scale by solid diffusion, local melting, or solutions.

recumbent fold: A fold in which the axial plane is more or less horizontal.

red and yellow podzols: Modern red and yellow soils developed on ancient, deeply weathered parent materials.

red shift: The apparent increase in the wavelength of radiation given off by a receding star as a consequence of the *Doppler effect.*

regional metamorphism: Metamorphism occurring over tens or scores of kilometers.

rejuvenation: A change in conditions of erosion that causes a stream to begin more active erosion and entrenching of its channel.

relative time: Dating of events by means of their place in a chronologic order of occurrence rather than in terms of years. Compare *absolute time.*

reserve of minerals: See *mineral reserve.*

reservoir rock: A porous (usually permeable) rock that will yield water, oil, or gas.

residence time: The average time that a quantity of a material spends in a container, with the rate at which the material is added or removed from the container taken into account.

resonance: In physics, a vibration of large amplitude that results from the application of vibrations to a system that have the same period as a natural vibration of the system.

resource (of minerals): See *mineral resource.*

retrograde orbit: Of a moon, motion in a direction opposite to that of a planet being orbited.

reverse fault: A fault along which the hanging wall appears to have moved upward relative to the footwall. Compare *normal fault.*

rhyolite: A fine-grained igneous rock with the composition of granite.

ridge–ridge transform fault: A lateral fault at the side of a tectonic plate connecting two midocean ridges (spreading centers).

ridge–trench transform fault: A lateral fault at the side of a tectonic plate and connecting a midocean ridge with an oceanic trench.

rift zone: A system of roughly parallel fractures in the earth's crust. Often associated with extrusion of lava.

right-lateral fault: A strike-slip fault along which the ground on the opposite side appears, to an observer facing it across the fault, to have moved rightward.

rigid: Resisting internal deformation.

rill: A miniature stream channel.

ripple marks: Small waves produced in unconsolidated material by wind or water.

rock: An aggregate of minerals of different kinds in varying proportions.

rock flour: Finely divided rock material pulverized by a glacier and carried by streams fed by melting ice.

rock-forming silicate minerals: Minerals built around frameworks of silicon–oxygen tetrahedra. Examples: *olivine, augite, hornblende, biotite, muscovite, orthoclase, albite, anorthite, quartz.*

rock glacier: A tongue of rock waste found in the valleys of certain mountainous regions. Characteristically lobate and marked by a series of arcurate, rounded ridges that give it the aspect of having flowed as a viscous mass.

rock slide: A sudden and rapid slide of bedrock along planes of weakness.

runoff: Water that flows off the land.

S wave: See *secondary wave.*

salt: In geology this term usually refers to halite, or rock salt, NaCl, particularly in such combinations as salt water and salt dome.

salt dome: A mass of halite generally of roughly cylindrical shape and with a diameter of about 1–2 kilometers near the top. These masses have pushed through surrounding sediments into their present positions from below. Reservoir rocks above and alongside salt domes may trap oil and gas.

saltwater wedge: A body of salt water, found in some estuaries, that thins toward the head of the estuary and is overridden by fresh river water.

sand: Particles of *sand size,* commonly, but not always, composed of mineral quartz.

sand size: A volume with a diameter between 0.0625 millimeter and 2 millimeters.

sand wave: A body of sand that has a wavelike form and moves as a wave in response to wind or water currents. Generally submarine, they range from roughly 1 centimeter to 1 kilometer in wavelength.

sandstone: A detrital sedimentary rock formed by the cementation of individual grains of sand size and commonly composed of quartz.

satellite, artificial: An artificial moon. Some of those that orbit the earth take photographs, sense variations in heat and light at the surface of the earth, and make other environmental measurements. Others transmit messages, enable ships to locate themselves accurately, and perform many other services.

savanna: A flat, treeless grassland of tropical or subtropical regions.

schist: A metamorphic rock dominated by fibrous or platy minerals. Has schistose cleavage and is a product of regional metamorphism.

schistose cleavage: Rock cleavage in which grains and flakes are clearly visible and cleavage surfaces are rougher than in slaty or phyllitic cleavage.

scuba: An acronym for self-contained underwater breathing apparatus.

sea-floor spreading: The concept that ocean floors spread laterally relative to the crests of the mid-ocean ridges. As material moves laterally from the ridge, new material replaces it along the ridge crest by welling up from below.

seamount: An isolated, steep-sloped peak rising from the deep-ocean floor but not reaching the surface. Most are extinct volcanoes.

secondary wave: An earthquake wave that travels at a lower velocity than a *primary wave*.

sediment: A comprehensive term for all materials deposited by water, wind, or ice. Includes *pebbles, sand, silt, mud,* and some, but not all, *clay*.

sedimentary cycle: The sequence of geological phenomena in which continental rock is weathered, eroded and transported, ultimately, to the deep-sea floor. Sediment is recycled when the tectonic and igneous–metamorphic cycles restore it to the continents.

sedimentary facies: An accumulation of deposits that exhibits specific characteristics and grades laterally into other sedimentary accumulations that were formed at the same time but that exhibit different characteristics.

sedimentary ore: An ore that is deposited by sedimentary processes. Includes alluvial and precipitated deposits.

sedimentary rock: Rock formed from accumulations of sediment, which may consist of rock fragments of various sizes, the remains or products of animals or plants, the products of chemical action or evaporation, or mixtures of these.

sedimentation: The process by which mineral matter and organic matter are laid down.

seismogram: The record obtained on a *seismograph*.

seismograph: An instrument for recording earthquake waves and other vibrations. Compare *seismometer*.

seismology: The scientific study of earthquakes and other earth vibrations.

seismometer: An instrument for detecting earthquake waves. Compare *seismograph*.

semiconductor: A solid crystalline substance having electrical conductivity greater than insulators but less than good conductors.

serpentine: A hydrated ultramafic rock consisting largely of the oxides of magnesium and silicon.

shale: A fissile, fine-grained, detrital, sedimentary rock made up of silt- and clay-sized particles. Contains clay minerals as well as particles of quartz, feldspar, calcite, dolomite, and other minerals.

shallow wave: In oceanography, a wave whose wavelength is more than twice as great as the water depth.

shatter cone: A conical fracture form in rock caused by the high energies associated with meteoritic impacts.

shear: A change in shape without change in volume.

shearing stress: A stress causing parts of a solid to slip past one another like cards in a deck.

sheeting: Joints that are essentially parallel to the ground surface. They are more closely spaced near the surface and become progressively farther apart with depth. Particularly well developed in granitic rocks, but sometimes in other massive rocks as well.

shield volcano: A volcano built up almost entirely of lava, with subaerial slopes seldom as great as 10° at the summit and 2° at the base. Submarine slopes average about 20°. Examples: the five volcanoes on the island of Hawaii.

shock wave: A large-amplitude compressional wave caused by supersonic motion of a body in a medium. Shockwaves occur when large meteorites hit the earth, or when aircraft fly at supersonic speeds.

siderite: A mineral, iron carbonate, $FeCO_3$. An ore of iron.

silicate minerals: Minerals with crystal structures containing SiO_4 tetrahedra.

silicon–oxygen tetrahedron: A complex ion composed of a silicon ion surrounded by four oxygen ions. It has a negative charge of 4 units and is represented by the symbol $(SiO_4)^{4-}$. It is the diagnostic unit of silicate minerals.

silt: Particles of *silt size*.

silt size: A volume with a diameter between 0.0039 millimeter and 0.0625 millimeter.

sinkhole: A depression in the surface of the ground caused by the collapse of the roof over a *solution cavern*.

slate: A fine-grained metamorphic rock with well-developed slaty cleavage. Formed by low-grade regional metamorphism of shale.

slaty cleavage: Cleavage exhibited by rocks with parallel cleavage planes separated only by microscopic distances.

slip vector: In seismology, the direction of relative motion along a fault during an earthquake.

slump: The downward and outward movement of rock or unconsolidated material as a unit or as a series of units.

snowline: The lower limit of perennial snow.

sodic feldspar: Albite, $Na(AlSi_3O_8)$.

soil: The superficial detrital material that forms at the earth's surface as a result of organic and inorganic processes. Soil varies with climate, plant and animal life, time, slope of the land, and parent material.

soil horizon: A subdivision of a layered soil. Approximately parallel to the land surface. Commonly designated from top to bottom by the letters A, B, and C.

soil mechanics: The study of the mechanical properties of sediment.

solar constant: The average rate at which radiant energy from the sun is received by the earth, measured on a plane perpendicular to the sun's rays at the outer edge of the atmosphere when the earth is at a mean distance from the sun.

solar system: Generally, a sun with a group of celestial bodies held by the sun's gravitational attraction and revolving around it. Specifically, the system in which the earth is one of the revolving bodies.

solid: Matter with a definite shape and volume and some fundamental strength. May be crystalline, glassy, or amorphous.

solifluction: Mass movement of soil affected by alternate freezing and thawing. Characteristic of saturated soils in high latitudes.

solution: (1) The state of being dissolved. (2) A homogeneous mixture of two substances.

solution cavern: A cavern formed in relatively soluble rock, such as limestone, by the action of ground water.

sorting: The degree of uniformity of grain sizes in a sediment.

space lattice: In the crystalline structure of a mineral, a three-dimensional array of points representing the pattern of locations of identical atoms or groups of atoms that constitute a mineral's *unit cell*. There are 230 pattern types.

specific gravity: The ratio between the weight of a given volume of a material and the weight of an equal volume of water at 4°C.

specific heat: The amount of heat necessary to raise the temperature of 1 gram of a material 1°C.

spelunker: An explorer of caves.

sphalerite: A mineral, zinc sulfide, ZnS. Nearly always contains iron, in varying proportions, as an impurity: (Zn,Fe)S. The principal ore of zinc.

spheroidal weathering: The spalling off of concentric shells from rock masses of various sizes as a result of pressure built up during chemical weathering.

spit: A linear body of sand extending from the shore. Intermediate in attachment between a *beach* and a *barrier island*.

spreading center: In tectonics, the line or narrow zone between two lithospheric plates that are moving apart. Generally perpendicular to fracture zones. *Midocean ridges* occur over spreading centers.

spreading rate: In plate tectonics, the rate at which two lithospheric plates are moving apart. Usually expressed in centimeters per year. The rate at which each plate is moving relative to the spreading center is half the spreading rate.

spring: A place where the water table intersects the surface of the ground and where water flows out more or less continuously.

spring tide: A tide of highest range, produced when sun, earth, and moon are in line.

stalactite: An icicle-shaped accumulation of *dripstone* hanging from a cave roof.

stalagmite: A post of *dripstone* growing upward from the floor of a cave.

stishovite: A high-pressure form of SiO_2 with a density of 4.34 grams per cubic centimeter.

storm surge: A broad elevation of the sea surface caused by the inflowing winds and the low atmospheric pressure in a hurricane or typhoon. In conjunction with hurricane winds, they are among the worst destroyers of life and property of all natural phenomena.

strain: A change in the dimensions of matter in response to *stress*.

stratification: The structure produced by the deposition of sediments in layers or beds.

stratigraphic trap: A structure that traps petroleum or natural gas because of variation in permeability of the reservoir rock, or the termination of an inclined reservoir formation on the up-dip side.

stratovolcano: A volcano built of layers of volcanic ash intermixed with solidified lava flows.

stream capture: The process whereby a stream, rapidly eroding headward, cuts into the divide separating it from another drainage basin and provides an outlet for a section of a stream in the adjoining valley.

strength: Resistance to failure.

stress: Force per unit area applied to material that tends to change its dimensions. Compare *strain*.

striation: A scratch or small gouged channel. Commonly produced when rocks trapped in glacial ice are ground against bedrock or other rocks. Striations along a bedrock surface are oriented in the direction of ice flow across that surface.

strike: The direction of the line formed by the intersection of a rock surface with a horizontal plane. The strike is always perpendicular to the direction of the *dip*.

strike-slip fault: A fault in which movement is almost in the direction of the fault's strike.

strip mining: Surface mining in which soil and rock covering the sought-for commodity are moved to one side.

stromatolite: Biogenic sedimentary structures produced by deposition of calcium carbonate by successive mats of algae.

subduction zone: In plate tectonics, a linear zone where two lithospheric plates push together and, generally, one plunges under the other into the mantle. Oceanic trenches, lines of volcanoes, and intense faulting and folding occur in such zones.

sublimation: The process by which solid material passes into the gaseous state without first becoming a liquid.

subpolar convergence: In meteorology, a zone at about 60°N or 60°S latitude where cool air from the poles meets warm air from midlatitudes and the mixed air rises and cools.

subtropical divergence: In meteorology, a zone at about 30°N or 30°S latitude where warm air from the tropics meets cooler air from midlatitudes. They meet aloft and plunge to the surface bringing dry air and creating deserts on the land below.

sulfate mineral: A mineral formed by the combination of the complex ion $(SO_4)^{2-}$ with a positive ion. Common example: gypsum, $CaSO_4 \cdot 2H_2O$.

sulfide mineral: A mineral formed by the direct union of an element with sulfur. Examples: *argentite, chalcocite, galena, sphalerite, pyrite,* and *cinnabar.*

sunspot: A dark marking on the sun's surface.

sunspot cycle: A regular variation in the number, size, and location of sunspots whose period is about eleven years. Many attempts have been made to correlate this cycle with natural phenomena on earth.

superposition, law of: If a series of sedimentary rocks has not been overturned, the topmost layer is always the youngest and the lowermost is always the oldest.

surf depth: The depth at which surf begins to break and most wave energy is expended on the shore.

surface wave: A wave that travels along the free surface of a medium.

surge: (1) a rapid advance of glacial ice. (2) A broad elevation of the surface of the sea under a center of low atmospheric pressure.

suspension: The process by which material is buoyed up in a gas or liquid and moved from one place to another without making contact with the bottom while in transit. Compare *traction.*

synchronous orbit: An orbit in which a satellite remains fixed over a point on a planet as it rotates.

syncline: A troughlike down-bending of stratified rocks. The reverse of an *anticline.*

talc: A silicate of magnesium common among metamorphic minerals. Its crystalline structure is based on tetrahedra arranged in sheets. Greasy and extremely soft. Sometimes called *soapstone.*

talus: A slope established by an accumulation of rock fragments at the foot of a cliff or ridge. The rock fragments that form the talus may be rock waste, sliderock, or pieces broken by frost action. However, the term "talus" is widely used to refer to the rock debris itself.

tectonic cycle: The sequence of geological phenomena in which the lithosphere is broken into drifting plates whose interactions deform the earth. In this cycle, oceanic crust forms and is destroyed; continents are split and moved about; and sediment is returned to the continents at subduction zones.

tectonic plate: One of the large, rigid or relatively rigid, mobile fragments of the lithosphere at whose boundaries occur spreading centers, transform faults, and subduction zones.

tektites: Small, black, glassy stones with distinctive shapes. Apparently, they are solidified droplets from a great splash of molten rock.

temperature: A measure of the activity of atoms; degree of heat.

temperature inversion: In the atmosphere, a temperature inversion occurs when air becomes warmer with increasing altitude and then inverts and becomes cooler with a further increase in altitude.

tension: Stretching stress that tends to increase the volume of a material.

terminal moraine: A ridge or belt of till marking the farthest advance of a glacier. Sometimes called *end moraine.*

terminal velocity: The constant rate of fall eventually attained by a body when the acceleration caused by gravity is balanced by the resistance of the fluid through which it falls.

terrace: A nearly level surface, relatively narrow, bordering a stream or body of water, and terminating in a steep bank. Commonly, the term is modified to indicate origin, as in stream terrace and wave-cut terrace.

terrigenous deposit: Material derived from above sea level and deposited in the deep ocean. Example: volcanic ash.

tethys sea: An ancient sea that once separated Eurasia from India and Africa.

tetrahedron: A regular four-sided solid (*pl.* tetrahedra).

texture: The general physical appearance of a material as shown by the size, shape, and arrangement of the particles that constitute it.

thermal gradient: In the earth, the rate at which temperature increases with depth below the surface. A general average seems to be around 30°C increase per kilometer of depth.

thermal pollution: Increase in the normal temperatures of natural waters caused by human activities.

thermal spring: A spring that brings warm or hot water to the surface. Sometimes called warm spring or hot spring.

thrust fault: A low-angle fault in which the hanging wall appears to have moved upward relative to the footwall.

tidal bore: A steep-faced wave that moves up some bays as the tide rises.

tidal bulge: With regard to the earth, a change in its shape that results because different parts are at different distances from the moon and sun and, thus, are more or less attracted toward them. Occurs in all pairs of bodies.

tidal current: A water current generated by the tide-producing forces of the sun and the moon.

tidal inlet: A waterway from open water into a lagoon.

tidal wave: Popular but confusing designation for *tsunami.*

tide: Alternate rising and falling of the surface of the ocean, other bodies of water, or the earth itself, in response to the gravitational attraction of the moon and sun.

till: Unstratified and unsorted glacial drift deposited directly by glacier ice.

tillite: Rock formed by the lithification of till.

tiltmeter: An instrument for measuring tilt of the ground.

tombolo: A sand bar connecting an island to the mainland, or joining two islands.

topography: The shape and physical features of the land.

tornado: An extremely intense whirlwind.

traction: The process by which material is transported in a stream by being dragged along the bottom.

transform fault: A lateral fault at the side of a tectonic plate. Such faults connect spreading centers and subduction zones.

transistor: A three-terminal semiconductor device used for altering electrical currents.

transpiration: The process by which water vapor escapes from the leaf surfaces of a living plant and enters the atmosphere.

transportation: In the sedimentary cycle, the generally horizontal movement of sediment between erosion and deposition.

travertine: A form of calcium carbonate, $CaCO_3$, that forms stalactites, stalagmites, other deposits in limestone caves, and incrustations around the mouths of hot and cold calcareous springs. Sometimes known as *tufa*, or *dripstone.*

trench: See *deep-sea trench.*

trench–trench transform fault: A transform fault connecting two trenches.

triple junction: In plate tectonics, a point where three plate boundaries meet.

tritium: A short-lived radioactive isotope of hydrogen with atomic mass 3 that is created by cosmic rays in the upper atmosphere and circulates in air and water.

tsunami: A large wave in the ocean generated at the time of an earthquake or, rarely, a volcanic eruption. Popularly, but confusingly, called *tidal wave.*

tufa: Calcium carbonate, $CaCO_3$, formed in stalactites, stalagmites, and other deposits in limestone caves, as incrustations around the mouths of hot and cold calcareous springs, or along streams carrying large amounts of calcium carbonate in solution. Sometimes known as *travertine*, or *dripstone.*

tuff: A rock consolidated from volcanic ash.

tundra: A stretch of arctic swampland developed on top of permanently frozen ground. Extensive tundra regions have developed in parts of North America, Europe, and Asia.

turbidites: Sedimentary deposits that have settled out of turbidity currents.

turbidity current: A current in which a limited volume of turbid or muddy water moves relative to the surrounding water because of its greater density.

turbulent flow: The motion of a fluid having local velocities and pressures that fluctuate randomly. It is the motion by which a fluid such as water moves over or past a rough surface. Fluid not in contact with the irregular boundary outruns that which is slowed by friction or deflected by the uneven surface. Fluid particles move in a series of eddies or whirls. Compare *laminar flow.*

typhoon: The eastern-hemisphere equivalent of a *hurricane*, originating in the western Pacific or the China Sea.

ultimate strength: The greatest stress difference that a material can stand under given conditions.

ultramafic rock: An igneous rock consisting almost entirely of ferromagnesian minerals.

unconformity: A buried erosion surface separating two rock masses, the older of which was exposed to erosion before the deposition of the younger.

uniformitarianism: The concept that the present is the key to the past. This means that the processes now operating to modify the earth's surface have also operated in the geologic past, that there is a uniformity of processes past and present.

unit cell: The simplest polyhedron that, repeated indefinitely, constitutes the lattice of a crystal and embodies all of the characteristics of its structure. See *space lattice*.

varve: A pair of thin sedimentary beds, one coarse and one fine. This couplet of beds has been interpreted as representing a cycle of one year, or an interval of thaw followed by an interval of freezing, in lakes fringing a glacier.

velocity: The rate and direction of motion of a body. The *speed* of a car traveling eastward may be "60 kilometers per hour," but its velocity is "60 kilometers per hour toward the east."

vesicle: A small cavity in an igneous rock formed by the expansion of a bubble of gas during the solidification of the rock.

viscosity: An internal property of a material that offers resistance to flow. The ratio of the deforming force to the rate at which the material changes in shape.

volcanic ash: Material consisting of pyroclastic particles with diameters of 0.06 millimeter or less.

volcanic bomb: A rounded mass of newly congealed magma blown out in an eruption.

volcanic breccia: Rock formed from relatively large blocks of congealed lava embedded in a mass of ash.

volcanic dust: Pyroclastic detritus consisting of particles of dust size.

volcanic eruption: The explosive or quiet emission of lava, pyroclastics, or volcanic gases at the earth's surface, usually from a volcano but rarely from fissures.

volcanic layer: In seismology, the layer of the oceanic crust beneath the sediment.

volcanic neck: The solidified material filling a vent or pipe of a dead volcano.

volcano: A landform developed by the accumulation of magmatic products near a central vent.

vulcanism: The aggregate of processes associated with the transfer of materials from the interior to the surface of the earth.

warm front: In meteorology, the gently sloping boundary along which a warm air mass flows over a cold air mass.

water cycle: The general movement of water from the sea by evaporation, through the air to the land by rain and snow, and back to the sea by rivers.

water table: The upper surface of the zone of saturation for ground water. It is an irregular surface with a slope or shape determined by the quantity of ground water and the permeability of the earth materials. In general, it is highest beneath hills and lowest beneath valleys.

watershed: (1) A drainage divide. (2) An area that collects water that flows to a single river or lake.

wave: A configuration of matter that transmits energy from one point to another.

wave height: The vertical distance from the top of a wave crest to the bottom of a trough.

wave period: The time required for a wave to advance the distance of one wavelength.

wave train: A group of waves generated by some phenomenon such as a storm at sea.

wavelength: The horizontal distance between two successive wave crests.

weather: Such atmospheric phenomena as rain, snow, wind velocity, and temperature.

weathering: The response of materials that were once in equilibrium within the earth's crust to new conditions in contact with water, air, or living matter.

westerlies: In meteorology, the winds in intermediate latitudes that come from the west as a result of the global circulation of the air.

western boundary current: In oceanography, a current that is concentrated on the western side of an ocean. The Gulf Stream is an example.

wetted perimeter: The sides and bottom that are in contact with water in an open channel.

wobble of earth's axis: A motion of the pole along an elliptical path with a mean diameter of 6 meters and a period of 14 months. Also called Chandler wobble.

work: Force times the distance in which it acts.

work hardening: The increase in strength that may occur during plastic deformation.

yield stress: The stress at which permanent *strain* begins.

zonal circulation: In meteorology, the component of the global atmospheric circulation that varies in longitude. It mixes polar and tropical air by the latitudinal movement of storms along the surface of the earth.

zone of aeration: A zone immediately below the surface of the ground in which the openings are partially filled with air and partially with water trapped by molecular attraction. Subdivided into (a) belt of soil moisture, (b) intermediate belt, and (c) capillary fringe.

zone of saturation: Underground region within which all openings are filled with water. The top of the zone of saturation is called the *water table*. The water contained within the zone of saturation is called *ground water*.

INDEX